VOLUME 3
1450-1699

Science and Its Times

Understanding the

Social Significance of

Scientific Discovery

VOLUME 3
1450-1699

Science and Its Times

Understanding the

Social Significance of

Scientific Discovery

Neil Schlager, Editor

Josh Lauer, Associate Editor

Produced by Schlager Information Group

GALE GROUP

Detroit
New York
San Francisco
London
Boston
Woodbridge, CT

Science and Its Times

VOLUME 3

1450-1699

NEIL SCHLAGER, *Editor*
JOSH LAUER, *Associate Editor*

GALE GROUP STAFF

Amy Loerch Strumolo, *Project Coordinator*
Christine B. Jeryan, *Contributing Editor*

Mary K. Fyke, *Editorial Technical Specialist*

Maria Franklin, *Permissions Manager*
Margaret A. Chamberlain, *Permissions Specialist*
Shalice Shah-Caldwell, *Permissions Associate*

Mary Beth Trimper, *Production Director*
Evi Seoud, *Assistant Production Manager*
Stacy L. Melson, *Buyer*

Cynthia D. Baldwin, *Product Design Manager*
Tracey Rowens, *Senior Art Director*
Barbara Yarrow, *Graphic Services Manager*
Randy Bassett, *Image Database Supervisor*
Mike Logusz, *Imaging Specialist*
Pamela A. Reed, *Photography Coordinator*
Leitha Etheridge-Sims *Junior Image Cataloger*

Library of Congress Cataloging-in-Publication Data

Science and its times : understanding the social significance of scientific discovery / Neil Schlager, editor.
    p.cm.
    Includes bibliographical references and index.
    ISBN 0-7876-3933-8 (vol. 1 : alk. paper) — ISBN 0-7876-3934-6 (vol. 2 : alk. paper) —
    ISBN 0-7876-3935-4 (vol. 3 : alk. paper) — ISBN 0-7876-3936-2 (vol. 4 : alk. paper) —
    ISBN 0-7876-3937-0 (vol. 5 : alk. paper) — ISBN 0-7876-3938-9 (vol. 6 : alk. paper) —
    ISBN 0-7876-3939-7 (vol. 7 : alk. paper) — ISBN 0-7876-3932-X (set : hardcover)
      1. Science—Social aspects—History. I. Schlager, Neil, 1966-
Q175.46 .S35 2001
509—dc21
                                         00-037542

Contents

Exploration and Discovery

Life Sciences and Medicine

Contents

Technology and Invention

Preface

~

The interaction of science and society is increasingly a focal point of high school studies, and with good reason: by exploring the achievements of science within their historical context, students can better understand a given event, era, or culture. This cross-disciplinary approach to science is at the heart of *Science and Its Times*.

Readers of *Science and Its Times* will find a comprehensive treatment of the history of science, including specific events, issues, and trends through history as well as the scientists who set in motion—or who were influenced by—those events. From the ancient world's invention of the plowshare and development of seafaring vessels; to the Renaissance-era conflict between the Catholic Church and scientists advocating a sun-centered solar system; to the development of modern surgery in the nineteenth century; and to the mass migration of European scientists to the United States as a result of Adolf Hitler's Nazi regime in Germany during the 1930s and 1940s, science's involvement in human progress—and sometimes brutality—is indisputable.

While science has had an enormous impact on society, that impact has often worked in the opposite direction, with social norms greatly influencing the course of scientific achievement through the ages. In the same way, just as history can not be viewed as an unbroken line of ever-expanding progress, neither can science be seen as a string of ever-more amazing triumphs. *Science and Its Times* aims to present the history of science within its historical context—a context marked not only by genius and stunning invention but also by war, disease, bigotry, and persecution.

Format of the Series

Science and Its Times is divided into seven volumes, each covering a distinct time period:

Volume 1: 2000 B.C.-699 A.D.

Volume 2: 700-1449

Volume 3: 1450-1699

Volume 4: 1700-1799

Volume 5: 1800-1899

Volume 6: 1900-1949

Volume 7: 1950-present

Dividing the history of science according to such strict chronological subsets has its own drawbacks. Many scientific events—and scientists themselves—overlap two different time periods. Also, throughout history it has been common for the impact of a certain scientific advancement to fall much later than the advancement itself. Readers looking for information about a topic should begin their search by checking the index at the back of each volume. Readers perusing more than one volume may find the same scientist featured in two different volumes.

Readers should also be aware that many scientists worked in more than one discipline during their lives. In such cases, scientists may be featured in two different chapters in the same volume. To facilitate searches for a specific person or subject, main entries on a given person or subject are indicated by bold-faced page numbers in the index.

Within each volume, material is divided into chapters according to subject area. For volumes 5, 6, and 7, these areas are: Exploration and Discovery, Life Sciences, Mathematics, Medicine, Physical Sciences, and Technology and Invention. For volumes 1, 2, 3, and 4, readers will find that the Life Sciences and Medicine chapters have been combined into a single section, reflecting the historical union of these disciplines before 1800.

Arrangement of Volume 3: 1450-1699

Volume 3 begins with two notable sections in the frontmatter: a general introduction to science and society during the period, and a general chronology that presents key scientific events during the period alongside key world historical events.

The volume is then organized into five chapters, corresponding to the five subject areas listed above in "Format of the Series." Within each chapter, readers will find the following entry types:

Chronology of Key Events: Notable events in the subject area during the period are featured in this section.

Overview: This essay provides an overview of important trends, issues, and scientists in the subject area during the period.

Topical Essays: Ranging between 1,500 and 2,000 words, these essays discuss notable events, issues, and trends in a given subject area. Each essay includes a Further Reading section that points users to additional sources of information on the topic, including books, articles, and web sites.

Biographical Sketches: Key scientists during the era are featured in entries ranging between 500 and 1,000 words in length.

Biographical Mentions: Additional brief biographical entries on notable scientists during the era.

Bibliography of Primary Source Documents: These annotated bibliographic listings feature key books and articles pertaining to the subject area.

Following the final chapter are two additional sections: a general bibliography of sources related to the history of science, and a general subject index. Readers are urged to make heavy use of the index, because many scientists and topics are discussed in several different entries.

A note should be made about the arrangement of individual entries within each chapter: while the long and short biographical sketches are arranged alphabetically according to the scientist's surname, the topical essays lend themselves to no such easy arrangement. Again, readers looking for a specific topic should consult the index. Readers wanting to browse the list of essays in a given subject area can refer to the table of contents in the book's frontmatter.

Additional Features

Throughout each volume readers will find sidebars whose purpose is to feature interesting events or issues that otherwise might be overlooked. These sidebars add an engaging element to the more straightforward presentation of science and its times in the rest of the entries. In addition, each volume contains photographs, illustrations, and maps scattered throughout the chapters.

Comments and Suggestions

Your comments on this series and suggestions for future editions are welcome. Please write: The Editor, *Science and Its Times*, Gale Group, 27500 Drake Road, Farmington Hills, MI 48331.

Advisory Board

Amir Alexander
Research Fellow
Center for 17th and 18th Century Studies
UCLA

Amy Sue Bix
Associate Professor of History
Iowa State University

Elizabeth Fee
Chief, History of Medicine Division
National Library of Medicine

Lois N. Magner
Professor Emerita
Purdue University

Henry Petroski
A.S. Vesic Professor of Civil Engineering and
*   Professor of History*
Duke University

F. Jamil Ragep
Associate Professor of the History of Science
University of Oklahoma

David L. Roberts
Post-Doctoral Fellow, National Academy of
*   Education*

Morton L. Schagrin
Emeritus Professor of Philosophy and History of
*   Science*
SUNY College at Fredonia

Hilda K. Weisburg
Library Media Specialist
Morristown High School, Morristown, NJ

Contributors

Mark H. Allenbaugh
Lecturer
George Washington University

James A. Altena
The University of Chicago

Peter J. Andrews
Freelance Writer

Kenneth E. Barber
Professor of Biology
Western Oklahoma State College

Bob Batchelor
Writer
Arter & Hadden LLP

Kristy Wilson Bowers
University of Maryland

Sherri Chasin Calvo
Freelance Writer

Matt Dowd
Graduate Student
University of Notre Dame

Thomas Drucker
Graduate Student, Department of Philosophy
University of Wisconsin

H. J. Eisenman
Professor of History
University of Missouri-Rolla

Ellen Elghobashi
Freelance Writer

Loren Butler Feffer
Independent Scholar

Keith Ferrell
Freelance Writer

Randolph Fillmore
Freelance Science Writer

Richard Fitzgerald
Freelance Writer

Maura C. Flannery
Professor of Biology
St. John's University, New York

Katrina Ford
Post-graduate Student
Victoria University of Wellington, New Zealand

Donald R. Franceschetti
Distinguished Service Professor of Physics and
*   Chemistry*
The University of Memphis

Jean-François Gauvin
Historian of Science
Musée Stewart au Fort de l'ile Sainte-Hélène,
*   Montréal*

Brook Ellen Hall
Professor of Biology
California State University at Sacramento

Diane K. Hawkins
Head, Reference Services—Health Sciences Library
SUNY Upstate Medical University

Robert Hendrick
Professor of History
St. John's University, New York

James J. Hoffmann
Diablo Valley College

Leslie Hutchinson
Freelance Writer

Matt Kadane
Ph.D. Candidate
Brown University

P. Andrew Karam
Environmental Medicine Department
University of Rochester

Evelyn B. Kelly
Professor of Education
Saint Leo University, Florida

Judson Knight
Freelance Writer

Lyndall Landauer
Professor of History
Lake Tahoe Community College

Josh Lauer
Editor and Writer
President, Lauer InfoText Inc.

Adrienne Wilmoth Lerner
Department of History
Vanderbilt University

Brenda Wilmoth Lerner
Science Correspondent

K. Lee Lerner
Prof. Fellow (r), Science Research & Policy Institute
Advanced Physics, Chemistry and Mathematics,
Shaw School

Eric v. d. Luft
Curator of Historical Collections
SUNY Upstate Medical University

Lois N. Magner
Professor Emerita
Purdue University

Amy Lewis Marquis
Freelance Writer

Ann T. Marsden
Writer

Kyla Maslaniec
Freelance Writer

William McPeak
Independent Scholar
Institute for Historical Study (San Francisco)

Lolly Merrell
Freelance Writer

Leslie Mertz
Biologist and Freelance Science Writer

Kelli Miller
Freelance Writer

J. William Moncrief
Professor of Chemistry
Lyon College

Stacey R. Murray
Freelance Writer

Lisa Nocks
Historian of Technology and Culture

Stephen D. Norton
Committee on the History & Philosophy of Science
University of Maryland, College Park

Glyn Parry
Sr. Lecturer in History
Victoria University of Wellington, New Zealand

Michelle Rose
Freelance Science Writer

Neil Schlager
Editor and Writer
President, Schlager Information Group

Keir B. Sterling
Historian, U.S. Army Combined Arms Support
Command
Fort Lee, Virginia

Gary S. Stoudt
Professor of Mathematics
Indiana University of Pennsylvania

Zeno G. Swijtink
Professor of Philosophy
Sonoma State University

Dean Swinford
Ph.D. Candidate
University of Florida

Lana Thompson
Freelance Writer

Philippa Tucker
Post-graduate Student
Victoria University of Wellington, New Zealand

David Tulloch
Graduate Student
Victoria University of Wellington, New Zealand

Roger Turner
Brown University

Stephanie Watson
Freelance Writer

Giselle Weiss
Freelance Writer

Michael T. Yancey
Freelance Writer

Introduction: 1450–1699

Overview

The years between 1450 and 1699 were a time of worldwide upheaval and change, of discovery and rediscovery, of exploration and invention. During this period the boundaries of man's physical world expanded, intellectual horizons broadened almost beyond belief, and a technological explosion put into motion an ongoing wave of learning, advancement, and innovation that has continued, albeit fitfully and chaotically at times, to this very day.

During these two and a half centuries, science itself, particularly in the West, underwent a dramatic evolution, becoming evermore central to human endeavor, and expanding its scope to encompass a more accurate view of the world and the universe in which it is located. The age-old belief that both man and the earth were the center of the universe crumbled, though not without resistance, as scientists employed new tools and techniques to explore the skies above and the interior of the human body. Moving virtually hand in hand with science were advances in mathematics, which gave scientists new tools to measure and calculate the forces that shape the world.

Technology, the application of science to practical ends, made greater progress during these centuries than during all the preceding centuries of human existence. Key to it all was the development of the printing press, which provided near-universal access to learning. Knowledge had been made available to everyone who could read, and the effectiveness of printing for capturing and disseminating information insured that it would continue to spread throughout the world.

The spread of learning proved a great threat to religious and political power, and much effort was expended to prohibit "improper" investigations or speculations. The effort proved fruitless—the march of science against ignorance could not be stopped, and the social upheavals that accompanied scientific and technological advance would transform society at every level. While theoretical science altered fundamental beliefs, technological advances brought a higher standard of living, advances in medicine, progress in hygiene and creature comfort, and an array of new products and capabilities. As always, technological advances were also applied to warfare, often with devastating effectiveness.

In short, this period encompassed one of the great shifts in human perspective, the Scientific Revolution, and laid most of the groundwork for another major change, the Industrial Revolution of the 1700s and 1800s.

The Renaissance Expands

The Renaissance, that stunning period of rebirth and renewal that began roughly around 1400, gathered force in the latter half of the fifteenth century. What had been a slow climb out of the Dark Ages 500 years before now became a race toward enlightenment, and the acquisition of knowledge became one of the great undertakings of mankind. Scientists, who had previously worked independently, or for patrons who sought to control their knowledge, began to work cooperatively in the first suggestions of scientific societies, the initial impulses toward a community of science that transcended national boundaries.

The ability of explorers—and increasingly traders and settlers—to transcend those borders in the centuries before 1450 proved one of the great spurs to scientific, technical, and cultural advance. During the twelfth century both Chinese and Europeans used their knowledge of magnetism to produce the first crude compasses; later incarnations would make possible the voyages of exploration to the unknown. The Chinese were the first to invent gunpowder, which increased the capacity of nations make war on one another, lifting combat to previously unimaginable levels of destructiveness.

Pure knowledge traveled from nation to nation as well during those years between the Dark Ages and the Renaissance. Perhaps most significant bit of knowledge to make the journey was the use of numerals, which Europeans acquired from Arabs, who had borrowed them from Hindu mathematicians. Knowledge traveled through time as well: As the Dark Ages receded, scholars began to rediscover the great works of ancient scholars, scientists, and historians, and translated them for the modern world.

By 1450, especially in Europe, the recreation of the past, the expansion of borders in the present, the rise of the scientific method, and the roots of higher mathematics came together, lighting a fuse that ignited a period of ferocious progress unlike anything that had gone before.

The Greatest Invention

Knowledge that cannot be shared is almost meaningless. Disseminating information in an age of handwritten manuscripts, however, was laborious. In 1450 Johann Gutenberg (c. 1398-1468) changed the world forever when he invented movable type. Gutenberg's printing press enabled the rapid duplication of pages of text (and numbers and symbols). No longer would knowledge be restricted to those who had access to rare, hand-copied manuscripts. Books could now be mass-produced and mass-distributed. Knowledge could travel wherever people went.

Gutenberg's revolution was immediate and overwhelming. In 1454 he printed 300 copies of the Bible (an edition many still consider the most beautiful book ever published). By the end of the century the number of books available had exploded, and the price had plummeted. This technological revolution was also an educational revolution, so that as the number of books increased, so did the number of people able to read them. Inexpensive, widely available books were the key to progress in the next two centuries, and they continue to affect the world even in our modern, electronic age. Five and a half centuries after the debut of movable type, Gutenberg's invention can still be called the most influential in all of history.

The Greatest Discovery

From the very beginnings of human history, the night skies exerted a phenomenal influence.

Myths and legends grew up about the stars, and central among them was the concept that man and the Earth were the center of the universe. That changed in 1543, barely a hundred years after Gutenberg, when Polish astronomer Nicolaus Copernicus (1473-1543) cast aside thousands of years of human centrality. The Earth revolved around the Sun, Copernicus said. Many did not want to hear him. One of his supporters, Italian astronomer Galileo Galilei (1564-1642), was forced by the Catholic Church to recant the Copernican view despite evidence of its accuracy.

It was the nature of observational astronomy, however, that while such recantations served political and social ends, they could not withstand the steady accretion of proof. For this is the essence of the Scientific Revolution: evidence, observation, and experiment produce verifiable results that, even if they conflict with long-held articles of faith, are demonstrably true. Copernicus set in motion the greatest of all revolutions, the shift from acceptance based on faith and tradition, to acceptance based on objective, rational proof.

The workings of the universe themselves rapidly became the focus of much scientific effort. Galileo himself applied the scientific method—observation, experimentation, analysis, verification—to the workings of gravity. (The Scientific Method itself would not be codified until 1620, by English philosopher Francis Bacon [1561-1626].) Astronomers throughout the world began using new and improved tools—telescopes (invented in 1698) equipped with lenses that were themselves the product of improvements and refinements in glassmaking—to discover much of the richness of our solar system. Galileo found moons orbiting Jupiter and explored the vast starfield of the Milky Way. Astronomers including Tycho Brahe (1546-1601) and Johannes Kepler (1571-1630) married observational astronomy to higher mathematics and began determining the nature of planetary orbits. The universe itself had been opened to our explorations.

Realm of Numbers

The universe of numbers likewise expanded during this period. If observation is the essence of science, then mathematics is its heart. Mathematical proofs of observed phenomena became vital to scientific consensus—agreement that experimental or observational results were accurate. For mathematics to approach the new complexities that observers reported, however, new methods were needed, beginning with the great effort to develop equations that could solve problems in which some values are unknown or variable.

Virtually all of modern mathematics rests upon advances made during the period between 1440 and 1699. After a period in which ancient mathematics were consolidated, an explosion of knowledge continued almost unabated for more than a century. Negative numbers were introduced in 1545, and trigonometric tables just six years later. Decimal fractions arrived in 1586 as a result of the work of Dutch mathematician Simon Stevin (1548-1620). By 1591 algebraic symbols were being introduced. In 1614 logarithms simplified the calculations of complex numbers; eight years later lograrithmic tables were built into a mechanical device called a slide rule, an early precursor of the calculator and computer. The first mechanical adding machine was built by French mathematician Blaise Pascal (1623-1662) in 1642.

Mathematics's analytical power took a large leap forward in 1637 with the development of analytic geometry, which married algebra to geometry. This development, in turn, led to the greatest of all mathematical advances, the simultaneous development by Isaac Newton (1642-1727) and Gottfried Wilhelm Leibniz (1646-1716) of calculus in the late 1660s. The true beginning of modern higher mathematics, calculus proved a supple tool for constantly varying elements, such as the positions of bodies in motion. Calculus also proved essential to approaching questions of planetary orbits and gravity over distance.

Matters of Gravity

The relationship between astronomy and mathematics was especially apparent in the many scientific studies of gravity and bodies in motion. Galileo himself applied his observations of gravity to the workings of the pendulum, and in 1581 began to measure the time it took a pendulum to complete its arc. (Decades later, further pendulum experiments would result in dramatic advances in timekeeping and the first accurate clocks—themselves among the most revolutionary of all inventions.)

More directly related to gravity itself were Galileo's famous experiments with falling and rolling objects, experiments that established the constant attraction of gravitational force. In 1657 English physicist Robert Hooke (1635-1703) conducted similar experiments, performing some of them in vacuum and proving that,

without air resistance to affect the results, all bodies fall at the same rate. From these experiments and others came English mathematician John Wallis's (1616-1703) 1668 revelation of the law of conservation of momentum: momentum can neither be created nor destroyed.

By 1687 Newton's studies of gravity and bodies in motion had produced his three laws of motion, defining the rules that govern inertia, force as the product of mass and acceleration, and the nature of actions and equal and opposite reactions.

Modern physics was born.

The Universe Within

Even as scores of scientists and scholars cast their interests outward to the larger universe, others looked inward, to the worlds within our bodies. In 1543 (the same year Copernicus upset notions of the universe) Flemish anatomist Andreas Vesalius (1514-1564) radically revised and improved human knowledge of human anatomy. Two years later the French barber Ambroise Paré (1510-1590) published an account of new surgical methods, including tying off rather than cauterizing (burning) severed arteries to stop them from bleeding, and other improvements that would alter the face of medical care.

In 1590 the infinitesimally small became visible when the first microscope was invented. In 1665 Robert Hooke revealed that he had found tiny chambers in a piece of cork examined under a microscope. He called these self-contained chambers "cells." In 1628 English physician William Harvey (1578-1657) explored the nature of the circulatory system in an influential book, *Exercitatio Anatomica de Motu Cordis et Sanguinis in Animalibus* (An Anatomical Exercise Concerning the Motion of the Heart and Blood in Animals). By 1658 corpuscles had been discovered, and capillaries were identified just two years later. In 1668 Italian physician Francesco Redi (1626-1697) disproved long-held beliefs about spontaneous generation—the ability of life to rise from nonliving matter.

Dutch scientist Anton van Leeuwenhoek (1632-1723) made perhaps the most startling discovery of all when he used the microscope to reveal the existence of protozoans, which he called *animalcules*. He also used his microscope to view different types of bacteria, although he did not recognize their importance. His discoveries launched a campaign of microscopic exploration that continues today.

The Chemical World

Chemistry, the combination of elements to form new materials, likewise came of age during this time. Irish physicist and chemist Robert Boyle (1627-1691) rejected the superstitions and half-truths of ancient science, arguing that the four Aristotelian elements or earth, air, fire, and water could not be the building blocks of the physical world. He proposed instead that all matter was made up of "primary particles," which could combine to form compounds, which he called "corpuscles." This systematic approach eventually led to the discovery of chemical elements.

Throughout this period, advances were made in identifying and understanding the different forms elements could take, and the different uses to which those forms could be put. As early as 1592 the fact that some materials expand or contract with temperature changes was used to create primitive thermometers. By 1624 experimentation showed how materials could change from liquids to gases. In 1643 the first barometer was developed, leading to further experiments with air pressure. Better understanding of differences in pressure and the nature of gases led the development of air pumps in the mid-1600s.

Air pumps made vacuum experiments possible, and they, coupled with science's increased understanding of liquids and gases, particularly steam, led by 1698 to the development of the first water pumps. These would prove to be the key invention that led to the Industrial Revolution of the next century.

Exploring and Expanding

Even as scholars explored the scientific world, others explored the physical world. By the end of the fifteenth century Christopher Columbus (1451-1506) had traveled from Europe to the New World, Vasco da Gama (c. 1460-1524) had sailed from Lisbon around the Cape of Good Hope to India, and Amerigo Vespucci (1454-1512) had begun mapping the coast of South America. By 1513 Vasco Núñez de Balboa (1475-1519) had crossed Panama and found the Pacific Ocean, and Juan Ponce de Léon (1460-1521) had begun the settlement of Florida. At roughly the same time a Portuguese ship reached China and established an outpost there. By 1519 Hernán Cortés (1485-1547) had launched his brutal conquest of Mexico.

In 1519 the greatest of all voyages was undertaken when Ferdinand Magellan (c. 1480-1521) undertook a the first circumnavigation of the world, taking five ships and 270 men with him. Although Magellan was killed in the Philippines, four of the ships were lost, and only 17 men returned to Spain in 1522, the voyage was undeniably historic. Never again would geographical barriers limit human expansion. The voyage also confirmed the ancient Greek Eratosthones's calculation of Earth's circumference as 25,000 miles (40,234 km).

Exploration was followed by settlement. Europeans eventually colonized the New World and set in motion a cycle of trade and further exploration that would lead over the next two centuries to the emergence of North America as the richest land on the planet. The explorers, traders, merchants, and settlers brought books with them—knowledge every bit as valuable a cargo as people or materials. The Scientific Revolution, like those who engendered it, knew no boundaries.

The Modern Age Begins

No brief survey can hope to encompass all the scientific, technological, and social progress that occurred between 1450 and 1699. The Scientific Revolution gave birth to an unparalleled expansion of technological capability, which in turn elevated the lives of all. Machines enabled more work to be done, and the results of that work were distributed—slowly, and against much social resistance—to more and more people. The arts were likewise affected, with great paintings, works of music, and above all drama reflecting our new understanding of ourselves and our place in the universe.

Hardship accompanied advance as ignorance, slavery, and warfare continued. But they were also opposed: The Scientific Revolution deposed ancient ignorance and superstition and replaced them with reason, giving rise to new schools of thought, a heightened understanding of humanity's place in the universe, and the importance of the individual within humanity.

Newton himself, acknowledging the scholars who had come before him, said "If I have seen further it is by standing on ye shoulders of Giants." It is no overstatement to say that the century and a half between 1450 and 1699 were an age of giants—in the sciences, in the technologies, and indeed in all of human endeavor.

Chronology: 1450–1699

1450 Johann Gutenberg invents a printing press with movable type, an event that will lead to an explosion of knowledge as new ideas become much easier to disseminate.

1453 Constantinople falls to the Turks, bringing an end to more than 1,100 years of Byzantine rule.

1492 Christopher Columbus encounters the New World.

1500 Hindu-Arabic numerals come into general use in Europe, replacing Roman numerals.

1500-20 During the High Renaissance, numerous artists—among them Michelangelo, Leonardo da Vinci, and Raphael—create their most memorable works.

1517 Martin Luther posts his 95 theses on the door of Wittenberg's Castle Church, a seminal event in the Reformation.

1519-22 Ferdinand Magellan leads the first circumnavigation of the globe and discovers the Strait of Magellan at the southern tip of South America.

1532 Niccolo Machiavelli writes *The Prince*, which provides rulers with a model for achieving and maintaining power.

1534 King Henry VIII officially breaks with Rome, establishing the Church of England.

1543 Nicolaus Copernicus's publication of *De Revolutionibus Orbium,* in which he proposes a heliocentric or Sun-centered universe, sparks the beginnings of the Scientific Revolution.

1588 The English fleet destroys the Spanish Armada, establishing English naval supremacy.

1603 Japan is pacified and united under the Tokugawa Shogunate, which takes measures to isolate the country from European influences.

1618-48 The Thirty Years' War involves most of Europe in a protracted political and religious struggle, fought mainly in Germany; hostilities conclude with the Holy Roman Empire virtually destroyed, Hapsburg power eclipsed, and France the chief power on the continent.

1628 English physician William Harvey, considered the founder of modern physiology, first demonstrates the correct theory of blood circulation in *De Motu Cordis et Circulatione Sanguinis.*

1637 French philosopher René Descartes's *Discours de la méthode* applies a mechanistic view to science and medicine, establishing a worldview that dominates the study of man for some time.

1642-48 Civil war in England results in the establishment of a dictatorship under Oliver Cromwell, but ultimately leads to increased power for the middle class and Parliament.

1644 China's last imperial dynasty, the Ch'ing or Manchu, assumes power.

1669 Isaac Newton circulates a paper, "De Analysi per Aequationes Numero Terminorum Infinitas," in which he lays the foundations for differential and integral calculus; four years later, and completely independent of Newton, G. W. Leibniz in Germany also develops calculus.

1681 France builds the Languedoc Canal, also known as the Canal du Midi, a 150-mile (241-km) waterway considered the greatest feat of civil engineering between Roman times and the nineteenth century.

1683 The Ottoman Empire invades Hapsburg lands in Eastern Europe and lays siege to Vienna.

1687 Isaac Newton publishes *Philosophiae Naturalis Principia Mathematica,* generally considered the greatest scientific work ever written, in which he outlines his three laws of motion and offers an equation that becomes the law of universal gravitation.

1688 England's Glorious Revolution establishes constitutional government under the joint rule of King William and Queen Mary.

Exploration and Discovery

Chronology

1488 Portuguese navigator Bartolomeu Dias first sails around the Cape of Good Hope; a decade later, Vasco da Gama will use this route to become the first European to travel by sea to India.

1492 Christopher Columbus discovers the New World.

1494 The Treaty of Tordesillas divides the New World between Spain and Portugal.

1497 John Cabot is the first European, other than Norse adventures some 500 years before, to set foot on North America.

1507 Cartographer Martin Waldseemüller becomes the first to call the New World "America," after explorer Amerigo Vespucci.

1513 Vasco Núñez de Balboa becomes the first European to see the Pacific Ocean.

1519-22 Ferdinand Magellan leads the first circumnavigation of the globe, and discovers the Strait of Magellan at the southern tip of South America.

1595 Flemish geographer Gerardus Mercator introduces the use of cylindrical projection—later dubbed Mercator projection—to depict Earth's spherical surface on flat paper.

1605 Willem Jansz is the first European to set foot on the Australian mainland.

1607 John Smith leads the establishment of the first permanent English colony in the New World, at Jamestown in Virginia.

1616 While searching for the Northwest Passage, William Baffin explores Greenland and Baffin Island, and ventures further north—within some 800 miles (1,287 km) of the North Pole—than any explorer will until the nineteenth century.

1648 Semyon Dezhnev is the first to sail through the sea channel between Siberia and Alaska, proving that Asia and North America are not connected; however, because Dezhnev's records are not found until much later, Vitus Bering—for whom the strait is named—receives credit for the discovery.

Overview:
Exploration and Discovery 1450-1699

As the civilizations of the world developed and expanded, so, too, did man's desire to explore and conquer new lands and peoples. The Vikings were prime examples of this need to discover and conquer, first with their raids throughout Europe from the middle of the eighth century, and later with their epic voyages for the adventure of discovering new lands, which they did in North America in the late tenth century. Other civilizations that turned to exploration for the purpose of expanding their empires included Genghis Khan's (1162?-1227) Mongols, whose vast empire stretched across Asia. The Crusades of the eleventh, twelfth, and thirteenth centuries brought European military expeditions to the Holy Land, introducing Islamic culture (and the science of cartography) to the West. With the journeys of Marco Polo (1254-1324) in the thirteenth and fourteenth centuries, the European spirit of exploration was further inspired. Polo's tales of the Great Khan and the wealth of the East—its silks and spices—spurred the nations of Europe into a period known as the Age of Discovery, with a focus on quests to seek out new lands and trade routes by sea.

Apart from the Norse voyagers, all official early European explorers had one goal: the discovery of a route to China and the Indies. The two main objectives of this goal were the riches of the Indies and the conversion of native "infidels" to Christianity. So began a tradition of European maritime discovery. The person who most encouraged fifteenth century sea exploration was Portuguese Prince Henry (1394-1460), known as Henry the Navigator, who established a "navigational" school at Sagres, near Cabo de São, Portugal. By the time of his death, expeditions under his sponsorship had explored southward along the coast of Africa as far as Gambia.

Explorer Bartolomeu Dias (1450?-1500) was the next great Portuguese navigator; he discovered the Cape of Good Hope in 1488 after he was blown around it while outrunning a storm. Others soon followed in his wake: Vasco da Gama (1460?-1524) rounded Africa and reached India in 1498, opening the Indian Ocean to trade; Christopher Columbus (1451?-1506) discovered the "New World" in 1492, but was convinced he had reached Asia; Amerigo Vespucci

(1454-1512) rediscovered North America on his return voyage from Brazil; John Cabot (1450?-1499?), the first European since the Vikings to make landfall in the northern reaches of North America—Newfoundland and Nova Scotia—around 1497; and Ferdinand Magellan (1480?-1521) led the first circumnavigation of the globe (1519-1522), though was killed by natives in the Philippines before he could return. These expeditions rapidly added details to maps of the world, as did others in the Pacific Ocean in the late sixteenth and early seventeenth century when Australia, New Zealand, and the Fiji Islands were discovered by Dutch sailors looking to expand lines of commerce for their nation.

The oceanic exploration begun with Henry the Navigator led to a quest for wealth and adventure and to the Spanish tradition of conquistadors—adventurers, part soldier, part sailor, interested in the myths and legends of gold, spices, and new lands to conquer. Their desire for conquest and the building of empires also concealed the objective of converting those they conquered to Christianity, thus beginning an era of colonization and commerce in the New World. The first Spaniard to disrupt an established New World civilization was Hernando Cortés (1485-1547), who conquered the Aztec empire in Mexico (1518). Eventually the Spanish occupied Mexico, sending out expeditions to the southwestern parts of North America, such as that of Francisco de Coronado (1510?-1554), who discovered the Grand Canyon (1540).

Overland exploration by the conquistadors led to critical geographic discoveries. Alvar Nunez Cabeza de Vaca (1490?-1560?) and Hernando de Soto (1496?-1542) led expeditions to the southeastern sections of North America. Vasco Nuñez de Balboa (1475-1519) and Diego de Almagro (1474?-1538) led expeditions to South America, where Francisco Pizarro (1475-1541) eventually conquered the Incas (1532). Two significant expeditions were conducted along the Amazon River, the first by Francisco de Orellana (1511?-1546), who made a 4,000-mile (6,437-km) journey along the Amazon to the Atlantic Ocean (1539-1541), and the second by Pedro Teixeira (1570?-1640), who in 1639 spent 10 months surveying the river. In addition to their contributions to geography and political expansion, the Spanish conquistadors also helped establish the first permanent European settlement on Cuba in 1512. In 1565 the first permanent European settlement in North America was founded in St. Augustine by the Spanish.

With the Spanish and Portuguese establishing settlements in South America and Mexico, the French, British, and Dutch looked to North America. As settlements were planned, explorers such as Henry Hudson (?-1611) ventured north and then inland, where Etienne Brulé (1592?-1632?) discovered and explored the Great Lakes—Huron (1611), Ontario (1615), and Superior (1621). In 1603 French explorer Samuel de Champlain (1567?-1635) founded the first settlement in Canada at Montreal. A few years later, the British founded Jamestown in Virginia (1607), followed by New Plymouth (1620), Salem (1628), and Boston (1630). The Dutch founded New Amsterdam, site of present-day New York City, in 1626. With the colonies came opportunities for trade and commerce, as new crops such as tobacco and sugar cane impacted the economies of Europe.

In North America attempts were made to reach the Pacific Ocean by traveling overland, including that of René Robert Cavelier, Sieur de La Salle (1643-1687), who reached the Mississippi (1681) and followed it to the Gulf of Mexico to lay claim to Louisiana. Frenchman Jean Nicollet de Belleborne (1598?-1642) explored between the St. Lawrence and Mississippi rivers, and Belgian friar Louis Hennepin (1626-1705?) explored the upper Mississippi (1679) and was the first European to see Niagara Falls. At the end of the seventeenth century, the unknown territory west of the Mississippi would remain for later explorers to discover.

While explorers were beginning to establish settlements in North America, others were making the first attempts to explore the cold northern Atlantic—in search of a passage to the East. The first recorded attempt to discover the Northwest Passage was made by Italian Sebastian Cabot (1476?-1557) for British investors. While unsuccessful, his voyage spurred others such as Jacques Cartier (1491-1557), who discovered the Gulf of St. Lawrence (1534) and the St. Lawrence River (1535) during his search; Martin Frobisher (1540-1594), who sailed up the coast of Greenland toward what is now Baffin Island and shared the first meeting between Englishmen and Eskimos (1576); and John Davis (1550?-1650), who led three voyages to discover the Northwest Passage between 1585-87 and reached just over 1,100 miles (1,770 km) from the North Pole. In 1616 William Baffin (1586?-1622) came within 800 miles (1,287 km) of the North Pole on an expedition that resulted in the discovery of Baffin Bay and Baffin Island. Other expeditions searched for a Northeast sea route to China—and were as unsuccessful, though they resulted in trade routes between England and Russia. As with explorations further west in North America, it would remain for later explorers to discover answers to the mysteries of the Arctic.

In the 1700s European explorers had expanded their knowledge of the world, defining its boundaries and cataloging its natural shape. With much of the Atlantic Ocean and its coastlines surveyed, explorers turned to the larger Pacific Ocean and began to survey and lay claim to islands in its waters and adjoining lands. Expeditions ventured further into the interiors of North America and Africa. Others made great strides in compiling more accurate geographic and meteorological data and maps of the world. Exciting developments were made in the fields of archaeology, geology, anthropology, ethnology, and other natural sciences. By the end of the eighteenth century the world seemed smaller due to the knowledge gained by its explorers.

ANN T. MARSDEN

Spanish Exploration and Colonization

Overview

Beginning in 1492 with the first voyage of Christopher Columbus (1451?-1506), Spanish explorers and conquistadors built a colonial empire that turned Spain into one of the great European powers. Spanish fleets returned from the New World with holds full of gold, silver, and precious gemstones while Spanish priests traveled the world to convert and save the souls of the native populations. However, Spain's time of dominance was to be relatively short-lived; only two centuries later, Spain's European power was in decline, and a century after that, virtually all her colonies were in open revolt. Much of the reason for this sequence of events, and for the subsequent history of former Spanish territories can be traced back to the reasons for and the nature of Spanish imperialism.

Background

For almost 800 years, Arabs occupied and ruled the Iberian Peninsula. For over a century, a succession of Spanish rulers fought the Moors, gradually pushing them back and reestablishing Spain as a Christian nation. This goal was finally achieved in 1492, when the Moorish bastion of Granada finally surrendered after a decade of siege. In that same year, Spain expelled thousands of Jews, a Spaniard was elected Pope, and another Spaniard published the first formal grammar of any European language. And Genoan navigator Christopher Columbus sailed on a voyage of discovery to find a more direct route to the Orient. All of these factors turned out to have great importance for the next 300 years of Spanish history, and for all subsequent Latin American history.

Columbus returned to Spain, convinced he had succeeded in finding the Orient and not realizing his discovery was, instead, much greater. He was quickly followed by others: Francisco Pizzaro (1475-1541), Vasco Núñez de Balboa (1475-1519), Hernan Cortés (1485-1547), and others. Within a few decades, Spain had explored most of South and Central America, and had found the Americas to be rich with precious metals and stones. Meanwhile, Spanish priests discovered a new continent full of, in their opinion, savages whose souls needed to be saved. So Spain descended on the Americas with a cross in one hand and a gun in the other, determined to convert the natives while stripping their lands to fill the Spanish treasury.

While this description may sound unnecessarily harsh, Spain's actions are understandable to some degree. Spain had just emerged from centuries of domination by a foreign power and (by their lights) heathen religion. They earned their liberty by force of arms and, they believed, divine help. This belief seemed vindicated when a Spaniard became Pope in the very year the last Moors were defeated, cementing in the national consciousness the link between religion and military power. This, plus Spain's late emergence from medieval feudalism, helped mold the national character that was to have such a pro-found influence in Spain's management of her overseas possessions.

Spain's religious fervor was no less understandable than was her elevation of the military to a position of prominence in society. Spain's recent emergence from seven centuries of Moorish rule had only served to emphasize to her the importance of the Christian Church (this was before the Protestant Reformation), and religious belief was an important fact of daily life. Then, in 1517, Martin Luther (1483-1546) tacked his famous 95 theses to the door of a church in Germany, launching the Reformation, which was to subject Europe to centuries of religious bloodshed as Protestants and Catholics battled for supremacy. Against this backdrop, Spain's desire to spread the Catholic Church overseas is entirely understandable, especially given Protestant England's later colonization of North America.

The Spanish did not treat their New World possessions kindly. The conquistadors came to conquer new territories for power and riches. They overthrew the Inca and the Aztecs, plus a host of less-advanced civilizations. Spanish settlers came to make a fortune and return to Spain, not to stay in a new home. They felt that many chores were beneath their dignity, so they employed or enslaved the native populations to till the land, mine precious metals, and do the other menial work of empire. In this, they were a microcosm of the Spanish government, and their colonial style was to have significant ramifications for both the Spanish colonies and for Spain herself.

Impact

During the Age of Exploration and subsequent years, there were five major colonial powers: England, Spain, France, Portugal, and Holland. Each of these nations had a different motivation for establishing overseas colonies, and each treated her colonies differently. Most of their former colonies still bear an unmistakable imprint of their colonial heritage, made of equal parts of the motivations of their parent country in establishing colonies and the manner in which they were treated before independence.

In general, the Dutch came to trade, the Portuguese to explore and to trade, the English to expand, the French to counter English maneuvers, and the Spanish to get rich. Another generalization is that the English and French settlers came looking for freedom and opportunity in a new home, the Portuguese and Dutch settlers came to work what was, in effect, an "overseas assignment" be-

Fifteenth-century woodcut depicting a Spanish ship in Hispaniola.*(Corbis Corporation. Reproduced with permission.)*

fore returning home again, and the Spanish came to take what they could to advance themselves, their families, their religion, and their nation.

During their centuries of domination, the Spanish colonies returned an incredible amount of wealth to Spain, making Spain one of the most powerful and most feared nations in Europe. However, this money was not used wisely, in part because Spain was not expecting it and her government was not ready for it, similar to how a child is not ready to inherit and manage a million dollars. So Spain spent her wealth building up a large army and larger navy, waging wars, subduing a continent, and defending her colonies against opportunistic attack. At the same time, Spain's European ambitions led to her dominating large sections of Europe, only to lose them in later years through war or political maneuvering.

Because she spent her money unwisely, Spain almost immediately went into debt, if that can be believed. She began borrowing against future treasure, primarily from foreign governments because Spain's Catholics were not permitted to lend money, and she had expelled her Jews, who had no Biblical injunction against lending money. So most of Spain's New World revenues passed through Spain and ended up in France, Switzerland, and the other nations of Europe while the Spanish economy and people benefited little. In effect, Spain's mismanagement of her great wealth drove her into bankruptcy, and Spanish power began to decline. In 1588 the seemingly invincible Spanish Armada failed to defeat the English navy, while at the same time, her New World possessions had been repeatedly attacked by English ships led, more often than not, by Sir Francis Drake (1540?-1596). Although Spanish power would continue to be feared for more than a century longer, by the start of the seventeenth century it was already apparent that Spanish power would not last forever.

Spain's colonies were perhaps most dramatically influenced by Spanish practices. As noted above, they were settled largely by men who came to the New World simply to conquer, convert, or become rich. This was a direct outgrowth of the period in which Spain found herself at that time. By the time of the Latin American revolutions in the last part of the eighteenth century and the first part of the nineteenth, these characteristics were deeply ingrained into the national psyches of virtually all Latin American nations, and they remain visible today. Most Latin American nations are devoutly Roman Catholic. The military has a prominence in most of them that is almost unique among the world's democracies, and Latin American politics and government are still strongly reminiscent of the Spanish feudal heritage, in which a strong leader dominated the nation's political machinery. This was seen in Chile and Argentina in the 1970s and 1980s, also in Panama, Nicaragua, and El Salvador during this same time frame, and continues to be the case in Peru, Venezuela, Mexico, Cuba, and other nations today. Some of these nations, in particular Venezuela and Mexico, continued their progenitor's profligate ways with national wealth; in both cases, vast amounts of revenue from petroleum and mineral deposits has been either squandered or vanished.

Although Spain's power was broken in the wake of the Armada's defeat, she remained a power to be reckoned with until her defeat in the Spanish-American War in 1898-99. During this time, she continued to play a role in European politics and wars, including the Napoleonic Wars, though usually in a supporting role.

It is also noteworthy that the treasure brought back from the New World, while it did not often benefit Spain, did benefit Spain's European lenders. In spite of the incredible imported wealth, Spain defaulted on loans several times in the late 1500s and early 1600s, and some of her military defeats were due to army mutinies over lack of pay. In particular, the Dutch, the Swiss, and the French held Spanish loans, but the Spanish borrowed from just about any government with which they were not actively at war. This money, in turn, was often put to good use by the recipient nations, helping to build their economies.

It is probably safe to say that Spanish aims in exploring and colonizing Latin America were not bad, but they turned out badly. Arriving with the near-absolutism of the zealot, Spanish missionaries were determined to convert native populations to Catholicism, in part to combat the spread of Protestantism in Europe. And, recently emerged from a long and bloody religious war against the Moors, Spanish settlers were more than willing to believe in the advantages of a powerful central government, a strong military, and the necessity of military conquest to tame a new continent. In addition, a strongly patriarchal society gave familial lands to the oldest son, leaving younger sons often destitute and eager to spend a few years in the Americas to make their fortune, which they tended to do with the labor of native populations. This almost inevitably led to the establishment of strong central governments presiding over largely Catholic nations and supported by a large, strong military—exactly the pattern seen in many Latin American nations for nearly two centuries. In addition, Spain's mismanagement of her imported wealth led just as inevitably to her economic and military downturn, taking Spain from a prominent position in European power to that of a second-class power within just a few centuries.

P. ANDREW KARAM

Exploration
& Discovery

1450-1699

Further Reading

Copeland, John, Ralph Kite, and Lynne Sandstedt. *Civilización y Cultura*. New York: Holt, Rinehart, and Winston, 1989.

Crow, John. *The Epic of Latin America*. Berkeley: University of California Press, 1992.

Kennedy, Paul. *The Rise and Fall of the Great Powers*. New York: Random House, 1987.

Manchester, William. *A World Lit Only by Fire: Portrait of an Age*. Boston: Little, Brown, 1992.

Wood, Peter. *The Spanish Main*. Alexandria, VA: Time-Life Books, 1979.

Portugal Launches Age of Discovery

Overview

Over a period of about 150 years, the tiny nation of Portugal founded Brazil, discovered the sea route around Africa to India, and established colonies and trading posts in Tangiers, Angola, the Congo, the Gulf of Ormuz, India, the Spice Islands, and China. For most of that time, Portugal dominated trade between Asia and Western Europe, undercutting the economies of flourishing trading cities, including Naples and Genoa. Henry the Navigator (1394-1460) set up a prototypical research center in Sagres where maps were systematically charted and both sailing vessels and techniques that made exploration possible were invented. With these tools, Portugal was able to secure luxury goods from the East, to spread Christianity, and to increase its wealth, influence, and power. When the power shifted, it went to other European countries that followed Portugal's successful lead. Holland, England, France, and Spain joined in a scramble to discover, explore and claim new lands that lasted all the way to Captain James Cook's (1728-1779) final voyage.

Background

Portugal's geography, politics, and personality came together to encourage it to become a nautical power. The country faces outward to the Atlantic, with 1,118 miles (1,800 km) of coastline. But, looking eastward toward the most vibrant trading centers, Portugal found itself relatively far away, with difficult land routes and no coastline on the Mediterranean. While to the west, Portugal had navigable rivers and deep, natural ports, including Lisbon and Setubal, to provide safe harbor, to the east were disadvantages of cost, time, and hazard.

One such hazard was enemies, including the kingdoms that became Spain on its own peninsula and the powerful Moors to the south.

The Moors, in fact, still held the territory south of Lisbon as late as 1200.

Despite foreign conflicts, the people of Portugal were relatively tolerant. Portugal's population originated from a variety of different tribes, including Celtic, African, Iberian, English, and Germanic. This encouraged a cultural broad-mindedness that gave the Portuguese critical access to tools like the compass (from Islamic countries) and maps (from Jews).

By the fifteenth century, Portugal was united internally and at peace with Spain. At the same time, it was hemmed in by the Moors to the south. The possibilities for expansion and trade were revealed when Henry the Navigator went on a crusade that seized the city of Ceuta in 1415. This trading center was filled with shops, precious metals, jewels, and spices. However, the captured city's trade stopped with the departure of the Moors, and Portugal was left with a hollow victory. If Portugal could find a route around the Moors to the East, it could participate in this rich trade directly.

Their primary trading need was pepper, which both helped preserve food and made heavily salted meat palatable. Because of Portugal's location, goods from the East went through many middlemen, and the costs to the Portuguese were high. With direct access to the East, Portugal hoped to lower prices and capture a portion of the wealth of trading. But trade was not the only reason exploration became a national goal for the Portuguese. In fact, it was 20 years before the acquisition of African slaves brought the first returns on their investments. There was another reason—conversions.

Portugal was at the forefront of the struggle between Christian and Islamic religion. Like Spain, many of its territories had been held by Islamic powers. Islamic strongholds were just across the Gulf of Cadiz, and the Popes were for-

Seventeenth-century Portuguese colonialists in East Africa. *(Corbis Corporation. Reproduced with permission.)*

mally blessing crusades against the Moors. Though there was a political basis for the enmity, there was also a rising tide of religious fervor within Portugal that led to forced conversions and trials of inquisition. Within this context, the zeal for gaining religious converts rose, and the spread of Christianity became an important motivation for exploration.

To this was added the curious legend of Prester John, a wise and powerful Christian leader located in the East. The story probably originated from misinformation about the Mongol Empire, a bogus letter from Prester John to European rulers, and wishful thinking. But the Portuguese accepted the existence of Prester John as fact, and pursued a strategy to link up with this Christian ally and outflank the followers of Islam. Rather than being contained and controlled by the Moors, Portugal would contain and control its rival.

Besides trade and conversions, curiosity was also a powerful motive for exploration that should not be underestimated. Henry the Navigator had seen the economic stakes in Ceuta and had sacrificed a ransomed brother to the cause of the spread of Christianity. But he was also hungry for new knowledge, and Portugal's adventures in exploration really began with his leadership and his financial backing of a center for exploration in Sagres. It was there that better maps were drawn, navigational instruments were adopted,

and a new kind of ship, the caravel, was developed. Quick, lightweight, and able to sail windward, the caravel become the key vehicle for discovery. Christopher Columbus's (1451?-1506) *Nina* and *Pinta* were both caravels. Most significantly, Henry systematically sent voyage after voyage down along the coast of Africa. This was unprecedented. He persisted even when the only benefit to Portugal was increase in the extent of known geography. Progress came to a halt when Henry's captains came to a bump on the coastline known as Cape Bojador. This was purportedly a point of no return; to pass it meant being killed or lost forever. Fifteen times over the course of 10 years captains were sent to take on this challenge for king and country, and 15 times they came back with word that it was impossible. Finally, Henry made Gil Eannes (?-1435?) swear that he would not return unless he had gone south of the Cape and, in 1435, Eannes rounded Cape Bojador, opening up territories south for further exploration. By Henry's death in 1460, the Portuguese had gone all the way to what would become Liberia, 1,864 miles (3,000 km) into unknown territory; by 1482, the Portuguese had gone as far as the Congo; by 1485, Bartholomeu Dias (1450?-1500) had rounded the Cape of Good Hope; and by 1499 Vasco da Gama (1460?-1524) had completed his trip to India. For the next hundred years, Portugal dominated the spice trade and was a world power.

Impact

With its many voyages, Portugal initiated the Age of Discovery. The Portuguese brought knowledge as well as wealth. They dispelled superstition and changed the political balance within Europe. For the Portuguese themselves, the most important geopolitical legacy is the nation of Brazil, the largest, most powerful country in South America. But the indirect results are of even greater significance. Portugal's success encouraged others, most notably the Dutch, the English, the French, and the Spanish, to engage in exploration and colonization. Portugal's competition with Spain led to a Papal decree, the Treaty of Tordesillas (1494), that secured Portugal's claims in Africa and the East and brought Spanish culture to most of Latin America. In fact, the political and cultural map of the Western Hemisphere was drawn during this era, and its shape is largely the result of forces set loose by Portugal.

Unfortunately, discovery included slavery and colonization. The slave trade that the Portuguese initiated in Africa grew quickly. One thousand slaves had already been brought to Portuguese territory by 1448, and the Portuguese continued to deal in slaves for two centuries. The Portuguese base at Elmina (Ghana) became an infamous link in the chain that brought millions of Africans to the Americas. Built in 1482, it was captured by the Dutch in 1637 and taken by the British in 1664. By the 1700s, 30,000 slaves were passing through Elmina each year.

Trading centers from Angola to India to China were established at the point of a gun. Though the Portuguese were generally not as thorough-going as the Spanish conquistadors, they did establish patterns of violence and distrust that persist today in Portuguese former colonies, such as East Timor and Angola, as well as in the former colonies of their imitators.

Portugal can be credited with more than just political and economic leadership in the Age of Discovery. The Portuguese also developed the tools and processes for exploration. Prince Henry the Navigator's center in Sagres improved ship design, transformed mapmaking into a rigorous discipline, and spurred the adoption of key navigational tools, including the compass (which others had superstitiously avoided) and the sextant. The Portuguese had a program of exploration that was systematic, objective, cumulative, patient, and determined. This approach, which was adopted by other nations, produced success over and over again. It also created a model for scientific exploration; Sagres had many of the same values and procedures that are part of the culture of today's research centers.

Though the Portuguese never linked up with Prester John, their plot to outflank the Moors succeeded. This went beyond a short-term trading advantage and political security. Thanks to communication with the East, the rise of scientific techniques, access to classical manuscripts, and the wealth of the New World, the dominance of Portugal, and more broadly Western Europe, in worldwide culture began. Islamic culture, lacking newer technology and relatively weaker in trade, went into decline and receded as a global presence and as a threat to Europe.

PETER J. ANDREWS

Further Reading

Asimov, Isaac. *Isaac Asimov's Biographical Encyclopedia of Science & Technology*. New York: Doubleday, 1976.

Boorstin, Daniel J. *The Discoverers*. New York: Vintage, 1985.

Cuyvers, Luc. *Into the Rising Sun: Vasco Da Gama and the Search for the Sea Route to the East*. New York: TV Books, 1999.

Dutch Exploration and Colonization

Overview

In the sixteenth century the United Provinces of the Netherlands rose from the status of a Spanish possession to a great European power. Dutch ships carried goods throughout the world for virtually every European nation, Dutch merchants and bankers made Amsterdam the economic center of Europe, and the Dutch navy was a power to be reckoned with. The Dutch empire was built on industry and trade, and Dutch merchants were remarkably pragmatic in political and economic matters. As a result, Dutch power

Seventeenth-century Dutch explorers encounter Chilean natives at Cape Horn. (*Corbis Corbis. Reproduced with permission.*)

grew more rapidly than English or French and, when Holland's power had peaked, it did not decline as precipitously as did Spain's. These same traits have helped make the Netherlands one of the world's most prosperous and egalitarian nations, a country that remains an economic powerhouse today.

Background

When Charles V of Spain was crowned the Holy Roman Emperor in 1519, among his holdings was the territory of the Netherlands, which he had inherited through his paternal grandmother, Mary of Burgundy. Apparently this arrangement did not sit well with the Dutch who, by century's end, had successfully freed themselves from Spanish domination and had become a formidable military and economic power.

Dutch success was due to a number of political, economic, and military factors. Politically, the Dutch were the only European nation at that time with a republican government, rather than an absolute monarchy. This gave each citizen a greater stake in the nation's success, and a greater responsibility for helping the country to do well. This also gave more power to the Dutch merchants, whose shrewd business sense and pragmatism led them to a position of prominence in Europe. The success of Dutch mer-

chants provided ample tax revenues from which the Dutch government could wage war, protect its borders, establish colonies, and care for its citizens. It also provided a large supply of money for lending at favorable interest rates, which, in turn, helped the Dutch government finance its activities when tax revenues were not sufficient. These three factors reinforced each other and enabled the Netherlands to achieve a prominence that belied its relatively small size and population.

During the seventeenth and eighteenth centuries Europe was in a nearly constant state of war. Alliances developed and shifted continually between England, France, Spain, the Netherlands, Sweden, and smaller states as the European nations first built themselves and then jockeyed for power and dominance. The Dutch and English fought three wars before allying against a French-Spanish force trying to reunite the Netherlands with Spain. Other alliances were made and broken over the years as nations sought the most advantageous situation for themselves in the shifting European political scene.

Against this backdrop the Dutch were busy defending their borders and carefully building their trade empire. Sturdy Dutch merchant ships carried most of Europe's trade, even trading with their enemy, the Spanish, if the potential profit outweighed their risks (and, ironically, helping deplete Spain's treasury, which helped contribute to Spain's downfall). As Dutch merchants and shipbuilders grew more confident in their respective crafts, Dutch ships began to sail further afield, and the Dutch saw economic advantage in establishing their own colonies, rather than simply carrying goods for others.

Although the Dutch colonial empire did not come close to matching the scope of English, French, or Spanish possessions, Dutch colonies were carefully selected and tenaciously defended. After abandoning their North American colonies (in what is now New York), the Dutch established outposts in the Caribbean, South America (what is now Suriname), South Africa, and what is now Indonesia. Holland also established a trading center in Japan, one of only a few European nations to do so. Between 1598 and 1605, 150 Dutch ships sailed to the Caribbean each year. Another 25 ships carried goods to and from Africa, 20 left for Brazil, and 10 plied trade routes to the East Indies. Some of these ships served Dutch colonies, some the colonies of other nations. All added to Dutch wealth and power.

Impact

The Dutch were not explorers in the same sense as other European nations. Unlike England, Portugal, and Spain, they were not prone to sailing forth on voyages of discovery, planting their flag wherever they set foot, and claiming lands for the Dutch crown. They were, at heart, shrewd and pragmatic businessmen, expanding cautiously and carefully, reluctant to commit themselves to the large investment a colony entailed unless the potential financial gain warranted the risk. This is not to say that every single Dutch move was carefully considered and weighed, but in general the Dutch sailed for profit and not for glory. This caution left an indelible mark on Dutch colonies, Dutch power, and the current Dutch nation.

Dutch aims in colonizing new territories were primarily commercial: maximize profit and minimize financial risk. Unlike the English in North America and (later) in South Africa, they had little interest in establishing colonies with a high degree of political autonomy. Instead, their preference was to establish colonial governments that would help organize the efforts of the native populations and the colonists so that the colonies could ship raw materials back to the Netherlands on a regular and continuing basis. This, however, helped make the Dutch poor colonial masters, as they tended to place great demands on Dutch colonists and native populations. At the same time, the Dutch tended to demolish the existing tribal or political structure, ruling almost entirely with Dutch nationals. This combination tended to not only anger the native populations, but also left them in a disadvantaged position when Dutch colonial rule ended. This is most obvious in Indonesia, which, since Dutch rule ended in the mid-twentieth century, has been subject to an endless succession of corrupt governments.

Unlike the Spanish, the Dutch did expect their colonies to produce goods on a relatively sustainable basis, and the Dutch colonists expected that a great deal of hard work would be involved. In addition, the Dutch were never as adamantly religious as the Spanish, and religious proselytizing and conversion was not a primary focus of Dutch overseas efforts. So, although the Dutch were not ideal colonial masters, they were better than the Spanish, and they did not plunder their possessions as the Spanish did.

The Dutch focus on commerce led to huge revenues that poured into the Dutch economy and government coffers, and in a short period of time the Netherlands was one of the wealthiest nations in Europe. In addition to carrying cargo for most European nations, the Dutch also imported raw materials, turning them into finished goods that were subsequently exported at a tidy profit. And Holland's role in trade helped make Amsterdam one of Europe's financial centers, further adding to Dutch revenue.

All of this income enabled them to fortify their borders and hire foreign mercenaries to protect against the attempted depredations of their neighbors. With all their shipbuilding experience, the Dutch shipyards built an impressive navy that helped with national defense, escorted Dutch merchant vessels, and protected Dutch colonies from foreign incursions. For a time the Dutch navy was the world's most powerful, and the Dutch army was more than adequate to defend its borders against any European power. There is little doubt that none of this would have been possible without the steady stream of revenue from Dutch commerce, including that from its overseas possessions.

Although Dutch military power was rarely sufficient to dominate European politics, it was enough to guarantee the nation's security against both land and sea attack by any great power. And, as all the great powers of the time discovered, the Netherlands's entry into a contest was often sufficient to tip the balance of power against its foes. This gave the Netherlands political "muscle" that was belied by its small size and population.

As their overt political and military power was eclipsed by that of England and France, the Dutch seem to have settled (not entirely willingly) into a different role in European politics. Although the term "power broker" is not entirely apt, it is also not entirely inappropriate because Dutch involvement in any close issue could be sufficient to decide the matter. From this, the Dutch seem to have grown into a philosophy of judicious international involvement which, in conjunction with their still-considerable economic might, gives them a continuing prominent role in many international organizations, including NATO and the United Nations.

As noted above, the Dutch tended to manage their colonies for long-term profitability rather than short-term gain. Part of this no doubt stemmed from their having established colonies largely in areas that did not appear to have great mineral wealth, but in which spices or tropical hardwoods could be harvested. This forced them to manage their resources with an eye towards some degree of sustainability, for if they harvested every single spice plant, their revenue source would disappear. In turn, this assured the Dutch a long-term source of income, and this income helped cushion the Dutch when they were militarily overtaken by other great European powers. This is also one of the reasons that the Netherlands remains economically strong and politically influential to this day.

Finally, all of these events had a distinct impact on the Dutch people, which still reverberates. The Netherlands remains one of the most egalitarian and affluent nations on Earth, and still wields what seems a disproportionate amount of influence in European and world affairs. A great deal of this stems from the Dutch policy of engagement with foreign nations, either through treaties, membership in international organizations, or foreign aid. All of this helps to make the Netherlands a very cosmopolitan nation in which a large number of citizens have an active interest in world affairs.

In summary, the Dutch left their shores to establish the trade and commerce that helped make them a respected European power. Dutch traders were more interested in financial return than exploration or national glory, so they were as happy to be ferrying French trade goods as they were establishing their own colonies, and their explorations were never as extensive as those of other European powers. As colonial masters, they were better than some and not as good as others, but they left their colonies largely unready for self-rule. As a result, though the Netherlands remains economically and politically strong today, its former colonies have not fared as well.

P. ANDREW KARAM

Further Reading

Kennedy, Paul. *The Rise and Fall of the Great Powers.* New York: Random House, 1987.

Manchester, William. *A World Lit Only by Fire: Portrait of an Age.* Little, Brown, 1992.

Overview of English Exploration

Overview

Until the mid-sixteenth century Spain and Portugal were the two main European seapowers; the English had little interest in overseas exploration. Yet, by the end of the seventeenth century, England had become a powerful presence on the seas with a sphere of influence that had expanded to include settlements in North America, the West Indies, and India. While individual motives for exploration were mixed, the main impetus was economic—the search for riches. The English were not interested in discovery for its own sake, but sought the opportunities for trade that were opened up by new markets and new routes to existing markets. Accordingly, English merchants, not the British crown, were the driving force behind many of England's overseas ventures. English exploration, however, was also shaped by political considerations and was often proposed and supported under the guise of religious motives.

Background

European demand for goods from the East spurred the first voyages of discovery. Imports of silk from China, cotton cloth from India, and

"spices," which referred to dyes and perfumes as well as condiments such as pepper, cloves, nutmeg, mace, cinnamon, and ginger, were highly prized. Europe had been trading with the East for these items since medieval times, but the trade had been conducted through the merchants of the Ottoman Empire. European merchants wanted to improve their profits by eliminating the middlemen and trading directly with the Orient. Portugal found a route to the Indian Ocean by sailing around Africa that enabled them to trade directly with the East; Spain's attempt to reach Asia from the West resulted instead in the their dominance of Central and South America.

English merchants and explorers sought their own sea routes to Asia via the northeast and the northwest. The first of these set sail In 1497, when John Cabot (c. 1450-c. 1500) set out to discover a Northwest Passage, similar to Christopher Columbus's quest a few years earlier. He reached Newfoundland, but believed that he had arrived in northeast Asia. (His mistake was soon corrected.) England's interest in exploration waned during the rule of Henry VIII (1491-1547), and resumed in earnest during the 1550s, thanks, ironically, to Spanish support.

Queen Elizabeth knights English explorer Francis Drake. *(Baldwin H. Ward & Kathryn C. Ward/Corbis. Reproduced with permission.)*

Philip II of Spain (1527-1598), husband of England's Queen Mary I (1516-1558), arranged for Stephen Borough (1525-1584) to be trained in Atlantic navigation at the Spanish maritime academy at Seville, and he taught his newly acquired skills to other English sailors.

Because English exploration focused on the north, they contributed greatly to Europe's emerging knowledge of world geography. Although they didn't reach the Orient, English westward forays established trade interests and settlements in the West Indies and along the east coast of North America in the early seventeenth century. English merchants remained interested in Asia, as well. In 1600 the English East India Company was formed as Portuguese dominance of Asian trade began to decline. After the Dutch won the struggle for the East Indies and their spices, the English shifted their focus to China and India. Their presence on the subcontinent allowed them to increase their presence in India when the ruling Mughal Empire began to collapse in 1707.

Profit was not the only motive for exploration. Religious goals—particularly the desire to convert indigenous peoples to Christianity—often prompted those who planned or advocated voyages. English explorers and adventurers, however, were generally more interested in trading with the people they encountered than in converting them. Following England's religious reformation in the sixteenth century, the desire to export Protestantism through overseas exploration was mainly a consequence of rivalry with Catholic Spain and Portugal.

Still another motive for exploration and expansion was an emerging sense of national pride and interest. In particular, Francis Drake's (c. 1540-1596) circumnavigation of the globe (1577-1580) fueled English confidence in the quest for mastery of the seas both to the East and the West. Influential individuals, notably John Dee (1527-1609) and Richard Hakluyt (c. 1552-1616), began to envision a vast sea-based empire as the nation's destiny. The defeat of the Spanish Armada in 1588 further reinforced England's sense of national pride in seafaring.

English voyages of exploration were strongly influenced by the crown's diplomatic policies toward other European powers, and those policies increasingly recognized the importance of trade. The prevailing economic philosophy of the era, called mercantilism, encouraged this. According to this doctrine, the world's store of wealth (such as precious metals) was finite and measurable; the expansion of one nation's trade volume was thought invariably to diminish that of other nations. Therefore, power and trade were inextricably linked, with nations jealously guarding their own trade routes and bases while trying to encroach upon or diminish those of others.

This was particularly evident in England's Atlantic ventures. When the crown wanted to appease Spain, exploration through Spanish territory was curtailed. After relations with Spain deteriorated, however, territorial claims were ignored: English buccaneers, such as Francis Drake, preyed on Spanish ships and seized their cargo in a literal trade war. These raiders also helped to pave the way for English colonization of the West Indies in the 1620s by undermining Spanish control of the region.

Impact

English overseas ventures had a significant economic effect. During this period, merchants organized and financed voyages. The crown granted them licenses to explore and trade, and benefited by taxing the profits. By the latter half of the sixteenth century, however, voyages became too complex for one individual or even a small group to finance. To obtain the necessary resources a new type of organization emerged: the joint-stock company, which allowed many investors to pool their resources. The first of these ventures was formed in 1553, when a group of merchants funded an expedition to search for a Northeast Passage to China. Although one of the group's two ships was lost, the other managed to reach Russia, and set up trade with Moscow. Two years later the group formed the Muscovy Company, and was given sole rights to trade with Russia.

At the beginning of this period England's manufacture and export of woolen cloth to Europe dominated the economy, and foreign merchants controlled much of England's trade. By the end of the seventeenth century, however, England increasingly exploited the new resources made available by exploration, particularly tobacco and sugar.

First cultivated in the Caribbean in the sixteenth century, tobacco farming in Virginia began in 1612. The crop was produced and exported back to England in such quantities that by the mid-seventeenth century it became significantly cheaper. What had formerly been expensive indulgence of the wealthy became a widespread habit.

The development of sugar production followed a similar route. Following its introduction to Barbados around 1640, sugar grown on British plantations in the West Indies quickly became the dominant crop. British colonials, many of them loyalists fleeing the civil war in England, bought large tracts of land and established huge sugar farms. So many workers were needed to man the growing plantations that a slave society was soon in place, vastly outnumbering the whites who owned and worked them. This transition to a slave- and sugar-based economy is known as the *sugar revolution.*

English merchants, investors, and colonists reaped the benefits of England's tobacco and sugar trade, importing them from North America and the West Indies, and selling them to the rest of Europe. The government benefited from customs duties on this trade, and the overseas settlements themselves were a growing market for goods produced in England.

Exploration and subsequent colonization also enabled religious dissidents to emigrate and establish settlements where they could live and worship according to their beliefs. New England was settled by Protestants in the early seventeenth century, and Maryland welcomed many persecuted Catholics and other Christians after its charter was granted in 1632. While emigration for religious freedom was not a new concept, previous dissidents had gone to other parts of Europe. By establishing themselves in North America they were able to retain their English culture while achieving a measure of self-governance. This relative independence made England's colonies unique; other European powers preferred to retain much more direct control over their colonies.

English colonization and the introduction of new crops took place alongside a wider process now termed the *Columbian exchange,* the exchange of plants, animals, microbes, and people between Europe and the Americas. The process transformed the diets, economies, and cultures of both continents. One especially devastating effect of this exchange, unfortunately, was the ravaging of America's indigenous populations by new diseases, particularly smallpox.

While England's overseas ventures need to be seen in the context of European discovery as a whole, its specific contribution was in northern exploration, part of a commercial enterprise to reach the wealthy markets of the Orient. Although these voyages failed to discover a passage to the East, the English focused instead on new opportunities for trade and colonization in the Americas. In addition, their unsuccessful bid to gain control of the spice trade in the East in the seventeenth century resulted in their entry into India instead. Ironically, these "failures" enabled England to emerge as a major European seapower by the end of this period.

PHILIPPA TUCKER

Further Reading

Andrews, Kenneth R. *Trade, Plunder and Settlement: Maritime Enterprise and the Genesis of the British Empire 1480-1630.* Cambridge: Cambridge University Press, 1984.

Crosby, Alfred W. Jr. *The Columbian Exchange: Biological and Cultural Consequences of 1492.* Westport, CT: Greenwood Publishing Company, 1972.

James, Lawrence. *The Rise and Fall of the British Empire.* London: Little, Brown and Company, 1994.

Lloyd, T.O. *The British Empire 1558-1995.* 2nd ed. Oxford: Oxford University Press, 1996.

Scammell, G.V. *The First Imperial Age: European Overseas Expansion c.1400-1715.* London: Unwin Hyman, 1989.

The Voyages of Christopher Columbus: European Contact with the New World and the Age of Exploration

Overview

In A.D. 1000, Viking Norsemen commanded by Leif Eriksson (fl. eleventh century) landed on the shores of Newfoundland and established temporary settlements there. Four hundred years would pass before another generation of explorers, equipped with the navigational and technological innovations of the Renaissance, rediscovered the New World. Christopher Columbus (1451-1506), an Italian-born Spanish explorer, is popularly held to be the first European to cross the Atlantic Ocean and make landfall in the Americas. Historical myth asserts that Columbus discovered the New World inadvertently while attempting to find a more expedient and safe sea passage to the trading ports of Asia and that the greatest fruit of his voyage was disproving the theory that Earth is flat. Neither of these popular legends associated with Columbus is entirely accurate. The initial impetus for his voyages remains widely disputed, and even prior to his voyages to the New World, the "flat-Earth" myth was criticized not only by the leading scholars of the sciences but on a more practical level by sailors, navigators, and astronomers.

Background

Christopher Columbus began his career as a mariner in the Portuguese merchant fleet. First employed as a chartmaker, Columbus quickly climbed through the ranks and became an agent for a mercantile and luxury goods firm in Genoa, Italy. Between 1477 and 1485, Columbus's trade voyages ranged from Iceland to the Gold Coast of equatorial West Africa. During these years, he learned the business of trade, studied the Atlantic wind systems, and gained a reputation as a master navigator. His voyages, however, rarely strayed from well-known coastlines, and Columbus himself is rumored to have bragged about his scant use of navigational instruments and his reliance on intuition.

Columbus moved his residence to Spain in 1486 and began to lobby the Spanish monarchs for a commission. His stated objectives were to find more expedient trade routes to the East and carry the banners of both Spain and Christianity, but he acknowledged early on the possibility of finding the long-legendary antipodal continent *terra australis incognito.* After two failed attempts to gain the support of Spanish monarchs King Ferdinand and Queen Isabella, Columbus was finally granted patronage in 1492 and given a commission of three ships with which to pursue his quest for transatlantic trade routes.

The first voyage commenced at Palos, Spain, where Columbus's three ships, the *Niña, Pinta,* and *Santa María,* were fitted. Contrary to legend, Queen Isabella did not sell her jewels to fund the expedition; in fact, Columbus himself put up a third of the venture's cost. Columbus left Spain on August 3, 1492, and sailed south to catch the northeast tradewinds, with which he had become familiar on previous merchant voyages to the Canary Islands. On October 12 land was sighted from the deck of the *Pinta.* (The locations of this original sighting and Columbus's subsequent first landfall remain uncertain.) The fleet pressed on and within a fortnight landed in Cuba, which Columbus convinced himself was the mainland of Cathay (China) despite the notable absence of great cities described by earlier travelers to the East.

Setting sail again, Columbus decided to turn south, thereby missing the North American

Christopher Columbus arrives in the New World. *(Corbis Corporation. Reproduced with permission.)*

mainland by the narrowest of margins. He landed next in present-day Haiti, naming the island *La Isla Española,* (Hispaniola) and claiming it for Spain. There Columbus plundered enough gold and silver from the indigenous Taino people (whom he called Indians) to save both his repu-

tation and his commission upon returning to Europe. When the *Santa María,* ran aground in December, Columbus used its salvaged wood and provisions to construct a crude fort, which he named La Navidad. To secure Spain's claim to the island, Columbus garrisoned the fort with

39 men, who were instructed to hold it until his return. Because the Taino seemed friendly and regarded the Europeans as gods, Columbus was sure there would be no problem. "[T]hey are the most timorous creatures there are in the world," he wrote, and the sailors should be in no danger "if they know how to behave themselves."

With Columbus in the *Niña*, the remaining two ships began the voyage home. They rode the westerlies to the Azores, but were then caught in a storm and separated. Columbus was forced to land in Portugal, and made the rest of the journey back to Spanish court over land, bringing with him the somewhat meager spoils of his journey. The *Pinta* arrived in Spain only hours after Columbus.

Despite his limited success, the crown was sufficiently impressed with Columbus to extend his commission and outfit him for successive voyages. His next fleet, comprised of 17 ships and as many as 1,500 personnel, left Cádiz on September 25, 1493. They made landfall in Dominica in the Lesser Antilles in November 1493. He expertly directed the fleet to return to Hispaniola, demonstrating his prowess as a navigator. When the men went ashore at La Navidad, however, they found that the fort had been destroyed and the men killed. Despite Columbus's warnings, they had not "behaved themselves," and in retaliation for their abuse and cruelty had been slaughtered by the Taino.

Realizing that the native population was now hostile to the European presence and strengthening in their defiance, he exacted a harsh revenge on the Taino for the Navidad massacre, taking many captives. He then launched a ruthless campaign of conquest for the entire island of Hispaniola, established a brutal governorship, and built several more forts on the island.

Determined to make this voyage more visibly successful than the last, Columbus sent 12 of his ships back to Spain, conveying small samples of the riches of Hispaniola and some captured Taino (most of whom did not survive the voyage) to the king and queen. The ships also brought news of the massacre, along with grumblings about what could most charitably be called Columbus's "management style."

Leaving his brothers Bartholomeo and Giacomo in charge, Columbus went back to Spain in 1496, and immediately urged the Spanish monarchs to fund another voyage to the New World. His request was granted, and Columbus once again set out as an explorer. He ventured south again, this time with six ships, landing on the Island of Trinidad and the Coast of Venezuela—his only contact with the continental New World. Landing on the Paria Peninsula, he claimed the land for Spain, then sent some of his men to investigate the northern branches of the Orinoco River. Columbus noted that the great influx of freshwater into the gulf signaled that he had indeed landed upon an uncharted continent. However, he found neither a passage

WAS THE WORLD EVER FLAT?

Modern legend has it that Christopher Columbus risked life and limb, sailing off into the unknown in defiance of the day's conventional wisdom that held the world was flat. According to this story, Columbus was nearly alone in believing that if he sailed west he would find Asia—not sail off the edge of the world. In fact, this story is far from the truth because, when Columbus sailed, people had understood for centuries that the world was round. Among the first to suggest that Earth was round was the Pythagoran school in ancient Greece, sometime around 500 B.C. The Pythagorans made several observations, including the fact that Earth's shadow on the Moon during a lunar eclipse is round, not straight. They also noted that when a ship sails out of sight, the hull disappears first followed by the sails, instead of the ship simply growing increasingly smaller. By about 240 B.C., Eratosthenes not only accepted the roundness of Earth, but calculated its diameter at about 28,500 miles (45,866 km), not far from what we now know to be accurate. When Columbus sailed, the true debate was not about the shape of Earth but about its size. Thinking Earth to be only about 17,000 miles (27,359 km) around, Columbus calculated he could travel across the Atlantic, reaching Asia in only a few months or less. He saw land at about the right time. What he didn't realize is that a new continent and another ocean still lay between him and the Orient.

P. ANDREW KARAM

to India nor gold, both of which he had expected at latitudes that far south. Columbus returned to Hispaniola, only to find the colony in dire straits.

His brothers' rule had by now become intolerable, especially to the Taino populace, who were rapidly being enslaved. Even the island's Spanish settlers were bristling with hostility for the pair. Their discontent eventually burgeoned into open rebellion and pleas to the Spanish court for intercession. When Columbus arrived, he attempted to restore order with his usual harsh tactics, including hanging. Soon, however, the Spanish chief justice arrived, and the results of his investigation did not flatter the Columbus family. The brothers were shackled and shipped unceremoniously back to Spain, where Columbus was stripped of his governorship of Hispaniola in 1499. He was permitted, however, to keep his title "Admiral of the Sea" as well as the privileges bestowed upon him after the first voyages. Columbus was given a token commission of four ships and barred from returning to Hispaniola. Undeterred, and suffering from a variety of ailments (some of which may have been psychological), he again sailed for the Caribbean in 1502.

Despite the royal edict keeping him from the island, he demanded entrance to Hispaniola, but was refused by the governor. He then turned his attentions to a transcaribbean crossing, a difficult task that enabled him to chart the region as a whole. Columbus then probed the eastern Panamanian coastline for a passage to India. Disappointed and riddled with hardships, Columbus turned his fleet, which by now consisted of only two ships, back to Hispaniola. Disregarding Columbus's advice, the navigator plotted the wrong course, beaching the ships and stranding Columbus and his crewmen in Jamaica for a year. After their rescue in June of 1504, Columbus returned to Spain. He died in 1506, before he could make another voyage. His remains were eventually interred in the Cathedral of Santo Domingo, Hispaniola.

Impact

The impact of Columbus's travels seems self-evident. His landing in America ushered in the Age of Exploration, sparking a frenzy of European exploration and colonization. As more explorers took to the sea, improvements in sailing vessels, navigation, cartography, and geography rapidly followed. Permanent settlements in the New World revolutionized trade—and the European economy. The quest for gold gave way to agriculture and the cultivation of luxury goods such as cocoa, coffee, corn, cotton, tobacco, and sugar. The rapid and relentless expansion of these markets also expanded the African slave trade. In Europe, the procurement, import, and export of trade goods and slaves spawned the rise of merchant companies, stock ventures, and banking. The ramifications of Columbus's venture are manifold, and almost impossible to evaluate fully.

Contrary to popular belief, Columbus's voyages did not debunk the notion of a flat Earth. Long before Columbus, mathematicians and cosmologists of ancient Greece had proposed a spherical Earth. In the European Middle Ages, the epistemologies of St. Isidore of Seville also suggested a spheroid model of Earth. Moorish mathematicians and astronomers confirmed the work of Isidore until the resurgence of classical scholarship in the thirteenth and fourteenth centuries revived the ancient idea. Columbus's reasoning, which led him to believe that a transatlantic passage to India existed, was based upon the knowledge that the surface of Earth is curved. He suspected, based on the travels of Marco Polo (c. 1254-1324), that the lands of the East were vast enough to wrap around a significant portion of the globe. Columbus's earliest calculations put the eastern coastline of these lands 1,500 miles (2,414 km) off the coast of the Azores—a distance shorter than that from the Azores to the present-day Virgin Islands.

The ultimate legacy of the voyages of Columbus—the rediscovery of the New World—was the product of a series of miscalculations. Even without knowing which lands were across the Atlantic, Columbus grievously miscalculated the distance to Cathay and the span of the Atlantic Ocean. According to his calculations and early charts, the islands upon which he landed exactly matched his projections for the locations of Cipango (Japan) and Cathay. This miscalculation in distance may have been initially willful on his part in order to gain the support of the crown and his crewman. Afterward, however, he refused to alter his assertion that he had reached Cipango and Cathay. Despite overwhelming evidence that he had discovered new lands, Columbus's steadfast public denial of that possibility was perhaps his greatest miscalculation. The voyages of other explorers, and evidence in Columbus's personal writings, established firmly that Columbus had indeed made contact with the New World.

Researching the material remains of Columbus's voyage is difficult; few direct remnants of Columbus's voyages exist beyond his personal logs and scant contemporary accounts. Archaeological remains are equally scarce. Though native sites contemporary to Columbus's first voyage have been excavated in Haiti, his original landing site has yet to be located. Evidence suggests that another site that founded by Columbus, Concepción de la Vega, might be the present town of La Vega Vieja, Dominican Republic. There is, however, some proof of Columbus's fourth voyage and his landing in Jamaica. Excavations at Sevilla la Nueva have even yielded evidence of his beached caravels. As excavations continue in the Caribbean, the material record of Columbus's travels will perhaps supply new insight into existing historical accounts.

Five hundred years after Columbus's discovery (or rediscovery) of America, there is still great debate over his ultimate legacy to the New World. Recent work by archaeologists and anthropologists and heightened political and social regard for the roles of Native Americans and Africans Americans in the shaping of the history of the New World has altered the portrayal of the European explorer as hero. Once focused largely upon the progress of European conquest and colonization, scholarship surrounding the arrival of Europeans in the New World now addresses not only the plight of the European explorers and settlers, but also their effect upon indigenous peoples and landscape. Many historians and archaeologists prefer the terms "encounter" and "European contact" to "discovery," recognizing that Columbus and his contemporaries interacted with native cultures that long predated their arrival.

Columbus's voyages opened the New World to colonization and trade—and disease. The study of trade between the New World and the Old also encompasses its human impact. In their surveys of contact-era sites, archaeologists now often study of the evidence of disease and trauma on human remains, a technique called *pale-*

opathological analysis, to assess more accurately the effect of European diseases upon indigenous populations. There is also a growing interest in the transport of New World diseases to Europe, and the introduction of African diseases through the slave trade.

There exists a great temptation to ascribe the faults of European contact with the New World to Columbus himself. However, Columbus was a product of his time, whose brutality and religious zealotry was most likely garnered from the political and social climate that surrounded him. The ongoing war against the Spanish Moors and their defeat in 1492, the relentless persecution of Spanish Jews (whose confiscated estates almost certainly helped fund Columbus's voyages), and the turmoil that surrounded the unification of Aragon, Castile, and Leon all surely influenced Columbus. In the debate surrounding the legacy of his voyages, history must recognize the full scope of the impact of European contact with the New World, as well as the personal accomplishments of Columbus as sailor and brilliant navigator.

ADRIENNE WILMOTH LERNER

Further Reading

Bedini, S.A., and David Buisseret. *Christopher Columbus and the Age of Exploration: An Encyclopedia*. Da Capo Press, 1998.

Paiewonsky, Michael. *Conquest of Eden (1493-1515)*. Rome: Mapes Monde, 1990.

Wright, Ronald. *Stolen Continents: The "New World" through Indian Eyes*. Boston: Houghton Mifflin, 1993.

Juan Ponce de León Explores Florida and the Bahama Channel

Overview

Unknown to the indigenous people of the New World, their destiny was being determined by political and economic forces taking place across the Atlantic Ocean in Europe. Toward the end of the fifteenth century, thousands of daring adventurers would be crossing the ocean to conquer within a few centuries what had taken the Indians thousands to years to inhabit. This "Age of Exploration" was fostered by technological advancements in maritime practices, the belief in an eco-

nomic philosophy called mercantilism, and an interest in converting the religious beliefs of native populations. Mercantilism was the idea that if a nation was not self-sufficient in its affairs, then its neighbors would dominate it. The two areas that seemed ripe for establishing this ideal were the Middle East and the Americas. Many of the Spanish conquistadors headed for the New World seeking wealth and adventure. One such man was Don Juan Ponce de León (1460?-1521), commonly referred to as simply Ponce de León.

Background

Ponce de León was a Spanish conqueror and explorer. He was born in Spain around 1460. He is well known for claiming and naming what is now Florida, being the first European to discover Mexico, conquering and governing Puerto Rico, and searching endlessly for the mythical Fountain of Youth. While there are some authorities who dispute the claim that he was indeed searching for the Fountain of Youth, Ponce de León's name has been associated with this endeavor more often than with anything else.

While details involving Ponce de León's family background are sketchy, it is believed that he was born into a noble family. He was an experienced soldier, having fought against the Moors; he later traveled to the New World in 1493 as part of Christopher Columbus's (1451-1506) second voyage. In 1502, while in the West Indies serving as a captain under the governor of Hispaniola, Ponce de León suppressed an Indian uprising and was rewarded by being named the provincial governor of the eastern part of Hispaniola. However, he was dissatisfied with political life and looked for further conquests in Puerto Rico. After exploring and settling that island, he was named governor but was displaced by the political maneuverings of his rivals. Though Ponce de León needed little encouragement, the Spanish crown implored him to seek out new lands and opportunities, which led to his exploration of Florida.

As legend has it, Ponce de León learned of a miraculous spring that could rejuvenate those who drank from it. While the Indian who told him about it had never seen it, he indicated that a number of his comrades had left to seek it and had never returned. The Indian reasoned that they must have found the Fountain of Youth. Ponce de León was quite interested in finding this place, so he led a privately outfitted expedition from Puerto Rico in March of 1513. In April of that year, after investigating various islands, he landed on the coast of Florida near the site of modern-day Daytona Beach. He claimed the land for his king. Ponce de León initially assumed that he had landed on an island, not a large continent. When he first sighted land it was during the Easter season known as *pascua florida*. Because of the flowers that he saw and in the spirit of the season, he named the newly discovered land *la florida*. He mapped a part of the Florida coast, but never ventured to the interior because he was under constant attack from Indians. Ponce de León was never even given a chance to find his Fountain of Youth. He eventually returned to Spain where he secured the title of governor of Florida with permission to colonize the area.

Indian insurrections prevented Ponce de León from returning to Florida until 1521, when he attempted to establish a colony there. Upon landing, he was struck by a Seminole arrow during an Indian attack, and the colonists were repelled. He was rushed back to Cuba in order to seek medical help, but died soon after his arrival. It took many years and countless numbers of lives before Europeans were able to colonize the area.

Impact

Ponce de León is credited as being the first European to discover both Mexico and the United States. Specifically, he named Florida and took possession of it in the name of Spain. However, there is ample evidence that Europeans had previously been to Florida on slave-trading missions. Because the people enslaved on Hispanola and other islands were dying due to disease and inhumane treatment, expeditions were formed to gather replacements. It is believed that some of these made it to the Florida coast. This would, at least in part, explain why the native population in this area was so aggressive. They had experienced previous interactions with Europeans that result in disaster, so they vehemently defended themselves.

Expeditions similar to those conducted by Ponce de León in Florida served to motivate thousands of Spanish peasants to join the military. The discovery of riches and wealth enticed these peasants to travel to the New World in search of a new life. A successful colonial mission could possibly lead to a governorship or a pension for the participants. If one were particularly lucky, he could procure untold riches. Other men were drawn to the New World by promises of adventure. They looked for quick advancement in the military and diplomatic careers. Still others came on a mission of God. These men wanted to convert the native population to Catholicism. By converting the Americas to God, they believed they would receive eternal blessings.

The discovery of Florida did not initially prove to be a huge downfall for the natives. They fought well and resisted early efforts to colonize the land. Five years after Ponce de León's ill-fated attempt in 1521, Spanish explorer Lucas Vázquez de Ayllon (1475?-1526) sought to es-

Spanish explorers attack native inhabitants of Florida. *(Corbis Corporation. Reproduced with permission.)*

tablish a colony in Florida. In addition to 600 colonists, he brought a contingent of African slaves with him. This is the first record of slaves being used in the United States. The settlement lasted for less than two months when an uprising of the slaves killed the majority of the population and just 150 survivors made it safely away. The next conquistador to test himself and his men in Florida was Pánfilo de Narváez (1480?-1528), who landed near Tampa Bay with 300 men and 40 horses in 1528. His expedition has become famous because it was chronicled by one of the five surviving members, Álvar Núñez Cabeza de Vaca, in what is regarded as one of the greatest stories of survival ever written. Cabeza de Vaca's descriptions are the first surviving documents from a European regarding the interior of Florida. According to Cabeza de Vaca,

the expedition was first attacked by Indians, then the Spanish missed a connection with their ships. Building rafts in an attempt to sail to Mexico, they were beset by a hurricane, which killed their leader; only 80 men made it safely to the Texan coastline. The death rate continued to climb until 1536, when the remaining five of the expedition arrived safely in Mexico, more than eight years after they had landed in Florida. After many other failed attempts at colonization, it was reported that Florida would be too difficult to colonize, and there was nothing of value to be had. Furthermore, there should be no fear that any other country would try to colonize it because of the previously stated conditions. This stood as the official Spanish position until the French attempted to establish a settlement in Florida.

Eventually, modern weaponry and unfamiliar disease overwhelmed the Native Americans, and like most other indigenous populations, they were overrun by the Europeans. Ponce de León had opened the door for explorers like Spaniard Hernando de Soto (1500?-1542), who marched throughout the southeastern portion of the United States looking for treasure and exploring the countryside. The most significant result of de Soto's march was the devastation of several native populations. Many native warriors were severely injured or killed following confrontations with the Spanish, and entire villages were wiped out, though not as the result of warfare, but from the introduction of European diseases against which the Indians had no natural immunity. These included such diseases as smallpox, measles, and the flu.

Ponce de León also popularized the use of ferocious dogs as warriors against native populations. These fierce dogs would terrorize the natives, as they were not accustomed to such attacks. His most famous dog was one that he owned personally, named Berezillo. His dog was so valued and renowned throughout the Caribbean that Ponce de León even awarded him soldier's pay.

Another important discovery associated with Florida and Ponce de León is that he was the first to describe the Gulf Stream (the world's strongest ocean current). While he was trying to sail south with the prevailing wind, his vessel was in a current so strong that he was actually going backwards. He was able to extricate himself from the current and found a countercurrent running south closer to the coast. The Gulf Stream is part of a general clockwise-rotating system of currents in the North Atlantic. It is fed by the westward-flowing North Equatorial Current moving from North Africa to the West Indies. In the region Ponce de León discovered, it flows roughly parallel to the eastern coast of the United States in a northerly direction. This current made Florida a valuable asset because the Gulf Stream could be used to help propel ships from North America to Europe.

JAMES J. HOFFMANN

Further Reading

Berger, Josef. *Discoverers of the New World.* New York: American Heritage Publishing Co., Inc., 1960.

Faber, Harold. *The Discoverers of America.* New York: Macmillan Publishing Company, 1992.

Quinn, David. *North America: From Earliest Discovery to First Settlements.* New York: Harper Row, 1977.

Alonso Alvarez de Piñeda explores the Gulf of Mexico and Is the First European to See the Mississippi River

Overview

The Gulf of Mexico was the first real entry point to the North American mainland, but by the time Christopher Columbus (c. 1451-1506) came to America in 1492, it was still unexplored. The sixteenth century, however, saw a rapid increase in Spanish exploration of the Americas, with Vasco Núñez de Balboa (1475-1519), Juan Ponce de León (c. 1460-1521), and Hernán Cortés (1485-1547) headlining the major conquests of Cuba, Mexico, and South America. But it was a lesser known and less-fabled explorer, Alonso Alvarez de Piñeda (d. 1520), who first sailed the entire Gulf of Mexico coastline, spotting the Mississippi River and confirming that Florida was not an island, as was previously believed. Piñeda's observations contributed to the exploration of the Mississippi later by Hernando de Soto (c. 1496-1542), opening North America to its era of European discovery.

Background

In 1492 Christopher Columbus landed on the northeastern shore of Cuba, and unable to explore the Gulf side of the island successfully, claimed that Cuba was actually a peninsula, and that no body of water—the Gulf of Mexico—existed. In fact, all of the great bodies of water were thought by the Spanish to be one sea, called "el mar oceana," or "the Ocean Sea."

Columbus, like many of the explorers of the early sixteenth century, was looking for gold, habitable islands, and strategic trading ports for Spain, who had "split" the world with Portugal to explore and exploit. The Spanish were prolific: Juan Ponce de León, in 1508, landed in Puerto Rico, and eventually became governor of the island. By 1510, Diego Velásquez de Cuéllar (1465-1524) had invaded Cuba, planning to conquer the island, and engaged in bloody combat with the Arawak tribe, eventually defeating them and taking control the island. In the mean time, Ponce de León explored the coast of Florida, supposedly in search of the "Fountain of Youth" among other objectives, and declared the peninsula an island. Ponce de León theorized that a channel of water ran across the state, and it would be the impetus for Alvarez de Piñeda's explorations a decade later.

Hernán Cortés, a violent, impulsive man, set out from Cuba to explore the mainland of Mexico in order to confirm reports of the existence of large, native civilizations in the interior. His plan, however, was to conquer these tribes and search for gold, and he eventually would terrorize the Aztecs throughout Mexico. While planning his strategy, he heard about a plan to send four ships from Jamaica to explore the unknown northern coast. The man responsible for the ships was Governor Francisco de Garay of Jamaica, a Spanish business man of sorts who had lucked into his appointed position in Jamaica after the previous governor had died.

Garay, who had sailed with Columbus on his journey to the West Indies, had been in and out of debt before he left for Jamaica and intended to employ the conquered native population there to develop sugar cane and cotton and build the island's economy. Much to his disappointment, most of the natives had left, and his agriculture plans failed. Having used a good portion of his money, and facing debt again, he finally got permission to obtain four ships for exploration of the lands north of him. He was to search between Ponce de León's discoveries and those of Velásquez for the purported channel across Florida that connected what was the Gulf of Mexico to the "South Sea," or the Atlantic Ocean.

While Garay was organizing his expedition, he elected Alonso Alvarez de Piñeda to lead the ships along the coast. Alvarez left Jamaica in March of 1519. By now, Cortés had begun his attacks on the Aztecs, and was thick into the interior of Mexico. Alvarez left through the Yucatan channel and sailed north until he spotted what is now western Florida. Assuming it was an island, as Ponce de León had claimed, he

traveled to its tip, never encountering the alleged channel that would send him across to the South Sea. He may not have found the passage, but Alvarez de Piñeda discovered that Florida was indeed a peninsula.

The ships turned back to the west, following along the coast, recording observations of Alabama's coastal islands and the exit of the Mississippi River, which purged a wide, strong current of muddy water into the Gulf. All along the coast Alvarez noted small native villages, fertile land, ports, and rivers. By observing the jewels worn by the Native Americans along his route, he determined that the rivers held "fine gold." He also noted that the people he encountered were friendly, and would therefore be easily converted to the Catholic faith, a habit of the Spanish conquistadors. Alvarez de Piñeda also claimed to have seen giants and dwarfs among the Indians he met.

At this point, after sailing along Texas and claiming it for Spain, Alvarez encountered Cortés in the beginnings of his conquest of the Aztec Empire. He sailed up the Veracruz coast and the Rio Panuco about 20 miles (32 km). Here the ships stayed for 40 days, where the crew resupplied and fortified the vessels. Alvarez then returned to Jamaica and presented Garay with his rough sketch of the Gulf of Mexico, the first of its kind that didn't speculate on the land forms and bodies of water en route, but actually confirmed them. Alvarez had proven Florida's geography and discovered the greatest river in America.

Impact

The officials representing King Charles V of Spain received the map, and claimed that Garay, Velásquez, and Ponce de León had collectively solved the geographic questions of the Gulf of Mexico. Garay was given permission to colonize the land that Alvarez de Piñeda had observed, which was the Texas gulf coast, and call it "Amichel." He sent three ships back, with 240 soldiers, horses, and musketeers with Alvarez de Piñeda as captain. Although the Huasteca Indians were friendly on Alvarez's first journey, they turned violent when he landed with his men. Although no records have proven the crew's fate, it is believed that the Huasteca killed all of the soldiers, horses, and Alvarez de Piñeda.

Garay, in Jamaica, had no idea of the massacre by the Huasteca. In fact, he dispatched two additional ships with more supplies for the new colonizers. The captain of this effort, Miguel

Diaz de Aux, arrived on the coast but found no indication that Piñeda had settled. While sailing down the coast, still searching, his ships were caught in a squall, and he brought them ashore, only to have his men join Cortés's ongoing conquest of the Aztecs.

It wasn't until 1539, when Hernando de Soto sailed back to Florida, that the Mississippi River was more thoroughly explored. In 1541 he arrived at the river south of what is now Memphis, Tennessee. While de Soto had hoped to conquer the Indians and find gold in their river, he and his crew were repeatedly attacked by tribes along the shore. In the end it was the Mississippi floods that beat down the Spanish, and de Soto, who was buried there, never established his empire. Nevertheless, the initial discoveries by Alvarez along the Gulf Coast were the first steps in the long and turbulent history of the European colonization of the Americas.

LOLLY MERRELL

Further Reading

Weddle, Robert S. *Spanish Sea*. College Station, TX: Texas A&M University Press, 1985.

Hernando de Soto and the Spanish Exploration of the American Southeast, 1539-1542

Overview

By the end of the first third of the sixteenth century, Spanish conquistadors and explorers had already claimed substantial lands in the New World. These ventures had yielded the "discovery" of new fruits, exotic spices, and whole civilizations. In Spain both the Crown and some individuals had already begun to profit from plundering gold and luxury trade items from newly claimed lands. However, vast tracts of land claimed under the banner of Spain had yet to be fully explored. One such region was Spanish Florida and the American Southeast. Both tactical advantage, namely the conquest of more territory than rival European nations, and the widely spun legend of "cities of gold" pushed Spain to invest in the exploration of its claims in this region. Following the initial voyage of Juan Ponce de León (1460-1521), young, veteran explorer Hernando de Soto (1496-1542) was chosen to return to Florida and solidify Spain's claim and expand the territory. De Soto had accompanied Francisco Pizarro (c. 1475-1541) on earlier voyages to South America and had grown rich from trade with—and exploitation of—the Inca. Hoping to gain the same wealth and renown from his venture to North America, de Soto embarked on an ambitious sea and land venture. The resulting expedition was one of the most devastating episodes in the history of European contact with the New World.

Background

In 1537 de Soto appealed to the King of Spain to be granted control of the New World territorial province that stretched from Rio de Las Palmas in South America to Florida. De Soto won his claim and was also granted the governorship of Cuba. However, his appointment stipulated that, within a year, he had to personally re-conquer and occupy Spanish Florida at his own expense. Previous ventures to South America with Pizarro had earned de Soto tremendous wealth and prestige; as a result, he found several willing financial partners for the venture, some of who accompanied de Soto on the actual voyage. He assembled and armada of 10 ships and 600 men. In April of 1538 his fleet departed from the port of San Lucar, Spain, for the shores of the New World. He landed in Cuba, remaining on the island for a few months to gather supplies, rest his men, and plan his expedition in Florida.

De Soto landed in Florida in May of 1539 and claimed formal possession of the land on June 3 despite ongoing hostility between his men and some of the neighboring Indian tribes. Welcomed by one local Native American chief, de Soto and his crew wintered in the village of Apalache before beginning their expedition. De Soto supposed that great indigenous civilizations, like those he encountered on voyages to South America, lay in the region's interior. Determined to garner further plunder for both his own interests and for the Spanish court, de Soto and his men headed northward through present-day Georgia. Once reaching the Piedmont, or the Appalachian foothills, de Soto turned his forces westward, exploring the Carolinas and Tennessee. Though he located the Tennessee River, de Soto had failed to find the material wealth and plunder after which he sought.

Disappointed and weary, in 1540 de Soto attempted to head south to Mobile Bay in Alabama to rendezvous with his ships. Two hundred miles (322 km) south of the Tennessee River, de Soto and his men encountered a warrior band led by Chief Tuscaloosa. The Native American forces were ill equipped to fight the Spaniards, and the ensuring battle proved disastrous for Tuscaloosa's men. The clash was perhaps the bloodiest single encounter between Native Americans and whites in American history. Crippled by the encounter with Tuscaloosa and running short on supplies, de Soto continued to head south, believing that he would not meet with further resistance. A few miles from the headwaters of Mobile Bay, however, the indigenous peoples at Mauvilia (Mobile) confronted de Soto's men. The local Native Americans were decimated, and the Spanish forces were weakened severely. Losing most of his men, supplies, and plunder, de Soto rashly decided to extend his expedition and recoup his losses instead of immediately returning to Spain.

After regrouping with some of his fleet and resting for a month, de Soto again pushed northward—though this time the decision would prove fatal. His expedition was plagued by Indian attacks as they made their way through western Alabama and Mississippi. On May 21, 1541, de Soto became the first European to sight the Mississippi River. He encountered the river south of Memphis, Tennessee, and instead of following the river and charting its path to the Gulf of Mexico, de Soto crossed the river into Arkansas in search of more wealth. The expedition was fruitless and de Soto lost more of his already diminished crew to fatigue and disease. Resolved to finally reunite with his fleet and return to Spain, de Soto decided to turn back and follow the Mississippi River southward. De Soto fell ill, most likely with Yellow Fever, and died in Louisiana on May 21, 1542, exactly one year after first sighting the Mississippi River.

The surviving members of de Soto's crew endured perhaps the most trying part of their travels after de Soto's demise. Continuing their way southward, they were unable to return to the remnants of de Soto's fleet. They made their way to Mexico via handmade rafts and eventually caught passage back to Spain. De Soto's second in command, Luis de Moscoso, arrived at the Spanish court over a year and half after de Soto's death.

Impact

De Soto's exploration of the Southeast was monumental in scope. He covered territory from the Gulf of Mexico coast north to the Appalachian Mountains, from the Florida shores of the Atlantic to slightly west of the Mississippi River. He discovered major coastal inlets and inland waterways, such as the Tennessee and Mississippi rivers, which paved the way not only for future exploration of the American eastern interior but its eventual settlement as well.

The records that de Soto kept of his expedition through the interior of the American Southeast are renowned for their detailed descriptions of landmarks, geographic locations, and the various indigenous peoples that he and his crew encountered. The work is thought to have been entrusted to one of the expedition officers after de Soto's death. From the chronicles, the first charts of the interior of the Southeast were devised. Future expedition not only relied on the geographic information provided in the work, but also utilized information on de Soto's dealings with different Indian groups.

In the 1930s de Soto's detailed records of his travels in the Southeast became the subject of not only historical, but also scientific, study. The chronicles were studied, with careful attention paid to the distances and landmarks they described, in conjunction with old maps, other records, and reports from several known archaeological sites in order to determine de Soto's precise path thorough North America. Surveyors charted positions using old Spanish units of measure. Archaeologists attempted to locate artifact assemblages that reflected Spanish contact with local tribes. The result was the unveiling of the de Soto Trail—a detailed and mostly accurate retracing of the de Soto expedition. Active archaeological survey continues along the trail today.

As de Soto pushed his way through the Southeast in search of gold, he abducted Native guides to lead his expedition. However, the Native peoples of the Southeast did not possess the gold wealth of the highly advanced Incan civilizations de Soto had encountered on his earlier ventures in Peru. Reports from the de Soto expedition, when the surviving members finally returned to Spain, changed the nature of European involvement in America. The failure of the de Soto expedition to locate gold and other precious metals in the Southeast made evident that the value of Spanish Florida was not in plunder, but in the actual land itself. Future expeditions to Spanish Florida largely focused on the establishment of various settlements, missions, and ports of trade. Furthermore, de Soto's expedition shaped the geographical boundaries of Spanish territories in the American Southeast. Violent encounters with indigenous tribes in Alabama convinced de Soto to abandon plans to establish Mobile as the chief city of the Spanish territories in the region. Future Spanish expeditions paid little attention to the area, which was eventually claimed and settled by the French.

The de Soto expedition left a legacy of decimation and destruction. The expedition proved disastrous and costly, its redeeming and ultimate value not recognized until years later. The few surviving men from de Soto's crew, who made their way first to Mexico and then to Spain, returned to a Spanish Crown leery of their accomplishments in the New World and angry about the loss of money and human lives. In the New World the inadvertent consequence of exploration was the introduction of European disease that swept through Native populations that were not able to fend off foreign contagions. The grand sweep of de Soto's venture, as well as travel among the Indians themselves, drastically increased the number of people who were exposed to bubonic plague, smallpox, and various fevers. The onset of foreign diseases aided in the fragmentation of large Indian towns as people fled to escape illness, and in several decades, the great mound-building chiefdoms of the American Southeast all but vanished. In the two centuries after de Soto's travels, an estimated 90% of the Indian population that existed before European contact was decimated.

ADRIENNE WILMOTH LERNER

Further Reading

Clayton, Lawrence, Vernon James Knight, Jr., and Edward C. Moore. *The de Soto Chronicles: The Expedition of Hernando de Soto to North America, 1519-1543.* Tuscaloosa: University of Alabama Press, 1996.

Worth, John. *The Timucuan Chiefdoms of Spanish Florida.* Vols. 1-2. Gainesville: University of Florida Press, 1998.

Wright, Ronald. *Stolen Continents: The New World through Indian Eyes.* Boston: Houghton Mifflin, 1993.

Coronado's Search for the Seven Cities of Gold Leads to Spanish Dominion over Southwestern North America

Overview

The year 1542 was the great climax of the Spanish age of discovery—a year in which Spain had expeditions under way stretching halfway around the globe. Soon after the making of the Spanish empire in the New World with the discoveries of Christopher Columbus (1451-1506), the great colonial effort moved towards establishing roots on the northern and southern continents of North America. Francisco Vázquez de Coronado led the last of these expeditions in search of new lands in North America for Spain. In 1540 he led a two-year epic journey that gave him and his companions the distinction of being the first Europeans to explore California, to see the Grand Canyon, to live among Pueblo Indians, and to explore the Great Plains homeland of the Quivira Indians in central Kansas. Coronado made one of the most significant expeditions of the remarkable era of the opening of the Western Hemisphere by Europeans. Coronado's expedition gave Spain what is now known as the southwestern United States.

Background

Although Francisco Vázquez de Coronado (c. 1510-1554) is generally credited with being the first European to arrive in what is now known as the southwestern United States, other Spaniards preceded this expedition by 13 years. In 1527 a Spanish ship carrying 400 people sank off the coast of Florida (possibly Texas), and four survivors spent nine years traveling west across the continent. These men were members of a Spanish expedition dispatched to explore the inner lands of the North American continent. One of the survivors, Cabeza de Vaca (c. 1490-1556), wrote about how he and the others adapted and developed an unusual sensitivity to the native people, accepting their help and surviving for 13 years in an unknown land.

The four men eventually reached the Gulf of California and then headed south, until they arrived in the village of Culiacan in southern Mexico. They told stories of what they had seen (or heard about, or imagined)—seven huge cities whose houses were made of turquoise and gold, the fabled "Seven Cities of Cibola." The odyssey of Cabeza de Vaca and his men set in motion the rumors of "opulent countries" to the north, and four years later the mammoth expedition led by Francisco Vázquez de Coronado was authorized.

In early 1540 Coronado led some 300 soldiers, up from Compostela, New Spain. An eager and well-appointed army, if a somewhat inexperienced one, it consisted not only of Spaniards but of Portuguese, Italians, and a Frenchman, a German, a Scot, and three women. On foot in the front ranks were Fray Marcos and four Franciscan padres, and bringing up the rear were 700 "Indian allies" who went along as servants, wranglers, and herdsman of the sheep, horses, and cattle brought along for food and transport. The expedition would follow the coast of the Gulf of California northward towards the state of Sonora, entering present-day Arizona and into New Mexico.

In July 1540 Coronado and his advance party of soldiers encountered the Zuni pueblo Hawikuh, which already had experienced an encounter with the Spanish the previous year; one of the survivors of Cabeza de Vaca's expedition was killed by Zuni warriors. Coronado arrived at the Zuni pueblo with the hope that he had finally "found" one of the famed cities of gold. Arriving at the high point of Zuni summer ceremonies, the Zuni people were not receptive to Coronado's declaration of the *requirimiento*—the standard Spanish speech to native peoples, which informed them that the Catholic Church was " the ruler and superior of the whole world." Coronado warned the Zuni that if they failed to obey orders, "with the help of God we shall make war against you and take you and your wives and children and shall make slaves of them."

Coronado ordered his men to attack the pueblo. The Zuni warriors fought bravely, but they could not stop the Spaniards. They were terrified by the sound of the loud guns and the charging horses. Their arrows and spears bounced off of the Spaniards' metal armor. The better-armed and mounted Spaniards entered the pueblo and finally the Zuni retreated, leaving Coronado and his men standing in an empty village.

Following this encounter, Coronado and his men discovered no gold in the Zuni pueblos. However, they did find ample food, producing fields, and a social system that was based on sharing and working together. From their base at Zuni, in hopes of redeeming the expedition, Coronado sent out scouting parties to investigate and hopefully find the illusive gold that had sent them into new unexplored lands.

Pedro de Tovar was sent to the Hopi pueblo at Tuysayan near the Grand Canyon. Meanwhile, Garcia Lopez de Cardenas was sent out to find the great river that Tovar had heard about from the Hopi people. Cardenas retraced Tovar's route to the Hopi mesas and there acquired some Indian guides. They eventually arrived at the south rim of the Grand Canyon. At first the discovers' eyes were deceived by the scale of the canyon. The Colorado River "looked like a brook" even though the Hopi guides told them that it was very wide and swift. Cardenas sent three of his most agile men to climb down to the river. These men spent a full day inching along a ridge and got "a third of the way down" before they had to turn back.

Melchor Diaz, another of Coronado's men, had instructions to travel west into an unexplored desert, find the flotilla of another Spanish ship, and collect supplies. Diaz negotiated a route to the Colorado River, and there he found a message stating that the ship had sailed back to Mexico. He then decided to travel upstream and traversed an area of sand dunes in the Mojave Desert of present-day California.

Winter quarters were set up in Tiguex, a large pueblo along the Rio Grande River. During this tumultuous winter of a near full-scale war against the Indians, Coronado met an Indian, called the Turk, who informed him of Quivera, a city rich with silver and gold. Coronado, in an attempt to salvage the expedition, decided to look for Quivera, taking the Turk on as his guide. He traversed the Texas panhandle and marched on further north. The Turk led Coronado on a wild goose chase that ultimately led to his death. When Quivera was found it was yet another disappointment, as the Quivera Indians were not rich in gold and silver, and the village consisted mostly of thatched huts. However, as at the Zuni pueblos, there was an abundance of cultivated crops, buffalo, and a sophisticated social system. Coronado's men were the first Europeans to see the great herds of buffalo inhabiting the vast plains.

The Spaniards returned to Tiguex, where they spent another winter. In 1542 Coronado went back to Mexico, roughly following the same route he had come. Fewer than 100 of the original 300 men returned. Many of the soldiers—weary, disgruntled, and fearing punishment because of their failure to find any treasure—deserted the expedition, as did the Catholic friars, who decided to stay in New Mexico to convert the Native American tribes to Christianity.

Impact

Initially for Spain, it made no great difference that Coronado had discovered vast fertile territories and staked the claim to the entire southwestern quadrant of the North American continent. Gold and silver had not been found, and that alone condemned the journey to failure and a pointless endeavor. However, exploration was a necessary first step to the colonization, exploitation, and social development of new lands in the New World. To the geographic map Coronado added Cibola, Tusayan, Tigeux, the Llanos del Ciloba, and Quivera, regions that became known as the Southwest or the Spanish Borderlands. Historical tradition in this vast area, from Nebraska to California, traces its lineage to the expedition of Coronado.

A notable contribution of the Coronado expedition to North American geography was the discovery of the Continental Divide—the watershed between the Pacific and the Atlantic Oceans from which two river systems run in opposite directions. It was also Coronado who first acquired a relatively accurate knowledge of the immensity of the southwestern part of the North American continent. European maps at the time showed the oceans north of Mexico as close together.

However, Coronado dispelled this idea and established the idea of a vast land mass separating the Gulf of California from the Gulf of Mexico.

By a strange misunderstanding, the European mapmakers reversed the direction of Coronado's route. The location of the pueblos were reversed, and the province of Quivera was shifted to the shores of the Pacific ocean, where for several decades it roamed up and down the map. The Rio Grande, the great river that today marks the boundary between Mexico and the southwestern states, was shown as flowing west into the Pacific Ocean in northern California. Relatively unimportant though these curious mistakes of the mapmakers may have been, they make it clear that Coronado contributed more to North American geography than Europeans could easily digest.

There has been a widespread misconception that Coronado introduced the horse to the Plains Indians. In the available written records of the expedition there are few notations of the disappearance of horses. In fact, in the accounts of the Spanish expeditions in the later sixteenth and early seventeenth century, no mention is made of horses or mounted Indians. It seems likely that the horses were descended from stock that strayed or were obtained from Spanish settlements after the permanent colonization of New Mexico and Texas in the seventeenth century.

The expedition of Coronado left a rich legacy for Spain and opened up the settlement of southwestern North America by European colonizers.

LESLIE HUTCHINSON

Further Reading

Bolton, Herbert. *Coronado: Knight of Pueblos and Plains.* Albuquerque: University of New Mexico Press, 1949.

Horgan, Paul. *Conquistadors in North American History.* New York: Farrar, Straus and Company, New York, 1963.

Udall, Stewart L. *Majestic Journey: Coronado's Inland Empire.* Albuquerque: Museum of New Mexico Press, 1987.

Spanish Florida
and the Founding of St. Augustine

Overview

In search of the legendary Fountain of Youth, Juan Ponce de León (1460-1521) landed on the

shores of present-day northern Florida on Easter, March 27, 1513. He claimed the territory for his native Spain, but did not leave a lasting set-

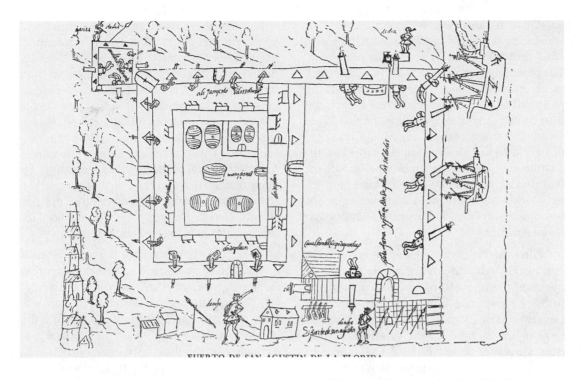

Sixteenth-century diagram of the Spanish fort at St. Augustine, established by Pedro Menéndez de Avilés. *(Bettmann/ Corbis. Reproduced with permission.)*

tlement at his point of first contact. The Spanish Crown sent six subsequent expeditions back to Florida to relocate the area of Ponce de León's landing and establish a settlement, but none were successful. Nearly 50 years passed until St. Augustine, Florida, was founded by a new generation of Spanish explorers, Christian missionaries, and European settlers. From its inception, St. Augustine was plagued by siege, Indian uprising, disease, and territorial boarder disputes. However, the small Spanish settlement, which predated the British settlement at Jamestown (1607) by 42 years, thrived under the stewardship of three nations to become the oldest continuously inhabited settlement in America.

Background

Pedro Menéndez de Avilés (1519-1574) was commissioned by King Philip II of Spain to resecure Spanish possessions near present-day Jacksonville. In July of 1565 Menéndez de Avilés led a fleet of 11 ships and 1,900 men to Florida. On August 28, the Feast of St. Augustine, he entered a bay near the delta of the St. Johns River. Upon making landfall 11 days later, the explorer rededicated the land to Spain and ordered his men to build a fort, which he named St. Augustine after the Catholic holy day. The fort was built on the site of the local indigenous village of Seloy. The existing village was hastily fortified,

but no initial plans for a more permanent fort structure were made.

The location chosen for Spanish fort and settlement was not selected by accident. The Spanish were not the first to settle the area around St. Augustine. French Protestants, known as Huguenots, established Ft. Caroline in 1564. The group, originally led by Jean Ribault, was plagued by problems of disease and supply shortages. Nonetheless, the French colony survived for over a year. Menéndez de Avilés's commission entailed ensuring that Spain's coastlines in the New World were free from interfering settlements from rival European nations—most especially France. Spain perceived the small colony as a threat to Spanish shipping interests between their colonies in the Caribbean and Europe. Control of northern Florida was thought to be vital. In 1565 soldiers under the command of Menéndez de Avilés seized control of Ft. Caroline. Menéndez de Avilés ordered most of the colony to be massacred, hanging the bodies of victims of in trees with the inscription "Not as Frenchmen, but as heretics." Incorporating the settlement that he had founded and the former French settlement on the St. Johns River, the Spanish secured their dominion over northern Florida.

Menéndez de Avilés continued to fulfill his obligations to the King of Spain by establishing a string of Spanish forts along the Northern Florida, Georgia, and South Carolina coastlines. Menéndez de Avilés was recalled to Spain in 1567, but his colony at St. Augustine thrived. The other outposts, however, did not fare as well as St. Augustine and within the span of 150 years had all been incorporated into French and then British territories. St. Augustine was left as the northernmost of these original outposts, and in 1672 with the construction of a stone fort, the Castillo de San Marcos, it became one of the most heavily armed and guarded forts in Spanish Florida, its strategic location acting as the first line of defense for Spanish territories in North America.

After the British established colonies to the north in Georgia and South Carolina, St. Augustine was the site of ongoing attempts of the British to gain land from rival Spain. In 1702 Governor James Moore of South Carolina laid siege to the city. Following closely two years later, Governor James Edward Oglethorpe of Georgia attacked St. Augustine. Neither attempt to take the city was successful.

In 1763 Spain ceded Florida to England in exchange for regaining control over the capital of Cuba. The British ruled St. Augustine for a 20-year period that coincided with the American Revolution. Florida, as a British possession, remained loyal to the British Crown. When the Revolution ended, Florida was granted back to Spain until the United States purchased the territory in 1821. The Castillo de San Marcos was renamed Ft. Marion, and St. Augustine was firmly under American control. The period after the American take-over of Florida marked a difficult time for St. Augustine. An epidemic of Yellow Fever swept the town in 1821, and a Native uprising in 1836 culminated in the Seminole War. No longer an active mission, trade hub, or large garrison town, the often-troubled economy of the city ground to a halt.

Impact

Though St. Augustine was precariously close to borders with rival colonies, the ongoing struggle among the Spanish, French, British, and later the Americans, for dominion over the area were due mainly to the town's prized location on shipping and trade routes. The Atlantic currents near St. Augustine allowed ships that disembarked from the town to make more expedient voyages to Europe than those that left similar New World ports just 100 miles (161 km) away. Control of the town granted the owner dominion over the Atlantic access waters to this trade

route—and the possibility for substantial economic prosperity. In 1586 English buccaneer and mariner Sir Frances Drake (c. 1540-1596) landed in St. Augustine and burned the town in an attempt to gain control of the region. Despite the immense damage caused to the town, Drake was ultimately unsuccessful. In 1668 pirate captain John Davis attacked the city, killing 64 inhabitants. These periodic raids by privateers were detrimental to the economic growth of St. Augustine. Davis's raid not only encouraged the Spanish to reroute some trade to other ports in Spanish Florida, but also pointed to the need for a more stalwart defense system for the city.

The massive stone fort that the Spanish constructed to guard St. Augustine, the Castillo de San Marcos, was itself an engineering marvel. Though the pointed design of the fort was fairly standard, the Spanish could not find the building materials with which they were accustomed to using in the construction of such massive defense works. Quarrying hard stones and moving them to St. Augustine proved to be impractical. Thus, the builders of the Castillo de San Marcos utilized coquina (a compact, densely packed, concrete-like material of shell and hardened sediment), which could be locally quarried on nearby Anastasia Island and ferried across to St. Augustine. The material was solid enough to be used like stone, but it had a unique ability to "swallow" enemy cannon fire with little or no damage to the integrity of the fort's walls. This use of indigenous coquina, a strategic location near the confluence of the San Sebastian and St. John's rivers, and a strong military presence allowed Castillo de San Marcos to withstand repeated enemy attacks. As a Spanish, British, and United States outpost, the fort never fell into enemy hands until 1862 during the American Civil War.

Several archaeological investigations, over a span of decades, have helped to locate Ponce de León's landing area, the French colony, the settlement established by Menéndez de Avilés, and series of city fortifications, walls, and structures from various other periods. Recently, there has been a renewed interest in investigating the precontact and proto-historic (the earliest years of European contact with the New World) sites in the St. Augustine area. Archaeological investigations have yielded valuable information about the structure and nature of the society of the indigenous peoples who settled the area before Europeans. Studying these early settlements will facilitate archaeologists and historians in interpreting how Native American, and later African slave, societies changed as a result of European conquest. Paleopathology research, the study of the effects of disease on ancient remains, will aid in understanding the transfer and dissemination of various foreign diseases, and their devastating impact on indigenous, slave, and immigrant populations.

ADRIENNE WILMOTH LERNER

Further Reading

Landers, Jane G., and Peter H. Wood. *Black Society in Spanish Florida.* Urbana: University of Illinois Press, 1999.

Various. *Oldest City.* St. Augustine, FL: St. Augustine Historical Society, 1983.

The English Establish a Colony in Jamestown, Virginia

Overview

After a few failed expeditions across the Atlantic, English colonists established the settlement of Jamestown, Virginia, in 1607. The representative legislature instituted there in 1624 was the first of its kind in America. Captain John Smith (1580?-1631), an early leader of the colony who had returned to England in 1609, published the first major English account of the New World in 1624. As the first successful English colony in the Americas, Jamestown helped establish Virginia as a major player in the history of the United States.

Background

Although Europeans had for centuries regarded themselves as having discovered the Americas, the continents were not empty of human inhabitants when the first sailing ships arrived from across the Atlantic. Virginia's coastal regions were populated by the Powhatan Confederacy, a part of the Algonquin language group. Other parts of Virginia were inhabited by tribes that spoke Siouan and Iroquoian languages. Altogether, there were about 18,000 Native Americans in Virginia at that time; half lived in the areas around the Chesapeake Bay.

Relations between the Native Americans and Europeans were often difficult. The Europeans saw the natives, whom they called Indians, as savage heathens to be Christianized, and considered the land unexploited and ripe for colonization. The indigenous people naturally saw things differently. The first European settlement in Virginia, established by Jesuits from Spain, was destroyed in an Indian raid a few months later.

In 1584 Queen Elizabeth I granted Sir Walter Raleigh (1554-1618) authority to establish

John Smith. *(National Portrait Gallery. Reproduced with permission.)*

colonies in America. Expeditions were sent to the eastern United States, and Raleigh named the area Virginia in honor of Elizabeth, who was known as the Virgin Queen. However, without enough supplies to make a proper start, the earliest attempts failed. The Virginia Company of London was chartered by King James I in 1606 to make a fresh start on colonization.

Impact

On December 20 of that year, Captain Christopher Newport set out from England with three ships, the *Susan Constant,* the *Godspeed,* and the *Discovery.* On board the ships were 144 men and boys, recruited with the goals of converting the Indians to Christianity, searching for gold and silver, looking for a waterway to the Pacific Ocean, and growing agricultural products to send back to England. They were ill prepared for a hard life; most were from the upper class, had few practical skills, and little experience with manual labor. On May 14, 1607, they set up an encampment on a peninsula about 60 mile (96.6 km) upriver from the Chesapeake Bay on Virginia's coast. This settlement, which was to become the first permanent English colony in the New World, was named Jamestown after the King.

The site of the colony was chosen because it seemed as if it would be easy to defend, and the water around it was deep enough for the ships to drop anchor near the shore. But the area turned out to be swampy and mosquito-infested, and the drinking water from the James River was brackish and muddy. Both contributed to disease, and lack of sufficient food weakened the colonists further. Two-thirds died of malnutrition, malaria, pneumonia, or dysentery.

Captain John Smith, a veteran of military exploits in the Netherlands, Hungary, and Turkey, took over as leader of the settlement in 1608. He held the situation together by trading for food with the Powhatans and forcing the reluctant men to take part in the physical labor needed to establish the colony and build up its defenses.

Another setback for the colony was Smith's return to England in 1609 for medical treatment after his gunpowder bag caught fire. The winter after his departure nearly put an end to Jamestown. Conflict with the Indians meant both bloody attacks and an end to trade. Supplies ran low. Fire, drought, and disease took their toll. Many more settlers died in what became known as the "starving time."

That spring, the remaining colonists prepared to give up and return to England. Just in time, at Hampton Roads, they met the incoming ships of Governor Thomas West (1577-1618), bearing additional settlers and replenished supplies. Like Smith, West was an effective leader. Without these two men, it is unlikely that Jamestown would have survived.

One of the Jamestown colonists, John Rolfe (1585-1622), began raising tobacco in 1612. He grew a type of tobacco from Trinidad that was sweeter than the native Virginian plant, and developed a method of curing the tobacco so that it could be exported. Once the colony had a way

to produce income, its prospects were much brighter. Corn and hogs were also grown successfully at Jamestown. These remain important agricultural products in Virginia today.

Rolfe also contributed to the welfare of the colony by marrying Pocahontas (1595?-1617), daughter of Powhatan, a powerful chief, in 1614. The marriage inaugurated a period of peace between the Native Americans and the English. However, after Powhatan died four years later, relations deteriorated between the settlers and the new chief, Opechancanough. Opechancanough perceived that as the colony prospered and grew, his people were being squeezed out of their territory. He led an attack in 1622 in which 347 colonists were massacred.

Still, by this time the colony was fairly well established. Settlers had been granted land of their own. A ship full of young women had been sent over from England for the colonists to marry. Having become householders with farms and families, the settlers were far more likely to stay. In 1619 Dutch traders brought the first African slaves to Virginia, at the same time increasing the colony's prosperity and opening a sad chapter in the history of the region and of the future United States.

That same year, the first representative legislature in America was formed on instructions from the Virginia Company. Called the House of Burgesses, it met with the governor and his council in a lawmaking body called the General Assembly of Virginia. The House of Burgesses served as the model for many of the colonial and state legislatures that were to follow.

Meanwhile, Captain John Smith had recovered from his wounds and returned to America in 1614, exploring the region of New England, a name he coined. In his later years, he lived in London and wrote about his adventurous life. His principal work, *The Generall Historie of Virginia, New England, and the Summer Isles,* was published in 1624. It was the first significant account of the New World in English.

In 1626 James I revoked the charter of the Virginia Company and declared Virginia a royal colony. Some of the royal governors dispatched from England had quarrelsome relationships with the colonists. Sir William Berkeley (1606-1677), who arrived in 1642, was an exception. However, he was ousted 10 years later, after King Charles I was overthrown by Oliver Cromwell in the English Civil War.

During the Cromwellian period, the colony was generally left to run its own affairs. Its population was augmented by a number of royalists fleeing England. With the restoration of the monarchy in 1660, Berkeley was re-appointed, but this time his tenure was not as successful. The population center was moving west to the edge of Virginia's Piedmont region, and many objected to Berkeley's allowing wealthy coastal families, known as the Tidewater aristocracy, to rule the colony. The Piedmont settlers wanted protection from Native Americans, fewer regulations imposed upon them from the east, and freedom from British restrictions on colonial trade. In 1676 a group of colonists rebelled, led by a young planter named Nathaniel Bacon (1647-1676). Jamestown was burned to the ground. The settlement was rebuilt, but the statehouse was again destroyed, this time by an accidental fire, in 1698.

As a consequence, the capital of the Virginia colony was moved from Jamestown to Williamsburg, and the earlier settlement was deserted. It had served its purpose as the English foothold in the New World. At the start of the eighteenth century, Virginia was the largest North American colony, with a population of about 58,000.

Virginia continued to play a pivotal role in American history. Four of the first five presidents—George Washington, Thomas Jefferson, James Madison, and James Monroe—were born there. Its large territory included regions that eventually became all or part of eight other states: Illinois, Indiana, Michigan, Wisconsin, Minnesota, Ohio, Kentucky, and West Virginia. Virginia's current capital, Richmond, was the capital of the Confederacy during the Civil War.

The peninsula on which Jamestown was established eventually became an island, as the tides eroded the neck of land. For many years it was believed that the remains of the 1607 fort were underwater, but in 1996 archaeologists discovered evidence of the fort on the island as well as artifacts from the settlement. Today the island is part of the Colonial National Historical Park, visited by more than a million people every year. Meanwhile, archaeological research continues to shed light on this important early colony.

SHERRI CHASIN CALVO

Further Reading

Bridenbaugh, Carl. *Jamestown, 1544-1699.* New York: Oxford University Press, 1980.

Friddell, Guy. *We Began at Jamestown.* Richmond, VA: Dietz Press, 1968.

Hume, Ivor Noel. *The Virginia Adventure: Roanoke to James Towne.* New York: Alfred A. Knopf, 1994.

Vaughan, Alden T. *American Genesis: Captain John Smith and the Founding of Virginia.* Boston: Little, Brown and Company, 1975.

John Cabot's Exploration of North America

Overview

In 1497 John Cabot (1450?-1499?), an Italian explorer sailing for England, reached land somewhere in the northern part of North America. Although unsuccessful in his attempt to reach Asia, his landfall gave England a territorial claim in the New World that would be the basis for her eventual colonization of parts of that continent. In addition, Cabot's son, Sebastian (1476?-1557), became the first of many explorers who attempted to sail across the top of the world in an effort to find a Northwest Passage from Europe to the wealth of Asia.

Background

In 1494 John Cabot (born Giovanni Caboto) moved his family to England from Valencia, Spain. He moved to England for the same reason that he had moved to Spain—to be a part of the exploration of the Atlantic Ocean and the lands on the other side, presumed at that time to be parts of Asia. He had been unsuccessful in convincing the Spanish and Portuguese to hire him, so he hoped to improve his luck in England.

Approaching the English king, Henry VII, Cabot offered to find a northern route to the Orient, challenging the Spanish route blazed by Christopher Columbus (1451?-1506) a few years earlier. Henry, who had barely missed the opportunity to sponsor Columbus's trip, jumped at the chance, provided Cabot could find some financial backing. In exchange for promises of an import monopoly from the Crown, a group of Bristol merchants underwrote Cabot's voyage, and the King issued a letter of patent authorizing

John Cabot. *(Corbis Corporation. Reproduced with permission.)*

Cabot to claim any lands he found for England. What none knew was that Cabot had made the same error as Columbus; he believed what he read, and his reading said that the world was actually only 17,000 miles (27,359 km) around, not the 24,000 miles (38,624 km) we now know it to be. Between England and Asia lay not only one ocean, but two, and a continent as well.

Cabot set sail for Asia in the spring of 1497, making landfall on June 24 of that year. The exact place of his landing is not known; convincing cases have been put forth for virtually every reasonable location between Maine and Labrador. What is known for certain is that he realized he had discovered a new continent, that he was the first European to land in North America since the Vikings, and that he claimed the territories he found for England.

Upon his return to England, Cabot announced his discoveries and immediately found backing for a larger expedition the next year. With five ships this time, he again set sail and again landed on the shores of the New World. This time, however, he traveled some way to the south in search of Japan or China. Failing to find them again, he returned to England. On his way, his small flotilla crossed a part of ocean swarming with fish, what is now known as the Grand Banks. Although he did not realize it at the time,

this discovery alone justified his trip, for the Grand Banks has been one of the world's most productive fishing grounds for several centuries.

When he returned to England, Cabot reported his failure to prove he had landed in Asia and, from there, faded from history. In fact, it is not even certain when or where he died. However, his discoveries were to have a profound and lasting impact on England and the world.

Impact

Although Cabot, like Columbus and so many others, was unsuccessful in discovering a short and easy route to Asia, his voyages were significant for a number of reasons. Among these are:

1. His territorial claims for the English Crown gave England a toehold in the New World.

2. His son, Sebastian, was encouraged by his father to continue exploring, beginning the quest for the elusive Northwest Passage.

3. His work helped prove that North America was a new continent, one that proved every bit as rich as Asia.

The letter of patent Cabot received from Henry VII allowed him to take possession of lands "which before that time were unknown to all Christians" for the English Crown. North America certainly qualified, and Cabot claimed everything he could for England. The next claim, based on Jacques Cartier's (1491-1557) 1535-1536 explorations of the Saint Lawrence River, established France as a colonial power along this river and into what is now Quebec. Thus, the future of Canada was set, even though England did not pursue its claim to these new lands for nearly a century. One result of these competing claims was the French and Indian War in which George Washington (1732-1799) gained most of his early military experience. Another outcome is still seen today, in Canada's perennial conflict between the French-speaking Quebecois and the English-speaking populations that settled the rest of Canada. There is some irony, too, that this territorial claim not only helped give England its North American colonies, but at the same time set in motion events that would eventually lead to training the man who would help take them away. However, at the time of Cabot's landing, none of this could even be envisioned.

Although it is interesting to speculate about how world history might have changed had

Cabot not landed in North America, this is more properly the realm of speculative fiction. It is entirely likely that England would have found some pretext to launch her settlements in precisely the same locations as in fact occurred, though possibly without the strategic advantage granted by also possessing the port of Halifax in Nova Scotia. Although these claims may have been contested more vigorously, the history of North America probably would not have changed much because England's rivals were largely interested in different parts of the Americas. Spain had already claimed almost the entirety of South and Central America, plus a large part of the American Southwest and Florida, while the French had staked a claim to the lion's share of what is now Canada. This left the Dutch as England's only serious rivals in North America, primarily for what is now New York, and they elected to withdraw and concentrate on their possessions in South Africa and the Dutch East Indies rather than to contest English claims in North America.

Also important was the effect that Cabot's voyages had on his son, Sebastian. Instead of being discouraged by the perception that his father had failed in his voyages, Sebastian went on to try to discover a Northwest Passage, across the "top" of the new continent and leading to Asia. To this end, in 1508 Sebastian Cabot set out on a self-financed expedition to find a passage to Asia. This time, he was well aware of the fact that North America was, in fact, not Asia but a new continent entirely, and he seems to have purposely set out to find a northern route to Asia across this new continent. Although he failed in this attempt, he set in motion countless attempts by those who were to follow him for nearly four more centuries. Success would not come until 1906, when Roald Amundsen (1872-1928) was able to successfully thread his way through the tortuous channels that comprise the Canadian Arctic islands. Between Cabot and Amundsen, the Northwest Passage claimed scores of lives and many ships. In that time, too, the Northwest Passage went from being an important goal to a near-afterthought because, in the interim, other routes to Asia were pioneered and made both routine and profitable.

The other major result of Cabot's discoveries—the realization that North America was a continent in its own right—was made somewhat gradually. Many, including Sebastian Cabot, seemed to grasp this more quickly than others, but maps until the 1700s continued to show North America connected to Siberia. However, in spite of this purported land connection, by the middle of the sixteenth century, North America's status as a full-fledged continent was fairly well established. However, it was to be some time before the English saw it as anything other than an impediment hampering their ready access to Asia.

In fact, it is hard today to fathom England's near-obsession with ignoring America in favor of trade with Asia. However, at the time Asia represented a known commodity. Beginning with Marco Polo (1254?-1324), some fortunate European nations and city-states had grown wealthy on Asian trade, and the British had been left out because of their status as a relatively weak and poor nation on the outskirts of Europe. At that time, European power and civilization was centered on the Mediterranean and, without a strong navy, England simply could not open trade routes of its own, or compete with the established Mediterranean powers. So, rather than try, the English opted instead to seek alternate routes to these riches, and looked to the West. England was hoping to find a new route to the Orient that would be faster and safer than existing routes. By doing this, the English hoped to share in the wealth generated in China, Japan, and the Spice Islands.

It was only after a century or so of fruitless effort that England finally realized that the North American continent could be the key to wealth, too. With the establishment of its North American colonies, England began profiting from tobacco, timber, fish, and other goods. At the same time, British sea power was in the ascendancy, and England began to project this power throughout the world. It is also worth remembering that, in spite of the American Revolution, British influence in North America continued until the passage of the British North America Act in 1867. This gave Britain nearly three centuries of domination over the American colonies, and nearly a century more of colonial influence in North America. The North American colonies were among England's first overseas possessions, starting England on the path to global empire in the nineteenth century, and enriching its treasury at the same time.

P. ANDREW KARAM

Further Reading

Lehane, Brendan. *The Northwest Passage.* Alexandria, VA: Time-Life Books, 1981.

Morison, Samuel E. *The European Discovery of America.* New York: Oxford University Press, 1971.

Morison, Samuel E. *The Great Explorers.* New York: Oxford University Press, 1978.

The Search for a Northwest Passage

Overview

Jacques Cartier (1491-1557) explored northern North America with hopes of finding gold and precious metals, and possibly a cross-continental passage connecting the Atlantic and Pacific oceans. He never realized those goals, but he did find something of great value: a water passage into Canada's interior. His discovery in the 1530s of the Gulf of St. Lawrence and the St. Lawrence River gave the Europeans, and specifically the French, access to the interior of the continent for the first time. This also provided the first views of the areas that would eventually become the sites of such major Canadian cities as Montreal and Quebec.

Background

Before Cartier's voyages to Canada, North America was known to the Europeans mostly from its eastern coastline. Fishermen sailed its ocean waters, but had neither need nor desire to venture far onto land or even to explore inland waterways. Descriptions of the coast were poor and maps were rudimentary and used mainly to direct them across the Atlantic and to a desired location off of North America, but not to skirt along the rugged shoreline, into its bays or around its many islands. The fishermen of the day were satisfied to know only as much as needed to safely navigate the coastal waters, to locate a safe harbor when necessary, and to find the prime fishing grounds.

The Frenchman Cartier sought something more. He went to Canada with a different idea, and approached the coastline with the eye of a navigator, geographer, and explorer.

Trained as a navigator in Dieppe, northwest of Paris, Cartier made his first well-documented voyage in 1534 as the leader of an expedition to find precious metals in North America. Although he found no gold on the six-month journey, he did discover the Magdalen and Prince Edward islands, and spent considerable time investigating the Gulf of St. Lawrence. His records indicate that he was close to the mouth of the St. Lawrence River, but he was either slightly too far north of it or the weather prevented him from spotting the waterway. While in Canada he befriended an Iroquois chief and brought two of the chief's sons with him to France. Cartier's ad-

ventures brought him fame in his hometown, but even before the accolades had died down, he was preparing for his next trip.

In May of 1535, he again crossed the Atlantic for North America. He led three ships carrying a crew of 110 men, plus the two native Americans who had learned the French language well enough during their overseas stay to serve as interpreters. Cartier immediately led the ships back to the Gulf of St. Lawrence where he hoped to find the waterway to Canada's interior and possibly across the continent. The interpreters guided him to the St. Lawrence estuary in August. With the discovery made, Cartier sailed his ships more than 200 miles (322 km) into the waterway and up to the Iroquois village of Stadacona. From there, they continued by longboat for another 100-plus miles (161 km) to the Iroquois village of Hochelaga. From a hill near Hochelaga, which he named Mont Réal, Cartier could see impassable rapids on the river beyond. At that point, he decided to return to Stadacona to spend the winter months.

Cartier returned to France in May of 1536 with news of the St. Lawrence River and its potential as a route to China. He also brought back to France another 10 Indians, including the chief whose sons had made the trip a year earlier. Based on stories told to him by his Indian interpreters, Cartier believed that if he traveled along the river just a bit farther, he might find that the waterway traversed the whole continent or at the very least revealed a wealth of precious metals. King François I was excited by the prospect and had planned to send the explorer back to Canada as soon as possible. While plans were under way, however, a war broke out between France and the Habsburg Empire, and the return trip was postponed until 1541. While waiting for the voyage back to their homeland, all but one of the 10 visiting Indians died in France.

When Cartier did set sail again for the New World, he was billed as the chief pilot under expedition commander and French nobleman Jean François de la Rocque de Roberval. The expedition included 1,500 men, including settlers for Canadian outposts, and eight ships. Cartier took the first ship out and arrived in Canada earlier than Roberval. He waited out the winter in an area north of Stadacona. The remainder of the fleet overwintered in Newfoundland. During the

Jacques Cartier encounters native people of the St. Lawrence River region. *(The Granger Collection, Ltd. Reproduced with permission.)*

winter months, Cartier discovered what he thought to be gold and diamonds. (As it turned out, they were mimics of little or no value.) He decided to return to France with his finds, and was on his way back when he ran into Roberval in Newfoundland. Roberval ordered him to return to the mainland to help him set up a colony. Cartier disobeyed, however, and made his way back to France, where he took a home in the country. Cartier never left France again.

Impact

Cartier's major achievements were the discoveries of the Gulf of St. Lawrence, the estuary leading into the St. Lawrence River and the river itself. The St. Lawrence waterway opened the first passage to take explorers and traders deep into Canada's interior. In fact, his voyage along the river brought him farther inland than any North American explorer had yet attained. Along the way, Cartier stopped at the Iroquois villages of Stadacona and Hochelaga, both of which would eventually become the sites of major Canadian cities. The city of Quebec now straddles Stadacona, and Montreal encompasses Hochelaga and the Cartier-named Mont Réal.

The impact of his discovery and his meticulous descriptions of the entire St. Lawrence water system was far-reaching. The geographical information he collected helped to create the most accurate maps of the period. Atlases of the day soon included details of the entire Gulf of St. Lawrence and the great river. In addition, his explorations of the eastern coast of northern North America led to discoveries of Prince Edward and Anticosti islands, Gaspé Bay, and other sites.

It was his discovery of the St. Lawrence River, however, that was the most influential. Once discovered and charted, it became for the French the most important and well-used entrance point into North America. As the fur trade blossomed into a major financial institution, the route garnered additional importance. The river became the major trade route for the French, and cemented France's major role in the development of Canada. As long as the French controlled the waterway, it also served as a political barrier to the expansion of other nations into the North American interior. As a result, France for many years held title to the major explorations of Canada and what would become the Midwestern portion of the United States. Even today, the influence of France and its explorers are still visible in many of the names of Canadian and Midwestern U.S. towns and cities. Had Cartier been able to bypass the rapids he saw outside of modern-day Montreal, the river would have taken him to Lake Ontario, the first in the chain of Great Lakes.

Beyond his geographical achievements, Cartier influenced future explorers in another way. His stay in Stadacona from late fall of 1535 to spring of 1536 represented the first time that Europeans had overwintered in Canada. The cold was more than he and his crew expected, but the most severe consequence of the winter months was the scurvy suffered by more than 90% of his crew. Fortunately for Cartier's men, the Iroquois had a remedy. They brewed a drink made from the bark of a cedar tree and served it to the crew. Laden with vitamin C, the curative beverage brought all but 25 crewmen back from the brink of death. Based on Cartier's winter experience, other Europeans who came to Canada were able to prepare better for the long winter days and nights.

Cartier's experiences as an expedition leader, his skills as a navigator and geographer, and his ability to form relationships with the native people helped not only to open Canada to exploration, but to create the climate for France to become highly influential in the development of Canada and the Midwestern United States.

LESLIE A. MERTZ

Further Reading

Baker, Daniel B., ed. *Explorers and Discovers of the World, first edition.* Detroit: Gale Research, 1993.

Biggar, H. *The Voyages of Jacques Cartier.* Ottawa: Public Archives of Canada, 1924.

Byers, Paula K. *Encyclopedia of World Biography, second edition.* Detroit: Gale Research, 1998.

Edmonds, J., commissioning ed. *Oxford Atlas of Exploration.* New York: Oxford University Press, 1996.

Morison, S. *The European Discovery of America: The Northern Voyages.* New York: Oxford University Press, 1971.

North America's
First Permanent European Colony

Overview

Although Samuel de Champlain (1567-1635) was not the first European to explore the coastline of Canada, he was arguably the most influential. Noted as the geographer who generated a set of the most complete maps of the coast, Champlain charted the seaboard from Nova Scotia to Rhode Island. He also explored what would become major trade routes along the inland rivers of Canada, was likely the first European to see the Great Lakes, and helped establish colonies at Quebec and Annapolis Royal.

Background

Champlain began his explorations of New France, or Canada, in 1603 as a member of an expedition led by François Grave Du Pont. Du Pont's charge was to sail up the St. Lawrence River system. The expedition made its way to what is now Montreal in the summer of 1603. That voyage lasted less than a year, but Champlain was so taken by the Canadian expanse that he wrote a book about his experiences in the wilderness that same year. The book was titled *Des Sauvages, ou, Voyage de Samuel Champlain.* In the book, he credited the native people, particu-larly the Hurons and Montagnais, for assisting his explorations of the countryside, and noted information that they had provided about the existence of expansive inland waters. He mistakenly believed that these waters, now known to be the Great Lakes, might provide a cross-continental passage from the Atlantic Ocean to the Pacific. Discovery of such a passage was desired for its potential to open up a shorter trade route to and from China.

Champlain's interest in exploring the lands of North America heightened when he returned to Canada less than a year later as geographer aboard an expedition led by Pierre du Gua de Monts. During this expedition, and the subsequent two years that Champlain spent living in the largely unknown continent, he spent a great deal of time touring the area and taking meticulous notes. From these, he created a general map of St. Croix and Annapolis Royal in Nova Scotia. Although not perfect, this map was much more accurate and contained more detail than previous maps of the area.

After a short return trip to France in 1607, Champlain went back to Canada. He took on the monumental task of establishing a fort at Quebec (near the site of an already existing Iroquois village) in 1608, serving as its administrator, and within two years generating enough interest in the site to have it acknowledged as his country's North American fur-trading center. Despite his duties in Quebec, Champlain continued his explorations. In 1615, he traveled up the Ottawa River and, with the help of another Indian guide—a tribal chief this time—he saw Lake Huron for the first time. He ventured along the shoreline of this "Fresh Water Sea," meeting with other small groups of Indians and learning about their cultures.

Although he conducted other explorations afterward, he turned a great deal of attention to his responsibilities in Quebec. There, he hired younger explorers, including such men as Etienne Brûlé (1592?-1633) and Jean Nicollet (1598-1642), to become interpreters between the French and native American populations, and to help expand France's trading circle into Canada's interior.

A war between England and France resulted in Champlain's ouster from not only Quebec, but from Canada as well, for about four years. As soon as the two countries struck a deal and France regained Quebec, Champlain returned to live out his life in his adopted home. He died on December 25, 1635.

Impact

Champlain's explorations of the eastern seaboard of Canada and the northern United States, along with his attention to detail and his skills as a geographer, provided the most accurate maps to that time of the coastline from Nova Scotia well into New England. His voyages along the St. Lawrence and Ottawa rivers also generated the maps necessary to set up primary trading routes through the waterways, and helped to assure France's place as one of the foremost trading nations of the world. His geographical expertise became widely respected through the publication of his 1632 map of the web of waterways connecting the Atlantic Ocean with the Great Lakes.

France's name in trading was further elevated when Champlain established and governed Quebec. Through his insistence about the potential benefits of the fort as a trading center, the explorer founded what was to become one of the largest and most influential cities in all of Canada.

Beyond his geographical contributions, Champlain was one of the first explorers to not only form working relationships with the native people, but to try to understand their cultures. His book, *Des Sauvages, ou, Voyage de Samuel Champlain*, contains never-before-printed ethnographic detail about the American Indians. Unlike many other explorers of his time who feared—and often slaughtered—the Indians they encountered, Champlain understood the importance of creating alliances. It was through such friendships that he was able to safely travel through the waterways and forests of the North American interior, and how he learned of the Great Lakes.

Champlain's ability to relate well with the Indians also allowed him to locate the native population's long-standing trading routes. Those trading routes became the pathways of the European explorers as they began to trek farther onto the North American continent.

His expeditions into Canada's interior opened the door to the exploration of America's Midwest, as well. His discovery of Lake Huron, and his belief that it might be the cross-continental passageway to the Pacific Ocean and China, combined to fuel the exploration of Michigan and Wisconsin. One of his hired interpreters/explorers, Etienne Brûlé, became the first European to set foot in what is now Michigan. Jean Nicollet pressed on to Lake Michigan and into Green Bay. During his trip, Nicollet learned about the existence of a great river, the Mississippi, and that discovery led to additional exploration of the area in ensuing years.

LESLIE A. MERTZ

Further Reading

Baker, Daniel B., ed. *Explorers and Discovers of the World, first edition.* Detroit: Gale Research, 1993.

Biggar, H., ed. *The Works of Samuel de Champlain.* 6 vols. Toronto: The Champlain Society, 1927-35.

Byers, Paula K. *Encyclopedia of World Biography, second edition.* Detroit: Gale Research, 1998.

Edmonds, J., commissioning ed. *Oxford Atlas of Exploration.* New York: Oxford University Press, 1996.

Morison, S. *Samuel de Champlain: Father of New France.* Boston: Little, Brown & Co., 1972.

Pedro Cabral and the Portuguese Settlement of Brazil

Overview

The exploration of Brazil by Pedro Cabral (1467-1520) established the nation of Portugal as a major power on the continent of South America. This would have a profound effect on the Native American people in the region and would eventually establish the modern culture of Latin America.

Background

Pedro Cabral is accepted as the first European to recognize the vast potential of what is now Brazil. This was part of the vast expansion of Europe around the globe, which began in the fifteenth century. Events had been pushing Europe outward for at least five centuries. By the turn of the millennium, Western Civilization had begun to recover from the destruction of the fall of the Roman Empire. Advances were made in agriculture, communication, and transportation that allowed Europeans to develop a sound economy.

By the end of the eleventh century, Pope Urban II (1035-1099) called for a holy crusade to free Jerusalem from Islamic domination. These holy wars would be a turning point in Western Civilization. Although they were tragic military disasters, the Crusades reintroduced Europe to the products of the East, especially perfumes, spice, and silk.

The Islamic world system controlled the movement of these precious items first through the Indian Ocean and then across Southwest Asia by caravan. Monopolies were established with certain Italian city-states, and these were so successful that they provided the financial basis for the great Southern Italian Renaissance. It was also during this period that the great inland Asian empire of the Mongols began to collapse. This disrupted trade along the great Eurasian Silk Road, which reduced the flow of goods into Europe. The capture of Constantinople in 1453 established the Ottoman Empire as a major force in the world. This new strategic reality forced many of the Italian city-states to establish military and economic alliances with the Ottomans. These treaties provided the nations of Western Europe with the incentive to find an all-water route to the East and to break the monopoly of this Islamic/Italian connection.

Europe had also experienced centuries of religious and dynastic wars. One lasting effect of these struggles was the growth of military technology. On the eve of Western expansion, the major nations of the Atlantic coast began to expand their military capabilities beyond that of their potential rivals. On the Iberian Peninsula, Spain and Portugal began to compete against each other for control of the lucrative spice trade. In 1492, the Spanish monarchy decided to fund an expedition based upon the belief that India could be reached by sailing west. This voyage of exploration was headed by Christopher Columbus (1456-1501) and resulted in the discovery of the Western Hemisphere. In 1498, Vasco da Gama (1460-1524) sailed around the Cape of Good Hope, reached India, and returned home with a cargo worth sixty times the cost of his voyage. His great successes pushed the Portuguese government to fund a second trip in 1500. Exhausted from the physical and emotional stress of the first voyage, da Gama recommended that Pedro Cabral lead the second expedition.

Impact

In 1500, Cabral began his voyage to India. His ships were blown off course by strong westerly winds, and he ended up off the coast of Brazil. The dense vegetation along the coast proved a formidable obstacle to the early Portuguese explorers. The first years of contact were known as the "factory period." During this time, fortified warehouses or factories were constructed to house and protect the products extracted from the Brazilian forests. The most important resource in the early years was Brazilwood. Over the centuries Europeans had decimated their forests, first for fuel, and then for their massive shipbuilding programs. Brazilwood provided the Portuguese with much needed lumber for the construction and repair of their maritime fleet. The climate and soil of Brazil was also compatible to the cultivation of sugar, which was an important commodity because it could be stored for long periods and shipped great distances without spoiling. The government initially turned this program over to a group of merchants from Lisbon. However, they proved unable to handle the task, and the production of sugar reverted back to the crown.

In addition, Brazil became a strategic problem for the Lisbon government. The great potential wealth of the Western hemisphere created intense competition among Spain, France, and Portugal. Both countries challenged the Portuguese for the ultimate control of Brazil's resources. This competition drastically changed the way the Lisbon Government viewed its colony. The crown decided that a permanent settlement would have to be established to counteract any potential attempts by Spain or France to gain control of Brazil. The government sent four hundred settlers to create this defensive force. The colony not only had permanent working settlements but also patrolled the coast and undertook the exploration of the Amazon and LaPlatta rivers. The impact of these settlements led to serious changes to the entire region.

In the early sixteenth century, Portugal and the rest of Europe were still recovering from the effects of the Bubonic Plague on their populations. The Black Death created a labor shortage throughout Europe, thus there was no incentive for people to emigrate to the New World. The government had to look to its prisons to provide settlers for the new colony; so a vast majority of the inhabitants were either criminals or political prisoners. These were obviously not the type of people that insured a successful colonial experience. This was especially true in the production of sugar. The cultivation and harvesting of sugar required large numbers of people who were willing to perform the strenuous tasks required to send the crop to the factory. This work was also rigorous and required long, uninterrupted hours of labor to process the cane into refined sugar. It became quite obvious that a new a source of labor had to be found. Initially the Portuguese tried to use Indian slaves. The source of these slaves was the dominant Native American tribe in the region. They had practiced their own form of slavery for generations, and they were most willing to become the providers for the Portuguese. Initially slavery posed an ethical problem for the Lisbon government. Christianity had always preached the equality of all people in the eyes of God, and Jesuit missionaries were sent to Brazil to convert the native population. The Portuguese adopted the same principles they had used in their reconquest against the Moslems. If the Indians accepted Christianity, they would be protected from forced labor, but if they refused this new religion they were forced into bondage. The cultural shock of bondage broke the spirit of many Native American slaves. In response to the terrible working conditions many took their own

Pedro Álvares Cabral. *(Corbis Corporation. Reproduced with permission.)*

lives; others perished as a result of European diseases. This was part of what has come to be known as the "Great Dying." It soon became obvious that another source of labor was needed to keep the colony in operation. Since the Portuguese had a long established relationship with the tribes of West Africa, they decided to use African slaves from the western region of the continent. These tribes had an extensive slave trade network that coincided with the movement of Islam into the continent. The crown took advantage of the situation to establish a slave trading system to supply their settlements in Brazil with a much needed labor force. African slaves were strong, intelligent, and used to working in a subtropical climate. The success of the Portuguese sugar industry rested upon the labor of these West Africans.

The massive relocation of Atlantic peoples was part of a larger historical phenomenon known as the "Columbian Exchange." This was the most extensive biological and demographic redistribution in the history of the world. Since the Western Hemisphere had been isolated from the Old World, the Native American population had not been exposed to many of the world's deadliest diseases. When the Portuguese established permanent settlements in Brazil, they exposed the indigenous population to smallpox and measles. With the introduction of slavery from West Africa, malaria was also unleashed against an unsuspecting population. Demographers suspect that the Native American population in Brazil was reduced from about 2.5 million to under 500,000. This demographic disaster affected Brazil in two ways: it reduced the Indian population by 80%, which prohibited the people of Brazil from organizing a successful resistance to European settlement; and it created an entirely new Latin American culture. This new civilization is a hybrid of three different continents, Europe, Africa, and South America. In addition, the new European-dominated class structure that was created remains in place to the present day.

RICHARD D. FITZGERALD

Further Reading

Crosby, Alfred. *The Columbian Exchange: The Biological and Cultural Consequences of 1492*. Westport, CT: Greenwood Press, 1972.

Crosby, Alfred. *Ecological Imperialism: The Biological Expansion of Europe ,900-1900*. Cambridge, MA: Harvard University Press, 1986.

Curtin, Philip. D. *Cross-Cultural Trade In World History*. New York: Cambridge University Press, 1984.

Frank, Andre Gunder. *ReORIENT: Global Economy in the Asian Age*. Berkeley: University of California Press, 1998.

Morison, Samuel Eliot. *The Great Explorers: The European Discovery of America*. New York: Oxford University Press, 1978.

Vasco Núñez de Balboa
Reaches the Pacific Ocean

Overview

Unknown to the indigenous people of the New World, their destiny was being determined by political and economic forces taking place across the Atlantic Ocean in Europe. Toward the end of the fifteenth century, thousands of daring adventurers would be crossing the ocean to conquer within a few centuries what had taken the Indians thousands to years to inhabit. This "Age of Exploration" was fostered by technological advancements in maritime practices, the belief in an economic philosophy called mercantilism, and an interest in converting the religious beliefs of native populations. Mercantilism was the idea that if a nation was not self-sufficient in its affairs, then its neighbors would dominate it. The two areas that seemed ripe for establishing this ideal were the Middle East and the Americas. Many of the Spanish conquistadors headed for the New World seeking wealth and adventure. One such man was Vasco Núñez de Balboa (1475-1519).

Background

Balboa came from the ranks of that lower nobility whose sons often sought their fortunes in the West Indies. In 1500 he was part of an expedition led by Rodrigo de Bastidas (b. 1460?), which explored the coast of present-day Colombia. Balboa then settled in Hispaniola and was given a farm to tend. Balboa did not enjoy the agrarian lifestyle and ac-

cumulated much debt. He wished to leave the country and seek his fortune elsewhere but was told he could not leave the island with outstanding debts. He decided to bribe some men getting ready to leave on an expedition so that he and his faithful dog could stowaway in a barrel. The voyage was organized in 1510 by Martín Fernández de Enciso (1470?-1528) to bring aid and reinforcements to a colony off the coast of Uraba (present-day Colombia). When they arrived, the colony was in ruins and there were few survivors. The Indians in the area were hostile and used arrows with tips that were soaked in poison. On the advice of Balboa the settlers moved across the Gulf of Uraba to an area known as Darien. This area was much less hostile, and they founded the town of Antigua. Balboa began to accumulate wealth from the Indians by befriending them or, if that was not successful, by going to war with them. Eventually Balboa was elected as the comagistrate of the settlement. He was later named by the king as interim governor and captain general of Darien.

Balboa meanwhile had organized a series of expeditions to hunt for gold and slaves. His Indian policy combined the use of barter, every kind of force, including torture, to extract information, and the tactic of divide and conquer by forming alliances with certain tribes against others. He was able to do this because of his vast knowledge of the area. The Indians of Darien were more timid that those of Uraba, so they were easily subdued.

One day, in a fit of rage over the Spanish love of gold, an angry Indian told of both a land to the south by a sea and a province infinitely rich in gold. It is thought that these references were to the Pacific Ocean and perhaps to the Inca Empire. The conquest of that land, their informants declared, would require 1,000 men. Balboa dispatched men to request reinforcements; the news they brought created much excitement, and a large expedition was promptly organized. But Balboa was not given command of the expedition because he had fallen out of favor with King Ferdinand II. Instead, that position went to an elderly, powerful nobleman, Pedrarias (1440?-1531). The expedition, numbering over 2,000 persons, left Spain in April 1514.

Balboa decided to move ahead without reinforcements and sailed on September 1, 1513, to Acla, at the narrowest part of the Panama isthmus. His troop numbered nearly 200 Spaniards and hundreds of Indian carriers. They marched across the isthmus through dense jungles, rivers, and swamps. Finally on September 27, 1513, after ascending a hill by himself, Balboa sighted the South Sea, or the Pacific Ocean. Some days later he reached the shore of the Pacific at the Gulf of San Miguel and took possession of the South Sea and the adjacent lands for his king. He then retraced his steps and returned in January of 1514. Once the king was informed of Balboa's feat, he immediately appointed Balboa the governor of the South Sea and Panama, but Balboa remained subject to the authority of Pedrarias.

When Pedrarias finally arrived in Darien in June of 1514, relations between the men were strained. As a show of good faith, Pedrarias betrothed his daughter Maria in Spain to Balboa. But the underlying causes of friction remained. Highly suspicious and jealous of Balboa, Pedrarias implemented policies that were meant to impede Balboa. After much effort, he granted Balboa permission to explore the Gulf of San Miguel. Soon thereafter, the king decided to have a judicial review of Pedrarias, as it was believed he was unfit to govern. One of the chief witnesses against Pedrarias would be Balboa. Pedrarias feared that Balboa's presence and testimony would contribute to his demise, so he decided to eliminate his rival. Summoned home, Balboa was seized and charged with rebellion, high treason, and mistreatment of Indians. After a mock trial, Balboa was found guilty, condemned to death, and decapitated in January of 1519.

Impact

Expeditions similar to those conducted by Balboa on the Isthmus of Panama served to motivate thousands of Spanish peasants to join the military. The discovery of riches and wealth enticed these peasants to travel to the New World in search of a new life. A successful colonial mission could possibly lead to a governorship or a pension for the participants. If one were extremely lucky, he could amass untold riches. Other men were drawn to the New World by promises of adventure. They looked for quick advancement in the military and for diplomatic careers. Still others came on a mission of God. These men wanted to convert the native population to Catholicism. By converting the Americas to God, they believed they would receive eternal blessings.

One legacy that Balboa tried to leave was his treatment of the indigenous people. Balboa had a reputation for treating the natives with respect, fostering relationships and keeping promises that he had made. He respected the native governments and societies and listened to them in order to increase his knowledge of the land. He helped to settle disputes between various native factions and gained the trust of most. This did not mean that he would not be swift and cruel if he felt it necessary. He often used torture to extract information not readily revealed and had numerous dogs in his command to use as executioners to tear Indian victims to pieces. One was his own Leoncico, who was such a respected warrior that he was given a soldier's full rate of pay. However, Balboa's style of governing was largely ignored by most people who came to the New World, and the native populations were treated for the most part as nonentities. Despite Balboa's treatment, many of the Indians in the New World were eventually overwhelmed with modern weaponry and unfamiliar diseases. While many natives were destroyed during confrontations with Europeans, even those under the rule of Balboa could not withstand the onslaught of disease. Entire villages were wiped out with the introduction of European diseases against which the Indians had no natural immunity. These included smallpox, measles, and the flu. Thus, one unintended legacy of Balboa is the destruction of entire populations of indigenous people by the introduction of disease.

Balboa was the first European to see the eastern shore of the great South Sea (the Pacific Ocean), on September 13, 1513. While he is often erroneously credited for naming it, it was actually named by the Portuguese explorer Ferdinand Magellan (1480?-1521) during his circumnavigation of the globe. He named it as such because its waters seemed so calm. He named the body of water *Pacifica* (meaning peaceful). Balboa claimed the Pacific Ocean and all its shores for Spain. This one act opened the way for Spanish exploration and conquest along the western coast of South America, giving Spain a solid foothold in this region of the world. It was through Balboa's conquest of this region and the information he gained through exploration that conquests further south could be made, such as that over the Incas.

The conquistadors of Spain were generally single-minded and brutal in their obsession with gold and riches in this part of the world. Balboa was mayor of the first profitable settlement in the Americas, but his type of rule was seldom seen in the New World. Most of the conquistadors were driven by their greed and lust for gold, often turning on each other to gain a share. They quickly decimated large Indian popula-

tions and at the same time relieved them of much of their riches. At the same time, Portugal was becoming rich from its newly established sea trade routes to India. Thus Portugal and Spain had taken the early lead in the race for riches from far-away lands. The English, Dutch, and French, who argued that the seas should be open and that possession of land should depend on occupation, would soon challenge this position. Before long all five of these countries would vie for supremacy of these lands.

JAMES J. HOFFMANN

Further Reading

Berger, Josef. *Discoverers of the New World*. New York: American Heritage Publishing Co., Inc., 1960.

Faber, Harold. *The Discoverers of America*. New York: Macmillan Publishing Company, 1992.

Lomask, Milton. *Exploration: Great Lives*. New York: Macmillan Publishing Company, 1988.

The First Maritime Circumnavigation of the Globe

Overview

Fewer than three decades after Christopher Columbus (1451-1506) made his voyage to the New World, Ferdinand Magellan (c. 1480-1521) set sail in 1519 with nearly 600 men and five ships on a voyage to the Spice Islands (East Indies) via a westward route from Spain. Magellan, undervalued by the Portuguese crown, made the trip under the Spanish flag. They crossed the Atlantic, sailed down the eastern coast of South America, rounded the southern tip of the continent through the shortcut now called the Strait of Magellan and named the Pacific Ocean before reaching the eastern shores of Asia. Although Magellan died partway through the trip, one of the five ships in the fleet completed what became the first circumnavigation of the globe in 1522.

Background

At the time of the Magellan voyage around the world, Europeans had known for less than two decades that South America existed, let alone the Pacific Ocean on the other side of the continent. A Portuguese mariner first discovered

South America while following a standard maritime practice of sailing the currents of the Atlantic Ocean far to the west before circling back to Africa. In 1500, Pedro Álvares Cabral (c. 1467-1520) made first landfall on the continent near Salvador, Brazil. Although the land was already occupied with native people, the Europeans as a whole did not view them as having any rights to the continent, and Cabral claimed the land for the Portuguese crown in accordance with a recent treaty between the rivals of Portugal and Spain.

When Magellan set sail in 1519, only six years had passed since the first European crew had actually viewed the Pacific Ocean. Vasco Núñez de Balboa (1475-1519) saw the ocean first in 1513 when he passed through the Darien Isthmus in Central America. He named the ocean the South Sea.

Magellan approached the King of Spain with a proposal for a voyage to the Spice Islands by way of the southern tip of South America. Although he was Portuguese and had sailed under the Portuguese crown for more than a decade,

Magellan was under-appreciated in his home country. Denied promotion and falsely accused of corruption, Magellan turned to his country's rival to fund the expedition. The king of Spain agreed. Magellan spent a year preparing, then began the trip on September 8, 1519, with a 560-man crew aboard five ships. The excursion from Spain to South America took three months. Magellan then led the fleet down the east coast, exploring inlets and bays along the way in search of a shortcut across the continent. Finding none, he continued down toward the tip of the continent.

The fleet encountered difficulties almost as soon as it reached South America. In addition to harsh weather conditions that forced the men to overwinter in the southern reaches of the continent, the voyage was fraught with internal bickering, plotting and attempted mutinies. One unsuccessful attempt to overthrow Magellan's command ended with the marooning of numerous crewmembers on the eastern coast of what is now Venezuela along with the beheading of the ship captains behind the plot. The winter season also claimed one of the five ships in the fleet.

When spring came, the fleet started out again. Magellan continued to take the ships in and out of the bays dotting the coast in his quest to find a shortcut across the continent. When his crew became tired of the lost time with these side trips, Magellan fabricated a story about a map describing a shortcut and convinced the crew to try one last time. It was on this exploratory trip that they discovered what is now known as the Strait of Magellan, a channel that, it was later learned, shaved several hundred miles from their voyage around the tip of the continent. While exploring the strait, the crew of one of the four remaining ships mutinied, turned the ship around and set off for home.

The now-three-ship fleet pressed on, entering the Pacific Ocean on November 20. Magellan and his crew dubbed the watery expanse the Pacific, because it was so calm and peaceful in comparison with the Atlantic they had just left.

Magellan wrongly assumed that the Pacific was a small ocean and that they would be in the Spice Islands in a matter of days. Days turned to weeks and to months. Rations dwindled, and sickness followed. Before they struck land on Guam more than three months after they had entered the waters of the Pacific, 19 of the men had died from scurvy.

Their troubles didn't end there. After landing in Guam, some of the native people took one

of the ship's boats. It was only after a bloody battle that Magellan was able to take back the boat and set off again. Their next stop was Cebu in the Philippines. Here, Magellan formed what was to become a fatal relationship with the king of the island. The explorer agreed to take part in the king's attack on a nearby island, and died after sustaining wounds from a poisoned arrow, two spears, and at least one lance.

The crew continued without him, making the decision to return home by continuing their westward route into Portuguese waters. Apparently, their desire to return to Spain overrode their fear of the Portuguese. The crew abandoned one of the three ships. Of the two remaining ships, the *Trinidad* tried to return across the Pacific against prevailing winds, but was forced back. It was captured by the Portuguese and its crew jailed. In 1525, after being released from the Portuguese jail, four members of the ill-fated Trinidad crew found their way back to Spain.

Juan Sebastián de Elcano took command of the other ship, the *Victoria*. He rounded the Cape of Good Hope and successfully brought the ship back to Spain on September 8, 1522, three years to the day since the voyage set out. The final crew numbered only 18 men. De Elcano received numerous accolades, including honors from the king and a monument in his hometown. The monument consists of a globe with the inscription: "The first one to circle me."

Impact

The Magellan-led journey was important for a number of reasons. Magellan's insistence on exploring the numerous bays along the eastern coast of South America allowed for the creation of detailed maps that allowed future expeditions to find refuge from rough waters and storms. His discovery of the Strait of Magellan became a widely used navigational route around the southern edge of South America. The strait offered shelter from the southern waters of the Atlantic, and also eliminated approximately 500 miles (805 km) from the trip around the continent.

The long voyage across the Pacific Ocean gave Europeans a sense of the sea's massive size. Before Magellan's trip, Europeans were under the mistaken impression that the Pacific was a small body of water that would allow quick passage to the East Indies. Magellan's more-than-three-month voyage helped European mapmakers to not only reveal its true extent (covering a third of

Earth's surface), but to understand much more about the relative sizes and locations of Earth's land masses. It also provided evidence of the reaches of the human race. At nearly every site where Magellan landed, he met native people. Almost all of even the most remote islands inevitably carried human populations. These discoveries often proved to be detrimental to the local people, however, as many lost their freedoms and many others lost their lives to the egotistical-thinking members of European nations.

On an economical level, Magellan's voyage opened the doors to trade with the East Indies. Although the route around South America was too long and dangerous to make sense financially, the circumnavigation whetted the European appetite for expeditions and explorations into and beyond the Pacific. Nations stepped up efforts to travel back and forth across the Pacific by way of the Darien Isthmus in Central America or via other shortcut routes. More than four decades after the circumnavigation of the globe, the Spanish discovered a route—known as the Westerlies—to carry their sailing ships back from the East Indies and to the Americas. Although some historians believe that Magellan might have learned of the west-to-east route across the Pacific and was heading northward toward that route when he landed the fleet in

Cebu, his intentions died with him on the Philippine island. The discovery of the Westerlies opened the trade route between Europe and the East Indies.

England made the second circumnavigation of the globe in 1577-1580. Francis Drake (c. 1540-1596) led the expedition, which had the purpose of locating a cross-continental strait somewhere around the area now known as northern California. Without luck, Drake rounded the Americas at the southern tip of South America, sailed past the Strait of Magellan and confirmed Magellan's notion that the strait separated the main continent from a large island, now called Tierra del Fuego.

LESLIE A. MERTZ

Further Reading

Baker, Daniel B., ed. *Explorers and Discovers of the World, first edition.* Detroit: Gale Research, 1993.

Byers, Paula K. *Encyclopedia of World Biography, second edition.* Detroit: Gale Research, 1998.

Edmonds, J., commissioning ed. *Oxford Atlas of Exploration.* New York: Oxford University Press, 1996.

Hildebrand, A. *Magellan.* New York: Harcourt, Brace and Co., 1924.

Nowell, C., ed. *Magellan's Voyage Around the World: Three Contemporary Accounts.* Evanston, Ill.: Northwestern University Press, 1962.

European Contact Overwhelms the Inca Empire: Francisco Pizarro's Conquest of Peru

Overview

Unknown to the indigenous people of the New World, their destiny was being determined by political and economic forces taking place across the Atlantic Ocean in Europe. Toward the end of the fifteenth century, thousands of daring adventurers would be crossing the ocean to conquer within a few centuries what had taken the Indians thousands to years to inhabit. This "Age of Exploration" was fostered by technological advancements in maritime practices, the belief in an economic philosophy called mercantilism, and an interest in converting the religious beliefs of native populations. Mercantilism was the idea that if a nation was not self-sufficient in its affairs, then its neighbors would dominate it. The two areas that seemed ripe for establishing this ideal were

the Middle East and the Americas. Many of the Spanish conquistadors headed for the New World seeking wealth and adventure. One such conquistador was Francisco Pizarro (1470?-1541).

Background

Spanish interest in the west coast of South America grew after Vasco Núñez de Balboa (1475-1519) discovered the Pacific Ocean in 1513 and brought back tales of untold riches. In 1523, Pizarro, Diego de Almagro (1475?-1538), and Hernando de Luque undertook the initial exploration of Peru that eventually led to its conquest. Their initial contract called for them to divide their shares equally. By 1527 they were convinced of the wealth of the Inca Empire. Failing to secure help

Spanish conquistadors battle native peoples in the New World. *(Historical Picture Archive/Corbis. Reproduced with permission.)*

in the New World, Pizarro returned to Spain, where he received authorization from emperor Charles V to conquer and govern the area extending 600 miles south from Panama. When Pizarro returned with his brothers to deliver the news, Almagro became incensed at Pizarro, claiming that Pizarro was trying to cheat him out of his fair share of the spoils. Despite this conflict Almagro continued to collaborate with Pizarro, and they were working together when the expedition embarked for Peru in late 1530 with 180 men.

When Pizarro arrived in Peru, he established a base at San Miguel on the north coast of Peru. He crossed the mountains to seek an interview with Atahuallpa, the Inca Sun King who had been victorious in the recent civil war. It is clear that the Spanish understood the implications of that war as they were dealing with emissaries from both factions. The actions of the Spanish were to just cover all the bases, but this may have been puzzling to Atahuallpa. At the same time Pizarro was deposing leaders who were loyal to Atahuallpa, he was sending messages that recognized him as the legitimate ruler. Prior to the meeting set up in Cajamarca, the Spaniards indicated that they would come to the aid of Atahuallpa against any group that opposed his rule. Atahuallpa clearly underestimated the Spanish when he agreed to meet with them on November 16, 1532.

Atahuallpa arrived at Cajamarca with an army of about 30,000 men. Confident that the Spanish would not attempt an attack, he saw no reason not to accept their invitation into the village square. He entered the village square carried on a litter and surrounded by 5,000 men. He was approached by a priest who asked him if he would accept God and the King of Spain. He was then given a Bible. According to scholars, he promptly glanced at a few pages and then cast the book to the ground. The priest then implored Pizarro to strike down the heathens; with that, Pizarro lunched a surprise attack. Soldiers came bursting out of buildings, horsemen came flying through doors, and canons and arquebuses opened fire, cutting down the Inca by the hundreds. The Inca were so overwhelmed that many tried to flee the square only to find themselves being suffocated by the weight of the others trying to escape. Despite the overwhelming numbers of the Inca, they were not prepared for such a fight. The battle lasted less than an hour, and the Inca lost 5,000 men and had just as many injuries. But most importantly, Atahuallpa was captured by Pizarro himself.

While in prison, Atahuallpa was allowed to carry on much of his daily existence. Realizing that it was gold that the Spanish were after, he agreed to fill a room with gold and two others with silver as a ransom. The enormous ransom was raised, but Pizarro feared that to release Atahuallpa would mean certain death for the Incan ruler, so they had him executed after a mock trial. Realizing that Atahuallpa's death was a mistake because it weakened their position, the Spaniards approved the coronation of a new leader, Topa Huallpa. He was subsequently poisoned, so the Spanish placed Manco Inca on the thrown. The real Spanish conquest of Peru then began in earnest. The Spanish prevented Manco Inca from having any real power, and he soon realized that the Inca needed to fight for their freedom or at least die trying. He led a year-long rebellion against the city of Cuzco, which had been occupied by the Spanish. In the end, the Spanish weaponry and war tactics were too advanced for the Inca to overcome with just sheer numbers. The Spanish won, and the Incan people were subjected to the perils of slavery, many of them literally being worked to death mining their own precious metals.

Impact

Expeditions similar to those conducted by Pizarro in Peru served to motivate thousands of Spanish peasants to join the military. The discovery of riches and wealth enticed these peasants to travel to the New World in search of a new life. A successful colonial mission could possibly lead to a governorship, wealth, or a pension for the participants. If one were particularly lucky, he could procure riches beyond his wildest imagination. Other men were drawn to the New World by promises of adventure. They looked for quick advancement in the military and diplomatic careers. Still others came on a mission of God. These men wanted to convert the native population to Catholicism. By converting the Americas to God, they believed they would receive eternal blessings.

The discovery of the Inca Civilization in Peru proved to be a huge downfall for the natives. In what would be their first contact with Europeans, nearly 5,000 were killed in just over 30 minutes. With their leader captured, the populace did not know what to do. They were intelligent, loyal subjects willing to do anything for their human god, but they were not trained to think and act for themselves. They offered little resistance to the Spanish onslaught. In fact, many people lament that that was the day the Inca civilization died. While that is probably an overstatement, things changed drastically from that day forward. The Spanish had superior technology with their weapons and were much more en-

lightened when it came to military strategy. Many warriors were severely injured or killed following confrontations with the Spanish, and entire villages were wiped out, not as the result of warfare, but from the introduction of European diseases against which the Indians had no natural immunity. These included such diseases as smallpox, measles, and the flu. The native population had a difficult enough time defending themselves against a known enemy like the Spanish, but it was impossible to protect themselves from the invisible attacks from these diseases.

Another factor that greatly favored the Spanish over the Inca was the constant struggle between neighboring factions. The civilization seemed to always be at war with someone, and therefore the indigenous populations did not fight for common goals. This pattern of dispersed regional groups that frequently were at war with one another may have facilitated the relatively effortless Spanish victory because the people would not or could not band together.

Spain was obsessed with its quest for gold and riches from the New World. The Spanish were single-minded and brutal in their efforts to obtain their prize. Villages were often taken by force until Spain had moved the Americas from a conglomeration of thousands of separate tribes to hundreds of scattered remnants, leaving Spain as the ruling power. One could not have been predicted how quickly the Spanish would rise to power and how far they would extend their influence. Within 50 years they had become the richest, most influential nation in the world. The pushed their way inland and established footholds to expand their territory with ruthless efficiency. They believed it was their divine right to expand their empire and even had papal approval for their conquests. They were ruthless and unyielding in their endeavors and built themselves the enviable position of controlling much of the known gold and silver in the New World. At the same time, Portugal was getting rich through the sea trade routes to India and the Orient. Thus Portugal and Spain had taken the early lead in the race for riches from far-away lands. The English, Dutch, and French, who argued that the seas should be open and that possession of land should depend on occupation, would soon challenge this position. Soon all five of these countries would vie for supremacy of these lands. Because Spain had such a vast area to defend, it could not adequately protect its interests in some areas. This paved the way for the English, Dutch, and French to step in and seize command of much of the trade in that area.

Thus, although Spain pioneered the way and showed tremendous immediate profit, in some instances, it was other countries that would reap the long-term benefits.

JAMES J. HOFFMANN

Further Reading

Berger, Josef. *Discoverers of the New World.* New York: American Heritage Publishing Co., Inc., 1960.

Faber, Harold. *The Discoverers of America.* New York: Macmillan Publishing Company, 1992.

Marrin, Albert. *Inca & Spaniard: Pizarro and the Conquest of Peru.* New York: Macmillan Publishing Company, 1989.

Wilcox, Desmond. *The Ten Who Dared.* Boston, MA: Little, Brown and Company, 1977.

Exploring the Amazon River

Overview

Portuguese captain Pedro de Teixeira (1587-1641) led the first full-length upstream exploration of the Amazon River in 1637. Taking almost two years to complete, the expedition covered more than 3,500 land and nautical miles (5,633 km). Teixeira's voyage was the first to systematically document the Amazon from its silt-laden outlet in Belém, northern Brazil, to the headwaters of its source in the Andes Mountains. The expedition established Portuguese dominance in the vast Amazon basin area of South America, and brought knowledge of the river that Portuguese explorers called the "Rio Mar," or river sea, to the world. Political impact of the expedition reached across Western Europe, Brazil, and the vast sought-after lands and riches of South America. The human impact of Teixeira's expedition included comprised both the Portuguese royalty (and, ultimately, their subjects), as well as the indigenous peoples of South America and Africa.

Background

Portugal was slow to enter the race for exploration of the New World, as most Portuguese assets were committed to solidifying and maintaining Portugal's interests in Asia. In 1494 the Treaty of Tordesillas was engineered by Pope Alexander VI, who feared conflict between two Catholic nations committed to expansion. The treaty divided the unexplored New World, including South America, between Spain and Portugal. Portugal was prohibited to explore beyond a meridian drawn 1,000 miles (1,609 km) west of the Cape Verde Islands. Essentially, the eastern half of South America was reserved for Portugal. Peru, (along with what is now part of Ecuador, Bolivia, Columbia, Brazil, and Venezuela) was reserved for Spain. By 1530 the first Portuguese colonists were sent to Brazil to cultivate indigenous crops and to introduce sugarcane to the region.

In 1540 the first European expedition down the Amazon was led by Spaniard Francisco de Orellana (c. 1511-1546), starting down the Napo River in what is now Ecuador, and finally reaching the Atlantic Ocean in northern Brazil, territory reserved for Portuguese exploration. The Papal decree was difficult to interpret with the technology of the day, as Spain and Portugal both held different interpretations of where the line crossed the coasts of the New World. Other European nations also did not accept the Papal ruling, which prohibited them from any conquest of the territory divided among Portugal and Spain. The French formed a colony in what is now French Guyana, the Dutch in what is now Surinam. Both colonies were near the mouth of the Amazon. Soon, the English and Irish also colonized areas within the Amazon basin and built commercial outposts for trading.

In 1580 Portugal was annexed by Spain. Spain's King Philip IV became monarch over both of the rival nations. By the early 1600s Spain was entrenched in a drawn-out war against France and summoned the help of Portugal to rid the Amazon region from foreign merchants and colonizers. Pedro de Teixeira served in the ruthless campaign, as the Portuguese attacked Dutch and English trading outposts along the Amazon. European colonists were killed or imprisoned, and their primitive homes and businesses were destroyed.

In 1637 Teixeira was chosen to lead an expedition to explore the Amazon from its mouth at Pará (now Belém) to its source. Officially, the mission was prompted by a group of starving Jesuits who had arrived near Belém from deep in the Amazon basin. The Jesuits relayed accounts of their attempts to return hampered by endless, confusing river tributaries and hostile indigenous peoples. Teixeira mounted an organized expedition with over 2,000 men in 47 boats and canoes. Accompanying Teixeira were men skilled in cartography (mapmaking) and navigation. The expedition left Pará in October 1637, and began its slow, deliberate journey documenting the Amazon and its tributaries. After only a few months Teixeira squelched a possible mutiny of his crew. While surveying a particular tributary, the crew observed a village of native people wearing bracelets and holding objects that appeared to be made of gold. The crew expressed a desire to stay for a time at the village they named Aldeia do Ouro (village of gold), and consented to continue the mission only after Teixeira's assurance that they would return later on the downriver portion of the excursion.

Teixeira and his men reached the upper Amazon basin in early 1638 and continued in a northwestern direction up the Napo River, an Amazon tributary that reaches into Peru. When the party reached the junction of the Aguarico River, Teixeira formed a small party to journey overland to Quito, a colonial Peruvian capital in what is now Ecuador. In Peru Teixeira presented the report of his journey along with maps to the surprised Spanish colonial officials. The Spanish Viceroy in Peru, Chinchon, ordered Teixeira and his Portuguese crew to return to Belém via the same route they ascended the Amazon. Chinchon also ordered the Spanish Jesuit Cristobal de Acuna (1597?-1676) to accompany Teixeira on his return journey, and to record a vivid, analytical account of the expedition downstream for Spanish authorities in Seville.

On the return journey, reportedly near the junction of the Napo and Aguarico Rivers, Teixeira held a ceremony claiming the western Amazon region for Portugal. This act of possession defied the Papal ruling of over a hundred years prior by more than 1,000 miles (1,609 km). Reportedly, Teixeira had orders from the Portuguese governor Noronha to claim the territory before the start of the expedition. The Spanish Jesuit Acuna did not record Teixeira's act of possession in his diary of the journey. As the expedition made its way downriver, the crew, anxious to return home, chose not to delay at the village where the gold was observed. The successful expedition returned to Pará in late 1639.

Teixeira was welcomed with an appointment as governor of the state of Pará. He did not have the opportunity to serve a lengthy term, however, as he died within a year of returning from his voyage up and down the Amazon.

Impact

Teixeira's act of possession caused controversy among the two Iberian nations for over a hundred years. In 1640 Portugal broke its alliance with Spain and declared its independence. The Portuguese monarchy was restored with the crowning of King Joao IV in late 1640, breaking its alliance with Spain's King Phillip IV. During the 60-year alliance with Spain, Portugal's empire greatly deteriorated, as the Portuguese aided Spain in its wars with England and Holland. The two countries reciprocated by attacking Portugal's holdings in Asia. Portugal lost its commercial monopoly in the Far East to Holland. Portugal also lost its burgeoning commercial interest in India to England. With the Portuguese empire greatly reduced, holdings in Brazil became more important. Teixeira's claim encompassed almost the entire Amazon basin and greatly enlarged Portugal's influence in Brazil. Acuna, upon reporting to his superiors in Seville, advised Spain to defray ambitions in Brazil, lest they fight the Portuguese as well as the Dutch. Even after the Dutch attacked Rio de Janeiro and Baia, the resolute Portuguese held Brazil.

The accurate surveys and maps made by Teixeira's party aided the Portuguese in their negotiations with the Spanish, and helped open Brazil for colonization. After Teixeira, many expeditions were sent to the interior during the seventeenth century. By the late 1600s and into the 1700s, when gold and diamonds were discovered in villages along the Amazon, a gold rush from all over the world to Brazil was sparked. The Portuguese crown lavished the new-found riches on the restored monarchy, building baroque palaces, and buying expensive imported goods. The ideas of social reform and of building the middle class that were a priority in Portuguese politics before the Iberian union were set aside as the Portuguese crown enjoyed Brazil's riches.

Amazon gold also encouraged the world to update its commercial relations with Portugal. The English allowed Portugal a preferential tariff on wine imported from Portugal. In return, Portugal subsidized English exports and paid for the resulting trade imbalance with Brazilian gold. Portugal intensified the cultivation of sugar, cotton, and spices in Brazil for export. Port towns and trading establishments boomed along the coast of Brazil and near the mouth of the Amazon. The riches from the interior of the Amazon continued to pose challenges to bring to market, and Teixeira's expedition maps served as guides.

The human cost of Brazilian colonization enhanced by Teixeira's expedition is often considered by historians. The expansion of agriculture required a greatly increased labor force. This need culminated in the importation and enslavement of Africans, mostly from Angola. The Spanish and Portuguese Jesuits defended the native peoples of the Amazon against enslavement. Nevertheless, many of the native peoples of the Amazon were forever changed by the presence of the first European explorers. The Spanish and Portuguese introduced modern weapons, Christianity, and European diseases to native Amazonians. Smallpox,

measles, and unfamiliar cold-type viruses decimated whole tribes. Syphilis spread quickly among indigenous peoples. Native culture was altered by the introduction of Christianity, often conflicting with engrained native customs allowing for living in harmony with the environment. Many indigenous tribes, however, refused the counsel or aid of the early Jesuit missionaries.

BRENDA WILMOTH LERNER

Further Reading

de Bare, Capistrano. *Chapters of Brazil's Colonial History, 1500-1800.* New York: Oxford university Press, 1997.

McAlister, Lye. *Spain and Portugal in the New World 1492-1700.* Minneapolis: University of Minnesota Press, 1984.

Smith, Anthony. *Exploration of the Amazon.* Chicago: University of Chicago Press, 1994.

Willem Barents Searches for the Northeast Passage and Finds Svalbard Instead

Overview

The sixteenth century saw the rise of two new Western European powers, England and Holland, each of which had hopes of building international trading empires. Both, however, recognized that Spanish and Portuguese dominance prevented them from plying the routes to the Americas, Africa, and Asia already claimed by the Iberian powers; thus was born the idea of finding a northern passage to Cathay or China. England was the first to send expeditions, both along the northeastern and later the northwestern routes. Each of these efforts was doomed to failure, and finally England gave up the quest. It was at that point that Holland stepped in, sending a captain named Willem Barents (1550-1597) on a voyage to find the Northeast Passage.

Background

Spain and Portugal inaugurated the great era of European exploration, and the first century of that age belonged almost exclusively to them. Portugal's Bartolomeu Dias (c. 1450-1500) rounded the Cape of Good Hope at the southern tip of Africa in 1487-88, opening the way for Vasco da Gama's (c. 1460-1524) historic voyage

to India a decade later. Meanwhile Christopher Columbus (1451-1506) had planted the Spanish flag in the New World, and after him came hordes of Spanish explorers and adventures. While Spain and Portugal prospered from their colonies, two other emerging powers of Western Europe—England and Holland—cast about for ways to develop their own international trade routes.

"There is one way [left] to discover," wrote English merchant Robert Thorne in 1527, "which is into the North. For out of Spain they have discovered all the Indies and seas occidental, and out of Portugal all the Indies and seas oriental." This need became increasingly pressing by the mid-sixteenth century, as Turkish pirates threatened Mediterranean sea lanes and the Iberian powers sought to strengthen their control over the routes they had discovered. Thus was born the idea of a northern passage, a route to Cathay either by the northwest—along the islands to the north of what is now Canada—or by the northeast, around Siberia to China.

The first efforts to find the northern passage fell to the English, whose monarchs (unlike their counterparts in Lisbon and Madrid) had yet to see the value of overseas exploration. Instead,

Willem Barents and his crew prepare to winter at Ice Haven, Novaya Zemlya. *(The Granger Collection, Ltd. Reproduced with permission.)*

these early attempts were almost entirely the result of investment by private individuals, with the guidance of scientific minds who saw the enterprise as one that would benefit knowledge as much as commerce. Initially the English favored the northeastern route, and in 1553 Sir Hugh Willoughby (d. 1554) set sail past Norway's North Cape with three ships bound for Cathay. They made it as far as Novaya Zemlya, a group of islands to the north of Russia, where they all died from a combination of cold and scurvy. A 1556 expedition by Stephen Burrough (1525-1584) got as far as the Kara Sea, but Burrough was luckier than Willoughby: he and his crew managed to winter in the White Sea before returning home the following spring.

In the years that followed, various Englishmen debated the advisability of the northwestern and northeastern routes, and for a time the former gained the upper hand. Unsuccessful expeditions by Sir Martin Frobisher (c. 1535-1594), however, combined with an effort by the Turks to cut off land routes through Persia, influenced a return to the Northeast Passage. The result was a 1580 expedition by Charles Jackman and Arthur Pet, a disastrous effort that took the life of the former. So England again devoted itself to finding the Northwest Passage, sending expeditions such as the one in which Henry

Hudson (d. 1611) perished. The 1631 voyage of Luke Fox (1586-1635) and Thomas James marked the last pre-twentieth-century attempt to find the Northwest Passage, and the last English effort to find any northerly route at all. Now it was the turn of the Dutch.

In 1594 a group of merchants in Amsterdam commissioned the first Dutch effort to find the Northeast Passage, an expedition led by Willem Barents. A native of Tar Schelling, an island off the coast of the northern Netherlands, Barents had spent much of his earlier career sailing in the Mediterranean, and published a travel guide on that subject in 1595. In the meantime, on June 5, 1594, he and his crew set sail from the Dutch island of Texel on what would be the first of three attempts to find the Northeast Passage. The expedition soon encountered treacherous ice floes, and returned to Holland.

In 1595 the Dutch parliamentary body, the States General, financed a second expedition. This time there were two boats, one commanded by Barents and the other by Jan Huyghen van Linschoten (1563-1611), who had earlier distinguished himself in voyages to the East Indies. The expedition set off late, in July, and got no further than the earlier one had. As a result, the States General lost interest in funding a third effort, but the City of Amsterdam stepped into the breach, and in 1596 commissioned yet another attempt to find the Northeast Passage.

Barents was appointed as commander of the expedition, with two captains named Heemskerk and Rijp as his immediate subordinates. The two ships set sail on May 15, and initially the journey seemed to go well. They even discovered new lands: Svalbard, a group of islands comprising some 24,000 square miles (about 65,000 square kilometers) to the north of Norway. (The archipelago is sometimes mistakenly called Spitsbergen, which is actually the name of the largest of its island groups.) Despite this promising beginning, however, disputes between Barents and Rijp soon led to a parting of the ways. Rijp sailed northward, where he ran into icefields and decided to return to Holland, while Barents, Heemskerk, and the other ship continued eastward.

They landed on Novaya Zemlya, circled the island, and found that the ice would not allow them to go any further. Nor could they leave, and they were forced to winter in a bay on the east coast of the largest island. Historical accounts vary on the subject of where the men found the wood with which they built the dwelling where they spent the winter: some writers maintain that they found driftwood, while others hold that they broke up parts of the ship. If the latter was true, it would not have made much difference in the long run, because the gathering ice exerted such pressure on the vessel that it eventually cracked, and was useless to them when the spring thaw came.

In the meantime, Barents and his crew passed a winter of almost inconceivable misery and hardship in their cabin, ironically named the "safe home" or "het Behouden Huys." Lacking knowledge of igloo-building, which would have provided them with better insulation than the wood cabin, the men were literally freezing in their beds even as a roaring fire polluted the air of the dwelling and made breathing difficult. At one point they opened a crate containing linen, cargo intended for Cathay, to give themselves a change of underwear; but they the made mistake of trying to wash their clothes, and the latter froze like stiff boards.

Men were dying of cold and scurvy, and even for those in the best of health, the sound of constantly shifting ice outside made sleep difficult at best. Watches froze, and they had only an hourglass to keep track of the time as the long night of winter settled over the Arctic. Gerrit de Veer, a sailor who kept a journal during this time, lamented the passing of the Sun, "the most beautiful creation by God," in December. Yet there were small blessings in this barren landscape: the surrounding area yielded foxes and bears for food and skins, and plenty of wood for fuel. Early in their time there, the men killed a polar bear and set its frozen body upright in front of their cabin to ward off other creatures.

When spring finally came, the group prepared for the return trip. By then many of them had died, and many more—Barents included—were sick. Lacking the option of using their ship, they set out in two open boats, but Barents and another man died before they had gotten far from Novaya Zemlya. The survivors eventually reached the mainland, where they met a native (probably a member of the nomadic Samoyed people) with whom they communicated by using sign language. They learned that the man had seen another boat in the area, and this gave them hope, which turned out to be justified: after 11 more weeks in the open boat, they were discovered by Rijp, who had returned to the area.

Impact

The Barents expedition effectively ended all attempts to find a northern passage either to the east or the west. Only in 1878-79 would Baron Nils Nordenskiöld (1832-1901) of Sweden finally traverse the Northeast Passage in the *Vega,* and not until the twentieth century would Roald Amundsen (1872-1928) succeed in making the Northwest Passage. Even then, during a three-year journey that ended in 1906, Amundsen his crew were trapped on several occasions, and might have perished if they had not possessed more modern technology and knowledge than that to which Barents had access.

By then the quest for both northern passages had been revealed as futile; driven by fantasies that could never find fulfillment, and resulting in enormous cost of human lives. But Barents's efforts, in a larger sense, were far from a dead end. This series of expeditions, despite their failure, only increased Dutch determination to build a trading empire on the high seas, and as it turned out, this readiness coincided with opportunity. Spain and Portugal, once overwhelmingly dominant in the realm of exploration and foreign trade, began to turn inward, beset by problems such as the defeat of Spanish Armada by the English in 1588. England founded its East India Company in 1600, and Holland the Dutch East India Company two years later. In the decades that followed, Dutch mariners would build colonies in the East Indies, fattening the coffers of trading houses in Amsterdam and Rotterdam, and spawning the era of prosperity so memorably evoked in the canvases of Rembrandt van Rijn (1606-1669).

Eventually the sea to the west of Novaya Zemlya would be named for Barents—who with his crew is remembered as the first European to spend an entire winter in the Arctic—and Svalbard would become an important (if sparsely populated) island group. A number of European nations, eager for whaling and later mining rights, vied for possession of it in the period from the seventeenth to the early twentieth centuries, and after a bitter battle, Norway claimed the area in 1925. The area was the site of heavy Nazi bombing during World War II, and following the war the Soviet Union took advantage of an international treaty governing Spitsbergen to establish mining rights on part of the island. Today the region, described firsthand by English journalist Tim Moore in his 1999 bestseller *Frost on My Moustache,* is of primary interest for environmental studies and research involving extreme cold.

In the nineteenth century a Norwegian seal hunter discovered the encampment where Barents and his men endured the harrowing winter of 1596-1597. Among other things he found a pitiful note from Barents, explaining that he had been detained in his efforts to find a route to

Cathay. During the 1990s Dutch and Russian archaeologists conducted research on the remains of "het Behouden Huys," while scholars began to reconsider Barents's ill-fated voyage as the beginning of a "golden age" of trade between Russia and Western Europe.

An ironic footnote to the Barents story came in the summer of 2000, when a group of Russian sailors was trapped in a submarine deep beneath the Barents Sea. Unlike the man for whom their watery grave had been named, the 118 sailors aboard the *Kursk* did not die in obscurity; they perished as the world watched, through satellite television and Internet updates. Yet in the end, they were as helpless as Barents and his crew had been four centuries before.

JUDSON KNIGHT

Further Reading

Heide, Albert van der. "Dutch Explorer Sought Northerly Route to the Indies." http://www.godutch.com/herald/Feature/barentsz.htm (August 17, 2000).

Moore, Tim. *Frost on My Moustache: The Arctic Adventures of a Lord and a Loafer.* New York: St. Martin's Press, 1999.

"A Voyage through Time: The Story of Barents's Wintering Hut." http://icarus.cc.uic.edu/~jzeebe1/barents.htm (August 17, 2000).

The Discovery of Baffin Bay

Overview

William Baffin (c. 1584-1622) was one of many explorers who searched the waters of northern North America for a cross-continental Northwest Passage linking the Atlantic and Pacific oceans. Baffin is noted as a highly skilled navigator and ship pilot who discovered Baffin Bay, traveled to the northernmost reaches of the continent, and became the first to determine longitude at sea. Although he never found the passage, he came close by charting the Lancaster Sound. More than two centuries later, explorers identified the sound as an entrance to the Northwest Passage.

Background

By the time of Baffin's voyages in the early 1600s, the Spanish and Portuguese had already

found paths to the treasured lands of China and southeast Asia by routes around South America and Africa. Their control of these southern passages left other European nations, including France and England, to place their focus on finding a passage through the northern reaches of either North America or Asia. With the great riches of eastern Asia at stake, the desire to find Northeast or Northwest passages was intense.

The search began in earnest in the late 1500s and early 1600s with the voyages of Englishman Martin Frobisher (c. 1535-1594) in 1576, John Davis (c. 1550-1605) in 1585-1587, and Henry Hudson (1565?-1611) in 1610. Through these expeditions, the explorers collectively discovered straits, sounds, and bays, some of which retain their names today. Baffin joined the quest to find the Northwest passage in 1615 as the chief pilot on the *Discovery*, the ship made famous by Hudson's historic voyage of 1610.

Baffin joined the *Discovery* under the command of explorer Robert Bylot, and the men set sail from England on March 15, 1615. They had hoped to scour the coast of Hudson Bay for entry to the Northwest Passage, but when they neared Hudson Bay in the autumn, they found it packed in by ice. Instead, Baffin made numerous observations, kept meticulous notes and charted as far as possible along the Hudson Strait. He also collected enough information to conclude that the passage wasn't accessible via the Hudson Bay. He was later found to have been correct.

After returning to England long enough to put together another expedition, Baffin and Bylot returned to North America in 1616 to search farther north for a passage. They sailed to the west of Greenland into what is now known as Baffin Bay. Exploring its entire coastline, Baffin again made detailed geographical descriptions of the surrounding area. The navigator noted and charted the presence of the large Lancaster Sound on the west coast of the bay, but failed to identify it as an entry point for the Northwest Passage. While exploring the coast of the bay, he also kept precise records of his astronomical observations, tidal changes and compass readings. Before it was over, the voyage took the expedition to within 800 miles (1,287 km) of the North Pole, a latitude farther north than any other European explorer had ventured or would venture for more than 200 years. The five-month expedition returned to England in August.

In 1617, Baffin signed on with an expedition run by the East India Company. He left with hopes of finding the Northwest Passage from the Pacific side rather than the Atlantic, but was disappointed. The expedition headed eastward as far as India, but never even entered the Pacific Ocean before its return in 1619. Undaunted, he set sail on a company ship again in 1620. This expedition had a fatal ending for Baffin, who was killed during a battle with a Portuguese stronghold in the Persian Gulf on January 20, 1622.

Impact

With each expedition into northern North America, European explorers provided a glimpse into the unknown reaches of Canada and the Canadian Arctic. The stories of their adventures, along with ship logs, navigational notes and geographical descriptions continued to build upon one another. Each expedition paved the way for

the next, allowing explorers to follow previous men's paths into Canada's northern waters and along her inland straits, and then press farther.

Baffin's first recorded voyage in 1612 took him to the west coast of Greenland. During the next two years, he took part in two whaling trips to islands east of Greenland. With a familiarity for the area, this inquisitive navigator was prepared to serve as chief pilot of the 1615 and 1616 voyages with Bylot. The 1615 trip up the Hudson Strait helped verify Hudson's important discovery of the bay named in his honor, and provided critical detail of the Hudson Strait. Perhaps most important, it also discounted the Hudson Bay as a potential entrance to the Northwest Passage, and encouraged further exploration of the Canadian Arctic.

While his trip to Hudson Bay was noteworthy, his 1616 voyage to Baffin Bay brought him fame. Although explorers and fishermen had rounded the massive expanse of Greenland and viewed its adjacent southwestern waters, the expedition of Baffin and Bylot penetrated much farther north into the remote Baffin Bay. Baffin was careful in his descriptions of the geography of the coast. The results of the circumnavigation of the bay were extremely accurate maps of its coastline, including explicit notations of Lancaster Sound. Despite Baffin's reputation as a skilled recorder of geographical features, however, Baffin Bay was eventually removed from some maps because future generations of mapmakers doubted its existence. The next major explorer to tour the northwestern waters off Greenland verified the bay's presence, but not until 1818.

Baffin discounted Baffin Bay as an entry point to the Northwest Passage. Incorrect in this conclusion, his report still had an impact on future exploration by turning attention away from the isolated bay and to other northern areas of both North America and Asia. In the meantime, the Portuguese and Spanish were able to maintain their hold on trade with the nations of eastern Asia.

On the scientific front, Baffin's keen observational and proficient record-keeping skills provided important insights into navigation. Until the 1616 voyage, ship captains and navigators had to rely on imprecise estimates and educated guesses to determine their exact east-west location. Baffin solved the problem by watching how the Moon tracked across the night sky. Specifically, he measured the distance of the Moon in degrees from some other celestial body that remained more fixed in the sky, such as a star. Taking into account the Moon's sweeping arc across the sky each night, and the distance between the Moon and the fixed object, he was able to calculate the near-exact location of the ship. This calculation made by Baffin is often heralded as the first time that longitude was determined at sea.

His records of deviations in his compass readings also proved to be significant as scientists began to study Earth's magnetic field and particularly variations in its magnetic north. Studies now indicate that magnetic north and true north can vary by several degrees, and the amount of that variation can change from year to year. The information Baffin collected made future Arctic explorers aware that their compasses might not always point to true north, instead fading off by a few degrees. Such a variation could potentially veer a ship off-course by many miles.

Overall, Baffin made many contributions to navigation and exploration. His mastery in observation, navigation and record-keeping opened up the Canadian Arctic to exploration, generated exceptional detail for future maps, solved the puzzle of longitude determination at sea, and provided information that helped scientists understand more about the Earth's magnetic field.

LESLIE A. MERTZ

Further Reading

Baker, Daniel B., ed. *Explorers and Discovers of the World, first edition*. Detroit: Gale Research, 1993.

Byers, Paula K. *Encyclopedia of World Biography, second edition*. Detroit: Gale Research, 1998.

Crouse, N. *The Search for the Northwest Passage*. New York: Columbia University Press, 1934.

Edmonds, J., commissioning ed. *Oxford Atlas of Exploration*. New York: Oxford University Press, 1996.

Markham, C., ed. *The Voyages of William Baffin, 1612-1622*. London: Hakluyt Society, 1881.

Semyon Dezhnyov Finds the Bering Strait— Eighty Years before Bering

Overview

In 1728 Vitus Bering (1681-1741) discovered the strait that bears his name, a body of water just 53 miles (85 kilometers) wide at its narrowest point, which separates the Asian and North American land masses. But Bering was not the first European to pass through the Bering Strait: Semyon Ivanov Dezhnyov (c. 1605-1673), a Cossack whose surname is sometimes rendered as Dezhnev, had done so 80 years before, in 1648. Dezhnyov, however, did not know what he had accomplished; nor, thanks to a number of factors—not least of which was czarist secrecy concerning Russian exploration efforts—did the rest of the world.

Background

In an attempt to compete with Spain and Portugal as trading powers during the sixteenth century, both England and Holland launched efforts to locate the Northeast Passage, a sea route from Europe through the Arctic Ocean to East Asia. These attempts would meet with disaster, and in fact it would not be until the nineteenth century that anyone managed to successfully traverse the icy seas above Siberia. By then sailors had long since recognized that the Northeast Passage was only for adventures, and as a trade route had no value. But a number of unintended effects resulted from the effort to discover the passage, among them the growth of trade with Russia and subsequent Russian efforts at exploration.

A 1553 English expedition led by Hugh Willoughby (d. 1554) had proved a failure, but in the course of it his chief pilot, Richard Chancellor (d. 1556), had landed at the port now known as Archangel and traveled over land to Moscow some 1,500 miles (2,400 kilometers). The result of this contact was the formation in 1555 of the Muscovy Company, an English enterprise aimed at Russian trade. The Muscovy Company prospered for nearly a century, but in 1649, Russia's czar ended its trading privileges.

By then Russia itself had become heavily involved in trade and exploration, and no doubt the czar's action resulted from a desire to keep more of the profits in Russian hands. From the late sixteenth century, Russians had begun seeking routes eastward, through the largely unexplored regions of Siberia, but here again government control proved an impediment to exploration—only this time the exploration was being conducted by Russians. Thus in 1616 and 1619, the czar closed an Arctic trade route via the Gulf of Ob.

Meanwhile, in 1581-1582, the Cossack leader Yermak Timofeyevich (d. 1584 or 1585) had crossed the Urals, conquering the Tatar khanate of Sibir and thus opening the region to Russian fur traders. In the years that followed, a number of Russian adventurers explored river-ine routes, though because most rivers in Siberia flow generally north-south rather than east-west, these could only take them so far in their quest to reach the Pacific. By 1633 Cossacks were using the Lena and Kolyma rivers, which they bridged by overland travel, to ply the route between the Arctic and Pacific oceans.

It is important to note that at this point, no one knew where the northeastern corner of the Asian land mass ended, and where the northwestern portion of the North American one began. For all anyone knew, in fact, the two could be connected—as indeed they were 20,000 to 35,000 years ago, when the Ice Age caused a drop in sea level, and permitted the migration of the Siberian tribes who later became known as Native Americans. This knowledge, too, lay far in the future when Dezhnyov set off on his voyage in 1648.

By then in his early forties, Dezhnyov had spent much of his career in Siberia, where he served the czar in posts at Tobolsk and Yeniseysk. In 1638 he moved to Yakutsk, the principal Russian post along the Lena in eastern Siberia, and it may have been during this time that he took a native Yakut wife, with whom he had a son. He moved still further east, to the Yana River, in 1640-1641, and during the following winter took part in an expedition along the upper Indigirka River led by Mikhail Stadukhin. In 1643 he followed the river to its mouth on the Arctic, then sailed east to the Alazeya River. A year later, he was on the lower Kolyma.

Up to this point, Dezhnyov had followed the established path of Russian explorers, traversing north-south rivers to the Arctic, then sailing a little further east to the next river. In 1647, Fyodor Alekseyev Popov invited him to take part in a voyage from the mouth of the Kolyma to that of the Anadyr. Since the former river empties into the Arctic and the latter into the Pacific, this meant that they would have to round the eastern tip of Siberia. The attempt failed, however, as heavy ice in the region prevented them from completing the voyage.

The two men set out again in June 1648 with seven boats and more than 100 men. They reached the mouth of the Kolyma in July, and soon afterward rounded what is sometimes called the East Cape. The latter is also known as Mys Dezhneva, or Cape Dezhnev, and though they did not know it, they had just passed the easternmost tip of the Asian continent. Nor did they realize that they had crossed from the Arctic to the Pacific Ocean, thus proving that Asia and North America are two separate land masses.

At the time, the men had far more pressing concerns on their minds. They had already lost four of the boats, and after entering the Pacific, another was lost. The remaining two boats landed, and were promptly attacked by native Chukchis. As a result, Popov was wounded—his boat was later lost as well—and Dezhnyov became commander of the expedition. Finally Dezhnyov, his crew now reduced to just 25 men, landed south of the Anadyr River. More men died in an attempt to travel up the Anadyr during the winter, and only when summer came was Dezhnyov able, with his 12 remaining men, to make the journey.

Halfway up the river, Dezhnyov and his crew built a fort, which became Anadyrsk, the focal point of later Russian exploration in eastern Siberia. They were finally met by Stadukhin, who had reached the Anadyr overland from the Kolyma, in 1650. The meeting was not, however, a happy one: by then Dezhnyov had begun collecting tribute from the local tribes—a practice typical of Cossacks in Siberia—and Stadukhin was jealous of his profits.

Two years later, in 1662, Dezhnyov sailed down the Anadyr to the Gulf of Anadyr, where he found a large pile of walrus tusks. He returned to Moscow in 1664 with tales of large treasures of ivory to be gained in the Far East, and this spawned further exploration efforts. By 1666 he was back in Yakutsk, but eventually returned to Moscow, where he died in 1672 or 1673. Later his son served Vladimir Atlasov in the conquest of Kamchatka.

Impact

Though he was celebrated in his time, word of Dezhnyov's findings gradually assumed the status of legend rather than fact. Only in 1736 did German historian Gerhard Friedrich Müller, studying archives at Yakutsk, uncover evidence of the groundbreaking expedition. Much had happened in the meantime: Czar Peter the Great

(1672-1725) had taken an interest in eastward exploration, and in the year of his death commissioned Bering to make his historic voyage. In 1733 Russia launched one of the greatest efforts in the history of Arctic exploration, the decade-long Great Northern Expedition. The latter, in which Bering himself perished, resulted in the mapping of virtually all of the Arctic and northern Pacific coastline.

By the time of the next notable venture into the Bering Strait, by Captain James Cook (1728-1779) during his crew's last voyage (1776-1780), Dezhnyov's role in discovering the Bering Strait had been recognized. In 1898 the Russian government named the easternmost point of Asia Mys Dezhneva in his honor, but he was never accorded full worldwide recognition for his efforts.

During the Cold War between the Soviet Union and the United States in the latter half of the twentieth century, the Bering Strait acquired new significance as a strategic barrier between the two superpowers. Some observers noted a physical irony in the existence of the Diomede Islands, two tiny spots of land discovered by Bering in the strait that bears his name. In this place, the Soviet Union and the United States, so widely separated by ideology, were geographically at their closest point: just 2 miles (3 kilometers) separates the Russian Big Diomede from the American Little Diomede.

With the end of the Cold War, the waters of the Bering Strait again became peaceful, with disputes confined chiefly to questions over fishing rights. By the end of the twentieth century, an international group with a site on the Internet called for the construction of a tunnel under the Bering Strait, which would once again link the Asian and North American land masses.

JUDSON KNIGHT

Further Reading

"Bering Strait Tunnel Project." http://www.arctic.net/ ~snnr/tunnel/ (August 17, 2000.)

Fisher, Raymond H., ed. *The Voyage of Semen Dezhnev in 1648: Bering's Precursor.* London: Hakluyt Society, 1981.

Lantzeff, George V., and Richard A. Pierce. *Eastward to Empire: Exploration and Conquest on the Russian Open Frontier, to 1750.* Montreal: McGill-Queen's University Press, 1973.

Diogo Cão and the Portuguese in West Africa

Overview

The voyage of Diogo Cão (1450-1487) up the Congo River established the Portuguese as a major power in West Africa, and especially in the Congo. As a result of this journey, political and economic alliances would be created that would change the history of both West Africa and South America, the most important of which centered on plantation agriculture and the use of slave labor. This would eventually result in the establishment of new cultures on both continents.

Background

By the late fifteenth century, the Portuguese established themselves as a naval presence in the Atlantic Ocean and had developed a highly successful fishing industry that extended into Northern Europe. The Reconquest or wars against the Islamic presence on the Iberian Peninsula also played a major role in their maritime policy. The initial success against the Muslims in Europe gave the Lisbon government the confidence to extend the battle to North Africa. Landings against Islamic strongholds were carried out for two reasons. Initially they were conducted to reestablish Christianity in the area, but eventually the Portuguese began fighting a fifteenth-century "Cold War" against the Islamic empire. Their long-range goals were to contain any future spread of Islam and eventually to roll back Muslim presence in North Africa. In time the Portuguese would come to recognize their presence in the area as a way to gain control of the lucrative African gold trade. After years of failing to penetrate the interior of the continent, the monarchy hoped it could acquire gold by controlling the caravan routes of North Africa. The Portuguese also found a new and growing market for pepper in Northern Europe. By the 1450s, the cattle herds of the continent were so large that the farmers were unable to keep them fed throughout the winter. Every autumn hundreds of livestock were butchered and pepper was used to preserve the meat. The attempt to control the flow of this precious substance became the third reason for the aggressive expansionist policy of the Lisbon government.

The Portuguese were also at the forefront of research and development in navigational and marine technology. Prince Henry the Navigator (1394-1460) created the first modern think tank and invited experts from all over Europe to come to Lisbon to work and research under optimum conditions. His scholars collected a wealth of information about the winds and currents of the Atlantic. He also created an extensive library dedicated to the science of cartography. Although Henry was a fierce nationalist and anti-Muslim, he did respect the scientific accomplishments of other civilizations. He encouraged his scholars to research in the fields of both Chinese and Islamic marine technology. This enabled the European Community to become proficient with both the compass and the astrolabe. These two instruments gave Portuguese navigators the ability to acquire accurate information concerning direction and location north or south of the equator, which gave Portuguese captains greater opportunity to successfully sail the world's oceans. Members of Henry's group also helped perfect the most important advancement in fifteenth century marine engineering, the caravel. This ship was the state-of-the-art vessel of exploration; it was both fast and reliable. The triangular lateen sail allowed the pilot of the ship to take advantage of the wind no matter what direction it was blowing. Improvements in the construction of the hull, including the ability to manufacture the keel from one piece of lumber, enabled the caravel to successfully sail the rough waters of the Atlantic. Finally, the size of the ship allowed it to navigate very narrow and shallow harbors without running aground. These advancements gave the Portuguese the technological advantage they needed to dominate the world's sea-lanes in the middle of the fifteenth century. Their influence was extended first by Bartolomeu Dias (1450-1500) when he rounded the Cape of Good Hope and again by Vasco da Gama (1460-1524) when he reached India.

Based upon the success of da Gama, the Portuguese intended to dominate trade in the Indian Ocean. This body of water provided a route for goods moving from China, South Asia, Southeast Asia, Southwest Asia, and Africa. It was truly the world's first international sea-lane.

The country or civilization that dominated the area would control a significant portion of the world's trade. This also allowed the Portuguese to strike another blow for Christianity in its struggle against Islam. Until the arrival of da Gama the Indian Ocean was alternatively dominated by the Chinese and Moslem civilizations, with Islam the most recent. The Moslem navy was no match for Portuguese technology, therefore the balance of power in the area shifted to the Lisbon government. This new geostragetic situation transferred the control of the trade in spice and other commodities to the Portuguese, and they quickly established themselves as the dominant power in the region. They created a series of strategic fortifications to control this trade. Some were constructed to control the flow of goods from the interior of Asia and Africa to ports on the Indian Ocean. Among the most important were Macao in China, Goa in India, and Mombassa in Africa. The Portuguese also gained control of strategic "choke points" where the flow of trade could be shut down by the nation that dominated these waterways. The Strait of Hormuz at the eastern end of the Persian Gulf and the Strait of Malacca at the tip of the Mayla Peninsula were two of the most critical points. The Portuguese soon discovered that despite their technological superiority, the task of controlling such a wide expanse of territory so far from the Iberian Peninsula was impossible to maintain. The government in Lisbon decided to concentrate its attention closer to home.

Impact

Brazil and West Africa were perceived to be more promising areas of economic growth and would eventually become linked to Portuguese imperialistic ambitions. The turning point in Africa came when Diogo Cão began to explore the Congo River in 1482 and eventually came into contact with the tribal state of Kongo. The leaders of this African political unit were strong and confident, thus they were able to deal with the Portuguese on an equal footing. The power of the government was based upon its control of the flow of important goods from across Africa. A single ruler, referred to as a "Big Man," attempted to regulate this important business. He was far from an absolute ruler, and like his European counterparts he had to deal with many challenges to his power. Two groups, which could prove to be a particular danger, were the secular elite running the government and the religious leadership. When the Portuguese arrived in 1485 the Kongo was locked in a struggle over succession. They were soon perceived as potential political allies, and when Cão threw his support behind the successful pretender Portugal's position in the area was solidified.

An alliance was established between the crown and Kongo government. The Portuguese supported the new leadership with military assistance and in turn they received slaves. The success of plantation agriculture in Brazil created a market for slave labor. The work was so strenuous that the lifespan of the slaves was very short. The work was performed predominately by men, so there were few women to provide replacements for the plantation. This would significantly impact the cultural and historical development of both South America and West Africa. Disease from both Europe and Africa reduced the indigenous population by as much as 80%. A new Latin American culture was produced from the influx of people from Europe and Africa. West Africa on the other hand was completely torn apart by the slave trade. The "Big Men" from the Kongo extended their slave raids deeper and deeper into the interior. Year after year young, strong, and intelligent African males would be taken to satisfy the needs of plantation agriculture. This produced history's greatest population drain and deprived Africa of its youth and vitality. In time it also helped to destabilize the Kongo government. So profitable was the slave trade that individual merchants tried to undercut the Portuguese government's monopoly. They provided anti-government forces with weapons that

could be used to overthrow the existing power structure. This established a culture of rebellion that destroyed the peace and security of the region. In the end, the populations of both Africa and Brazil suffered greatly from the imperialistic drive of the Portuguese.

These problems would extend into the late nineteenth and twentieth centuries. When Europe's second age of imperialism began in the 1870s, Africa was still recovering from the disruption of this fifteenth century invasion. The vast damage that was caused to the continent as a result of slavery, disease, and political destabilization prevented the African people from defending themselves against another wave of colonization. Much of the current political, social, and economic turmoil found on the continent today can be traced historically to this first wave of imperialism.

RICHARD D. FITZGERALD

Further Reading

Abu-Lughod, Janet. *Before European Hegemony: The World System A.D. 1250-1350.* New York: Oxford University Press, 1889.

Crosby, Alfred. *The Columbian Exchange: The Biological and Cultural Consequences of 1492.* Westport, CT: Greenwood Press, 1972.

Curtin, Philip. *Cross-Cultural Trade In World History.* New York: Cambridge University Press, 1984.

Morison, Samuel Eliot. *The Great Explorers: The European Discovery of America.* New York: Oxford University Press, 1978.

Bartolomeu Dias and the Opening of the Indian Ocean Trade Route to India, 1487-88

Overview

The Portuguese Bartolomeu Dias (c. 1450-1500) lies at a crossroad in the history of exploration. For more than 50 years before he set sail to what would become the Cape of Good Hope, Portugal had explored to its own profit along most of the western coast of Africa. When Dias reached the Cape in 1487, he triggered a completely new series of explorations in the Indian Ocean. His achievement should thus be seen as the end of one epoch in the history of European explo-

ration and colonization of the world and the beginning of another.

Background

By going beyond the southern tip of Africa, Bartolomeu Dias fulfilled a hope of many centuries—circumnavigating that great continent. His exploit, however, was not something that came out of the blue, the result of a lone buccaneer's ship in search of great treasures. Rather, it was part of a grand orchestrated strategy that would give Portugal complete control of the eastward trading routes to India before the turn of the sixteenth century. What Dias actually accomplished was to lead the tiny Iberic nation to the threshold of the Indian Ocean—which was crossed ten years later by his countryman Vasco da Gama (c. 1469-1524). Although Dias was following in the footsteps of skilled and daring Portuguese seamen, his achievement was only made possible because of innovative breakthroughs in seamanship.

As a matter of fact, the second half of the fifteenth century saw the art of navigation radically transformed. Prince Henry the Navigator (1394-1460), architect and patron of the epoch-making explorations along the coast of Africa, questioned mathematicians and astronomers (Jews, Arabs, and Christians alike) to resolve a number of problems involving navigation on the high seas, first and foremost establishing one's latitude south of the Equator where one loses sight of the Pole Star—the latter always a key navigation tool for mariners. These scholars provided seamen with new theoretical and practical tools that enabled them to calculate their latitude anywhere south of the Equator by measuring the altitude of the Sun at noon. And so, armed with a cross-staff and mathematical tables to calculate the declination of the Sun (the so-called "Regiments of the Sun), Prince Henry's sailors were capable of finding their way along the west coast of the African continent. Shipbuilding also changed considerably; without the strength and maneuverability of the newly designed caravels, Dias's discovery would have been virtually impossible due in most part to unfavorable currents and winds (whirling, that is, counterclockwise south of the Equator).

When Bartolomeu Dias set sail on a journey that lead him past the Cape of Good Hope, Portugal had already discovered and conquered most of western Africa. There is no question that political, economic, and religious motives were at the foundation of such territorial expansion. The outcome for science, though, was somewhat unexpected. The Portuguese—by sailing beyond Cape Bojador (Gil Eannes, 1434) and into the mysterious and treacherous "Sea of Darkness"—discovered new lands and new stars, and some unheard-of plants and animals. Never before in history had the scientific authority of the ancient scientists, and most of all the geography of Ptolemy, been challenged by such a wealth of observed facts. For some historians of science this chain of geographical discoveries triggered nothing less than the seventeenth-century Scientific Revolution and the rise of modern science.

Impact

By reaching what Diogo Cāo (fl. 1480-1486) before him missed by many leagues, Dias opened an entire new vista of exploration. In fact, when his weather-beaten caravels landed in Lisbon harbor in 1488 after more than a year at sea, the news of his rounding the Cape of Good Hope disappointed an explorer whose dream was also to get to Asia, but by going westward across the Atlantic. His name was Christopher Columbus (c. 1451-1506). Contemporary with Dias, Columbus was at that moment in Portugal trying for a second time to convince King John II (1455-1495) of the viability of his westward expedition. Even though the Italian impressed the King with his "industry and good talent," it became rather apparent to the former—since a sea route to the Indies around Africa was now found to be practicable—that his project was superfluous. Columbus left Portugal to find again his good fortune under the aegis of the Spanish crown. Thus, Dias's achievement delayed once more Columbus's own discovery.

It took a few years of convincing, but Columbus finally accomplished his life-long dream and came back in 1493 with the news that he discovered an alternate route to the Orient (Columbus did not know at that moment that he had landed on a new continent). Spain claimed these new discovered lands for itself, but so did Portugal, saying, for instance, that they were not far enough away from the Azores (islands belonging to Portugal) to be out of their jurisdiction. The Pope had to settle the difference, which resulted in the signing of the famous Treaty of Tordesillas (June 7, 1494). Since the discovery of Dias, it was pivotal for the Portuguese crown to keep intact the gate to the Indian Ocean by the circumnavigation of Africa. For Spain, it was important to lay claim on these new lands (whether or not they were part of a New World or Asia) to ensure that Portugal would not be alone to profit from these new discoveries. Hence it was ruled that a meridian line, drawn from pole to pole 370 leagues (1,185 miles or 1,907 km) west of the Cape Verde Islands, would separate the world between the two Iberic countries, leaving out all the other European nations. Spain was given exclusive rights to all newly discovered and undiscovered lands in the region west of the line, while Portuguese expeditions were to keep to the east of the line. Neither power, of course, was to occupy any territory already in the hands of a Christian ruler. The treaty thus affirmed Portugal's exclusive rights to Dias's discovery and eastward sea route to the Indies.

The running disputes between Spain and Portugal, however, postponed for a full decade the fulfillment of Prince Henry's cherished project: to colonize, Christianize, and take control of the economic trade between Europe and the empire of silk and spices. During his journey, when Dias realized that he had passed the southern tip of Africa, he had wanted to pursue the exploration further. But his crew was becoming restless and longing for home. It was Vasco da Gama, 10 years later, who was chosen by the new king Manuel I (1469-1521) to reach India. When he left in 1497, Dias escorted him as far as Cape Verde Islands, but in a subordinate position. Da Gama was then left on his own to further Dias's previous discoveries. In later years the success of da Gama's oriental mission was considered to be so significant for Portugal that Luis da Camões (1524?-1580) composed an epic poem, *Os Lusiadas* (The Lusiads," 1572), narrating the voyage.

Bartolomeu Dias was involved in one final important geographical discovery, owing in good part to the sea route promptly adopted to reach the Cape of Good Hope. Indeed, because of the strong opposing currents and winds found along the coast of Africa in the South Atlantic, it was easier to sail far to the southwest of the Azores and afterward veer to the east (in order to catch the now favorable currents and winds) than to follow the said coast all the way down to the Cape. Under the leadership of Pedro Álvares Cabral (c. 1460-1526), an armada of 13 ships (composed of seamen, priests, soldiers, and merchants) left Portugal in 1500 en route to Calicut, India, to civilize, Christianize, and trade. One of the caravel's captains was Dias. On their way to the Cape, and mostly because they miscalculated the longitude, they went so far to the southwest that they saw land and forests unnoticed before. The trees were bright red, like glowing embers, hence the name given to the new territory: Brazil. Since the Treaty of Tordesillas was still enforced, Portugal could claim these newfound lands. Some time later, in the vicinity of his epoch-making discovery of the Cape of Good Hope, Dias would meet his destiny.

João de Barros (c. 1496-1570), a sixteenth-century Portuguese historian, gave an account of that fatal day of May 29, 1500: "This happened suddenly: the wind burst down in an instant so furiously that there was no time for the seamen to work the sails, and four vessels were overwhelmed, one of which was that of Bartolomeu Dias; he who had passed so many dangers at sea in the discoveries he had made, principally of the Cabo de Boa Esperanqa. But this fury of the wind ended his life and those of other fellow mariners, casting them into the great abyss of that ocean sea ... giving human bodies as food for the fishes of those waters."

Bartolomeu Dias's explorations are often overlooked in comparison to the fame earned by such explorers as Columbus or Ferdinand Magellan (c. 1480-1521). Dias has the great merit of having found the gates to the sea-route to India even though it was da Gama who forced them open. But most of all let us not forget that, regardless of the great achievements of other explorers before and after Dias, it was as a result of these earliest ocean voyages that scientific instruments (and later technology) were made vital to scientific knowledge and progress. The development of increasingly accurate tools of science to measure time and space went hand in hand with the new geographical discoveries, and hence to ever clearer depictions of the universe.

JEAN-FRANÇOIS GAUVIN

Further Reading

Books

Axelson, Eric. *Congo to Cape: Early Portuguese Explorers.* New York: Barnes and Noble, 1973.

Axelson, Eric. *Portuguese in South-East Africa, 1488-1600.* Johannesburg: C. Struik, 1973.

Hooykaas, Reyer. "The Portuguese Discoveries and the Rise of Modern Science." In *Selected Studies in the History of Science.* Coimbra, 1983: 579-98.

Lamb, Ursula, ed. *The Globe Encircled and the World Revealed.* Aldershot, Hampshire, UK: Variorum, 1995.

Internert Sites

"The European Voyages of Exploration: The Fifteenth and Sixteenth Centuries." http://www.acs.ucalgary.ca/HIST/tutor/eurvoya/index.html.

Vasco da Gama Establishes the First Ocean Trade Route from Europe to India and Asia

Overview

Prince Henry the Navigator (1394-1460) of Portugal is often credited with initiating the "Age of Discovery." This was the term given to the quest of European countries to seek out new lands and trade routes by sea. Prince Henry had multiple motives for this endeavor. He wanted to establish new routes of trade, find a possible route to attack the Moors from the rear, test new advances in shipbuilding and navigational aids, and fulfill his curiosity regarding the world. He commissioned numerous expeditions throughout the fifteenth century to explore and chart the African coast. Although Prince Henry died in 1460, his legacy was firmly established and exploration continued.

King John II of Portugal sought to establish both a land route and a sea route to India. The sea route was to go around the southern tip of Africa, which was not even believed to exist by some at that time. In 1487 Portuguese navigator Bartholomeu Dias (1450?-1500) rounded the cape of Africa in stormy seas and began sailing in a northeast direction to reach what is now South Africa. Dias had shown that there was indeed a possible route to India via the southern tip of Africa. Upon his return voyage, he set up a pillar on the Cape to commemorate its discovery.

Although Spanish explorer Christopher Columbus (1451-1506) claimed to have reached India by a much easier westerly route in 1492, interest in the route around Africa was renewed when the validity of his contention came under close scrutiny. Although the land that Columbus discovered would prove to be very important—he had discovered America—he had not found the rich land of India that was so desperately sought by many. In 1497 a Portuguese captain named Vasco da Gama (1469?-1524) put together an expedition in an attempt to sail around the southern edge of Africa to the port of Calicut, located on the west coast of India.

Background

Da Gama sailed from Lisbon on July 8, 1497, with a four-vessel fleet consisting of two medium-sized sailing ships, a caravel (a small, fast ship), and a large storeship. Because of previous voyages, da Gama knew that the currents along the African coast would impede his progress, so he boldly set a course that took him far from land, sailing in uncharted waters. The explorers rounded the southern tip of Africa, which da Gama named the Cape of Good Hope, on November 22. At this point they no longer need the storeship, so it was broken up and burned. They continued sailing up the eastern African coast but stopped because many of the crews were sick with scurvy. The expedition rested a month so that the men could heal and the ships could be repaired.

On March 2 the fleet reached the island of Mozambique. They were treated friendly because the inhabitants believed the Portuguese sailors were Muslims like themselves. While in port, da Gama learned that the natives traded with Arab merchants, and the Sultan of Mozambique supplied da Gama with a pilot to help guide them. The expedition reached Malindi (present-day Kenya) on April 14, and another pilot who knew the route to Calicut was taken aboard. He proved to be very skilled, and they safely made the treacherous crossing to Calicut in less than a month. They were now in the most important trading center in Southern India at that time and were initially welcomed by Zamorin, the Hindu ruler. Da Gama could not persuade him to make a trade agreement, though, partly because of the cheap gifts that da Gama had brought and partly because the Muslim merchants were extremely hostile to the Christian sailors. As tensions mounted, da Gama left for Malindi in late August with little to show for his efforts.

The crossing to Malindi was extremely harsh. The pilot had abandoned them in Calicut, so they were forced to traverse their way back on their own. The weather was uncooperative, and they were not prepared for a return trip that would last three times as long as the initial voyage. Most of the men contracted scurvy and many died, while the others were nearing death when they finally arrived in Malindi. There, the Sultan helped the crew by providing life-giving oranges and allowing the crew a chance to heal. He also gave them gifts to bring back home. Because da Gama had lost so many men, he ordered one of the ships burned before they began their long journey home.

The remaining two ships set out for home and skirted the Cape of Good Hope on March 20.

They were later separated by a storm, with one ship eventually reaching Portugal on July 10. Da Gama's ship continued on to the Azores, and he reached his original starting point, Lisbon, on September 9. While he was hailed as a hero and his achievement was remarkable—for he had traveled some 27,000 miles (43,452 km) by sea—the trip had taken a great toll. He returned with only half of his ships and less than half of his men. One casualty was especially hard on him: he had lost his brother, Paulo, to sickness on the last leg of the journey. Although da Gama had returned with a small amount of tradable goods, he brought back a much more valuable commodity, a sea route to India. The door was open, and the Portuguese intended to use it to their fullest.

Impact

The discovery of a sea route to India proved to be extremely valuable to the Portuguese. They amassed huge profits in a limited amount of time due to their exclusive hold on commerce in that area. Da Gama opened a passage for his countrymen to follow, and the route was one of the most closely guarded secrets of that era. Pedro Álvares Cabral (1460?-1526) traveled to the Orient within a year of da Gama's return and established a trade treaty with Calicut, which gave the Portuguese a foothold on the commerce of that area. Things did not go smoothly, however, and after some serious fighting, Cabral moved on to another port, Cochin. The ruler of this port was the bitter rival of those at Calicut, so Cabral was able to trade quite easily by playing the ruling parties against one another. He also set up a depot, where trading could take place and where ships could unload. This pattern set the precedent for trading in that area. The Portuguese would play the various mercantile factions against each other to get what they wanted, and if that was not successful, they would use force. Although more than half the fleet and men were lost, Cabral returned ladened with spice, which was sold at a huge profit. Expeditions became annual events, and Portugal profited greatly from these trips. Portugal was now at the forefront of maritime European commerce.

These expeditions to India and the Far East had tremendous impact on the European population. The supply of exotic goods such as spices was higher than it had ever been. While still expensive, these types of items were now available. In addition, many people profited from the sale of such items. These voyages set all of Europe ablaze with thoughts of exotic lands just waiting to have treasure plucked from them.

The Portuguese discovered very important sea routes that had not previously been used by other people. In their quest, they had significantly advanced maritime knowledge, including fine-tuning systematic nautical practices, furthering scientific and technical innovations, increasing cartographical and navigational skills, and fostering a spirit of adventure and endurance. These achievements are important contributions to world history and had a significant influence on other nations of Europe.

Although the Portuguese benefited tremendously from their expeditions, they did so to the detriment of the people living both in that area and even some living far from it. Much of the area and trade that the Portuguese controlled came at the expense of Moslem merchants who had been trading in that area for centuries. Additionally, Venice had a virtual monopoly on trade in that area from land routes, and much of their wealth was derived from it. Although there were other factors that helped its downfall, ultimately it was Portugal's establishment of trade in India that led to its demise.

Portugal was single-minded and brutal in its obsession for this part of the world. Ports and towns were taken by force until Portugal had moved India from a mere trading post to a full-fledged colony. It could not have been predicted how quickly the Portuguese would rise to power and how far they would extend their influence. Within 50 years they had established trade with ports as far away as Japan and all points in between. They pushed their way inland and established footholds to expand their territory. They believed it was their divine right to expand their empire and even had papal approval for their conquests. They were ruthless and unyielding in their endeavors and built themselves the enviable position of controlling the trade routes to India and the Orient. At the same time, Spain was becoming rich from their conquests in America. Thus Portugal and Spain had taken the early lead in the race for riches from far-away lands. The English, Dutch, and French, who argued that the seas should be open and that possession of land should depend on occupation, would soon challenge this position. Soon all five of these countries would vie for supremacy of these lands. And because Portugal was so single-minded and had set up such a wide-ranging trading network, it was soon overextended and could not adequately protect its interests. This paved the way for the English, Dutch, and French to step in and seize command of much of the trade in that area. Thus, although Portugal pioneered the way and showed tremendous immediate profit, it was other countries that would reap the long-term benefits.

JAMES J. HOFFMANN

Further Reading

Berger, Josef. *Discovers of the New World.* New York: American Heritage Publishing Co., Inc., 1960.

Lomask, Milton. *Exploration: Great Lives.* New York: Macmillan Publishing Company, 1988.

Syme, Ronald. *Vasco de Gama. Sailor Toward the Sunrise.* New York: Morrow Publishing, 1960.

Willem Jansz Lands on the Australian Mainland and Sets Off a Century of Dutch Exploration of the Region

Overview

In 1606 Dutchman Willem Jansz (1570-?) arrived on the Australian mainland, becoming perhaps the first European to do so. His achievement did not lead to Dutch rule of the area, as the Dutch were not interested in colonizing it. Nevertheless, his voyage was a milestone because it launched almost a century of successful Dutch exploration of Australia.

Background

Scholars, including the ancient Greeks and Romans, had long contended that a continent must exist in the Southern Hemisphere to balance the large land areas in the Northern Hemisphere. Ptolemy's (fl. A.D. 127-145) world map in the second century and later Renaissance maps depicted a Pacific *terra australis*, Latin for "southern land." Gerardus Mercator's (1512-1594) 1541 map of the world referred similarly to a territory south of Indonesia.

Some sixteenth-century Portuguese maps clearly depict the outline of the northern part of Australia. The most important are the Dieppe maps, so-named after the then-famous cartographic center located in Dieppe, France. The Dieppe maps were well known to eminent geographers of France and England until the mid-nineteenth century; moreover, they were accepted as proof that Portugal had seen and charted the coast of Australia some 60 years before Jansz.

Portugal's impressive maritime history can be traced to Prince Henry the Navigator (1394-1460) and to the great Portuguese explorers of the late fifteenth century, such as Vasco da Gama (c. 1460-1524) and Bartolomeu Dias (c. 1450-1500). Inspired by Prince Henry's dream that

Portugal should promote trade and spread the Christian faith to India, Portuguese mariners ventured into the Atlantic Ocean and around Africa's Cape of Good Hope in search of a passage to India and the Spice Islands. Within a century, Portugal had established a colonial empire in South America and in the Pacific Far East.

On their voyages to the Dutch East Indies, Portuguese sailors were sometimes blown off course by treacherous winds in the Indian Ocean and found themselves along the shoreline of an unknown land somewhere southeast of the Dutch East Indies. The mariners charted part of the northern and eastern coasts of the territory and named this region "Java La Grande." Sixteenth-century cartographers were certain that Java La Grande was the southern continent that had long been sought. One of the Dieppe maps, the 1536 Dauphin Map, shows a rough outline of northeastern Australia as charted by Portuguese sailors. Decades later, the narrow waterway between New Guinea and northwestern Australia was named Torres Strait after Portuguese navigator Luis Vaez de Torres (d. 1613); de Torres sailed under the Spanish flag through the passage north of Cape York only a few months after Willem Jansz. At the time, Torres made no special mention of the landmass he must have seen to the south because, as some historians claim, he was already aware of *terra australis* from earlier Portuguese explorers and maps.

Besides the Dieppe maps, however, there is simply no other tangible evidence that points to a Portuguese discovery of Australia. Therefore, historians still credit the Netherlands and Willem Jansz with the first documented sighting of Australia in 1606.

In November 1605 Jansz set out to explore the area southeast of the Spice Islands. Jansz received his orders from Jan Willem Verschoor, a director of the Dutch East India Company in Bantam on the island of Java (today Indonesia). Jansz's mission was to explore the new region and to determine its trade possibilities with the Netherlands. Commanding the *Duifken* (Dove), Jansz sailed in March 1606 across what was later named the Torres Strait. He and his crew then continued for 200 miles (322 km) along an uncharted coastline. After losing a man in a clash with Aborigines (native Australians), Jansz spotted a land projection that he named Cape Keer-Weer (Turn Again). Unbeknownst to them, Jansz and his crew had actually discovered what is today known as Cape York, the northeastern tip of Australia. Jansz eventually lost nine men in battles with the Aborigines.

After the *Duifken* returned to New Guinea, Dutch cartographer Hessel Gerritsz traced Jansz's voyage and in 1622 drew a map of Australia that showed the continent as a tiny bit of land in the midst of the surrounding sea. Gerritsz's map, although distorted like previous maps of Australia, was important because it charted the earliest documented exploration of the continent. A more accurate map would not be drawn for more than a hundred years, when British seaman James Cook (1728-1779) circumnavigated Australia.

Impact

Jansz's expedition proved to be the catalyst for a concerted Dutch effort to explore the region. In fact, Dutch explorers who followed Jansz were so numerous that Australia was called New Holland for almost 200 years. However, because the Dutch had no interest in colonizing the territory, their interest in the region was gradually eclipsed by the British.

Following Jansz's 1606 expedition, the next notable Dutch landing on Australia occurred a decade later. In 1616 Dutchman Dirck Hartog (fl. 1610s), commanding the *Eendracht*, was accidentally blown off course as he was sailing from the Cape of Good Hope to the Dutch East Indies. For three days he explored the western coast of Australia from 35° to 2° south latitude. At a place that is today called Shark Bay, Hartog left a tin (or perhaps pewter) plate inscribed with handwriting that described his ship's arrival at the Bay on October 25, 1616. Until the eighteenth century a stretch of land running parallel to Shark Bay was called Eendrachtsland, in

honor of Hartog's ship. Today a small island in the bay is still called Dirk Hartog Island. The Dutchman returned several times, in 1618, 1619, and 1620, for further exploration of Australia's western coast. In 1696 Willem de Vlamingh landed in the same area as Hartog had. He found Hartog's plate, copied the words onto a new plate and added his own, describing his visit. (Hartog's original plate now resides in the Rijksmuseum in Amsterdam.) Vlamingh continued with his expedition and discovered a river, later named the Black Swan River. He took with him live black swans, a species unusual enough to have caused quite a sensation when he returned with them to Batavia (present-day Jakarta, Indonesia).

In addition to Dirk Hartog Island, another small island off Australia's west coast is named after a Dutch explorer. In 1619 Frederik de Houtman (1571-1627) and Jacob Dedel sailed the vessels *Dordrecht* and *Amsterdam* along Australia's western shore and described the dangerous shoals off the coast, a site of numerous shipwrecks. There, Houtman Island was named in memory of the captain of the *Dordrecht*.

In January 1623 Dutch East Indies Governor General Jan Pietersz Coen ordered Jan Carstens to command an expedition to explore New Guinea. The *Arnhem*, captained by Willem van Colster, and the *Pera*, commanded by Carstens, landed at Cape York on January 21, 1623. In April Carstens explored the area inland from the Cape and became the first white man to penetrate the interior part of Australia. However, natives killed him and some of his crew. Part of Australia's Northern Territory is named Arnhem Land to commemorate Carstens's expedition in 1623 that explored the western shore of the Gulf of Carpentaria.

In October 1628 Francisco Pelsaert sailed the *Batavia* from Texel, on the northwestern coast of the Netherlands, to the Dutch East Indies. The vessel shipwrecked June 4, 1629, on the Abrolos reefs, the same dangerous shoals southwest of Dirk Hartog Island that had been described by Frederik de Houtman. While Pelsaert explored the shore for water and food, some of his crew mutinied. Pelsaert punished the instigators and after returning to Europe, wrote about his ordeal. Pelsaert's gripping tale of the shipwreck, mutiny, and his days spent on the mainland is one of the earliest about the discovery of Australia and includes the following of the kangaroo: "...a species of cat, which are very strange creatures—the forepaws are very short—

and its hindlegs are upwards of half an ell, and it walks on these alone." Pelsaert's is the first known description of the Australian marsupial.

Finally, by 1636, with the information gathered from the many Dutch sightings of Australian territory, including the *Leeuwin* (1622), the *Gulden Zeepard* (1627), and William De Witt's voyage in 1628, cartographers were able to draw more accurate maps of Australia. In 1642 Anthony van Diemen (1593-1645), governor-general of the Dutch East Indies, dispatched fellow Dutchman, the brilliant Abel Tasman (1603-1659), on an expedition to Australia. Tasman's voyages took him to the southern coast of Australia and on to New Zealand. The body of water between the latter two countries was named the Tasman Sea; the island off the southwestern coast was called Van Diemen's Land and then later received its present-day name, Tasmania.

Despite their extensive exploration of Australia, the Dutch showed no interest in colonizing New Holland. When Britain became the world's supreme naval power in the eighteenth century, James Cook claimed Australia for Britain in 1770. Subsequently settled by the English, Australia became part of the British Empire. Proud of their English heritage, Australians came to think of Cook as the true discoverer of Australia. This pro-British bias was especially evident by the early twentieth century, when some scholars belittled the Dutch role in Australian history and decried any claim that Portugal could have discovered Australia. Several Australian historians engaged in an academic battle over the veracity of the Dieppe maps. Hampered by being unfamiliar with the Portuguese language and blinded by a deep-seated pro-British prejudice, a few historians claimed that the maps were a sixteenth-century hoax. Otherwise brilliant scholars went to extremes to discredit the maps as plausible evidence that Portugal had discovered Australia. These misconceptions were gradually rectified and the authenticity of the Dieppe maps acknowledged. Historians now not only agree that Willem Jansz sighted Australia in 1606, but by the late twentieth century they had recognized that the intrepid Portuguese navigators of the sixteenth century most likely did, too.

ELLEN ELGHOBASHI

Further Reading

Books

Eisler, William. *The Furthest Shore*. New York: Cambridge University Press, 1995.

McIntyre, Kenneth Gordon. *The Secret Discovery of Australia*. Souvenir Press, 1977.

Internet Sites

"European Discovery of Australia." http://www.finalword.com/Touring_Australia/gta_www/gta_011a.htm.

The Voyages of Abel Janszoon Tasman

Overview

It was a long-held belief prior to the seventeenth century that there existed a huge continent in the Southern Hemisphere that would balance the large continents of the Northern Hemisphere. It was commonly known as the great unknown southern continent and was called either Terra Australia Incognita or Nondum Cognita. It was boldly drawn by cartographers, even though there was no evidence of its existence. The discovery of North and South America further fueled the conjecture that below the equator was a huge continent, which had yet to be discovered and explored.

Anthony van Diemen (1593-1645), as governor-general at Batavia in Dutch East Indies, was intent on the exploration of the Southern Hemisphere in order to expand commerce and accumulate wealth. The discovery of land up to this point had been merely coincidental, despite the fact that there was some idea of what was to be found in the Indian Ocean and the South Pacific Ocean. The western and northern borders of New Holland (present-day Australia) were known to exist, but it was not known what laid beyond these. In fact, it was speculated by many that these were actually the coastal regions of the theoretical great southern continent. Others felt that this was actually part of a large island and could be circumnavigated. However, these and many other questions on the geography of the area were unanswered and this is what drove most of the exploration of this area at that time. After one abortive expedition where the leader died shortly after the start, van Diemen decided to send a expedition north toward Japan in order to find the rumored "shores of gold and silver." While this adventure proved also to be unsuccessful, it was noteworthy on three major points. First, the captain on all of the ships was Dutch explorer Abel Janszoon Tasman (1603?-1659?). Second, nearly half of the men were lost due to the disease scurvy. Third, although the expedition was deemed a failure by the Dutch East India Company, it was acknowledged that the region needed further exploration.

Frans Jacobszoon Visscher was a noted geographer in the employ of the Company. He reviewed the known regions and competently designed the scope for expeditions to solve the great southland problem and help support Dutch interests in the region. Visscher envisioned a search for the great southland which would initially track eastward, then turn northward, reverse the initial direction back to the west, and then return to the point of origin by going south. Thus, if no land was found, the journey would take the explorers in a great square. Tasman was chosen to head the expedition, which was slated to begin in August of 1642.

Background

Tasman was given two ships (the *Heemskerk* and the *Zeehaen*) and was told to sail initially to Mauritius. From there, he was instructed to sail south in search of the southern continent. It should be noted that the main focus of this expedition was not exploration, but rather to find better trade routes and search for new sources of wealth and commerce. Tasman was given explicit directions on the route he should take, keeping in mind the objectives of the expedition. As an example, if possible, Tasman was supposed to at one point head for the Chilean coast in an attempt to discover an advantageous route by which Dutch interests could snatch trade from the Spanish in this region of the world. Tasman was not to disclose the importance placed on silver and gold should he encounter possible trading sources, and he was to treat all natives in the most friendly manner possible.

With Visscher on board, Tasman sailed for Mauritius from Batavia on August 14, 1642. In Mauritius, he refitted his ships and set off on the intended course on October 8th. Weather forced a change in plans, and Tasman came upon Van Diemen's Land (presently Tasmania) in late November. Again the weather was a problem, and it was difficult to explore the eastern coast of van

Diemen's land. Eventually the explorers gave up and continued east.

On the 13th of December Tasman saw land again, having reached the shore of South Island, New Zealand. He anchored in an area he termed Murderer's Bay, because three Dutchmen were killed there when their small boat was rammed by a native canoe. Tasman determined that he would not be able to befriend the local native population and continued on to the North Island.

Tasman could not find suitable fresh water in New Zealand, so the explorers turned northeast, discovering Tonga in early 1643. There they found fresh water, and Tasman was treated well by the natives. He now turned northwest and soon discovered the Fiji islands. These islands were uncharted, and Tasman questioned whether their position was calculated correctly. Tasman suggested returning to Batavia; the ship's committee agreed.

They chose a northerly route home, and after some anxious weeks during which they were unable to determine their position due to weather, they eventually were able to deduce their correct position. They took some time to explore the coast of New Guinea and determined that there was no passage through it. Tasman returned to Batavia in June, completing a ten-month voyage. During this time, he had completely circumnavigated the continent of Australia without ever catching a glimpse of its land.

Van Diemen was not especially pleased with the results of the voyage since no new trade routes or sources of wealth had been discovered. However, he was still interested in resolving several issues around New Guinea and New Holland, so he outfitted Tasman with three ships, the *Limmen*, the *Zeemeuw*, and the *Bracq* for another expedition in February of 1644. There is little information on this voyage except for the fact that the explorers charted the northern coast of Australia and returned in October of the same year. While Tasman noted the presence of natives, he did not seek to trade with them. The Dutch East Indies Company, which had sponsored Tasman's voyage, again was not happy because there was no return on their investment. Thus Tasman's efforts were not highly regarded by the company.

Impact

Tasman had set a new standard for Dutch exploration on his voyages, but his expeditions were deemed to have been relatively fruitless and created little excitement for the Dutch East India Company. In fact, he was looked upon with some scorn by the stockholders and many officials. The voyages resulted in no new trading partnerships, and there were no major resources or wealth uncovered, which would be of obvious benefit to the company. Furthermore, in his first voyage, Tasman had not proven there was a passage through the south ocean to Chile, which was one of the original objectives of the expedition. One positive of his first voyage was that he had only lost ten men on the entire trip and over half of those were due to natural causes. This was remarkable in a day and age where many expeditions would return with only a handful of men. There were a few promising details from the first voyage, and Tasman was thought of highly enough to figure prominently in a second expedition.

The second voyage proved to be even a bigger disappointment than the first. In a dispatch to the company's council, van Diemen expressed his disappointment and discontent that the expedition had not discovered a passage through New Guinea. He further stated that Tasman had done nothing but sail along the coasts and had gained no knowledge of the country and its productions; according to van Diemen, Tasman claimed that the explorers did not have enough manpower to venture onto land in the face of the savages. Van Diemen went on to state that Tasman in his two voyages had circumnavigated the hitherto unknown South Land, which was calculated to have an extent of 8,000 miles (12,875 km) of coast, yet in so great a country, with such a variety of climates, he had found nothing of great importance and profit for the company. It was plain that van Diemen was dis-

satisfied with Tasman. He had looked for immediate results in the extension of trade, or at least for the finding of the New Guinea strait, and, disappointed in this, he could not appreciate the importance of the discoveries from a geographical standpoint. As a direct result of Tasman's voyages, the Dutch were reluctant to undertake any costly expeditions unless it could readily be proven that there would be immediate and substantial profits. Because of this attitude, Tasman had completed the last great Dutch exploration.

The legacy that Tasman left is a much-improved understanding of the geography of the South Seas. While not all of the assumptions proved to be correct, Tasman had proved that New Holland did not extend indefinitely to the east, but that it was an extremely large island. He now separated the real land from the legend. He had also shown that, in fact, there was no large continent in that area of the world. He had discovered many islands and charted that region of the world better than anyone had previously. This stimulated a keen interest in the area and helped to drive the exploration of this area in the future.

JAMES J. HOFFMANN

Further Reading

Allen, Oliver E. *The Pacific Navigators*. Alexandria, VA: Time Life Books, 1980.

Beaglehole, J. C. *The Exploration of the Pacific*. London: Adam & Charles Black, 1966.

Sharp, Andrew. *The Voyages of Abel Janszoon Tasman*. Oxford: Clarendon Press, 1968.

Introduction of the Mercator World Map Revolutionizes Nautical Navigation

Overview

In 1569, Flemish cartographer Gerardus Mercator (1512-1594) broke away from the teachings of Greek mathematician and astronomer Ptolemy (90-168) and published a world map, which introduced a new system of projection for marine charts featuring true bearings, or rhumblines, between any two points. His system presented a revolutionary cylindrical projection where a straight line between any two points

forms the same angle with all the meridians, and became the basis for modern day navigational charts.

Background

The practice of cartography, or map-making, can be traced back to early examples from Babylon, Egypt, and China, where the first maps were printed. European map-making can

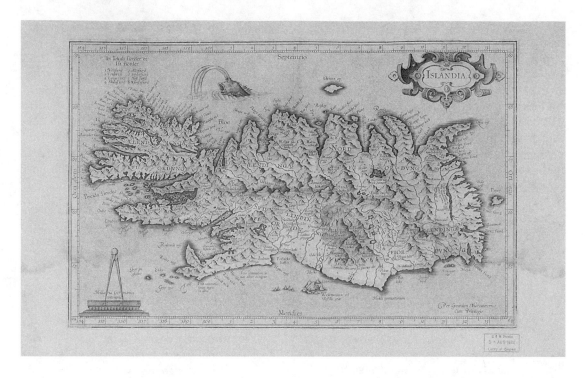

A map of Iceland by Gerardus Mercator. *(Corbis Corporation. Reproduced with permission.)*

be traced to early Greek culture and the most significant contributions to the study of geography were made by Claudius Ptolemaeus known as Ptolemy, a Greek astronomer and mathematician. Ptolemy created an eight-volume cartographical work entitled *Geographike Hhegesis or Guide to Geography*, which featured his own research as well as that of his predecessors. These included Eratosthenes of Cyrene (c. 276-c. 194 B.C.), who developed an accurate measure of the circumference of the globe; Crates of Mallus (fl. early second century B.C.), who formalized the concept of a globe; Hipparchus (fl. late second century B.C.), who worked out a grid of Earth; and Poseidonius (c. 135-c. 51 B.C.), who developed a "corrected" figure of Earth's circumference, which was smaller and less accurate than that of Eratosthenes and, unfortunately, was the figure chosen by Ptolemy to represent the circumference of the globe.

The cartographic principles established by Ptolemy became the fundamental elements of geography for centuries. His *Geography* was translated to Latin in the early fifteenth century and maps of his world were constructed and became the model for Renaissance exploration. Before that, however, two developments of the twelfth and thirteenth century had an impact on navigation and cartography: the magnetic compass, in use in Western Europe by the late 1100s, and Portolan charts. Based on the *por-*

tolano, a harbor-finding manual of the Middle Ages, Portolan charts, printed on goat and sheepskin and very rare and expensive, were the first charts developed exclusively for nautical use. Common in the Mediterranean, the Portolan chart eventually covered the upper coasts of Africa and those of northern Europe.

About the same time as Portolan charts were being enhanced to include Africa, Prince Henry of Portugal (1394-1460), known as Prince Henry the Navigator, established his "navigational" school at Sagres near Cabo de São, Portugal. The school employed cartographers, including Portolan chart-makers, to develop new maps based on data collected during expeditions sponsored by Prince Henry. By the time of his death, expeditions under his sponsorship had rediscovered Madeira and the Azores in the Atlantic and had explored southward along the coast of Africa as far as Gambia. As information from these discoveries were added to the cartographic record, more European countries became interested in maritime exploration and Prince Henry's lasting contribution (influenced by his accomplishments and the tales of the thirteenth century explorer Marco Polo (c.1254-1324)) was to instigate a tradition of European maritime discovery.

Men like Bartolomeu Dias (c.1450-1500), who discovered the Cape of Good Hope; Vasco da Gama (c.1460-1524), who rounded Africa and reached India in 1498; Christopher Columbus (1451-1506), who discovered North America but was convinced he had reached Asia; and Amerigo Vespucci (1454-1512), who rediscovered North America on his return from Brazil, all contributed to events which would radically alter the world map of the time. The 1519-1522 circumnavigation of the globe by Ferdinand Magellan brought back better understanding of the expanse of the Americas and the Pacific Ocean. These momentous discoveries by European explorers were matched by the invention of printing in Europe and the first European printed map. By 1507, much of the geographic knowledge and recent discoveries appeared on a world map, one of the first ever printed, published by Martin Waldseemüller (c.1475-1522) of Germany. His map depicted the new continent discovered across the Atlantic as a separate entity and named *America* in honor of Vespucci.

As men explored the oceans and coastlines of the world, they found that the Portolan charts were inadequate for navigation over the expanses of oceans. The need for a chart of latitudes and longitudes instead of directions and distances prompted Renaissance mathematicians to experiment with various map projections to accommodate both the new geographical data and the problem with navigation. This new scientific approach to cartography stimulated one Flemish cartographer to abandon the teachings of Ptolemy and develop a new system for navigation charts.

Impact

The growing volume of new geographical data persuaded Gerardus Mercator (1512-1594) to abandon the time-honored theories of Ptolemy and other cartographers and to construct his maps and globes to reflect the latest observations and geographic knowledge. Using his background in mathematics, Mercator sought a practical solution to represent the curved meridians and parallels of the globe on the flat surface of a chart. This problem was especially significant to mariners who were beginning to explore the vast expanse of the world's oceans after centuries of sticking to the coast lines.

Compiling data from Spanish and Portuguese charts with his mathematical system of projection, Mercator published his world map for navigation at sea in 1569. The cylindrical projection enabled navigators to compute a compass bearing in a straight line, enabling the sailing of the shortest distance between two points. Because the compass bearings could be plotted as straight segments on Mercator's map, a ship's course could be found by laying a straight rule across it. The map, of which only four copies survived, was made up of 24 sheets with dimensions being 131 x 208 cm. Intended for navigators, it also represented the land surfaces of the globe as accurately as possible based on current geographical data gathered on expeditions around the world. Mercator regarded his world map as one part of a coordinated scheme of cartographical research that culminated in the posthumous publication of his greatest work, his *Atlas* (1595), a collection of maps which included 27 originally prepared by Ptolemy (with corrections and commentary by Mercator), and his own maps of the countries of Europe and the world. Mercator is credited with coining the phrase *atlas* for a collection of maps.

Although sailors were slow to adopt the new Mercator mapping device, the projection was eventually used by some of the most important explorers of the time. The concept of Mercator's projection was embraced by scientists who were determined to discover the mathematical reasoning behind it. Because Mercator did not explain his map—he merely included a graphical device showing how it could be used to solve the nautical triangle—it was not practical for chart-makers and sailors. His solution was not perfect; it approximated longitude because neither the knowledge of the true size and shape (many still, despite vast evidence to the contrary, believed the world to be flat) of the globe nor the mathematical resources of his time were adequate to permit great accuracy.

The problem of determining longitude delayed the use of Mercator's charts until two English mathematicians developed a solution using trigonometric tables. Thomas Harriot (1560-1621), working privately for Sir Walter Raleigh (c. 1554-1618), and Edward Wright (1558-1615), constructed tables of meridional parts, by which lines of latitude on a Mercator chart could be spaced. The compass bearing, or plot, on a Mercator chart is thus determined using the difference in meridional parts and longitude. (Wright published his table in 1599 and is often given sole credit for the solution.)

Despite the table of meridional parts, determining longitude was still problematic for sailors and would require the invention of an accurate chronometer, which was accomplished in 1759 by English inventor John Harrison (1693-

1776). Harrison's seagoing chronometer was employed by James Cook (1728-1779) during his circumnavigation of the globe. The charts Cook compiled during his voyage were so accurate and detailed that they changed the nature of navigation and cartography forever. In 1884, the countries of the world agreed to adopt the meridian of Greenwich, England, as the Prime Meridian (0°), making longitude constant on all future navigational charts around the globe.

Today, nearly all navigational charts are constructed using the Mercator projection with the exception of maps of large areas, such as the entire Pacific Ocean, or charts of the Polar regions. However, modern cartography owes much to the inventions of other scientists, such as photography, the airplane, the computer, and space technology. The advent of aerial photography, first from balloon, then airplane, and now satellite, has had a tremendous impact on the development of more accurate and infinitely detailed maps of the globe. The use of modern technologies in the maritime realm, such as global positioning systems and highly sophisticated radar, has significantly changed the work of a navigator on the seas and oceans of the world. The Mercator map is still used, but it has been overshadowed by these commonplace modern inventions.

ANN T. MARSDEN

Further Reading

Books

Crone, Gerald Roe. *Maps and Their Makers.* Folkestone, UK.: Wm Dawson & Sons Ltd, 1978.

May, W.E. *A History of Marine Navigation.* New York: W.W. Norton & Company, Inc., 1973.

Tooley, Ronald Vere. *Maps and Map-makers.* London: B.T. Batsford, 1987.

Internet Sites

"The History of Navigation." http://boatsafe.com/kids/navigation.htm.

Biographical Sketches

Alonso Alvarez de Piñeda
1494-1519
Spanish Navigator

According to meager historical references, Captain Alonso Alvarez de Piñeda's life was short but exceedingly eventful and productive. The only actual mention of his date of birth appears in a biography of his immediate superior, Francisco de Garay. It states that Piñeda was born in Spain in 1494 in the village of Centernera.

His adventures began when Garay (who was then governor of Jamaica in the Caribbean) commissioned Alvarez de Piñeda to command a flotilla of four vessels with the express purpose of finding the imagined Southwest Passage water route to the Orient and the subsequent treasures of China and other Asian civilizations. It was a huge responsibility for a captain who was only 25 years of age.

Alvarez de Piñeda took his assignment seriously. He spent the next nine months sailing along the coast of present-day western Florida and all around the gulf to Vera Cruz, Mexico. The remarkable talent of this young explorer was his attention to detail. He not only stopped numerous times along the unfamiliar coastline, but made comprehensive maps of each area he encountered with notes about the settlers and their cultures.

Before leaving the Caribbean, Francisco Garay gave Alvarez de Piñeda instructions to intercept the flotilla commanded by Hérnan Cortés (1485-1547) in Vera Cruz. His plan was to claim that portion of Mexico for Spain and to oust Cortés. This mistake in military judgment was soon evident when Alvarez de Piñeda anchored his ships and sent men ashore to take command. The plan backfired when Cortés captured the landing force and sent Alvarez de Piñeda and his flotilla packing.

They resumed their voyage along the coast of Texas and Alvarez de Piñeda is further credited with the discovery of what is now Corpus Christi. He gave the settlement this name to honor the Catholic feast day of Corpus Christi. He also went ashore at another settlement in south Texas and was instrumental in colonizing the area that is now Brownsville.

Because his vessels were all badly in need of repairs, Captain Alvarez de Piñeda anchored his fleet at the mouth of what he called Rio de las Palmas, now believed to have been the Rio Grande. He remained there 40 days while his men secured the supplies needed for repairing the damaged vessels.

Alvarez de Piñeda used this sailing break to travel inland and discover any inhabitants who might aid him in his mapping work. From archive records of the time, it appears that he traveled about 18 miles (29 km) inland along the river and reported encountering approximately 40 different Indian tribes, many of whom were wearing ornaments made of gold. This encouraged him to make written recommendations to Governor Garay that colonists be sent to the area and, in the name of Spain, to claim the entire coastal landmass from Texas to Florida under the name "Amichel."

Although written accounts of these events are lacking, there was an unexpected discovery that turned speculation into historical fact. In 1974 the Harlingen Naval Reserve Unit was excavating for Civil War artifacts on behalf of the Rio Grande Valley Museum. While digging for these treasures, they came upon a clay tablet with a Spanish inscription that translated into English as: "Here [. . .] Capt. Alonso Alvarez de Piñeda in 1519 with 270 men and 4 of Garay's Ships."

According to Bernal Diaz del Castillo (one of Piñeda's officers), Captain Alvarez de Piñeda died as a result of wounds he received while fighting Indians on the Panuco River. He was 25 years of age when he was reported dead.

Although Piñeda's career was very short, it was of great importance to all who followed him in that he proved—without a doubt—that the coast of the Gulf of Mexico was a solid landmass with no possibility of a water passage to anywhere else on the continent.

Scholars are indebted to Clotilde P. Garcia, M.D., of Corpus Christi, Texas, for much of the known historical information about Alvarez de Piñeda. Her years of research were conducted in order to prepare a paper to qualify for the issuance of a historical marker for Captain Alonso Alvarez de Piñeda in Corpus Christi.

BROOK HALL

William Baffin
1584-1622
English Navigator, Ship Pilot, and Explorer

While searching for the elusive Northwest Passage, William Baffin made important

discoveries about the geography of the northern reaches of the New World. During two of his most noteworthy expeditions—both with explorer Robert Bylot as commander—Baffin searched the waters now known as Hudson Bay or Baffin Bay for entrances to the Northwest Passage, which would connect the Atlantic Ocean to the Pacific. One of his expeditions with Bylot led the men and their crew to within 800 miles (1,287 km) of the North Pole—the northernmost point ever reached in the Canadian Arctic. The record stood for more than two centuries. Baffin's exploits in the Canadian Arctic are now immortalized in the names of Baffin Bay and Baffin Island.

Baffin was born around 1584 in or near London, but little else is known of him until he took the position of chief pilot for a Greenland-bound ship in 1612. He continued his adventures by taking positions in 1613 and 1614 aboard ships funded by the whaling outfit named the Muscovy Company. During these voyages, the English pilot was able to learn about the coasts of the Spitsbergen Islands in the icy waters about 500 miles (805 km) east of Greenland.

Baffin's experience in the Arctic helped him attain the title of chief pilot for an expedition commanded by Robert Bylot. The men set sail on March 15, 1615, aboard the *Discovery*, a ship made famous by explorer Henry Hudson (c. 1565-1611) when he discovered what is now known as Hudson Bay in 1610-1611. Baffin and Bylot planned to lead the *Discovery* back to Hudson Bay with the express purpose of determining whether a Northwest Passage originated from its waters. They got as far as the Hudson Strait, but were forced back by thick ice before they could enter Hudson Bay. Nonetheless, Baffin's observations during the trip allowed him to conclude that the Northwest Passage did not connect with Hudson Bay. He was right.

Baffin and Bylot returned to England for a short time before leaving on the *Discovery* for the Canadian Arctic on March 26, 1616. During this expedition, they ventured into a large expanse of water west of Greenland to a point farther north than any other North American explorers had gone. Baffin charted the serpentine coastline of the bay and the large adjacent island, both of which now carry his name. He also charted and named the Lancaster Sound, which connects to the bay on the west. The sound was later shown to be an entry point for the Northwest Passage, Baffin never identified it as such.

Although unsuccessful in the discovery of a Northwest Passage, Baffin's charts of the Hudson Strait and Baffin Bay provided important new information. He is also credited with making a significant navigational finding during the voyage by determining longitude at sea. His method involved calculating the distance of the moon from another more fixed celestial object. In addition, he maintained careful records on his observations of the Moon and stars, the tides and even the variations in his compass readings as they neared the Earth's magnetic pole. The latter helped future scientists to learn more about the pole's variations from year to year.

After he returned to England, Baffin continued his quest to find a Northwest Passage. He began a two-year voyage on February 4, 1617, in hopes of finding the passage from the Pacific side of the continent. The ship, commissioned by the East India Company, never traveled as far as the Pacific, however. Undeterred, he set out on another East India Company expedition in 1620. Almost two years into the voyage, the fleet engaged in a battle with Portuguese adversaries in the Persian Gulf. Baffin died in combat there on January 20, 1622.

LESLIE A. MERTZ

Vasco Núñez de Balboa
1475-1519
Spanish Conquistador

Vasco Núñez de Balboa was a Spanish conquistador who explored Central America, was the first to establish a permanent settlement in Central America, and was the first European to see the Pacific Ocean.

Balboa was a descendent of the Galician family of nobles in Castile; he began his life at Jerez de los Caballeros in the province of Estremadura. He grew up in a time when many from his social class were sailing to the New World to seek their fortune; he set out on his own in 1500. He sailed to present-day Colombia with Rodrigo de Bastidas (b. 1460?) but eventually moved to Hispaniola (present-day Haiti) to try his hand at pioneer farming. Unfortunately, Balboa experienced financial troubles, and in an effort to evade his creditors, he stowed away in a provisions cask aboard an expedition headed by Martín Fernández de Enciso (1470?-1528). This expedition took Balboa to a struggling Spanish colony in present-day Colombia. There, using his knowledge of the area along with his intelli-

Vasco Núñez de Balboa. *(Library of Congress. Reproduced with permission.)*

gence and sheer willpower, he persuaded the remaining members of the colony to relocate across the Gulf of Uraba to Darien on the Isthmus of present-day Panama. Once there, he established the town of Santa Maria de la Antigua, the first permanent settlement in Central America. The town elected two magistrates, one of whom was Balboa. With the departure of Enciso to Hispanola, Balboa quickly moved to become the leader of the settlement. King Ferdinand of Spain declared Balboa the interim governor and captain general of the area in December of 1511; Balboa was 36.

Balboa began to explore and ultimately dominate the area, subjugating the Indians to slavery and sometimes torture to extract information about other Indian tribes. His treatment of the Indians was marked by force and by a policy designed to make the tribes war with themselves, making Balboa's task of domination all the easier.

Balboa and the Spaniards were told by the Indians of a sea that was to the south and of a gold-rich culture of Indians; he set about immediately to gain support for the expedition. Unknown to Balboa, an expedition was set out from Spain, but he ultimately was not to be the commander. Pedro Arias Davila, an aging nobleman, with 2,000 personnel, left Spain in April of 1514 with the objective of taking over for Balboa.

Balboa, impatient with waiting for the support from Spain, moved on from the settlement in Santa Maria de la Antigua to the narrowest part of the isthmus with 190 Spaniards and Indian support. On September 25 (or 27), 1513, Balboa became the first European to see the Pacific Ocean, which they called the South Sea. Balboa claimed the land and the sea in the name of the king of Spain; he was made governor of the *Mar del Sur* (South Sea) and the provinces of Panama but was to be under the authority of Pedro Arias Dávila.

Balboa and Dávila had a relationship marked by distrust and jealousy. Even under these conditions, Balboa was given authority to explore the South Sea, or the Pacific Ocean. Balboa oversaw the tremendous effort to build a fleet of ships on the Atlantic Ocean side, disassemble them, and then transport them across the isthmus, over mountains and through swamps, to the Pacific side where they were reassembled and used to explore the Gulf of San Miguel. During this time, Balboa's claims of incompetence leveled at Dávila succeeded—the king replaced Dávila with another governor. Dávila, in an effort to save his career and possibly his life, ordered Balboa home to discuss matters of mutual concern. Once Balboa arrived, he was charged with rebellion and after a mock trial was beheaded along with four accomplices in January 1519.

MICHAEL T. YANCEY

Willem Barents
1550?-1597
Dutch Navigator

The name Willem Barents is almost as well known to Dutch children as Hans Brinker, hero of the famous finger-in-the-dyke folk story. Born around 1550, Barents went on to a naval career that brought him a permanent place in history for his deeds and heroism, for which the Barents Sea is named after him.

In the early 1500s both the Dutch and the English were interested in finding a northeast passage to China and the Indies to facilitate trade and commerce in these promising, fruitful areas. The newly established Muscovy Trading Company in London funded the first expedition in 1553 but, after 25 years without tangible results, the British settled for profitable trading with the northern Russians. This left the field wide open for the ambitious merchants in the Netherlands.

Because of his successful voyages to and from Spain, as well as numerous ports on the Mediterranean, Barents worked in Amsterdam with Dutch geographer Peter Plancius to create a navigational guide for those voyages. Barents was a cartographer and provided future historians with a now-famous introduction to the art of cartography as well as competent seamanship.

The initial voyage was sponsored by the Estates of Holland and left the island of Texel on June 5, 1594, to explore the possibilities of a northern passage to the Indies. They continued for a relatively short time before encountering a daunting sea full of ice floes and bergs of all sizes. Satisfied, for the time being, they returned home and made their reports.

The Estates decided to try again the following year and appointed another officer to command the expedition. Because of a later departure in July 1595, they found the ice fields even more treacherous and seas that had been previously navigable were now impossible to cross. In addition to these disappointments, several men were lost while trying to return to Amsterdam.

The third venture was undertaken and financed by the City of Amsterdam with Willem Barents in command. He was able to depart on May 15, 1596, along with two other ships. Surviving records show that Barents and one of his captains, Rijp, had a disagreement, which resulted in Captain Rijp's changing course, running into formidable ice fields, and returning home. Captain Heemskerk remained with his commander and both ships were caught in the deadly grip of the hardening ice that surrounded them. Eventually, both vessels were forced upward, out of the ice and were broken up by the inexorable forces that surrounded them.

By this time they had reached the icy shores of Nova Zembla, an island sometimes called Novaya Zemyla, off the Russian coast. Realizing they would be spending at least six months in the harshest of circumstances, the crew members and their officers began at once to salvage the ships' lumber to build a longhouse or cabin to house them and to store whatever they could recover from the wreckage.

Journals kept and brought back by survivors recount a tale of unbelievable hardship that they endured on Nova Zembla. Even with a wood fire kept burning at all times, the sheets on their makeshift beds would be frozen solid along with whatever they would cook and try to eat or drink. They gave up trying to wash any clothing since it would start to freeze as soon as it left the warm water and could never be dried or worn again.

Since there was no outlet for the smoke from the fire, it settled in the cabin and made breathing not only unhealthy but almost impossible. It got so cold that their watches stopped and they used a 12-hour glass to keep time.

When the worst was over, they realized they had to try to leave the island or perish if they remained. They provisioned several longboats as best they could and started out on the 1,600-mile (2,575-km) journey home. Unfortunately Willem Barents did not survive the harrowing trip and died at sea. Those who made it back told of his inspiring leadership along with other accounts of the adventure, which are still told around Dutch fireplaces and remain relevant today. Willem Barents was close to 47 when he perished in 1597.

Confirming evidence of their incredible story was found in 1871, when another explorer discovered the remains of the Arctic dwelling they had built, along with the tools, instruments, and other artifacts they had left behind. These relics have been preserved and can be viewed in The Hague, Netherlands.

BROOK HALL

Jacques Cartier
1491-1557
French Navigator and Explorer

Jacques Cartier is the adventurer who is often credited with discovering Canada. He was the first European to locate the Gulf of St. Lawrence and the St. Lawrence River, and was the first to venture deep into the northern wilderness of the North American continent. The accounts he made of his journeys became the basis for the first maps of the area. Those maps opened the major routes followed by later French explorers into Canada.

Born in the port of Saint-Malo, in the French province of Brittany in 1491, Cartier began his seafaring career with a number of voyages to Brazil, Newfoundland and perhaps even to America as a member of Giovanni da Verrazzano's crew in the 1520s. His first recorded journey, however, was his expedition to the New World in 1534. He received command of the voyage after the bishop of Saint-Malo recom-

Jacques Cartier. *(The Granger Collection, Ltd. Reproduced with permission.)*

mended Cartier to King François I. Cartier set sail with two ships and five dozen men in April 1534. Their mission was to find gold and other valuable minerals. Less than three weeks after they left Saint-Malo, Cartier's ships reached Newfoundland at a point only 11 miles (17.7 km) from their set destination. They continued south to the Gulf of St. Lawrence, recording the first European accounts of many sites along the way, including Prince Edward and Anticosti islands, and the Bay of Gaspe. Although he was near the St. Lawrence River, Cartier either didn't reach it or missed it due to poor weather conditions. He returned from his six-month journey to a hero's welcome in Saint-Malo.

With the success of the 1534 voyage, Cartier embarked on another expedition in the spring of 1535. He took three ships this time, along with a full crew plus two Native Americans he had brought back with him to France on his previous voyage. The Native Americans, who had learned the French language during their stay overseas, served as interpreters for Cartier on his 1535 visit. With information garnered from the Native Americans, Cartier was able to locate the St. Lawrence estuary and by September to travel up the river to a village named Stadacona, located at the present-day Quebec. A month later, by longboat rather than in his sailing ship, he arrived in a village named Hochelaga. Cartier named the hill near the village Mont Réal. The area is now known as the major Canadian city of Montreal.

Cartier set sail for his next voyage to the New World in 1541. He served as chief pilot for the expedition under Jean François de la Rocque de Roberval, who commanded the eight-ship, 1,500-man voyage. Cartier's ship arrived in Canada early and overwintered apart from the other ships north of Quebec. Cartier rejoined Roberval in June, but disobeyed Roberval's orders to return to Quebec and instead made his way back to France. There, Cartier lived the rest of his life outside Saint-Malo. He died on September 1, 1557.

LESLIE A. MERTZ

Samuel de Champlain
1567-1635
French Geographer and Explorer

Samuel de Champlain was among the first explorers to travel along the east coast of North America and into its interior. He is often credited as the founder of New France, which is known today as Canada. He also established or helped to establish colonies at Montreal, Annapolis Royal in Nova Scotia, and Quebec. By founding Quebec, he created the first permanent European settlement in North America. He also helped to create much more accurate maps of northeastern North America, providing needed geographical information to later expeditions.

Born around 1567 in the town of Brouage on the coast of France, Champlain soon began a career at sea. He learned about navigation, map-making and chart-reading, and took part in his first major expeditions to the West Indies from 1601-1603. His interest spurred, he set sail again, this time aboard a ship that was headed to Canada, or New France. Led by François Gravé du Pont, the 1603 expedition toured Tadoussac and ventured north to Montreal. Through an interpreter who translated between Champlain and the native people, he learned about the existence of the Great Lakes.

Champlain's trip with du Pont kindled a desire to explore more of northern North America, and in particular, an area known as Acadia (Newfoundland and surrounding regions). Back from the du Pont expedition for less than a year, Champlain joined another voyage to Canada. Lieutenant General Pierre du Gua de Monts led this trip, which Champlain made as the ship's geographer. Champlain spent the fall of 1604

Samuel de Champlain. *(AP/Wide World Photos. Reproduced with permission.)*

exploring the coasts of Nova Scotia, the Bay of Fundy and the St. John River as part of the crew, and then spent many days on his own touring and writing meticulous accounts of the coastline and adjoining inland areas. The expedition overwintered on what is now known as Douchet Island in the St. Croix River, then made its way down the coast, traveling south to Cape Cod. Champlain continued his highly detailed description of the areas he visited, and generated needed information for maps of the Atlantic coastline of northern North America.

De Monts returned to France, but Champlain and other Frenchmen remained in Acadia for two years. Champlain spent his time exploring and further charting the eastern seaboard from Nova Scotia to Rhode Island. He and the other Frenchmen returned to France in 1607, but Champlain quickly landed a position as lieutenant for another expedition led by de Monts. By June 1608, Champlain was again in Canada. There, he led the construction of a fort in what is now Quebec. Within two years, the fort had become France's North American center for fur trading.

Champlain's interests in exploring this new wilderness became particularly evident during his 1615 journey into Canada's interior, where his eyes fell for the first time upon Lake Huron. In fact, he was likely the first European to ever see the freshwater expanse. He relied on native people to guide him through the forests and to assist with safe passage through the Indian-inhabited countryside. Intertribal violence persisted, however. One skirmish with a warring group of Indians left Champlain with a severe knee injury that required several months of recovery time.

After his explorations continued, Champlain not only was able to learn about the inland geography of Canada, but was successful in documenting the lives and lifestyles of the native people. His account was one of the first thorough descriptions of the Canadian native populations.

Champlain went to France periodically over the years, but always returned to Canada. He had even attained the title of commander of the colony at Quebec. A war between France and England forced him to surrender Quebec in 1629 and return to France, but his country regained both Quebec and Annapolis Royal in 1632. By then nearing 70 years old, Champlain went back to Quebec for the last time. His health deteriorated, and he died there on December 25, 1635.

LESLIE A. MERTZ

Christopher Columbus
1451-1506
Italian Explorer

If there is any explorer who, in the eyes of most Americans, seems to need no introduction, it is Christopher Columbus. Yet few figures in history have been the subject of so much myth. Old-fashioned political correctness maintained that Columbus was a sort of savior for discovering the New World, whereas modern political correctness—manifested particularly in 1992, during the 500th anniversary of his discovery—condemns him as a murderer of Native Americans and destroyer of the environment. In fact, both views miss the point that Columbus ultimately had no idea what he was doing: though he was right in surmising that it was possible to reach Asia by sea, he went to his grave believing (incorrectly) that he had done so.

He was born Cristoforo Colombo (Columbus" is an Anglicized version) in Genoa at some time between August and September 1451. His parents, Domenico and Suzanna Fontanarossa Colombo, were humble people: Domenico was a weaver, and what little education their son received was primarily a result of his own efforts. Young Columbus read, and was fascinated by, Marco Polo's (1254-1324) account of his

Christopher Columbus. *(Library of Congress. Reproduced with permission.)*

odyssey on the ancient Silk Road to China. By Columbus's time, however, the Turks' destruction of the Byzantine Empire had virtually sealed off the eastward land route; thus explorers, beginning with those sent out by Portugal's Prince Henry the Navigator (1394-1460), had attempted to find a sea route.

Columbus, who first went to sea as a nine- or ten-year-old, gained considerable experience sailing the relatively safe Mediterranean. After being wounded in a battle off the coast of Portugal in 1476, he settled in that country, where he and his brother Bartholomeu worked as mapmakers. During this time, he married Felipa Perestrelo, who gave him one son, Diego, before dying in 1483. The loss of his wife seemed to spark a restlessness in Columbus, now in his early thirties, that led him into the events that would make him an immortal.

Portuguese efforts at eastward exploration had concentrated on attempts to round the coast of Africa and reach Asia via the Indian Ocean; Columbus, by contrast, presented King John II with the idea of a westward expedition to achieve the same goal. John turned a deaf ear, so Columbus went instead to the court of Queen Isabella and King Ferdinand of Spain. The latter did not agree to support the expedition, but took enough of an interest in Columbus to grant him a small annuity. He would wait for the better part of seven years to begin his voyage, during which time he had an affair with Beatriz Enriquez, with whom he had a son named Ferdinand. Then suddenly in 1492, a Spanish priest acted as broker in an agreement between the monarchs and Columbus, who promised them vast riches to be gained from the expedition.

On August 3, 1492, Columbus set sail with some 100 men aboard the *Niña,* the *Pinta,* and the *Santa María.* After 37 perilous days' voyage, the crew sighted land, and on October 12, set foot on what is now the island of San Salvador in the Bahamas. There they were greeted by the aboriginal Arawaks, who Columbus—believing he had reached Asia—dubbed "Indians." After some time on San Salvador, the crew explored the islands of Cuba and Hispaniola. On the latter, they built a fort called Santo Domingo, today the capital of the Dominican Republic and the oldest continuous European settlement in the Americas. Frustrated in his attempts to find either treasure or clear confirmation that he had reached Asia, Columbus departed for Spain in January 1493 with a pair of captured Indians, a few trinkets, and a small quantity of gold he had managed to obtain from the Arawaks. He left behind a group of 40 men, and one of the ships, at Santo Domingo.

Columbus received a hero's welcome in Spain, and his rising fortunes were signified by the size of his second expedition: 17 ships, some 1,200 men, and six months' worth of supplies. Yet things began to turn sour upon their return to Hispaniola in November 1493: as it turned out, tensions between the Indians and the greedy Europeans had resulted in the slaughter of all 40 Spaniards. A number of Columbus's men began succumbing to New World illnesses, and with supplies dwindling, he sent a dozen ships back to Spain. He and the remaining group explored parts of Cuba and Jamaica, but their demands for treasure again put them into conflict with the Indians.

Returning without significant treasure in 1496, Columbus found that his standing with the royal couple had diminished considerably, and this was reflected in the size of the third expedition: just eight ships. This time Columbus, desperate to find the Asian mainland, sailed southward to Trinidad before returning to Hispaniola in August 1498. In Santo Domingo, he found a full-scale mutiny, and when returning sailors brought this news to Ferdinand and Isabella, they sent an official named Bobadilla to investigate. On Bobadilla's orders, Columbus was brought back to Spain in chains in October 1500.

Within a few weeks of his arrival, however, Columbus managed to talk his way back into the royal couple's good graces. Finally they authorized what would be his last voyage, in May 1502, this time with just four ships. The situation in the New World was even worse than before: a new governor in Hispaniola prevented Columbus from landing on the island, and after his crew survived a hurricane, he had to wait a year before the colonial governor sent him help. By November 1504 he was back in Spain, a virtually forgotten man.

Within days of his arrival, Columbus lost his chief supporter, Queen Isabella. He himself would not live more than 18 months, during which time he continually beseeched King Ferdinand for the rewards that had been promised him in their 1492 agreement. He died in the town of Valladolid on May 20, 1506.

JUDSON KNIGHT

Francisco Vasquez de Coronado
1510?-1554
Spanish Explorer

Born around 1510 in Salamanca, Spain, Francisco Vasquez, better known as Coronado, explored large regions of the American Southwest, including Arizona, New Mexico, Texas, Oklahoma, and Kansas. His name would forever link Mexico with the United States.

As one of the younger sons of a wealthy father, he was raised and educated in good circumstances but without hope of inheriting any portion of the family estate. The daughters received dowries to ensure suitable marriages but the younger sons were sent out to make their own ways in the world. Coronado chose Mexico; his brother, Juan, opted for Costa Rica.

Coronado received a warm welcome from Viceroy Mendoza and began his career with the government of Mexico. He was said to be attractive, popular, and competent, and within two years he married the beautiful (and extremely wealthy) Dona Beatriz, daughter of Alonso de Estrada.

He moved up the political ladder quickly. By the time he was 28 years of age, he was appointed Governor of Nueva Galicia with all its responsibilities and the need to distinguish himself. The opportunity arose when Fray Marcos

Francisco Vasquez de Coronado. *(Library of Congress. Reproduced with permission.)*

de Niza (?-1558) returned from an exploratory mission to New Mexico and told eager listeners of a wealthy, golden city named Cíbola. The story generated the same kind of unreasonable excitement that inspired Jason to chase the Golden Fleece, thousands of Americans to the goldfields of California, and millions of today's population to the lottery machines.

Coronado saw this as his opportunity to replenish the government coffers and to secure his position in his adopted country. He assembled 340 Spanish, 300 Indians, and 1,000 Native American and African slaves and left Mexico City with the entire force in 1540. They headed north toward the territory that was west of today's New Mexico. When the impressive body of explorers reached Cíbola, they found nothing to resemble the golden city described by Fray Marcos. Instead, they met and easily overcame a modest pueblo of Zuni Indians. Further searches revealed six other Zuni encampments, but nothing to indicate the existence of any treasures to be taken home.

After sending Fray Marcos back to Mexico in disgrace, Coronado sent out various expedition parties to map the source of the Colorado River, to confirm the rumor of another great river in the west and to pick up supplies he was expecting from Mexico City. The search for the second "great river" led to the discovery of the Grand Canyon in Arizona.

When most of the journals, reports, maps, and other records were finally assembled, it appeared that Coronado accomplished far more than what was originally believed. He acquainted himself with many Indian tribes besides the Zunis and was the first foreigner to see the famous City in the Sky, built by and inhabited by the Acoma tribe near present-day Albuquerque. Reportedly, Coronado and his men were the first explorers to pass through the Texas panhandle, then Oklahoma, finding the Cimarron and Arkansas rivers on their way to eastern Kansas. Most of the territories they passed through and mapped appear in American atlases as the south central and far southwest portions of the United States. As a matter of record, though Juan Rodriguez Cabrillo (1498?-1543) is credited with discovering California, Coronado preceded him by two full years.

When Coronado returned to Mexico in 1542, he was accompanied by only 100 of his original force. The quest for the City of Gold was a dismal failure but Coronado retained his post as Governor of Nueva Galacia for another two years. He died in 1554, only 44 years of age but living in retirement.

Today, the name "Coronado" is in widespread use throughout the United States. Everything from state monuments, forests, bridges, parks, cities, and schools, to beaches, automobiles, industries, and hotels bear the name that unites the United States with its southern neighbor, Mexico.

BROOK HALL

Hernán Cortés
1485-1547
Spanish Conquistador

Hernán Cortés was a Spanish conquistador who succeeded in claiming most of present-day Mexico for Spain by conquering the Aztec people in their capital city of Tenochtitlan.

Cortés was born in the region of Medellin, Spain, to parents of good social standing. Cortés was sent to the University of Salamenca at the age of 14 where, among other subjects, he studied Latin for two years. Following his studies, Cortés set out to make his way in life. His imagination ignited with the possibilities that awaited in the Indies recently discovered by Christopher Columbus (1451-1506), Cortés sailed, at age 19, to Hispaniola (present-day Santo Domingo) in 1504.

Cortés spent approximately six years as a farmer in Hispaniola, then in 1511 joined Diego Velazquez (1465?-1522), with whom he became an intimate friend, on an expedition to Cuba. When Velazquez was appointed governor of Cuba in 1511, Cortés was made clerk to the treasurer.

Velazquez made plans to send an expedition to what is now Mexico; Cortés was placed in charge of the expedition. With his experience as a leader and his position in politics, Cortés was able to quickly recruit 300 men and acquire six ships. Cortés soon became aware that Velazquez was intending to name another leader of the expedition, and in an effort to preserve his efforts he slipped away and headed along the coast of Cuba, recruiting more men. When Cortés finally left for Mexico on February 18, 1519, he was the leader of over 600 soldiers and sailors, 11 ships, 200 Indians for support, and 16 horses.

Cortés went first to Yucatan and then along the coast of Mexico, where he founded the town of Villa Rica de Veracruz. He had himself elected, by his soldiers, as captain general as well as chief justice, making himself the sole authority for the expedition. He also burned their ships, a tactic designed to raise the level of commitment in his men to the conquest of Mexico.

Cortés led his men into the interior of Mexico, relying on information from Indians with whom he had become friendly; the most signifi-

Hernán Cortés.

cant of these were the Tlascala, who were at war with the Aztecs. With the information gathered, Cortés set out for Tenochtitlan, the island-city capital of the Aztec empire in Lake Texcoco, ruled by Montezuma II. In early November of 1519 Cortés and his men reached Lake Texcoco. Accompanied by 1,000 Tlascala Indians, they entered the city on November 8, 1519. Montezuma initially thought that Cortés was the Aztec god Quetzalcoatl and as a result opened the city to him. Cortés then took Montezuma hostage in an effort to conquer the entire Aztec empire with one action.

Soon thereafter, Cortés learned that an expedition, sent by Velazquez and led by Pánfilo de Narváez (1480?-1528), had left Cuba and had arrived at the coast of Mexico with the intent of relieving Cortés of his command of Tenochtitlan. Cortés left the Aztec city and, after a surprise attack and victory over Narvaez and his men, returned to Tenochtitlan. In the meantime, the Aztecs had revolted against the Spaniards that Cortés had left behind. So under cover of night on June 30, 1520, Cortés and his men left the city and found refuge with the Tlascala Indians.

Cortés reassembled his army and with Indian support marched on the Aztec capital in December of 1520. The attack on the city began in May of 1521 and by August 13 the city was in the control of the Spaniards; the Aztec empire was officially defeated. On October 15, 1522, Charles V named Cortés the governor of New Spain, or the territories in Mexico conquered by the Spanish; he rebuilt the city of Tenochtitlan. However, Cortés was only in power for a few years, and by 1526 he was removed as governor; from 1528-1530 he tried to regain his position in Mexico. He did not succeed in this effort and returned to Mexico where he retired to his estate about 30 miles (48 km) south of Mexico City (Tenochtitlan). He spent his time building a castle on his estate and leading expeditions, but he was not to acquire the position he had held before. In 1540 he returned to Spain, where he died in 1547; his body was returned to Mexico for burial.

MICHAEL T. YANCEY

Bartolomeu Dias
c. 1450-1500
Portuguese Mariner and Navigator

Bartolomeu Dias was the first mariner to round Africa's Cape of Good Hope, opening up a coveted sea route to the West Indies for Portugal. In the latter part of his life, he took part in the Portuguese discovery of Brazil.

Little is known about Dias's early life. Although his surname was a common one, historians believe that Dias came from a long line of navigators that may have included Dinis Dias, who rounded the Cape Verde in 1455, and Joao Dias, who rounded Cape Bojador in 1437. In 1486, on the appointment of King John II, Dias was instructed to find "Polly Prism Promontorium," the southernmost extremity of Africa, and hopefully discover the coveted sea route to India. Finding this route was important to Portugal because unrest in the Mongol Empire had closed the overland trade routes to the Far East. Commanding a fleet of three ships, Dias left Lisbon in 1487 after 10 months of preparation. Using a new strategy, one of the three ships was designated as a supply vessel, allowing the expedition to stay at sea longer. Dias's brother, Pedro, captained the supply ship. Dias also took with him several African interpreters who had lived in Europe and would assist in establishing trade with the native Africans.

The expedition sailed south for four months, stopping along the way to trade. The explorers passed the stone pillar left by Diogo Cão (1450-1486) near present-day Namibia to mark the southernmost point the Portuguese had reached thus far. By the end of December

Bartolomeu Dias. *(Archive Photos, Inc. Reproduced with permission.)*

Dias's ships passed the Orange River, just to the north of present-day South Africa. Shortly after passing Cape Volts, named for its strong winds, the ships encountered a massive storm that blew them on a southerly course for nearly 13 days. When the wind died down, Dias sailed east, expecting to encounter the west coast of Africa. When no land appeared, he turned north, reaching Bahia de Vaquieros, roughly 200 miles (322 km) east of the Cape of Good Hope, not realizing at this point that they had rounded the cape.

Dias was convinced by his crew to turn back to Portugal. It was on the return journey that Dias spotted the much-looked-for cape as he sailed west. Although some controversy surrounds who named the Cape of Good Hope, historians generally credit it to Dias rather than King John II, who may have named it Cape Tormentoso, in recognition of the storms encountered there.

Maps drawn by the Greek cartographer Ptolemy had shown the Indian Ocean as a great landlocked sea, and land from the east reaching around and touching Africa's west coast. Dias's expedition proved this incorrect, but it was nearly 10 years before Portugal was able to take advantage of this important geographic discovery. Within Portugal, Dias was not granted the credit due as the person responsible for locating and mapping the Cape of Good Hope. Nevertheless, in 1494 he was appointed to oversee the construction and outfitting of a fleet of ships for an expedition to reach India by way of the Cape of Good Hope. Vasco da Gama (c. 1460-1524) was to lead the expedition with Dias accompanying the voyage as far as the Cape Verde Islands. After reaching the Cape Verde Islands, the new Portuguese King Manuel sent Dias to establish trading posts in present-day Mozambique.

In March 1500 Dias embarked on his final voyage of discovery. One of 13 ships under the command of Pedro Álvares Cabral (c. 1467-c. 1520), the expedition was to duplicate da Gama's voyage. The expedition embarked south from the Cape Verde Islands and crossed the equator. When the fleet encountered the trade winds they were blown off course and as a result made the first recorded European landing on the coast of Brazil. After leaving Brazil, the fleet encountered a fierce storm and four ships were lost in the vicinity of the Cape of Good Hope, including that of Dias.

LESLIE HUTCHINSON

Sir Francis Drake
1543?-1596
English Navigator

Sir Francis Drake was an English seaman and explorer who made the second circumnavigation around the world and later led the English in defeating the Spanish Armada in an historic battle in 1588. At the time of his death, Drake was an internationally acclaimed figure known throughout the world for his privateering, pirating, and geographical knowledge.

Francis Drake was born in the town of Tavistock in Devonshire, England. He was one of 12 sons born into a modest family, almost all of who became seaman. In his teens Drake was apprenticed to a shipmaster, and in 1566 he was employed as a seaman on English ships engaged in the slave trade between Africa and the Caribbean.

On his return to England in 1569, Drake reported Spanish and Portequese attacks on English ships to Queen Elizabeth I. Due to the great rivalry between England, Spain, and Portugal, the Queen officially gave Drake a privateering commission to cruise the Panamanian coast. A "privateer" was someone authorized by the government of one country to attack the ships of another country and then retain a portion of the proceeds. From these privateering expeditions, Drake learned the art of sailing and the geography of the Western Hemisphere, and he developed a life-long hatred of the Spanish. For the remainder of his life he conducted a personal war against Spain.

In 1577 Queen Elizabeth commissioned Drake to sail around the world. The expedition would sail to the South Seas through the Strait of Magellan, something no Englishman had ever done. English trading posts were to be established throughout the Pacific. Queen Elizabeth also assumed that a small force of about 200 men could severely disrupt the flow of gold and silver to Spain, giving England an advantage over her arch rival.

In 1577 the circumnavigation expedition departed with a crew of 166 men—primarily hardened seamen who could handle weapons. The *Pelican* (commanded by Drake), the *Elizabeth*, and three smaller vessels, heavily armed, set sail from Plymouth, England, en route to Brazil, the mouth of the Rio de la Plata in Argentina, and Patagonia and on through the Straits of Magellan in 1578. Once through the treacherous straits, only three ships remained as two ships had burned. Once in the Pacific the little fleet was hit by a fierce storm that lasted 52 days and drove Drake's ship far off course to the south. One of the remaining ships was sunk with the crew onboard, and the other returned home. Drake renamed his ship the *Golden Hind* and continued the journey.

Drake and the *Golden Hind* sailed up the coast of South America, provisioning themselves with supplies captured from Spanish storehouses in Chile and several Spanish ships captured en route. In 1579 he anchored near Coos Bay, Oregon, and then repaired his ship in Drake's Bay, California. From there he headed out into the Pacific and did not sight land for 68 days until he reached Palau in southwestern Micronesia. He refitted his ship in Java, sailed to the Cape of Good Hope, and arrived back in England in 1580, where Queen Elizabeth knighted him in 1581. He was the second person to lead an expedition around the world and the first Englishman to do so.

By this time the Spanish were determined to put an end to the English raids and mounted the Great Armada of 1588. Drake was put in charge of the English fleet. After a series of fierce battles, the Spanish were forced to retreat, and the remainder of their ships were destroyed by storms. At this defining moment, England replaced Spain as the most influential sea power in the world.

Drake died in 1596 off of the coast of Panama.

LESLIE HUTCHINSON

Sir Martin Frobisher
1539-1594
English Explorer

Sir Martin Frobisher was one of the first Englishmen to search for the Northwest Passage, and he personally led three expeditions to the Canadian Arctic. He is known as one of Queen Elizabeth I's most aggressive and enterprising seamen, feared by Europeans for his privateering and for his role in the battle against the Spanish Armada in 1588.

Martin Frobisher was born into an influential family in England in 1539—his father was the director of the English mint. In his teen years, instead of pursuing higher education, he joined two trading voyages to West Africa. After his adventures in Africa, Frobisher fought in the English army in Ireland and then became a privateer—a pirate commissioned by the Queen to attack enemy ships and then keep part of the proceeds. During his privateer days, he became very interested in the tales of the Northwest Passage, a supposed route far to the north of Canada that would link Europe to Asia.

He discussed this idea with many leading scientists in England at the time. Well known was Sir Humphrey Gilbert's (c. 1539-1583) book, *Discourse of a Discoverie for a New Passage to Cataia*, a speculative geography suggesting the probability of a passage above North America, mirroring the passage around South America at the Cape of Good Hope. This work intrigued him and he decided to raise an expedition and investigate the hypothesis.

In 1575 Frobisher convinced the shareholders of a Russian company to invest in a voyage to search for the Northwest Passage. Funding procured, the expedition embarked with three ships and 35 crew members. Sailing northwestward, they sighted the coast of Greenland. However, they were caught in a storm off the coast of Greenland, and the ships were separated. One returned to England and another with Frobisher on board was reported missing. He, however, continued to sail westward, sighted Resolution Island, discovered the great inlet on Baffin Bay, and encountered a group of native Inuit people. After collecting ore samples and an Inuit Indian,

he returned to England. Most of the samples were considered pyrite or "fool's gold," however, some individuals believed that they indicated deposits of real gold, and he was able to get funding for another expedition. Queen Elizabeth I herself invested money in this company and furnished a ship in hopes of Frobisher finding gold.

In 1577 Frobisher, under his new company name, Cathay, set out with three ships and 120 crew members to travel back to Baffin Bay (now called Frobisher Bay) and look for gold. He loaded approximately 200 tons of ore onto his ships and captured an Inuit man, woman, and child to take back to England. On his return, the ore assay showed the mineral deposits to be worthless, but Frobisher decided to send out an even larger expedition to bring back more extensive samples.

In 1578 Frobisher left England as the head of 15 ships with the mission to mine for ore and look for a Northwest Passage that might lead to the Orient. While sailing south of Baffin Island, they entered what is now known as Hudson Strait. He was convinced this was the passage they had been looking for. However, the massive ice, freezing wind, and erratic currents would not allow the ships to proceed. He returned to Frobisher Bay and built a stone house whose remains were found nearly 300 years later.

After the expedition's return to England, metal workers were unable to refine the ore into gold. Abandoning this venture, Frobisher continued on in the British navy and as a privateer. In 1588 he served as a commander in the battle against the Spanish Armada, during which he was wounded; he eventually died of complications in 1594. Although he did not find the Northwest Passage, he is credited with adding vital information to the geographical understanding of the northern latitudes.

LESLIE HUTCHINSON

Vasco da Gama
1460-1524
Portuguese Explorer

In the last years of the fifteenth century, an explorer set off from the Iberian Peninsula, full of grand illusions and hoping to reach India by going where no European had ever gone before. Though that statement would seem to describe the 1492 voyage of Christopher Columbus (1451-1506) to the New World, it is equally true of a less famous expedition—from an American perspective, at least—that set sail five years later.

This one was led by Vasco da Gama, who sailed under the Portuguese flag and rounded the southern tip of Africa to become the first European to reach the Indian subcontinent by sea.

Da Gama was born in Sines, Portugal, where his father was governor. As a member of the nobility, he led a Portuguese attack on French ships in 1492, and later served as a gentleman at the court of King Manuel I. Under the leadership of Manuel, the Portuguese continued the tradition, begun by Prince Henry the Navigator (1394-1460) and maintained sporadically ever since, of exploring the African coast. This had been done by bits and pieces, with each subsequent probe venturing just a bit further south, until Bartolomeu Dias (c. 1450-1500) had rounded the Cape of Good Hope at the continent's southern tip in 1487-1488. Now Manuel was prepared to take the bold step of passing the Cape by and sailing across thousands of miles of open sea to India. Therefore on July 7, 1498, da Gama and his crew set sail from Lisbon aboard four ships.

Their goal was the city of Calicut (not to be confused with Calcutta) on the Malabar, or southwestern, coast of India, and da Gama took with him letters of introduction both to the ruler of Calicut and to Prester John. The latter, supposedly the ruler of a Christian kingdom, is now known to have been an utterly fictitious character, created by a sort of early urban legend around 1150; but people in da Gama's time did not know that, and Manuel was convinced that Christian Portugal would find an ally in India.

Sailing well west of Africa, the crew rounded the Cape on November 22, then began tracing the continent's east coast. This put them in contact with coastal trading cities, which served as ports for Arab and Persian vessels plying the Indian Ocean route da Gama intended to cross. The Portuguese did battle with the Muslims in Mozambique and Mombasa (now part of Kenya), but found a better reception in the city of Malindi, whose sultan provided them with an Indian pilot to guide them across the ocean. Thanks in part to this help, da Gama landed in Calicut on May 20, 1498.

At first the Portuguese were sure they had found Prester John's land, because they mistook a temple to a Hindu goddess as a shrine to the Virgin Mary. Disappointment followed when the zamorin, the local ruler, examined the treasures Manuel had sent him as examples of Portugal's economic might. From the standpoint of India, wealthy in natural resources, these were cheap trinkets, and though the zamorin sent back sam-

Vasco da Gama. *(New York Public Library Picture Collection. Reproduced with permission.)*

ples of treasure and spices when da Gama set sail again in August 1498, this was probably more from courtesy than from a genuine belief that trade with Europe would prove profitable. The zamorin could not have known that the ragtag band of sailors were the advance party for waves of European colonization that would not end until the nation of India annexed the Portuguese colony of Goa in 1961.

As for da Gama, his crew ran into considerable hardships on the return voyage, which claimed the life of his brother Paulo (captain of one of the ships), along with many other crew members. He arrived in Lisbon on September 9, 1499, and would spend most of his remaining years enjoying the wealth and titles he had accrued through his pioneering voyage. Da Gama returned to India twice: first in 1502, then in 1524, when he served as viceroy of the Portuguese colony before dying on December 24 in the city of Cochin.

JUDSON KNIGHT

Richard Hakluyt
c. 1552-1616
English Geographer

Though he never personally took part in any expeditions, Richard Hakluyt greatly advanced the cause of English exploration in North America. One of England's first geographers, he collected and disseminated information, and promoted the colonization efforts of Sir Walter Raleigh (1554-1618) and others.

Hakluyt's father, a skinner in London, was named Richard Hakluyt; so too was his cousin (c. 1535-1591), a geographer who influenced Hakluyt's interest in the subject. Hakluyt attended first Westminster School, then Christ Church, Oxford, where he earned a B.A. in 1575 and an M.A. in 1577. Soon afterward, he was ordained to the priesthood, but continued to study and lecture on geography at Oxford. During this time, he made the acquaintance of Sir Francis Drake (1540?-1596), and sponsored a translation of two accounts of voyages by Jacques Cartier (1491-1557).

The first geographical work by Hakluyt appeared in 1582 as *Divers Voyages Touching the Discovery of America.* The book, which greatly encouraged colonization efforts, contained an account of previous English voyages to North America, a list of America's known resources, and a report on the Northwest Passage. During the following year, Hakluyt promoted an expedition led by Sir Humphrey Gilbert (c. 1539-1583), and began a stint as chaplain to the English ambassador in France. The latter position gave him the opportunity to study French, Portuguese, and Spanish geographical information. In 1584, he visited England long enough to beseech Queen Elizabeth I to encourage the planting of crops in the New World.

The Principal Navigations, Voyages, and Discoveries of the English Nation, a history of English overseas exploration, was published in 1589. Hakluyt married soon afterward, and between 1598 and 1600 published a second and greatly expanded version of the book. He died in 1616, the same year as his fellow Englishman William Shakespeare.

JUDSON KNIGHT

Henry Hudson
c. 1565-1611
English Explorer

Both a small but significant river in New York and an immense bay—by far the world's largest—in Canada are named after Henry Hudson. The great sea distance between these two bodies of water is a tribute to his wide-ranging explorations, and to his bold but ultimately trag-

Henry Hudson. *(American Museum of Natural History. Reproduced with permission.)*

ic attempts toward finding the elusive Northwest Passage.

Ironically, Hudson spent the early part of his career searching for the North*east* Passage, the sea route via the northern coast of Russia and Siberia to China. In 1607, the Muscovy Company in his native England hired him for this purpose, but though he explored the forbidding regions of Jay Mayen Island and Svalbard between Scandinavia and Greenland, he did not find the passage. Another voyage the following year was cut short due to heavy ice.

In 1609, Hudson set sail yet again, this time under the aegis of the Dutch East India Company, aboard the *Half Moon.* The expedition had gotten no farther than the North Cape of Norway before running into heavy ice, and the crew refused to venture any further. Instead of returning to Holland, however, Hudson set his sights on finding the Northwest Passage, the sea route to Asia via the northern coast of North America.

By July 1609, the expedition had reached Nova Scotia, then turned south to Chesapeake Bay. They then explored Delaware Bay before entering New York harbor—none of these places bore these names at the time, of course—on September 12. They followed the harbor to the mouth of what is now known as the Hudson River, then sailed up the Hudson to the site of present-day Albany. They then sailed back down, stopping at the island called "Manna-hata" by the Indians. At first the latter threatened them, even killing one of the crew members, but when the sailors offered them European goods, relations improved.

On his return trip to Europe, Hudson docked at Dartmouth, England, where he was seized by the authorities and forbidden to sail for any foreign powers. Nonetheless, he was able to pass on to his Dutch employers what he had observed on his trip: that the land around the harbor and river he had explored were rich and promising. This would influence the Dutch founding of New Netherland, a colony comprising what is now New York and New Jersey, in 1614.

Back in England, Hudson raised support for another voyage to find the Northwest Passage, and set sail aboard the *Discovery* on April 17, 1610. It appears that Hudson was never a very good manager of men, and long before they caught sight of land—Resolution Island, to the southeast of Baffin Island—discord had arisen among the crew members. Nonetheless, Hudson sailed onward.

The *Discovery* plied what is now called Hudson Strait, between Baffin Island and the northern coast of Quebec. It seemed to Hudson that he was on the verge of finding the Northwest Passage, and he proceeded with optimism. In fact they were sailing into Hudson Bay, which would be more properly identified as a sea—a sea half as large as the Mediterranean, but not a tenth as inviting to mariners.

By October, the expedition had reached a dead-end, a shallow inlet called James Bay at the southeastern corner of Hudson Bay. Cold weather was beginning to set in, and with it ice, so that the crew was forced to spend the winter there. Lacking adequate provisions—another mistake on Hudson's part—they suffered terribly during the cold months, and tensions grew.

Only on June 12, 1611 were they finally able to set sail again, but after just 12 days the crew mutinied against Hudson. On June 22, they set the captain adrift in a small boat with his 19-year-old son John and six of the less physically fit crew members. The eight men were never heard from again.

JUDSON KNIGHT

Willem Jansz
1570-1629
Dutch Explorer

Willem Jansz, a Dutch sea captain, was the first European to catch sight of Australia.

But like another man far more famous for being the first Caucasian to glimpse a continent—Christopher Columbus (1451-1506)—Jansz had no idea what he had found.

Virtually nothing is known about Jansz's life either before or after the point when he played his role in history. His voyage took place in 1605, four years after Holland founded the East India Company for the purpose of exploring and colonizing the East Indies, or modern-day Indonesia and surrounding islands. Jansz's boat, a relatively small craft called the *Duifken,* had earlier distinguished itself in a sea battle with the Portuguese over control of Java, and on November 28 it sailed from the port of Bantam with orders to explore the southern coast of New Guinea.

Jansz and his crew crossed the Banda Sea to reach southwestern New Guinea at Dolak Island. They then entered what is now known as the Torres Strait before running into shallows leading to the Great Barrier Reef. Fearing that they might run aground, Jansz turned the prow southward, where they again sighted land. Though they thought this was simply another part of New Guinea, in fact what they were seeing was the west coast of the Cape York Peninsula in northeastern Australia—an entirely new continent.

The crew of the *Duifken* were also the first Europeans to glimpse Australian aborigines—and nine of them were the first to be killed by these natives. Between the hostility of the aborigines and the dryness of the region he saw, Jansz concluded that the land offered no promise for future exploration or trade, and he explained as much to his superiors when he returned to Java. In the years that followed, the Dutch made tentative efforts to explore the continent, which they claimed for their own under the name New Holland. They established so little presence in Australia, however, that it was easy for the British to take control in the eighteenth century.

JUDSON KNIGHT

Ferdinand Magellan
1480-1521
Portuguese Explorer

Ferdinand Magellan initiated, organized and led what was to become the first circumnavigation of the globe. Under his leadership, five ships set sail from Spain in 1519. Although Magellan died during the three-year voyage—the victim of a conflict between warring island nations—and only one of the five ships completed the expedition and returned to Spain, Magellan is credited as the man behind the first trip around the world.

Magellan was born Ferñao de Magalhaes into Portuguese nobility in 1480. Following a youth spent as a page in the home of the queen of Portugal, he took a position with the nation's fleet, which sought to expand the spice trade—often by way of bloody battles. Through these

Ferdinand Magellan.

MAGELLAN'S CREW DISCOVERS THE INTERNATIONAL DATE LINE

In 1519 Ferdinand Magellan started what was to become the first circumnavigation of the world. Leaving Spain with a small fleet of five ships on August 10, Magellan would not survive the journey. In fact, only one ship and nineteen men returned to Spain after more than three years at sea. Upon their return, the acting captain, Juan Sebastián de Elcano, was surprised to find his calendar off by a day. Through storms, battles, overhauls, and all other adversity, the ship's log had been meticulously maintained, as had the expedition's calendar. In spite of everything, however, there was no denying the discrepancy in dates; the expedition's logs showed they had returned to Spain on September 6, 1522, when in fact they returned on September 7. What had happened, of course, was that the fleet unknowingly crossed what is now the International Date Line on their voyage. This line is more than an abstraction or a simple line on the map. In traveling around the world, the voyagers crossed from time zone to time zone, always moving back by an hour. Upon their return to Spain, they had traveled through all 24 time zones, rolling back their clocks by a whole day. Since they did not advance their calendar by a day when they crossed the date line, they effectively "lost" 24 hours, one for each time zone they crossed in their travels.

P. ANDREW KARAM

trips, including his first in 1505 to the East Indies, he learned how to sail and how to fight. It was during a skirmish in Morocco nearly a decade later that he sustained a leg wound, which affected him for the rest of his life.

Despite his years of service to his country, Magellan met a disappointing reception in his homeland. He not only fell under suspicion for corruption, a charge that was proven false, but

received notice from the Portuguese crown to begin looking for work elsewhere. In 1517, he and friend Ruy de Falero (Faleiro), an astronomer, left together to seek opportunities in Spain, a longtime enemy of Portugal. In Seville, Magellan approached the Spanish court, the advisors of King Charles V, and finally the king with a proposition to explore uncharted waters and search for spice-rich islands in East Asia. King Charles approved the expedition. In exchange Magellan and Falero received decade-long, exclusive rights to the new trade routes they developed.

Falero eventually bowed out of the expedition, but Magellan pressed forward during the next year by planning the voyage and outfitting the five ships the king had allotted for the trip. The expedition set sail on September 8, 1519, with a crew of 560 men aboard the *Trinidad, San Antonio, Concepción, Victoria,* and *Santiago.* The ships took a southwesterly route from Seville across the Atlantic to Rio de Janeiro, Brazil, where they arrived on December 13. From there, the five ships continued on along the coast of South America, spending time to investigate coves and inlets for a possible shortcut across the continent to the waters on the other side. Without luck, they sailed down the coast.

Magellan decided to overwinter in the Bay of San Julián before continuing the journey.

While there for four months, the captains of four of the five ships tried to organize a mutiny. Magellan and those loyal to him quashed the attempt and killed at least three of the opposing captains. The expedition also lost one of its ships to heavy damage over the winter.

After the expedition again set sail, Magellan led three of the four ships through what became known as the Strait of Magellan at the southern tip of South America and into the Pacific Ocean. The captain of the fourth ship broke ranks while in the strait, turned around and left for home. Magellan and his remaining fleet went on, heading north to Guam and then to the Philippines, where they landed at Cebu. There, Magellan made the fatal mistake of taking sides in a local war and died in battle on April 27, 1521.

More than a year later, the voyage that Magellan had initiated finally ended when one of the original five ships completed the worldwide circuit. The battered *Victoria,* carrying 18 adventure-weary, sick, and starving crew members, made shore in Spain on September 8, 1522.

LESLIE A. MERTZ

Gerardus Mercator
1512-1594
Flemish Geographer and Cartographer

The word "atlas" to define a collection of maps was coined by Gerardus Mercator, who is best known for his 1569 invention of a new system of projection for marine charts, called the Mercator projection, which revolutionized cartography as well as nautical navigation.

The son of a shoemaker, Mercator was born Gerhard de Kremer on March 5, 1512, in Rupelmonde, Flanders (now Belgium). His name was later latinized to Gerardus Mercator. Mercator began his education in Hertogenbosch in the Netherlands, where he studied Christian doctrine, dialectics, and Latin. From there, he continued his education at the University of Louvain, studying philosophy and the humanities and graduating in 1532 in geography, geometry, and astronomy. While at university, Mercator was a mathematical pupil and assistant to Gemma Frisius (1508-1555), a physician, astronomer, and mathematician for whom he worked as an instrument-maker and globe-maker after graduating.

Before going to work with Frisius, however, young Mercator spent two years of religious study in Antwerp and Mechelen to reconcile his doubts regarding the biblical account of the origin of the

world versus the teachings of the philosopher Aristotle (384-322 B.C.). During this time, he developed a passion for geography and subsequently began pursuing a career as a professional cartographer. In 1534 he was married to Barbara Schellekens (with whom he had six children).

Mercator returned to Louvain and from 1535 to 1536 worked with Frisius and Gaspar à Myrica to construct a terrestrial globe. In 1537 the team completed a celestial globe. The same year, Mercator created his first map, of Palestine, and one year later, had completed his first map of the world, a unique double heart-shaped projection. He began gaining a well-deserved reputation as a cartographer and talented engraver and, in 1540, published his first book, a manual on the italic lettering he had introduced on his maps entitled *Literatum Latinarum quas Italicas cursoriasque vocant scribende ratio,* for which he created woodblock engravings depicting the lettering.

From 1541 to 1551 Mercator spent much of his time creating globes, interrupted only by a seven-month stint in prison for Lutheran heresy (1544). In 1552 he embarked on a new career as a mapmaker and lecturer at the University of Duisburg in the Duchy of Cleve in Germany. Two years later he published a map of Europe, which perfected earlier maps such as those of Ptolemy (c. 100-c. 170). He followed it, in 1564, with a map of the British Isles. That same year, he became the court cosmographer to Duke Wilhelm of Cleve.

In 1569 Mercator introduced, with the publication of his world map, a new system of projection for marine charts featuring true bearings, or rhumb-lines, between any two points (a ship's course could be found by laying a ruler across the map). His system, mathematically derived in an early form of calculus, allowed for a more accurate representation of the world's continents and resulted in a revolutionary cylindrical projection where a straight line between any two points forms the same angle with all the meridians, which appear as straight, or longitude, lines that are perpendicular to the Equator and parallels, which appear as straight, or latitude, lines that are parallel to the Equator.

Mercator's 1569 world map was one part of his plan of publications that began the same year with the publishing of a chronology of the world from its creation to 1568, followed, in 1578, by the publication of 27 maps originally prepared by the Greek geographer Ptolemy with corrections and commentary by Mercator. He finished the whole work, but it wasn't published until 1595, shortly after his December 2, 1594, death; it carried the title *Atlas*, the name chosen by Mercator to represent a collection of maps. The *Atlas* was preceded by the publication of a series of maps—in 1585 Mercator issued new maps of France, Germany, and the Netherlands; in 1589 he published new maps of Italy, Sclavonia (the Balkans), and Greece; and in 1595 the posthumously published *Atlas* included all his previous maps as well as a new series of maps on the British Isles. After Mercator's death, his *Atlas* had a new printing (in 1602), and a 1606 edition including new maps by Jodocus Hondius (1563-1612) became known as the *Mercator-Hondius Atlas*.

ANN T. MARSDEN

Francisco Pizarro
1470?-1541
Spanish Soldier and Explorer

Francisco Pizarro was a Spanish soldier and explorer who conquered the Incan empire and founded the city of Lima, Peru, in 1535. He was also a member of Balboa's expedition that discovered the Pacific Ocean in 1513. Like many of the Spanish conquistadors, Pizarro led a dangerous life dedicated primarily to the accumulation of wealth through the conquering of new lands and people.

Pizarro was born in Trujillo, Spain, around 1470. He was the illegitimate son of Gonzalo Pizarro, who was a captain of infantry. Pizarro was not interested in education while growing up and did not learn to read or write. He was, however, intrigued by adventure. This was especially true of the stories he heard regarding the exploits of his countrymen in America. Determined to live that style of life, he traveled to Hispanola (presently Haiti and the Dominican Republic) in 1502. Bored by colonial life, Pizarro joined an expedition to Columbia with Alonso de Ojeda (1465?-1515) in 1509. He began to develop a reputation as a man who could be trusted in difficult situations; therefore, he was made a captain on the expedition led by Spanish explorer Vasco Núñez de Balboa (1475-1519) that was credited with the European discovery of the Pacific Ocean in 1513. Pizarro later served from 1519-1523 as the mayor of Panama, where he accumulated a small fortune. However, it was not until he was almost 50 years old that the events that he is most famous for took place.

In 1523 Pizarro embarked on an expedition to the west coast of South America in partnership

Francisco Pizarro. (*Corbis Corporation. Reproduced with permission.*)

with a soldier, Diego de Almagro (1475?-1538), and a priest, Hernando de Luque. Travel and conditions were extremely difficult, and many men perished in the harsh conditions. Almagro was sent back to Panama to obtain reinforcements, but the new governor of Panama ordered the expedition to be halted. Legend has it that at hearing this, Pizarro drew a line in the sand with his sword and invited anyone who was interested in wealth and glory to step over to his side. The men who crossed the line and continued the exploration were known as the "famous thirteen." They continued on to present-day Peru, obtaining first-hand accounts of the Inca Empire. Pizarro traveled back to Spain to gain permission from the emperor Charles V to continue his exploits.

The king agreed with Pizarro and placed him in charge of New Castle, a province 600 miles (966 km) south of Panama. While all of the "famous thirteen" were granted significant privileges in the new land, both Almagro and Luque were given positions subordinate to Pizarro. Pizarro and his four brothers eventually arrived in Peru with a relatively small contingent of men compared to the 30,000 men in the Incan army. Pizarro set up a meeting with the leader of the Incas, Atahuallpa. He asked the Incan leader to submit to Christianity and to Spain, but the king refused. Pizarro immediately ordered an attack, seizing Atahuallpa and demoralizing the Incan army. Atahuallpa was held for ransom, while great rooms were filled by the Incas with gold and silver. Atahuallpa, however, was never released; he was put to death in 1533 on charges of plotting against the Spanish government. With their leader dead, the Incas offered little resistance while Pizarro took over the entire empire.

Pizarro now had the difficult task of defending his hold on the Incan empire. Almagro had grown jealous of Pizarro and his power from the king of Spain. Almagro demanded an equal share of the spoils of the expedition, so an agreement was established that gave him a large portion of Chile. After finding that country poverty stricken, Almagro returned to Peru, where he was captured and executed by Pizarro's brother, Hernando (1475?-1578). Almagro's allies were rounded up and sent to Lima so they could be watched. They realized they were in immediate danger and attacked the Pizarro palace on June 26, 1541. Francisco Pizarro is said to have fought gallantly, but was killed in a sword fight at the hands of Almagro's allies.

JAMES J. HOFFMANN

Don Juan Ponce de León
1460-1521
Spanish Explorer and Soldier

Juan Ponce de León was one of the leading early Spanish explorers and colonizers of the Western Hemisphere. He is best remembered for his discovery of Florida and for conquering and settling Puerto Rico.

Little is known of Ponce de León's early life. His family was part of the Spanish nobility, and he served as a page in the court of King Ferdinand and Queen Isabella. In 1492, he participated in a military campaign against the Moors in Granada that successfully expelled the Muslims from their last foothold in Spain.

The time in which Ponce de León lived was an age of exploration and Spanish expansion, and he took full advantage of the opportunities this afforded. His countryman Christopher Columbus (1451-1506) discovered America in 1492, thinking it to be the Far East, and the following year, Ponce de León sailed with him on his second expedition to the New World. Upon arrival, Ponce de León apparently responded eagerly and successfully to the many challenges of the colonization enterprise. During 1502-1504, he commanded the effort to subdue the natives on the island of Hispaniola, which the Spanish were attempting to colonize. As a reward for his success, he was made provincial governor of northeastern Hispaniola, a post he held until 1508.

Attracted by reports of gold, he left Hispaniola in 1508, leading an expedition to Puerto Rico. He quickly conquered the island and established a settlement there. He was made governor of Puerto Rico in 1509, retaining the position until 1511. In this case, the stories of gold proved to be true: a large supply was discovered, and he became one of the wealthiest men in the New World.

The natives throughout the Caribbean spoke of a spring located on an island, that they called Bimini, located somewhere north of Cuba. According to the legend, the water of this spring would make anyone who drank from it young again. This story corresponded with a European legend concerning such a Fountain of Youth located at the site of the Garden of Eden. In 1513, the king commanded Ponce de León to search for Bimini and its legendary Fountain of Youth. He obeyed, setting sail as soon as an expedition could be organized. On this voyage, he explored the Bahamas and discovered Florida. Landing near the present site of St. Augustine, he claimed what he thought was a large island for Spain, unaware that he had landed on the mainland of a large continent. He then sailed south along the coast around the isthmus of Florida, exploring as he went. On the return trip to Cuba, he discovered the Yucatan peninsula of Mexico. In 1514, he was named military governor of Florida and told to conquer and settle it. In 1515, he returned to the Caribbean under orders to subdue a tribe of cannibals that was causing considerable problems for the new Spanish settlements. Again he was successful.

In 1521, he returned to Florida with a large armed force, intent on conquering and settling what was still thought to be a large island. Landing near either present-day Sanibel Island or Charlotte Harbor, the Spanish force was attacked and defeated by the natives. In the fighting, Ponce de León received a wound from a poison-tipped arrow and died soon after the survivors returned to Cuba.

Although Ponce de León's explorations and his attempts to form colonies in the New World appear to have been motivated by a desire for personal wealth and glory, they were an important part of a tenacious and successful effort which resulted in opening the Western Hemisphere to European influence and, ultimately, control.

J. WILLIAM MONCRIEF

Sir Walter Raleigh
1552-1618
English Adventurer and Writer

Sir Walter Raleigh. *(Library of Congress. Reproduced with permission.)*

Walter Raleigh's place in history results primarily from his eccentric character and his ambiguous relationship with Queen Elizabeth I rather than his actual accomplishments. The latter include his unsuccessful attempts to found a settlement in North Carolina and his exploits in South America.

His family was Protestant, and in 1569, at an early age, he fought in the French Wars of Religion in support of the Protestant Huguenots. He studied at Oriel College of Oxford University and at the Middle Temple (law college) in London.

Raleigh's participation in the successful suppression of an uprising in Ireland in 1580 brought him to the attention of Queen Elizabeth I. By 1582, he was clearly the favorite of Elizabeth. She knighted him in 1585, and he rapidly grew wealthy and influential as the result of gifts of monopolies, properties, and positions from the Queen. One of these grants was the right to establish a colony on land claimed by the English in the area of present-day North Carolina and Virginia. Although the Queen, wishing to keep him nearby, forbade him to travel to America, he financed a number of groups of settlers who attempted, during the period 1584-1589, to form a settlement near Roanoke Island. All these attempts were unsuccessful, the last group disappearing mysteriously.

In 1587, he was named Captain of the queen's guard, and the following year financed one of the ships that fought the Spanish Armada. Some time later, however, he began a romantic relationship with Elizabeth Throckmorton. They managed to keep their marriage a secret from the Queen, who was jealous of Raleigh's attention and affection, until the birth of their son in 1592. In her jealous anger, Queen Elizabeth imprisoned the pair in the Tower of London. Raleigh, however, bought his way out with profits from a voyage he'd funded, and moved away from the court.

In 1595, he financed and led an expedition to present-day Guyana in South America. Since Guyana was in the center of territory claimed by Spain, this journey can be viewed as a continuation of Raleigh's antipathy for the Spanish as well as of his dream of establishing English colonies in the New World. In 1596, he participated in an attack on Cadiz, and he commanded a successful assault on Fayal in 1597. These anti-Spanish exploits regained some of the Queen's favor. He served in Parliament in 1597 and 1601 and was named Governor of Jersey in 1600.

His return to a position of influence came to an abrupt end when Queen Elizabeth died in 1603 and James I, the son of Mary Queen of Scots, became king. Raleigh's enemies quickly convinced the King that Raleigh was plotting to overthrow him. Raleigh was arrested in 1603 and convicted of treason. The death sentence, however, was changed to life imprisonment in the Tower of London where he lived comfortably for twelve years with his family and their servants. During this time, he studied chemistry and mathematics, and performed scientific experiments. He also continued his writing of poetry and prose. His work, *The History of the World*, was published in 1614.

He was released from prison, but not pardoned, in 1616 with the condition that he finance and lead a second expedition to Guyana in search of gold but to do so without angering the Spanish. Raleigh's troops, however, burned a Spanish settlement, and his son was killed in the fighting. In addition, no gold was found. When he returned to England, the King invoked the suspended sentence, and in 1618 he was beheaded. His wife carried his embalmed head with her until her death 29 years later.

J. WILLIAM MONCRIEF

Hernando de Soto
1496-1542
Spanish Conquistador, Explorer, and Mariner

Hernando, known also as Fernando, de Soto was born in Jerez de los Caballeros, Spain. He spent nearly all of his youth at his family manor house, but at 17 he told his father of his desire to go to Seville and attempt to secure employment in a merchant fleet that traded in the West Indies. Though his father desired for him to study the law, de Soto was eventually permitted to pursue his interests in Seville. The young de Soto did quickly garner a position on a ship in 1514, but as a member of Pedro Arias Dávila's exploratory expedition to the West Indies and not specifically as a merchant. A skilled horseman and trader, de Soto quickly became known for accomplishing daring feats to gain high profits from his ventures.

De Soto's renown helped him create several successful partnerships with fellow explorers, such as Francisco Campañón and Juan Ponce de León. The influence of these men, however, led de Soto to abandon his mercantile interests in favor of conquistador pursuits. Adopting a military-like approach, de Soto commanded his fleet to vie for control of Nicaragua against fellow Spaniard Gil González de Ávila. He defeated his rival in Central America in 1527, and then plundered his new territory for precious metals and slaves. De Soto captured the bulk of his capital through slave trading—mostly by capturing natives.

After the death of his patron, de Soto allied himself with explorer Francisco Pizarro. After confirmed reports of a civilization in South America that possessed great wealth in gold, the two men planned an expedition to Peru in 1532. De Soto lent the fellow explorer two ships in return for being named Pizarro's Chief Lieutenant and the expedition's "Captain of Horse." The expedition led to the conquer of the Incan Empire. De Soto and the men under his command were instrumental in defeating the Inca at Cajamarca—a devastating battle for the Inca. Shortly thereafter, he became the first European to make contact with Atahuallpa, the Incan emperor. De Soto formed an amicable political alliance with the Incan ruler after the Spanish defeated the Incan capital at Cuzco. Pizarro undermined this alliance and held Atahuallpa for enormous ransom. Though the sum demanded was met and offered to Pizarro, he grew suspicious of the Incan ruler's power and murdered him. Dissatisfied with Pizarro's actions, de Soto left South

Hernando de Soto. *(Corbis Corporation. Reproduced with permission.)*

America to return to Spain in 1536, taking with him enough plunder to make him one of the wealthiest men in Europe.

Though his feats in Peru gained him power and accolades from the Spanish court, de Soto was soon anxious to return to the New World. He petitioned the crown in 1537 to grant him permission to lead an expedition to conquer Equador but was refused. Instead he was made governor of Cuba and charged with the conquest of Spanish territory in North America. The following year, de Soto took 10 ships and 700 men to Cuba. In 1539, his Spanish forces landed in Florida, near present-day Tampa. What ensued was one of the most far reaching and devastating episodes in the history of European contact with the populations of the New World.

De Soto pushed his way through not only Florida, but Alabama, Arkansas, Georgia, Louisiana, Mississippi, South Carolina, and Tennessee. He abducted native guides to lead his expedition through the southeast, in search of gold. However, the native peoples of the southeast did not possess the gold wealth of the highly advanced Incan civilizations de Soto had encountered in Peru. Disappointed, in 1540 de Soto attempted to head to Mobile Bay in Alabama to rendezvous with his ships. He was met with resistance from the natives at Mauvilia (Mobile). The local Native Americans were decimated, but the Spanish forces were weakened severely. Losing most of his men, supplies, and plunder, de Soto decided to extend his expedition and recoup his losses instead of returning to Spain.

De Soto again pushed northward, though this time the decision would prove fatal. His expedition was plagued by Indian attacks as they made their way through Alabama and Mississippi. On May 21, 1541, de Soto became the first European to sight the Mississippi River. However, he encountered the river south of Memphis, Tennessee, and instead of following the river and charting its path to the Gulf of Mexico, de Soto crossed the river into Arkansas in search of more wealth. The expedition was fruitless. De Soto decided to turn back and follow the Mississippi River southward. De Soto fell ill—most likely with Yellow Fever. He died in Louisiana, exactly one year after first sighting the Mississippi River, and was given a mariner's burial in that river.

ADRIENNE WILMOTH LERNER

Abel Janszoon Tasman
1603?-1659?
Dutch Explorer

Abel Janszoon Tasman was a Dutch navigator and explorer who discovered Tasmania, New Zealand, Tonga, and the Fiji Islands. Tasman made two important voyages (1642 and 1644) through both the Indian and South Pacific Oceans that helped to map the southern hemisphere. With exploration a secondary goal of his voyages, he was primarily interested in establishing trade and finding sources of wealth for his employer, the Dutch East India Company. Because he failed in both respects with the newly discovered lands, his voyages were initially considered to be disappointments. However, with the passage of time, his voyages have been recognized as important contributions to the knowledge of that part of the world.

Little is known about the life of Tasman outside of his service with the Dutch East India Company. While the details of his birth are unknown, it is generally believed that he was born in 1603 at Lutjegast in the Netherlands. In 1632 or 1633 he joined the Dutch East India Company and made his first exploratory voyage to Indonesia as the captain of the *Mocha* in 1634. Five years later he served on an expedition that futilely searched for the "islands of gold and silver" in the seas surrounding Japan. He later made a series of trading voyages to the coastal areas of Asia. Dur-

Abel Tasman. *(Corbis Corporation. Reproduced with permission.)*

ing this time he proved to be an excellent seaman and was subsequently chosen for an ambitious exploration of the Southern Hemisphere.

In 1642 Anthony van Diemen (1593-1645), the governor-general of Dutch East Indies, selected Tasman to command an exploratory voyage to the Southern Hemisphere in an attempt to locate new sources of wealth and commerce. In addition, although stretches of the Australian coast had been previously discovered, it was not known if these were part of a large continent or if they were unconnected masses of land. Relying heavily on the memoir of chief pilot Frans Jacobszoon Visscher, Tasman was instructed to explore the Indian Ocean in an easterly direction and then sail into the Pacific Ocean to search for a passage to Chile.

Tasman sailed from Batavia (present-day Jakarta) to Mauritius on August 14, 1642, with two ships, the *Heemskerk* and *Zeehaen*. From there, he sailed southeast until he discovered land on November 24, which he named Van Diemen's Land (present-day Tasmania). He later discovered the coast of South Island, New Zealand. Continuing his voyage, Tasman became convinced that there was a passage to Chile, so he turned in a northeast direction. On January 21 he discovered Tonga and later the Fiji Islands. Tasman then directed the crew northwest and returned to Batavia on June 14, 1643. In his ten-month voyage, Tasman only lost 10 men to illness and had actually circumnavigated the entire continent of Australia without ever sighting land, thus showing that it was not attached to any continent.

Despite this seeming success, the council of the Dutch East India Company was not pleased with the voyage. They ordered Tasman to make another voyage to establish trading relationships with the areas he had discovered. His second voyage took him to the south coast of New Guinea and then to many of the coastal portions of New Holland (present-day Australia). Once again, he failed to provide a significant amount of wealth or commerce for his company, and this voyage was also considered to be a failure from a financial point of view. Despite this, Tasman was rewarded with the rank of commander and was even made a council member. He later commanded trading and war fleets for the company before he left the service of the Dutch East India Company in 1653. Tasman is believed to have died on October 22, 1659.

JAMES J. HOFFMANN

Pedro de Teixeira
1587-1641
Portuguese Explorer

The fact that 150 million Brazilians—and thus a large portion of South America's population—speak Portuguese rather than Spanish owes much to Pedro de Teixeira. In the course of a 1637-1639 expedition, he became the first European to travel up the Amazon River, and claimed the entire river valley for Portugal.

Born in 1587, Teixeira was 30 years old when he first arrived in Brazil, where he would spend most of his remaining 34 years. After helping to drive out French would-be colonists from the coastal town of Sao Luis (now a major city) in 1615, he helped establish Fort Presépio in the Amazon delta in 1616. The fort later became Belém, destined to become the largest city in the entire river valley.

More than 20 years later, a group of Spanish soldiers and priests arrived in Belém after travelling down the Amazon from Ecuador. Even though Spain and Portugal were at that point ruled by the same king, Portugal feared Spanish dominance in what it perceived as an unequal partnership. It seemed quite possible that Spain might seek to extend its influence over the Amazon; therefore the Portuguese governor of

Maranhao commissioned Teixeira to lead an expedition upriver to stake Portuguese claims.

On October 28, 1637, Teixeira set out from the village of Cametá on the Tocantins River, an Amazon tributary in eastern Brazil. He led an enormous expedition, composed of 70 soldiers, some 1,200 Indian men, and many more women and children, who travelled in a fleet of more than 70 large canoes. The party sailed down the Tocantins, which flows northward to the Amazon, and from there they began the arduous task of moving upstream—that is, westward—along the world's largest river. Other Europeans had come *down* the Amazon before, but Teixeira's was the first to fight the mighty river's current.

After eight hard months of travel, first on the Amazon and then along a tributary, the group reached Spanish territories along the Andes. It took another four months of overland travel to reach the city of Quito, today the capital of Ecuador, where they met with the Spanish governor. Communications between the two groups were tense, but the governor received them with a show of cordiality. The Portuguese stayed for more than three months, and when they left, the governor sent with them his brother, a Jesuit priest. While this appeared like a gesture of friendship, in fact it made it possible for the Spanish to keep track of what their alleged allies were doing.

The group left Quito on February 16, 1639. When they reached the place where the Napo and Aguarico rivers come together—now in eastern Ecuador, near the Peruvian border—Teixeira claimed all lands and rivers "that enter from the east" for Portugal. On reaching Belém on December 12, Teixeira received a hero's welcome, but he did not have long to enjoy his new success. Appointed governor of Belém on February 28, 1640, he soon had to retire due to poor health, and died on June 4. During that year, Spain and Portugal dissolved their uneasy alliance—and thanks to Teixeira's efforts, the Amazon basin remained under the control of Portugal.

JUDSON KNIGHT

Amerigo Vespucci
1454-1512
Italian Merchant and Geographer

Amerigo Vespucci was one of the most important personalities of the European Age of Exploration. His vast knowledge of geography would set the stage for the European colonization of the Western hemisphere.

Amerigo Vespucci. *(Library of Congress. Reproduced with permission.)*

Amerigo Vespucci was a child of the Renaissance and the Scientific Revolution. Nicholas Copernicus (1473-1543) would move the Earth from the center of creation to the position of third satellite orbiting the sun. Galileo (1564-1642) and his telescope proved the heliocentric theory, and a questioning attitude would define this early modern period. The development of the scientific method created a process through which humankind could decipher God's "Book of Nature."

Italy was at the center of this knowledge explosion. By the late fourteenth century, certain Italian city-states had seized control of the flow of spice, perfumes, and silk from the East. Florence was the most powerful of these city-states. Into this intellectually vibrant environment, in 1454, Amerigo Vespucci was born. The Vespucci family had been prominent for over a century, with family members holding important positions in the city's government. These family connections enabled Amerigo to receive an exceptional education, including an introduction to the latest geographic theories, and very early in his education he decided to make geography his intellectual focus. The turning point in his formation as a geographer came when he began an intellectual relationship with Paolo Toscanelli

(1397-1492). Toscanelli was regarded as Florence's greatest intellectual, and he always stressed the importance of experience over authority. He believed that in the modern world one should reject all knowledge that did not stand the test of empirical examination.

In 1492 Christopher Columbus (1456-1501) declared that he had reached India by sailing west. As this information became public Vespucci began to question the veracity of Columbus's claims. The length of his voyage was less than a month, and Vespucci believed that was too short a period of time to travel such a great distance. Most experienced geographers believed that a degree on the surface of the Earth was equal to $66\frac{2}{3}$ miles (107.3 km). Columbus argued that his voyage was shorter than expected because in fact a degree was only equal to $56\frac{2}{3}$ miles (91.2 km), thus making the circumference of the earth much smaller than previously thought. Vespucci's second problem was based upon the fact that Columbus had sailed directly west from Spain. It was common knowledge that Bartholomeu Dias's (1450-1500) voyage to the Cape of Good Hope not only had taken much longer than that of Columbus, but he also had to sail south of the equator. These two facts were in direct conflict with the information put forth by Columbus.

Following the training he received from Toscanelli, Vespucci set out to gather his own empirical data and signed on as an expert astronomer for the next expedition funded by the Spanish monarchy. Of the five ships assigned to this voyage, Vespucci was in charge of two. Both ships sailed westward and reached the coast of what is now Brazil. Along with mapping the entire coastline, he also charted territory, which consists of present-day Colombia, Uruguay, and Argentina. He then explored parts of the Amazon, the Para, and the La Plata rivers. The information from these detailed expeditions convinced European scholars that Columbus had not reached India but had found a vast uncharted territory. Vespucci's accurate maps would eventually be used for further exploration of the Western hemisphere, setting the stage for Europe's colonization of the New World. Amerigo Vespucci was held in such high esteem that in 1507 the German cartographer Martin Waldseemüller (1470-1521) named this new region "America" to honor Vespucci's achievements as a geographer.

RICHARD D. FITZGERALD

Biographical Mentions

Cristóbal de Acuña
1597?-1676

Spanish explorer and missionary who is known for navigating and exploring the far reaches of the Amazon river with Portuguese explorer Pedro Teixeira (1575?-1640) from 1637-39.

Anton de Alaminos
1478-?

Spanish navigator who piloted explorers to and around the New World. Alaminos served as pilot on Christopher Columbus's fourth voyage. Alaminos also accompanied Spanish explorer Ponce de León (1460?-1521) as pilot on his voyage of discovery to Florida. Alaminos is credited as being the first to note the existence of the Gulf Stream while piloting for Ponce de León. In addition, Alaminos piloted for Hernán Cortés (1485-1547) as well as Hernández de Córdoba (d. 1518) on their respective explorations of Mexico.

Afonso de Albuquerque, the Great
1453-1515

Portuguese military officer and strategist who contributed significantly to Portuguese efforts to control the main trade routes to Asia. He conquered India and established a colony in Goa (India) in 1510 and a colony in Melaka (Malay Peninsula) in 1511. He gained additional Portuguese control by having his men marry the widows of the defeated Indian Muslims. His successful efforts provided a base for the expansion of Christianity into India and East Asia.

Diego de Almagro
1474?-1538

Spanish explorer who, in a partnership formed with Spanish explorer Francisco Pizzaro (1470?-1541) in 1522, explored the Inca territories of Mexico for wealth. Almagro made exploratory trips into Columbia and was the first European to enter Chile by land. The rivalry of the conquistadors turned ugly for Almagro when Francisco Pizzaro's brother, Hernando (1475?-1578), had Almagro beheaded in July of 1538 after the latter was charged with rebellion against Spain.

Pedro de Alvarado
1486-1541

Spanish military officer who played a major role in conquering Mexico and Central America and establishing the rule of Spain in America. In 1519, as Hernán Cortes's chief lieutenant, he participated in the subjugation of Mexico. He conquered Guatemala in 1523-1524 and served as Governor of much of Central America from 1527 until his death in a campaign against rebellious natives in Mexico in 1541. He, like other Spanish conquistadors, was known for his brutality.

Francisco Alvares
c. 1465-c. 1541

Portuguese explorer in Ethiopia who visited the remains of ancient Aksum and Kush, as well as the rock-hewn churches of Lalibala. A Jesuit lay missionary, he published a book translated into English as *Narrative of the Portuguese Embassy to Abyssinia, During the Years 1520-1527*. Alvares helped to spur European interest in the Christian kingdom of Ethiopia and its queen, Candace (actually a title and not a name).

Duarte Barbosa
c. 1480-1521

Spanish explorer who visited Africa and India before meeting his death in the Philippines alongside Ferdinand Magellan (c. 1480-1521). Among Barbosa's writings are a description of the East African coastal trading city of Mombasa, as well as the Indian practice of sati or suttee, the ritual suicide of a widow on her late husband's funeral pyre. Barbosa's uncle Diego, warden of the castle of Seville, was Magellan's father-in-law, and Barbosa himself accompanied the explorer in his famous voyage around the world. Both met their deaths in a battle with Filipino tribesmen on the island of Cebu.

Rodrigo de Bastidas
1460-1526

Spanish explorer of what is now Colombia. Working variously with Vasco Núñez de Balboa (1475-1519) and Juan de la Cosa (1460?-1510), Bastidas travelled along the South American coast from Trinidad to the isthmus of Panama during the years 1500-1502. He discovered the mouths of the Magdalena River near present-day Baranquilla, Colombia, and in 1525 founded the colony of Santa Marta nearby. Today Santa Marta is Colombia's oldest city.

Martin Behaim
1459-1507

German-Portuguese cartographer and navigator who created the first world globe. Behaim was born to a well-to-do merchant family in Nuremberg, Germany. After studying with the famed astronomer Regiomontanus, he traveled to Lisbon, Portugal, where he was appointed to the "junta dos mathmaticos" commission of King John II. The purpose of this commission was to find a better method for determining latitude, which Behaim accomplished by ascertaining the position of the Sun, Moon and stars. Behaim gained the respect of the Portuguese elite, and was offered the chance to set sail with Diogo Cão to the west coast of Africa (1485-1486). On this journey, Cão discovered the mouth of the Congo River. Upon his return to Nuremberg in 1490, Behaim constructed the very first globe, with the assistance of painter Georg Glockendon. On it is represented the equator, one meridian, the tropics and the constellations of the zodiac.

Sebastián de Belalcázar
1495-1551

Spanish explorer and conquistador, also known as Sebastián de Benalcázar, who founded settlements throughout Nicaragua, Ecuador and southwestern Colombia. Belalcázar accompanied Christopher Columbus on his third voyage, then assisted Francisco Pizarro in conquering Peru. In 1533, he set out from his base in Piura, Peru and defeated the Incas, taking over command of Quito. There, he founded the settlement of Quayaquil. Two years later, he went in search of the fabled city of gold, El Dorado. Along his journey, he established the Colombian cities of Pasto, Cali, and the Popayán province, over which he became governor in 1541. In his latter years, Belalcázar was involved in struggles with several Spanish leaders, and was eventually convicted of murdering Jorge Robledo, the head of a neighboring province.

Etienne Brûlé
1592?-1633

French explorer who is believed to be the first European to reach many of the Great Lakes, including Lakes Ontario, Erie, Superior, and Huron. He is also credited as the first European to set foot in what is now Michigan. Under orders from discoverer Samuel de Champlain, Brûlé learned Native American languages, became an interpreter, and traveled into the unknown interior of North America to live among various tribes. After more than two decades of exploration, Brûlé was murdered in the land of the Huron Indians.

Sir Thomas Button
?-1634

Welsh explorer who was one of a group of men searching for a water passage from the Atlantic to the Pacific. While the Northwest Passage was not discovered until more than two centuries later, the expeditions of Button and such explorers as Henry Hudson and William Baffin provided the earliest descriptions and charts of North America's Atlantic coast. On Button's best-known journey, he led an expedition aboard the famed Arctic ship *Discovery* to rescue Henry Hudson in 1612-1613. He never found Hudson. He was knighted in 1616.

Robert Bylot

English explorer who was one of the first Europeans to travel to the Canadian Arctic. Bylot was one of a number of explorers who scoured the Arctic for a Northwest Passage to Asia. During the period of 1610-1616, Bylot made a name for himself as a competent navigator and/or commander of four Arctic voyages. During one expedition, he and fellow explorer William Baffin discovered the Lancaster Sound. Although they were unaware of it, the sound was proved some two centuries later to be an entry point for the Northwest Passage.

Álvar Núñez Cabeza de Vaca
c. 1490-c. 1560

Spanish explorer in the region of what is now Texas, whose claims regarding legendary cities of gold influenced later exploration efforts by Hernando de Soto (c. 1500-1542) and Francisco de Coronado (c. 1510-1554). In 1528 Núñez landed near the site of modern-day Galveston, and spent eight years wandering among Native American tribes, during which time most of his men died. When found by fellow Spaniards in northern Mexico in 1536, he was full of wild tales concerning the Seven Cities of Cibola, about which he had heard but which he did not claim to have visited. Later, as governor of Rio de la Plata in South America (1541-1545), he created a route from Santos, Brazil, to Asunción, Paraguay. He recorded his North American journeys in *Naufragios* (1542), and his South American ones in *La Relación y Comentarios* (1555).

John Cabot
c. 1450-c. 1500

Italian navigator, also known as Giovanni Caboto, who was the first European to discover the North American mainland. In 1497 he set out with his three sons on a voyage for Henry VII of England to seek a western route to Asia. On June 24 he discovered Cape Breton Island but believed that he had reached northeast Asia. He undertook another voyage the following year and probably died at sea.

Sebastian Cabot
1476?-1557

Italian navigator and cartographer who sailed (for British investors) to North America in the first recorded attempt to find the Northwest Passage (1508-1509). Cabot served as royal cartographer for both England (1509-1512) and Spain (1512-1518). From 1518 to the 1540s, Cabot sailed for Spain as chief pilot and explored parts of South America, including the Río de la Plata, the Paraná River and the Paraguay River (1526-1529). Retiring from the sea, he turned to mapmaking and, in 1544, completed a world map based, in part, on his voyages and exploration. In 1551, Cabot began his last career as governor of the Muscovy Company, an English trading organization that initiated trade between Russia and England.

Pedro Álvares Cabral
c. 1467-1520

Portuguese navigator who is credited with the discovery of Brazil. Cabral was born into a family of great privilege, held in high esteem among the Portuguese royal family. He spent many years in the service of King Manuel I, who in 1500 entrusted him to lead an expedition of 13 ships to India. Cabral was to follow the route taken by Vasco da Gama, and was charged with cementing trade alliances between India and Portugal. After sailing far westward of his course, Cabral landed in the country he called Island of the True Cross, which would later be renamed Brazil. He took possession of the country for Portugal, and dispatched one of his ships to send word of his conquest to the king. Cabral spent only 10 days in Brazil, before setting sail once again for India. After a voyage fraught with disaster, his fleet cast anchor at Calicut, India, where he entered into a fierce battle against Muslim soldiers. Many of Cabral's crew were killed in the struggle, but the Portuguese eventually prevailed, seizing 10 Muslim ships. He then sailed for the port of Cochin, and there established successful trade relations with local merchants. Cabral returned to Portugal in 1501.

Juan Rodriguez Cabrillo
1498?-1543

Portuguese explorer who discovered the West Coast of the United States. Little is known of Cabrillo's early life, only that he appears to have fought in the army of Hernán Cortés against the Aztecs of Mexico. Later, he was one of the conquistadors of the countries now known as Guatemala, El Salvador, and Nicaragua. By the mid-1530s, he settled in Guatemala, where he encouraged a flourishing trade with Spain and other parts of the New World. In 1542, Cabrillo departed from the port of Navidad, Mexico on an expedition of the coasts north and west of Mexico. After several months, he sailed into San Diego Bay, which he called San Miguel. His fleet of ships reached as far north as Monterey Bay, discovering eight islands off the California coast along the way.

Alvise Cadamosto
1432-1488

Venetian trader who discovered the Cape Verde Islands and explored coastal West Africa. Cadamosto (sometimes rendered as Ca' da Mosto), sailed under the Portuguese flag at the behest of Prince Henry the Navigator (1394-1460). In his first significant voyage (1455), he landed on the Madeira and Canary islands, then traversed the African coastline to a point just south of the Senegal River delta. He attempted to sail up the Gambia River, but when he ran into trouble with the native peoples there, he turned his craft around. On a voyage during the following year, he visited two of the Cape Verde Islands, which were uninhabited.

Diogo Cão
1450-c. 1486

Portuguese navigator and explorer who was the first European to reach the Congo River. In his quest to discover a sea route around Africa to India, King João of Portugal commissioned Diogo Cão to explore the region. Cão left Portugal in 1482 and sailed along the West Coast of Africa until he reached the mouth of the Congo. There he erected a stone pillar dedicated to Portuguese sovereignty of the area. He sailed further, to Cape Santa Maria along what is now the coast of Angola, and erected a second pillar. Upon his return to Lisbon in 1484, King John II honored him with a title of nobility in honor of his discoveries. Cão made his second voyage in around 1485, reaching Cape Cross in what is present-day Namibia.

Thomas Cavendish
1560-1592

British navigator and buccaneer who led the third expedition to circumnavigate the globe (1586-1588). A member of Parliament, Cavendish began his maritime career around 1585 when he joined British admiral Sir Richard Grenville (1541?-1591) on a voyage to the colony of Virginia. Then, following in the path of Sir Francis Drake (1543?-1596), he began his historic circumnavigation in July 1586, reaching the Pacific in February 1587. He then sailed up the South American coast, attacking Spanish settlements and capturing Spanish ships, including treasure galleon *Santa Ana* (November 1587). In October 1592, while attempting a second circumnavigation, Cavendish died at sea.

Richard Chancellor
?-1556

British navigator who opened the White Sea trading route between England and Russia. In 1553, Chancellor functioned as pilot-general of a small fleet of ships sailing for the Company of Merchant Adventurers (later renamed the Muscovy Company) on its first expedition, led by Sir Hugh Willoughby. The ships were separated in a storm and Chancellor's ship reached the White Sea coast. He established a trading post and traveled inland at the invitation of the Russian czar Ivan IV in Moscow. In 1555, he led a second commercial expedition to Russia, obtaining formal trade agreements from the czar and bringing back the first Russian ambassador to London, Ossip Gregorevitch Nepeja. On the return voyage (in 1556), the ship was wrecked off the coast of Scotland and, although Chancellor was killed, the Russian ambassador survived and eventually reached London.

Martin Cortes de Albacar
1532-1589

Spanish navigational scientist who wrote an influential book on navigation, *Breve compendio de la Sphera y del arte de navegar: con nuevos instrumentos y reglas, exemplificado con muy subtiles demonstraciones* (Short Compendium of the Sphere and of the Art of Navigation: With New Instruments and Rules, Exemplified with Very Simple Demonstrations). This work discusses the variability of compass readings in different parts of the world, and offers explanations counter to generalizations in other works of the era. Upon its publication, Cortes's book was considered very important, particularly to the English. Indeed, the English translation went through six editions in the sixteenth century alone.

Juan de la Cosa
c. 1460-1510

Spanish navigator and cartographer who participated in the earliest explorations of America. He was captain of the *Santa María* on Christopher Columbus's first voyage to America in 1492 and returned with Columbus on his second American journey in 1493. In 1499 he participated in the exploration of the northern coast of South America. Cosa drew the first map of the world containing the new American continent in 1500. He was killed by natives in South America in 1510.

Pêro da Covilhã
c. 1460-after 1526

Portuguese explorer who helped establish diplomatic relations between Portugal and Abyssinia (Ethiopia). Determined to find Prester John, a mythical Christian king in the East, Portugal's John II in 1487 sent Covilhã to India and Afonso de Paiva to Ethiopia. The two men agreed to meet in Cairo, but when Covilhã returned there in late 1490, de Paiva was dead. He then proceeded to Abyssinia, on the way making a detour to the Middle East, where he visited the Muslim holy city of Mecca disguised as a Muslim. Covilhã arrived in Abyssinia in 1493, and though he was well-treated, he was not allowed to leave. He married and raised a family, and when the first Portuguese embassy arrived in Abyssinia in 1520, Covilhã served as an interpreter.

William Cecil Dampier
c. 1652-1715

English buccaneer and explorer who achieved fame writing about his voyages. He circumnavigated the globe, visiting Africa, the Philippines, and Australia. Returning to England, he published *A New Voyage Round the World* (1697), a vivid account of his adventures and the places he had visited. In 1699 he commanded a voyage of exploration in the South Seas. He explored Australia and New Guinea, discovering Dampier Archipelago and Strait.

John Davis
1550?-1605

British navigator and Arctic explorer who made three voyages in search of a Northwest Passage from Europe to the Indies (1585, 1586, and 1587), visiting Greenland and Baffin Island. As pilot and navigator for Thomas Cavendish's second privateering circumnavigation expedition in 1591, Davis became separated from the fleet near the Straits of Magellan and he journeyed back to England, discovering the Falkland Islands on his return voyage (1592). Davis authored two books on navigation, *The Seaman's Secrets* (1594) and *The World's Hydrographical Description* (1595); and he invented the Davis quadrant, a double quadrant that was the principle instrument of navigation until the early 1700s. Beginning in 1598, he served as pilot on three voyages to the East Indies, including the first successful expedition of the East India

Company. Davis was killed by Japanese pirates off the Malaysian coast on his third voyage. His *Traverse Book* from his final voyage became the model for ships' log books.

Semyon Ivanov Dezhnyov
c. 1605-1673

Russian explorer who discovered the Bering Strait, but whose discovery did not become widely known until after the time of Vitus Bering (1681-1741). A Cossack, Dezhnyov—sometimes rendered as Dezhnev—explored Siberia in the 1640s, and in 1648 reached what is now known as the Bering Strait. Thus he confirmed that the Asian and North American continents are separated, but his findings were lost until German historian Gerhard Friedrich Müller uncovered them in 1736. By then Bering, whose 1728 voyage attracted far more notice than Dezhnyov's 80 years before, had been credited with the discovery.

Sir Robert Dudley
c. 1574-1649

English sailor, engineer and cartographer who wrote the model for sailing and navigation for his day. Dudley was the son of the elder Robert Dudley, earl of Leicester. He would eventually go on to hold his own titles, as Duke of Northumberland and Earl of Warwick. In 1594, Dudley sailed to Trinidad on his way to explore Guiana and traveled a great distance up the Orinoco River. Two years later he was knighted for his service to England's navy. In 1605, he traveled to Italy, where he served the Grand Duke of Tuscany, assisting in the construction of the port of Leghorn. Late in life, he published his three-volume *Dell'Arcano del mare* (Concerning the Secret of the Sea), a treatise on modern shipbuilding, naval operations and cartography.

Sieur Daniel Greysolon Dulhut
1639?-1710

French soldier and discoverer who led explorations of the expansive Lake Superior. Born into French nobility, Dulhut set sail for New France (Canada) in 1674 and settled in Montreal the following year. In 1678, he formed an expedition to try to form alliances between the French and the Native American populations living along the Lake Superior shoreline. This and ensuing travels and negotiations helped to ensure France's role as the primary European force in the fur trade within New France. Dulhut died in Montreal in 1710.

Juan Sebastián de Elcano
c. 1476-1526

Spanish Basque mariner who completed the historic circumnavigation of the globe begun by Ferdinand Magellan (c. 1480-1521). Elcano sailed from Spain in 1519 as captain of the *Concepción*, but after Magellan perished in the Philippines in April 1521, he assumed command of the entire expedition. The group faced enormous hardships on the return journey, and by the time they landed in Spain in September 1522, only one ship (the *Victoria*), 18 sailors including Elcano, and four captured Indians remained. Elcano died in 1525 during an expedition to seize the Moluccas on behalf of Spain's Charles V.

Nikolaus Federmann
1501?-1542

German conquistador and explorer who led two expeditions (1530-1532 and 1533-1539) into South America for the Weslers, a powerful German banking family. In 1530, the Wesler family sent Federmann to South America to explore the interior of a new colony granted to them by King Charles I of Spain (in present-day Venezuela), seeking the fabled city of gold, El Dorado. In 1533, after a brief return to Europe, Federmann returned to Venezuela to begin another search for El Dorado, during which he explored present-day Colombia, making the first east-west crossing of the Andes, to Bogota. In 1539, he, along with two Spanish explorers, sailed to Spain to submit claims to the land in the region, which eventually was granted to Spain by the Council of the Indes. Federmann died in Madrid in 1542.

Humphrey Gilbert
1539?-1583

British navigator and explorer who played a significant role in early British colonization, setting up the first British colony in North America in 1583 at St. John's, Newfoundland. After an early career in the military, Sir Gilbert (knighted in 1570 for his military achievements), a half-brother of Sir Walter Raleigh (1552?-1618), was the first to petition Queen Elizabeth I of England for exploration seeking a Northwest Passage to the Orient. As a result, in 1578 he was granted a royal charter for the privileges of exploration and colonization in North America. After an unsuccessful first voyage, a second expedition in 1583 reached Newfoundland, where Gilbert founded the first English colony in America. On the return voyage, the ship carrying the explorer was lost at sea north of the Azores in the Atlantic.

Bento de Goes
1562-1607

Portuguese Jesuit who discovered that Cathay and China were the same place. Born in the Azores, Goes was first a soldier before joining the Jesuits in the Portuguese province of Goa, India, where he was befriended by the Mogul emperor Akbar. Goes became intrigued by the question of whether Cathay, as described by Marco Polo (1254-1324), was the same as the land of China that had recently been visited by Italian missionary Matteo Ricci (1552-1610). In 1602-1607, he traveled northward from India to the Silk Road, which he followed east into China. He died at the Great Wall near Beijing, his question answered.

Médard Chouart des Groseilliers
1618?-1696?

French-born Canadian fur trader and explorer who was the first to recognize the possibilities for fur trade and, subsequently, open up that trade in the western Great Lakes and Hudson Bay regions. After serving at a Jesuit mission for several years after arriving in New France (present-day Canada), Groseilliers used his knowledge of Algonguin, Huron, and Iroquois languages to begin trading in furs with Indians. With his second wife's half-brother, Pierre-Esprit Radisson (1636-1710), as his companion, Groseilliers established an active fur trade for the French colony. His successes helped bring about the founding of the Hudson's Bay Company by English investors (in 1670), for whom he worked until 1674. He retired from fur trading in 1683 and died at Trois-Rivières in Quebec around 1696.

John Hadley
1682-1744

English mathematician and inventor who built the first working reflector telescope. Hadley built his Gregorian reflector telescope in 1721. The device contained a 6-inch (15.2-cm) mirror, and proved accurate enough to be successfully used by astronomers. The response to Hadley's invention was so favorable, he later built a larger, more sophisticated version. In 1730 he invented a quadrant, known as Hadley's quadrant, which was used to measure the altitude of the Sun or a star in order to ascertain a ship's geographic position while out at sea. His design eventually evolved into a sextant, a navigational

instrument that measured the distance between the horizon and either the Sun or stars.

Dirck Hartog
fl. 1610s

Dutch sea captain who explored part of the west coast of Australia. While attempting to follow a new route for trade with the East Indies, he reached Australia by mistake in October 1616 in his ship, the *Eendracht*. He left a record of his visit inscribed on a pewter plate on the island known today as Dirk Hartog Island. The surrounding region was subsequently labeled "Eendrachtsland" by the Dutch.

Louis Hennepin
1626-1705?

Belgian explorer and Franciscan missionary priest who is presumed to be the first European to have seen and described Niagara Falls (in his book *A New Discovery of a Vast Country in America*, published in 1697). In 1675, Father Hennepin traveled to Quebec as a missionary among the Iroquois Indians. In 1678, he joined the expedition of French explorer Robert Cavelier, Sieur de La Salle, traveling through the Great Lakes to the Illinois River. Hennepin was sent by La Salle on a voyage via canoe to explore the upper Mississippi River, where in 1680 he discovered the Falls of Saint Anthony (in present-day Minnesota) while a captive of Sioux Indians. Known for exaggerating his explorations in writings such as his book *A New Voyage* (1698), the priest fell into disgrace in his later years and the end of his life remains a mystery.

Pierre Le Moyne d' Iberville
1661-1706

French-Canadian naval captain and discoverer who led explorations of the southeastern United States. Born to a wealthy fur trader in Montreal, Quebec, Iberville spent a good deal of his youth at sea on his father's ship. As an adult, he joined French expeditions that captured English-held Hudson's Bay Company posts in James Bay. His military career eventually took him to the Gulf Coast where he established forts in what are now the states of Mississippi and Louisiana. He died while attacking the English on the islands of the West Indies.

Domingo Martinez de Irala
1487-1557

Spanish explorer who established the first colonial settlements in what is now Paraguay. In 1537, Irala, along with fellow lieutenant Juan de Ayolas, sailed up the Plata and Paraguay Rivers.

Ayolas vanished on an expedition, but Irala carried on, eventually founding Asunción, (now in Paraguay). Asunción soon became the first permanent Spanish settlement in southeastern South America, and would later help Spain in its quest to conquer northern Argentina.

Anthony Jenkinson
1525?-1611

British merchant and explorer who made significant contributions to English trade in Russian and Central Asia. In 1557, Jenkinson, an explorer interested in establishing new markets for English merchants, traveled to Russia for the Muscovy Company to build new trade opportunities in China using company contacts in Moscow. With letters of introduction from czar Ivan IV, he continued on to Central Asia to the Caspian Sea. Jenkinson's party, who at one point joined a camel caravan in the desert, were the first Englishmen to reach the trading city of Bukhara (now in Uzbekistan) in December 1558. He left Bukhara in March 1559 on another trade caravan, eventually reaching the Caspian Sea then Moscow and finally London. In 1561, Jenkinson returned to Persia seeking further trade opportunities and made two more trips to Russia (1566 and 1571) before retiring to England, where he died in 1611.

Gonzalo Jiménez de Quesada
c. 1499-1579

Spanish conquistador who claimed New Granada (now Colombia) for Spain. Jiménez de Quesada was trained as a lawyer in Granada, then traveled to the colony of Santa Marta on the northern coast of South America to serve as its chief justice. In 1536, he was commissioned by Pedro Fernandez de Lugo to lead an expedition into the center of New Granada. After a difficult journey, in which his men were repeatedly attacked by Native Americans, the group eventually reached and conquered the region, founding Bogota as the capital of New Granada. In 1538, two rival conquistadors arrived in the area, and challenged Jiménez de Quesada for authority. Jiménez de Quesada returned to Spain, and exerted his claim to the region. He was named honorary governor, and returned to New Granada in 1549. In 1569, he went on a quest for the mythical city of gold, El Dorado, but returned defeated after only two years.

Louis Jolliet
1645-1700

French-Canadian fur trader who became the first European to make a voyage down the Mis-sissippi River. Jolliet switched careers from priest to fur trader and eventually to explorer when he accepted the leadership of an expedition to discover whether the Mississippi flowed south to the Gulf Coast or west to the Pacific Ocean. Jolliet and fellow traveler Jacques Marquette followed the river only as far as present-day Louisiana, but they learned enough to know it emptied into the Gulf of Mexico.

Henry Kelsey
c. 1667-1724

English mariner who explored the Canadian plains under the employ of the Hudson's Bay Company. Kelsey began his apprenticeship with the Hudson's Bay Company at the age of 17, and continued to work there for nearly 40 years. His first expedition with the company was in 1684, along the western shore of Hudson Bay. During his travels, Kelsey became proficient in the native languages, and in 1690 journeyed to the Saskatchewan River to promote trade with the Indians. During this trip, he is believed to have become the first white man to explore Canada's central plains. From 1718 to 1722, he served as overseas governor for the Hudson's Bay Company.

Eusebio Francisco Kino
1645-1711

Spanish priest and explorer who introduced Christianity and Spanish influence into northern Mexico and present-day Arizona. Sent to Mexico in 1681, he established missions in many native villages and made at least 40 expeditions in Arizona, exploring extensive areas including the Rio Grande, Colorado, and Gila rivers. Father Kino was a compassionate man. He opposed the use of natives as forced labor in the mines, and he improved native agriculture by introducing wheat and cattle farming.

René-Robert Cavelier, sieur de La Salle
1643-1687

French explorer who was the first European to follow the Mississippi south to the Gulf of Mexico. La Salle left France for New France (Canada) in 1667. He conducted many expeditions, including an exploration of the Great Lakes aboard the *Griffon* in 1679. Although many of his voyages wound up unsuccessful, he achieved what no European had before: a voyage south on the Mississippi River to the Gulf Coast in 1682. Five years and several calamitous adventures later, La Salle's men became so enraged with their leader that they shot him dead and left his body where it fell in the wilderness of what is now Texas.

Miguel López de Legazpi
1510-1572

Spanish explorer who claimed Spanish control over the Philippines. Legazpi left Spain for Mexico (then called New Spain) in 1545, where he served as a clerk in its local government. In 1564, Luis de Velasco, the viceroy of New Spain, sent Legazpi to the Philippine archipelago to establish the first Spanish settlement there, which he accomplished the following year. Legazpi served as the first governor of the Philippines from 1565 until his death. In 1571, he journeyed to the northern Philippine island of Luzon, and there set up the capital of the new Spanish colony, Manila.

Jan Huyghen van Linschoten
1563-1611

Dutch explorer who went in search of a northeast passage to India. From 1583 to 1589, Linschoten lived in India, working as a bookkeeper for the archbishop of Goa. Seeking a shorter route from the Netherlands to India, he sailed with the Dutch navigator Willem Barents in search of a northeast passage through the Arctic. He was eventually forced to turn back due to bad weather in this, as well as a subsequent voyage. In 1601, Linschoten published the journal describing his expedition, which inspired later Dutch and English explorers to continue his search.

Jerónimo Lobo
1595-1678

Portuguese Jesuit missionary to India and Ethiopia who wrote an account of his visit to the latter, translated as *Voyage to Abyssinia*. Lobo first went to India in 1621, and in 1624 was sent to convert the Abyssinians from Coptic Christianity to Catholicism. Initially he enjoyed the protection of the Emperor Segued, but with the latter's death, he fell into a series of misadventures and ultimately departed to India. Lobo, who was said to have traveled 38,000 leagues (about 133,000 miles or 214,700 kilometers) in his career, died in Lisbon. His memoirs of Abyssinia first saw publication in 1728, with the French translation *Voyage historique d'Abissinie*.

Diogo Lopes de Sequeira
?-c. 1520

Portuguese sea captain who led the first European expedition to Melaka (modern-day Malaysia). He set sail from Lisbon on April 5, 1508, with four ships, and arrived in Cochin, India, more than a year later. After four months in Cochin, he departed for Melaka in August 1509, with Ferdinand Magellan (c. 1480-1521) among the men under his command. Landing on the island of Sumatra in what is now Indonesia, he erected two stone pillars claiming those lands for Portugal, then sailed for Melaka, arriving on September 11, 1509. There the Portuguese were met by hostile Muslim forces, and Lopes de Sequeira lost some 60 of his men in battle before sailing back to Cochin. This encounter led to the Portuguese conquest of Melaka under Afonso de Albuquerque (1453-1515) in 1511.

Jacques Marquette
1637-1675

French Jesuit missionary who accompanied French-Canadian explorer Louis Jolliet on the first voyage down the Mississippi River. In 1673, The two men traveled far enough to determine that the river emptied into the Gulf of Mexico, rather than west into the Pacific as had been proposed. Marquette's primary calling was as a missionary, however. He established two missions at Sault Ste. Marie and St. Ignace in what is now the Upper Peninsula of Michigan in 1668 and 1671, respectively. A year after the Mississippi voyage, Marquette set out for Illinois to set up another mission, but died en route where the river now known as Pere Marquette flows into Lake Michigan.

Alvaro de Mendaña de Nehra
1541-1595

Spanish explorer who made discoveries in the South Pacific. As the nephew of the viceroy of Peru, in 1567 he was appointed commander of an expedition in search of the mythical great southern continent, terra australis. This expedition failed to find the southern continent, but did discover the Solomon Islands. In 1595 he was sent on a second expedition in search of terra australis but died at sea.

Pedro Menéndez de Avilés
1519-1574

Spanish mariner who founded what has since become one of the oldest European settlements in North America. Menéndez's career as a seaman began at the age of 14 when he joined a ship's crew. He gained fame at the age of 30 when he engaged in a duel with the notorious French corsair Jean Alphonse, and struck him down. Five years later, Menéndez accepted the title of captain-general of the Fleet of the Indies and began leading voyages to America. On his first trip to America in 1565, he established a settlement at St. Augustine, Florida. That settlement is now considered the oldest white settlement in the nation.

Sebastian Münster
1488-1552

German cartographer, cosmographer, and Hebrew scholar whose work revived European interest in geography. Professor of Hebrew at the University of Basel from 1527, he translated a number of ancient works, including Ptolemy's *Geographia*. His description of the world, published as *Cosmographia* in 1544, was widely read. It was translated into five languages and printed in forty editions. His map of the New World was the most widely used for a number of years.

Pánfilo de Narváez
1470?-1528

Spanish conquistador and colonial official who explored Cuba, Mexico, and Florida in search of gold and glory for himself and Spain. In 1498, Narváez emigrated to Hispaniola to seek adventure and fortune as a soldier. He participated in the conquest of Jamaica (1509) before commanding part of an expeditionary army sent to survey and conquer the island of Cuba (1511-1518). After an unsuccessful attempt in 1520 to wrestle control of New Spain (present-day Mexico) from Hernán Cortés during which he lost an eye and was imprisoned, Narváez eventually returned to Spain. In 1526, he was appointed governor of Florida by King Charles I of Spain and was sent to explore and conquer the lands between Florida and northeastern Mexico. He was lost at sea in the Gulf of Mexico in 1528.

Jean Nicollet
1598-1642

Frenchman who is credited as the first European to explore Lake Michigan and reach the American Midwest. At the age of 20, Nicollet joined a French trading company and set out for Canada to live among the Native Americans and become a company interpreter. With direction from Samuel de Champlain, Nicollet began his journey through the Great Lakes in 1633. Nicollet had hoped not only to negotiate trading arrangements with Native Americans, but also to find the rumored transcontinental route to China. Nicollet made history when he arrived onshore in what is now Wisconsin, dressed in a fine Chinese robe. After this voyage, he returned to Quebec, where he held the positions of colonial interpreter and merchant.

Afanasy Nikitin
fl. 1466-1472

Earliest known Russian visitor to India. During the period 1466-1472, Nikitin traveled south from his homeland, through the Indian subcontinent and Persia. His account of his travels, *Khozhdeniye za tri morya* (Journey Beyond Three Seas), became one of the early classics in the then-nascent Russian literary tradition. Known as the "Russian Marco Polo," Nikitin was celebrated with a monument in his hometown of Tver on the Volga River, as well as a commemorative coin issued by the Russian government in 1997.

Abraham Oertel
1527-1598

Flemish cartographer who published the first modern atlas. Born in Antwerp, where he resided until his death, he was involved in the map trade from a young age. In 1570 he published his great work, *Theatrum Orbis Terrarum*, a collection of 70 maps by different cartographers printed in a uniform format. It appeared in more than 40 editions between 1570 and 1612 and was translated into Dutch, German, French, Spanish, Italian, and English.

Alonso de Ojeda
1466-1510

Spanish explorer who participated in the early exploration and conquest of America by Spain. He accompanied Christopher Columbus on his second American voyage in 1493, and in 1496 he brought natives back to Spain as slaves from another voyage of exploration. In 1502 and 1510, his attempts to establish colonies in America failed. Most of his men were killed by natives in the second attempt, and Ojeda died soon after his return to Hispaniola, the Spanish outpost in the West Indies.

Juan de Oñate
1550?-1630

Spanish conquistador who established the colony of New Mexico and explored parts of western North America. Oñate, who married the granddaughter of Hernán Cortés (1485-1548), was born and raised in the New World. In 1598, he founded New Mexico with some 400 settlers, but his actions soon revealed that his principal interest was gold, and he dealt ruthlessly with Spaniards and Native Americans who got in the way of his ambition. During his 1601 expedition to what is now central Kansas, where he hoped to find the legendary city of gold at Quivira, most of the colonists in New Mexico escaped. Oñate led one final attempt to find gold, this time along the Colorado River to the Gulf of California. He was later tried and punished for his acts of cruelty, but appealed and received a reversal of his sentence.

Francisco de Orellana
c. 1490-c. 1546

Spanish soldier who became the first European to explore the Amazon River. Orellana, who served with Francisco Pizarro (c. 1475-1541) during the latter's conquest of Peru in 1535, in 1541 led the advance party for an expedition under Pizarro's half-brother Gonzalo (1502?-1548) to explore regions east of what is now Ecuador. He ended up drifting down the Amazon with 50 soldiers, and after reaching the mouth of the river (1542) and returning to Spain, told stories of attacks by armed women. The latter he compared to the Amazons of Greek mythology—hence the river's name. A second expedition to the region was a monstrous failure, and Orellana drowned as his ship attempted to enter the mouth of the river.

Pedro Páez
1564-1622

Spanish Jesuit priest and explorer who was the first European to locate the source of the Blue Nile near Lake Tana, Ethiopia, in 1618. Another Jesuit priest, Father Jeronymo Lobo, wrote *Voyage Historique,* a book about Páez's discovery. Father Páez was a prisoner in Arabia for several years before he traveled to Ethiopia, where he converted the emperor to Roman Catholicism. After the emperor's death, Father Páez and many other Catholic priests were executed.

Antonio Pigafetta
1491?-1534

Italian sailor who wrote a classic account of his participation in the historic global circumnavigation begun by Ferdinand Magellan (c. 1480-1521). In 1519, the expedition sailed from Spain with five ships, but its leader was killed in 1521, and the following year only one ship—the *Victoria*—returned to Spain bearing a handful of men. Pigafetta was among the survivors, and at the behest of King Charles V, he wrote about his experiences in *Primo viaggio inforno al globo terraqueo.* Colombian novelist Gabriel Garcia Marquez, in his 1982 Nobel Prize lecture, praised Pigafetta for "a strictly accurate account that nonetheless resembles a venture into fantasy."

Martín Alonso Pinzón
c. 1441-1493

Spanish navigator who, with his brothers Vicente (c. 1460-c. 1523) and Francisco Martín (1440?-1493?) took part in Christopher Columbus's (1451-1506) first voyage to the New World. Commander of the *Pinta,* Martín owned a partial share in both that ship and the *Niña,* and was responsible for a change in course on October 7, 1492, which led to the expedition reaching landfall in what is now the Bahamas on October 12. He later separated from the rest of the fleet with the *Pinta,* hoping to find treasure, and rushed back to Spain in an unsuccessful attempt to precede Columbus with the news of the expedition's discovery.

Vicente Yáñez Pinzón
c. 1460-c. 1523

Spanish mariner who commanded the *Niña* during the first New World voyage of Christopher Columbus (1451-1506), and later explored the coast of South and Central America. In 1499 Vicente, younger brother of Martín Alonso Pinzón (c. 1441-1493), discovered a Brazilian cape that he named Santa María de la Consolación. He also explored the region around the Amazon estuary and what is now northeastern Venezuela, and took part in three more voyages to the New World. On the last one, he sailed along the coast of Central America with Juan Díaz de Solís (1470?-1516) before the two had a falling-out and returned to Spain in 1509.

Samuel Purchas
c. 1577-1626

English clergyman and editor who compiled several collections of writings on travel and exploration. Much of his original material has disappeared, and thus his books are the only source of important information about travel and exploration during this period. He became interested in reports of travel and, while serving in several pastorates, collected and edited a large number of these stories. His principal work is *Purchas His Pilgrims* which was published in four volumes in 1625.

Pedro Fernandez de Quirós
1565-1614

Portuguese navigator who explored the South Pacific in a vain attempt to find the elusive "southern continent," or Terra Australis. In 1603, King Philip III of Spain granted him three vessels for a southward voyage, and on December 21, 1605, the expedition departed from Callao, Peru. They moved gradually across the southern Pacific, through the island regions now designated as Pitcairn and French Polynesia and beyond, far to the west. Finally they arrived in Espiritu Santo in what is now Vanuatu—only about 1,200 miles (1,931 km) from Australia—where Quirós established the colony of Nova Jerusalem on May 14, 1606. Due to infighting among the crew, Quirós was forced to turn back soon afterward.

Pierre-Esprit Radisson
1636?-1710

French-born Canadian adventurer who with brother-in-law Médard Chouart des Groseilliers journeyed into the Canadian wilderness north of Lake Superior, became fur traders and eventually set up the English Hudson's Bay Company. After their early success as fur traders, Radisson and Groseilliers traveled to France to sell their pelts. The French authorities seized many of their furs, and the two men turned their attention to England where Radisson received financial support along with a charter to establish Hudson's Bay Company in 1670.

Antonio Raposo de Tavares
1598-1659

Portuguese slave trader who reputedly made one of the first trips across South America. Raposo Tavares led *bandeiras,* or slaving raids in which soldiers of mixed Portuguese, African, and Native American extraction attacked villages, often doing battle with Jesuit priests as well as Indian defenders. His first *bandeira* took place in 1629, the second in 1636-1638. On the third *bandeira,* which began in 1648, Raposo Tavaraes supposedly travelled north from what is now the border between Brazil and Paraguay all the way to the city of Quito in present-day Ecuador, and thence along the Amazon River to Belém at is mouth. If this account is true, then he traversed some 8,000 miles (12,875 km) before returning to Sao Paolo in 1652.

Matteo Ricci
1552-1610

Italian Jesuit missionary who brought European mathematics, geography and Christian teachings to the Chinese, who called him Li Ma-tou. After attending a Jesuit college, Ricci volunteered for missionary work in the Far East and voyaged to the Portuguese colonies at Goa (1578) and the island of Macao near Canton (1582), before being chosen to establish a Christian mission in mainland China. He settled in Chao-ch'ing in 1583, studying the Chinese language and culture and introducing the locals to the culture of Europe. Ricci also introduced the Chinese to Western geography, creating several influential maps, including a large world map with extensive geographical annotations (1584). From 1601, when he established a new mission in Peking, until his death in 1610, Ricci published

several books in Chinese and conducted explorations of China's interiors.

Johannes Schöner
1477-1547

German astronomer and geographer. Ordained a priest in 1515, Schöner later converted to Lutheranism and married. From his days in the priesthood, he operated a printing shop in his house, and also produced globes—including the first using the word "America," which he made in 1515. Schöner wrote a number of works on mathematical, astronomical, and geographical subjects.

Willem Corneliszoon Schouten
c. 1567-1625

Dutch mariner who discovered the Drake Passage around the southern tip of South America. Schouten and Jakob Le Maire (1585-1616) sailed in two vessels from Holland in May 1615, en route to the East Indies. Rather than pass through the perilous Strait of Magellan at the tip of South America, they navigated a much more logical and safer path between Tierra del Fuego and Estados Island, a route that came to be known as the Drake Passage. Arriving in Batavia, Java (now Jakarta, Indonesia) in October 1616, they were arrested on charges of attempting to infringe on the Dutch East India Company monopoly over East Indies trade. Eventually they were cleared of all charges, and Schouten's memoirs, when published, proved highly valuable to mariners and explorers.

Pedro de Sintra
c. 1446-after 1462

Portuguese explorer who became the first European to explore the coast of West Africa in the region of what is now Liberia and Sierra Leone. Operating under the aegis of the navigation "school" established by Prince Henry the Navigator (1394-1460), Sintra voyaged to the West African coast in 1461. He named Cabo do Monte, a large rock promontory on the Liberian coast, and in 1462 named Sierra Leone, or "lion mountain."

John Smith
1579-1631

English soldier, explorer, and adventurer. The son of a Lincolnshire farmer, he completed his elementary school education before being apprenticed to a merchant in Cambridgeshire. Beginning in 1599, he fought against the Spaniards in the Netherlands and later with the Hungarians against the Turks, rising to the rank of captain in the Imperial Army. Wounded, captured, and sold into slavery by the Turks, he escaped, later serving on a French privateer off the Moroccan coast. In 1606 Smith sailed with the Jamestown Expedition to Virginia, serving for a time as president of the colony. He explored the Potomac River region, learned much about Indian society, and became an enthusiastic promoter of English colonization in the New World. Returning to England in 1609, he briefly explored and mapped New England (which he first named) in 1614. Most of his remaining years were spent in England, writing at least eight books, including *The Generall Historie of Virginia, New England, and the Summer Isles* (1624) and *The True Travels, Adventures, and Observations of Captain John Smith* (1630). His maps and writings have much historical value.

Luis Vaez de Torres
?-1613

Spanish navigator who became the first to sail through the strait later named after him, which separates New Guinea from Australia. A ship's captain in the fruitless 1605-1606 South Seas expedition led by Pedro Fernandez de Quirós (1565-1614), Torres and his men were abandoned in what is now Vanuatu by their leader, who inexplicably sailed back to Mexico without them. At that time point Torres opened sealed orders from the Viceroy of Peru, directing him to head for the Philippines. In so doing, he sailed along the southern coast of New Guinea, a route made dangerous by the many reefs and shifting currents along the 90-mile-wide (145 km) passage. During this time, Torres may have glimpsed Australia, making him only the second European to do so, after Willem Jansz (b. 1570) a few months earlier. Due to Spanish security concerns, the strait was kept a secret for more than a century, but in 1762 a British hydrographer named it after Torres.

Andres de Urdaneta
1498-1568

Spanish monk and navigator who discovered a viable route across the Pacific Ocean from west to east, facilitating the Spanish conquest of the Philippines. He was a military officer, then spent 11 years (1525-1536) in the Spice Islands as an adventurer. In 1553 he became an Augustinian monk in Mexico. In 1564-1565 he led an expedition to the Philippines from Mexico and, on his return trip, discovered a northerly return route with favorable winds.

Lodovico de Varthema
c. 1465-1517

Italian adventurer, also known as Ludovico di Varthema, who made significant discoveries and observations as the first European to visit many areas of the Middle East and Asia. He traveled extensively from 1502-1507, visiting Egypt, Syria, the Arabian and Malay peninsulas, Aden, Yemen, Ceylon, Burma, and India. He risked his life as the first non-Muslim to visit the holy city of Mecca. He joined the Portuguese army in India and was knighted. The report of his travels was published in 1510.

Diego de Velásquez
1460?-1524

Spanish conquistador and colonial official who was commissioned by the Spanish Crown to colonize Cuba, which he conquered in 1512. After sailing to the West Indies with Christopher Columbus on his second voyage of exploration (1493), Velásquez settled on Hispaniola, where he helped establish several towns. In 1511, he commanded the Spanish fleet that conquered Cuba. Velásquez founded Baracoa, the first permanent European settlement on the island, and later that year became governor general. From 1517 to 1520, historic expeditions were sent by the new governor from Cuba to the Yucatan and Mexico under the commands of Francisco Hernandez de Cordoba, Juan de Grijalva, Hernán Cortés, and Pánfilo de Narváez (1470?-1528), firmly establishing the Spanish colonies in the New World.

Giovanni da Verrazzano
1485-1528

Italian discoverer who was the first European to explore and chart the Atlantic coastline from Newfoundland to as far south as North Carolina, including what is now known as the New York harbor. In the service of the King of France, Verrazzano began the voyage with the goal of finding a waterway that would connect the Atlantic Ocean to the Pacific, and ultimately Asia. While he didn't find a passage, he was successful in describing the eastern coast of North America. He led later voyages to Brazil and to the West Indies, where he died in 1528 at the hands of a Carib tribe.

Martin Waldseemüller
c. 1470-c. 1521

German clergyman and cartographer who coined the name *America* for the New World. Waldseemüller read of Amerigo Vespucci's explorations in a 1504 letter. Adept at inventing names, including his own, he named the newly discovered continents in honor of Vespucci in *Cosmographiae Introductio* and on a world map, both in 1507. His map was widely distributed, and thus two continents came to be known by the name of a relatively minor Italian navigator.

Hugh Willoughby
?-1554

English ship commander who was one of the many explorers of the 1500s in search of a water passage from Europe to the lucrative markets of southeast Asia. Sir Hugh Willoughby is perhaps most remembered for his attempt to find a Northeast Passage across the northern reaches of Europe and Asia. He set sail in 1553 into the cold waters north of Norway. His ship was separated from the other two in the expedition, however, and he died off Lapland. The expedition continued east to Moscow without him, and initiated trade between England and Russia.

Edward Wright
1558?-1615

English mathematician who improved on the cartographic system devised by Gerhard Mercator (1512-1594), and thus ensured the widespread acceptance of the Mercator projection. In 1599, Wright published *The Correction of Certaine Errors in Navigation*, which he revised as *Certaine Errors in Navigation, Detected and Corrected* (1610). These books offered modifications on Mercator, as well as mathematical tables and practical information for mariners plotting straight-line courses on Mercator-projection maps. The British Navy adopted Wright's system, which enabled it to gain an advantage over other navies.

Saint Francis Xavier
1506-1552

Spanish missionary who brought Christianity to India, Malaysia, and Japan. Of noble birth, he studied in Paris where he helped found the Society of Jesus (Jesuits). Ordained in 1537, he went to Portuguese Goa, in India, as a missionary in 1542. His mission expanded into Malaysia (1545) and Japan (1549). He advocated the use of local language and culture by missionaries and the utilization of natives as clergy. Xavier was canonized in 1622.

Timofeyevich Yermak
?-1584

Russian adventurer who brought Siberia under Russian control. Yermak (or Ermak) was the leader of a band of outlaw Cossacks who pirated boats on the Volga River. In 1579 they were hired by a wealthy Russian family to defend their property against raids by Siberian tribes. He led a force of less than 1,000 Cossacks over the Ural Mountains and, in 1582, captured the Siberian capital, opening Siberia for eventual absorption into Czarist Russia.

Bibliography of Primary Sources

Cortes de Albacar, Martin. *Breve compendio de la Sphera y del arte de navegar: con nuevos instrumentos y reglas, exemplificado con muy subtiles demonstraciones* (Short Compendium of the Sphere and of the Art of Navigation: With New Instruments and Rules, Exemplified with Very Simple Demonstrations). 1556. An influential work on navigation, including discussion of the variability of compass readings in different parts of the world, and explanations counter to generalizations in other works of the era. The book was particularly important for the English, with its English translation going through six editions in the sixteenth century alone.

Dampier, William. *A New Voyage Round the World.* 1697. A vivid account of Dampier's adventures and the places he visited, including Africa, the Philippines, and Australia, while circumnavigating the globe.

Hakluyt, Richard. *Divers Voyages Touching the Discovery of America.* 1582. A geographical work that greatly encouraged colonization efforts. It contained an account of previous English voyages to North America, a list of America's known resources, and a report on the Northwest Passage.

Hakluyt, Richard. *The Principal Navigations, Voyages, and Discoveries of the English Nation.* 1589. A comprehensive three-volume history of English overseas exploration, considered a classic of English literature.

Hennepin, Louis. *A New Discovery of a Vast Country in America.* 1697. Contains the first European description of Niagara Falls.

Lobo, Jeronymo. *A Short Relation of the River Nile.* 1669. Contains an account of Pedro Páez's discovery of the source of the Blue Nile near Lake Tana, Ethiopia, in 1618. Lobo was the second European to arrive at this location.

Münster, Sebastian. *Cosmographia.* 1544. A description of the world that revived European interest in geography. It was widely read, translated into five languages, and printed in forty editions.

Oertel, Abraham. *Theatrum Orbis Terrarum.* 1570. A collection of 70 maps by different cartographers, printed in a uniform format. It appeared in more than 40 editions between 1570 and 1612, and was translated into Dutch, German, French, Spanish, Italian, and English.

Purchas, Samuel. *Hakluytus Posthumus or Purchas His Pilgrimes*. 1625. A four-volume collection of writings on travel and exploration, based in part on Richard Hakluyt's accounts. Purchas's work is an important source of information about travel and exploration during this period, as it is one of the few reliable sources to have survived.

Smith, John. *The Generall Historie of Virginia, New England, and the Summer Isles*. 1624. The first significant account of the New World in English.

Waldseemüller, Martin. *Cosmographiae Introductio*. 1507. An atlas containing a map of the New World under the name "America" a designation he coined in honor of explorer Amerigo Vespucci. The map was widely distributed, and thus two continents came to be known by the name of a relatively minor Italian navigator.

Wright, Edward. *The Correction of Certaine Errors in Navigation*. 1599. Provided modifications of Mercator's maps, as well as mathematical tables and practical information for mariners plotting straight-line courses on Mercator-projection maps. Revised as *Certaine Errors in Navigation, Detected and Corrected* (1610).

JOSH LAUER

Life Sciences and Medicine

Chronology

1451 Bernardo di Rapallo of Italy devises a perineal operation for kidney stones.

1521 The first anatomical drawings made from nature, by Giacomo Berengario da Carpi, are published; those of Leonardo da Vinci, made earlier, will not be published for centuries.

1536 *Chirurgia Magna*, a surgical treatise by Paracelsus, stresses the importance of minerals in treating diseases.

1543 Andreas Vesalius publishes his stunningly illustrated book of anatomy, *De humani corporis fabrica libri septem* (Seven books on the structure of the human body), one of the most important books in medical history. The illustrations were probably done by Jan Stephan van Calcar.

1551 Konrad von Gesner publishes the first volume of *Historia Animalium*, the first modern, scientific study of animal life.

1555 By depicting the homologies between the skeleton of a bird and that of a human, Pierre Belon in *L'Histoire de la Nature des Oyseaux* establishes comparative anatomy as a discipline.

1602 Swiss anatomist Felix Plater publishes *Praxis Medica*, the first modern attempt at the classification of diseases.

1628 English physician William Harvey, considered the founder of modern physiology, first demonstrates the correct theory of blood circulation in *De Motu Cordis et Sanguinis in Animalibus*.

1637 French philosopher René Descartes's *Discours de la Méthode* applies a mechanistic view to science and medicine, establishing a worldview that dominates the study of man for some time.

1664 Thomas Willis, an English physician, writes his *Cerebri Anatome,* which gives a complete and accurate account of the nervous system.

1668 The first successful intravenous injections on humans are made independently by Johann Major and Johann Elscholtz, both German physicians.

1674 French physician Morel de Villiers invents the tourniquet for stopping major hemorrhages.

1677 Antoni van Leeuwenhoek, a Dutch biologist and microscopist, discovers and describes spermatozoa.

Overview:
Life Sciences and Medicine 1450-1699

Previous Period

There was relatively little interest expressed in the life sciences during the Middle Ages. However, several factors developed in the later medieval period that led to a renewed interest in the careful observation of nature on the part of Europeans. These factors were: the writings of the Greek philosophers whose work had been rediscovered in the late Middle Ages; the learning of Arab philosophers and physicians, which became known after the crusades; and the work of such European scholars as Roger Bacon (1220?-1292?) and Albertus Magnus (1200?-1280). All three of these groups had stressed the importance of investigating the natural world. With the close examination of nature and the striving for realism that marked the Renaissance, sciences such as anatomy and botany began to develop their modern forms. The invention of the printing press in the mid-fifteenth century made it possible to transmit this learning much more easily, thus setting the stage for significant developments in the life sciences and medicine during the Renaissance.

Anatomy

Unillustrated medieval anatomical texts simply copied the writings of classical scholars such as Galen (c. 130-c. 200), who based his work on the dissection of animals, not humans. By the fourteenth century, human dissections were becoming common in Italian universities and it became clear that Galen's descriptions were not always accurate. One of the most significant events in the study of anatomy was the publication in 1543 of Andreas Vesalius's (1514-1564) book on human anatomy, *De humani corporis fabrica,* which combined exquisite illustrations with a text that was a significant improvement on the anatomies passed down from ancient times.

Historians have suggested that the artist who illustrated Vesalius's text was influenced by the unpublished drawings in Leonardo da Vinci's (1452-1519) famous notebooks, drawings based on actual dissections. Many artists of the time, including da Vinci and Albrecht Dürer (1471-1528), were extremely interested in anatomy, as well as other aspects of nature; they painted and drew much more realistic and accurate images of plants and animals than had been created in the previous centuries.

The philosopher and natural scientist René Descartes (1596-1650) developed the influential idea that the human body was like a machine and that the body and the mind, which he thought was seated in the brain's pineal gland, were quite separate from each other. A number of other anatomists also studied the nervous system in the seventeenth century; for instance, Thomas Willis (1621-1675) investigated blood circulation in the brain. One of the most significant anatomical works of the seventeenth century was done by William Harvey (1578-1657), the first to accurately describe the human circulatory system, including the circulation of blood through the heart. This work led to further interest in circulation and to the first blood transfusions, which were attempted in the 1660s. There were also considerable advances made in other areas of anatomy. Bartolommeo Eustachio (1520?-1574) investigated the structure of the ear, Harvey and his teacher Girolamo Fabrici (1537-1619) worked on the developing embryo, and Vesalius's pupil Gabriele Falloppio (1523-1562) studied the female reproductive system.

Medical Theories and Practices

In the sixteenth and seventeenth centuries medical practice remained a mixture of ancient ideas and new observations, with the latter leading to improved therapies. One of the individuals whose ideas represent such a mixture of old and new was Paracelsus (1493-1541), a Swiss alchemist and physician who argued against the barbaric, medieval techniques in medicine and called for radical change in medical practice, while at the same time holding to the ancient ideas of alchemy. These ideas led Paracelsus to encourage the use of chemicals such as sulfur and copper compounds as medicines, rather than the traditional plant-derived remedies.

Paracelsus was not the only physician developing new treatments at this time. Because there was little reliable information, there were numerous theories of disease. Many physicians, including some with very questionable remedies, preyed on people who were being assaulted by a large number of medical problems. Bubonic plague was still present in Europe, as was yellow fever, sweating sickness, leprosy, and a number of other infectious diseases. At the end of the fifteenth century, syphilis appeared in armies fighting in Italy and continued to spread throughout Europe in epidemic proportions for a long period of time.

Barber-surgeons were prominent dispensers of medical treatment, but during this time the field of surgery became more professionalized. The French surgeon Ambroise Paré (1510-1590) made many notable contributions to the field, including the practice of ligature, or tying off severed blood vessels. He also condemned the use of boiling oil as a treatment for gunshot wounds. In addition, there were changes in procedures related to childbirth. Traditionally babies had been delivered by female midwives, but with the increasing influence of surgeons, childbirth became the domain of these physicians, who were called "men-midwives."

All areas of medicine were enriched by the development of printing. Books called herbals gave information on the medicinal uses of plants, and many illustrated works of anatomy followed those of Vesalius. Printing also made possible wider distribution of ancient texts, and this led to more questioning of old ideas and to the search for better information. Medical manuals began to be produced in languages other than Latin, making knowledge more accessible to those with less education.

New Worlds

During the period 1450-1699, much of the activity in the life sciences resulted from the opening up of new worlds. The age of exploration was just beginning, and as explorers returned from voyages throughout the world, they often brought back both living and dead specimens of new plant and animal species, some of which were very unusual and had characteristics that had never been seen before by Europeans. Some plants soon became economically significant crops in Europe, including maize or corn and tobacco as well as potatoes and tomatoes. This wealth of new species brought questions of organization to the fore, and several classification schemes were developed, one of the most significant being that of the English botanist John Ray (1627-1705).

There was also an unintentional movement of species in the other direction as well, with those of the Old World being brought to the New World. For example, rats on European ships were responsible for the extinction of many species, particularly on islands where animals were not adapted to coping with rodent predators. Europeans also brought diseases to the lands they colonized, causing large-scale epidemics among native populations. Smallpox and other infectious diseases had arrived in the Caribbean by 1518 and they soon spread to Mexico and South America.

Another new world that opened up in the seventeenth century was the microscopic world, when magnifying lenses were perfected to the point that the tiny organisms in pond water and other fluids could be seen for the first time. In 1665 Robert Hooke (1635-1703) published *Micrographia,* a book which contained many drawings of specimens he had examined under the microscope, including a flea and the cells in a piece of cork. Jan Swammerdam (1637-1680) and Anton van Leeuwenhoek (1632-1723) were two Dutch naturalists who were particularly important in opening up this new microscopic world.

Botany and Zoology

Once the microscope was invented, it was used by a number of botanists, including Nehemiah Grew (1641-1712) and Marcello Malpighi (1628-1694), to investigate plant anatomy. In zoology, Konrad von Gesner (1516-1565) inaugurated the modern age of this science with a four-volume work on animals, published in the 1550s. The renewed interest in botany and zoology led to the establishment of a number of zoological and botanical gardens in Europe, which allowed the general public, for the first time, to be able to view organisms gathered from around the world.

The Future

By the time the seventeenth century ended, the life sciences were poised for the tremendous growth of knowledge that occurred in the eighteenth century. Work on classification, microscopic investigations, and anatomical studies all continued to bear fruit. There would also be more emphasis placed on experimentation, particularly in physiology, the study of plant and animal function.

ROBERT HENDRICK

Philosophy of Science:
Baconian and Cartesian Approaches

Overview

The Renaissance and Scientific Revolution encompassed the transformation of art, science, medicine, and philosophy, as well as the social, economic, and political life of Europe. Ancient concepts were challenged by new ideas and facts generated by the exploration of the world, the heavens, and the human body. Natural philosophers, physicians, and surgeons were confronted with plants, animals, and diseases unknown to the ancient authorities. Although Francis Bacon (1561-1639) and René Descartes (1596-1650) developed different methodologies, these two seventeenth century philosophers helped to guide and systematize the new sciences and define the modern scientific method.

Background

Although he made no direct contributions to scientific knowledge, Francis Bacon is remembered as Britain's major seventeenth-century British philosopher of science. A keen observer of the great events of his time, Bacon said that of all the products of human ingenuity the three most significant were the compass, gunpowder, and printing. Through their combined effects, Bacon argued, these inventions had "changed the appearance and state of the whole world." Bacon himself became the guiding spirit of the new experimental science and the scientific societies that nurtured it. His impact on the sciences came about through his emphasis on defining the methodology of science, suggesting means of insuring its application, and providing encouragement and direction for the new scientific enterprises he predicted. Bacon planned an encyclopedia of the crafts and experimental facts, a review of all branches of human knowledge, and new scientific institutions that would improve human welfare, comfort, and prosperity. Ultimately, according to Bacon, science would increase human knowledge, power, and control over nature. Bacon rejected the scholasticism of the universities and launched open attacks on Aristotle and Plato. He insisted that fact gathering and experiment must replace the sterile burden of deductive logic so that naturalists could produce new scientific knowledge.

Bacon's *Advancement of Learning* (1605) proposed a new science of observation and experiment to replace traditional Aristotelian science. The Baconian method, also known as the inductive method, involves the exhaustive collection of particular instances or facts and the elimination of factors, which do not accompany the phenomenon under investigation. Generally suspicious of mathematics, deductive logic, and intuitive thinking, Bacon believed that valid hypotheses should be derived from the assembly and analysis of "Tables and Arrangements of Instances." Rather than passively collecting facts, the scientist must be actively involved in putting questions to nature. Scientists would analyze experience "as if by a machine" to arrive at true conclusions by proceeding from less to more general propositions. The result of applying this scientific method, Bacon assured his readers, would be a great new synthesis of all human knowledge, a true and lawful marriage between the empirical and the rational faculty. Nevertheless, Bacon apparently appreciated the significance of what is now known as the falsifiability principle, which is usually associated with the twentieth-century philosopher Karl Popper (1902-1994). Despite his enthusiasm for the collection of facts, Bacon realized that it is impossible to provide absolute proof of inductive generalizations based on a finite number of observations. Because a few "negative instances" have the power to falsify an induction, experimental results that contradict a general theory may reveal more about nature than another bit of data that appears to support the theory.

Like Bacon, the French philosopher René Descartes believed that a new science would lead to knowledge and inventions that would promote human welfare. Unlike Bacon, Descartes was a gifted mathematician, honored as the inventor of analytic geometry, and the advocate of a deductive, mathematical approach to the sciences. Descartes believed that his approach to science would allow human beings to master and possess nature's abundance and establish a new medical science capable of eliminating disease and extending the human life span. Unlike Bacon, whose work he had studied and criticized, Descartes placed *a priori* principles first and subordinated his observations and experimental findings to them. Nevertheless, he too had a grand scheme and task for natural philosophy. An examination of method was a primary part of his plan to use the mathematical method in developing a general mechanical model of the workings of nature. Although experimentation had a role in Descartes's system, it was a subordinate one. He believed that experiments should serve as illustrations of ideas that had been deduced from primary principles or should help decide between alternative possibilities when the consequences of intuitive deduction were ambiguous.

Following Bacon's example, Descartes also opposed scholastic Aristotelianism and called for new approaches to science and philosophical inquiry. Applying his methods to science, philosophy, or any other rational inquiry, Descartes asserted, would not only resolve problems but would lead to the discovery of useful philosophical knowledge. Descartes began by methodically doubting knowledge based on authority, the senses, and reason. Ultimately, he found certainty in the intuitive knowledge that he was thinking, and, therefore, he must exist. He expressed this insight in his famous declaration: "I think, therefore I am." Descartes developed a dualistic system that separated mind, the essence of which is thinking, from matter, the essence of which is extension in three dimensions.

Descartes's metaphysical system is intuitionist, derived by reason from innate ideas, but his physics and physiology, based on sensory knowledge, are mechanistic and empiricist. The mechanistic philosophy asserts that all life phenomena can be completely explained in terms of the physical-chemical laws that govern the inanimate world. Vitalist philosophy claims that the real entity of life is the soul or vital force and that the body exists for and through the soul, which is incomprehensible in strictly scientific terms. The writings of Descartes provided the most influential philosophical framework for a mechanistic approach to physiology. Descartes's own physiological experiments and texts provided his followers with a complete and satisfying mechanistic system, embedded in a general system of philosophy. The fundamental platform of Descartes's mechanical philosophy was that all natural phenomena could be explained solely by matter and motion. In his *Treatise of Man*, Descartes extended his concept of the universe as a machine to the explanation of human beings as machines working in accordance with physical laws.

Impact

Cartesian doctrine essentially treated animals as machines whose activities were explained in purely mechanical terms as the motions of material corpuscles and the heat generated by the heart. Descartes systematized the mechanical philosophy and provided a rationale for describing the human body as a machine. Even human beings could be investigated as earthly machines that differed from animals only because they possessed a rational soul that governed their actions. Serving as the agent of thought, will, conscious perception, memory, imagination, and reason, the rational soul was the only entity exempted from a purely mechanical explanation. Except for thought processes, all physiological functions of the human body were as mechanical as the workings of a clock.

Descartes challenged scientists to treat the physical and mental aspects of human beings in the same manner as all other scientific problems. According to Descartes, a human being is a union of mind and body, two dissimilar substances that interact only in the pineal gland. He reasoned that the pineal gland must be the uniting point because it is the only nondouble organ in the brain, and double reports, as from two eyes, must have one place to merge. He argued that each action on a person's sense organs causes subtle matter to move through tubular nerves to the pineal gland, causing it to vibrate distinctively. These vibrations give rise to emotions and passions and cause the body to act. Bodily action is thus the outcome of a reflex arc that begins with external stimuli and involves first an internal response, as, for example, when a soldier sees the enemy, feels fear, and flees. The mind cannot change bodily reactions directly—for example, it cannot will the body to fight—but it can change the pineal vibrations from those that cause fear and fleeing to those that cause courage and fighting.

Even though the nervous system carried out the commands of the rational soul, Descartes provided a mechanical explanation for the nervous system. Direct interaction between the rational soul and the earthly machine occurred in the pineal gland, an unpaired organ that was erroneously thought to be present only in humans. Through conduits in the brain, the animal spirits were able to enter the nerves, which were hollow tubes that incorporated hypothetical valves governing the flow of nervous fluid. Delicate threads along the length of the interior of the nerves connected the brain to the sense organs. The tiniest motion along the thread tugged at the site of the brain where the thread originated and opened pores that allowed the animal spirits to flow into the muscles. Bodily action was, therefore, the result of a reflex arc that began with external stimuli and involved an internal response. Movement of the subtle fluid through the nerves in response to stimulation of the sense organs caused the pineal gland to vibrate, resulting in changes in the emotions and passions. Although the mind could not change bodily reactions to external stimuli directly, it could affect the distinctive pineal vibrations. Thus, external stimuli could cause fear, but the mind could determine whether the reaction would be flight or fight.

Descartes's work was widely read, imitated, and honored. He challenged scientists to treat the physical and mental aspects of human beings in the same manner as all other scientific problems. His disciples saw him as the first philosopher to dare to explain all the functions of human beings, even the brain, in a purely mechanical manner. Guided by Descartes, many seventeenth century physiologists tried to force all vital phenomena to fit mechanical analogies. Such physiologists were known as iatromechanists, because they believed that all functions of the living body could be explained on physical and mathematical principles. In contrast, iatrochemists attempted to explain vital phenomena as chemical events. The mechanical philosophy allowed naturalists to investigate nature without relying on the vitalistic "soul" and "spirits" that had characterized ancient and Renaissance science. Only the rational soul of human beings remained.

Descartes's influence on philosophy, literature, and French culture was both profound and subtle. Eventually, however, it was the "Baconian method" that became virtually synonymous with the "scientific method." Nevertheless, neither Descartes nor Bacon alone could have served as a complete guide for the development of experimental and theoretical science. Even some of their contemporaries recognized the deficiencies of the pure Baconian system and the pure Cartesian system. The great mathematician and physicist Christiaan Huygens (1629-1687) recognized this when he remarked that Descartes had ignored the role of experimentation, while Bacon had failed to appreciate the role of mathematics in scientific method. A synthesis of the two approaches was needed, or the admission that there is no one scientific method sufficient for posing and solving all possible problems. Mechanical fact-finding, daydreams, and flashes of intuition have played a role in science, no matter what formal doctrine or method scientists professed to follow.

LOIS N. MAGNER

Further Reading

Blasius W., Boylan, J. W,. and K. Kramer, eds. *Founders of Experimental Physiology*. Munich: Lehmanns, 1971.

Carter, Richard B. *Descartes' Medical Philosophy: The Organic Solution to the Mind-Body Problem*. Baltimore, MD: Johns Hopkins University Press, 1983.

Descartes, René. *Treatise of Man (1622)*. Translated by T. S. Hall. Cambridge, MA: Harvard University Press, 1972.

Farrington, Benjamin. *Francis Bacon, Philosopher of Industrial Science*. New York: Schuman, 1949.

Hall, Thomas Steele. *History of General Physiology 600 B.C. to A.D. 1900*. 2 vols. Chicago, IL: University of Chicago Press, 1975.

Rothschuh, Karl E. *History of Physiology*. New York: Robert E. Krieger, 1973.

Shea, William R. *The Magic of Numbers and Motion: The Scientific Career of René Descartes*. Canton, MA: Science History Publications. 1991.

Whitney, Charles. *Francis Bacon and Modernity*. New Haven, CT: Yale University Press.

Zagorin, Perez. *Francis Bacon*. Princeton, NJ: Princeton University Press, 1998.

Theory and Experiment Redefine Medical Practice and Philosophy

Overview

In the sixteenth and seventeenth centuries there was considerable debate among physicians about the appropriate philosophical basis for their art. The humoral theory of Galen and Avicenna remained influential while the iatro-chemists advocated a more aggressive application of new chemical remedies. Drawing on Descartes, Galileo, Newton, and Harvey, the iatromechanist school began to think of the human body as a machine. In contrast, the early vitalists believed in a vital force that differentiated living matter from nonliving. The boundaries between the different systems were not rigid. The majority of medical practitioners, not holding university degrees, relied more on first-hand experience and traditional medical knowledge than on a philosophical framework.

Background

With the appearance of universities and their faculties of medicine in the twelfth through fourteenth centuries, physicians educated at universities became eager to distinguish themselves from other medical practitioners, who they disparagingly described as "empirics." The traditional medical curriculum was devoted to the study of the works of the Greek medical writer Galen (c. A.D. 130-c. 200), and of the great Persian physician Abu-Ali Al-Husain Ibn Abdulla Ibn Sina (c. 980-1037), known in the Western world by the Latinized name Avicenna, who extended Galen's system.

Galen based his medical system on the theory of four humors, which like the theory of four elements originated in the teaching of the Greek philosopher Empedocles (490-c. 430 B.C.) and had been embraced by the philosopher Aristotle. In the Galenic view, health resulted from the balance between blood, phlegm, yellow bile, and black bile. Disease was a matter of internal imbalance, which could be restored through diet or through purgation or bleeding.

By the sixteenth century medical scholars had joined in the controversy between "ancients" and "moderns" that also affected the study of literature and theology. While some scholars sought to restore the teachings of the ancient writers in their original purity, others felt that too much respect had been paid to the teachers of antiquity. In scientific areas there was a new emphasis on experiment and observation as well as new theories of health and healing. For physicians the need for new approaches was also seen as necessary to deal with new diseases, particularly syphilis, which made its first appearance in Europe after the return of Christopher Columbus (1451-1506) from the New World.

The alternative system known as iatrochemistry is usually considered the creation of the controversial Swiss physician Theophrastus Bombast von Hohenheim (1493-1541), generally known by the name he adopted, Paracelsus. Scholars agree that Paracelsus traveled widely and sought knowledge from alchemists, barber-surgeons, midwives, and miners. They disagree

as to the extent of his study at universities. There is evidence he received a doctoral degree at Ferrara in Italy. In his practice of medicine he was credited with a number of nearly miraculous cures, and in 1526 he was appointed town physician in Basel, Switzerland, where he publicly burned the books of Galen as a protest against the traditional teachings. His supporters called him, among other things, the Luther of the Physicians, seeing in him a needed reformer. His opponents were in the majority, however, and had him run out of town.

As alchemists, Paracelsus and his followers wanted to replace the four elements of Empedocles, and the corresponding system of four humors, by a set of three fundamental "principles": salt, sulfur, and mercury. This proposal grew out of alchemical experimentation, which increasingly involved the application of heat to separate materials into their volatile part (mercury), fluid part (sulfur), and solid remainder (salt). More important than this new theory, perhaps, was the Paracelsans' insistence that the proper purpose of chemistry was not the changing of base metals into gold but the creation of medicines. Paracelsus gained some followers among the empirics but loved controversy too much to be accepted by the majority of university-educated physicians. His followers frequently gained support from royal families. In France the King supported Paracelsan physicians against the medical faculty at Paris, which remained faithful to Galen.

New discoveries in the physical sciences stimulated an alternative school, based more on physics than chemistry, known as iatromechanism, or sometimes iatromathematics. The Italian mathematician and physiologist Giovanni Alfonso Borelli (1608-1679), a friend of Galileo (1564-1642), is generally considered the founder of this school. In his book *De Motu Animalum*, published shortly after his death, he describes the skeletal system of various animals as a system of mechanical levers controlled by the muscles. Iatromechanists were quick to seize on the idea of the English physician William Harvey (1578-1673) that the heart acts as a mechanical pump. They also claimed the great French philosopher René Descartes (1596-1650) as one of their number. While Descartes is known primarily for his contributions to philosophy and mathematics, he wrote also on physics and, living near the butcher's quarter in Amsterdam, conducted many dissections of animals. Descartes concluded that animals were in essence machines while humans were a combination of mechanism and a nonmaterial mind.

The success of the British mathematician Sir Isaac Newton (1642-1727) in providing a mathematical framework for physics through his three laws of motion also served to strengthen the iatromechanist school.

Vitalism is the belief that living organisms possess qualities that are not present in inorganic matter. The vitalist approach to medicine appears in the work of the Flemish physician and alchemist Jan Baptista van Helmont (c. 1750-1644), a follower of Paracelsus. Van Helmont and his followers believed that only living organisms could produce the "ferments" that permitted the digesting of foodstuffs and their incorporation into the organism, and thus that the chemistry of living or once-living matter was forever separated from that of inanimate matter.

Impact

The increased emphasis on observation and experiment in the seventeenth century would change the way in which university-educated physicians viewed their calling. In 1600 such physicians considered themselves to be students of nature. Many, including William Gilbert (1544-1603), physician to Queen Elizabeth I, engaged in physical experiments. "Physic," from the Greek word meaning nature, was understood to include recommending good diet and other health practices as well as trying to cure the sick. By 1700 the physician was mainly concerned with treating the sick, and clinical experience gradually replaced much of the traditional philosophical training.

Each of the major medical theories current in the seventeenth century continued to influence the practice of medicine up to the nineteenth century and to some extent even the twentieth. Blood-letting, and the application of medicinal leeches, continued through most of the nineteenth century. While a young ship's surgeon, Julius Mayer (1814-1878) made the observation that sailors bled in the tropics showed less use of oxygen than people in temperate climates. These observations were the catalyst that led him to formulate the law of energy conservation. Notions of restoring the equilibrium of the patient may have played a role in the experimentation with "antagonist therapies" in the 1920s, such as the deliberate infection of patients with malaria to treat advanced syphilis.

Iatrochemistry might claim credit for the expansion of chemical therapies. A flurry of new chemical therapies followed the late-nineteenth-century discovery of dyes that could stain, and thus attach themselves to, specific tissues. The influence of iatrochemical theory on the diet of Europeans was also profound. The diet of the wealthy, the only people who could afford a variety of foods, in 1600 emphasized heavy use of sugar, cooked vegetables, and warmed wine, all of which were expected to insure the balance of the Galenic humors. A hundred years later there had been a shift to fresh vegetables or salad, minimal use of sugar, and cold beverages, much as in modern practice, justified by the three Paracelsan principles and the idea of digestion as fermentation.

The iatromechanist tradition has continued also, perhaps with the least fundamental change. The diagnosing physician's reliance on blood pressure, temperature, and timing the pulse to assess the state of health of the patient has not changed in principle since the seventeenth century.

Vitalism has had a more checkered career. Vitalists are behind the original separation of chemistry into organic and inorganic categories. The debate about the spontaneous generation of organisms from nonliving matter appeared to be finally resolved by the demonstration in 1860 by the French chemist Louis Pasteur (1822-1895) that meat broth can be preserved indefinitely if microbes are excluded, an apparent victory for vitalism. In 1828, however, the German chemist Friedrich Wöhler (1800-1882) demonstrated that urea, a simple compound found in animal waste, can be made in the laboratory from ammonium cyanate, considered to be an inorganic compound. In modern science, the vitalist position is no longer accepted, at least outside the area of mental phenomena, and the difference between living organisms and nonliving is considered to be a matter of vastly greater complexity, much of it to be found in the molecular structure of enzymes, the modern counterpart of van Helmont's ferments.

Assessments of Paracelsus have varied through the centuries. He has been considered something of a Swiss national hero. The young romantic poet Robert Browning wrote a play in verse about him in 1835. The Swiss psychoanalyst Carl Gustav Jung (1875-1961) wrote two essays, "Paracelsus the Physician" and "Paracelsus as a Spiritual Phenomenon," praising the controversial physician as an innovator in medicine. His collected works have been published, and several scholarly papers about his role in the evolution of modern medicine appear each year.

DONALD R. FRANCESCHETTI

Further Reading

Books

Cumston, Charles Greene. *An Introduction to the History of Medicine*. London: Dawsons of Pall Mall, 1968.

King, Lester S. *The Philosophy of Medicine. The Early Eighteenth Century*. Cambridge, MA: Harvard University Press, 1978.

Porter, Roy. *The Greatest Benefit to Mankind: A Medical History of Humanity*. New York: Norton, 1997.

Articles

Cook, Harold J. "The New Philosophy and Medicine in Seventeenth-Century England." In D. C. Lindberg and R. S. Westman, eds., *Reappraisals of the Scientific Revolution*. New York: Cambridge University Press, 1990.

Laudan, Rachel. "Birth of the Modern Diet." *Scientific American* 283 (August 2000): 76-81.

Advances in Midwifery and Obstetrics

Overview

The routine delivery of a healthy child by skilled and knowledgeable birth attendants is a relatively modern phenomenon. Centuries ago, ignorance about anatomy, particularly female anatomy, combined with a lack of knowledge about pregnancy, labor, and delivery, made childbirth a life-threatening process for both mother and child. Before the fourteenth century, pregnancy and childbirth were of little interest to the medical community. Female midwives, whose actions fell under the jurisdiction of the Church, attended laboring women. The admission of medical men and their technologies to the birth chamber, which became increasingly common during the fifteenth century, signaled the beginning of obstetrics as a scientific discipline and marked the decline of midwifery. The inclusion of physicians, unfortunately, did little to improve a woman's chances of safe delivery during a difficult labor.

Background

Until the beginning of the fourteenth century, care for women during pregnancy and labor was the exclusive province of midwives. Despite a lack of formal training in anatomy, they learned their craft through apprenticeships with experienced midwives. Thus, both knowledge and superstition passed from one generation to the next. The medical community on the other hand, viewed pregnancy as an illness and treated it as such. Bleeding was a common treatment for the discomforts and complications of expectant women; it was intended to restore the balance of humors in the body.

Midwives attended both normal and difficult births. A normal birth was (and still is) one in which the child presented head first and facing the mother's back. During a normal delivery the midwife's role was to assist the mother without interfering with labor, becoming impatient, or trying to hurry the birth. Wise midwives allowed labor and delivery to follow a natural course whenever possible, because intervention, haste, or incomplete removal of the placenta was often fatal for the mother. Following the birth, the midwife tended to both mother and child, usually for several days.

During a difficult birth, a midwife had few genuine options. With a poor understanding of anatomy and birth, midwives resorted to several different strategies in their attempts to help a mother deliver a child. These included pressing on the mother's abdomen and having the mother inhale smells thought to encourage active labor. Sometimes symbolic rituals thought to aid labor were performed: removing the patient's hairpins, or opening anything in the house that was closed: doors, drawers, cabinets—even bottles and jugs. In the event of the mother's death, a midwife needed to act quickly to either save the child or baptize it before it died. To accomplish this, a midwife had to perform a caesarean section (the surgical removal of a fetus through an incision made in the abdomen) on the dead woman.

Impact

The dawn of the Renaissance (c. 1453) marked a new beginning in the annals of medicine, childbirth, and midwifery. The invention of the printing press spurred a new interest in learning and inquiry, making documents and information newly available to a wide audience, including midwives and doctors. During this period some local ordinances began to limit midwives' traditional functions, requiring them to send for the assistance of a physician or barber-surgeon in dif-

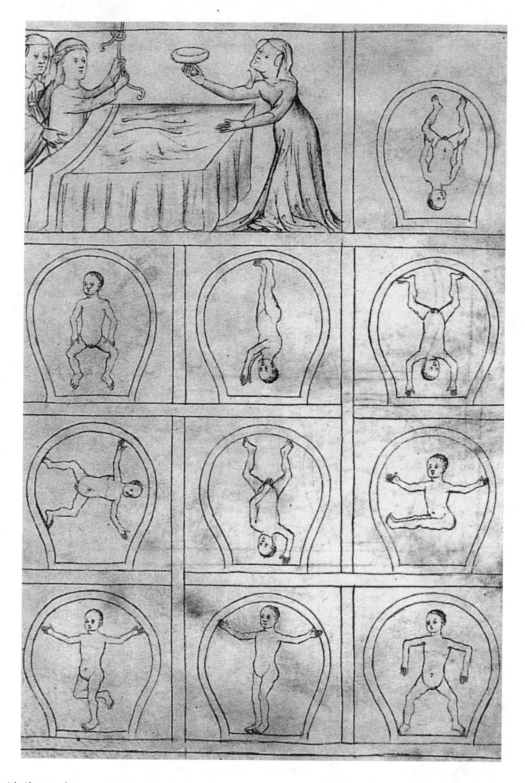

A midwife attending a patient, from an English manuscript from around 1400. The diagrams show the normal fetal presentation (top right) as well as those that could complicate delivery. *(Bodleian Library, Oxford University. Reproduced with permission.)*

ficult births, and forbidding them to use sharp instruments (such as hooks and knives) to extract a dead fetus. The Guild of Surgeons further solidified the segregation of midwives from surgical duties (including cesarean sections) in the 1540 statute that stated "no carpenter, smith, weaver, or woman shall practice surgery." The passage of these regulations and statutes gave the medical community a toehold that allowed them to enter a realm about which they knew very little.

As the sixteenth century began, the need to improve the education of midwives became apparent. The first English work on midwifery—a translation of a German text written by Eucharius Rösslin (1490?-1526) in 1513—was published in 1540. In this text Rösslin praised midwives for their positive contributions and at the same time criticized the overuse of medical instruments and surgical interventions. He advocated attendance of a birth that was neither hurried nor interfered with. Rösslin's text proved of little practical use because while it did contain a few nuggets of wisdom, it also contained much of the old and often inaccurate information published in ancient medical texts. If midwives were to improve their knowledge and skills, it would have to be through the their own efforts.

There was no shortage of information in medical circles. Several notable obstetricians built their reputations on their studies of the birth process and innovation of new techniques. Ambroise Paré (1510-1590), surgeon to the King of France, pioneered the field of operative obstetrics while studying the mechanics of labor and delivery. His nonsurgical method for assisting in certain difficult births, which involved manually turning the fetus to deliver the head first, was considered a major medical advance and subsequently became part of the professional repertoire of man-midwives. By the early 1600s, although not widely accepted by women, the man-midwife was a regular feature in birth chambers during difficult deliveries. In the latter part of the same century, François Mauriceau (1637-1709) followed in Paré's footsteps; he developed a technique for delivering babies in the breech position and wrote extensively on other malpresentations and difficult labors.

François Rousset (1535-1590?), a contemporary of Paré, was far ahead of his time in advocating cesarean sections for living women. His insistence that this procedure could be performed without killing the mother or affecting future pregnancies was highly (and rightly) criticized by his colleagues, particularly Paré. Despite Rousset's insistence, the few cesareans performed in this period were almost invariably fatal to the mother, either because of hemorrhage or overwhelming infection; they were attempted only in desperation. Not until the late-nineteenth century, when surgical and antiseptic procedures improved, did mortality rates fall.

At the same time, publications by now renowned midwives such as Louise Bourgeois (1563-1636), French midwife to Marie de Medici, and English midwife Jane Sharp (fl. 1700) sought to provide midwives with sound anatomical information, midwifery techniques, and encouragement. In 1688, Elizabeth Cellier submitted a proposal for a midwives' college that would instruct and supervise midwives. Although her proposal was refused, Cellier continued to press her cause. In France the prestigious Hôtel Dieu trained a handful of midwives each year. For the time being, normal births and deliveries still belonged to female midwives. That all changed, however, with the next great development in obstetric technology.

For more than 100 years, the Chamberlen family employed a secret weapon in their fight to be the "miracle workers" who could deliver babies during both normal and difficult deliveries. Their secret was so closely guarded that the only physicians allowed to see the instrument were those who had bought the rights to use it. Even the mothers on whom it was used never saw the instrument. In 1720, the Chamberlen family publicly revealed the wondrous instrument that had saved so many lives. In the skilled hands of a trained obstetrician or man-midwife, the Chamberlen forceps significantly reduced the incidence of fetal injury and death and also decreased the rates of maternal disease, injury, and death. Because forceps were classified as a mechanical instrument, midwives were prohibited from using them to assist during childbirth.

The Chamberlen forceps was the final blow to dominance of midwives in obstetrics. Male physicians were increasingly called even for normal labors and deliveries; the establishment of lying-in hospitals that sprung up around Europe further aided the medicalization of childbirth. By the middle of the eighteenth century the marginalization of midwives was in full swing.

MICHELLE ROSE

Further Reading

Arney, William R. *Power and the Profession of Obstetrics.* Chicago: The University of Chicago Press, 1982.

Blumenfeld-Kosinski, Renate. *Not of Woman Born.* Ithaca, NY: Cornell University Press, 1990.

Donnison, Jean. *Midwives and Medical Men: A History of Inter-Professional Rivalries and Women's Rights.* Schocken Books, 1977.

Gelis, Jacques. *History of Childbirth: Fertility, Pregnancy and Birth in Early Modern Europe.* Northeastern University Press, 1991.

O'Dowd, Michael J., and Elliot E. Phillipp. *The History of Obstetrics and Gynecology.* The Parthenon Publishing Group, 1994.

Advances in Understanding the Female Reproductive System

Overview

The Renaissance, a period of immense cultural change during the fourteenth, fifteenth, and sixteenth centuries, contributed greatly to learning. During this time, the knowledge of anatomy grew significantly, an advance that contributed more to basic understanding than to improving health. The working of the female reproductive system was particularly shrouded in mystery and superstition. Anatomists generally assumed that male and female anatomy were exactly alike except for the childbearing function. When taboos against dissection began to be lifted, anatomists were still slow to recognize the significance or the function of the various discoveries.

Gabriele Falloppio (1523-1566), a student of Andreas Vesalius (1514-1564), explored the fields of urology and the female reproductive system. While the fallopian tubes, which he discovered, bear his name, he did not realize how they function. Two centuries would pass before it was known how the eggs released in the ovaries pass through them into the uterus.

As the seventeenth century unfolded, bitter disputes arose about the nature of reproduction and problems of regeneration. Around 1665 many writers addressed the subject, but the one who stood above the rest was Dutch anatomist Regnier de Graaf (1641-1673). His treatise on the female reproductive system was an important step in the history of biology.

Several other writers also described gynecological problems and obstetric skills, among them William Chamberlen (1540-1596), who developed forceps for delivery. But healing therapies still were shrouded in tradition and new discoveries in anatomy did not result in better treatment for patients.

Background

The largest early treatise on the female reproductive system was by Soranus, a Greek physician who practiced at Ephesus during the second century A.D. While he worked in the Hippocratic tradition studying the diseases of women, he maintained many male prejudices. The female constitution was assumed to be an imperfect version of the male, and the "wandering womb" was blamed for certain hysteria-like illnesses. His book *Gynaecology* (literally "study of women.") circulated widely and included sections on both delivery and "womb-caused" diseases.

Galen (c. 130-c. 200), the main authority in Western medicine for hundreds of years, spoke little about the female reproductive system. Because of the prohibitions on dissecting human cadavers, the few studies that were done were conducted on animals. This often led to erroneous conclusions. Galen, for example, thought the uterus had two horn-like projections. He believed that male and female children developed in the right and left horn, respectively.

During the Middle Ages childbirth was strictly the domain of women. Midwives and others passed on their knowledge via oral tradition, since few of them could read or write. In the eleventh century, a woman named Trotula taught at the medical school in Salerno, Italy. She wrote an important treatise, *Passionibus Mulierum Curandorum* (The Diseases of Women), which discusses menstrual disorders, and the problems of women after childbirth, but little about birth itself. Unfortunately, the book's illustrations of the fetus, which depict miniature people with adult characteristics floating in the uterus, show little actual knowledge about its development.

Impact

Understanding of the female reproductive system was not a high priority for physicians during the Middle Ages and Renaissance. Although there were a few bright lights, basic knowledge, along with attitude, progressed little during this period. The prevailing beliefs about women's anatomy were based on the writings of Aristotle (384-322 B.C.), Galen, Soranus, and the Bible. The standard view was that men and women share a common physiology, although the male was the perfect version, and the female a flawed imitation. In this view, the male and female reproductive organs were exactly alike, except that those of the female were inverted and inferior. From animal observations, they assumed the female vagina was an inverted, immature penis.

In general, the females were considered faulty and weak. Since the womb was an unstable organ (and linked to female hysteria), women were thought to be mentally less balanced than men. Menstruation and tearfulness were believed to make female flesh flabby and moist, whereas men were hard and muscular. The female psyche was also deemed weak because of all the oozing female body fluids. Because the female did not have adequate "vital heat," women could not make semen like the men, but produced milk instead. Women, in short, were leaky vessels full of holes, as evidenced by crying, milk production, and menstruation. The anatomists of this period who revived interest in the structure of the human body did not develop an understanding of the physiology of the female reproductive system and did not translate the new knowledge into practice.

Falloppio added to the knowledge of dissection by not only using adult human cadavers, but also those of fetuses, newborn infants, and children. Although he was interested in bone and muscle development, some of his most important work was in urology, especially the kidneys. Criticizing Vesalius for using the dog to illustrate these structures, he was then naturally drawn to the structures of reproduction.

He first described the fallopian tubes as small trumpets. The accuracy of the shape, which he called *tuba*, loses some of the meaning when translated into English. But his descriptions were accurate and warrant the structure bearing his name. He first used the term vagina and showed it was a separate structure from the uterus. He also described structures that were filled with a watery fluid and others that have a yellow humor or fluid. These were possibly the ovaries with the egg encased in the follicle and fluid. Falloppio's works were published in *Opera omnia* in Venice in 1584.

Studying the reproductive system, or generativity, as it was called, was not a top priority among anatomists because childbirth, or generation, was still the domain of women and midwives. Medical writings of the time were intended to assist labor, and little effort was made to understand the process until well into the seventeenth century.

One of the great creators of experimental physiology was Regnier de Graaf, who studied and published on many topics. He took the field of gynecology from simple observation of the structures to understanding how they worked. De Graaf is credited with naming the ovaries, a term also claimed by van Horn and Jan Swammerdam (1637-1680). (Swammerdam and de Graaf had been classmates and close friends until they became embroiled in disputes.) De

Graaf was a great illustrator and examined the ovaries of numerous mammals. He isolated ovarian vesicles with their envelopes, and described how bovine (cow) ovaries changed after mating. He was the first to describe changes in the ovaries and other body parts during the menstrual cycle.

De Graaf was also the first to describe the corpus luteum (yellow body), a structure that develops within a ruptured ovarian follicle and secretes the hormone progesterone. The function of these structures was not definitely established until around 1900. In the tradition of naming structures after their discoverer, the follicle around the egg is now called the graafian follicle. His one error was in supposing the entire follicle was the egg, which then was expelled into the fallopian tube.

De Graaf followed the progress of pregnancy in the rabbit, from mating until birth, and left several plates illustrating each stage. He was aware that the egg was smaller than the follicle but was unable to explain this since he could not observe the eruption of the egg from the follicle. Actually, the mammalian egg was not discovered until 1827 by Karl Ernst von Baer (1792-1876) and its role was not clarified until the early twentieth century.

Compared to the writings and works of his contemporaries, de Graaf was ahead of his time. While the drawings of others were inexact and full of speculation, de Graaf's were precise and detailed. He established a level of scientific inquiry into the reproductive systems that would not be perfected until centuries later.

Throughout this period women remained in charge of childbirth, only calling in a surgeon to extract a dead fetus. However, in the last half of the seventeenth century gynecology and obstetric problems were discussed in a stream of books. Hendrik van Roonhuyze (1622-1672) of Amsterdam published a 1661 work describing cesarian section, ectopic or tubal pregnancy, and certain fistulas, and François Mariceau (1637-1709) deposed the myth that pelvic bones separated during labor to enable birth.

Another advance involved Chamberlen's obstetrical forceps. The instrument included a curved metal blade that could be inserted to grasp the baby's head. The Chamberlen family of male midwives kept the forceps a secret, passing it down through several generations. Since a sheet covered the woman in labor at all times, it was easy to hide the instruments.

The Renaissance was a period of great learning and advances in anatomy and the rudiments of physiology. It was not until much later, however, that this new understanding of the structure and function of female reproduction improved patient care, lowered infant morality, and prevented maternal deaths.

EVELYN B. KELLY

Further Reading

Garcia-Ballester, Luis. *Practical Medicine from Salerno to the Black Death.* Cambridge: Cambridge University Press, 1994.

Porter, Roy. *The Greatest Benefit to Mankind: A Medical History of Humanity.* New York: W.W. Norton, 1997.

Rowland, Beryl. *Medieval Woman's Guide to Health: The First English Gynecological Handbook.* Kent, Ohio: Kent State University Press, 1981.

The Medical Role of Women: Women as Patients and Practitioners

Overview

Throughout history more than half of the people involved in health care and healing have been women. More than half of the patients have also been women. Historically, the disproportionate fame and recognition given to male practitioners is largely due to the fact that surviving manuscripts from earlier times were written by men, and because women generally were not accepted into medical schools. Women have long practiced medicine, but dealt primarily with childbirth and conditions of the female reproductive system—women took care of women.

During the Renaissance, the period of intellectual and cultural revival that marks the end of the Middle Ages in the fifteenth and sixteenth centuries, women had a freer life, but living conditions were still poor and life was cheap. The use of herbs and potions for healing led to associations between healing and witchcraft, in some cases resulting in unjust trials and the execution of alleged sorcerers—many of whom were women.

The sixteenth century brought about the beginning of a new era in geography, religion, the arts, and science. The Renaissance in medicine began in 1543 when Andreas Vesalius (1514-1564) published his anatomy works. Several unsung women also participated in the medical renaissance.

However, as the medical profession became more regulated, women did not fare as well. By the end of the 1600s the role of women had not improved and ultimately gave way to complete male dominance in obstetrical care in the next centuries.

Background

For many centuries women had significant roles as physicians. For example, in the Egyptian medical schools at Helipolis and Sais, the golden-haired Agamede was skilled in medicine and herbal lore. Philistra (318-372 B.C.) lectured so well that pupils flocked to her. She was also so attractive she had to lecture from behind a curtain. Women physicians were numerous during Roman times. Roman records attest to gynecological work done by both obstetrices, or midwives, and *medicae,* female doctors. According to Tacitus, the practice of medicine was common among the German barbarians.

During medieval times the scholars of both genders kept learning alive in the Christian monasteries. Many herbal and diagnostic skills were continued. The best known nun was Hildegard von Bingen (1098-1179), who wrote two medical manuscripts on plant, animal, and mineral medicines and on physiology and the nature of disease. Her remedies were partly herbal and partly spiritual or magical.

In the eleventh century there emerged at Salerno treatises attributed to Trotula. She gained a great reputation as a physician and obstetrician and wrote on many topics. One manuscript reports that Trotula was so loved that at her funeral in 1099 her procession was two miles long. Some scholars argue that the woman Trotula did not do the writing. Regardless of who wrote it, many of the remedies and procedures were practiced for centuries. The books, which had to be first hand-copied, appeared in print in the fifteenth century.

Much of the obstetric care was given by midwives who passed down herbal remedies via

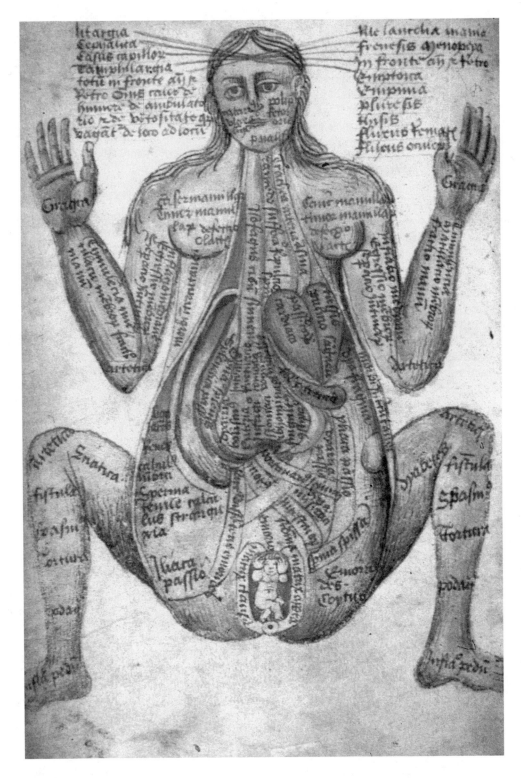

An illustration of female anatomy from the late Middle Ages. This image reflects the typical and deliberate distortion of medieval medical diagrams, which were used to illustrate both anatomy and disease. Zodiacal influences, which were also thought to affect health, are also indicated. *(Wellcome Institute Library. Reproduced with permission.)*

oral tradition. The herbals became closely related to spiritual and magic, and consequently led to the presumption of witchcraft and sorcery. From the fifteenth century the Catholic Church was active in the persecution of witches, many of whom were women who were herbalists and healers. In 1486 *The Hammer of Witches* was the black bible of this movement and led to the death of thousands of innocent people. The last witch was supposedly burned in Germany in 1775.

Impact

To understand the status of women, one must look at the traditions passed down by Galen (c. 130-c. 200), Soranus, and others. Women were thought to be inferior models of men. Men were strong and muscular; women were weak and flabby. Physiologically, they were viewed as the same except the uterus was an inverted version of the male penis. Women were considered leaky vessels compared to men. Menstruation, tears, and producing milk ostensibly proved this. Although women were generally considered to be inferior mortals, some girls of the upper classes were educated like the men by tutors.

Medicine in general in the fifteenth century reflected the times—terrible. Personal hygiene was unknown and plagues and war were rampant. For example, during the plague of 1478 one-third of the population of Europe died. In 1484 Pope Sixtus IV (1471-1484) issued an edict proclaiming that only university graduates could practice medicine. These laws were continuously broken because there was such a need.

Medical schools were generally unpopular, and it was only in Italy that women were admitted to the university. Beatrix Galindo (1473-1535) was educated in Italy and went on to be professor of Latin, philosophy, and medicine at the University of Salamanca in Spain. At that time professors taught several subjects that were part of the liberal arts curriculum, which included medicine, though it was not medical training as we think of today.

The Medicis of Florence were great collectors of books and their researchers found copies of Galen, Aristotle (384-322 B.C.), and Hippocrates (460?-377? B.C.) in Greek. They hired Cassandra Fidelis, a scholar known for her knowledge of medicine, to translate. She wrote a book on the natural sciences and treatment of diseases in 1484.

Italy's famous medical families and the university women were crucial in Italy's acceptance and advancement of medicine.

The fifteenth century also witnessed the widespread belief in witchcraft for the first time. The public was convinced that a large number of women conspired with the devil to injure others. Many of these innocent women were tortured into confessions, then publicly killed. With the ban against women physicians and the witchcraft mania, many women were reluctant to even become midwives. If the baby or mother died, the midwife or doctor might be blamed.

The education of midwives was passed on via oral tradition, but toward the end of the fifteenth century certain midwives began to demand books in their vernacular languages. The earliest treatise for midwives, *Das Frauen Buchlein* (The little book for wives) was printed in 1500. In 1513 Eucharius Roesslin (1490?-1526), a city physician at Frankfurt-am-Main, published *A Rose Garden for Pregnant Women and Midwives*. In 1545 Thomas Raynalde published *The Byrth of Mankynd*. This book included an illustration of "The Woman's Stool" and adult-like figures of fetuses floating in an inverted light-bulb shaped uterus. In general, the state of obstetrics was stunted by superstition and quackery. However, this century has been dubbed as the end of the Middle Ages.

The sixteenth century was the beginning of the new era of freethinking and the Renaissance. The great age of discovery of Christopher Columbus (1451-1506) and the Cabots opened geography and astronomy, art, music, and drama. The renaissance of medicine began in 1543 with the publication of the anatomical text *De Humani Corporis Fabrica* by Vesalius.

In France the position of medical women was worse than ever, but women could attend births, nurse charity patients, and care for their own families. The most noted French obstetrician of the sixteenth century was Louise Bourgeois, friend and pupil of Ambroise Paré (1510-1590), the famous French surgeon. Born in 1563, Bourgeois, like most women of her age, married young. She had three children and was widowed. Deciding that lace-making was not enough to provide for her family, she begged to be taught medicine. Acquiring great skill, she published books on midwifery in 1608 and 1653.

By the end of the century Bourgeois had formed an association to make rules for the protection of midwives and to elevate midwifery in general. The association sought to have their practitioners attend dissections of female bodies at medical schools to learn about anatomy.

The most famous medical woman in Switzerland was Marie Colinet of Bern, who was married to the renowned surgeon Geronimo Fabricius (1537-1619). He taught his wife surgery and admitted that she was a better surgeon than he was. He praised her skill as a bonesetter and told of a case where she wired the ribs of an injured man then placed an effective dressing containing oil of roses. She also performed obstetrical operations and was skilled in cesarean sections.

Medical books of the sixteenth century give insight into the conditions of the times. Roesslin's *Rose Garden* had many later editions with woodcuts that illustrate the care of mother and child. Caspar Wolff also printed Trotula's work. The books of this period indicate that gynecology was becoming a medical specialty.

Since the sixteenth century was a transitional period of thought, there were extreme and puzzling contradictions. Medicine, however, was not changed as much as other fields because it was still closely tied to religion. Astrology held tight, but people were beginning to realize the suffering of the plague could not be helped with heavenly bodies. Education was spreading slowly among women, and the Catholic church no longer ruled in Protestant countries.

Medicine in the seventeenth century advanced slowly. While the work of Bourgeois and others had spread some hope in a small part of the world, the practice of obstetrics in Europe and the American colonies went back to medieval practices before Trotula. Doctors performed vaginal examinations with unclean hands and broke the bag of waters with their long dirty fingernails. The rate of puerperal fever ran high among both rich and poor. There also emerged a new problem for women healers—aggressive witchcraft persecution.

It was unfair that medical women, denied a university education, except in Italy, should be blamed for their lack of success. Men were frequently called by midwives for help. Francois Mauriceau (1637-1709) delivered charity patients at Hotel Dieu as well as noblewomen. In 1668 he wrote and illustrated a book for midwives that was used for 150 years, until it was replaced by Madame Boivin's book in the nineteenth century.

Slowly men began to emerge as professional leaders. The Chamberlen family of male midwives developed a type of short obstetrical forceps for the delivery of babies, an instrument they kept secret for four generations. Obstetrics and gynecology eventually became a male-dominated profession, and remained so until the twentieth century.

EVELYN B. KELLY

Further Reading

Hurd-Mead, Kate Cappella. *A History of Women in Medicine*. Ahead, CT: Ahead Press, 1938.

Porter, Roe. *The Greatest Benefit to Mankind: A Medical History of Humanity*. New York: W.W. Norton, 1997.

Rowland, Beryl. *Medieval Woman's Guide to Health: The First English Gynecological Handbook*. Kent, Ohio: Kent State University Press, 1981.

Wear, A., R. K. French, and I. M. Lonieed, eds. *The Medical Renaissance of the Sixteenth Century*. Cambridge: Cambridge University Press, 1985.

Mechanical Printing and Its Impact on Medicine

Overview

Before the fifteenth century, medical practitioners relied on texts that were laboriously handwritten and recopied through the centuries. This method of distributing medical information was slow, limited to only a few translations, and frequently altered the content and illustrations to the point of inaccuracy. After the invention of the printing press around 1450, however, medical texts, especially classical works by Hippocrates, Galen, and Aristotle, among others, experienced a new life and reproductions closer to the original text. Distribution of medical literature increased, translations in several languages stretched across the world, and subsequently, medical science and practice progressed rapidly.

Background

For nearly 5,000 years preceding the invention of the printing press, medical books were written by hand. Prescriptions, incantations, and spiritually inspired healing rituals were exchanged orally, but were difficult to remember. Eventually these complicated treatments were written down, transcribed in painstaking, elaborate manuscript on expensive parchment. This method of transcription changed little through the fifth century B.C., when Hippocrates (c. 460-c. 377 B.C.), the "father of medicine," wrote several texts that treated medicine with a more realistic philosophy, free from superstition and magic. But the early Hippocratic works, as well as many other early medical texts, were written in Ionic dialect and limited to the number of copies that could be hand copied.

By the Middle Ages, the few books published were mainly religious records or biblical transcriptions. The act of transcribing and bounding books became an art form of sorts, and flourished in monasteries. During this era, the number of monasteries increased and book production subsequently grew, but within the censorship of the church. The availability of medical information changed at the whim of the current religious tide, and many medical books were banned and even destroyed if deemed heretical or obscene. Soon, however, as literacy spread, the demand for books outgrew the production capabilities of the monasteries. Professional scribes, unassociated with the church, became widespread, and censorship was somewhat sidestepped.

RARE MEDICAL BOOKS

Classics in the history of medicine are among the books most highly prized by collectors. Several prominent libraries, including the College of Physicians of Philadelphia, the Wellcome Trust (London), and the largest medical library in the world, the National Library of Medicine of the United States (Bethesda, Maryland), have built their reputations for excellence by acquiring and protecting works of this kind.

By the thirteenth century, cheaper, more easily produced paper became plentiful, and many authors were now their own scriveners. Doctors who had copies of the most used medical texts of the day—Avicenna's all-purpose *Canon of Medicine*, Guy de Chauliac's surgical manual *Chirurgia*, and Galen's and Hippocrates' classic medical works—would add their own observations and corrections in the notes, much as a chef would alter a recipe. Without widely distributed, identical texts, or easily modified and critiqued information, medical professionals had no baseline with which to work. Translations were frequently limited to Latin, and as demand for medical books continued, scribes would often hasten through details, omitting information and altering illustrations.

By the fourteenth century, most copied books suffered from poor handwriting and illus-

trations that were so stylized that they were totally inaccurate. Surgeons, in many cases, were improvising through procedures since text and illustrations were only barely coherent. There was a great need for more accuracy, wider dissemination and critique of existing manuals, and corrected copies of the classical medical tomes in a wide variety of languages.

Impact

It is believed that the first mechanical printing press was developed in Germany sometime near 1450. Soon thereafter, books were being printed around the world at a fast clip. For medicine, one of the most immediate and important additions to medical writing were more illustrations, clearer diagrams of surgical procedures, instruments, and disease identification. Obviously, identifying parts of the anatomy was critical to successfully treating a patient, and poor reproductions from over-copied and hastily illustrated manuals had decreased the value of these before the printing press. Now that a book could be reproduced identically to its original, great care was taken to keep anatomical illustrations accurate. In addition, skin diseases, medicinal plants, and instruments were identified more clearly and consistently than ever before.

Avicenna (980-1037), an Arab physician, philosopher, poet, and politician also known as Ibn Sina, wrote *Canon of Medicine* in the eleventh century. It was translated into Latin and republished in 1473. The work combined the teachings of Galen, Hippocrates, and Aristotle, and became the single most authoritative text for universities. The *Canon* included Avicenna's own observations on everything from nervous ailments to skin diseases and disease distribution through water and soil, and he recorded 760 different kinds of medicines. The text was translated into several different languages, and remained the foremost medical text through the seventeenth century. While Avicenna's work was well regarded and had a tremendous influence on the spread of medical knowledge, it also had a negative effect. The author demonstrated an aversion to surgery, instead prescribing medicinal cures, implying that surgery was an inferior approach to medicine. Because of the expansive printing and subsequent influence of his work, surgical progress suffered a severe setback.

Nevertheless, the standard surgical manual of the day, Guy de Chauliac's *Chirurgia*, written in 1363, was printed mechanically in France in 1478 and had a tremendous impact. Chauliac

(c. 1300-1368) recorded his own observations of the Black Death (the plague, either bubonic or pneumonic) and the patients he treated. He promoted the excision of abnormal growths, and described what was probably the earliest anesthesia. The dissemination of *Chirurgia*, especially under the influence of Avicenna's works, invoked a wide critique of his procedures, a positive step in medicine, in that previously, timely improvements were rarely included in subsequent copies of texts. By 1683 *Chirurgia* had gone through 68 editions published in many different languages, with critical forewords and bibliographic references.

Included with nearly every medical text were elaborate illustrations. In the fifteenth and sixteenth centuries anatomical renderings were especially in vogue. Leonardo da Vinci (1452-1519) and Michelangelo (1475-1564), for example, both illustrated anatomical manuals, and would frequently dissect cadavers to understand their subjects more completely. But while anatomical renderings seemed to clarify the workings of the human body, the approaches to understanding the symbiotic systems of the body were in question. Thanks to the revival of the classical works of Hippocrates, who emphasized a humanist and practical approach to medicine, and Galen (c. 130-c. 200), who stressed the importance of observation and application in his works, medical learning shifted toward diagnosis and curing of disease. Medical literature now described every fine detail, including the color, smell, and structure of diseases, and doctors had a philosophical guide from which to work.

Combining both the fine artistic renderings and the new detail of medical texts, William Harvey (1578-1657), an English physician, eventually used the wealth of texts produced in the fifteenth and sixteenth centuries to formulate a theory about the circulation of the blood. Harvey, in his *De Motu Cordis*, explained the collective function of the veins, blood, heart, lungs, and liver, and was the first person to attempt a blood transfusion.

Another result of the invention of the mechanical press was that medicine was brought to the masses. Even women, who during the Black Plague medicated patients in the home, used the studied medical practices that were suddenly available in "vulgar" languages, rather than the Latin of tradition. That tradition continues: The influence of Hippocrates is as vital today as it was in its creation. Galen's works are frequently drawn from, and Avicenna's *Canon* is still an important text in some Arab countries.

The two most important books in the history of medicine are *De humani corporis fabrica libri septem* (Seven books on the structure of the human body) (1543) by Andreas Vesalius and *Exercitatio anatomica de motu cordis et sanguinis in animalibus* (Anatomical exercise on the motion of the heart and blood in animals) (1628) by William Harvey. The copy of the *Fabrica* that Vesalius himself presented to Holy Roman Emperor Charles V eventually found its way into the collection of Haskell F. Norman. When Christie's auction house sold Norman's collection in New York in 1998, that *Fabrica* sold for $1.5 million. Norman also owned the copy of *De motu cordis* that once belonged to Johann Friedrich Blumenbach. It fetched $480,000 at the same auction. In 1989 the firm of Pickering and Chatto sold a slightly damaged copy for $100,000. In 1988 Christie's received $181,000 for the copy owned by cardiologist Myron Prinzmetal. In 1934 the New York Academy of Medicine and the University of Munich, Germany, jointly published *Andreae Vesalii Bruxellensis icones anatomicae* (Anatomical images of Andreas Vesalius of Brussels), using 227 of the original sixteenth-century woodblocks carved for the illustrations of Vesalius's *Fabrica* and *Epitome*. These woodblocks were accidentally destroyed by Allied bombs in World War II. First priced at only a hundred dollars, one of the 615 numbered copies of *Icones anatomicae* typically sells for $5,000-6,000.

LOLLY MERRELL

Further Reading

Carter, John and Percy Muir. *Printing and the Mind of Man*. London: Oxford University Press, 1967.

Loudon, Irvine, ed. *Western Medicine: An Illustrated History*. New York: Oxford University Press, 1997.

Smith, W.D. *The Hippocratic Tradition*. Ithaca, NY: Cornell University Press, 1979.

The Development and Impact of Medical Illustrations

Overview

Artists were in many ways the driving force behind the study of medicine. Seeking to perfect their skills in realism and accuracy, artists, rather than professional anatomists, studied the bodies of animals and men. Not content to just observe animals, they gained first hand knowledge of anatomy through dissection.

Another great contributing force was the development of the printing press, which enabled not only textual information but also drawings to be replicated. The manufacture of paper, as well as improvements in wood engraving, made possible a growing number of illustrated books. As in all areas, the power of visual images attests to the adage that "a picture is worth a thousand words." The impact of art and visual representations was shattering.

Several illustrators of the period were responsible for this development. Johannes de Ketham (fl. 1460?) published the first illustrated medical work. Hans von Gerssdorff (1455-1529) wrote and illustrated work from the battlefield. Giacomo Berengario da Carpi (1470-1530) was a serious student of anatomy and showed bodies with the skin removed as part of a landscape. Giovanni Battista Canano (1513-1579) refined his work using copper plates. However, the greatest illustrator of the period was Andreas Vesalius (1514-1564). The publication of his *De humani corporis fabrica* in 1543 marked the beginning of the renaissance of medicine.

Background

Life in the fifteenth and sixteenth centuries was not a pleasant affair. Epidemics of bubonic plague, typhus, smallpox, and sexually transmitted diseases killed and maimed large numbers of the general population. Add warfare and famine, and misery was rampant.

The heritage of a thousand years of the Dark Ages was superstition and ignorance. The classical medical works interpreted by word of mouth and influenced by tradition had become a standard. But the deplorable condition among the people forced scholars to consider new interpretations and ideas. However, few findings would be translated into change for daily life for many years.

In the thirteenth century, human dissection began at Bologna to determine the causes of deaths. The dissections piqued interest in anatomy, and Mondino de Luzzi (1270-1326) wrote the first modern work on the subject. The universities began to teach anatomy and medicine as part of the natural science of the liberal arts. Other universities at Padua, Florence, Pisa, and Venice adopted this approach and were dedicated to supporting and proving the ideas of the Greek physician Galen (c. 130-c. 200).

Throughout the Middle Ages medical illustrations had appeared, but the drawings were childish, not realistic. Texts were necessarily copied by hand, and the drawings, even in medical centers, were teaching aids similar to sketches, used to represent general truths rather than being exact.

The most common type of illustration was the "Zodiac Man." This male figure is marked with points for bloodletting correlated with the zodiac signs to assist the barber-surgeons. For example, Taurus controlled and cured diseases of the neck and throat, and Scorpio controlled the genitals. The moon and constellations controlled the right way and place to bloodlet. Charts also described how to examine urine, called uroscopy. Just by looking at the color of the urine, the physician could supposedly tell what was wrong.

Medieval painters also depicted the figure of Death with grotesque grins, calling for peasants, merchants, and princes alike. Typical was the famous painting called *The Dance of Death* by Hans Holbein the Younger. However, with the new zeal and fervor for realism among medical scholars, such medieval caricatures of the body and its afflictions gave way to new forms of anatomical illustration.

Impact

The most startling development to affect illustration was the invention of the printing press. In 1450 Johann Gutenberg (1390?-1468) had developed movable type that could be set to make many copies. The process of making woodcuts for engraving was also refined for illustrations. Access to paper made books available to many.

The first medical work that included illustrations was Ketham's *Fasciculus medicinae,* published in Venice in 1491. The Latin version had six illustrations—a uroscopy chart in red and black, phlebotomy figure, zodiac man, the female viscera, wound man, and disease man. In 1493 an edition with 10 illustrations was translated into Italian. The most notable addition is a dissection scene in which a professor presides over a group of barber-surgeons cutting open a human cadaver. The illustration also reveals the transition to the formal unity of Renaissance art; the lecturer is vertical and the body is presented horizontally so that everything is symmetrical.

Hippocrates (460?-377? B.C.) had said that he who wished to become a surgeon should go to war. In the fifteenth and sixteenth centuries there were ample opportunities because wars were prevalent. In addition, there were new problems concerning the care of the wounded. No longer were the soldiers just shot with bows and arrows, but the use of gun powder, cannon balls, and lead shot would pierce the flesh, leaving gaping wounds and major infection. In 1497 Heironymus Brunschwig (1450-1533) wrote the *Buch der Wund Artzney* (Book of Wound Dressing), which includes the earliest printed illustration of surgical instruments. His book showed how shot wounds were poisoned by gunpowder and needed cauterizing. Hans von Gersdorff (1455?-1529) wrote and illustrated from field experience. His *Feldbuch der Wundartzney* (Fieldbook of Wound Dressing), describes how to extract bullets with special instruments and how to dress wounds with hot oil. He also showed how to enclose amputated stumps with animal bladders.

Jacopo Berengario da Carpi was a surgeon and anatomist at Bologna from 1502-27. It was during this time that serious inquiry into human anatomy began, and the drawings took on a unique feature. The bodies, portrayed with skin removed, were positioned to show their dissected muscles while standing and observing a landscape. One drawing from *Isagogae breves* (1523) shows a figure, with skin stripped, leaning on an axe with clouds, trees, and hills in the background. The landscape is set in the hills around Bologna. Another figure is sitting like a figure S on a rock.

Da Carpi's skeletal figures are also involved in the environment. One skeleton stands in front of a grave, from which he had obviously been taken, holding two other skulls in his uplifted hands. Another convention of da Carpi's medical illustration is the use of multiple views of a single part of

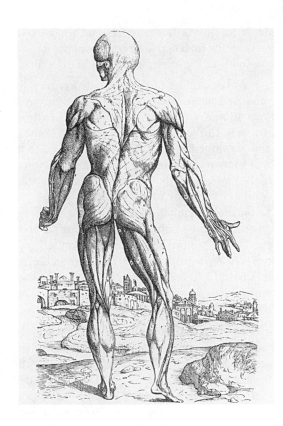

An illustration from Vesalius's 1543 book, *De humani corporis fabrica,* which revolutionized anatomical drawing. *(Bettmann/Corbis. Reproduced with permission.)*

a muscle. Leonardo da Vinci (1452-1519) also sketched multiple views in his notebooks and probably influenced these drawings.

Giovanni Battista Canano (1515-1579), while a student at the University of Ferrara, performed private dissections in his home. In 1541 he published *Musculorum humani corporis pictuata dissectio,* or *Picture of the Dissection of the Muscles of the Human Body.* The drawings, which feature the muscles of the arm, were made by the Ferranese painter Girolamo da Carpi using 27 copperplates.

Another book, *Picturata dissectio,* is based directly on structures of the human body and of living animals, and not on the dissection of the ape as performed by Galen. The works of Canano have two innovations. Copperplates were used for the first time, which allowed finer details than the woodcuts used by Berengario and Vesalius. Canano also featured the fine muscles of the hand. While he followed the same order as Galen, he pointed out omissions and errors. He also drew the valves of the deep veins and described their function in controlling blood flow. His illustrations showed a novel approach to myology (the study of muscles), as he depicted only a few muscles and their movement of the fingers. However, this approach was not followed by subsequent medical illustrators.

As the middle of the sixteenth century approached, medical illustrations came into their own in the works of Vesalius. Vesalius from Brussels studied at the University of Louvain and continued his education in Paris, where he was known occasionally to rob a grave to get body parts to study. Commissioned to write a comparative study of the works of Galen, who had made his conclusions using the anatomy of an ape, Vesalius pointed out more than 200 errors by the famous physician. This caused controversy, especially between Vesalius and his former friend and teacher Jacobus Sylvius, a Galen enthusiast. Vesalius published *De humani corporis fabrica* in Basel at the age of 28.

The text had 663 illustrations and was divided into seven books. The contrast with the conservative scenes of Ketham was obvious. The frontispiece illustration depicted Vesalius with a mass of students pressing around. He is portrayed as the dissector, and the barber-surgeons crouch under the table. A skeleton sits in the professor's chair, and a monkey and dog are vying to get in the picture. The identity of the artist who actually designed the plates is subject to debate. For years it was believed to be de-signed by the famous Venetian painter Titian, but more than likely it was his Dutch assistant Calcar. However, scientists agree that Vesalius's 1543 text is a pivotal work that marks the beginning of the renaissance in medicine.

The Renaissance saw an emergence of realism in art and new ideals that supported the direct and factual representation of natural phenomena. The rules of perspective and mathematics prevailed. Art had gone scientific, and by the fifteenth and sixteenth centuries the new theory was firmly established and fully accepted in the painting of masters such as Albrecht Dürer, Michelangelo, and Raphael. Likewise, science had gone artistic, and modern ideas owe much to the efforts of the theorizing artists.

EVELYN B. KELLY

Further Reading

Andreas Vesalius of Brussels. New York: World Publishing, 1950.

Garcia-Ballester, Luis. *Practical Medicine from Salerno to the Black Death.* Cambridge: Cambridge University Press, 1994.

Porter, Roy. *The Greatest Benefit to Mankind: A Medical History of Humanity.* New York: W.W. Norton, 1997.

Schultz, Bernard. *Art and Anatomy in Renaissance Italy.* Ann Arbor: UMI Research, 1985.

The Invention of the Microscope

Overview

Historical records indicate that around the time of Christ the ancient Assyrians first realized that glass spheres could be used as magnifying devices. Claudius Ptolemy, a second-century mathematician and astronomer in Alexandria, wrote a paper on the optical properties of lenses. He discussed how glass spheres filled with water could be used for magnification and refraction. However, despite this knowledge, glass lenses were not extensively used for over a millennium. Around 1300, spectacles were invented to improve vision. This innovation served as the springboard to strong interest and research into the properties of magnifying lenses. Several treatises were published in the sixteenth century as a result. Near the end of the sixteenth century, it was found that if certain lenses were joined to-gether by a cylinder, they would become what is called either a Galilean telescope or a Galilean microscope, depending on which end is used to view objects. Italian mathematician and astronomer Galileo (1564-1642) used this device as a telescope to observe the stars and planets but did little to advance its use as a microscope for biological purposes.

The earliest simple microscopes (containing only a single lens) used drops of water confined to a small hole that functioned as a magnifying lens. Eventually glass replaced water as the medium, but it is not clear when glass lenses first began to be used. It has been well established that by the seventeenth century Antoni van Leeuwenhoek, a Dutch scientist, had developed techniques for making high-quality ground lenses for simple microscopes. While these were limited in power, he used them in combination with both light and his keen eyesight to observe specimens just a few micrometers in size.

The compound microscope, which uses a multiple lens system, was first described in the sixteenth century but had little practical use at that time due to the arrangement of the lenses, the blurring of images from improper grinding, and chromatic aberrations due to problems with light. The first useful compound microscope was constructed in the Netherlands sometime between 1590 and 1608. Three different people, all of the optometrists, have been credited with the invention at one time or another: Hans Jansen, his son Zacharias Jansen, and Hans Lippershey (d. 1619?). It would be over 200 years before these problems were completely resolved, making the compound microscope an important biological tool.

Four microscopists are considered to have influenced the development and use of microscopes in biology and medicine. These individuals made significant improvements either in the technology involving microscopes or in accumulating the body of knowledge of microscopic structure. They were Marcello Malpighi (1628-1694), Anton van Leeuwenhoek (1632-1723), Jan Swammerdam (1637-1680), and Robert Hooke (1635-1703).

Background

Marcello Malpighi was an Italian biologist and physician who conducted extensive studies in animal anatomy. He was one of the first scientists to use a microscope to study the structure, composition, and function of tissues, so he is often known as the father of histology (the microscopic study of tissues). Among his many accomplishments dealing with the human body was the first description of capillaries, the inner layer (dermis) of the skin, the papillae of the tongue, the outer portion of the brain (cerebral cortex), and red blood cells. He wrote detailed treatises on animals and insects, including descriptions of the development of the chick embryo and the lifecycle of the silkworm, and demonstrated that pests such as the flea and weevil reproduce through ordinary insect means and not by spontaneous generation (the concept that living organisms could be created from nonliving matter). In addition, Malpighi made detailed investigations into plant anatomy. He systematically described the various parts of plants, such as bark, stem, roots, and seeds, and discussed such processes as germination (the beginning of growth). Although Malpighi did little

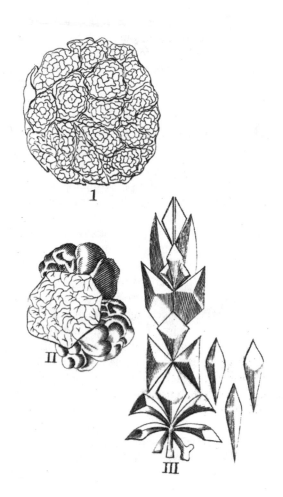

A page from Marcello Malpighi's 1686 book, *Opera Omnia*. Malpighi's pioneering studies were made possible by the invention of the microscope. *(Wellcome Institute Library. Reproduced with permission.)*

to expand the technical aspect of microscopes, he did have a significant impact on the advancement of the body of knowledge in biology.

Anton van Leeuwenhoek, a Dutchman who spent most of his life in Delft, sold cloth for a living. Although he had little formal schooling, as a young man he became interested in making magnifying lenses and recording his observations. This soon became an obsession. Leeuwenhoek used a small single-lens system and achieved magnifications that allowed him to see in much greater detail than was allowed by any microscope of the time. Unfortunately, Leeuwenhoek closely guarded his lens-making technique, so his improvements died with him. Although he was not willing to share his methods, Leeuwenhoek was more than willing to share his observations. In fact, he became somewhat of a celebrity because of his publications regarding his research. He is most famous for his discovery of *animalcules* (one-celled animals, now known as *protozoa*) in stagnant water, which he reported in the mid-1670s. He made significant contributions in the areas of capillaries, the structure of muscle, the lens of the eye, the reproductive system, and teeth. He studied bacteria from the mouth and recognized the various shapes, postulating correctly on their relative size to red blood cells.

Jan Swammerdam was a contemporary and countryman of Leeuwenhoek. In contrast to Leeuwenhoek, Swammerdam was a well educated and highly systematic scientist who confined his attention to studying relatively few organisms in great detail. He was responsible for many highly innovative techniques, such as injecting wax into objects to hold them firm, dissecting fragile objects under water, and using micropipettes to inject organisms under the microscope. Swammerdam concentrated his research on what he considered to be insects based on their mode of development. These included such organisms as spiders, snails, scorpions, fishes, and worms. Unfortunately, Swammerdam was subject to fits of mental instability and had financial difficulties that led to periods of depression. He died at the early age of 43, having contributed a significant amount of research to biology.

English physicist Robert Hooke may be the most famous of the early microscopists. He certainly had the widest array of interests. He was curator of instruments at the Royal Society of London, which allowed him to remain abreast of all new scientific developments. He made significant contributions to many areas of science and has been credited with coining the term *cells* while looking at cork under a microscope. In 1665 Hooke published his book *Micrographia*, which is primarily a review of a series of observations that he had made while following the development and improvement of the microscope. This book had tremendous influence at that time. Hooke described in detail the structure of feathers, the stinger of a bee, and the foot of the fly. He also noted similar structures to cells in the tissue of trees and plants and discerned that in some tissues the cells were filled with a liquid while in others they were empty. He therefore supposed that the function of the cells was to transport substances throughout the plant. Hooke, like the others previously mentioned, had a significant impact on biology by utilizing the microscope.

Impact

Although the work of any of the classical microscopists seems to lack a definite objective, they made significant strides by using the techniques of observation and experimentation to their fullest. It is remarkable that so few men, working independently, should have made so many fundamental observations of significant importance. Their work revealed for the first time the incredible complexity of living organisms. At the same time, it helped debunk many prevailing and unquestioned theories that had existed since the time of antiquity. The ideas of the Greek scientist Galen (A.D. 130?-200?) persisted for more than 1,000 years because they were rarely questioned, even when evidence was shown to the contrary. Thus, each of these men had to fight the preconceptions of the time and contradict prevailing wisdom to make their ideas public. They played a significant role in starting the scientific revolution.

The significance of these advances may be difficult to understand today. The microscope was relatively new, and it was not clear back then that it would ever be useful in making scientific discoveries. There was some thought that it was more of a curiosity than a scientific tool. In fact, at that time it was more a recreational device for noblemen than a tool for research. However, each of these scientists appreciated the fact that by looking at something close up, they could view things in a significantly different way. And they believed that the perspective they gained was of scientific importance. They made many important discoveries that helped to clarify some of the thinking of their time.

As an example, Malpighi made a discovery of monumental importance in physiology. William Harvey (1578-1657), a famous British physiologist, stunned the academic and medical environment with the publication of his book *De motu cordis et sanguinis in animalibus* (On the Motion of the Heart and Blood in Animals), in which he presented experimental and logical proof that the long-held theories of Galen were wrong. In the work Harvey proposed that blood actually travels in a circuit from the heart, around the body to the tissues, and then back to the heart. The current thinking was that blood was produced from the intestines, traveled to the liver and to the heart, and was then distributed to the body by both veins and arteries, where the tissue consumed it. Harvey's theory was a radical idea, and if it had not been for his stature, he might have been imprisoned. While some considered Harvey's idea, there was no proof that blood circulated because there was no evidence of the connection between arteries and veins. In 1660, three years after Harvey's death, Malpighi used a microscope to see the capillaries, the extremely thin blood vessels, which formed the needed connection between the arteries and veins that could not be seen with the naked eye. Other examples such as Leeuwenhoek's *animalcules* also raised some disquieting thoughts in the minds of his contemporaries. This helped to provide some initial evidence against the theory of spontaneous generation, held by the ancient world and passed on unquestioned up to that time. Thus, these men helped to not only provide information, but to also change the prevailing ideas of entire societies.

JAMES J. HOFFMANN

Further Reading

Croft, William J. *Under the Microscope: A Brief History of Microscopy (Series in Popular Science)*. River Edge, NJ: World Scientific Publishing Company, 2000.

Ford, Brian, J. *The Leeuwenhoek Legacy*. New York: Lubrecht & Cramer Ltd., 1991.

Ruestow, Edward G. *The Microscope in the Dutch Republic: The Shaping of Discovery*. Cambridge, MA: Cambridge University Press, 1996.

The Alliance of Science and Art in Early Modern Europe

Overview

Throughout the Middle Ages, European artists concentrated on religious subjects. Whether a piece of sculpture on the facade of a cathedral, a mural on a monastery wall, an altarpiece, or an illustration in a prayer book, most medieval art served a religious function: to focus people's attention on attaining salvation. The development of the Renaissance in fifteenth-century Italy and its spread to Northern Europe dramatically changed this. Since Renaissance humanistic thought emphasized nature and the beauty of the human body, artists now attempted to duplicate the natural world in their work. They enthusiastically incorporated this new naturalism into their art and in doing so, played a major role in creating new sciences, particularly those of anatomy and botany, the earliest of the modern life sciences. Artists such as Leonardo da Vinci (1452-1519) and Albrecht Dürer were as crucial to the development of modern science as they were to the formation of postmedieval art.

Background

The period after 1450 was marked by constant upheavals as the millennium-long culture of medieval Europe disintegrated. Religious unity was broken by the Protestant Reformation. Feudal society was radically changed by the rise of cities, the beginnings of capitalism, and the emergence of the middle class. The great age of exploration began—an expansion that would have momentous conse-quences for both the Old and New Worlds. There were also dangers before which the medieval consensus seemed powerless: bitter power struggles between nobles and monarchs; outbreaks of the Black Death (bubonic plague), which repeatedly swept through Europe (syphilis also appeared in epidemic proportions); and in 1453 Constantinople fell to the Turks, a new Moslem power that began pushing relentlessly into Europe.

Many intellectuals responded by seeking new ways to view the world and mankind's place in it. Paradoxically, they found the basis of their modern thought in writings from the ancient world. For centuries, texts from ancient Greece and Rome had been known in Europe, many having been preserved in the Arabic world. They had been largely ignored until the collapse of the Byzantine Empire, which culminated in the fall of Constantinople, bringing a flood of Greek texts to Italy as Byzantine scholars fled Turkish rule. As these texts were translated into Latin, an intellectual movement called *humanism* arose. Like the ancient writings themselves, humanists stressed that the proper study of scholars should be mankind, not heaven. Thus, humanism was a shift from theology to philosophy. Since the ancient texts also expressed a strong interest in science, a major factor in Renaissance humanism was the desire for more knowledge of the human body and of the world of nature.

One result of this emphasis on the body was a growing interest in anatomy, the study of the structures beneath the skin. The pre-Renaissance authority on human anatomy was Galen (130?-200?), a Greek physician in the Roman Empire. However, Galen had not dissected humans, since the Romans opposed that practice. He had instead carefully dissected animals, mainly pigs and monkeys, and assumed that humans were the same. Since there are numerous differences between human and animal anatomy, his work was filled with errors. Galen's texts had been preserved in Arabic translations but had never been corrected, since there were also prohibitions against human dissection in the Islamic world. When early Renaissance scholars did note a difference between Galen's descriptions and a human cadaver, they assumed that the text they were using had been erroneously copied.

The duplication of texts and illustrations was crucial to the development of Renaissance science. Hand-copied texts usually contained errors and hand-copied illustrations were virtually worthless because repeatedly copying a drawing necessarily degrades what the original artist intended. The adoption of printing from moveable type (around 1450) and the incorporation of woodcut (after 1500) and then engraved (after 1550) drawings permitted accurate texts and detailed, informative illustrations to be disseminated throughout Europe. Artists, of course, were essential to producing these accurate, detailed illustrations. This was particularly important in the sciences of anatomy and botany, where illustrations are of greater value than the text in conveying information.

Impact

Since 1300, Italian universities were permitted to carry out dissections of humans in their medical schools. Although not frequent events because the only legal cadavers were those of executed criminals, by about 1500 scholars began to notice that Galen's descriptions were often different from what was seen in the bodies on the dissection table. They realized that if the science of anatomy was to progress, these discoveries had to be illustrated and disseminated. At the same time, based on the revival of Greek and Roman texts, Renaissance artists sought to depict the human body more realistically, particularly in situations of dramatic movement. The few medieval nudes that existed (Adam and Eve, Christ's crucifixion) gave artists no naturalistic tradition to follow.

Artists began to examine carefully the body as it moved. This concentration on the exterior muscles of the body reached its culmination in Anto-

nio Pollaiuolo's (1432-1498) large engraving, *The Battle of Ten Naked Men* (1460s), in which he depicted male nudes in the violent action of mortal combat. However, he crammed so many contracted muscles and sinews on each figure that the drawings were not realistic. Artists realized that for more accurate knowledge of the human body they would have to attend dissections. Indeed, many artists taught their pupils that only by doing dissections themselves could they attain the knowledge they sought. By the mid-1500s, dissections had become part of the artist's routine.

No one was more responsible for this development than Leonardo da Vinci. He insisted that artists could avoid the errors made by Pollaiuolo only by probing beneath the muscles deep into the body's structure to explore the organs, the nervous system, arteries, and so forth. Da Vinci himself dissected at least thirty bodies, including one of a pregnant woman and one of an old man. His meticulous drawings were the first accurate and informative anatomical illustrations ever made. He compared human anatomy with that of animals and explored pathological (diseased) anatomy such as hardening of the arteries. He also originated a number of illustration techniques that became commonplace in the subsequent history of anatomical images.

One crucial innovation was that of rotation, of looking at the same body part from a number of different angles. This allowed da Vinci to accurately portray a three-dimensional object in a two-dimensional drawing. Other techniques first used by Leonardo were that of transparency (making overlying tissue transparent in order to reveal underlying structures) and traverse sections, or drawings that revealed what would be seen if the body was sliced crosswise.

Da Vinci's drawings and their accompanying text were not published until the late nineteenth century. However, they had some immediate influence because numerous contemporary artists saw his notebooks and copied his work. More importantly, Leonardo widely publicized the fact that words were inadequate to describe the complexities of anatomy and that the science could only advance through the use of careful, artistic illustrations: "The more detail you write," he noted, "the more you will confuse the mind of the reader." The truth of this observation can be seen in the most famous of all anatomies, Andreas Vesalius's (1514-1564) *De humani corporis fabrica* (1543). While Vesalius made numerous corrections of Galen's work, the more than 200 accompanying woodcut illustrations gave his text intelligible meaning. Their unknown artist (perhaps Joannes Stephanus of Calcar, of whom very little is known) deserved as much credit for the book's success as did Vesalius himself. Over the next century and a half, anatomical knowledge was conveyed more by illustrations than by texts. Their artists and engravers, virtually all of whom are unknown to us, developed their skills by doing dissections themselves or by watching anatomists dissect. Vesalius complained that so many artists crowded around his dissecting table, they interfered with his work.

One of the men influenced by Leonardo da Vinci and Italian humanism was the German artist Albrecht Dürer (1471-1528). He made several protracted visits to Italy, where he saw Leonardo's notebooks. Dürer's interest in the human body is obvious in his clinically accurate portraits, including several self-portraits. His anatomical drawings are still admired today, especially his famous *Praying Hands* (1508). His nudes perhaps most clearly reflect his devotion to capturing reality. His ink drawing *The Women's Bath* (1496), depicting a group of nude women of all ages in a public bath, was an attempt to show what the aging process does to the human body. He even drew a nude self-portrait. He spent much of his energy trying to find the "laws" of anatomical proportions, and was working on his *Treatise on Human Proportions* at the time of his death. In all of this, Dürer was motivated by his belief that beauty could only be attained by the exact copying of the natural world.

Dürer's alliance of artistic observation and scientific detachment is even clearer in his many drawings and paintings of animals, birds, and plants. He wrote that "life in nature reveals the truth of things." Ironically, he contracted the sickness that eventually killed him on a trip to sketch a beached whale. Like Leonardo, Dürer took great pride in the accuracy of his plant drawings. Undoubtedly the most important of these nature paintings was his *The Great Piece of Turf* (1503), a life-sized representation of grasses and dandelions in a clod of earth. No one as famous as Dürer had ever painted anything so apparently insignificant before, yet this single painting had a tremendous artistic and scientific impact in Europe.

The period from about 1470 to 1670 was one when herbals (compilations of plants focusing on their medicinal uses) were being replaced by books that dealt with the structure and classification of plants. An impetus for this trend was the thousands of new plant specimens brought to Europe from around the world by the return-

ing adventurers of the Age of Exploration. As the modern science of botany evolved, Dürer's *Turf* played a pivotal role at a crucial period. Most publishers had been content to keep using the old, inaccurate woodcuts from medieval herbals because they were cheaper than hiring artists to produce new, accurate illustrations. The influence of Dürer's insistence on accuracy made that practice impossible, at least in Northern Europe.

In the 1530s, Otto Brunfels (1489?-1534) published a three-volume *Living Images of Plants*. While his text was unexceptional, the book's more than two hundred illustrations by Hans Weiditz were revolutionary. Unlike his predecessors, Weiditz had clearly used real specimens for his models, even including leaves partially eaten by insects. In 1542 another German, Leonhard Fuchs (1501-1566), published *De historia stirpium*. Its 500 illustrations were drawn by three artists of exceptional skill. Unlike Weiditz's work, they drew ideal plants with no flaws. Subsequent botanical artists used this approach because the purpose of the illustrations was to represent the species as accurately as possible. Fuchs's artists also produced botanical illustrations of great beauty, a trend that would continue throughout the early modern period and culminate in the remarkable plant illustrations of Maria-Sybilla Merian (1647-1717).

ROBERT HENDRICK

Further Reading
Belt, Elmer. *Leonardo the Anatomist.* New York: Greenwood Press, 1969.

Braham, Allan. *Dürer.* London: Spring Books, 1965.

Cazort, Mimi, Monique Kornell, and K.B. Roberts. *The Ingenious Machine of Nature: Four Centuries of Art and Anatomy.* Ottawa: National Gallery of Canada, 1996.

Clayton, Martin. *Leonardo da Vinci: The Anatomy of Man.* Boston: Little, Brown, 1992.

Eisler, Colin. *Dürer's Animals.* Washington, DC: Smithsonian Institution Press, 1991.

Emboden, William A. *Leonardo da Vinci on Plants and Gardens.* Portland, OR: Dioscorides Press, 1987.

Mayor, A. Hyatt. *Artists and Anatomists.* New York: Metropolitan Museum of Art, 1984.

Panofsky, Erwin. *The Life and Art of Albrecht Dürer.* 4th ed. Princeton, NJ: Princeton University Press, 1955.

Pinault, Madeleine. *The Painter as Naturalist: From Dürer to Redouté.* Trans. by Philip Sturgess. Paris: Flammarion, 1991.

Roberts, K.B. and J.D.W. Tomlinson. *The Fabric of the Body: European Traditions of Anatomical Illustration.* Oxford: Clarendon Press, 1992.

Russell, Francis. *The World of Dürer, 1471-1528.* New York: Time, 1967.

Advancements in Surgery

Overview

During the Renaissance, between 1450 and 1699, surgery was a mix of art, science, and myth. The art of caring for a soldier's battle wounds, the myth of blood-letting to cure or prevent disease, and the advances in scientific surgery for breast cancer, hernias, and bladder stones were all common to surgery at this time. It became a period for advancing the science of surgery as new ways to control bleeding were developed, plastic surgery was invented, complex surgeries to remove stones in the bladder were beginning to be performed, and surgeons tried their steady hands at cesarean sections with living mothers.

While significant improvements in surgical techniques and publications came from Italian, German, and French surgeons, such as Fabricius Hildanus (1560-1634) and Ambroise Paré (1510?-1590)—who dominated the field in the 1500s—British physicians such as William Harvey (1578-1657) made significant discoveries in anatomy during the 1600s that would make surgeries in the 1700s more successful.

Background

In the fifteenth century, blood letting (or phlebotomy) already had a long surgical history. The barber's red and white pole became a symbol of bleeding and bandaging and the practice of bleeding persisted well into the eighteenth century, as surgeons and barbers promoted health by periodically "thinning" the blood by bleeding. Its use proved more myth than science, however.

The turn of the sixteenth century saw the beginning of more scientific approaches to cutting to cure, however, and professional surgeons performed a wider range of services during the Renaissance. Military surgeons predominated. The advent of the use of gunpowder during the Middle Ages meant that military surgeons had to deal with gunshot wounds and wounds inflicted by canons. Cleaning combat wounds using cautery—searing remaining flesh with a hot instrument—was common practice. Amputation became the preferred method of treatment and, slowly, became more art and science than butchery.

The period from 1450 to 1699 also saw advancements in surgery for cancers, hernias, and cesarean sections, as well as the advent of plastic surgery, and new and more efficient ways to control bleeding. Thanks to the anatomists who solved many of the mysteries of the circulatory system, more surgical successes resulted. The greatest surgeons of the times studied in Italy at the University of Padua. They went on to serve the kings of Europe while uneducated but hardworking itinerant surgeons traveled from village to village in Europe cutting and healing and teaching others in their footsteps. Both groups made considerable contributions.

Impact

War provided a testing ground for Renaissance surgeons. Amputation, most often the result of wounds turned gangrenous, was the most frequent call for the military surgeon. Theoretically divided, military surgeons debated over whether or not to amputate through healthy tissue or amputate just to the limits of the gangrenous side of the limb. Questions over when and where to amputate were also tied to steps to control bleeding. Until better methods of controlling bleeding were developed, surgeons primarily amputated legs below the knee because bleeding from the major artery above the knee was very difficult to control.

Cleaning and closing wounds were major concerns for military and nonmilitary surgeons alike in this period. German surgeon Fabricius Hildanus used a red-hot knife blade to control bleeding and cauterize the wound during amputations. He and others cleaned wounds by applying boiling oil.

Ambroise Paré, a French surgeon, disapproved of both cautery and boiling oil. In 1537, Paré became an army surgeon around the age of 26. On the battlefield he became a leading authority on military wounds, finding that, contrary to current opinion, gunshot wounds were not poisonous and did not need cautery. Rather than use boiling oil, he cleaned gunshot wounds with a concoction of turpentine, rose oil, and egg yolk. One of Paré's major contributions was the ligature—a method of tying off a bleeding blood vessel—instead of cauterizing. With the ligature, he could control bleeding and amputate above the knee, where large blood vessels and arteries were previously likely to hemorrhage.

In contrast to Paré, who served monarchs, French surgeon Pierre Franco (1500-1561) occupied a professional niche somewhere between a barber-surgeon and an itinerant "cutter." It is said that his greatest contribution to surgery was to retrieve it from the quacks, bonesetters, and charlatans practicing in Europe by requiring surgical supervision from physicians. Franco specialized in hernia surgery. Whereas prior surgeons always removed the testicle to repair inguinal hernias, rather than incise at the level of the pubis, which he considered dangerous, Franco invented a low incision at the base of the scrotum to release a strangulated hernia. He performed his first successful operation using this technique in 1556.

Aurelous Philip Theophratus Bombastus von Hohenheim, who lived from about 1493-1541 and called himself "Paracelsus," was a major figure in sixteenth-century medicine. He encouraged medicine and surgery to be joined in the same art and science and dedicated his craft to wound management, traveling extensively in Europe practicing medicine, performing surgery, and teaching.

Many surgical "firsts" came in the mid and late sixteenth century. The first account of a successful tracheotomy occurred in 1546 when Italian surgeon Antonio Bravasola opened the trachea of a patient who was near death from a respiratory obstruction caused by an abscess in his windpipe. The first reported splenectomy was in Naples in 1549, performed by Zaccarelli. Georg Bartisch of Dresden performed the first eye removal for cancer in 1583. During this period British surgery lagged behind the Italian and French masters, but Englishman William Clowes was the first to perform a successful thigh amputation for gangrene in 1588. He wrote extensively on the art of surgery.

Cesarean section—the surgical removal of a baby from the mother's womb—had a long history of both fact and legend by the time of the Renaissance. Roman emperor Julius Caesar was fabled to have been delivered by the surgical technique still bearing his name. French surgeon Paré, noting that most cesarean sections on living mothers ended with the mother's death, recommended waiting until the mother died and then removing the baby as quickly as possible. However, in 1581, French surgeon François Rousset (1535-c. 1590), physician to the Duke of Savoy, published *Hysterotomotokie*, reporting successful cesarean sections on mothers who continued to live following their surgery.

Hieronymous Fabricius (1537-1619) was perhaps the most prominent Italian surgeon of

Ambroise Paré. *(Library of Congress. Reproduced with permission.)*

the era. He was first a student, then a professor of anatomy at the University of Padua in Italy. He was concerned with anatomical research but also practiced surgery, improving the practice of tracheostomy by showing surgeons how to move, rather than cut muscles in the neck, to expose the trachea. In addition to his institutional contribution of building the first permanent surgical amphitheater (1594), he also improved treatment for urethral blockages and strictures.

The leading German surgeon at the end of the sixteenth century was Guilhelmus Fabricius (Fabry) Hildanus. As a young man, Fabry was too poor to receive formal medical training, so he became apprenticed in surgery at a low level. He rose in the profession, however, and traveled widely in Europe practicing surgery and publishing on surgery. He favored amputating through healthy tissue to fight gangrene and devised a tourniquet. He operated on cancers on the breast and described in his works the removal of the lymph glands. Fabry said that all the "sprouts" had to be removed with the breast tumor. He invented a device that pinched the base of the breast as a blade swept it off.

Gaspare Tagliacozzi (1545-1599) created the field of plastic surgery when he took skin from a patient and fashioned him a new nose. After removing the damaged nose, Tagliacozzi made a paper model of the nose he intended to build, used it as a pattern for cutting skin from the arm, and sutured the new nose to the face. He treated harelips by cutting away the deformity and suturing the ends back together. Tagliacozzi's fame spread until he was banned from plastic surgery by the Church, which claimed that he was undoing "God's handicraft." In Bologna, a statue was later erected of Tagliacozzi. It depicts him holding an artificial nose. His son, Antonio, advanced his work by restoring lips and ears by taking skin from the arm by a pedicled-flap.

Many surgical firsts also greeted the new seventeenth century. Gastric surgery to remove a foreign object made its debut in 1602 when barber-surgeon Florian Mattjis removed a knife from a

SURGICAL FIRSTS

Many surgical firsts came in the mid and late sixteenth century. The first account of a successful tracheotomy occurred in 1546 when Italian surgeon Antonio Bravascola opened the trachea of a patient who was near death from a respiratory obstruction caused by an abscess in his windpipe. The first reported splenectomy was in Naples in 1549, performed by Zaccarelli. Georg Bartisch of Dresden performed the first eye removal for cancer in 1583. During this period British surgery lagged behind the Italian and French masters, but Englishman William Clowes was the first to perform a successful thigh amputation for gangrene in 1588. He wrote extensively on the art of surgery.

peasant knife-swallower in Prague—during a show the patient had swallowed the knife too far, and it slid into his stomach when he drank a stein of ale. In 1635, the first reported surgery for cancer of the tongue was followed by a hemiglossectomy (partial tongue removal) in 1638. In 1664, a professor of surgery at Padua demonstrated the resectioning of ulcers and cancers of the lip.

British surgery began to catch up with its European counterparts in the early 1600s. British physician-surgeon William Harvey, who studied at Cambridge, spent several years in Europe, primarily in Italy at the University of Padua where he was influenced by Fabricius. Harvey received a medical degree in 1602 and returned to England, where he became professor

of anatomy and surgery at St. Bartholomew's Hospital and Medical School. He is regarded as the father of our understanding of the circulatory system, which he wrote about in *Concerning the Motion of the Heart and Blood* in 1628. Detailing the contractions of the heart and how its valves operate, Harvey's discovery of the working of the circulatory system has been hailed as a landmark in medical progress that benefited, if not paved the way for, modern surgery.

British naval surgeon John Woodall, who wrote *The Surgeon's Mate* (1639), favored leaving healthy tissue intact during amputations and only cutting the gangrenous tissue. Woodall then recommended whittling away the remaining dead tissue little by little as the healing process progressed. Richard Wiseman (1622?-1676) of England operated on aneurysms, controlling bleeding and dissecting arteries. He published *Treatise on Wounds* in 1672 and *Several Chirurgical Treatises* in 1676.

A major advancement in surgery during the seventeenth century came with the development of a surgeon's ability to perform lithotomies—the removal of stones in the bladder. A new technique was developed during this period, which marked an improvement over running an object through the urethra to knock out a blockage. Perineal lithotomy required an incision at the base of the body, between the anus and the opening for the urinary tract. The surgeon, either with his hands or a scoop, entered the bladder and removed the stones. Jacques de Beaulieu (1651-1719), who called himself Brother Jacques, was a strolling Italian lithotomist who demonstrated the technique in Paris in 1697. While he and other lithotomists were often successful, survivors were often noted to lead sad lives of postsurgical dripping urine and fistulas. Lithotomists were considered specialists and often worked as itinerant surgeons moving from place to place to apply their skills. Many patients, however, died of the infections from the lithotomist's unclean hands and instruments.

As the seventeenth century closed, anatomists who discovered new and important information about how the human body worked greatly influenced the next century's surgeons.

RANDOLPH FILLMORE

Further Reading

Baas, J. Herman. *The History of Medicine*. Robert E. Krieger Publishing Co., Inc., 1971.

Cartwright, Frederick F. *The Development of Modern Surgery*. New York: Thomas Y. Crowell and Company, 1967.

Ingus, Brian. *A History of Medicine*. New York: The World Publishing Company, 1965.

Krummbhaar, E. B. *A History of Medicine*. New York: Alfred A. Knopf, 1947.

Meade, Richard H. *An Introduction to the History of General Surgery*. Philadelphia: W. B. Saunders Company, 1968.

Wangenstein, O. H. and Sarah Wangenstein. *The Rise of Surgery*. Minneapolis: The University of Minnesota Press, 1978.

Zimmerman, Leo M. and Ilza Veith. *Great Ideas in the History of Surgery*. New York: Dover Publications, 1967.

Empirics, Quacks, and Alternative Medical Practices

Overview

During the Middle Ages and through the Renaissance, physicians who were educated and trained in the scientific method—with a reliance on observation and experimentation—were few in number. Most of the population relied on a combination of alternative healers for their medical care—empirics (those outside the medical mainstream), barber-surgeons, and apothecaries. These healers learned their trade mostly through an apprenticeship, and used folklore, herbs, and guesswork to cure the sick. Disease was often attributed to supernatural causes, from evil spirits to punishment by God. Superstition and religious fervor dominated the philosophy of caring for the sick. In Europe the population embraced empirical healers along with the prevailing mysticism in desperate attempts to escape the great plagues of the late Middle Ages. By 1661, as London was in the grip of another plague, new explanations for disease were sought. Interest in science was rekindled and classical texts were rediscovered, as Renaissance thinking spread from Italy throughout Europe. With the rebirth of the scientific method based on observation and experimentation, the early foundations for modern medicine were laid. Slowly, empirics and folklore medicine declined, and the science of medicine assumed prominence.

Background

Most Renaissance villages were without a resident physician or surgeon. Often the local barber would perform surgeries, along with his traditional duties of trimming villagers' hair and beards. The barber-surgeon's repertoire of procedures, often performed with the same razor with which he cut hair, included bloodletting, lancing infections, and excising lesions. The barber-surgeon also set broken bones and extracted teeth. Bloodletting was seen as the appropriate treatment for many common illnesses, and attempted to aid the victim by cleansing the blood of the excess of the "sanguine humor" in the Galenic tradition. After the barber-surgeon made a small incision and sometimes placed a cannula (small hollow tube) into a vein, the patient was encouraged to drink large quantities of fluids thought to dilute the blood and have a cleansing effect. Up to a pint of blood was removed per session. The familiar barber's pole is a symbol of the barber-surgeon's role in the practice of bloodletting. The patient gripped a staff in order to make the veins appear prominent for the procedure, and afterward the barber-surgeon secured the blood soaked bandages to the staff to dry. The familiar red-striped effect was created by the wind blowing the bandage around the staff. Barber-surgeons also used the red-striped staff to advertise to the public that, according to the position of the stars, the time was optimal for bloodletting cures.

Until the seventeenth century surgery was a free-for-all trade that almost anyone could practice. A Papal edict during the thirteenth century forbade monks to draw blood, and therefore prohibited them from practicing medicine. Monks would often pass their knowledge of anatomy and medicine on to barbers, who often visited monasteries to help the monks maintain their smooth-shaven appearance. Thus, barbers acquired surgical knowledge, then practiced in public for others to observe. With few laws governing surgery, the result was an army of charlatans who wandered the countryside and filled the marketplaces and county fairs representing themselves as tooth-pullers, cataract-removers, and "cutters-for-the-stone," or lithotomists, who specialized in a quick surgical removal of stones

Bloodletting by a Dutch barber-surgeon. Draining blood from a sick person was considered a chief method to eliminate the illness. *(Bettmann/Corbis. Reproduced with permission.)*

of the urinary tract. The most notorious of these was the French lithotomist Jacques de Beaulieu. After an apprenticeship to a wandering Italian surgeon in 1690, Beaulieu donned monk's robes, named himself Frère Jacques (believed to be the inspiration for the familiar nursery rhyme), and

was eventually expelled from France due to the number of patients who died from his lithotomy techniques.

Apothecaries and herbalists were among the lowest in the social order of the Renaissance healers. Apothecaries dispensed medicines of the day upon the order of physicians and surgeons. Many also practiced alchemy, the ancient and futile search for a method of turning ordinary substances into extraordinary ones, notably gold. Herbalists, often known as "witch women," made potions of dried herbs and plants, parts of dead animals, and sometimes excreta. The efficacy of the potions was trial and error—if a sick or injured person recovered, the herbalist recorded the herbal potion as successful and added the concoction to her medicinal collection. Occasionally, an herbalist stumbled upon a plant with legitimate curative action. In the early eighteenth century an herbalist near Shropshire, England, reported great success in curing dropsy (congestive heart failure) with a mixture of brewed foxglove. Digitalis, an important drug still used today for its action on the heart, is derived from the foxglove plant. In 1638 the Countess of Chinchon, wife of the Spanish Viceroy of Peru, was cured of malaria by an extract of the bark of the Peruvian quina-quina tree. The news of the cure spread quickly throughout Europe, and the drug was named *chinchona* in honor of the countess. Now known as quinine, the drug remains an effective weapon in the prevention of malaria. More often, the herbalist and even the apothecary stocked medicines that had little effect on the diseases for which they were intended. Some curatives involved elaborate and expensive hoaxes. Mumia, a powder derived from dried Egyptian mummies, was one of the most valued commodities on the apothecaries' shelves. Believed to contain powerful, mystical cures for many diseases, the demand for mumia in seventeenth-century Europe was so great that it was impossible to supply it. As a result, medical charlatans grew wealthy selling fake mumia, made from the remains of recently executed prisoners.

Impact

With little understanding of the nature of disease, and little to offer their patients in the way of a cure, physicians as well as other healers often embraced the occult for an explanation of illness. This philosophy impeded the rediscovery of the scientific approach to medicine by hundreds of years. Especially in the fifteenth and sixteenth centuries, the belief of superstitions and demons as the cause of disease was widespread. Demons were said to invade persons of weak stature or character and thereby cause disease. Sore throats and pneumonia were believed to be caused by demons traveling with winds from a northern direction. Many persons cov-

ered their faces or stayed indoors to avoid north winds. Epilepsy and insanity were believed to be the result of particularly fierce demons taking residence in the unfortunate victim. Priests and other clergy encouraged the belief in demons as the cause of disease, as prayer and exorcism allowed them an active role in caring for the sick.

The officially sanctioned persecution of individuals for witchcraft began during the Renaissance period. Although many of those presumed to be witches were probably suffering from delusional mental illnesses, some empirical healers were also accused. Occasionally, healers were put to death if their potions or methods did not affect a cure. Others accused of witchcraft were first sent to an exorcist. If the demon did not depart, the demented person was blamed. He was then subjected to starvation, whippings, and imprisonment in a dungeon in an attempt to rid himself of the demon. If these tortures did not induce the demon to leave the body, the demented person was burned at the stake. Mental illness was not considered a reason for humane treatment via scientific means for over another hundred years, when French physician Philippe Pinel (1745-1826) wrote a treatise characterizing insanity as an illness, and its victims worthy of humane intervention.

Many healers earned their living by catering to the prevailing superstitions of the time, becoming traveling peddlers, selling amulets and other wares intended to ward off demons and disease. The modern custom of wearing lockets in the shape of a heart around the neck is a vestige of a belief that the heart possesses spiritual powers. Often garlic or herbs were worn in the locket to ward off disease. In Scotland, these amulets were known as witch brooches and were used to protect children from witchcraft. Christians wore prominent crosses to display their faith and to ward off demons. England's Queen Elizabeth wore an engraved gold ring suspended from her neck to dispel "bad airs" about her. Healers sold vast quantities of garlic and onions, both said to have medicinal value and powers against evil spirits. One of the most dramatic superstitions involved the ground powder from a unicorn's horn, believed to cure any poisoning and bring good fortune. When English surgeon Ambroise Paré (1510-1590) suggested that many apothecaries and quacks were growing rich substituting domestic animal horns for the imaginary unicorn, he was denounced by the dean of the Paris Medical College. Amid such superstition, sanctioned by religious and many academic leaders, scientific progress in medicine crept through the latter Middle Ages and first half of the Renaissance.

Beginning in about 1550, several factors diminished the ranks of empirics, barber-surgeons, herbalists, and quacks across Europe. The renewed interest in the classics brought about by Renaissance thinking reintroduced the Hippocratic method of careful observation of patients and their symptoms. Physician and philosopher René Descartes (1596-1650) demystified the human body with his mechanical conception of human physiology. When opposition to human dissection evaporated, Andreas Vesalius (1514-1565) saw first-hand the structure of the human body, and pioneered the study of modern anatomy and physiology. Anatomical observations were accurately illustrated by artists such as Leonardo da Vinci (1452-1519). Also, the invention of the printing press made possible the dissemination of medical literature and sharing of information across Europe. Several universities were built during the period that served to train learned physicians in the classical tradition and scientific method. Laws and regulations regarding practice were strengthened. By the eighteenth century those who practiced superstition and folklore remained, but were superceded by those physicians trained in the sciences at the dawn of modern medicine.

BRENDA WILMOTH LERNER

Further Reading

Cipolla, Carlo M. *Public Health and the Medical Profession in the Renaissance*. New York: Cambridge University Press, 1996.

Lyon, Sue, ed. *Exploring the Past: Shakespeare's England*. New York: Marshall Cavendish, 1989.

Taylor, Laurence and Angus McBride (illustrator). *Everyday Life: The Sixteenth Century*. Morristown, NJ: Silver Burdett Company, 1983.

William Harvey and the Discovery of the Human Circulatory System

Overview

William Harvey (1578-1657) is recognized as the man who discovered and published the first accurate description of the human circulatory system, based on his many years of experiments and observations as a scientist and physician. Harvey had accumulated a mass of irrefutable experimental evidence in support of his dramatic new view, knowing that a tremendous amount of criticism and disbelief would be mounted against his groundbreaking, revolutionary theory of the physiology of blood circulation. Although the majority of the physicians and scientists of his day refused to accept his research, Harvey's discovery and written description of the true functioning of the heart and circulatory system remains as one of the landmark medical textbooks and the foundation of modern physiology.

Background

Most physicians, scientists, and philosophers of seventeenth-century Europe were adherents of Galen's doctrine, which contained several significant errors regarding the movement of blood and the workings of the heart. These were actually quite ancient ideas and notions, still accepted more than 1,400 years after first being postulated by Galen (130?-200?), the Greek physician of Rome. Over time, the dogma of Galen became sacrosanct, even though most of his anatomical knowledge and physiological investigations were based on his studies of monkeys and pigs, because dissections of human bodies were typically not permitted. Galen recognized the usefulness of comparative anatomy for gaining understanding of the human body, and he studied the workings of animal bodies and various structures in some detail. He was a prolific writer and dedicated scientist, venerated for centuries, and long considered to be the authority on medicine and health.

Galen and his proponents believed that the circulation of blood began in the gastric and intestinal blood vessels, and was carried to the liver, where it was "elaborated" by the liver. This "venous blood" then entered the hepatic vein, which he believed to be the origin of the vena cava, and the "descending" vena cava transported blood to the lower body, while the "ascending" branch sent blood to the upper body. As blood entered the right side of the heart, it was thought

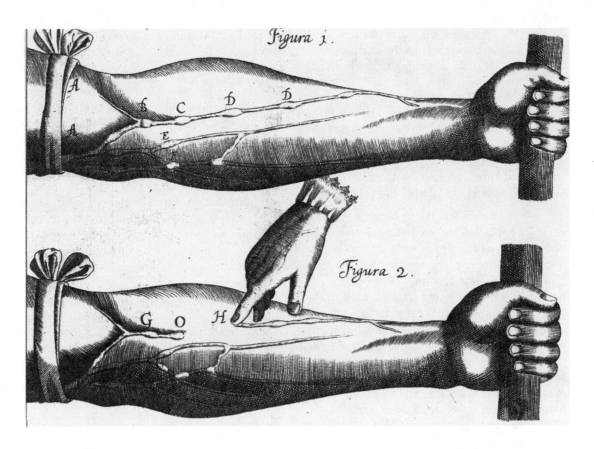

William Harvey's drawing of the veins of the forearm. *(Wellcome Institute Library. Reproduced with permission.)*

to pass through invisible pores in the septum that divided the heart, forced into the left ventricle, mixed with air brought in from the lungs by the pulmonary veins, and transformed into the "arterial blood." The heart was seen as a type of bellows, expanding when a small volume of blood in the left ventricle was greatly heated by the addition of "vital spirits," forcing the heart to expand and draw blood inside. In a similar fashion, the arteries carried this "boiled up" blood away from the heart to the body, but the blood did not return to the heart. According to Galenic doctrine, the liver was seen as the continual source of new blood, replenishing the blood that was vaporized and converted into waste material, and released from the lungs as "soot."

As the personal physician of King Charles I and the recipient of the best medical education possible, Harvey was perhaps the preeminent physician in England and perhaps all Europe, and he long doubted the accuracy of many of the "facts" that the medical profession espoused. Harvey finally published the results of his research in his text *Exercitatio anatomica de motu cordis sanguinis in animalius* (On the movement of the heart and blood in animals) in 1628. Harvey only accepted as facts those ideas that were supported with repeated experimental evidence, and, as was his nature, he methodically and forcefully exposed the errors of the long held misconceptions about the heart and blood circulation. His new system completely altered the Galenic concept of blood circulation, proving that the heart is a hollow muscle that contracts regularly to provide the single motive force of the blood's movement. He patiently exposed the other unacceptable aspects of Galen's erroneous system, using well-designed experiments that attempted to dispel various falsehoods.

Harvey was able to fully illustrate the actions of the heart, its chambers and valves, as well as clarify the long misunderstood pattern of pulmonary circulation. Harvey concluded that blood moved from the right ventricle into the lungs via the renamed pulmonary artery (correctly changed from pulmonary vein), which Galen thought carried only air and "soot" back and forth between the lungs and heart. Harvey properly stated that the blood then returned to the left side of the heart via the pulmonary veins. Harvey would not attempt to answer why the blood traveled to the lungs and back, as he did not have any knowledge about gas exchange during pulmonary respiration. Harvey also fully detailed the systematic circulatory system, tracing the flow of blood through the arteries coursing within the body, returning to the heart via the network of veins. He also artfully illustrated the workings of the valves in veins, proving the one-way circulation of venous blood towards the heart, and refuting the notion that the valves were actually reinforcement structures that prevented the over expansion of the veins as blood was forced through, as his university mentor Girolamo Fabrici (1537-1619) had taught.

When Harvey found that his experimental evidence could not provide an answer to a question, he did not attempt to evoke rational mysticism by way of explanation, as in the case of how and why blood in the arteries eventually passed into the veins and traveled back to the heart. Harvey could not see the capillaries found in tissues and had no way of addressing blood's metabolic function, but he did anticipate the presence of the "anastomoses" between arteries and veins and the possibility of blood providing nourishment or some other function. These blood-carrying structures were too small to be seen with the naked eye, but Harvey strongly believed that their existence would be detected eventually. Later, in the seventeenth century, both Marcello Malpighi (1628-1694) and Anton van Leeuwenhoek (1632-1723) would use the improved microscope to describe the presence of capillaries and blood cells in a wide variety of animals, including humans.

Impact

Harvey worked long and hard to create what became the starting point for modern mammalian physiology. His still impressive research is also seen as the first milestone of modern experimental science, and can be used as an example of how to perform experimental scientific research. Being the person to inaugurate two new scientific systems that condemned long-held beliefs, Galen's doctrine and the school of rationalism, Harvey must have recognized the likelihood of dire consequences. The derision and attack of the medical community was inevitable, and accusations and charges made by the Church and legal authority would not be without common precedence. New ideas that change entire systems of knowledge were always viewed with skepticism and apprehension, often evoking harsh criticisms and accusations of quackery. Harvey risked being rejected as foolishly misled or even acquiring the stigma of being labeled a quack. After his publication, his private practice

suffered a great decline as a result of the intense controversy he created, but Harvey steadfastly maintained himself and his convictions during the controversy.

As a professor and physician, Harvey advocated the use of comparative techniques to study anatomy and physiology, recognizing the advantages and practicality of using the animals that were available for study. Harvey worked with fish, amphibians and reptiles, birds, mammals, and humans, experimenting and comparing where ever possible, building his theory methodically and with great care. In the case of the action of the heart, he found that in many lower animals, the heart's movement was slower and could be seen more readily, and he used the slower heart rate of chilled fish and amphibians for analysis and comparison to the faster mammalian heart. Many of Harvey's experiments would later be described as direct, artfully simple, and beautifully designed.

Throughout his career, Harvey emphasized the experimental method of scientific research, which would become a basic tenet of modern science. Harvey would not accept any rationalism or mysticism as evidence for determining how or why something occurred in the body. Only experimental evidence that was repeated many times, using as many different animal examples as possible, could be considered in reaching any conclusions. Harvey avoided having any preconceived ideas about his experiments, rather, he gathered his evidence, analyzed the data, and then created a scientific hypothesis that he knew he could further test directly with more experiments. He built his new theory of blood circulation in a straightforward analysis of each step in the process, gathering extensive experimental data to confirm every aspect. He anticipated potential criticisms and designed more experiments to refute future controversies. His reliance on the experimental method was in contrast to many scientists and philosophers of his time, who instead employed rationalism or dialectics to essentially think their way through a question or problem, often following anecdotal or casual observational information, and using little to no experimental evidence. This type of analysis typically evoked the presence of unseen forces or "principles," usually a supernatural or divine phenomenon. Harvey tended to avoid this kind of philosophical reasoning, referred to as ratiocination.

The adherents of the Galenic doctrine did not surrender to the new physiology quietly, but rather a great controversy raged for many years and long after Harvey's death. Harvey, humble and dignified as a person and in his work, was patient and understanding when dealing with his critics and doubting contemporaries. Occasionally he would answer his critics with a direct letter or a publication that would add to or reiterate the existence of the relevant experimental evidence that confirmed his conclusions. Nevertheless, recognition of the truths that he illuminated did not come in his lifetime. Eventual acceptance came much later, when scientists developed new tools of investigation and better understanding of modern science. Harvey is remembered and revered both as the founder of modern physiology and a champion of modern experimental science.

KENNETH E. BARBER

Further Reading

Chauvois, Louis. *William Harvey, His Life and Times: His Discoveries, His Methods.* London: Hutchinson Medical Publications, 1957.

Harre, R. *Early Seventeenth Century Scientists.* Oxford: Pergamon Press, 1965.

Gardner, Eldon J. *History of Biology.* 3rd ed. Minneapolis, MN: Burgess Publishing Company, 1972.

Guthrie, Douglas. *A History of Medicine.* Philadelphia: J. B. Lippencott, 1946.

The Beginnings of Blood Transfusion

Overview

William Harvey's (1578-1657) discovery of the circulation of the blood, reported in *De motu cordis* (1628), resulted in attempts to inject various therapeutic agents, including blood itself, into the veins of animals and humans. The first significant experiments on blood transfusion were performed by Richard Lower (1631-1691) in England in 1666 and by Jean-Baptiste Denis (1640-1704) in Paris in 1667. Lower began with a series of experiments on animals in preparation for transfusion of blood into humans, but Denis was the first to perform the experiment on human beings. Interest in blood transfusion was high from 1660 until about 1680, when various countries began to outlaw this dangerous, experimental practice. After the deaths of several patients, this highly experimental form of therapy was abandoned for about 150 years.

Background

Long before 1628 when he assembled his evidence and published *De motu cordis* (Anatomical Exercises Concerning the Movement of the Heart and Blood in Animals), William Harvey seems to have arrived at an understanding of the motion of the heart and blood. The notes for his first Lumleian Lecture show that by 1616 he was performing demonstrations and conducting experiments to show that blood passes from the arteries into the veins. He concluded that the beat of the heart impels a continuous circular motion to the blood. His discovery that blood continuously circulates in a closed system, was an essential prerequisite to the concept that blood could be transplanted from one animal to another.

With arguments based on dissection, vivisection, clinical experience, and the works of Aristotle and Galen, Harvey proved that in the adult all the blood must go through the lungs to get from the right to the left side of the heart. Harvey proved that it was the beat of the heart that caused a continuous circular motion of the blood from the heart into the arteries and from the veins back to the heart. In support of the novel idea of a continuous circulation of the blood, Harvey turned to quantitative considerations. Although he did experiments aimed at getting an accurate measurement of the quantity of blood put out by the heart with each beat, he emphasized that exact measurement was unnecessary. Even the most cursory calculation proved that the amount of blood pumped out of the heart per hour was so great that it exceeded the weight of the entire body. That is, if the human heart pumps out two ounces of blood with each beat and beats about seventy times per minute, the heart must expel about 600 pounds (272 kg) per hour. Therefore, blood must move in a continuous circle through the body. So the purpose of the motion and contraction of the heart was to impart a continuous circular motion to the blood.

Once this basic principle was grasped, many observations fell neatly into place. Experience gained through phlebotomy and observations on the throbbing of arterial aneurysms, and the pulses of the wrists, temples and necks of patients similarly supported the central thesis of *De motu cordis*. Knowledge of the continuous circulation of the blood explained many puzzling clinical observations. Circulation explained why poisons or infections at one site could affect the whole system, as could bites from snakes and rabid animals. On the other hand, for traditional medical practice, Harvey's theory seemed to raise more questions than it answered. How could the new system explain how the parts of the body secured their proper nourishment if the blood was not continuously consumed? If the liver did not continuously synthesize the venous blood, what was the function of this organ? How did the body distribute the vital spirits and the innate heat? If the vital spirit was not produced in the lungs or the left ventricle, what was the function of respiration? If all of the blood moved in a circle, what was the difference between the arterial and venous blood? What principles would guide medical practice if the great Galenic synthesis were sacrificed for the theory of the circulation?

Arguing from experimental and quantitative data in biology was a novelty and opponents of Harvey's work presented what seemed to be quite logical alternatives, at least in light of Galenic theory. Harvey's work did not lead to the total rejection of Hippocratic and Galenic principles, or the rejection of venesection as a major therapeutic tool. Harvey's work opened up new fields of research, but not even Harvey seemed to reconsider the relationship between therapeutic bloodletting and a closed, continuous circulation. Indeed, Harvey defended venesection as a major therapeutic tool for the relief of diseases caused by plethora. Long after Harvey's theory had been accepted, physicians believed in the health-promoting virtues of bloodletting.

Impact

Paradoxically, while provoking new arguments about the selection of appropriate sites for venesection, the discovery of the circulation seemed to stimulate interest in bloodletting and other forms of depletion therapy. Bleeding was recommended in the treatment of inflammation, fevers, a multitude of disease states, and hemorrhage. Patients too weak for the lancet were candidates for milder methods, such as cupping and leeching. In addition to prescribing the amount of blood to be taken, physicians had to select the optimum site for bleeding. Arguments about site selection became ever more creative as knowledge of the circulatory system increased. Many physicians insisted on using distant sites on the side opposite the lesion. Others chose a site close to the source of inflammation in order to remove corrupt blood and attract good blood to repair the diseased area. Proper site selection was supposed to determine whether the primary effect of bloodletting would be evacuation (removal of blood), derivation (acceleration of the blood column upstream of the wound), or revulsion (acceleration of the blood column downstream of the wound).

As a scientist, Harvey demonstrated admirable skepticism towards dogma and superstition, but he was not especially innovative as a practitioner and he does not seem to have considered the possibility of therapeutic blood transfusions. Harvey's methods stimulated his disciples to work out the medical and physiological implications of his discovery. The questions raised by his work provided the Oxford physiologists—scientists such as Robert Boyle (1627-1691), Christopher Wren (1632-1723), Robert Hooke (1635-1703), John Mayow (1640-1679), and Richard Lower (1631-1691)—with a new research program, that is, the exploration of the systemic effects of injecting various drugs and fluids into the veins of animals and human beings. By the 1660s, British physiologists were performing ingenious experiments involving the injection of drugs, poisons, nutrients, pigments, and blood itself into animal and human veins. The transfusion and infusion of medicinal substances into the bloodstream did not become part of routine medical practice for many years, but seventeenth century experimentalists did raise many intriguing possibilities.

Competing claims for priority have created some confusion in the history of blood transfusion, but the body of evidence indicates that the first significant studies of blood transfusion were performed by Christopher Wren, Richard Lower, and Robert Boyle in England, and by Jean-Baptiste Denis (1640-1704) in Paris. According to Thomas Sprat's *History of the Royal Society* (1667), Christopher Wren was the first to carry out experiments on the injection of various fluids into the veins of animals. During experiments exhibited at meetings of the Royal Society, experimental animals were purged, vomited, intoxicated, killed, or revived by the intravenous injection of various fluids and drugs. Dogs, birds, and other animals were bled almost to death and sometimes revived by the injection of blood from another animal.

A late-seventeenth-century engraving depicting a blood transfusion from a sheep to a man. *(Corbis Corporation. Reproduced with permission.)*

Reasoning that the nature of blood must change after it was removed from the living body, the English physician and physiologist Richard Lower decided to transfer blood between living animals by connecting the artery of the donor to the vein of the recipient. During a demonstration performed at Oxford in February 1666, Lower removed blood from a medium sized dog until it was close to death. Blood taken via the cervical artery of a larger dog revived the experimental animal. Using additional donors, Lower was able to repeat this procedure several times. When the recipient's jugular vein was sewn up, it appeared to be in good condition. These experiments led observers to speculate that someday blood transfusions could cure the sick by replacing bad blood with healthy blood.

After learning about Lower's experiments performing a series of transfusions from dog to dog, Jean-Baptiste Denis, professor of philosophy and mathematics at Montpellier and physician to Louis XIV, conducted a series of experiments on the transfusion of blood from dog to dog. In March 1667, he transferred blood from a calf into a dog. Observing no immediate adverse effect, Denis concluded that animal blood could be used to treat human diseases. Denis argued that humans were able to assimilate the flesh of animals and that animal blood might be a better remedy than human blood, because animal blood would not be corrupted by human vices. Moreover, as a practical matter, animal blood could be transfused directly from the artery of the donor to the vein of the human recipient.

On June 15, 1667, with the help of Paul Emmerez, a surgeon and anatomist, Denis tested his methods on a fifteen-year-old boy who had suffered from a chronic fever and had endured numerous therapeutic bleedings. Emmerez drew off about three ounces of blood from a vein in the boy's arm and Denis injected about ten ounces of arterial blood from a lamb. The operation seemed to be a great success, except for the boy's complaint about the sensation of great heat in his arm. In another experiment, Denis injected about twenty ounces of lamb's blood into a healthy forty-five-year-old paid volunteer. Again, except for a sensation of warmth in the arm, no ill effects were reported. A man suffering from "frenzy" seemed to improve after a transfusion of calf's blood, but the patient experienced pains in the arm and back, rapid and irregular pulse, sweating, vomiting, diarrhea, and bloody urine. Given the patient's poor state of health and previous treatments, Denis saw no compelling reason to blame these problems on the transfusion. Thus, although Denis reported several apparently successful transfusions, his experiments also resulted in what was probably the first recorded account of the signs and symptoms of a hemolytic transfusion reaction. The death of a patient who had been given two injections of calf's blood as a treatment for recurring attacks of insanity precipitated a violent controversy, an avalanche of pamphlets, a lawsuit, and the arrest of Denis.

Although not found guilty of malpractice, Denis and Emmerez discontinued their experiments and returned to conventional careers. In 1668, the Chamber of Deputies declared that the transfusion of blood from animals into humans was prohibited, unless specifically approved by the Faculty of Medicine of Paris. Then years later, the British Parliament also prohibited blood transfusion. Little progress was made in the field in the next 150 years.

English scientists had been very critical of the experiments performed by Denis, but they too experienced mixed success in blood transfusions. About six months after Denis's first human transfusion, Richard Lower hired a "lunatic" named Arthur Coga, to serve as a test subject. Coga's condition seemed to improve after an injection of sheep's blood, but his condition deteriorated rapidly when he was given a second transfusion. The first transfusion experiments had briefly stimulated great expectations, but blood transfusion did not become routinely safe and effective until after World War I. Seventeenth century physiologists, who explained their world in terms of four elements and four humors, could not appreciate the immunological barriers between different species and individuals. Safe blood transfusions were made possible when the immunologist Karl Landsteiner (1868-1943) demonstrated the existence of distinct blood group types.

LOIS N. MAGNER

Further Reading

Books

Bylebyl, Jerome J., ed. *William Harvey and His Age: The Professional and Social Context of the Discovery of the Circulation.* Baltimore: Johns Hopkins University Press, 1978.

Dickinson, C. J. and J. Marks, eds. *Developments in Cardiovascular Medicine.* Lancaster, England: MTP Press, 1978.

Frank, R. G., Jr. *Harvey and the Oxford Physiologists. Scientific Ideas and Social Interactions.* Berkeley, CA: University of California Press, 1980.

Hackett, Earle. *Blood, The Paramount Humor.* London: Jonathan Cape, Ltd., 1973.

Harvey, W. *Anatomical Studies on the Motion of the Heart and Blood.* Trans. by G. Keynes. Birmingham, AL: Classics of Medicine Library, 1978.

Keynes, G. L., ed. *Blood Transfusion.* Bristol: John Wright and Sons, Ltd., 1949.

Whitteridge, G. *William Harvey and the Circulation of the Blood.* New York: American Elsevier, 1971.

Periodical Articles

Farr, A.D. "The First Human Blood Transfusion." *Medical History* 24 (1980): 143-62.

Hoff, Hebbel and Roger Guillemin. The First Experiments on Transfusion in France. *Journal of the History of Medicine and Allied Sciences* 18 (1963): 103-24.

Progress in Understanding Human Anatomy

Overview

The great European Renaissance, or revival of thinking, had a major impact on the study of human anatomy. The development of medicine established by the Greeks and Romans, and imbued with a spirit of inquiry, had long served as the unchallenged standard of medical practice and belief. During the Middle Ages knowledge of the human body was enveloped with ignorance and superstition. A religious and cultural taboo against human dissection limited anatomical knowledge to what could be gleaned from the study of animal specimens.

All of this changed with the development of universities in Italy. Responding to the need of the times to understand the causes of death, anatomists began to challenge ancient traditions by performing dissections of the human body. They were aided by an unusual group of professionals, artists who sought to understand the nature of the body to perfect their craft. Another great invention, printing, enabled duplication of the text, and wood engraving enabled drawings and illustrations to help people understand anatomy.

One of the first great anatomists was Andreas Vesalius (1514-1564), who had an amazing impact because he made anatomy acceptable and questioned long held traditions of the past. As anatomy developed in the great universities beginning in Bologna, others began a study in earnest. Antonio Benivieni (1440?-1502) dissected cadavers to study disease and the cause of death, and his findings were published in 1507.

A professor, Johannes Dryander (1500-1560), made the first illustrations directly from cadavers and published these in 1537. By the middle of the sixteenth century public dissections became a matter of curiosity among the populace in general. Many of these were done in the town square and drew crowds from miles around. Heironymus Fabricius (1583-1619) worked on the Aristotle Project, which sought not only knowledge of structure, but also comparative understanding of anatomy.

By the end of the period, and moving into the 1700s, the state of anatomy had so changed that it could hardly be compared to the simple beginnings in the mid-fifteenth century.

Background

Gross anatomy is the term used to describe a systematic knowledge of the structure of the body. From the earliest times several things prevented organized investigation of both human and animal bodies. Social ideas against touching and cutting a dead body developed. Tampering with the body was believed to be tampering with the soul. Religious beliefs became enveloped with myth about anatomy. For example, the belief prevailed that men had one less rib than women, as set forth in the biblical story of the creation in Genesis, when God took a rib to made Eve.

Aristotle (384-322 B.C.) applied logical and rational thinking to biology, and his precepts of anatomy and physiology were based on animal observation. Empedocles (490?-430 B.C.) of Sicily was the first to introduce the theory of the four humors, which influenced medical thought for 2,000 years. It was believed that when the humors were out of balance, the person was sick. Hippocrates (460?-377? B.C.) adopted these ideas for use in clinical medicine. Within this theory of knowledge, anatomy is not considered useful and is subordinate to practice. Galen (130?-200?), in the same tradition, prided himself on being a fine clinician and adapted his knowledge from the study of apes, pigs, sheep, goats, and even an elephant's heart—but not humans. He knew much about the skeletal anatomy, but human dissection was out of the question. Galen drew on his knowledge of animal anatomy, combined with Hippocratic medicine and Platonic reasoning, to establish a hold on thinking for hundreds of years.

Mondino di Luzzi (1275-1326) was a professor of anatomy at Bologna who first introduced dissections of humans instead of pigs. He wrote the first textbook on anatomy, which remained popular for two centuries and emerged in a 79-page printed version in 1515.

Human dissection began toward the end of the thirteenth century at the University of Bologna. Although frowned upon by the Church and tradition, dissection became a necessity to find out causes of death. The great universities supported the dissections, which extended to Padua, Florence, Pisa, and Venice. Anatomy was still influenced by Galen, and the efforts at illustration were initially to support his beliefs.

During the fifteenth century a great change occurred in the general culture, inspired by the rediscovery of Greek work. The Renaissance of learning that exploded in the sixteenth century included the development of anatomy.

Impact

The influence of Leonardo da Vinci (1452-1519) on anatomy was not felt by his contemporaries in the medical community, but he did establish a pattern for medical artists that began slowly to expand through Italy first, then to other areas. His notebooks included methods for the study of the arm and forehand, motor muscles of the hands and wings, and actions of the muscles in breathing. Da Vinci led a group of painters, philosophers, and poets to commend the beauty of the human form. Their emblem became the Vetruvian man, the human form superimposed on the cosmos. This interest in the human body led naturally into interest in its physical structure and workings.

Da Vinci spent hours in his basement dissecting and studying corpses. He created about 750 anatomical drawings, and in 1489 planned an anatomical atlas of the stages of man from the womb to the tomb, although this was never completed. Some of the drawings of early anatomists indicated exposure to da Vinci.

With interest in everything Greek, the first anatomy text was written by Allessandro Benedeti (d. 1512) who lived 16 years in Greece, then came to Padua in 1490 as a professor of anatomy. The discovery of Galen's *On Anatomical Procedures* showed how to carry out a dissection. Galen had encouraged people to find out for themselves about anatomy. The only flaw was that he used animals not humans.

Enthusiastic about the work of Galen, Vesalius took up the challenge to examine anatomy firsthand and, as a result, became one of Galen's great critics. Born in Brussels, Belgium, Vesalius studied under Jacob Sylvius, Galen's great champion. In later years, Sylvius would become Vesalius's adversary due to Vesalius's criticism of Galen, and even called him a mad man. When war in 1533 forced Vesalius to flee Paris, he went to Louvain, where he introduced human dissection. At the time, criminals were left on a gibbet (the place of hanging) for public scorn. Vesalius secretly stole these bodies, smuggled their bones home, and reconstructed the skeleton.

In 1537 he went to Padua, a great institution for dissecting, although it was not mandatory for physicians. In 1538 he designed for his students *Six Anatomical Pictures,* a treatise still influenced by Galen's legacy. But as he found out more about anatomy, it became more unsettling. He began to challenge Galen and came to the conclusion that human anatomy must be learned through dissections. Dead bodies, not dead language, were the key to learning.

To do this, he got a huge supply of dead bodies of executed criminals and began his great masterpiece *De humani corporis fabrica* (On the Structure of the Human Body). He took the book to Basel where it was published in 1543. This book was a great turning point in medicine, for it promoted the understanding of the structures of the human body. At the time the artwork was thought to be done by the famous Italian painter Titian, but was later attributed to one of his assistants, a Dutch artist named Jan Stephen van Calcar (1499-1546). The poses are astounding, showing a body without skin standing or posing in front of landscapes of various kinds. Vesalius showed the human body was the key to medical knowledge.

Another aspect of anatomy emerged from the work of Antonio Benivieni, that of pathological anatomy, or autopsy, to find the cause of disease. A native of Florence, Benivieni studied medicine at Pisa and became a respected physician in his city. While practicing, he became convinced that autopsy was the only way to discover causes of disease. He observed gallstones, cancer of the stomach, peritonitis, and many other conditions. He published several books that were lost at the time, but were later found and published. His *The Hidden Causes of Disease* presented a new method of thinking in medical science—that of the discovery of cause of death by autopsy.

Heironymus Fabricius became a professor at Padua in 1565. He was not only interested in anatomical structure, but also comparative approach. He stressed three aspects of anatomy: description, action, and use of body parts. His most significant work on the valves of the veins laid the foundation for William Harvey (1578-1657) to develop this theory of blood circulation.

An interesting trend also developed in anatomy—that of public dissections. Because of plagues and other diseases, people became very curious about how these were transmitted and about the body in general. Public dissections became quite a spectacle.

In the 1520s anatomical drawings and texts increased in number, whetting further interest and curiosity. Johannes Dryander, a professor at Marburg, Germany, carried out the first public dissection. Dryander also created a trend in anatomy—specializing on one part. He wrote a treatise detailing the anatomy of the head. Rolfinck, also in Germany, continued the trend of public dissection into the 1600s.

Later anatomists specialized in the study of body parts. Bartolomeo Eustachio (1500-1574) produced a detailed study of the kidney and ear. He criticized Vesalius for studying the kidneys of a dog instead of those of a human, and dog's ears instead of man's. A structure of the ear, the Eustachian tube, is named for him.

Most scientists recognize Vesalius's 1543 publication as the beginning of the renaissance in medicine and anatomical inquiry. The anatomists opened the floodgate of medicine. Their interest in the structure of the human body naturally led to studies of how these structures function. Human physiology would soon develop.

EVELYN B. KELLY

Further Reading

Clendening, Logan. *Source Book of Medical History.* New York: Dover, 1960.

Porter, Roy. *The Greatest Benefit to Mankind: A Medical History of Humanity.* New York: W.W. Norton, 1997.

Schultz, Bernard. *Art and Anatomy in Renaissance Italy.* Ann Arbor: University of Michigan Press, 1985.

Wear, A., R. K. French, and I. M. Lonieed, eds. *The Medical Renaissance of the Sixteenth Century.* Cambridge: Cambridge University Press, 1985.

Advances in Understanding the Nervous System

Overview

Neurology, the study of the brain along with the body's nervous system, had little structure in the medical community of late-Medieval and Renaissance Europe. The renewed intellectual enthusiasm of the Renaissance brought about an appreciation for classical study, and with it advances in the knowledge of the anatomy of the human brain. Much of the period, however, was devoted not to seeking practical medical knowledge about the nervous system, but to pondering the philosophies of its nature. Occasionally, empirics or quacks filled the void, performing surgeries on the scalp, administering drugs or herbs, or chanting for the benefit of a patient with headache, neuromuscular difficulty, or mental illness. By the end of the seventeenth century, however, the scientific and Hippocratic methods, based on direct observation and experimentation, were laying the foundation for the future study of neurology.

Background

Medieval understanding of the nervous system was basically limited to observations of animal anatomy, tempered by philosophies prevailing since antiquity. The influence of Greek physician Galen of Pergamum (c. 130-c. 200) on medical theory and practice was dominant in Europe throughout the Middle Ages and into the Renaissance. Galen thought that the best physicians were also philosophers, and that philosophy promoted medicine. Galenic tradition held that illness was a result of an imbalance of body fluids, or humors. While dissecting calves, Galen noticed a network of nerves and vessels at the base of the calf brain that he mistakenly assumed also existed in humans. Galen labeled this area the *rete mirabile*, and stated that this was the site where vital life spirits were transformed into man's animal spirits. After the advent of Christianity, these spirits were unified into the concept of a Christian soul, and physicians debated its base in the human body, presumably the heart or the brain.

By the mid-1600s scientists considered the structure and philosophy of the brain based upon observations on humans. English physician and anatomist Thomas Willis (1621-1675) was the first to name the fledgling science of neurology. Willis was professor of philosophy at Oxford University, and he built one of the largest and best-known medical practices of his day. Willis belonged to the new school of iatrochemists, who believed that animal physiological activity could best be explained by chemical interactions. By removing the brain from the cranium, Willis was able to clearly visualize its structures. Willis accurately described the hexagonal network of arteries (the circle of Willis) at the base of the brain that is responsible for ensuring the brain's blood supply as the body's highest priority. He identified the eleventh cranial nerve (the spinal accessory nerve) responsible for motor stimulation of major neck muscles. Willis also distinguished gray matter from white matter, and ventured that the gray matter held the animal spirits, while the white matter distributed the spirits throughout the body, giving rise to movement and sensation. Willis rejected the Galenic tradition of the spirits as fluids, however, when he noticed the speed with which sensations and muscular actions occur. Willis likened the spirits to rays of light, quickly filling the passages of nerves, thus resembling an elementary concept of nerve conduction. Along with French philosopher René Descartes (1594-1650), Willis was one of the last of the animal spirit theorists. Still, he accepted the idea of a soul unique to man, possessing the ability to reason. In 1664 Willis published *Cerebri Anatome*, the most accurate and complete account of the nervous system yet published, containing detailed illustrations by Christopher Wren (1632-1723).

In France Willis's work was carried on by the French physician Raymond Vieussens (1641-1715). Vieussens held the post of chief surgeon at Hotel Dieu at St. Eloi (near Montpellier) for most of his adult life. Inspired by both Descartes's mechanistic and the iatrochemical philosophies, Vieussens studied the white matter of the brain by tracing the path of its fibers. Because of his tendency to explain his observations with fantastic physiological explanations, Vieussens sometimes drew harsh criticism from the faculty of medicine at Montpellier University. Vieussens had an advocate in the Marquis de Castries, who shielded him from attacks of the faculty and probably provided financial patronage, allowing Vieussens to continue his clinical and experimental work in relative peace. In 1685 Vieussens published the well-received *Neurographica Universalis*, detailing the central and peripheral nervous systems.

Descartes, a French mathematician and philosopher, sought to separate the rational soul from the physiology of the body. Descartes's concept of the body was that of an intricate machine. Descartes belonged to the school of iatrophysics, in which the workings of the animal body were explained purely on mechanical

grounds. The rational soul, according to Descartes, interacted with the body via the pineal gland of the brain. The animal spirits flowed from the ventricles of the brain, carrying instructions to the muscles that could be overruled or modified by the rational soul. Descartes saw the rational soul as the seat of wisdom, and the pineal area of the brain as the center of all sensory input. Descartes also is credited with the founding of reflex theory. Descartes articulated a mechanism for an automatic reaction, beginning with an external motion displacing the peripheral ends of nerves. The central ends of the nerves were, in turn, displaced, which allowed the flow of animal spirits from the center of the nerves to the surrounding appropriate areas, creating an involuntary reaction. Although the mechanism for a reflex action advanced by Descartes was faulty, the fact that the reflex action was a predicted automatic response to a stimulus remains.

Impact

During the Renaissance progress was much slower in the practice of medicine than in its science. Typically, the medicine of the Renaissance benefited the doctor more than his patient, and many physicians occupied much of their time debating the numerous prevailing philosophies of the workings of the body, leaving the actual care of the sick to barbers, empirics, and folk healers. In no area of medicine did so many philosophies exist ready for discussion than the workings of the brain and nervous system.

Although Descartes was proved wrong in his belief that the pineal gland was the focus of all sensory input, his belief in the rational soul did not find disfavor with many religious leaders of the day. Christianity was consistent with a separate soul that departed the body after death, one that could be held accountable for the body's earthly actions. The Cartesian dualism of a separate body and rational soul allowed Descartes to justify the strident religious demands of the time (Descartes was aware of his contemporary, Galileo, and his persecution during the Inquisition) with his belief in the mechanical nature of the workings of the human body. Descartes's dualism was outlined in his *De Homine* (On Man), published posthumously in 1662.

The urge to apply philosophical principles to the practice of medicine continued throughout the Renaissance. This led to variations on systems to explain disease, sometimes with unusual methods of treatment. Friedrich Hoffmann (1660-1742), a chemist and physicist, envisioned the body as composed of individual fibers. The fibers dilated and contracted, according to Hoffmann, in response to a "nervous ether" substance in the brain, which traveled to all parts of the body via the spinal cord. The contractile property was *tonus*, and every change in tonus was thought to bring about a change in health. Hoffman attempted to regulate tonus with several drugs he labeled "tonics." Tonic became a catch-all term for questionable remedies up until the twentieth century.

Beginning in about 1650 several factors set the stage to nurture the legitimate, new science of neurology. The renewed interest in the classics brought about by Renaissance thinking reintroduced the Hippocratic method of careful observation of patients and their symptoms. Descartes demystified the human body with his mechanical conception of human physiology. When opposition to human dissection evaporated, Andreas Vesalius (1514-1565) saw first-hand the structure of the human body, and pioneered the study of modern anatomy and physiology. Removed from the hands of barber-surgeons, anatomy evolved to an exacting science viewed with respect among physicians and medical academics. With the invention of the microscope, nerve cells were first visualized in the early 1700s, and nerve cell function studies were initiated. Eventually, eighteenth-century scientists contemplated the spark of life itself, and debated its natural, spiritual, or electrical origin in the human brain.

BRENDA WILMOTH LERNER

Further Reading

Carter, R. B. *Descartes' Medical Philosophy: The Organic Solution to the Mind-Body Problem*. Baltimore: Johns Hopkins University Press, 1983.

Finger, S. *Origins of Neuroscience: A History of Explorations into Brain Function*. New York: Oxford University Press, 1994.

Spillane, J. *The Doctrine of the Nerves*. London: Oxford University Press, 1981.

Paracelsian Medicine Leads to a New Understanding of Therapy

Overview

Few issues stirred up debate in the sixteenth and seventeenth centuries as much as the ideas and writings of the medical reformer Theophrastus von Hohenheim, widely known as Paracelsus (c. 1493-1541). Paracelsus attacked the academic medical establishment of his time and offered an alternative way of thinking about pathology, the understanding of the processes through which diseases occur. For Paracelsus, diseases were specific entities or powers that attacked particular parts of the body. Medicine should therefore aim to attack diseases through the use of chemically prepared substances that have a correspondence with the disease. This approach led to a new understanding of therapy and encouraged many people to use chemically based remedies for diseases. Many Paracelsian ideas seem bizarre today, but they are significant because they provided an alternative to a medical system that many had come to see as stale, useless, and unable to meet people's medical needs.

Background

The medical system taught in Renaissance universities relied heavily on book learning and philosophical theory. Generally, health and disease was understood as a matter of balance and imbalance within the individual body. It was believed that the body contained substances called humors, which could become corrupted. Disease occurred when this corruption affected the functions of the body. Therapy worked by the principle of "cure by contraries," which aimed to restore the body to its normal balance. For example, the physician would attempt to cure a disease that was an imbalance of heat and moisture through the use of remedies with the contrary qualities of cold and dryness. The physician could also restore the body to balance by drawing the corrupt humor out of the body, through methods such as bleeding and purging. Every patient was a unique case, because the balance of humors and qualities that made up each person was unique. Therefore, therapy was supposed to be individually tailored to each separate case of illness, rather than based upon previous experience of the disease. Those nonacademic practitioners who treated people on the basis of experience alone were disparagingly referred to by physicians as "empirics" or "quacks."

During the sixteenth century, there were many signs of dissatisfaction with this system. The inability of orthodox medicine to adequately explain or treat some major diseases was a source of frustration for many people, both patients and practitioners. Plague continued to ravage most parts of Europe throughout the sixteenth and seventeenth centuries, and orthodox physicians appeared helpless in the face of the pestilence. In the late fifteenth century, Europe also experienced the first outbreaks of what came to be known as syphilis. Believed to be brought to Europe from the Americas by crew aboard Christopher Columbus's voyage of discovery, the pox caused a great deal of suffering and excited a moral panic similar to the twentieth-century response to AIDS. For many, the pox highlighted the inadequacies of orthodox medicine. Academic physicians seemed to expend more energy on arguing whether or not the disease was new than actually treating it. The medical approach that insisted that each illness be treated on the basis of the patient's individual characteristics did not fit in well with the treatment of widespread epidemics. It is in the discussion of these two diseases, plague and syphilis, that one is most likely to find innovation in the explanation and treatment of disease in the sixteenth century. Paracelsus focused upon the treatment of plague and syphilis in many of his writings.

As well as this medical background, there were many intellectual developments during the Renaissance that partially explain the development of Paracelsian medicine. Orthodox medicine was part of the intellectual system of university natural philosophy, which was based upon the thought of the classical philosopher Aristotle. However, during the Renaissance, other systems of intellectual thought, such as Neoplatonism, challenged the dominance of Aristotelianism. Those scholars who were interested in Neoplatonism looked at the universe in quite different ways. They explored the connections and harmonies between different aspects of the cosmos, and emphasized the way in which the wise man could manipulate these cosmic harmonies for his own benefit, including in the sphere of medicine. Many of these influences can be seen in Paracelsus's writings, particularly in the emphasis of his medicine upon man as a microcosm, containing within himself all the phenomena of the universe. Neoplatonism also influenced Paracelsus in his understanding of astrological influences upon health and disease. In Paracelsian thought, it was these connections between man and the universe that must be explored if the physician was to understand health and disease within the human body.

Paracelsus's career also coincides with the period of religious and social upheaval known as the Reformation. The history of Paracelsian medicine throughout Europe is closely involved with the spread of Protestantism. Paracelsus's condemnation of the authority and monopoly of the orthodox physician in administering medical care parallels Martin Luther's rejection of the Church's monopoly over access to God. Paracelsus's reform of medicine was a part of the reform of religion and society that was occurring throughout the German territories in which Paracelsus lived and worked. Though Paracelsus never identified himself within any of the various sects of the Reformation, nor officially renounced his Catholicism, the religious and philosophical content of his writings shows similarities with many of the mystical thinkers of the German Reformation.

In terms of the impact that it had upon medicine, the most significant insight of Paracelsus's thought was the way he took alchemical concepts and applied them to the understanding of the processes of nature. Alchemy was an ancient mystical tradition, which involved the quest for the secret of transforming metals into gold. It had practical medical significance because many alchemists were also involved in the search for substances that could purify the body and therefore prolong life. However, alchemy was a tradition that existed outside the bounds of the kinds of knowledge considered acceptable by establishment medicine. It was part of a manual, craft-based tradition, rather than the learned knowledge of the universities. Paracelsus insisted that alchemy was a crucial part of medicine. To develop remedies, the physician needed to gain access to the hidden powers involved in disease and its cure. Alchemical processes released these powers that existed within material objects, so that they could act upon the disease and drive it out of the body. In place of the elements and humors of orthodox medicine, Paracelsus offered a more chemically based analysis of matter, which rested upon what he termed the three principles: mercury, salt, and sulphur.

The effect of this approach was to change the nature of medical therapy. In the place of the changes in diet and the herbal concoctions that the orthodox physicians favored, Paracelsus advocated strong remedies, often metallic based, which attacked the diseases at their roots. These remedies did not work according to the orthodox principle of cure by contraries. Instead, Paracelsus believed that like cured like, and the physician had to identify a similarity between the disease in the body and the substance in the outside world that could be used to cure it. Seeing disease as an entity that attacked a part of the body, rather than a general imbalance of humors, encouraged the search for specific chemical remedies for these specific diseases. Paracelsian writings provided an important theoretical justification for those who believed in the worth of chemical medicines and encouraged many to develop chemical remedies for diseases.

Their use of metallic-based remedies meant that chemical physicians often had to answer the accusations that they were poisoners whose remedies killed more than they cured. But Paracelsus was well aware that many of the substances he used could be dangerous, and he criticized the rash use of mercury in the treatment of syphilis, while still maintaining that careful use of alchemically prepared mercury could cure syphilis. Exponents of Paracelsian medicine believed that by exposing some of these dangerous substances to alchemical processes, they could be rendered fit for human consumption and have dramatic results in the treatment of many diseases. An example of this is the ether substance Paracelsus called *spiritus vitrioli*, which he developed as a sedative for use in the treatment of epilepsy. This was first time the medicinal potential of ether was recognized.

Impact

In the years after Paracelsus's death, those who advocated chemical medicines became a serious challenge to the medical establishment. In the late sixteenth and early seventeenth centuries, the question of the worth of chemical medicines was one of the most fiercely debated issues among the intellectual elites of Europe. Not all who supported chemical medicine regarded themselves as Paracelsians, but most felt that Paracelsus had achieved some great insights in the development of chemical medicine. By the mid-seventeenth century, most physicians accepted that chemical remedies were part of the physician's arsenal against disease.

An important part of Paracelsus's attack upon orthodox medicine was his insistence that medical knowledge should rest upon experience, rather than book learning. This means that Paracelsus has often been seen as an early advocate of experimental science. This has been overstated, and it must be remembered that many of Paracelsus's claims are based upon a kind of mystical intuition rather than experiment or observation. However, Paracelsus's approach to the study of the cosmos certainly made a contribution to the major shift that occurred in this period in the way people viewed the world around them. His insistence that knowledge of health and disease in people required knowledge of the correspondences between the human body and the outside world was an encouragement to closer observation of the natural world. His belief that true knowledge was found not in the dead wisdom of books but in the study of the living physical world led him to make some bizarre claims, but it also stimulated new and profound insights.

Paracelsus's refusal to accept the limits that orthodox medicine placed upon medical knowledge was a crucial part of his appeal for later generations of scholars. His willingness to explore traditions of knowledge that were shunned by orthodox medicine, such as alchemy and therapies based in popular medicine, challenged the boundaries of what was accepted as proper knowledge. He believed that if a reform of knowledge could be instituted, then people would discover the cures that God had created for the treatment of all diseases. This optimistic belief in the human capacity for the discovery and improvement of wisdom was an encouragement to all those who were dissatisfied with the existing systems of knowledge. This is why, throughout the sixteenth and seventeenth centuries, Paracelsus's ideas were used by those who sought reform in a variety of spheres.

KATRINA FORD

Further Reading

Debus, Allen G. *The Chemical Philosophy: Paracelsian Medicine in the Sixteenth and Seventeenth Centuries*. 2 vols. New York: Science History Publications, 1977.

Pachter, Henry. *Paracelsus: Magic into Science*. New York: Schuman, 1951.

Pagel, Walter. *Paracelsus: An Introduction to Philosophical Medicine in the Age of the Renaissance*. Basel, New York: Karger, 1958, rev. ed., 1982.

Paracelsus: Essential Readings. Trans. by Nicholas Goodrick-Clarke. Berkeley: North Atlantic Books, 1999.

Webster, Charles. *From Paracelsus to Newton*. Cambridge: Cambridge University Press, 1987.

The Exchange of Plant and Animal Species Between the New World and Old World

Overview

When Europeans reached North America's shorelines in the late 1400s and began to explore the continent's interior in the 1500s, they saw the vast land as a source of new plants, animals, and minerals for them to use and to transport back to Europe. As they colonized this New World, they also brought with them many familiar plants and animals for food, farming, and other purposes. This exchange of species between the two continents had positive and negative effects, and they continue today. On the positive side, the exchange introduced what would become important agricultural crops and beneficial animals to both continents. It also, however, expanded the range of species that carried disease and competed with beneficial native species, and it also permanently changed the face of each continent.

Background

When the Europeans landed in the New World in the late 1400s and early 1500s, North America was an untamed wilderness filled with mysterious flowers, trees, birds, and mammals. Christopher Columbus wrote of the scent of a breeze from the shores of North America as "the sweetest thing in the world." Other explorers similarly reported of the biological wealth of this unknown continent: flocks of birds so thick they blocked out the noonday sun; fishes so large and numerous that they sometimes hampered river navigation; enormous stands of pines, oaks and chestnuts; and meadows so vast that their boundaries were beyond sight.

The explorers were in the New World to find items of economic benefit to their countries and saw the new land as a resource to be exploited and a wilderness to be tamed. The first explorers to North America came with hopes of finding a passageway to the East Indies, and eventually opening a lucrative trade route with the spice-rich islands. Although they were unsuccessful in this regard, they were able to discover many new and useful plants and animals, which they were only too happy to transport back to their homelands.

One successful transplant was corn. It was first cultivated in the Americas more than 5,000 years ago, by the Maya, Aztec, and Inca. Called *ma-hiz* by the Indians of Central America, its name was later corrupted to "maize" in Europe. By the time the New World was discovered in 1492, corn had spread across the continent as far north as Canada. Columbus is credited with bringing corn back to Europe from his first voyage, where it quickly became a popular crop.

It is believed that the native Indians also offered Columbus tobacco, but he discarded the gift as no more than fragrant dried leaves. Shortly thereafter, explorers Rodrigo de Jerez and Luis de Torres witnessed the actual smoking of tobacco while they were in Cuba. Jerez became a smoker himself, but when he returned to Europe and engaged in his habit, he was imprisoned by religious zealots for what was seen as an unholy activity. Nonetheless, smoking caught on in Spain while Jerez was serving his seven years of incarceration.

Potatoes were actually imported to both North America and to Europe from their native South America. In the late 1500s, European explorers discovered potatoes in South America and transported them to Spain, where the plants spread throughout Europe. Within about 50 years, Europeans transported the spuds back across the ocean to North America.

Other plants moved from Europe to the New World. On Columbus's second voyage to the Americas, he brought with him seeds for such plants as wheat, salad greens, grapes and sugarcane. Each grew well in the fertile American soil. Other explorers introduced additional agricultural plants, and settlers followed with European plants that served to remind them of their faraway homes.

Along with plant transportation, the Old World and New World exchanged many animal species. Europeans introduced such domestic animals as cattle, pigs, chickens, goats, and sheep to North America, with the intent of using the animal meat for food, and hides or wool for clothing. They also inadvertently brought pest animals and plants, such as rats and assorted weeds. Many of these pest species had disastrous consequences for endemic plants and animals in the New World.

Impact

Some of the plants introduced from the New World to the Old World had obvious beneficial aspects. When corn made its way to Europe, Spanish farmers began to plant the kernels. Their success of these crops prompted corn's quick spread throughout Europe and to other continents, where it became an important dietary component.

Potatoes likewise radiated throughout Europe. Following their initial journey from South America to Spain, the plant spread to England. English discoverers then shipped them back across the Atlantic to North America in the early seventeenth century. In both continents, potatoes thrived and became a nutritional staple on both sides of the Atlantic. The importance of the potato to European nations became dramatically apparent during the great potato famine of Ireland in the 1840s. Introduced to the country in the 1700s, potatoes had become so important—providing more than three-quarters of the calories in a commoner's diet—that when the crop fell to a fungus in the 1840s, more than one million people died of starvation and at least a million others left the country.

Tobacco's spread throughout Europe was similarly quick. Explorers introduced it to Spain in about 1500, smoking became popular shortly thereafter, and farmers began to cultivate the plant in 1531. By 1556, the plant appeared in France, and within a decade, it was present in England. Its use spread both as a recreational pastime and as an important medicinal herb. Its supposed curative properties ranged from headaches to toothaches, and lockjaw to cancer. By the end of the century, tobacco use had spread nearly around the globe despite mainly religious-based attempts to ban or control its cultivation and/or use.

The introduction of the new crop plants from Europe, along with the invasive European agricultural practices, changed the North American landscape. Native Americans planted corn and other crops sparingly and with little long-term effect on the environment. Europeans were more likely to plant large farm fields, often burning acres of forest and meadows to make way for agricultural plots. Where forests once stood and animals thrived, farm fields fragmented the land. Europeans systematically removed native plants to make way for the introduced crops. Animals that weren't killed outright by farmers often found the new cropland unsuitable, and were forced to move into smaller and smaller areas, facing increased competition and fewer resources. In addition, explorers and settlers unwittingly brought seeds from undesirable "weed" plants with them to the New World. These included dandelions, stinging nettle and crabgrass, which spread through fields and woodlands, and vied with the native vegetation for nutrients, sunlight and space to grow. Overall, the American ecosystem suffered, but the European settlers thrived.

Often the success of the European settlers also spelled misery for the American Indians. As Europeans came to rely more and more on the introduced crops and their productive but invasive agricultural practices, they began to see the Indians as either hindrances to the expansion of agriculture or as potential farm workers. The Indians were thus forced to work the land for the benefit of the Europeans or were pushed—sometimes violently—from land that Europeans began to claim as their own.

European explorers and settlers also obliviously or carelessly transported pest animals to the New World. An example is the European rat, which likely came to North America as a stowaway on Old World ships. The rats flourished. By 1609, for instance, records indicate the rats had become so numerous in Virginia that they devoured nearly the entire stored food supply in Jamestown, destroyed acres of crops and gnawed through the bark on the trunks of fruit trees.

Even the domestic animal introductions to North America weren't without detriment. On the positive side, settlers had a growing supply of meat, along with hides and wool for clothing. On the other hand, many of the domestic animals became feral. This new land had plenty of resources for them, but not the natural predators to keep their populations in check. These now-wild animals swiftly multiplied and competed with native animals, such as bear, deer and beaver, for dwindling resources. In turn, plant communities became devastated by the increased demand of more and more herbivores. In addition, indigenous animals found themselves fighting previously unknown diseases that were carried to the continent by the new domestic animals.

The exchange of plant and animal species brought change to the New World and the Old World. People on both continents gained much, including animal-produced clothing materials and bountiful new agricultural crops that would become dietary mainstays. They also lost a great deal. Author Frederick Turner (1993) wrote:

"There is something mythical about the New World described in the old travelers' tales. So much of it has vanished that it seems we are being told of some other, lost continent."

LESLIE A. MERTZ

Further Reading

Books

Crosby, A. *The Columbian Exchange.* Durham, NC: Duke University Press, 1967.

Cronon, W. *Changes in the Land.* New York, NY: Hill and Wang, 1983.

Drake, J., et al., eds. *Biological Invasions: A Global Perspective.* Chichester, U.K.: Wiley, 1989.

McKnight, B., ed. *Biological Pollution: The Control and Impact of Invasive Exotic Species.* Indianapolis: Indiana Academy of Science, 1993.

U.S. Congress, Office of Technology Assessment. *Harmful Non-Indigenous Species in the United States.* Washington, DC: U.S. Government Printing Office, 1993.

Other

Borio, G. Tobacco BBS (Internet bulletin board). http://www.tobacco.org/History/Tobacco_History.html.

Turner, F. "New World, Heartbreaking Beauty," (article). Reprinted in *Gale Environmental Almanac,* Detroit: Gale Research Inc., 1993: 3-10.

The Impact of European Diseases on Native Americans

Overview

Contact between Europeans and Native Americans led to a demographic disaster of unprecedented proportions. Many of the epidemic diseases that were well established in the Old World were absent from the Americas before the arrival of Christopher Columbus in 1492. The catastrophic epidemics that accompanied the European conquest of the New World decimated the indigenous population of the Americas. Influenza, smallpox, measles, and typhus fever were among the first European diseases imported to the Americas. During the first hundred years of contact with Europeans, Native Americans were trapped in a virtual web of new diseases. European diseases, seeds, weeds, and animals irreversibly transformed the original biological and social landscape of the Americas. By 1518, the Native American demographic catastrophe and the demands of Spanish settlers for labor led to the importation of slaves from Africa. Thus, the Americas quickly became the site of the mixing of the peoples and infectious agents of previously separate continents.

Background

Despite considerable progress in analyzing traces of the early migrations to the Americas, there is still doubt about the time of the arrival of the first humans. Some scholars believe that wandering bands of hunter-gatherers first crossed a land bridge from Asia to the New World about 10,000 years ago. Other evidence suggests that human beings might have arrived much earlier, but the earliest sites are very poorly preserved. In any case, migration from Siberia to Alaska might have served as a "cold filter" that screened out many Old World pathogens and insects. In addition, except for the late development of a few urban centers, primarily in Mesoamerica, population density in the New World rarely reached the levels needed to sustain epidemic diseases.

Centuries before Europeans arrived in the Western Hemisphere, advanced cultures and great cities had developed in Guatemala, Mexico, and the Andean Highlands. These areas were not free from disease, but accounts of pre-Conquest epidemics were generally associated with famines. Archeological evidence suggests that there were several periods of significant spurts of population growth and sudden declines in the Americas long before European contact. However, the impact of European diseases and military conquest was so profound and sudden that other patterns of possible development were abruptly transformed. Contact events involving the Aztecs, Mayans, and Inca civilizations were especially dramatic, primarily because Mexico and Peru had the highest population densities and the most extensive trade and transport networks in the Americas. Such factors provide ideal conditions for the spread of epidemic diseases.

Initial European reports about the New World speak of a veritable Eden, populated by healthy, long-lived people, who could cure illness with indigenous medicinal plants, and did not

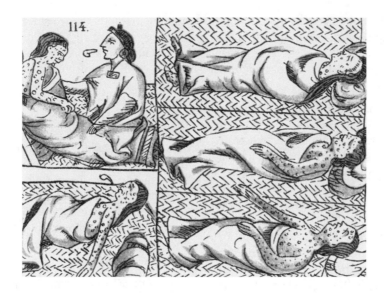

Smallpox victims as depicted by Fray Bernardino de Sahugún in a sixteenth-century book about Aztec history and culture. The disease devastated the indigenous peoples of the Americas. *(AKG London. Reproduced with permission.)*

know the diseases common in other parts of the world. Of course, the New World was not really a disease-free utopia. Diseases that were probably present in pre-Columbian America included American leishmaniasis, American trypanosomiasis (Chaga's disease), roundworms, pinworms, tapeworm, treponematosis, tuberculosis, arthritis, cancer, endocrine disorders, dysentery, pneumonia, rickettsial and viral fevers, goiter, and pinta.

Smallpox, measles, chicken pox, whooping cough, diphtheria, scarlet fever, trachoma, malaria, typhus fever, typhoid fever, influenza, cholera, bubonic plague, and probably gonorrhea and leprosy were unknown in the precontact period. The pre-Columbian distribution of certain diseases, especially syphilis and yellow fever, is still controversial, although some physicians believed that Europeans imported syphilis from the New World. Because yellow fever can be confused with malaria, dengue fever, or influenza, early accounts of such epidemics are unreliable. Modern immunological and entomological studies seem to have eliminated earlier claims that Mayan civilization was virtually destroyed by yellow fever, or that epidemics of this disease occurred in Vera Cruz and San Domingo between 1493 and 1496. Some epidemiologists contend that yellow fever was brought to the New World from Africa and that the first known epidemic occurred in Cuba in the seventeenth century.

Impact

Although a precise determination of the population of the Americas in 1492 is probably impossible, there is no doubt that contact with Euro-

peans resulted in a massive demographic collapse of the Native American population. The magnitude of the collapse and its causes remain controversial. Assessing the impact of European contact is a not simple matter because changes in population are the result of complex forces. Some scholars have argued that the devastating population decline in the New World was due primarily to imported diseases, while others have argued that the demographic catastrophe was the result of the chaos and exploitation that followed the Conquest. The rapid decline in the numbers of Native American peoples and the demands of Spanish settlers for labor, led to the establishment of the transatlantic slave trade by 1518. The Americas became the site of an unprecedented mixing of peoples and infectious agents from previously separate continents.

Although it is impossible to quantify with any certainty the impact of European contact on New World populations, estimates of the precontact population of the Americas have ranged from 8 to 30 million. Between 1492 and 1650 the Native American population may have declined by as much as 90% as the result of virgin-soil epidemics (outbreaks among populations that have not previously encountered the disease), compound epidemics, crop failures and food shortages.

The first Spaniards to reach the Caribbean islands found at least four distinct Indian cultures. Some recent estimates suggest that the pre-Columbian population of Hispaniola (modern Dominican Republic and Haiti) was close to 4 million. By 1508, fewer than 100,000 Indians remained. By 1570, almost all of the Caribbean Indians had disappeared, except for the Caribs in a fairly isolated area of the eastern Caribbean. A similar pattern occurred in Cuba, which was conquered in 1511.

Even before the first appearance of smallpox in the Caribbean, some epidemic disease seems to have swept through the islands and devastated the Indians of Hispaniola, Cuba, and the Bahamas. The first epidemic disease to attack the Caribbean Indians might have been swine influenza, brought to the West Indies in 1493 with pigs that Columbus had obtained from the Canary islands on his second voyage. Typhus may also have attacked the islands before the first known smallpox outbreaks in Hispaniola in 1518 and Cuba in 1519. Smallpox decimated the Arawaks of the West Indies, before making its way to Mexico with the Spaniards, and preceding them into the Inca Empire. The Spanish estimated that death rates among Native Americans from smallpox reached 25 to 50%. A similar death rate occurred in Europe, but the disease had essentially become one of the common childhood diseases. Therefore, most adults were immune to the disease. Other European diseases seem to have reached the islands before the measles epidemic of 1529. More recent examples of virgin soil outbreaks suggest that the mortality rate for swine influenza is about 25%, smallpox about 40%, measles about 25%, and typhus between 10 and 40% of the affected population.

With the establishment of the transatlantic slave trade by 1518, diseases from Africa were added to the epidemic burden imposed on Native Americans. The vector and virus for yellow fever probably appeared in San Juan, Puerto Rico by 1598. Better-documented outbreaks occurred on Barbados and Guadeloupe, Cuba, and the Gulf coasts of Mexico and Central America in 1647. Soon after the original human inhabitants of the islands were gone, the native plants and animals were forced to compete with Old World invaders. The peoples of the present day Caribbean trace their ancestry principally to Asia, Europe, and Africa. Slaves were imported as early as 1502, but by 1518 the decline in labor supply had become so acute that King Charles I of Spain approved the direct import of slaves from Africa. However, the Africanization of the islands was the result of the "sugar revolution" that began in the seventeenth century, along with the importation of epidemic yellow fever.

The Empire of the Aztecs was the first American civilization to encounter the Spanish and the first to be destroyed. Several factors, including devastating epidemics of smallpox, which killed many Aztec warriors and nobles, facilitated the Spanish capture of the Aztec capital in 1521. Native Americans came to see this smallpox epidemic as a true turning point in their history. The time before the arrival of the Spanish was remembered as a veritable paradise, free of fevers, smallpox, stomach pains, and tuberculosis. When the Spanish came, they brought fear and disease wherever they went. Mayan civilization had already experienced a long period of decline by the time it encountered European explorers and invaders, but the Inca Empire was at its peak when the Spaniards conquered it in 1532.

European diseases probably preceded European contact in the Andean region. A catastrophic epidemic, which might have been smallpox, swept the region in the mid-1520s, killing the Inca leader Huayna Capac and his son. Subsequent epidemics struck the region in the 1540s, 1558, and from the 1580s to 1590s. These waves of epidemic disease might have included smallpox, influenza, measles, mumps, dysentery, typhus, and pneumonia. The precise impact of smallpox and other European diseases throughout the Americas is difficult to document or comprehend. However, studies of more recent and limited virgin soil outbreaks clearly demonstrate how small a spark is needed to create a great conflagration in a native population.

LOIS N. MAGNER

Further Reading

Ashburn, Percey Moreau. *The Ranks of Death: A Medical History of the Conquest of America.* New York: Coward-McCann, Inc., 1947.

Cook, Nobel David. *Born to Die: Disease and New World Conquest, 1492-1650.* (New Approaches to the Americas.) New York: Cambridge University Press, 1998.

Crosby, Alfred W. *The Columbian Exchange: Biological and Cultural Consequences of 1492.* Westport, CT: Greenwood Press, 1972.

Crosby, Alfred W. *Germs, Seeds, and Animals: Studies in Ecological History.* Armonk, NY: M.E. Sharpe, 1993.

Denevan, William M. *The Native Population of the Americas in 1492.* Madison, WI: University of Wisconsin Press, 1976.

Dobyns, Henry F., with W.R Swagerty. *Their Number Became Thinned: Native American Population Dynamics in Eastern North America.* Knoxville, TN: University of Tennessee Press, 1983.

Henige, David. *Numbers from Nowhere: The American Indian Contact Population Debate.* Norman, OK: University of Oklahoma Press, 1998.

Kunitz, Stephen J. *Disease and Social Diversity: The European Impact on the Health of Non-Europeans.* New York: Oxford University Press, 1994.

Ramenofsky, Ann F. *Vectors of Death: The Archaeology of European Contact.* Albuquerque, NM: University of New Mexico Press, 1987.

Reff, Daniel T. *Disease, Depopulation, and Culture Changes in Northwestern New Spain, 1518-1764.* Salt Lake City, UT: University of Utah Press, 1991.

Life Sciences & Medicine

1450-1699

The Appearance of Syphilis in the 1490s

Overview

The earliest references to the disease now known as syphilis come from the 1490s, when it broke out among French troops besieging the city of Naples. Initially known as *morbus gallicus* (the French Disease), it soon became epidemic throughout Europe. The disease left visible and disfiguring signs of infection, which led to social stigmatization. Most damaging in its late stages, it often produced severe disabilities and even death. Believed to be a new disease imported from the Americas, syphilis helped challenge traditional ideas of disease causation and spread. While most early modern medical authorities believed syphilis was a new disease, scholars today continue to debate its origins and antiquity.

Background

In 1494, the Italian city-state of Milan appealed to King Charles VIII of France for military assistance. Seizing the opportunity, Charles VIII invaded with mercenary troops. Opposed by Italian forces from Florence, Venice, and the Papal States, as well as by those from Spain, Charles initiated what became known as the Italian Wars. The French succeeded in capturing Naples in 1495 after a long siege. Disease broke out among the troops in the midst of their subsequent celebrations, and when they were sent home many returned to their native lands bringing this new disease with them. By the end of 1495, all of Europe seemed to be infected with the French Disease, which caused painful aches and fevers, disfiguring sores, and often death. Sometimes called the Great Pox, to distinguish it from smallpox, the disease was also variously called the Spanish, Neapolitan, Polish, Russian, British or Portuguese disease, depending on the nationality of the speaker. Its common name today, syphilis, comes from a poem published in 1530 by the Italian physician and humanist, Girolamo Fracastoro (c. 1478-1553). Entitled

Syphilis sive morbus Gallicus (Syphilis or the French Disease), the poem tells the story of a young shepherd named Syphilis who was infected with a "pestilence unknown" after offending Apollo. It goes on to describe the course and symptoms of the disease, vividly describing events such as when "unsightly scabs break forth and foully defile the face and breast" and " a pustule resembling the top of an acorn, and rotting with thick phlegm, opens and soon splits apart flowing copiously with corrupted blood and matter." Although Fracastoro's term did not catch on immediately, by the nineteenth century *morbus gallicus* was known as syphilis.

Modern science has shown that syphilis is caused by a corkscrew-shaped bacterium, or spirochete, known as *Treponema pallidum*. It is one of four treponemes that infect humans, though the only one that is currently found worldwide. The other three cause the diseases pinta (*T. carateum*), a skin disease endemic today in Central and South America; yaws (*T. pertenue*), which affects both skin and bones and is found today in warm, humid climates including Africa, Southeast Asia, the Caribbean and South America; and endemic syphilis (*T. pallidum endemicum*), which is similar to yaws, but is found only in warm, arid climates such as Saharan Africa. Interestingly, despite causing four clinically distinct diseases, these treponemes are indistinguishable in the laboratory. All four are transmissible, though only venereal syphilis is passed through sexual contact (or through *in utero* infection of an unborn child by a syphilitic mother).

The first stage of venereal syphilis produces a painless lesion in the genitals, which will heal of its own accord within weeks. After a six-to-eight-week latent period, the second stage is generally marked by a rash, fever, and swollen lymph nodes. These symptoms also disappear spontaneously after a few weeks. The third stage of syphilis does not occur in all cases, and then only after a lengthy latent period, often as much as 20 years. During this latency, the disease continues to be transmissible. Tertiary syphilis is the most damaging stage of the disease. Most typically, it produces *gumma*, small rubbery lesions that can occur in all parts of the body, including on internal organs. The disease can affect the cardiovascular system, the nervous system, the spinal cord, and even the brain, producing a form of insanity known as *dementia paralytica*. Today, syphilis is treatable with antibiotics, and tertiary syphilis has largely disappeared.

Impact

The withdrawal of the French troops and dispersal of their multinational mercenaries is credited with aiding the quick spread of the disease throughout Europe. After this initial outbreak, syphilis remained epidemic throughout the sixteenth century. It produced violent symptoms and often resulted in death. By the seventeenth century, the disease seems to have become less virulent, settling down into the chronic disease still suffered in many areas today. While the emergence of syphilis as an epidemic disease did not have long lasting economic or demographic effects, it did have important social and medical repercussions.

Socially, syphilis led to increased stigmatization and fear. Joseph Grunpeck, a sufferer in this early epidemic, described it in his autobiography as "a disease which is so cruel, so distressing, so appalling that until now nothing so horrifying, nothing more terrible or disgusting, has ever been known on this earth." Quickly recognized as a venereal disease, so called because of its association with Venus, goddess of love, syphilis became a disease of vice, marking its sufferers as corrupt and licentious. This corruption was reflected in the very faces of the infected, including noses destroyed by both the disease and the mercury used to treat it. One result, then, was an increased demand for an early form of plastic surgery. Gasparo Tagliacozzi, a Bolognese surgeon in the late sixteenth century, acquired an international reputation for his technique of grafting skin from the upper arm onto the nose. His 1597 work, *De curtorum chirurgia per insitionem* (On the Surgery of the Mutilated by Grafting), provided a precise guide to the technique. A second repercussion of syphilis was that these visible signs reconfirmed the ancient association of sexual activity with sin, which in turn led to increased state regulation and restriction on licentious entertainment, such as brothels and steam baths.

Treatments for syphilis included bleeding, mercury treatments, and guaiac tree bark. Bleeding and mercury treatments had the same goal—to remove bad humors from the body. Mercury had long been used by Europeans in salves for skin diseases, and was soon applied for syphilis as well. Later used in baths or taken orally, mercury treatments prompted heavy salivation, which was believed to be a sign of the effectiveness of mercury in pulling poisons out of the body. In reality, it was a sign of mercury poisoning, along with other side effects such as loss of hair, bleeding gums and loose teeth. The third common treatment, guaiac wood, was from the New World. This tree (*Guaiacum officinalis*) grew only in the Americas and decoctions made from its bark or wood were used by natives to treat a variety of skin problems. Bolstering this trade was a belief, common at the time, that for every disease God had created a cure nearby. Many believed that the disease had been brought to Europe from the Americas, and therefore that the cure was sure to come from there also. Experience showed this treatment less effective than mercury, however, and by the late sixteenth century it had fallen out of favor.

The belief that syphilis was a new disease was common from the fifteenth century on, though there were some disagreements in the medical community over its origins. The earliest reference to a New World origin for syphilis comes from Gonzalo Fernández de Oviedo, who wrote of it in his 1526 work, *Summary of the Natural History of the Indies*. Another important support for this belief came from Ruy Díaz de Isla, who argued in his *Tractado contra el mal serpentino* (Treatise on the Serpentine Malady), published in 1539 but written as early as 1505, that the disease was brought back by Columbus and his men, and that he had treated one of them in 1493. As some of these men were known to be fighting at the siege of Naples, the spread of syphilis was blamed on them. Many other well-known and respected Spanish authors, such as Nicolás Monardes, Bartolomé de Las Casas, and Bernardino de Sahagún, all agreed that the disease had long been known in the Americas, and only recently brought to Europe. Arguments to the contrary, however, came as early as 1497, when Niccolò Leoniceno (1428-1524) argued in his *De epidemia quan vulgo morbum gallicum vocant* (On the Epidemic Vulgarly Called the French Disease) that ancient authors had indeed discussed this malady, but poor translations hindered modern understanding.

The idea that syphilis was a new disease also played a part in challenging some tradition-

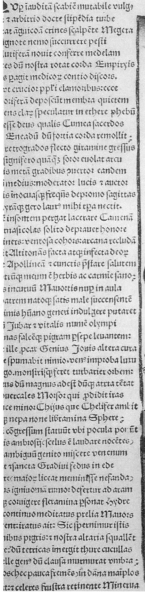

The Syphilitic by Albrecht Dürer, depicting the large impact the disease had on European society in the 1500s. (Wellcome Library, London, England. Reproduced with permission.)

al medical theories. The sixteenth century saw the beginnings of the scientific revolution, which may be summarized as a change from relying entirely on ancient authorities to relying on observation and experiment. The advent of recent new diseases, such as bubonic plague in the fourteenth century and syphilis in the fifteenth, prompted a reevaluation of traditional Galenic approaches to health and disease. Ancient authorities were found to be useless, as they had neither descriptions of these diseases nor treatments for them. Doctors were thus forced to adapt old theories or rely on their own experiences and observations. By the mid-sixteenth century, these diseases prompted a new theory of disease transmission, that of contagion, which was developed by the same Girolamo Fracastoro who had named syphilis. He argued for tiny causative agents, or "seeds", rather than internal humoral imbalance to explain disease. The process of changing or discarding ancient knowledge was a very slow one, and Fracastoro's contagion theory was not immediately embraced. Nonetheless, in hindsight we can see the emergence of syphilis, and reactions to it, as having a role in these larger ongoing changes.

Questions on the origins and antiquity of syphilis were revived during the Enlightenment, and the debate continues among many scholars today. Modern scholars have produced both skeletal and linguistic evidence to support the contention that syphilis existed in the Americas prior to contact with Europeans. What remains unclear is whether syphilis existed in Europe before Columbus's voyages. Archeologists have found what they believe to be skeletal evidence for syphilis in Europe prior to the age of discovery, but this evidence is not conclusive and remains subject to interpretation.

Literary evidence is also used by both sides. Modern proponents of the view that it was a new disease point to the large numbers of treatises written in the early sixteenth century. References to and descriptions of syphilis seem to suddenly explode from the 1490s onward, while earlier writings do not directly address a disease recognizable as syphilis. Certainly, *morbus gallicus* acted like a new disease—it spread quickly, caused extremely debilitating symptoms, and took a high toll in lives. All early chroniclers of the disease asserted it was new, and witnesses repeatedly declared it to be unlike anything they had previously seen.

Opponents of the American theory argue that syphilis existed in earlier centuries, but was simply not differentiated from other diseases. Most likely, it was encompassed by leprosy, another disfiguring disease often associated with corruption and uncleanliness. Leprosy was a common diagnosis in medieval Europe, one that has been shown to embrace more than just the medical condition implied by the term today. Most significantly, there are numerous references in medieval literature to "venereal leprosy", which was believed to occur from sexual contact with one already infected. True leprosy, however, is not sexually transmitted, and is transmissible only with years of close contact. In addition, leprosy was often treated with a mercury-based salve known as Saracen ointment. While mercury does little for true leprosy, it became the primary treatment for syphilis after the fifteenth century. Finally, at the same time that syphilis became epidemic, leprosy was on the wane and leper houses were being closed down, sending any remaining residents out into society. All this evidence is taken to show that leprosy simply masked the existence of syphilis in the Old World.

A third approach to the question, which focuses on the biology of syphilis, has been put forth in the last half century. Known as the unitarian theory, it takes into account the enormous similarities of the four human treponematoses, to argue that all four originated from a common ancestral spirochete and simply adapted to differing climates and societies, producing different diseases. Thus, these treponemes existed in both the Old and New Worlds, and simply evolved over time alongside humans. The oldest of the four diseases, pinta and yaws, are both transmissible through direct skin to skin contact, and flourish in warm humid climates where such contact is common. Endemic syphilis emerged later as a response to dryer regions, where the treponeme could not survive on the surface of the skin and was forced to use the body's mucosal tissues to remain moist. Venereal syphilis, according to this theory, was simply another adaptation to colder climates and heavier clothing, in which the spirochete found warmth and moisture by retreating further into the interior of the body, relying on sexual contact for transmission. The epidemic of venereal syphilis at the end of the fifteenth century is thus accounted for not by the introduction of a new causative agent, but by social changes, including population expansion, migration, and changing morality codes that increased sexual activity and enabled the treponeme to adapt and flourish. This theory has been well received by many modern scholars, and shows a great deal of promise for solving the riddle of syphilis's origins.

KRISTY WILSON BOWERS

Further Reading

Books

Arrizabalaga, John, John Henderson and Roger French. *The Great Pox: The French Disease in Renaissance Europe.* New Haven: Yale University Press, 1997.

Crosby, Alfred W. *The Columbian Exchange: Biological and Cultural Consequences of 1492.* Westport, CT: Greenwood Publishing Company, 1972.

Oriel, J.D. *The Scars of Venus: A History of Venereology.* London: Springer-Verlag, 1994.

Watts, Sheldon. *Epidemics and History: Disease, Power and Imperialism.* New Haven, CT: Yale University Press, 1997.

Periodical Articles

Baker, Brenda J. and George J. Armelagos. "The Origin and Antiquity of Syphilis." *Current Anthropology* 29 (1988); 703-734.

Guerra, Francisco. "The Dispute over Syphilis Europe versus America."*Clio Medica* 13 (1978): 39-61.

The Development of Zoology

Overview

Zoology, the branch of biology that studies animals, seeks to understand the sum total of all the properties of animals and animal populations. As a discipline, zoology is similar to others with major subdivisions that include anatomy, physiology, genetics, and interrelationships.

Much of the early information about human anatomy came from the dissection and study of animals, although some efforts were made to understand and classify animals. It was during the Renaissance that the study of zoology began to separate from human anatomy, as great artists who sought to understand the makeup of both men and animals emerged. Great natural scientists, such as Konrad Gesner (1516-1565), recognized as the father of zoology, developed the field as a scientific inquiry. Other investigators, such as Guillaume Rondelet (1507-1566) and Ulisse Aldrovandi (1522-1605), contributed accurate observations of animals. Natural philosopher and theologian John Ray (1627-1705) also sought to understand and classify all known animals.

The classification and physiological studies by these early naturalists provided the foundation upon which the zoology of the nineteenth century was unified by the theory of evolution. Comparison of animals allowed an understanding of how various animals might have developed. Zoology in the late twentieth century developed as a major force behind the understanding of total interrelationships and the ecology movement.

Background

Philosophers and thinkers throughout history have looked at animals and attempted to understand them by classifying and relating anatomy and physiology. Aristotle (384-322 B.C.) first attempted a comprehensive classification of animals. His organization and rational development of thought sought to include all things and established an area of natural philosophy that included living things. He was the first to establish some type of hierarchy of animals based on the logic of structure. Roman scholar Pliny the Elder (A.D. 23-79) wrote a major work on natural history. The great physician Galen (129-199?) dissected only animals for his studies of human anatomy, and his works became the standard for use in medicine throughout the Middle Ages.

In the twelfth century, a remarkable development took place in the southern part of Italy, beginning in the town of Salerno. The idea emerged that medical practice should be made a division of the "natural" part of philosophy, as Aristotle had done. By the mid-thirteenth century natural philosophy was one of the liberal arts that all scholars were required to study, and medicine was a part of this. In Italy major universities began to find in the ideas of Galen and Aristotle an encouragement to investigate. Their inquiries were limited to the intellectuals and wealthy, and did not impact the daily lives of people. These were merely seeds that would sprout at the end of the medieval period and bloom during the Renaissance in the fifteenth and sixteenth centuries.

The same driving forces that were part of the discovery period were present in many areas. Explorations of the New World by Christopher Columbus (1451-1506), the Cabots, and others told of new and unusual plants and animals. In 1453 Byzantium, or Constantinople, the capital of Eastern Christendom, fell to the Turks, forcing Greek scholars to move from the East to the West, bringing with them knowledge and access to ancient works. The explosion in the liberal arts with the discovery of Greek manuscripts brought new interest in all areas of classical thinking. Most important of all was the invention of the printing press with movable type by Johann Gutenberg (1398?-1468) around 1455. This enabled scholars to write about their findings and ideas, and using woodcut drawings, they could illustrate what they saw. Add the new availability of paper and the beginning of writing in the language of the people, and the interest in learning would spread rapidly.

Impact

The intellectual energy of the fifteenth and sixteenth centuries touched the field of zoology. The revival of classical works was evoked by artists seeking to realistically and correctly show the body. Leonardo da Vinci (1452-1519) was motivated to study animals, comparing them to the physical form of man. He was the first to describe the homology, or the arrangements of the bones and joints, of a horse. He noted how they were alike and how they differed from the human. Homology would become an important concept in classifying distinct units, and later play a part in the study of evolution. Although Andreas Vesalius (1514-1564), the great anatomist and illustrator, encouraged the new spirit of investigation by dissecting humans, he also used animal parts to show structures such as the kidney. These artists and early anatomists promoted knowledge through dissection and a new spirit of investigation. Francis Bacon (1561-1626) gave increasing emphasis to direct observation and experiment, which caught on in the seventeenth century.

Zurich, Switzerland, one of the centers of the Protestant Reformation, also emerged as a

THE LAST OF THE AUROCHS

Sixteen twenty-seven is the last year that anyone is known to have seen an aurochs in the wild. The last one was seen somewhere in Europe. In that year, cattle became fully domesticated animals; the aurochs was the wild ancestor of all domesticated cattle. The aurochs used to roam freely across large parts of Europe, Asia, and northern Africa. They were large and relatively docile herd animals that were well suited to domestication, and they were domesticated at some time in human prehistory, probably in what is now the Middle East. When domesticated, aurochs entered the service of humanity as draft animals, sources of milk and dairy products, and sources of meat. In fact, cattle today are among the more valuable food animals, providing us with butter, cheese, milk, and, of course, meat. To provide these benefits, we have made the aurochs over into a bevy of cattle breeds, each a specialist in its own way. And, in so doing, we have increased the productivity of our farms.

P. ANDREW KARAM

center for the study of the natural sciences or natural philosophy. In this town Konrad Gesner was born and became the godson of Ulrich Zwingli, the great Protestant reformer. Gesner developed as a person of many talents and interests.

When his godfather Zwingli was killed in the battle of Karpel in 1531, he went to Strasbourg to study theology, but soon became tired of this subject and turned to medicine. An eager scholar, he studied at Bourges and Paris, and received his doctorate at Basel in 1541. As a

scholar he was very interested in botany, which at the time was closely related to medicine because of the use of plants and herbs in the treatment of disease.

The search for plants led to the next step. Gesner became an ardent traveler and explored extensive areas. He climbed Mount Pilatus overlooking Lake Lucerne and brought back volumes of information on both alpine plants and animals. He found the love of his life in natural history.

Gesner was the first botanist to grasp the importance of floral structures to establish a systematic key of classification of plants. He drew over 1,500 plates himself for his *Opera botanica*, published in 1551 and 1571.

Several decades before, Aelian had conceived a work on the history of animals. Patterning his organization after the older document, Gesner worked on *Historia animalium* for years, ending with a 4,500-page volume. The work was immediately recognized by scholars. Using classification similar to his work with plants, Gesner used animal physiology and structure to group specimens. Some consider him to also be the father of veterinary medicine.

Since Gesner was the first naturalist to sketch fossils, he is considered to the first paleontologist. He also studied crystallography and was one of the first to include printed plates of crystals in his works. His drawings were a major addition to texts. He also had many interests. For example, while stories of sea serpents and monsters had been reported for centuries, he printed the first account of such a monsters—based on hearsay—but with a picture.

Guillaume Rondelet was a French naturalist and physician who contributed substantially to zoology through his description of marine animals, primarily those in the Mediterranean Sea. Rondelet published *Libri de Piscibus Marinis* (1554), or *Book of Marine Fish*, which contained detailed descriptions of 250 kinds of marine animals. The book had illustrations of each item. He also included whales, marine invertebrates, and seals. He was a professor of anatomy at the University of Montpellier and served the cardinal of that area.

Ulisse Aldrovandi was a nobleman from Bologna who studied mathematics, Latin, law, and philosophy, then went to Padua to study medicine. He became a professor at Bologna and presented very interesting lectures on natural history as a systematic study. He published numerous works, one of which included a detailed observation of day-to-day changes in the developing chick embryo. His museum of biological specimens was classified according to his own system and left to the city of Bologna at his death. He contributed to the development of animal taxonomy by using structure and formation.

The natural theologians of the period were part of the search for organization of zoology. A person dedicated to taxonomy is interested in establishing an orderly account of species. This fit well into the scheme of the natural theologians who believed in an orderly world. John Ray, a devout Puritan, was an academic scholar at Cambridge. With the restoration of the monarchy, Ray hit unfortunate times and was dismissed because he refused to sign the Act of Uniformity in 1662. Some prosperous friends supported him for 43 years while he continued to work as a naturalist. In 1660 Ray began to catalog plants growing around Cambridge and, after completion of that small area, explored the rest of Great Britain. A turning point in his life occurred when he met Francis Willughby, a fellow naturalist who convinced him to undertake a study of the complete natural history of all living things. Ray would investigate all the plants and Willughby, the animals.

The two searched Europe for flora (plants) and fauna (animals). In 1670 Ray had produced his *Catalog of English Plants* when Willughby suddenly died, leaving Ray to finish both projects.

He published *Ornithology* and *History of Fish,* giving all the credit and recognition to Willughby.

Ray correlated science with natural theology. He believed that it was obvious that form and function in organic nature demonstrated there must be an all-knowing God.

Many other people who made famous discoveries were interested in animals. William Harvey (1578-1657) demonstrated the circulation of the blood and function of the heart, arteries, and veins. The invention of the microscope by Anton van Leeuwenhoek (1632-1723) assisted in comparing fine structures that previously could not be seen with the unaided eye. Jan Swammerdam (1637-1680) and Marcello Malpighi (1628-1694), who discovered the role of the capillaries, added to the body of information about animals.

The work of the great naturalists culminated in the work of Carl Linnaeus (1707-1778). His binomial system of nomenclature (genus and species) and *Systema naturae* (1735) marked the beginning of the modern system of classification and helped define zoology as a distinct discipline of study.

EVELYN B. KELLY

Further Reading

Porter, Roy. *The Greatest Benefit to Mankind: A Medical History of Humanity.* New York: W.W. Norton, 1997.

Wear, A., R. K. French, and I. M. Lonieed, eds. *The Medical Renaissance of the Sixteenth Century.* Cambridge: Cambridge University Press, 1985.

Advances in Botany

Overview

With the beginning of the Renaissance in the fifteenth century came a new interest in the natural world, including the world of plants. This was manifested in closer observation of plant structures, the identification of new species, and the formation of botanical gardens and plant collections to serve as resources for botanists. In the seventeenth century, interest in experimentation led to early work on plant physiology, or function; and the development of the microscope prompted increased attention to the careful study of plant anatomy.

Background

In 1455 the printing press was a new invention, and it was one that had perhaps a greater influence on botany than on other sciences. While the printing press made it much easier to communicate all scientific ideas, it was particularly important in botany because it allowed for the accurate reproduction of images of plants. Until that time, illustrations had to be made by hand, and each time an illustration was copied there was the likelihood that inaccuracies would be introduced. Also, as images were copied repeatedly they tended to become simpler and carry less information. This meant that it was often impossible to identify a plant species from an illustration. Since two plant species may look very similar to each other, only accurate visual portrayal could make the differences apparent, since it was often difficult to present these differences concisely in words.

But it was not until 1530, almost a hundred years after the invention of the printing press, that a book on plants was published with accurate illustrations. This was Otto Brunfels's (1489?-1534) *Herbarum vivae eicones* (Living Images of Plants); it was followed 12 years later by Leohnard Fuchs's (1501-1566) *De historia stirpium*. These are considered among the best illustrated botanical books ever produced and set the standard for botanical illustration. That is why Brunfels and Fuchs, along with another author, Hieronymous Bock (1498-1554), are regarded as the "fathers of German botany." All three relied to some extent on the writings of ancient Greeks and Romans on plants, including Theophrastus (c. 370-285 B.C.), the great student of Aristotle (384-322 B.C.). It was only as the Renaissance progressed that botanists began to draw more on their own observations and less on the learning of the past.

During this time period interest in botany was directly related to attempts to improve the study of medicine because plants were the sources of most of the medications in use. This interest also led in 1545 to establishing the first botanical garden. Located in Padua, it was affiliated with the medical school at the university there. Gardens were also founded in other Italians cities, including Pisa where Luca Ghini became the first professor of botany. He established what many consider the first herbarium, a collection of dried plant specimens. This was an important innovation because such collections provided study materials for future botanists, and like book illustrations, were a way to document plant characteristics.

This documentation became important because by the beginning of the sixteenth century, the great age of exploration was in full swing, and plant specimens were arriving in Europe from around the world. The increased number of species, many of which had new and unusual characteristics, stimulated interest in classification as a way to organize all this new information. Andrea Cesalpino (1519-1603), a pupil of Ghini and one of the most important botanists of his time, was the first to develop a natural system of classification, that is, a system based not on some arbitrary characteristic such as the number of petals in a flower, but rather, on overall similarities. Cesalpino also produced much more detailed plant descriptions than had been common to that time.

Joachim Jung (1587-1657) built on Cesalpino's work and systematically presented the different parts of the plant in a much more organized way than had been done previously. He also made advances in the study of flower structure, emphasizing all the different variations that can be found in the parts of the flower. One of Jung's greatest contributions was the development of a precise vocabulary making it possible for botanists to communicate information about plant structures more accurately. In England, John Ray (1628-1705) provided the greatest advances in plant classification in the seventeenth century, and his work served as the basis for later natural classification systems. All this work on classification and the study of plant anatomy laid the groundwork for other types of botanical investigations, including experimentation.

Impact

In the seventeenth century, experimentation became a more common approach in scientific inquiry, and one of the most famous botanical studies of the age was done by Jean-Baptiste van Helmont (1577-1644). He planted a five- pound willow tree in 200 pounds of dry soil; all he did to the tree for the next five years was water it and collect the leaves that it shed each fall. At the end of the five years, he added up the weight of all the leaves, plus that of the tree, and it came to 169 pounds, three ounces. When he dried the earth, it had only lost two ounces. So about 164 pounds of tree weight seemed to be derived from water. What van Helmont did not know, and what would not be discovered for another hundred years, was that along with water, the tree was also absorbing carbon dioxide from the air. Using the energy of sunlight in the process of photosynthesis, carbon dioxide and water are converted to sugar, which is then used to provide the energy and building materials for growth. We now also know that the two ounces missing from the soil in which the tree was growing represented minerals that were also essential to growth. While van Helmont's experiment failed to explain at all aspects of photosynthesis, it was significant because it indicated that most of the tree's material did not derive from the soil and, perhaps most importantly, it emphasized the importance of careful measurement in scientific inquiry.

It was also in the seventeenth century that the microscope was developed. This instrument opened up a whole new world to investigation; and those interested in the living things, including botanists, soon took advantage of its power. It was Robert Hooke (1635-1703), who compared the structure of a piece of cork he observed under a microscope to cells in a prison, thus coining the word used to describe what later came to be known as the basic unit of living things.

Though the lenses in early microscopes were not nearly as powerful or distortion-free as those available today, they still made it possible for botanists to discover a great deal of structure within plant parts such as leaves, stems, and flowers. The two men who were responsible for the development of the science of plant anatomy, Marcello Malpighi (1628-1694) and Nehemiah Grew (1628-1711), both used magnifying lenses in their work.

Malpighi, who also did significant work in animal anatomy and discovered the circulation of blood through the tiny blood vessels called capillaries, investigated several parts of the plant. He made a thorough study of the stem of woody plants, including the bark. He found that the wood of trees increased by the yearly addition of layers of cells under the bark. He also discovered fibers in stems that he hypothesized carried water up to the leaves. His detailed descriptions called attention to the complex structure of the stem and led to the eventual discovery of two sets of conducting tubules. Malpighi also discovered the stomata, the microscopic pores on the surface of leaves, and he understood that air and water passed through them. In addition, he accurately described the layers of the seed coat as well as the process of germination by which the plant embryo within the seed enlarges, breaks through the seed coat, and develops into a seedling or young plant.

At the same time that Malpighi was working in Italy, Nehemiah Grew was studying plant anatomy in England, and though they were often investigating similar problems, they maintained very cordial relations. Their collaboration

is seen as a model of scientific cooperation at its best. Grew studied the structure of plant tissues and produced exceptionally clear and accurate drawings of the cellular organization found in these tissues. As did Malpighi, he investigated the development of buds. Grew also did extensive work comparing the structure of flowers in different species, but he did not understand the role of flower parts in sexual reproduction, nor did anyone else at that time.

It was at the very end of the seventeenth century that Rudolph Camerarius (1665-1721) identified the flower structures involved in reproduction. He observed that a female mulberry tree bore fruit even if there were no male trees in the vicinity, but that the fruit was seedless and thus wouldn't produce plants. He argued that the stamen (found on the male trees) contained the male sex organ and produced a yellow powder, the pollen, which was needed for seed production. He further observed that the seeds were produced in the carpel, the structure that contained the female sex organ on the female trees.

Camerarius's work spurred a great deal of further research on fertilization and seed production in the next century, just as great strides were also made in plant anatomy and in physiology, including the processes by which sap flows through stems and leaves absorb and release gases. The mid-eighteenth century also witnessed the work of Carolus Linnaeus (1707-1778), a botanist who developed a classification scheme that served as the basis for all future work in classification.

MAURA C. FLANNERY

Further Reading

Arber, Agnes. *Herbals: Their Origin and Evolution 1470-1670*. Cambridge: Cambridge University Press, 1986.

Blunt, Wilfred, and Sandra Raphael. *The Illustrated Herbal*. New York: Thames and Hudson, 1994.

Gabriel, Mordecai, and Seymour Fogel, eds. *Great Experiments in Biology.*. Englewood Cliffs, NJ: Prentice-Hall, 1955.

Iseley, Duane. *One Hundred and One Botanists*. Ames, IA: Iowa State University Press, 1994.

Magner, Lois. *A History of the Life Sciences*. 2d ed. New York: Marcel Dekker, 1994.

Renaissance Botanical and Zoological Gardens

Overview

Throughout the Middle Ages, Europeans regarded plants and animals from a very pragmatic viewpoint. Plants were seen as sources of food, medicines, and wood. Animals were valued as food and as aids to mankind by providing power (oxen and horses) and help in hunting (dogs). But neither plants nor animals were studied scientifically until the dawn of the Renaissance and the Age of Exploration. As thousands of previously unknown specimens poured into Europe from around the world, the science of botany began to evolve. Plants were carefully examined, classified, and exchanged between scholars. Scores of new animal species were discovered and brought back to Europe. One result of this activity was the founding of numerous botanical and zoological gardens, establishments that introduced these exotic species to the general public and helped break down many medieval myths.

Background

In medieval Europe, botany (the science of plants) was essentially limited to the study and copying of writings from the ancient Greek and Roman worlds. In particular, scholars relied heavily on the work of Pedanius Dioscorides, a Greek doctor who served in the Roman army, about whom virtually nothing more is known today. Sometime around A.D. 60 he published *De Materia Medica*, which was basically a summary of Greek pharmacological knowledge. In it, Dioscorides discussed approximately 1,000 drugs, three-fifths of which were derived from plants. Medieval interest in plants was so narrow that for the next fifteen centuries the *Materia Medica* remained the authoritative botanical text in Europe. Unfortunately, its descriptions of plants were quite short, often making it impossible for the reader to distinguish one species from another.

Materia Medica was extremely influential, however, because Dioscorides repeatedly stressed the importance of plants as sources of medicines. Throughout the entire Middle Ages, scholars studied plants solely for their medicinal properties. Their books, which are known as herbals, relied on Dioscorides as their primary source of knowledge. But as the *Materia Medica*

was repeatedly copied by hand, errors inevitably occurred and were incorporated into subsequent copies. Even more seriously, since botany relies on illustrations of plants to propagate information, copies of the original drawings degenerated to the point that they were replaced by formalized and increasingly fictitious pictures. The errors in medieval herbals seem ludicrous today. For example, they described trees whose leaves turned into birds if they fell on the ground, and into fish if they dropped into water. They also contained drawings of the Scythian lamb said to exist in Asia, which had the roots and trunk of a small tree but the body of a lamb in place of branches and leaves.

Medieval interest in animals was similarly restricted. While zoos existed in the Greek world as collections of animals for study as well as for exhibition, by the time of the Roman Empire animals were captured and brought to Europe either to be slaughtered by gladiators or as exotic food for the jaded tastes of epicureans. This declining interest continued in the millennium-long medieval period. The few animal collections that existed then were simply menageries, made with no scientific purpose in mind. Henry III (1207-1272) of England kept a menagerie in the Tower of London, which included camels, lions, leopards, and the first elephant ever seen in the country. The one medieval menagerie that did lead to some scientific knowledge was that of Frederick II (1194-1250), the king of Sicily and the Holy Roman Emperor. Frederick made several discoveries about animal behavior. For instance, he learned that the lead cranes flying in V formations switch places during their flight. Other than Frederick's work, however, the pre-Renaissance menagerie provided little scientific information.

The fifteenth century witnessed the Middle Ages finally coming to an end. One sign of this was the growing desire to explore the non-European world. A variety of motives lay behind the Renaissance's exploring zeal, among them: the search for new trade routes to Asia that would eliminate the Moslem middlemen; the desire to spread Catholicism outside of Europe after the Protestant Reformation in the early sixteenth century split the medieval church; and the strong sense of curiosity and adventure that permeated Renaissance thought. By the early 1500s, European navigators had sailed around Africa to India and were exploring the West Indies and the American continents. China and Japan were soon reached. When the Age of Exploration voyagers returned from these far-flung lands, they carried with them new species of plants and animals never before seen in Europe. These new species triggered a huge scientific enterprise, particularly in botany.

Impact

Among the new species flooding into Europe were plants such as maize, the potato and tomato, cassava, tobacco, vanilla, chocolate, pineapples, sunflowers, rhubarb, and tulips. Newly discovered animals included llamas, bisons, turkeys, iguanas, guinea pigs, toucans, and the anaconda. Even though the Italian city-states were not directly involved in the explorations, examples of these newly discovered species were quickly purchased by individuals in the commercial cities of northern Italy, such as Venice and Florence. The merchants and bankers in these cities were wealthy enough to be able to afford these specimens and their commitment to the new scholarship was very strong. They sought to make Italy the scientific leader of Europe, hence their support of the great universities located there.

Since wealthy Italian families also sought to collect unusual plants as status symbols, the first concentrated efforts to grow the nonindigenous specimens occurred in Italy. The Medici family in Florence experimented with growing potatoes and pineapples before the end of the fifteenth century. By 1550 Italians were regularly eating tomatoes, which they called "love-apples" for their supposed aphrodisiac powers; and peppers were a common food by 1580. As the cult of the private garden spread in Renaissance Italy, the search for plants with pharmacological properties was accelerated. Plants with reputed medicinal powers became, after gold and silver bullion and spices, the most eagerly sought of the New World products.

There were two crucial results of this revival of interest in botany in early Renaissance Italy. The first was the appointment in Italian universities of special professorships to teach botany. The most important of these appointments was that of Luca Ghini in 1534 at Bologna. Ghini perfected the method of preserving plants by drying them under pressure between sheets of paper. This marked the beginning of the herbarium, a basic tool in establishing botany as a science. Dried specimens could be preserved almost indefinitely; they could be stored and examined at will and could easily be exchanged between scholars. The second result of this fascination with botany was the founding of the first botanical gardens, which were established as branches of universities or other institutions of higher learning. They served the dual purpose of maintaining living plants for study and research, and providing educational tools for the public. Initially, both the new professors and the botanical gardens were tied to medical schools, reflecting the pervading interest in the medicinal properties of plants. But once established, these gardens were the cornerstone of the science of botany. Here plants were not only collected and studied, but botanists began to compare them to other plants and to devise systems of classification.

The first botanical garden was founded by the Venetian Senate in July 1545 at Padua. Almost immediately, a second one was set up in Pisa. Others rapidly followed, the most important being those of Florence and Ferrara (1550) and one in Bologna (1567). Soon botanical gardens were established outside of Italy such as those in Leipzig (1580), Leiden (1587), Montpellier (1593) Heidelberg (1593), Paris (1620), Jena (1629), Oxford (1632), Amsterdam (1646), Uppsala (1655), and Edinburgh (1670). Since the acclimatization of tropical plants was a very difficult undertaking, vast heated glass buildings (greenhouses) were constructed. By the end of the seventeenth century, the botanical gardens of Italy had been equalled by French gardens, especially by the one at Montpellier, where research was predominant, and at Paris (the *Jardin du Roi*, or King's Garden), where displays for the public took precedence. But the expansion of all these "living encyclopedias" was impressive. At Oxford, for example, the catalogue of 1658 listed about 2,000 plants, no more than 600 of which were indigenous to the British Isles, and this after only 26 years of the garden's existence.

The treatment of the newly discovered animals that were being imported into Europe was rather different from that accorded to plants. What was most dissimilar was that there was very little scholarly interest in the animals as compared to that expressed toward plants. There was, for instance, little attention paid to the establishment of zoological gardens by universities. The main reason for this was undoubtedly because while plants provided medicinal relief for many human ailments, animals did not. Foreign animals were valued for their curious appearances and differences from European animals, but they had no perceivable intrinsic value. They were expensive to buy and maintain and were often hard to keep alive; caring for plants was easy in comparison. In addition, plants were relatively simple to multiply while encouraging animals to propagate in captivity was often nearly impossible (this still remains the case with some species today). Further, a dead stuffed animal was regarded as having as much scientific value as one alive in a zoological garden. This attitude would retard the study of animal behavior for a very long time.

Although zoos did not fully assume their modern form until the nineteenth century, Renaissance nobles and monarchs did establish menageries as status symbols. These private collections were often referred to as zoological gardens, particularly when they were combined with collections of plants. They were located all over Europe; Leo X (a Medici and pope from 1513-1521) even set one up at the Vatican. One of the most notable of these early collections was founded at the Château de Chantilly in France by the Montmorency family. It often exhibited rare zoological specimens never before seen in Europe. It survived until 1792 when it was destroyed during the French Revolution. Another French zoological garden that came to an unhappy end was the royal menagerie in the Louvre in Paris, which was famous for its aviary. In January 1583, King Henry III (1551-1589) personally killed all of its animals after he had dreamed that they were going to eat him.

One of the most important of these early collections was the royal zoological garden established in Sweden in 1561. Over a period of two centuries, it evolved into a true research institution and had a strong influence on the work of Carolus Linnaeus (1707-1778), perhaps the greatest of all botanists. There was also an important zoological garden founded outside Paris at Versailles by Louis XIII (1601-1643) in 1624. Although it eventually fell victim to the French Revolution, among the species that had their first European exhibitions at Versailles were the condor, hummingbird, toucan, cardinal, tanager, lemur, and tapir. In central Europe, three outstanding menageries were founded by Maximilian II, the Holy Roman Emperor from 1564 to 1576: at Ebersdorf (1552) and at Prague and Nãugebãu (both in 1558). Descendants of the animals in these menageries eventually formed the first collection of the Schönbrunn in Vienna (1752), the oldest modern zoo in the world.

ROBERT HENDRICK

Further Reading

Croke, Vicki. *The Modern Ark: The Story of Zoos: Past, Present, and Future.* New York: Scribner, 1997.

Duval, Marguerite. *The King's Garden.* Trans. by Annette Tomarken and Claudine Cowen. Charlottesville: University Press of Virginia, 1982.

Fisher, James. *Zoos of the World: The Story of Animals in Captivity.* Garden City, NY: Natural History Press, 1967.

Minelli, Alessandro, ed. *The Botanical Garden of Padua, 1545-1995.* Venice: Marsilio, 1995.

Morton, A.G. *History of Botanical Science: An Account of the Development of Botany from Ancient Times to the Present Day.* London: Academic Press, 1981.

Prest, John. *The Garden of Eden: The Botanic Garden and the Re-Creation of Paradise.* New Haven, CT: Yale University Press, 1981.

Thacker, Christopher. *The History of Gardens.* Berkeley: University of California Press, 1979.

Biographical Sketches

Pierre Belon
1517-1564
French Naturalist

Pierre Belon was one of the first and most important naturalists to rely on his own exploration in order to further his research. He is considered the originator of comparative anatomy due to his systematic analysis of similarities be-

tween the skeletal systems of humans and birds. Likewise, his discussion of dolphin embryos signifies the emergence of modern embryology.

Pierre Belon was born in Soutiere, a small village near Le Mans, France, in 1517. He came from an obscure family and, before 1535, was apprenticed to René des Prez, who was the apothecary of Guillame du Prat, the bishop of Clermont. Later, Belon became the protégé of

René du Bellay, the bishop of Le Mans. Because of this alliance, Belon was able to begin his study of botany at the University of Wittenberg in 1540. In 1542 Belon went to Paris. There du Prat recommended him as an apothecary to François, the cardinal of Tournon, who was, throughout Belon's life, his most significant patron. Later, in 1542, Belon journeyed to Geneva, presumably because of a diplomatic mission for the cardinal. There he was imprisoned for six months following a violent altercation with two young Calvinists.

Between 1546-1548 Belon embarked on a series of diplomatic missions to Constantinople and the Middle East. On this extended tour he identified and described places, objects, animals, and plants mentioned by ancient writers. While Belon was engaged primarily on diplomatic missions, he recorded a detailed account of scientific curiosities that he discovered on his travels in *Les Observations de plusieurs singularitez et choses mémorables* (Observations of Several Curiosities and Memorable Things). The itinerary described by Belon was frequently followed by travelers interested in scientific discoveries for several centuries following its publication. Belon's account of his travels was so significant because of his detailed depiction of items of primarily scientific interest. In this sense, Belon established a pilgrimage route that was not exclusively religious.

The period between 1551-1555 formed a new stage in Belon's life. In this period he composed his principal works. *L'Histoire Naturelle des Etranges Poissons Marins* (The Natural History of Strange Marine Fish), published in 1551, contains descriptions and illustrations of fish and cetaceans, such as porpoises and whales, which Belon had dissected. Furthermore, this work established a classification system for marine fish. This classificatory system was based largely on Aristotle's (384-322 BC) animal classifications, and included both cetaceans and hippopotami under "fishes." However, in this work, Belon recognized that the milk glands of cetaceans were mammalian in type. Likewise, he noted that these creatures were air-breathing mammals, even though they lived underwater.

His last work, *L'Histoire de la Nature des Oyseaux* (The Natural History of Birds) was completed in 1555. This was the most significant of his works and brought him the greatest fame. In fact, it was so significant that King Henry II accepted Belon's dedication of the text to the crown. This dedication was significant because it was accompanied by the promise of a royal pension. Belon was also commissioned by the king to develop a botanical garden used to acclimate exotic plants to the French climate. However, Belon's payment was anything but prompt. A bit displeased by his unsuccessful attempts to collect his royal subsidy, Belon traveled to Switzerland and Italy and explored more of France in 1556. Upon returning, he received the money and, in 1558, obtained his medical license. Belon practiced medicine until 1564, when he was murdered in what was almost certainly a politically motivated crime.

DEAN SWINFORD

Antonio Benivieni
1433?-1502
Italian Physician and Anatomist

Antonio Benivieni is considered a late pre-Vesalian anatomist. Decades before Andreas Vesalius (1514-1564) published his landmark work of illustrated anatomy, *De Humani Corporis Fabrica* (1543), Benivieni was doing similar work.

The Renaissance was an exciting time because all arenas of knowledge acquisition were encouraged. Medieval dependence on authority was superceded by a need to observe and explore. Art, literature, philosophy, and science were cultivated and encouraged as a secular society increasingly grew away from theological explanations of existence and sought to find answers in the observable present. One of the most provocative goals in learning was the attempt to unravel the mystery of life. Confronting both theological taboos and pseudoscience, early anatomists sought to explain the interiority of living beings through dissection. What made people live, die, or become ill? Were there rational explanations for the cause of disease?

Although less well known than Vesalius, Benivieni's work was scientific in an age of early science and broke the tradition that forbade dissection. His published work was neither exclusively a dissection manual (like Vesalius's) nor a descriptive anatomy. He went one step further and attempted to describe pathology rather than normalcy. Obviously one has to know a great deal of anatomy in order to differentiate between normal variation and pathologic changes. He was a person clearly ahead of his time.

Benivieni practiced medicine in Florence, Italy, for more than 30 years, all the time documenting his cases and keeping notes on postmortem observations. His chief work, *De abditis nonnullis ac mirandis morborum et sanitorium causis* (On the secret causes of disease), was published posthumously in 1507. Its title may be a clue to the challenge of the worldview that held that all disease was caused by divine or diabolic intervention.

Although very little is known about his life, he must have been a disciplined rigorous person in order to maintain his study. He must also have been highly objective because during his lifetime, a patient was a neighbor and friend rather than a stranger. His cases were the stories of people he had known and treated in life. Among the descriptions of disease were: stones found in the tunic of the liver (gallstones), callous growth in the stomach (carcinoma of the stomach), death from difficulty in breathing (fibrinous pericarditis), degenerative hip joint disease, and ruptured intestine. He was able to observe symptoms in his living patients, and then search for "signatures" in organs after death.

LANA THOMPSON

Giovanni Alfonso Borelli
1608-1679
Italian Naturalist, Mathemetician and Physicist

Giovanni Borelli was born in Naples, Italy. Little is known about his early life, education, and family, other than his precocious mastery of mathematics. There is uncertainty about whether he ever studied medicine formally, but he did study mathematics with Benedetto Castelli (1577-1644) in Rome. Borelli was appointed to the chair of mathematics at the University of Messina by 1640. In 1642, inspired by the work of Galileo Galilei (1564-1642), Borelli obtained the consent of the University to take a prolonged leave from his professorship in order to study with Galileo and Evangelista Torricelli (1608-1647) in Florence. Unfortunately, Galileo died that year and Borelli returned to the University. In 1656 Ferdinand, Duke of Tuscany invited Borelli to serve as professor of mathematics at the University of Pisa. During the same year, Marcello Malpighi (1628-1694) joined the university as a professor of theoretical medicine. Malpighi held doctorates in both medicine and philosophy and was the founder of microscopic anatomy. Borelli and Malpighi became good friends and leaders of the Accademia del Cimento (Academy of Experiments), which was founded in Florence in 1657.

The Accademia del Cimento, one of the first scientific societies, helped establish Pisa as a center of research in mathematics and the natural sciences. After 12 years at Pisa, and many quarrels with his colleagues, Borelli left the university to seek quieter and healthier surroundings. In 1667 Borelli returned to the University of Messina, where he became involved in literary and antiquarian studies, investigated an eruption of Mount Etna, and continued to work on the problem of animal motion. In 1674 he was accused of being involved in a conspiracy to free Sicily from Spain. Forced into exile, he fled to Rome, where he lived under the protection of Christina, former queen of Sweden. Borelli died before his great work *De motu animalium* (On Motion in Animals, 1680-81) was published. An introduction was added by a church official who commended Borelli for upholding the authority of the Church in his lectures on astronomy. In physiological matters, too, Borelli remained a faithful son of the Church by acknowledging that, although he had attempted to explain the movements of the body in terms of purely mechanical principles, all the mechanical phenomena described in his book were ultimately governed by the soul.

The seventeenth century was a time of great ferment and change in scientific and medical thought. The Scientific Revolution created skepticism about ancient medical theories and resulted in the search for a simple system that could guide medical practice. René Descartes's (1596-1650) concept of the human body as a machine that functions in accordance with mechanical principles was very influential in medical thought. Physicians and scientists who accepted the mechanical model of physiology were called iatrophysicists. Those who thought of life as a series of chemical processes were called iatrochemists. Whereas Descartes was the founder of iatromechanism as a philosophy, Borelli was the founder of iatrophysics as an experimental science. Borelli was especially interested in the problem of muscle action.

De motu animalium was a sustained analysis of the mechanics of muscle contraction that dealt with the movements of individual muscles and groups of muscles treated geometrically in terms of mechanical principles. Animal movements were divided into external movements, such as those carried out by the skeletal muscles, and internal movements, such as those of the heart and viscera. Borelli's method of study involved a progression from the simplest element of the motor

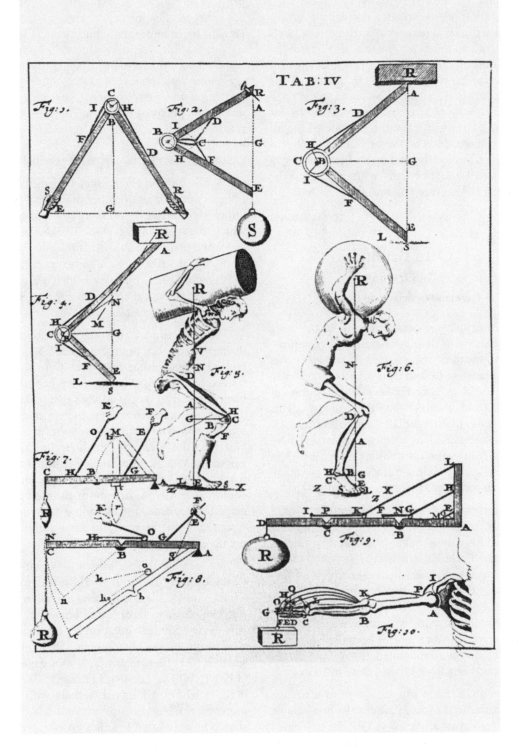

A page from Giovanni Borelli's 1680-81 book, *De motu animalium. (Wellcome Institute Library. Reproduced with permission.)*

system, the independent muscles, up to the more complicated organs and organ systems, and finally the power of movement of the organism considered as a whole. According to Borelli, the fleshy muscular fibers played a fundamental role in muscle contraction; the fibers of the tendons, in contrast, were merely passive agents, which did not take part in contraction. The action of the heart particularly intrigued Borelli. Unlike Descartes, he recognized that the heart was a muscular pump rather than a heat engine and confirmed this by simple experiments.

Although Borelli is primarily remembered for his attempts to explain muscular movement and other body functions according to the laws of statics and dynamics, he was also one of the early microscopists. Borelli carried out microscopic investigations of the circulation of the blood, nematodes, textile fibers, and spider eggs. Borelli also wrote many astronomical works, including a treatise in 1666 that considered the influence of attraction on the satellites of Jupiter. In a letter published in 1665 under the pseudonym Pier Maria Mutoli, he was the first to suggest the idea that comets travel in a parabolic path.

LOIS N. MAGNER

Otto Brunfels
1489?-1534
German Physician and Botanist

Otto Brunfels is considered one of the three "fathers" of German botany, in large part because of his three-volume *Herbarum vivae eicones* (Living Images of Plants), published in the 1530s. These books contained illustrated descriptions of plants, most of which had medical uses. This marked the first time that accurate and realistic images were used in a printed botanical text. Ironically, it is the illustrations drawn by Hans Weiditz, rather than Brunfels's quite mediocre text, that make the *Herbarum* noteworthy.

Brunfels was born about 1489 in the town of Braufels, near Mainz in what is now Germany. He received his education in Mainz and then entered a Roman Catholic monastery. He left there in 1521, changed his religion, and became a Lutheran preacher. In 1524 he settled in Strasbourg, where he got married and opened a small school. He then moved on to Switzerland, where in 1531 he received a medical degree from the University of Basel, and was appointed Town Physician of Berne two years later. He died there in 1534.

Brunfels's main interests were in medicine and botany. In the same year that the first volume of the *Herbarum* appeared (1530), he published *Catalogus,* a collection of older medical works he had edited and translated; it included one of the first medical bibliographies. The *Herbarum* is also related to medicine because it follows in the tradition of earlier herbals, or illustrated books on plants that provided information on the medicinal uses for plants. It was only later that interest developed in plants for their own sake.

The invention of movable type around 1440, and subsequent improvements in the printing press, made it possible to produce books more cheaply and in greater quantities. Perhaps more importantly for botanical science, it made feasible exact reproduction of images of plants. Before that time, artists had to copy images along with the text of herbals and extensive copying led to inaccuracies in the images. But despite the potential of printing to provide accurate images, the illustrations in the first printed herbals were of poor quality, which is what makes Brunfels's *Herbarum* so exceptional.

While most of the previous botanical illustrations were copies of earlier drawings, Weiditz obviously used real specimens as his models, which made his images so lifelike. In some cases, the plants are shown with leaves nibbled by insects or with other individual peculiarities. This is in marked contrast to the work of most subsequent botanical artists who portrayed ideal plants without blemishes.

In all, there are 238 woodcut illustrations in the *Herbarum;* they represent about 230 species, including over 40 species that were first described by Brunfels. The plants are not arranged in any organized way. It appears that it was Weiditz, rather than Brunfels, who determined the sequence in which the plants were presented, based on the order in which he finished the woodcuts. Weiditz also decided which plants he would draw, including several that Brunfels couldn't identify. At one point Brunfels apologized to the reader for including a picture of a plant that didn't have a Latin name and that was not used medicinally.

In contrast to the illustrations, which are extraordinarily better than those in previous books, Brunfels's text is not exceptional. It is just a rehash of the work of earlier botanists. This is why many historians question his designation as a father of German botany, giving that honor instead to two other botanists, Hieronymus Bock (1498-1554) and Leonhard Fuchs (1501-1566), both of whom published herbals with much more original and informative texts. But it should be noted that Brunfels encouraged Bock to write his own book.

MAURA C. FLANNERY

Realdo Matteo Colombo
1515-1559
Italian Anatomist, Physician, and Professor

Realdo Colombo was one of the first European scientists to clearly describe the pulmonary circulation, or passage of blood between the heart and the lungs. Columbo, the son of an apothecary, was born in Cremona, Italy. He served as an apprentice to a surgeon for seven years before be began his studies of medicine and anatomy at the University of Padua in 1540 with Andreas Vesalius (1514-1564), the founder of modern anatomy. When Vesalius left the university in 1543, Colombo became professor of surgery. Colombo was an excellent anatomist, as well as an influential scientist and teacher. Colombo carried out many systematic dissections on both living animals and human cadavers. However, he seems to have exaggerated the number of dissections that he performed and he was quite critical of the work of others, including Vesalius. Indeed, Columbo was one of the most vociferous critics of the Vesalius's great treatise *On the Fabric of the Human Body* (1543). Surprisingly, Colombo and Vesalius seem to have had a cordial collegial relationship before Colombo assumed his professorship. In response to Colombo's criticism, Vesalius called his former student a scoundrel, an ignoramus, and an "uncultivated smattering" whose general education was as deficient as his mastery of Latin. Later the anatomist Gabriele Fallopio (1523-1562) accused Colombo of plagiarizing the discoveries of other anatomists. In 1546 Colombo became the first professor of anatomy at the University of Pisa. Two years later he was appointed professor of anatomy at the Sapienza, or Papal University, in Rome. He also served as surgeon to Pope Julius III.

When he first became a professor, Colombo spoke of plans for the publication of a major illustrated anatomical treatise that would detail the errors made by Vesalius and others, but this ambitious enterprise never materialized. Colombo's only major anatomical treatise, *De re anatomica* (On Anatomy, 1559), was published posthumously by his heirs. Although the book is not illustrated, it was clearly written and organized and included several important original observations. Colombo carefully described the organs within the thoracic cavity, the membrane surrounding the lungs, and the membrane surrounding the abdominal organs. His most important contribution was his description of general heart action, which included his observations and experiments concerning the minor, or pulmonary circulation of the blood. However, Columbo may have been demonstrating the pulmonary circulation as early as 1545. Calling upon the reader to confirm his observations by dissection and vivisection, Columbo boasted that he alone had discovered the way in which the

lungs serve in the preparation and generation of the vital spirits. Unlike Michael Servetus (c. 1511-1553) and Ibn an Nafis, Colombo made it clear that his ideas on the pulmonary circulation were based on clinical observations, dissections, and experiments on animals. He described the anatomy of the heart in detail and corrected many ancient errors. Colombo explained that blood entered the ventricles of the heart during diastole (the relaxation phase of heart action), and is expelled from the ventricles during systole (the contraction of the heart muscle). Colombo clearly followed the movement of venous blood from the right ventricle, through the pulmonary artery to the lungs. He noted that blood was bright red when it left the lungs where it was mixed with a "spirit" present in the air. The blood then returned to the left ventricle through the pulmonary vein. In other words, air was received by the lungs where it mixed with blood brought in by the pulmonary artery from the right ventricle of the heart. Blood and air were taken up by the branches of the pulmonary vein and carried to the left ventricle of the heart to be distributed to all parts of the body.

Although pulmonary circulation had been described during the thirteenth century by the Egyptian physician Ibn an Nafis and by Servetus, these accounts were generally unknown. Therefore, sixteenth- and seventeenth-century scientists generally recognized Colombo as the discoverer of the pulmonary circulation. However, Colombo's claim to priority has been questioned on several accounts. Even if he begun his study of the pulmonary circulation as early as 1545, his ideas might have been stimulated or confirmed by learning about the speculations of Servetus. Despite his professions of originality and daring, Colombo was actually rather conservative concerning other aspects of the functions of the heart, blood, and respiration. Therefore, although Colombo's work proved that it was not necessary to invoke invisible Galenic pores in the septum of the heart in order to get blood from the right side to the left side, it did not explain the major, or systemic circulation of the blood.

LOIS N. MAGNER

René Descartes
1596-1650
French Philosopher, Physiologist, and Mathematician

René Descartes, who was born in La Haye (now Descartes), France and died in Stock-

René Descartes. *(Library of Congress. Reproduced with permission.)*

holm, Sweden, has been called the founder of modern philosophy, but he is also honored as a mathematician and physiologist. Descartes's father, Joachim, who owned farms and houses in Châtellerault and Poitiers, was a councilor in the Parliament of Brittany in Rennes. Descartes's mother died when he was only one year old. After his father remarried, Descartes remained in La Haye where his maternal grandmother and other relatives raised him. In 1606 Descartes was sent to the Jesuit college at La Flèche, which had been established in 1604 by Henry IV. The curriculum included classical studies, science, mathematics, metaphysics, music, poetry, dancing, riding, and fencing. Students were prepared for careers in military engineering, law, and government administration. Philosophy was taught in the scholastic Aristotelian manner that Descartes would eventually challenge.

In 1614 Descartes went to Poitiers where he earned a law degree two years later. In 1618 he went to Breda in the Netherlands to study mathematics and military architecture. From 1619 to 1628, Descartes traveled widely through Europe. By 1920, while he was serving in the army of Maximilian I, duke of Bavaria, Descartes discovered a method of deductive reasoning that he believed would be applicable to all the sciences. Descartes moved to Paris in 1622 where he associated with prominent writers, scholars, philoso-

phers, mathematicians, and scientists. During this period, Descartes wrote several treatises that have not survived. In 1628 Descartes presented a demonstration of his method for establishing truth, but he soon decided to move to the Netherlands, where he felt he could enjoy greater liberty and tolerance than anywhere else. Religious intolerance had grown in France and the Parliament of Paris had even passed a decree in 1624 that provided the death penalty for attacks on Aristotle.

In 1629 Descartes began writing his *Meditations*. In 1633 he was planning to publish a major work called *Le Monde* (The World), when he heard that Galileo Galilei (1564-1642) had been condemned by the Roman Catholic Church for publishing the view that the Earth revolves around the Sun. Although Descartes believed that eventually his physics would replace Aristotle's, he suppressed *Le Monde* because the Copernican theory was central to his own cosmology. *Le Monde* was finally published in 1664. *Discourse on Method* was published in 1637. It was one of the first significant modern philosophical works written in French instead of Latin so that all literate men and women could learn to use reason in the search for truth. Descartes's *Geometry*, which established analytic geometry, introduced modern algebraic notations. Descartes offered his readers four rules for reasoning: (1) Accept nothing as true that is not self-evident. (2) Divide problems into their simplest parts. (3) Solve problems by proceeding from simple to complex. (4) Recheck the reasoning.

Descartes began methodically doubting knowledge based on authority, the senses, and reason. For Descartes, the only certainty was that he was thinking and he must, therefore, exist: *Cogito, ergo sum* (I think, therefore I am). To escape the trap of solipsism (the view that nothing exists but one's individual self and thoughts), Descartes argued that all clear and distinct ideas must be true. Using his philosophical method, Descartes devoted the rest of his life to the exploration of mechanics, medicine, and morals. His theories of medicine and physiology are based on mechanics. Descartes believed that animal and human bodies operate on purely mechanical principles. In his physiological studies, he dissected animal bodies to show how their parts move. His ideas and experiments were described in *L'Homme, et un traité de la formation du foetus* (Man, and a Treatise on the Formation of the Foetus, 1664). Although Descartes argued that animals were purely mechanical and had no soul, he described human beings as a union of mind and body. The interaction between mind and body took place only in the pineal gland, the only unpaired organ in the brain.

Finally, in 1649, as intolerance grew in both France and the Netherlands, Descartes sought refuge with Queen Christina, of Sweden. Apparently as a result of being forced to give Queen Christina philosophy lessons at 5 A.M., Descartes contracted pneumonia and died in Stockholm in February 1650. The Roman Catholic Church put his works on the Index of Forbidden Books in 1667.

LOIS N. MAGNER

Bartolomeo Eustachio
1510?-1574
Italian Anatomist

Bartolomeo Eustachio, contemporary of Andreas Vesalius (1514-1564) and a major figure in the great flowering of knowledge in gross anatomy that occurred in Italian universities during the sixteenth century, is best known for his account of the auditory organ that bears his name, the Eustachian tube. His treatises on the kidney, venous system, and teeth were superior to anything yet produced, and his copperplate engravings, particularly of the sympathetic nervous system, are of such quality that they alone ensure Eustachio's place of eminence in anatomical history.

Eustachio was born in San Severino, Italy, between 1510-1520. The son of a physician, he received a good humanistic education and knew Greek, Hebrew, and Arabic. He studied medicine in Rome at the Archiginnasio della Sapienza and began to practice near his birthplace in about 1540. Thereafter he was invited to become physician to the Duke of Urbino and later to the Duke's brother, Cardinal della Rovere, whom he followed to Rome in 1549. There he joined the medical faculty of the Sapienza as the equivalent of a professor of anatomy, and was permitted to obtain cadavers for dissection from nearby hospitals. In later years Eustachio was so disabled with gout that he had to resign his chair, but continued to serve Cardinal della Rovere. In 1574 the Cardinal requested his services; Eustachio set out from Rome to Fassombrone to attend him, but met death along the way.

Eustachio's first works were directed against Vesalius and his challenges to Galenic theories. Although regarding himself as a follower of Galen (129-199), the renowned Greek physician who was quoted dogmatically for over 1,200 years, Eustachio nonetheless possessed a spirit of scientific inquiry and was an innovator in his anatomical investigations. Since the religious and civil prohibition against human dissection had somewhat eased, human anatomy as both science and art was poised to become a focus of excellence.

In 1562 and 1563 Eustachio produced a remarkable series of treatises on, among other structures, the kidney (*De renum structura*), the auditory organ (*De auditus organis*), and the teeth (*De dentibus*). These were collectively published in 1563 under the title *Opuscula anatomica*, which can loosely be translated as "Little Anatomical Works." The work on the kidney, the first to be dedicated to that organ, contains the first account of the suprarenal gland and the first detailed analysis of the concept of anatomical variation. The auditory tube, as was acknowledged by Eustachio, had been known to Aristotle (384-322 B.C.), and Alcmaeon (sixth century B.C.), but Eustachio expanded on this earlier knowledge with a careful description of the auditory tube and its anatomic relations. Eustachio's *De dentibus* was the first detailed study of the development of the teeth, describing the first and second dentitions, the structure of both the hard and soft tissues, and possible reasons for the sensitivity of the hard tissues of the tooth.

In 1552 Eustachio, with the help of the artist Pier Matteo Pini, prepared a series of 47 anatomical illustrations in copperplate engravings. Only eight were published during his lifetime, relating to the discussions of the kidney in the *Opuscula anatomica*. The others were lost after his death, reappearing only in the early eighteenth century in the possession of a descendent of Pini. They were then purchased by Pope Clement XI and presented to his physician Giovanni Maria Lancisi (1654-1720), a successor to Eustachio's chair of anatomy at Sapienza. Lancisi published the rediscovered plates, along with the previously published eight, in 1714 under the title *Tabulae anatomicae Bartholomaei Eustachi*. Tabula XVIII, on the sympathetic nervous system, is generally considered even today to be one of the best illustrations ever produced of that structure.

Eustachio's anatomical achievements were great, but much of his influence was posthumous. His work for the most part was not published during his lifetime, and nearly all the text is lost. The copperplate engravings of Eustachio are less beautiful than those of Vesalius, but are in many instances more accurate. They contain a number of discoveries that for originality rank him below only Leonardo da Vinci (1452-1519) and Vesalius. Had these engravings appeared when they were completed in 1552, knowledge of the nervous system and anatomical studies in general would have matured at least a century earlier.

DIANE K. HAWKINS

Girolamo Fabrici
1533-1610
Italian Physician

Girolamo Fabrici was born in Acquapendente, Italy, in 1533 and received his medical training (both an M.D. and Ph.D.) at the University of Padua in his home country. Although he preferred private practice and research to teaching, he is best remembered as the teacher and mentor of William Harvey (1578-1657), regarded as the father of modern medicine and physiology.

In addition to his famous student, Fabrici had a renowned teacher, Gabriele Falloppio (1523-1562), an anatomist who achieved a place in medical history by discovering the Fallopian tubes and other parts of the female reproductive system.

When Falloppio retired from teaching anatomy and surgery at the University of Padua, Fabrici was chosen to replace him. Although historical accounts vary in certain areas, it is generally accepted that Fabrici did not enjoy his teaching responsibilities and frequently avoided them by disappearing well before classes were completed. Since many of the students traveled long distances for their medical studies, they were often disappointed at their teacher's lack of interest in their education.

In Fabrici's defense, his consuming interest was research, primarily in anatomy, secondarily in surgery. He spent many years studying and publishing numerous volumes that were well received in the medical communities of his era. Prominent among these publications were books on anatomical observations containing detailed descriptions of the venous valves. It was this branch of research that led to Harvey's monumental discovery of the circulatory system.

Fabrici had an unusually interesting life outside of his university activities. In 1581 he became private physician to the Duke of Mantua as well as the Duke of Urbino in Florence. They latter summoned him in 1604 to treat his ailing son and, according to historical accounts, gave him two golden chains for his efforts. This led to an international reputation and Fabrici was even called upon by the King of Poland, who also believed in awarding gold chains and medals for medical help.

Fortunately for Fabrici, the sixteenth century was a time when patronage was in style and he made the most of it. Along with his general practice of medicine and surgery, he attended many who were celebrities of the time. He charged them nothing for his treatment and received unreasonably large gifts for his work. On the other side of the coin, he treated poor people for nothing and was well-regarded at all social levels. He lived in a magnificent villa and entertained prominent visitors in lavish style. On one of his visits to Venice, he was called upon to treat a wounded man named Paolo Sarpi. His treatment was so successful that the Republic of Venice honored him by naming him a Knight of St. Mark.

Following his publication of *On the Valves of the Veins* (1603), Fabrici not only continued his research on this subject, but began to work on embryological works. He published at least two known works on this topic: *Deformato foetu* (1604) and *De formatione* (1621).

Following the pattern of his medical work, he eventually became a pioneer in comparative anatomy and, throughout his life, continued to teach classes in this science, both privately and at the University of Padua.

In turn, the university recognized the genius of Fabrici and gave him life tenure on their staff along with the impressive title *Sopraordinario*.

His other accomplishments include the invention of both medical and dental instruments, memberships in the Medical Colleges of Padua and Venice and, from 1570 to 1584, he served on a board that examined and accredited surgeons for private practice.

Fabrici died in 1610 at the age of 77 while living in retirement at his villa outside of Padua.

BROOK HALL

Wilhelm Fabricius Hildanus
1560-1634
German Surgeon

Fabricius Hildanus was the "Father of German Surgery." He was the first to use magnets to extract iron slivers from the eye, the first to operate successfully for gallstones, and among the first to use tourniquets and ligatures to control bleeding. He improved amputation techniques and introduced many new surgical instruments.

Born Wilhelm Drees in Hilden, near Düsseldorf, Germany, on June 25, 1560, he is known variously as Wilhelm Fabry von Hilden, Wilhelm Fabricius von Hilden, Guilhelmus (or Guilielmus) Fabricius Hildanus, Wilhelm Fabricius Hildanus, or sometimes just as Fabricius Hildanus. He is not to be confused with the other great "Fabricius" of medicine, Italian anatomist Girolamo Fabrizio, known as Hieronymus Fabricius ab Aquapendente (1537-1619).

Although orphaned at a young age, Fabricius Hildanus received an excellent humanistic classical education, and became fluent in Latin. He was apprenticed to several barber-surgeons and military surgeons, including Jean Griffon in Geneva, Switzerland. After studying in Switzerland, France, and Italy, he practiced in Switzerland and Germany.

In 1587 he married Marie Colinet, the daughter of a printer in Geneva. She was already an accomplished midwife and surgeon. He taught her additional surgical skills and they practiced surgery together in Bern, Switzerland. If he was the Father of German Surgery, then she was its Mother. When a patient presented with a piece of steel in his eye in 1624, using a magnet to get it out was her idea. She wrote and published several books, all literary or religious, none medical or surgical. Yet her surgical insights permeated her husband's works, and he acknowledged that she surpassed him in orthopedics, ophthalmological surgery, cancer surgery, and several other kinds of surgery. She successfully performed caesarian section 40 times in an era when it was nearly always fatal to the pregnant woman.

Fabricius Hildanus was known for his skill and speed in operating, his conservative surgical theory, and his preference for ancient medical and surgical authors. He invented the otoscope, or aural speculum, a device for examining the entire ear canal, and improved the drug kit for battlefield surgery.

He either rejected or was ignorant of new techniques, advanced by Ambroise Paré (1510-1590), for the treatment of wounds. Fabricius Hildanus continued to rely on medieval treatments in military and trauma surgery, such as the so-called "weapon salve," a form of sympathetic magic in which lotions, herbs, or oils are applied to the weapon that caused the wound in the belief that it could heal the wound, even if the wounded person was miles away. Sympathetic magic is to perform an action on one object while intending another object to receive the effect of that action. Some examples are dancing to generate rain, shooting at mock animals to ensure a good hunt, and causing human illness by sticking pins in voodoo dolls.

In *De gangraena et sphacelo* (On gangrene and sloughing), published in 1593, Fabricius Hildanus recommended amputation for gangrene, high above the infected part. His greatest work was the six-volume *Observationum et curationum chirurgicarum centuriae* (Six hundred surgical observations and treatments), published in 1606-1641, to which Marie contributed immeasurably. In his 1607 *De combustionibus* (On burns), one of the earliest books to deal exclusively with injuries caused by heat and fire, he categorized burns in a practical, systematic way. His other works include: *De vulnere quodam gravissimo et periculoso* (On the most grave and dangerous wounds), 1614; *New Feldt Arztny Buch von Kranckheiten und Schäden* (New book of field medicine for illnesses and wounds), 1615; *De dysenteria* (On dysentery), 1616; *Anatomiae praestantia et utilitas* (The superiority and usefulness of anatomy), 1624; *Schatzkämmerlein der Gesundheit* (Little treasury of health), 1628; *Lithotomia vesicae* (Lithotomy in the bladder), 1628; *Consilium in quo de conservanda valetudine* (Consultation on the preservation of health), 1629; and *Cheirurgia militaris* (Military surgery), 1634.

ERIC V.D. LUFT

Gabriele Falloppio
1523-1562
Italian Anatomist

Gabriele Falloppio, an illustrious anatomist of the sixteenth century and one of the founders of modern anatomy, is best remembered for the first accurate description of human oviducts or "fallopian tubes," which he correctly described as resembling small trumpets. His 1561 work *Observations Anatomicae* included original observations on the eye, ear, teeth, and female reproductive system, introducing many anatomical terms including vagina, placenta, cochlea, labyrinth, and palate.

Born in 1523 in Modena, Italy, Falloppio was first educated in the classics and directed towards a career in the church. Later he studied medicine and then surgery, but after a series of fatal outcomes abandoned surgery and turned entirely to medical studies. He joined the medical school in Ferrara in 1545, accepted the chair of anatomy at the University of Pisa in 1549, and then in 1551 the famous chair of anatomy at Padua. While teaching at Padua he inspired many of his students, including Fabricius ab Aquapendente (1537-1619) and Volcher Coiter (1534-1576?), to further research in surgery and other fields of medicine. He was a respected physician as well as an anatomist, and was considered an authority in botany, mineral springs, and syphilis. He taught at Padua until his death from pulmonary tuberculosis at the age of 39.

Of the various works attributed to Falloppio, only the *Observations Anatomicae* was published during his lifetime and is known to be authentic. It is not a comprehensive textbook but rather a series of commentaries on the *De Humani Corporis Fabrica* of Andreas Vesalius (1514-1564). Although Falloppio regarded Vesalius with the highest respect, even referring to him as the "divine Vesalius" upon whose foundation he continued to build, he did not hesitate to point out shortcomings in the Vesalian text. For example, he criticizes Vesalius for describing and illustrating in the *Fabrica* the kidney of a dog instead of a human.

Falloppio's dissections included not only adults but also children, newborns, and even fetuses, and thus allowed him make observations of primary and secondary centers of ossification, i.e., bone formation. In his studies of the teeth he provided a clear description of primary dentition and the process of replacement of the primary by the secondary tooth. His description of the auditory apparatus gives the first clear account of several anatomical structures, including the round and oval windows of the ear, the cochlea, the semicircular canals, the scala vestibuli and the tympani.

Also important are Falloppio's contributions to the understanding of muscles of the scalp and face, most notably the muscles of the orbit. He made a major contribution to the knowledge of the nervous system through his clear description of the trochlear nerve, tracing it to its origin in the brain stem and its termination in the superior oblique muscle of the eye, and establishing it as a separate nerve root. In addition, he recognized and described 11 of the 12 cranial nerves.

Falloppio's most important contribution to urology is his account of the kidney, but it is unclear whether the priority belongs to him or to his contemporary Bartolomeo Eustachio (1510?-1574). He does provide the earliest account of bilateral duplication of the ureter and renal vessels, and was the first to describe the three muscle coats of the urinary bladder.

Falloppio was a very effective teacher and careful observer, who demonstrated courage in challenging the accepted medical authorities, especially those of Galen (129-199), whose sayings had been revered as laws for over 1,200 years. Falloppio's early death limited his output, but his influence can be traced in the work of his two able pupils, Coiter and Fabricius, founders of modern comparative anatomy and embryology. He may be considered a student of Vesalius because of his thorough analysis of Vesalius's works, and through his continuation of the Vesalian tradition of using independent judgment rather than adhering to previous authority.

DIANE K. HAWKINS

Girolamo Fracastoro
1478?-1553
Italian Physician

Girolamo Fracastoro was born in Verona, Italy, in or near 1478 (historical sources differ on this). Although his family was untitled by the crown, they were nonetheless an old family with a history of several prosperous generations. Fracastoro enjoyed a sheltered childhood, living on the family estate, Incaffi, which was situated on Lake Garda and located about 15 miles (24.1 km) from the city of Verona.

Although his greatest recognition came from his medical discoveries, Fracastoro had many other academic gifts. His father was his first instructor in philosophy and literature, but he went on to the University of Padua where he studied astronomy, mathematics, more philosophy and literature, and finally earned his M.D. degree in 1502.

In that same year he began teaching philosophy at the university, but engaged in the private practice of medicine in Verona. Eventually he became interested in medical research, specifically in the contagious disease we know today as syphilis. It had been known by various other names in different parts of the world, but Fracastoro (being also a poet and writer), created a story about a shepherd named Syphilis who had cursed the Sun and been punished by the gods who infected him with the loathsome disease that spread throughout the kingdom, even to its king.

Fracastoro's research led to other forms of contagion and how diseases were contracted.

Girolamo Fracastoro. *(Arte & Immagini srl/Corbis. Reproduced with permission.)*

His speculation about germs and their transfers was remarkable in that the microscope was not invented until the end of the sixteenth century. Despite this, he is considered the first person to advance the germ theory and its importance in medical advancement.

While conducting his lucrative medical practice in Verona, Fracastoro found time to study the medicinal properties of flowers and other plants. He wrote several treatises on botanical subjects and, in light of the widespread interest in herbs and natural medicines today, could certainly be considered an early founder of the movement.

Fracastoro led an interesting, multifaceted life. When the University of Padua was forced to close because of the war between the Holy Roman Emperor and League of Cambrai, a high-

ranking officer invited Fracastoro to Pordenone to stay with him and to be active in the short-lived Academia Friulana. Upon another occasion, another of his patrons, Bishop G. M. Giberti, presented Fracastoro with a house on Lake Garda in Malcasine, near his childhood home.

Another religious affiliation was enhanced when Fracastoro dedicated his syphilis research to Cardinal Pietro Bembo, Secretary of Briefs to Pope Leo X. The Cardinal had been an active assistant in the research and stated (in writing) that the dedication was the highpoint of his life and the most highly valued gift he had ever received.

While he lived in Verona, Fracastoro's residence was a gathering place for the elite intellectuals of the time. Many gifted students of philosophy and the sciences shared their knowledge with Fracastoro and corresponded with him frequently.

Even though he received many honors and the patronage of important officials, Fracastoro takes his place in the history of medicine for his work on the diagnosis and treatment (of that time) for syphilis. There were many theories of how the disease originated, one of them being that Christopher Columbus (1451?-1506) brought it back from his journeys to the New World, and another linking it to the importation of African slaves into Europe. The disease was known as "the calling card of civilized man," and was becoming more prevalent in all parts of the world.

Fracastoro began a series of treatments and believed that if treated in its earliest stages, syphilis could be cured. He prescribed a rigorous schedule for the afflicted that included exercising until heavy sweating occurred. In addition, he developed a mixture of mercury, sulfur, and hellebore that he smeared all over the patient, then encased the sufferer in wool fabric and confined him to bed to await profuse sweating and salivation. Fracastoro firmly believed that this procedure would cleanse the body of its infestation. Other treatments included purges, near-starvation food restrictions, and equally painful regimens. Fortunately, modern medicine has more effective and less painful treatments.

By any standard, Fracastoro was a remarkable man with numerous talents in both the arts and sciences. He lived to the age of 75 and died in 1553, near Verona and his first home on Lake Garda.

BROOK HALL

Leonhard Fuchs
1501-1566
German Physician and Botanist

Leonhard Fuchs is considered one of the "fathers" of German botany because of his book on plants called *De historia stirpium*. Published in 1542, it contains over 500 illustrations that accurately portray a wide variety of plants, most of which were useful in medicine. The book is particularly noteworthy because Fuchs chose superior artists to illustrate his text, which was both scholarly and accurate.

Fuchs was born in Wemding in the Bavarian region of Germany in 1501 and attended German universities, receiving his doctorate from the University of Ingolstadt in 1524. He briefly practiced medicine in Munich, and then went back to Ingolstadt in 1526 to teach medicine. Two years later be became court physician to the Margrave Georg von Brandenburg, returning to teach at Ingolstadt in 1533. But his travels were not over. Becoming a Lutheran, he had to leave the Catholic town of Ingolstadt for Tübingen, where he again became a professor of medicine, this time at the new Protestant University. When an epidemic of an infection called sweating sickness broke out, Fuchs provided an effective treatment. Word of his success spread and he was offered a number of positions in foreign countries, but he remained in Tübingen until his death in 1566.

It is not surprising that a physician was responsible for one of the most noted works in botany. At the time, the study of plants was an important part of medicine because plant substances were the primary source of remedies for disease. Since ancient times, books called herbals were produced which described plants and their medicinal uses. The development of the printing press in the mid-fifteenth century made possible the cheaper production of such books and the accurate reproduction of illustrations. But the images in these books remained crude until the publication of an herbal by another German physician-botanist, Otto Brunfels (1489?-1534), in the 1530s. While his text was not noteworthy, the illustrations done by the artist Hans Weiditz were both accurate and beautiful. Brunfels also inspired Hieronymous Bock (1498-1554), who produced an herbal with a much more original and informative text, but with inferior images. Because of their work, which set the stage for Fuchs's book, Brunfels and Bock are also considered founders of German botany.

Fuchs found three excellent artists—Albrecht Meyer, Heinrich Füllmaurer, and Veit Speckle—to create the drawings and woodcuts for his book. While Weiditz had drawn plants exactly as he saw them, flaws and all, these artists produced images of ideal plants, perfect specimens, and it is their approach that continues to be used in botanical illustration to this day. One particularly noteworthy image they produced was that of corn or maize, which had been brought back to Europe by Spanish explorers. It was the first recorded picture of this plant.

Along with the illustrations, Fuchs provided an informative text, though much of it was based on the plant descriptions handed down from ancient Greek and Roman texts, as was traditional in herbals. But Fuchs did present over 100 species that hadn't been described before, and his comments on firsthand experiences in the field indicate that he was a careful observer. The orga-

A page from Leonhard Fuchs's 1543 *New Plant Book. (Christel Gerstenberg/Corbis. Reproduced with permission.)*

nization of the book was not very informative. Instead of grouping species with similar characteristics, he presented them in alphabetical order. In the final analysis, it is the illustrations that made Fuchs's herbal so memorable. They served as the basis for pictures in many later herbals. Fuchs's book itself appeared in several editions, including ones that were smaller, less expensive, and easier to use for reference than the original large-scale edition. He planned two further volumes but they were never published.

The flowering plant, *Fuchsia*, was so named in honor of Fuchs and his contributions to botany.

MAURA C. FLANNERY

John Gerard
1542-1612
English Botanist

John Gerard was one of the most important English botanists. He is known primarily for his *Herball, or Generall Historie of Plantes* (1597), a book that combined medical knowledge of plants with poetic prose, personal observations, and elaborate illustrations. This book is significant for its content, but is also important because, due to its popularity, it demonstrates the extent to which publishing transformed Elizabethan English society.

John Gerard's botanical texts remain among the most significant and useful written in the English language. Gerard was born in Nantwick in 1545 and was trained as a barber-surgeon. While he never attended university, he was apprenticed to Alexander Mason, a London barber-surgeon with a large practice, from 1561-1568. From 1568 until some time in the 1570s Gerard was the surgeon of a merchant ship. It appears likely that he sailed on a ship of the Merchant Adventurers. The only details known of these voyages are those that Gerard himself recorded. He indicated travel to Moscow, Estonia, Poland, and throughout Scandinavia. These travels were significant in stimulating his interest in the collection of rare and unusual plants.

By 1577 he was married and had established a surgical practice in London. Around this time Gerard was appointed superintendent of the London and Hertfordshire gardens of William Cecil, Lord Burghley. Burghley was one of the most influential aristocrats in the court of Queen Elizabeth and helped Gerard through the course of his life. In fact, Gerard dedicated his first book, his catalogue, and the *Herball* to Burghley.

At this time it was fashionable for the aristocracy to maintain elaborate gardens filled with exotic plants, and to amass extensive collections of herbals in their libraries. As a norm these large, intensively illustrated volumes contained folklore, commentary gathered from antiquity, medicinal uses of plants, and descriptions of the plants themselves. Furthermore, at this point in European history, medicine and herb gardening were intrinsically related. Gerard's status as a surgeon necessitated his botanical knowledge.

Gerard was also responsible for his own garden, located on Fetter Lane, in Holborn, London. This area had, in 1600, long been a site of suburban gardens, and Gerard constantly expanded and enriched his personal plant collection. Indeed, the opening passage of Gerard's *First Booke of the Historie of Plants,* a catalogue of the plants in Gerard's own garden, depicts the extent of his enthusiasm:

Among the manifold creatures of God (right Honorable, and my singular good Lord) that all have in all ages diversly entertained many excellent wits, and drawne them to the contemplation of the divine wisdome, none have provoked mens studies more, satisfied their desires so much as Plants have done, and that upon just and worthy causes: For if delight may provoke mens labore, what greater delight is there than to behold the earth apparelled with plants, as with a robe of embroidered worke, set with Orient pearles, and garnished with great diversitie of rare and costly jewels?

For Gerard, the individual plant specimen invokes the wonder of an entire universe.

Gerard's most famous work, *The Herball, or Generall Historie of Plantes,* was an expanded version of this first book. Published in 1597, it became a landmark in botanical publishing, and is considered one of the monuments of the English language. Indeed, the fame of this book is the result of its felicitous prose and illustration. He used poetic language to vividly describe the natural world. To him, the lowly dandelion is "a floure . . . thick set together, of colour yellow, which is turned into a round downy blowbal that is carried away with the wind." His work appealed to a wide audience, and included information about both the medicinal qualities of plants and their decorative value. Because of the success of this book, he became the surgeon and herbalist of King James I and was granted a lease to a garden which adjoined the royally-owned Somerset House.

Though the *Herball* contains many errors in its identification of specimens and has been described by some as Dodoens's *Herball* for an English audience, Gerard's work dominated its time and is still the best known and most frequently quoted English herbal.

DEAN SWINFORD

Konrad Gesner
1516-1565
Swiss Physician, Zoologist, and Botanist

Gesner (also called "Gesnerus") was among the founders of modern zoology. He also made important contributions to botany, biology, natural history, and scientific bibliography.

Born on March 26, 1516, in Zürich, Switzerland, Gesner was the son of a poor Protestant furrier, Ursus Gesner. Because his father could not afford to care for all his children properly, Gesner was raised until 1527 by his uncle, Hans Frick, a chaplain who awakened Gesner's interest in plants and medicinal herbs, and after 1527 by Johann Jakob Ammann, a choirmaster. Gesner's father died alongside the Swiss Reformation leader Ulrich Zwingli in the Battle of Kappel on October 11, 1531, fighting against Zürich's Catholic neighbors.

At school in Zürich, the boy Gesner so impressed his teachers that they fostered his development in every way. They made it financially possible for him to continue his education. In 1533 he entered the University of Bourges, France, to study theology and languages. The next two years he attended the University of Paris. In 1536 he returned to Switzerland to avoid Catholic hostility toward Protestants and enrolled at the University of Basel as a medical student. Throughout the 1530s he taught school to help support himself. From 1537 to 1540 he was professor of Greek at the Lausanne Academy. After briefly studying medicine at the University of Montpellier in 1540, he received his M.D. from the University of Basel in 1541.

Gesner was a tireless worker, a quintessential Renaissance humanist, and a dedicated, unselfish scientist. He was fluent in Hebrew, Greek, Latin, Dutch, French, Italian, German, and Arabic. From 1541 until he died of the plague in Zürich on December 13, 1565, he wrote, edited, or contributed to dozens of books. Among his goals was to help revive ancient learning and use it to promote new research. In 1543 he published a Greek-Latin dictionary and in 1555 Latin excerpts of Greek medical encyclopedist Oribasius of Pergamon (325-403?).

Gesner's major work, *Historia animalium* (History of Animals), published in five volumes from 1551 to 1587, marked the beginning of modern zoology with a comprehensive catalog of all animals known up to that time. It contains over a thousand beautiful woodcut illustrations, including the famous, but inaccurate, depiction of the rhinoceros by Albrecht Dürer (1471-1528). Evidence of the strong reputation of *Historia animalium* is that both the text and the illustrations were still being plagiarized 150 years later.

Gesner cataloged plants in *Historia plantarum et vires* (History of Plants and Powers), published in 1541, and herbal cathartic and emetic agents in *Enumeratio medicamentorum purgantium* (List of Purgative Medicines), published in 1543.

Before the sixteenth century, the most authoritative catalog of plants and herbal medicines was by ancient Greek botanist Dioscorides Pedanius Anazarbeus (40?-90?). After the untimely death of the brilliant German medical botanist Valerius Cordus (1515-1544), Gesner edited and expanded Cordus's planned revision of Dioscorides, bringing the total of plants described from about 600 to about 1,100, and published it in 1561 as *Annotationes in Pedacij Dioscoridis Anazarbei de medica materia* (Notes on Dioscorides's materia medica).

Gesner was among the great medical bibliographers. After Symphorien Champier (1472-1539), he was very nearly the earliest. His appendix to *De chirurgia scriptores* (Authors on Surgery) in 1555 contained the first useful bibliography of surgery. His 1545-55 *Bibliotheca universalis* (Universal Library), though without a specific medical section, was a primary inspiration for Sir William Osler (1849-1919) to create his monumental *Bibliotheca Osleriana* (1929). Osler called Gesner the "Father of Bibliography."

In 1943 a new Swiss scholarly journal on the history of medicine and natural sciences was named *Gesnerus* in his honor.

ERIC V.D. LUFT

Francis Glisson
1597-1677
English Physician

Francis Glisson was an English physician who is known for, among other things, his medical work with infantile rickets, the liver and the mechanics of muscle contraction.

While the details of Glisson's childhood and upbringing are cloaked in obscurity, what is known is that he was born in 1597 in Rampisham, located in southwest England; he graduated with a Bachelors of Arts from the University of Cambridge in 1621 and obtained a Master of Arts in 1624. He was lecturer of Greek from 1625-1626 and he became a physician in 1634; that same year he was a candidate of the Royal College of Physicians of London. He was named regius professor of physics at Cambridge University in 1636 and he held this position until his death. In 1639 he was named anatomy reader at Cambridge University.

Glisson became involved with a group of physicians and scientists who began to meet on a regular basis in 1645. Out of this group, which became known as the Invisible College, emerged the Royal Society. It was while associated with this organization that Glisson, along with G. Bate and A. Regemorter, was appointed the task of writing a book on the disease rickets; this condition was becoming a noticeable affliction in England at the time. Glisson made such an impression on his collaborators that he became the primary contributor to the project. This work became one of the first books in English on a medical subject.

Titled *Tractatus de rachitide*, the document was concerned with the description and anatomy of the disease rickets which is a softening of the bones; Glisson wrote at that time that the condition was due primarily to a deficiency of nutrition. The bones become brittle and soft and deformities of the bones often result. Something else significant about this book, and other of Glisson's work, is that in them he tried to clearly lay out his empirical findings and arrange in them in a academic manner as well as argue his position and defend any questions or problems with his position. This manner of conveying scientific information in published form was to be a model for years to come.

Glisson's second book, *Anatomie hepatitis*, was concerned with the anatomy of the liver and in it he was the first to describe the layer of connective tissue that covers the liver now known as Glisson's Capsule. This book was chiefly the result of lectures he had made on the subject of the liver in 1640; the book was published in 1654.

It was in this same book that Glisson advanced a new and groundbreaking concept which he called irritability. Simply put, this idea was concerned with the stimulation of muscle and this phenomenon was independent of any external stimulus. Up to this point, the nervous system was poorly understood and muscle contraction had been explained in different ways. One prevailing idea was that on contraction the muscle was inflated much like a balloon only instead of air there was an unexplained "spirit" that inflated the muscle that emanated from the brain and spinal column.

Glisson's concept was more in line with a nervous stimulation, or irritability as he called it. He theorized that nerve signals were sent from the brain by way of a contraction which caused a vibration in the nerve that subsequently caused the "irritability" in the muscle resulting in contraction. It was later that the concept proposed by Glisson was carried on further to explain muscle contraction and the operation of the nervous system.

MICHAEL T. YANCEY

Regnier de Graaf
1641-1673
Dutch Anatomist

In his very short career, Regnier de Graaf contributed to knowledge of the pancreas and the female reproductive system. He became the first scientist to identify and describe the ovary, and discovered a structure within the latter known as the Graafian follicle.

De Graaf was born on July 30, 1641, in the Dutch town of Schoonhoven. During his academic career, he studied at a number of universities: Utrecht, Leiden, Paris, and Angers, where in 1665 he received his degree.

In 1664, de Graaf conducted the first of his important experiments—in this case, involving the pancreas. Using a fistula, a sort of tube, he figured out how to extract pancreatic juice from a living dog. As it turned out, however, the theory regarding pancreatic juice that he formed as a result of this experimentation was incorrect.

Four years later, in 1668, de Graaf began his studies in the reproductive system. He first examined the male system, but yielded no information that was not already known. With the female system, however, he identified the ovary, whose existence had earlier been posited by Johannes van Horne (1621-1670) and microscopist Jan Swammerdam (1637-1680).

Over the few years that followed for de Graaf, he dissected the ovaries of numerous animals and located the Graafian follicle. He even

identified changes in the ovarian structure following fertilization, and witnessed the release of fertilized ova. The latter had yet to be identified by science, however—in fact it would be many years before scientists understood the function of the ovum—and thus de Graaf did not really understand what he was seeing.

De Graaf was only 32 years old when he died on August 21, 1673, a victim of the plague. In his lifetime and afterward, his independent research—notable in part because it was not done under the aegis of any university—attracted the admiration of other anatomists.

JUDSON KNIGHT

William Harvey
1578-1657
English Physician

William Harvey was the first scientist to accurately describe the workings of the human blood circulatory system, thus establishing the modern science of physiology. Harvey based his research on extensive experiments and observations of animals and humans, rejecting ideas that were not confirmed by his experiments. Harvey's discoveries contradicted the long held beliefs of his contemporary physicians and scientists, and subjected him to great criticism and derision. When his work was finally acknowledged long after his death, Harvey's stature rose to that of England's most revered physician, as well as one of the founders of modern medical science.

Harvey was born in Folkestone, England, the eldest son of Thomas Harvey, a well-respected Levant merchant who eventually became mayor of Folkestone. After his first wife died in childbirth, Thomas married his second wife Joan, and they had seven sons and a daughter. William attended the King's Grammar School at Canterbury, benefiting from a proper English school education of academics and athletics. His extraordinary academic abilities became apparent and Harvey was enrolled in Gonville and Caius College at Cambridge University, where he received his B.A. in 1597. Harvey entered the University of Padua at the height of its influence as the most prestigious medical university in Europe, and he received his doctorate in 1602. William was then admitted to the College of Physicians and Surgeons, becoming a physician in 1609.

As part of his physician's training, Harvey served as an assistant surgeon at St. Barthol-

William Harvey. *(Library of Congress. Reproduced with permission.)*

omew's Hospital, where he gathered further clinical experience. After he received his physician's license, Harvey became a professor of anatomy at the College of Physicians and Surgeons and was awarded the Lumleian Lecturer position in 1615. This honored position included a series of weekly lectures over a six-year period, as well as six anatomies a year performed on executed criminals. Harvey's notes from this period indicate that he had begun to develop the ideas and concepts that would later lead to his monumental and critically important discovery of the true role of the heart and blood circulation. Harvey held his university post for 40 years, while serving as the personal physician of King Charles I, and maintaining his own medical practice.

In 1628 Harvey completed the publication that would be considered the most important medical textbook in history, *Exercitatio anatomica de motu cordis sanguinis in animalius* (On the movement of the heart and blood in animals). Harvey proved that blood flowed from the left ventricle of the heart, through the arteries of the body and then into the veins, which eventually returned the blood to the right side of the heart. Harvey confirmed that blood in the right ventricle went to the lungs and returned to the left side of the heart, as part of the pulmonary circulatory system. He also illustrated the functioning of the valves found in the heart and veins, and

assumed the presence of capillaries connecting arteries with veins, though final confirmation would come later in the century with the improvement of microscopic study.

Harvey's research corrected many long held erroneous beliefs about blood circulation and the heart, many originally postulated 1,400 years before by Galen (129-199?), the Greek physician of Rome. Harvey was able to cut through an immense accumulation of ignorance and incomprehension that had been held on to tenaciously by the physicians, scientists, and philosophers of Europe even during the seventeenth century. Harvey introduced both a new system of physiology of the heart and a new dependence on the experimental method of scientific research, a basic tenet of the era of modern science.

Harvey's practice suffered a serious decline and his work was largely rejected during his lifetime. When Charles I was dethroned, the aged Harvey retired into exile at his country home, but he did publish another textbook, *De generatione animale* (1651). In it he famously stated that "all living things come from an egg," and that the egg is a composite of both parents. He is credited with coining the term *epigenesis* to denote the developmental process of the embryo as it is gradually differentiated and emerges from the formless egg mass. This more accurate proposal was opposed by the more commonly held belief that all embryos existed as preformed, miniature individuals within the egg. Eventually a theory of epigenesis modified from that of Harvey was adopted and is currently accepted. Harvey is rightly remembered as the man who discovered the real workings of the human heart and circulatory system, thereby founding modern physiology.

KENNETH E. BARBER

Jan Baptista van Helmont
1580-1644
Belgian Alchemist/Chemist and Physician

Jan Baptista van Helmont played an important role in the transition from classical and medieval ideas and practices to those of modern science, especially in the fields of chemistry and biology. He is considered the founder of scientific pathology and the father of biochemistry.

Helmont received the M. D. degree in 1609 at Louvain. The ideas of the Swiss physician and alchemist Paracelsus (1493-1541) dominated the teaching of medicine at this time. Alchemy, a forerunner of chemistry, assumed the existence of a substance, known as the philosopher's stone, which could transform common metals into gold. Its practice combined mysticism (religious belief in reality beyond normal human perception) with pragmatic chemistry and astrology (belief that the stars and planets influence human action and health) to address various problems such as the prediction of events and the treatment of illnesses.

Helmont was firmly rooted in the alchemical tradition. For instance, he proposed that an agent, which he called archeus, functions as an alchemist within the body. This archeus, he believed, has a body and a soul and causes diseases by imagining them. Also, unlike Paracelsus, who believed that the human body was composed of salt, sulfur, and mercury, Helmont held that the body is made up of water, and that, indeed, water is the basic element from which all living matter is formed.

He, however, was able to move beyond this medieval context. He was an attentive observer and was the first to employ quantitative and experimental methods in biological and physiological problems. He was first to apply chemistry systematically to biological processes, studying such phenomena as digestion and nutrition from a chemical point of view, i.e. as resulting from the interaction of chemical substances. As an example, he used alkaline compounds to treat the pain caused by an excess of stomach acids. As a result of this groundbreaking work, he is regarded as the founder of biochemistry.

He also applied his method of deliberate observation to diseases, attempting to correlate diseases with their causes. His systematic approach placed such study on a scientific basis, and consequently he is considered to be the founder of the scientific discipline of pathology.

He was first to perceive fully that the substance known simply as air is not made up of a single entity but is composed of a number of substances. Noting that these substances have the ability to fill any space made available to them, he modified the Greek word for space, *chaos*, and invented the name *gas*. He also pointed out that gases are generated in various chemical processes. He discovered the gas carbon dioxide and showed that it is produced both in the burning of coal and in the fermentation process of winemaking. His ideas formed the basis of the work of Robert Boyle (1627-1691) who is regarded as the founder of modern chemistry.

In his effort to apply quantitative methods to his observations and experiments, Helmont was the first to use the boiling point and melting point of water as standard points for a temperature scale, thereby increasing the accuracy of temperature measurements.

During the period of history in which Helmont lived and worked, the world moved significantly away from the philosophy and the technology of the middle ages, a period also known appropriately as the dark ages. Although a product of the past and a participant in its beliefs and practices, Helmont's reliance on careful observation and quantitative experimentation led to significant contributions to the body of scientific methods and ideas which subsequently developed into the dramatic change in thought and practice known as the scientific revolution.

J. WILLIAM MONCRIEF

Anton van Leeuwenhoek
1632-1723
Dutch Microscopist and Scientist

Known as the father of microbiology, Anton van Leeuwenhoek was a Dutch scientist who was the first to use a microscope to observe bacteria and *protozoa* (one-celled animals). His researches on lower animals refuted the then-held doctrine of spontaneous generation (the idea that living organisms could be created from inanimate matter), and his observations helped lay the foundations for the sciences of bacteriology and protozoology. Leeuwenhoek was an unlikely scientist. He had no formal training, but with his skill, his diligence, his endless curiosity, and an open mind free of the scientific dogma of his day, he succeeded in making some of the most important discoveries in the history of biology. He routinely shared his research and in the process opened up the entire world of microscopic life to the scientific community.

Leeuwenhoek was born in Delft, Holland, on October 24, 1632. His relatives were primarily tradesmen, so although he was educated as a child in the town of Warmond, he did not receive any further educational training. He lived with his uncle for a time at Benthuizen and apprenticed in a linen-draper's shop. Around 1654 he returned to Delft, where he spent the rest of his life. He set himself up in business as a fabric merchant, but he is also known to have worked as a surveyor, a wine assayer, and as a minor city official. In 1660 he obtained a position as

Anton van Leeuwenhoek. *(Library of Congress. Reproduced with permission.)*

detailed descriptions of yeast. He also demonstrated that spermatozoa were necessary for fertilization of the egg. Because of his hobby, a friend put him in touch with the Royal Society of England; Leeuwenhoek was elected as a fellow in 1680. Many of his discoveries were made public by the society.

Leeuwenhoek's microscopic research on the life histories of various low forms of animal life was in opposition to the doctrine that they could be produced by spontaneous generation. He showed that both grain weevils and fleas were produced not from grain or sand, as was the prevailing wisdom at that time, but that they were bred in the regular way of insects. He also demonstrated similar ideas in other animals. Leeuwenhoek became famous because of the dramatic nature of his discoveries. He gladly demonstrated his microscope for dignitaries the world over and continued to do so until his death in 1723.

JAMES J. HOFFMANN

Li Shih Chen
1518-1593
Chinese Pharmacologist and Physician

Li Shih Chen was born in the Hupeh province in northeast China, into a family whose male ancestors were educated in medicine and practiced pharmacy. His father was a medical officer who wrote five treatises, one of them on smallpox. Li Shih Chen's major work is called the *Pen Ts'ao* or *Pen Ts'ao kang mu*.

Like European Renaissance notables, great thinkers in China were compiling knowledge of the past and creating encyclopedic works and collections. Li Shih Chen was born during the Ming Dynasty (1368-1644), a time known for its brilliance and prosperity. Yung Lo, the third emperor, built grand temples and palaces in Peking, architectural wonders that awed the rest of the world.

Li's father encouraged him to become a civil servant like himself, but Li was unable to progress. His interests lay in medicine, and his father permitted him to assist in observing and examining patients and learning what existed in Chinese *materia medica* (books on pharmacy).

Unfortunately, there is little information available about Li Sheh Chen's mother or female relatives because Chinese culture was patriarchal and elevated and rewarded the importance of males while subordinating women. This system went back to the philosophy of Confucius (551-

chamberlain to the sheriffs of Delft. This secured his income and enabled him to devote much of his time to his hobby of grinding lenses and using them to study tiny objects. He seems to have been inspired to take up microscopy by having seen a copy of English physicist Robert Hooke's (1635-1703) illustrated book *Micrographia* (1665), which depicts Hooke's own observations with the microscope. While Leeuwenhoek's microscopes were simple in design, he possessed tremendous skill in grinding lenses. While he is often incorrectly credited with inventing the microscope, his lenses are generally regarded as having the best resolution of any microscopes in the era. Although Leeuwenhoek's studies lacked the organization of formal scientific research, his powers of careful observation enabled him to make discoveries of fundamental importance.

In 1674 Leeuwenhoek began to observe bacteria and protozoa, which he isolated from different sources and named *animalcules*. He accurately calculated the size of these specimens. He was the first to describe spermatozoa from insects, dogs, and humans. Leeuwenhoek also studied the structure of the optic lens, striations in muscles, the mouthparts of insects, and the fine structure of plants. He also discovered parthenogenesis (the reproduction of an egg unfertilized by a sperm) in aphids. In 1680 he gave

479 B.C.) and his philosophy of filial piety and its hierarchical nature. Chinese people paid homage and obedience to the emperor, sons obeyed their fathers, and women obeyed and were subordinate to men.

The Chinese philosophy of health and healing is very different to that of European cultures. Humans were only a part of the great scheme of nature and all living things. Although the word "ecology" is new to Western culture, the link between agriculture, environment, and health was the essence of Chinese medicine. In contrast to the concept of "cure," prevention and health maintenance were the ideals in the time of Li Shih Chen and the medical tradition before him.

Chinese philosophy depends on the concept of two principles, yang and yin (male and female, principles of night and day), and five basic elements, five planets, five directions, five seasons, five colors, five sounds, and five organs of the human body. Disease was believed to be a disharmony between the five fundamental organs. And the restoration of the balance between yin and yang was viewed as the cure for any disease. Since the human being was inseparable from the universe and environment, all these factors had to be taken into account when treating a patient. In the medical world of Li Shih Chen, cosmic relationships with the 29 healing Buddhas were necessary. Hence pharmacy was a benevolent art.

Since Chinese medicine also depended on regional treatments and differences, the compendium that Li composed was detailed to a degree never before seen in China. As he gained skill as a physician and recognition for his abilities, he was able to read and research. He depended on prior versions of herbals from over 40 previous *Pen Ts'ao* and incorporated medical works into his writings. He made it his mission in life to write about, illustrate, and give directions to find the environment where the particular drug grows, enumerate its pharmacological qualities, state what it is used for, tell about its advantages and disadvantages, give directions for extracting the active ingredient, and determine the dosage for each medicine. There is no book in contemporary culture that approximates this work.

In 1578 the *Pen T'shao Kung Mu* was completed, a description of 1,892 drugs and 10,000 prescriptions. *Pen T'shao* means teaching based on an understanding of drugs. The *Pen Ts'ao Kang Mu* (Great Herbal) consisted of 52 volumes and took Li Shih Chen 27 years to complete.

LANA THOMPSON

Thomas Linacre
1460?-1524
British Physician

A classical scholar and physician, Thomas Linacre is remembered for being "the founder and prime mover of English medicine." Linacre served as personal physician to King Henry VIII. With permission from Henry VIII, he founded the Royal College of Physicians in London, an institution through which he and a body of educated physicians decided who could practice medicine in greater London. Members of the college were charged with the authority to examine and license physicians. Later in his life, Linacre published volumes of Latin grammar before leaving medicine in 1520 to become a Roman Catholic priest.

Linacre was born in Canterbury, Kent, England around 1460. He was educated at Oxford from 1480-1484, then traveled through Italy studying Greek and Latin classic literature. In Italy he studied medicine at the University of Padua, receiving a degree in 1496. Upon his return to England in 1500 he earned another degree in medicine, this one from Oxford. He was appointed tutor to Prince Arthur, son of King Henry VIII. He subsequently served as personal physician for Henry VIII from 1509-1520. Linacre also treated private patients, some of whom were among the most notable men in London: Cardinal Woolsey, Desiderus Erasmus (1466?-1536), and Sir Thomas More (1478-1535).

During this time, dissatisfied with a lack of government regulation over the practice of medicine—which could then be practiced by barbers, clergy, or anyone considering themselves a physician—Linacre sought and received (in the form of a letters patent) Henry VIII's permission in 1518 to institute the Royal College of Physicians. Cardinal Woolsey was instrumental in helping Linacre receive the patent of letters. As founder and president, Linacre, along with other formally educated London physicians, examined and licensed those who would be permitted to practice medicine in greater London. The Royal College of Physicians had the power to fine and imprison those practicing medicine without a license, with graduates of Cambridge and Oxford excepted.

Linacre biographer John Freind has claimed that Linacre sought to raise the profession of English medicine and subsequently gave "encouragement to men of reputation and learning" to become physicians. Under Linacre, British physicians through the Guild of Physicians consolidated against their less educated rivals of barbers, clerics, and apothecaries. They required graduation from Oxford or Cambridge for membership in the Guild. While the consolidated physicians were only partially successful at controlling medical practice, Linacre was considered "the founder and prime mover of English medicine."

Besides being a physician, Linacre was, of course, a man of letters. He continued studying the Greek and Latin classics and taught Greek and Latin to British scholars such as Sir Thomas More. His practice of medicine and his work as a grammarian "placed Linacre in the front rank of the medical humanists of the Renaissance," wrote his biographer. For Princess Mary, for example, Linacre translated into English Latin works on medicine written by the Greek physician Galen, the most important physician of pre-Renaissance Europe. He translated Galen's works on hygiene (1517), therapeutics (1519), temperament (1521), natural faculties (1523), the pulse (1523), and disease symptoms (1524).

During this same period, however, Linacre left his medical practice to become ordained as a Roman Catholic priest. He died in London of a stone in the bladder on October 20, 1524, at the age of 64. At the time of his death he had just completed a book on Latin syntax, which was published posthumously.

Linacre was so highly thought of as a grammarian that many consider the poem by British poet Robert Browning (1812-1889) entitled "The Grammarian's Funeral" to be a tribute to the physician.

RANDOLPH FILLMORE

Richard Lower
1631-1691
English Physician, Anatomist, and Physiologist

The son of a country gentleman, Lower was born on the family estate in Tremeer, Cornwall. From 1643 to 1649 he attended Westminster School of St. Peter's College, London, the most celebrated British preparatory school of the era. There he won the praise of the headmaster, Richard Busby, who sent him to Christ Church College, Oxford University, in 1649. Lower received his B.A. in 1653, his M.A. in 1655, and both his B.Med. and D.Med. in 1665, all from Oxford.

At Oxford he studied chemistry under Peter Stahl and became the protégé and later the assistant of Thomas Willis (1621-1675), who was then renowned as the greatest medical scientist in England. Lower was part of an informal group of researchers known as the "Oxford physiologists." Among his scientific collaborators at Oxford, besides Willis, were Ralph Bathurst (1620-1704), Robert Boyle (1627-1691), Robert Hooke (1635-1703), John Locke (1632-1704), John Mayow (1641-1679), Thomas Millington, Walter Needham, William Petty (1623-1687), Henry Stubbe, John Wallis (1616-1703), John Ward, and Christopher Wren (1632-1723). Lower, Millington, and Wren all helped Willis to produce the monumental *Cerebri anatome* (Anatomy of the Brain) in 1664.

Tensions in the mid-seventeenth century ran high between adherents of the ancient medical theories of Galen (130-200) and followers of the new physiology of William Harvey (1578-1657). Willis, Lower, and the rest of the Oxford physiologists were all Harveians. Harvey himself had been at Oxford from 1642 to 1646. One of the staunch Galenists, Edmund Meara, denounced Willis and the entire Harveian worldview in *Examen diatribae Thomae Willisii de febribus* (Examination of the discourse of Thomas Willis on fevers) in 1665. Considering himself Willis's disciple, Lower rushed to his master's defense. His first publication, *Diatribae Thomae Willisii de febribus vindicatio* (Vindication of the discourse of Thomas Willis on fevers), appeared only four months later. It was a vigorous polemic which not only counterattacked Meara, but also defended Boyle and Harvey and laid out a Harveian agenda for further research into the physiology of the blood. It contained some of the earliest correct observations about lung function, the interaction of the heart and lungs, the differences between arterial and venous blood, and the relation of respiration to blood color.

In 1666 Lower married Elizabeth Billing, by whom he had two daughters. When Willis moved to London in 1666, Lower followed him shortly thereafter, set up his own practice, but continued working with Willis on several research projects. Lower became a fellow of the Royal Society in 1667 and a fellow of the Royal College of Physicians in 1675. After Willis died in 1675, Lower was London's leading physician for a short time. His outspoken Whig politics drove most of his highborn patients away after 1678.

Lower was one of the first physicians to perform successful blood transfusions. He reported his experiments on dogs in *Philosophical Transactions* of the Royal Society in 1665 and described his transfusion of sheep's blood into a human in the same journal in 1667. In connection with this work, he greatly improved the design of the syringe.

Lower's 1669 *Tractatus de corde* (Treatise on the Heart) contained many noteworthy advances in cardiology, including the first accurate anatomical description of the structure of heart muscle. He improved Harvey's theory of blood circulation and speculated about why dark blood from a vein turns bright red when exposed to air. One particularly famous section of this book, "Dissertatio de origine catarrhi" (Dissertation on the origin of catarrh), disproved the traditional belief that nasal congestion was caused by mucus dripping down from the brain or the pituitary. Even Andreas Vesalius (1514-1564) had failed to refute Galen on this point.

ERIC V.D. LUFT

Marcello Malpighi
1628-1694
Italian Physician and Biologist

Marcello Malpighi was an Italian physician and biologist who pioneered experimental methods to study living organisms with the aid of the newly invented microscope, thereby founding the science of microscopic anatomy. After Malpighi's contributions, microscopic anatomy became essential for advancing the fields of physiology, embryology, and medicine. He is often called the father of histology (the microscopic study of tissues) because of his work with tissue and cell samples. He helped to change many of the antiquated ideas regarding medicine with his discoveries. As an example, he was the first to demonstrate that capillaries connect small arteries and veins, completing the circuit of blood at the tissue. This discovery provided the factual data to support English physician Willam Harvey's (1578-1657) groundbreaking and controversial theory of the circulation of blood (1628). For almost 40 years Malpighi used the microscope to describe the major types of plant and animal structures, having significant impact on future generations of biologists. Moreover, his lifework brought into question the prevailing concepts of body function. His enemies, who failed to see how his discoveries could possibly improve medical practice, were vigorously opposed to the work of Malpighi. However, he was correctly convinced that microscopic anatomy would prove to have significant value and influence on medicine.

Malpighi was born in Crevalcore on March 10, 1628. He entered the University of Bologna at the age of 17. He was the oldest of eight children and lost both of his parents at the age of 21, prior to completing his education. He placed his career on hold for two years while he settled the affairs for the family. When he returned to his studies, he received a degree in medicine in 1653. Three years later he became a professor at the University of Bologna.

Malpighi questioned the prevailing medical teachings at that time, especially the reliance on the writings of the ancient Greek doctor Galen (130?-200?). He performed experiments that attempted to explain anatomical, physiological, and medical problems of the day in a different light. He was one of the first scientists to recognize the importance and value of the microscope in medicine. In 1661 he identified and described capillaries, which was one of the major discoveries in the history of science. Malpighi's views evoked increasing controversy and dissent, mainly from envy, jealousy, and lack of understanding on the part of his colleagues. Because of this, Malpighi bounced between various institutions of higher learning throughout his entire lifetime.

In 1662, Malpighi accepted a professorship in medicine at the University of Messina in Sicily. It was during this time that he identified taste buds and described the minute structures of the brain, optic nerve, and fat reservoirs. In 1666 he was the first to identify red blood cells and to attribute the color of blood to them.

After four years at Messina, Malpighi returned in 1667 to Bologna, where, during his medical practice, he studied the microscopic subdivisions of specific living organs, such as the liver, brain, spleen, and kidneys, and of bone and the deeper layer of the skin that now bears his name (called the Malpighian layer). Malpighi's work at Messina attracted the attention of the Royal Society in London; in 1669 Malpighi was named an honorary member, the first such recognition given to an Italian. Malpighi continued to make huge contributions to the field of microscopy. Not just confining his work to medicine, he studied insect larvae and plants and published a historic work in 1673 on the embryology of the chick.

Malpighi's ideas were considered extremely controversial, and in 1684 his villa was burned, his apparatus and microscopes shattered, and his papers, books, and manuscripts destroyed. He accepted an invitation from Pope Innocent XII (1615-1700) in 1691 to become the papal archiater (personal physician), a position that he held until his death in 1694.

JAMES J. HOFFMANN

Paracelsus
1493?-1541
Swiss Physician, Pharmacologist and Alchemist

Paracelsus was arguably the most innovative medical mind of the Renaissance. Some denounced him as a charlatan (one who merely pretends to possess knowledge or skill) because of his devotion to magic and the occult. But other scholars agree that he accomplished too much in too many genuinely scientific fields for this accusation to make sense. In an age when authority was expected to remain unquestioned, Paracelsus rejected authority and conducted his own investigations. His iconoclasm inspired Canadian physician Sir William Osler (1849-1919) to dub him "the Luther of medicine."

Paracelsus was born in Einsiedeln, Switzerland, the only son of Wilhelm von Hohenheim, a poor country physician. His real name was Philipp Aureolus Theophrastus Bombast von Hohenheim. He created the pseudonym Paracelsus by combining the Greek prefix *para-*, meaning "beside" or "beyond," with the name of a great Roman physician, Aulus Aurelius Cornelius Celsus (25 B.C.-A.D. 50).

After the death of his mother, Theophrastus (as he was called) and his father moved in 1502 to Villach, Austria. Theophrastus attended the Bergschule in Villach, where his father taught chemistry and where students learned the properties of metals and the economics of mining. He served as an apprentice to his father in medicine and studied the works in his father's library. In 1507 he began his life of wandering. Eager for both knowledge and adventure, he traveled widely, briefly studying at several German universities. Around 1510 he may have received a bachelor's degree from the University of Vienna. In 1513 he enrolled at the University of Ferrara, Italy, where he may have received an M.D. degree in either 1515 or 1516. However, academic life and its pretensions disturbed him. He claimed that the true student should seek knowledge from sorcerers, nomads, thieves, and peasants, as well as from professors, and should travel in order to keep from stagnating. His journeys extended into England, Africa, and Asia. While in England, he claimed he could learn more in a Cornwall mine than at Oxford or Cambridge.

The alchemist Paracelsus. *(Library of Congress. Reproduced with permission.)*

Paracelsus discarded all previous medical systems and held Arabic medicine in particular contempt. To promote his lectures at the University of Basel, Switzerland, in 1527, he publicly burned the works of the acclaimed physicians Galen (129-c. 216) and Avicenna (980-1037). Through alchemy, he experimented with therapeutic applications of metallurgy and chemistry that would later develop into iatrochemistry and eventually into modern chemotherapy.

Some of the ancients believed that disease resulted from disturbances in the body's four humors: yellow bile, black bile, phlegm, and blood, which corresponded to the four elements, temperaments, and seasons. Choleric yellow bile was hot and dry like fire and summer; melancholic black bile was dry and cold like earth and autumn; impassive phlegm was cold and moist like water and winter; and sanguine blood was moist and hot like air and spring. It was believed that all four humors should be in balance in order to ensure good health. Paracelsus renounced this traditional humoral theory and instead attributed the onset of disease to environmental factors such as contagion, the pathogenicity of chemicals, and geographic location.

Paracelsus wrote much, but few of his writings were published during his lifetime. He left manuscripts behind him wherever he went. Thus many of his works were published posthumously and probably many more were lost. His *Grosse Wund Artzney* (Great Surgery) (1536), soon translated into Latin as *Chirurgia magna*, offers a detailed analysis of gunshot wounds and argues against treating them with hot oil, which was common among military surgeons before the work of the French surgeon Ambroise Paré (1510?-1590). (Paré discovered the therapeutic value of simple dressings and soothing ointments for wounds.) *Von der frantzüsischen Kranckheit* (1553) contains Paracelsus's studies of syphilis, which he called "French disease" or "French gonorrhea." He advocated mercury for its cure. In *De gradibus* (1562) he detailed most of his important improvements in drug therapy. One of the first books on occupational health hazards was *Von der Bergsucht oder Bergkrankheiten* (1567), which focuses on the diseases of miners. In *Von den Kranckheyten so die Vernunfft berauben als da sein* (1567), he rejected the popular notion that unwelcome mental states were caused by demons and described psychiatric disorders in terms of purely physical occurrences. In *De generatione stultorum* (1603), he revealed the association of cretinism with endemic goiter.

There is no reason to discount the standard view that Paracelsus was a coarse and brutish man. Sometime before 1524, he acquired a gigantic broadsword that he carried for the rest of his life, even sleeping with it. He supposedly hid his personal supply of laudanum in a secret compartment in its hilt. He died mysteriously in Salzburg, Austria, perhaps as the result of a bar fight.

ERIC V. D. LUFT

Ambroise Paré
1510-1590
French Surgeon

Ambroise Paré inaugurated modern military surgery and was the greatest military surgeon before Dominique Jean Larrey (1766-1842). He invented or introduced many surgical instruments and popularized the use of trusses, ligatures, artificial limbs, and dental implantations. His *Oeuvres* were first published in 1575 and had gone into five editions by 1598.

The son of an artisan in Laval, France, Paré served as apprentice to a barber-surgeon then studied surgery at the Hôtel Dieu hospital in Paris. He became a master barber-surgeon in 1536 and joined the army the same year.

In 1536 King François I made war on the Duke of Savoy and besieged Turin, Italy. Paré was a surgeon with the French army. Until this time, the standard surgical procedure for arrow, bullet, and similar puncture wounds was to cauterize them with hot oil. European doctors had used this ancient Arabic technique for over 500 years and no one questioned it. Paré used it too, until the Savoyard defenders of Turin shot so many French soldiers that he ran out of oil. Desperate because he could not cauterize, he wrapped the newer wounds in bandages soaked with egg yolk, rose oil, and turpentine. The next day he was surprised to discover that the bandaged wounds were healing better than the cauterized ones. He never cauterized again.

Paré experimented with various substances to soak the bandages and presented his results in his major work, *La méthode de traicter les playes faictes par hacquebutes et aultres bastons à feu, et de celles qui sont faictes par flèches, dardz, et semblables* (The method of treating wounds made by harquebuses and other firearms, and those made by arrows, darts, and the like), published in 1545. He was at first disbelieved because he did not know Latin. He wrote in the vernacular at a time when all learned treatises were supposed to be written in Latin. After 1552, when King Henri II appointed him as one of the royal surgeons, such criticism diminished.

While competing in a tournament in 1559, Henri II received a lance wound to the eye. He was attended by Paré, but died eleven days later. Because of this incident, Paré turned his research toward head wounds and published *La méthode curative des playes et fractures de la teste humaine* (The method of curing wounds and fractures of the human head) in 1561.

In his 1549 *Briefve collection de l'administration anatomique* (A short collection on governing the body), Paré improved upon podalic version, an obstetrical technique introduced in the second century by the Greek gynecologist Soranus. Podalic version is a means of extraction used in cases of difficult birth. The physician, with the right hand on the abdomen and the left hand inside the uterus, turns the fetus and extracts it by the feet. No further improvements of this technique were made until John Braxton Hicks (1823-1897) did so in the 1860s.

Paré's 1564 *Dix livres de la chirurgie* (Ten books on surgery) discuss dental and oral surgery, amputations, and several other topics. His 1573 *Deux livres de chirurgie* (Two books on surgery) contains both scientific and fanciful accounts of birth defects.

Ambroise Paré treating a soldier. *(Bettmann/Corbis. Reproduced with permission.)*

Among Paré's innovations were tying (ligating) blood vessels to control bleeding during operations, performing autopsies to determine the cause of death, and using bandages rather than stitches to close wounds so as to minimize scarring. His autopsies led him to think that syphilis might be the cause of some aneurysms. His colleague Jean François Fernel (1497-1558) held the same belief, but the relationship between syphilis and aneurysm was not proved until 1899 by Arnold Heller (1840-1913).

A lifelong devout Catholic, Paré's motto was "I treated him; God cured him."

ERIC V.D. LUFT

John Ray
1627-1705
British Botanist and Zoologist

A founding figure in British botany and zoology, John Ray made extensive classifications of flowering and nonflowering plants and laid the groundwork for the field of taxonomy and other evolutionary studies. The seventeenth-century naturalist is often referred to as the father of natural history in Britain.

Ray was born on November 29, 1627, in the village of Black Notley, Essex, England. His father was a village blacksmith. Many speculate that Ray gained his love of plants from his mother, who was an herbalist. After studying at Cambridge and Trinity universities, Ray traveled throughout England and once to Europe to collect plants, animals, and rocks. He began to document his samples and established specimens in his college garden. The naturalist conducted experimental work in embryology and plant physiology and proved that the wood of a living tree conducts water. Ray's fascination with living and extinct organisms would eventually help make sense of the chaotic mass of names used by the other naturalists of his time.

During this time, Ray also studied for the priesthood. He lectured regularly about natural theology—the doctrine that God's wisdom and power could be understood by studying the natural world He created. Ray was ordained a minister in the Anglican Church in 1660 after years of delay caused by the English Civil War. Ray's formal recognition as a priest, however, was short-lived. During the war a manifesto for church reform had been drafted. England's new king was displeased with the Covenant and in 1662 demanded every minister to swear an oath condemning the reformation. Ray disobeyed the king's order. His defiance cost him his university post, his house, and his treasured botanic garden.

After the Reformation, Ray joined naturalist Francis Willughby (1635-1672) on an expedition to Wales. The pair agreed to undertake the

John Ray. *(Corbis Corporation. Reproduced with permission.)*

malian class. Although a truly natural system of taxonomy would not be realized until the age of Darwin, Ray's system came closer than any of his contemporaries.

Ray's insight that fossils were the remains of living organisms was a significant advance over most other theories of his time. His ideas about the relationship of fossils and Earth's age would eventually be studied by generations of paleontologists.

Years of renowned research paved the way for Ray's induction into the newly formed Royal Society of London, one of the world's first scientific societies, in 1667. As poor health began to restrict his travels, Ray spent the last years of his life interacting with the leading scientists of his time, including zoologist Martin Lister and English scientist Robert Hooke (1635-1703).

After his death on January 17, 1705, at the age of 77, Ray's legacy endured. His book *Synopsis Methodica Avium et Piscium* was posthumously published in 1713, and natural theology remained an influential doctrine for well over a century.

KELLI A. MILLER

huge task of documenting the complete natural history of all living things, with Ray responsible for the plant kingdom and Willughby the animal. A three-year tour of the European continent greatly extended Ray's knowledge of flora and fauna. After Willughby's sudden death in 1672, Ray completed his portion of their project.

Ray's research slowly began to bring order to the study of species. His method of classification would become a powerful tool in evolutionary studies. In 1660 Ray published his *Catalogue of Cambridge Plants*, his first systematic work on plants, birds, mammals, fish, and insects. Ray's goal of a natural system of classification inspired generations of systematists, including Swedish botanist Carolus Linnaeus (1707-1778) and, eventually, Charles Darwin (1809-1882).

Because he was a natural theologian, Ray spent his time investigating the relationship of an organism's form to function. Both Ray and Linnaeus searched for a natural system of classifying organisms that would reflect God's order of creation. But unlike Linnaeus, who used the floral reproductive organs as the basis for classification, Ray classified plants by their overall form and structure, including internal anatomy. He was the first to divide flowering plants into monocots and dicots. His insistence on the importance of lungs and cardiac structure laid the groundwork for the establishment of the mam-

Santorio Santorio
1561-1636
Italian Physician

Santorio Santorio (Latinized as Sanctorius, or Santorius), is primarily remembered as the inventor of the clinical thermometer and the author of *De Statica Medicina* (*On Medical Measurement*, 1614). Santorio attempted to introduce quantitative experimental methods into medical research. Santorio Santorio was born in Justinopolis (now Koper). His mother was from a noble family in that region, and his father, Antonio Santorio, was a nobleman in the service of the Venetian republic. Santorio began his education in Justinopolis and continued his studies in Venice. In 1575 he entered the University of Padua and earned his M.D. degree in 1582. He served as the personal physician of a nobleman in Croatia from 1587 until 1599 when he established his medical practice in Venice, where he became a friend of Galileo (1564-1642). He was appointed to the chair of theoretical medicine at the University of Padua in 1611. As a practicing physician, Santorio appears to have relied on classical Hippocratic and Galenic methods. Instead of blindly following ancient authorities, however, Santorio urged his students to consider sense experience and reasoning, before accepting ancient authority. In 1624 his students charged him with negligence on the grounds that his private practice often took precedence over his teaching duties. Although he was found innocent, he retired from teaching and spent the rest of his life in Venice.

Santorio was a leading member of the iatrophysical school of medical thought and attempted to explain the workings of the animal body on purely mechanical grounds. In contrast to the traditional Aristotelian and Galenic qualities and essences that had been used as explanations for bodily functions, Santorio attempted to describe the fundamental properties of the body in mechanical and mathematical properties, such as number, position, and form. As an iatrophysicist, Santorio compared the body to a mechanical clock or machine. His ingenious inventions and instruments, including a wind gauge, a water current meter, a pulsilogium, and a thermoscope allowed him to describe various phenomena in numerical terms. Although there is some controversy over the invention of the thermoscope, Santorio was apparently the first to apply a numerical scale to the thermoscope and create the clinical thermometer. Galileo's experiments with pendulums probably inspired Santorio's adaptation of the pendulum to medical practice as an accurate way to determine pulse rate. The pulse clock was described by Santorio in a book that he published in 1603. Similarly, Santorio's invention of the clinical thermometer in 1612 might have been inspired by Galileo's thermoscope, a device used to measure hot and cold. The thermoscope devised by Galileo in 1597 consisted of a glass vessel about as large as an egg, with a long glass neck. When Galileo heard about Santorio's instrument, he appears to have complained and asserted his priority. Probably, Santorio deserves credit for adapting several of Galileo's inventions to medical practice and for inventing others. With his pulse clock, a clinical thermometer, and the large balance used in his metabolic experiments Santorio was a pioneer in establishing the importance of accurate physical measurements in medicine.

In order to test the ancient idea that a second kind of respiration occurs through the skin, Santorio constructed a large balance on which he often ate, worked, and slept, in order to measure fluctuations of his body weight in relation to his solid and liquid excretions. After 30 years of such measurements, he established that a large part of the food and drink that he had ingested was apparently lost from the body in the form of

Santorio Santoro performing a weighing experiment. *(Christel Gerstenberg/Corbis. Reproduced with permission.)*

perspiratio insensibilis, or "insensible perspiration." Santorio reported his observations in his landmark text *On Medical Measurement* (1614), which has been called the first systematic study of basal metabolism. The book was widely read and highly respected by his contemporaries.

LOIS N. MAGNER

Olivier de Serres
1539-1619
French Agronomist

Olivier de Serres is sometimes referred to as the father of French agriculture. His most famous work, the *Théatre d'agriculture* (1600), provided a complete guide to agricultural practices and helped to outline the ideals of French Protestant culture. Serres advocated agricultural reform, and explained land-management techniques of the utmost importance in an era threatened by nearly continuous famine, drought, and war.

Olivier de Serres was born in 1539 on his family manor, called Pradel. He spent his entire life on this estate, and established many experimental fields on its lands. His father, Jacques de

Serres, and his mother, Louis de Leyris, were both from well-established families of landowners in Vivarais. However, despite his affluent upbringing, there is little evidence to substantiate his level of education. He studied at the University of Valence but probably never graduated. Still, he was well versed in agronomy and studied all that was written concerning agricultural practices.

Serres's most important work was his *Théatre d'agriculture,* which was published in 1600. This book was immensely popular and appeared in more than 45 editions through the course of the century. Indeed, it remained the standard textbook of agriculture for an even greater period of time. The work is divided into sections that explain specific agronomic and horticultural practices. These sections focus on subjects as diverse as methods of tilling, dressing, and sowing, the nature of soils, the art of grafting, the maintenance of kitchen gardens, the planting of trees, and the comportment of the country gentleman.

In essence, this book operated as a guidebook for the French landowner. It discussed the domestication and cultivation of French plants and animals, and provided detailed instructions on how to maximize the productivity of the land. Serres advocated irrigation and provided useful tips for the maintenance of well-drained soils. Likewise, he was one of the first advocates of water conservation techniques. Furthermore, while he discussed the uses of native species, he also advocated the use of artificial grasses on fallow lands. Indeed, he introduced hops to France, an ingredient necessary for beer-brewing.

Serres dedicated his *Théatre d'agriculture* to Henry IV, who was a close friend. The two were so well acquainted that in 1599, Serres published a book on the art of silk collecting especially for the King. Because of this book, Serres was recognized as an expert on sericulture, the production of raw silk and the rearing of silkworms for this purpose.

The wealth of knowledge contained in Serres's *Théatre d'agriculture* was significant regardless of social circumstances. However, the social conditions of the time directly impacted Serres, and shaped his role as a public figure. His focus on agricultural practices was imperative during a time of constant famine, war, and drought. Increased harvests were necessary to accommodate the increased demands that were placed on the land. This era was particularly impacted by religious battles instigated by the uprisings of the Protestant reformation; Serres actively participated in some of these battles. A convert to Protestantism as a young man, Serres acted as a leader of the local Huguenots as early as 1561. Between 1560 and 1570, Serres commanded troops in local campaigns, and was driven from Pradel on many occasions.

Relatively recent developments in the printing and distribution of books, paired with a greater literacy rate, help to account for the extent of Serres's influence. His scientific methods and projects were based on the observation of his immediate surroundings. These methods, employed by many other scientists of the time, were developed in direct reaction to the emphasis on the cosmological and universal theories perpetuated by medieval Catholic scientists. Serres's agenda was both French and Protestant. His book outlined practical farming practices and also illustrated the means by which to attain a cultural and religious ideal.

DEAN SWINFORD

Michael Servetus
1511-1553
Spanish Physician and Theologian

Michael Servetus, or Miguel Serveto, was a person of many interests who is credited with the discovery of pulmonary circulation, the process of blood going to the lungs to pick up oxygen. Servetus's life was one of controversy—from the question of his place of birth, to the end of his life when he was burned at the stake for heresy.

The traditional site of his birth is Tudela, Navarre, in southern Spain, although some of his writings indicate he was born in Villaneuva, Spain, in 1511. Some of his statements lead others to think he was born in 1509. The son of a notary, he was sent to Toulouse, France, to study law but became interested in theology. His friend and mentor was a Franciscan monk Juan de Quintana, who took Servetus to the coronation of Emperor Charles V at Bologna. Disgusted with the extravagance and worldliness of the pope and church, he left Quintana and traveled to Lyons, Geneva, and Basel. These latter cities in Switzerland were the center of the Protestant reformation, with such leaders as John Oecolampadius, Martin Bucer, and John Calvin (1509-1564).

Through his biblical studies, Servetus concluded that the Trinity was not described in the Bible and angered both the Catholics and Protestants with his persistent arguments. But Servetus had a stubborn personality and was determined to voice and print his unpopular views.

Assuming the name Michel de Villeneuve and sometimes Villanovanus, he went to the University of Paris to study then moved to Lyons to work as an editor for the famous publishers, the Trechsel brothers. The editing duties led to his interest in medicine. There, while editing and reading hundreds of medical manuscripts, he met the medical humanist Symphorien Champier (1471-1539) who encouraged him go back to Paris to study medicine under several distinguished anatomists.

In 1537 he published a work supporting the use of syrups for curative purposes, the eating of "correct foods" including citrus, and maintained that sickness was the perversion of natural functions of the body organs, a basic contention of Galen (129-199?). He became an exciting and interesting teacher and lecturer, but Servetus's views on astrology led to his condemnation for teaching of medicine as a function of astrology. In 1538 he was charged and dismissed for lecturing on astrology.

Servetus then moved to Lyons, a port center more accepting of dissenting views, where he practiced medicine. For a while he lived at Vienne and served as personal physician to Archbishop Pierre Palmier. Establishing a general practice, he worked for the next 12 years and became a respected member of the medical community. He was elected by his colleagues to the Confraternity of Saint Luke, serving as supervisor to the apothecaries and overseeing work with indigent patients at the hospital.

In 1553 he wrote a book called *The Restoration of Christianity,* which discussed the pulmonary transit of the lungs within the framework of how the Holy Spirit entered man. According to the Bible, God breathed into man the breath of life or soul. Therefore, he reasoned that there must be a point of contact between the air and blood. Galen had surmised that the blood went through the septum, the dividers of the chambers of the heart. Challenging Galen's idea that this middle wall or septum was not suitable for such passing, he concluded that blood was pumped from the right side of the heart to the lungs through an artery and picked up the vital spirit or air. The "vital spirit" was then received in the left side of the heart, which then pumped blood into the arteries of the entire body.

This same document, along with letters to John Calvin, the Protestant reformer, was his downfall. His letters fell into the hands of the inquisitor in Lyons and his books were seized. He was tried and convicted of heresy but managed to escape with his life.

He decided Italy would be a safe place, but ended up going by way of Geneva, the hotbed of Protestant ideas. He was recognized and arrested. Calvin declared that Servetus must be put to death. He was given the opportunity to retract his ideas and on October 27, 1553, was burned at the stake while still declaring he was right and would never recant.

EVELYN B. KELLY

Jan Swammerdam
1637-1680
Dutch Anatomist and Microscopist

Jan Swammerdam, the son of a prosperous apothecary, was born in Amsterdam, Holland and died there only 43 years later. Even as a child, Swammerdam's main interest was the study of insects. His father was also interested in the natural sciences, but he demanded that his son take religious orders rather than study natural history. As a compromise, Swammerdam was allowed to study medicine at the University at Leiden. Although Swammerdam graduated as a doctor of medicine in 1667, he retained his interest in research on insects and never practiced medicine. Through his research on the natural history and anatomy of insects, Swammerdam established himself as one of the founders of modern comparative anatomy, entomology, and microscopy. Swammerdam saw the microscope as a tool rather than an end in itself and he used it to carry out systematic studies of entomology and comparative anatomy.

During the seventeenth century, naturalists began to use microscopes to examine the morphology of organisms that were difficult or impossible to see with the naked eye. Much of the early microscopical work was devoted to plant tissues, but Swammerdam was primarily interested in the fine structure of insects. In order to study the mouthparts and action of insects, Swammerdam subjected himself to the bite of lice and other insects. In addition, he systematically investigated the fine structure of plants and animals, and discovered the minute "seeds" of ferns. Through dissection and microscopic examinations of human cadavers, he made important discoveries about the uterus, spinal medulla, lymphatic system, and the organs of respiration. The studies of respiration included in his 1667 graduation thesis included original observations about the structure of the lung that had important implications for forensic science. Swammerdam reported that, before respiration had been established, the lungs of a newborn mammal would sink when placed in water. After respiration had been established, the lungs would float.

In 1669 Swammerdam published a general history of insects. The text was written in Dutch rather than Latin. He also carried out dissections of tadpoles, snails, marine invertebrates, and so forth. After contracting malaria, a mosquito-borne disease that plagued him for the rest of his life, Swammerdam was sent to the countryside to recuperate. Instead, he devoted himself to a painstaking study of the morphology and natural history of the mayfly. This research was also published in Dutch in 1675. The study of the mayfly reflected Swammerdam's growing interest in religion as well as his obsession with the details of insect life. He suggested that knowledge of the brief life of the mayfly might give humans beings an appreciation of the shortness of earthly existence and inspire them to a better life. Eventually, his health deteriorated and he became increasingly involved in religion and mysticism; especially after he became a disciple of Antoinette Bourignon, who was known as a mystic and religious fanatic.

More than fifty years after Swammerdam's death, his manuscripts were purchased by Hermann Boerhaave, who had them translated into Latin and published in two large volumes under the title *Biblia naturae* (Bible of Nature, 1737). The text, which included plates engraved from Swammerdam's own drawings, provided the first systematic account of insect microanatomy, classification, and metamorphosis. For each of the insects that had been studied, Swammerdam described the natural history and details of its anatomical structures. For example, Swammerdam provided the first accurate descriptions of the compound eyes, stinger, and mouthparts of the honeybee. Swammerdam discovered the nucleated, red blood corpuscles of the frog. He also described experiments that proved, contrary to traditional belief, muscles do not increase in volume when they contract. In addition, eighteenth century naturalists turned to Swammerdam's studies of insect metamorphosis in support of the theory of embryological development known as preformationism.

LOIS N. MAGNER

Thomas Sydenham
1624-1689
English Physician

Thomas Sydenham put British medical practice on a firm empirical foundation. He eschewed medical theorizing, discounted medieval medical traditions and Renaissance science, trusted no medical author except Hippocrates (460?-377? B.C.), and based his therapeutics on his own direct observation of each patient.

Sydenham was born the son of a country squire in Dorset. In 1642 he matriculated at Magdalen Hall, Oxford University, but the English Civil War interrupted his studies. He fought on the side of Parliament and achieved the rank of captain. He returned to Oxford in 1647 and received his bachelor of medicine there in 1648. This was not an earned degree, but a reward for his services to Cromwell. Much later, in 1676, he received an honorary M.D. from Pembroke College, Cambridge University. He became a licentiate of the Royal College of Physicians in 1663, but never was admitted as a fellow, probably because after the Restoration of Charles II in 1660, the British political climate was against those who had opposed the monarchy.

Although he remained at Oxford until 1655, he distrusted academic medicine and was skeptical of anything he was taught. He was more interested in curing diseases than in speculating about their causes. Nevertheless, he developed his own theories of the origins and transmissions of diseases and held to the humoral theory of Hippocrates that the healthy body is in balance with nature. He claimed that each disease must run its natural course, but that the physician, using nature, could ease the suffering of the patient along this course. His belief that each disease was a separate entity made him a godfather of the nosological movement in the eighteenth century.

In 1655 Sydenham opened the medical practice in Westminster, London, that occupied him for the rest of his life. The same year he married Mary Gee. They had three sons, one of whom, William, became a physician.

Sydenham's observations contributed much to the knowledge of dysentery, malaria, pneumonia, and several nervous conditions. He popularized using cinchona or Peruvian bark, the source of quinine, to treat malaria. He used opiates to treat nearly everything else and he invented laudanum. He was among the first to prescribe fresh air for convalescence, exercise for tuberculosis, and iron tonics for anemia. He formed close friendships with the chemist Robert Boyle (1627-1691) and the philosopher John Locke (1632-1704). The three mutually influenced each other's work for decades.

Sydenham's most important work was his 1683 *Tractatus de podagra et hydrope* (Treatise on gout and dropsy), in which he distinguished between gout and rheumatism. The first edition of his *Observationes medicae* (Medical observations), published in 1676, contained the best descriptions of measles and scarlet fever up to that time, and the first clear distinction between these two diseases. The fourth edition (1685) reported significant progress in smallpox research. His *Schedula monitoria de novae febris ingressu* (A warning essay on the emergence of a new fever), published in 1686, included his description of chorea minor or St. Vitus's dance, now called Sydenham's chorea or rheumatic chorea. Among his other books are: *Methodus curandi febres* (Method of curing fevers), 1666; *Epistolae responsoriae duae* (Two responsive letters), 1680; and *Dissertatio epistolaris ad . . . Guilielmum Cole* (Dissertation in the form of a letter to William Cole), 1682.

Sydenham had many followers, especially posthumously, and his positive influence lasted at least two centuries. His clinical reputation is based primarily on the fact that he made his patients feel better. He relied chiefly upon vegetable materia medica and noninvasive methods. He bled patients as often as did most other physicians of his time, but did not take such great quantities of blood. Even if he did not cure his patients, they were satisfied with his fatherly concern and gentle therapeutics.

The Sydenham Society, dedicated to the preservation of medical knowledge, was founded in London in 1844 and succeeded by the New Sydenham Society in 1858.

ERIC V.D. LUFT

Franciscus Sylvius
1614-1672
German Physician, Chemist, and Anatomist

Franciscus Sylvius was the founder of a school of medicine which proposed that all physical events of the body, including disease, are based on chemical reactions. This school of science later became known as "iatrochemistry," coming from the Greek work "iatro," which means to

heal. He helped shift the perspective of medicine from mystical speculation and superstition to a rational field based on the universal laws of physics and chemistry.

The Sylvius family was of southern Flemish extraction. His grandfather, a wealthy merchant, emigrated from Cambria in France to Frankfurt-am-Main. Born in Hanau, Prussia, which is now Hanover, Germany, Franciscus Sylvius received his education at Sedan, a Calvinist academy. Because of his ancestry and residence in several countries, Sylvius is also known as Franz Deleboe or Francois Du Bois, of which Franciscus Sylvius is the Latinized version.

He went to several great universities in Europe, including Leiden, Wittenburg, and Jena, and received his doctorate at Basel, Switzerland, in 1637. He went back to Hanau to practice medicine, but soon returned to Leiden to lecture on anatomy.

At first he just lectured using the book *Anatomicae intitutiones,* written by Caspar Bartholin (1585-1629). Soon he found himself demonstrating dissection and anatomy to a large audience in the botanical garden of the university. Later, he devised physiology experiments for the instruction of his students. William Harvey (1604-1649) had just proposed his new theory of blood circulation, and Sylvius became an enthusiastic supporter and used dogs to demonstrate his belief in the theory. Relating physiology and chemistry, he developed a theory of the interaction between acids and bases in the blood. Also he described the nature and use of body fluids, including blood, lymph, pancreatic juice, and saliva. He was in error to assume all fluids were either acid or bases and, in order to treat disease, the correct balance must be restored.

Sylvius seemed to be limited at Leiden and in 1641 moved to Amsterdam, where he set up a profitable medical practice and became a respected member of the community. A member of the Protestant Walloon Church, he was appointed physician in charge of relief to the poor and supervisor of the Amsterdam College of Physicians. Although a dedicated physician, he did not abandon his anatomy and physiology studies and devoted spare time to his experiments. He discovered the deep cleft separating the temporal area of the brain from the frontal and parietal lobes. The cleft or fissure is named the sylvian fissure.

In 1658 representatives from Leiden persuaded Sylvius to return to accept a professorship at twice the salary offered to other professors. He threw himself into the new task and attracted students from all over Europe. He remained at Leiden from 1658-72 and became one of Europe's outstanding teachers.

He persuaded the hospital to let him try a unique innovation—taking his students with him as he went to the hospitals. He was one of the first professors to instruct future physicians as they made their rounds through the wards.

He also performed autopsies himself. His students were enthusiastic about his teachings and defended them in public debates. He published his main work, *Praxeos medicae idea nova,* in 1670, but did not live to see the second volume in print. He died on November 16, 1672, at Leiden.

In 1647 Sylvius married Anna de Ligne, the daughter of a lawyer, who was 13 years younger than he. She died in 1657. In 1666, he married a 22-year-old woman, who died three years later. Only one of his children grew to adulthood.

Sometimes the Dutch Sylvius is confused with Jacobus Sylvius (1478-1555) of Paris, a skilled anatomist, teacher, and later opponent of Andreas Vesalius (1514-1564).

Franciscus Sylvius was able to work with the innovations of Harvey but kept them in the general framework of Galen's humoral system. However, in his therapies, he preferred chemical medicines to those of Galen (130-200), using mercury, antimony, and zinc. In this emphasis, his work was pivotal to a new outlook of scientific investigation. He taught many students who went on to be distinguished anatomists, including Jan Swammerdam (1637-1680) and Reinier de Graaf (1641-1673).

EVELYN B. KELLY

Andreas Vesalius
1514-1564
Belgian Anatomist, Physician, and Surgeon

Andreas Vesalius revolutionized the study of anatomy, contributed immeasurably toward making medicine a rigorous empirical science, and wrote one of the most important books in medical history.

Andries van Wesel is generally known by the Latin form of his name, Andreas Vesalius. He was born in Brussels and destined to serve the Holy Roman Emperors. His great-great-grandfather, Peter Witing, was physician to Emperor Freder-

ick III; his great-grandfather, Johannes Witing van Wesele (d. 1476), was physician to Frederick and Archduke Maximilian; his grandfather, Everart van Wesele (d. c. 1485), was physician to Maximilian; and his father, Andries van Wesele (c. 1479-1527), was apothecary to Emperor Maximilian I (the former Archduke) and Emperor Charles V. Frederick granted to Johannes a coat of arms with three weasels. This device is commonly seen in representations of Vesalius.

Vesalius received a Catholic education in Brussels, probably at the School of the Brothers of the Common Life, from about 1520 until 1529, when he entered the University of Louvain. He studied there until 1533, then at the University of Paris until 1536. He received his M.B. from Louvain and his M.D. from the University of Padua, Italy, both in 1537. From 1537 to 1544 he taught surgery and anatomy at Padua. In 1537 he published his paraphrase of nine books of the Arabic physician Rhazes (c. 854-c. 925) and in 1539 his first *Epistola* (The Venesection Letter).

When Vesalius was a student the dominant medical system in Western Europe derived from the Greek physician Galen (c. 130-c. 200). Vesalius had no reason to question Galen until, while at Paris, he began his lifelong habit of doing all his own dissections. The more he saw with his own eyes the less he believed in the anatomical textbooks. He became preoccupied with anatomical science, and even stole a corpse from a gibbet in Louvain in 1536. By 1540 he recognized that many fundamental Galenic presentations of human anatomy were wrong.

At Padua he hired artists to prepare accurate anatomical charts for his students. His first original publication, *Tabulae anatomicae sex* (Six anatomical charts) appeared in 1538. The illustrations were done by the Flemish artist Jan Stephan van Calcar (1499-1546?), a student of Titian (1488?-1576).

In 1540 at the University of Bologna, Vesalius publicly announced that Galenic anatomy was not human anatomy, but the anatomy of apes, pigs, and dogs transposed into human form. He had already proved this assertion to his own satisfaction, but to demonstrate it to the world, he began to prepare his monumental work, *De humani corporis fabrica libri septem* (Seven books on the structure of the human body), and an outline version to be published simultaneously, *Suorum de humani corporis fabrica librorum epitome* (Epitome of his books on the structure of the human body). Both appeared in

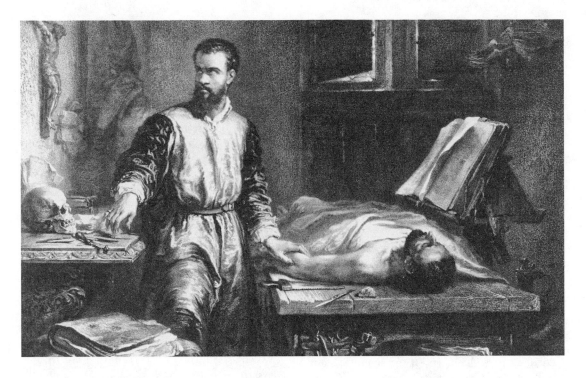

Andreas Vesalius, dissecting a cadaver. *(Corbis Corporation. Reproduced with permission.)*

1543. Vesalius visited Titian's studio in Venice in 1542 and Calcar probably did most of the illustrations for both the *Fabrica* and the *Epitome*.

Vesalius personally gave the uniquely colored dedication copy of the *Fabrica* to Charles V in Mainz, Germany, in 1543. Charles was impressed enough to appoint Vesalius his household physician. Vesalius abruptly abandoned the study of anatomy to become an imperial courtier, but remained active in other areas of medical research. In 1546 he published his second *Epistola* (The China Root Letter), about syphilis. He published a definitive edition of the *Fabrica* in 1555, but his primacy as an anatomist was usurped in his lifetime by his student Gabriele Falloppio (1523-1562).

In 1544 he married Anne van Hamme. Thereafter he spent his life mostly traveling with the emperor throughout Europe, sometimes as a military surgeon. When Charles V abdicated in 1556, he created Vesalius Count Palatine with a pension. With Ambroise Paré (c. 1509-1590), Vesalius attended the dying French King Henri II in 1559. The same year he moved to Madrid as court physician to Spanish King Philip II. He died on the Greek island of Zacynthus while returning from a pilgrimage to Jerusalem.

Vesalius inaugurated an important subgenre of medical publications: "artistic anatomy." His *Fabrica* and the hundreds of books that followed

it in this subgenre are renowned not only for their scientific accuracy, but also for their beauty and power as aesthetic objects.

ERIC V.D. LUFT

Raymond Vieussens
1641?-1715?
French Physician, Surgeon, and Anatomist

The son of François Vieussens, a townsman of Vigan, France, Raymond Vieussens was best known for his advancements in the knowledge of the brain and spinal cord. He received his medical degree from the University of Montpellier in 1670, then became chief physician at Hôtel Dieu St.-Eloi, the main hospital in Montpellier. Thereafter he divided his career between Montpellier and Paris, but held no university appointment. In the late 1670s he married Elisabeth Peyret, by whom he had twelve children. Two of his sons and two of his sons-in-law became physicians.

Favored and protected by the French aristocracy, Vieussens became wealthy through the patronage of royals and nobles. He was the personal physician of the Marquis de Castries, the Archbishop of Toulouse, and the Duchess of Montpensier. Even though he never treated King Louis XIV (1638-1715), he received the titles of Royal Physician in 1688 and State Councillor in 1707.

Throughout his career he was not on good terms with the medical faculty at Montpellier, primarily because of his long and bitter dispute with Professor Pierre Chirac (1650-1732) about which of the two had first discovered an acidic salt in the blood. Also involved in this public controversy were the English physician Richard Lower (1631-1691), Guy Crescent Fagon (1638- 1718), and William Briggs (1642-1704). The irony of the situation is that both Vieussens's and Chirac's results were incorrect.

As an anatomist, Vieussens was careful, observant, and generally accurate, but his physiological studies were suspect. Several dubious schools of physiological speculation held sway in the seventeenth century, and Vieussens was influenced by two of them—the dualistic iatromechanics of French scientist and philosopher René Descartes (1596-1650) and the mystical iatrochemistry of Franciscus de le Boe Sylvius (1614-1672). Twenty-first-century readers may discount Vieussens's accounts of bodily processes and functions but still find lasting value in his accounts of bodily structures.

Vieussens's two main influences in the study of anatomy were English anatomist and physician Thomas Willis (1621-1675) and Danish anatomist Nicolaus Steno (1638-1686). All three were interested in the anatomy of the brain. Vieussens concentrated his anatomical research on the nervous and vascular systems.

Many neurological and cardiological features are named after him, including Vieussens's centrum (the white oval core of each hemisphere of the brain); Vieussens's valve (a sheet of thin white tissue in the brain); Vieussens's ventricle (one of the fluid-filled spaces in the brain); Vieussens's ansa (a loop in the ganglia around the subclavian artery); Vieussens's ganglion (a network of nerves between the aorta and the stomach); Vieussens's anulus, isthmus, or limbus (a ring of muscle in the right atrium of the heart); Vieussens's foramina (tiny openings in the veins of the right atrium of the heart); and Vieussens's veins (small veins on the surface of the heart).

In his major work, *Neurographia universalis* (General neurography) (1684), Vieussens was the first to describe the centrum ovale precisely; thus it is sometimes called Vieussens's centrum. Because Félix Vicq d'Azyr (1748-1794) achieved a more refined understanding of its structure, it is more often called the centrum semiovale, or Vicq

d'Azyr's centrum. The beautifully executed copperplate illustrations make *Neurographia universalis* second in importance only to Willis's *Cerebri anatome* (Anatomy of the brain) (1664) among seventeenth-century books on neuroanatomy.

Vieussens's *Novum vasorum corporis humani systema* (New system of the vessels of the human body) (1705) is a classic of cardiology. It includes the earliest accurate descriptions of mitral stenosis, aortic insufficiency, aortic regurgitation, and several other heart diseases and circulatory disorders. Vieussens was also the first to describe the left ventricle and some of the blood vessels of the heart correctly.

The first part of Vieussens's *Tractatus duo* (Treatise on two subjects) (1688) discusses human anatomy; the second part discusses fermentation. Among his many other works are *Epistola de sanguinis humani* (A letter about human blood) (1698) and *Deux dissertations* (Two dissertations) (1698), both about blood; *Dissertatio anatomica de structura et usu uteri ac placentae muliebris* (Anatomical dissertation on the structure and use of the uterus and placenta in women) (1712); *Traité nouveau de la structure de l'oreille* (New treatise on the structure of the ear) (1714); *Traité nouveau des liqueurs du corps humain* (New treatise on human body fluids) (1715); and *Traité nouveau de la structure et des causes du mouvement naturel du coeur* (New treatise on the structure of the heart and the causes of its natural motion) (1715).

ERIC V. D. LUFT

Thomas Willis
1621-1675
British Physician

Thomas Willis was a British physician and the leader of the English iatrochemists, a group of scientists who strived to explain bodily functions and disease from a chemical standpoint. Willis's research laid the foundational text on the anatomy of the central nervous system.

Willis was born on January 27, 1621, in Great Bedwyn, Wiltshire, England. As did many men of his time, Willis chose to follow a career in the church. His early pursuit of the ministry, however, was thwarted when the English Civil War broke out. After deciding that such a career would be risky, he turned to medicine. He alternated between the classroom at Oxford University and the battlefield as he fought for the Royalist Army from 1643-1646.

For 15 years Willis served as an Oxford professor of natural philosophy. In 1660 he was appointed Sedleian Professor of Natural Philosophy. When the anatomist began his Sedleian lectures, he ignored the traditional Aristotelian science, instead emphasizing iatrochemistry and the correlation between chemical and physical interactions. With the help of his students, Christopher Wren (1632-1723), Robert Hooke (1635-1703), and John Locke (1632-1704), and his associate, Richard Lower (1631-1691), Willis actively engaged in anatomic and physiologic research. Lower's extraordinary skill in anatomical dissection allowed Willis to conduct his pioneering study of the central nervous system and cerebral circulation.

In 1664 Willis's careful studies of the nervous system and various diseases were outlined in his comprehensive work, *Cerebri Anatome, cui accessit Nervorum descriptio et usus.* His work was the most accurate account of the nervous system to date and was the first to clearly identify the distinct sub-cortical structures. Willis's work detailed the concept of circulation of the blood and introduced the world to the term "reflex action." He described the circle of arteries located at the base of the brain (which is still called the Circle of Willis) and explained its function.

Britain's new authority on the brain also documented the spinal accessory nerve, the nerve responsible for motor stimulation of the major neck muscles, also called the eleventh cranial nerve. In 1671 he first described myasthenia gravis, a chronic muscular fatigue marked by progressive paralysis and fever.

As a follower of the Paracelsian School of Iatrochemistry, Willis attempted to understand anatomy and physiology by studying the body's chemical interactions. Willis was the first to identify sugar in the urine of diabetics, a discovery that led to the classification of diabetes mellitus. His acute observations of various epidemics also resulted in the first clinical description of typhus fever and launched the English tradition of epidemiology.

In 1674 he published *Pharmaceutice rationalis,* a series of case histories, post mortems, and therapies that vied to establish anatomy and chemical experimentation as the basis of pharmacology. His book, however, was not considered a success.

Willis died a year later on November 11, 1675, in London, England.

KELLI A. MILLER

Francis Willughby
1635-1672
English Natural Historian

Francis Willughby was a gifted amateur observer and collector of natural history specimens. His work in zoology, particularly with insects and vertebrates, together with his moral and financial support, constituted a major contribution to the pioneering work of John Ray in biological systematics.

Francis Willughby was born the third child and only son of Sir Thomas Willughby and his wife, country gentry living in the county of Warwickshire. Willughby was educated at Sutton Coldfield School and at Trinity College, Cambridge. There he met John Ray (1628-1705), eight years his senior and a lecturer at the college, who later became a pioneering biological systematist. Following his graduation in 1656 Willughby continued his studies at Cambridge and through extensive private reading of natural history at the Bodleian Library, Oxford University. In the early 1660s Willughby and Ray undertook a series of collecting expeditions through various parts of England and Wales. This travel was largely underwritten by Willughby. The two men evidently decided to try and place the plant and animal worlds within some scientific system. In 1663 Willughby became one of the founding members of the Royal Society of London. In the spring of 1663 Willughby and Ray, in company with Phillip Skippon and Nathaniel Bacon, who had also been students of Ray's at Cambridge, traveled in the Netherlands, Germany, and Italy. They spent the winter of 1663-64 in Padua, where Willughby registered and studied anatomy at the university there. In 1664 Willughby left Ray and Skippon in Italy, returning to England via Spain. Willughby purchased many engravings of plants and animals and continued with certain botanical experiments, mainly having to do with sap rising in birch trees. When Ray returned to England, he and Willughby continued this activity, while Ray also helped arrange and classify Willughby's specimens. Willughby's father died in 1665, and in 1668 Willughby married Emma Barnard, by whom he had a daughter and several sons. He had been planning a trip to the American colonies to study the animal life there when he died at his home in Middleton, England, in the summer of 1672.

Ray had been living and working at Willughby's home as tutor to his friend's children; he continued there for several years in that capacity while simultaneously working on Willughby's collections. Ray was also a trustee of Willughby's estate, with Skippon and Francis Jessop, another colleague, as co-trustees. Willughby's will provided Ray an annual stipend of sixty pounds. Ray spent some time in editing and completing two books of Willughby's work and observations, *Ornithologia* (1676), and *Historia piscium* (1686). There was considerable debate in the nineteenth century concerning Willughby's share in the conclusions reached in these books and in Ray's own later publications. Willughby was an extremely able amateur naturalist whose observations informed much of Ray's work. He was also a warmly supportive friend and colleague. But, taken as a whole, his work did not match the stature of Ray's accomplishments. Willughby was a young man of many varied interests, some of which ran in other directions, including family history, foreign language vocabularies, and the recreational games played by people in late seventeenth-century England. Unfortunately, he died before accomplishing much of the work in natural history that he had planned. Ray's editions of Willughby's books incorporated conclusions drawn from Willughby's field work and experiments, his collections, and their many discussions concerning scientific method. With the aid of Skippon and Jessop, Ray, a very self-effacing person, placed all of these materials within an understandable scientific context. Ray's own later publications on animals probably benefited in some measure from the close association with his younger colleague. But Ray's conceptual and organizational genius, coupled with his own very extensive study and field work, merit his firm position as one of the fathers of modern biology.

KEIR B. STERLING

Biographical Mentions

Ulisse Aldrovandi
1522-1605

Italian physician and one of the founders of modern natural science. Aldrovani studied law, mathematics, philosophy, and medicine at the universities of Bologna, Pisa, and Padua, receiving his medical degree from Bologna in 1553. He did not practice medicine, but became a ju-

nior member of the university's medical faculty. The income from this post enabled him to collect natural history specimens throughout Italy, often with colleagues or students. Later appointed to teach the history of medicinal plants, he also established and directed a botanical garden, but soon was covering all natural history. Owing to increasing student interest in his subject, he was made a full professor of the natural sciences at Bologna in 1561. He did pioneering work in chick embryology and became an authority on both pharmacology and on civic hygiene for the City of Bologna. His dozen sumptuously illustrated published works, most of which appeared after his death, were published with financial assistance from Pope Gregory XIII. They included studies of mammals, birds, reptiles, fish, and invertebrates.

Prospero Alpini
1553-1616

Italian botanist and physician who first examined plants outside of their uses for medicine. Born in Marostica, he studied medicine at the University of Padua. His major contributions resulted from his travels, especially to Egypt, where he brought home exotic plants and sought to categorize them. Some of his descriptions were included in the writings of Linnaeus, the great Swedish organizer plants and animals, who named a genus after him.

Giulio Cesare Aranzio
1530-1589

Italian physician and surgeon who saw in the ancient surgical texts the foundations of a modern practice. A lecturer in anatomy for 22 years at Bologna, he sought to treat head wounds by adhering to the precepts of Hippocrates. A humanist surgeon, Aranzio sought to return to the masters of antiquity for practical advice. In 1564 he coined the term hippocampus, a structure in the brain, and described some of the blood vessels in the brain. He was also the first to do plastic surgery on people whose noses had deteriorated due to syphilis.

Giovanni Arcolani (Arculanus)
1390?-1484

Italian surgeon who was an early pioneer of dentistry. Born in Verona, Arcolani was the first to recommend the filling of teeth with gold fillings. As a surgeon, he specialized in dentistry in general and wrote about his findings in a book, *Practica,* published in 1483. He followed the teachings of Avicenna and was an opponent of Marliani, an avid supporter of Aristotle.

Gaspere Aselli
1581-1625

Italian physician who investigated the anatomy of the digestive system. As a professor of anatomy at Pavia, he cut open a recently fed dog and noted white structures though the mesentery and along the surface of the intestines that emitted a milky fluid when cut. He called these structures white veins or lacteals. Observing these in different species, he explored the function of the digestive juices. His investigations opened up the field of exploration of the digestive system, paving the way for others to overthrow Galen's liver-centered physiology.

Pedro Barba
1608-1671

Spanish physician who first wrote about the use of the bark of the cinchona tree to treat malaria. Until the early seventeenth century malaria was treated like other fevers by bleeding and purging. In 1632 Jesuit missionaries brought the bark of the cinchona tree back to Spain from Peru, where they had observed its effective use by the natives. Barba recognized the value of the tree, which produced quinine, an effective treatment against malaria.

Mariano Santo di Barletta
1488-1550

Italian surgeon who developed a method for the surgical removal of bladder stones. Earlier, in the first century A.D., Celsus described a risky method of inserting a finger into the rectum, causing the stones to bulge into the body cavity. Mariano developed a surgical procedure where the stone could be removed by an instrument, avoiding the complications of Celsus's procedure.

Caspar Bartholin
1585-1629

Danish anatomist and theologian who was a prodigy of his time. He went to grammar school at age three and lectured in Greek and Latin at age eleven. He became fascinated with anatomy and studied with Fabricius at Padua, where he wrote a manual on anatomy. His fame arose from his massive learning and reputation as a teacher. He first described the olfactory nerves and the suprarenal glands and published a textbook called *De studio medico* for his sons in 1628.

Thomas Bartholin
1616-1680

Danish physiologist who investigated the anatomy of the digestive system not only in animals but also in humans. He wrote about chyle, the milk-like contents of the lymph vessels of the intestines. His book, *Vasa lyphatica et hepatic exsequiae* (1653), or *The Lymphatic Vessels and the Secretion of the Liver,* introduced color plates into anatomical studies. He discovered the pancreatic duct in 1644 and the parotid glands in 1659. By the late seventeenth century the glands of digestion were known. This attacked and destroyed Galen's physiology, which centered on the liver as blood maker.

Gaspard Bauhin
1560-1624

Danish physiologist and botanist who described more than 6,000 medicinal plants. Encouraging both individual and collaborative efforts, he convinced the universities to work together to share discoveries in their botanical gardens. He wrote a large book called *Pinax theatri botani,* published in 1627. He evoked new interest in plants as the range of remedies increased and assisted in the renewed interest in classical drugs.

Basilius Besler
1561-1629

German naturalist whose botanical illustrations are popular for prints and posters even to this day. His magnificent engravings were the first large folio depictions of historical botanicals. He drew from the specimens in the garden of the Prince Bishop of Eichstatt. His artwork is vivid, realistic, and colorful.

Geradus Blasius
1626-1692

Dutch anatomist who studied the structure of the brain. He discovered a thin membrane that encloses the brain and spinal cord. Because of its resemblance to the web of a spider, he called it arachnoid, meaning "resembling a spider's web."

Theophile Bonet
1620-1689

Swiss anatomist who is known for writing the most complete record of its time regarding surgical, medical, and pharmacological knowledge both traditional and current at that time.

Jacobus Bontius
1598?-1631

Physician and naturalist, perhaps of Danish origin, who is known for supplying the first scientific documentation of the disease beriberi, is a disease known to Asia and the South Pacific and is caused by a nutritional deficiency of vitamin B_1; this disease is known in groups where there is a high intake of rice. He observed the disease firsthand in the South Pacific and published his findings in 1642.

Louise Bourgeois
1563-1636

French midwife who was the first woman to write a textbook on midwifery, *Observations Diverse sur la Sterilite, Perte de Fruict, Foecundite, Accouchements, et Maladies des Femmes, et Enfants Nouveaux Naiz* (Diverse Observations on Sterility, Loss of Fruit, Fecundity, Childbirth, and Diseases of Women, and Newborn Infants; Paris, 1609). Bourgeois, later Boursier, was a well-educated woman who was trained and licensed at the famous Paris hospital, Hôtel Dieu. She raised the respectability and social standing of midwives, eventually becoming official midwife to the court of Henry IV. Her book, based on the teachings of French surgeon Ambroise Paré, and written in the vernacular, was widely read in her time and established her reputation as a pioneer of scientific midwifery.

Robert Boyle
1627-1691

British chemist and natural philosopher who is primarily known for his experiments on the physical properties of gases. In *The Sceptical Chymist* (1661) Boyle proposed a corpuscular theory of matter. With the assistance of Robert Hooke, Boyle constructed an air pump with which he investigated the physical properties of air. The second edition of his landmark work *New Experiments Physio-Mechanicall, Touching the Spring of the Air and its Effects* (1662) established the relationship that is now known as Boyle's law: at a constant temperature the volume of a gas is inversely proportional to the pressure.

Hieronymus Brunschwig
1450?-1512?

German-Alsatian physician, pharmacologist, and military surgeon, also known as Brunschwygk, Braunschweig, or Bruynswijck. His *Buch der cirurgia* (Book of Surgery), published in 1497 and the first printed, illustrated surgical textbook, deals mainly with wounds and injuries, especially fractures, cuts, and gunshot wounds. He also wrote treatises in German and Latin on pharmaceutical distillation processes and on the plague. The 1512 edition of *Liber de arte distillandi* (Book on the Art of Distilling) is a classic of botanical medicine, demonstrating how to obtain essential oils and other chemicals from plants.

Robert Burton
1577-1640

British clergyman, author, and Oxford dean of divinity best known for his astute observations and descriptions of depressive disorders in *The Anatomy of Melancholy* (1621). This treatise, which influenced English writing, outlined the causes, symptoms, and cures of depression, although some of the etiologies were attributed to myth rather than fact. Similar in tone to Augustine's *Confessions,* Burton's self-reflective point of view provides insight into his own melancholia.

Andreas Caesalpinus
1519-1603

Italian physician and botanist whose attempts to create a philosophically grounded system of plant classification helped establish botany as an independent scientific discipline. His *De plantis libri XVI* (1583), the first textbook of botany, described and classified over 1,500 plants. Cesalpino also wrote about anatomy and practical medicine and some historians believe that his ideas about the heart and blood anticipated William Harvey's discovery of the circulation of the blood. Cesalpino believed that the heart was the most important organ of the body and that observations made in the course of bloodletting could lead to insights into the movement of the blood.

Rudolph Jacob Camerarius
1665-1721

German botanist who is primarily remembered for establishing the existence of plant sexuality. In addition to identifying and defining the male and female reproductive parts of plants Camerarius described the function of these parts and the role of pollen in fertilization. He conducted pioneering experiments on heredity in plants. His publications *On the Sex of Plants* (1694) and *Botanical Works* (1697) are landmarks in the history of botany.

Giovanni Battista Canano
1515-1579

Italian anatomist best known for his drawings of the muscles and their relationship to bones, particularly the upper extremity. Giovanni and Antonio Maria Canano studied muscles with Vesalius's brother and Bartolomeo Nigrisoli, whose father taught anatomy in Padua. Giovanni Canano's most famous work, published in 1542, had no title, no date, no name of publisher, and no place of publication given. A facsimile copy published in 1962 was entitled *Musculorum Humani Corporis Picturata Dissectio.* Although

Canano had found valves in veins, Vesalius and Eustachius insisted that they did not exist. However, four years before his death, Fabricius rediscovered them. In 1552 Canano was appointed physician to Pope Julius III at Rome.

Berengario da Carpi
1470-1550

Italian anatomist and surgeon who was the first to name the appendix (1521) and the vas deferens. Da Carpi was also one of the first to illustrate his texts with figures. Other original observations included description of the thymus and the action and existence of the heart valves. He showed that the kidney is not a sieve and that the bladder of an unborn child had urinary pores.

Julius Casserius (Giulio Casserio)
1552-1616

Italian teacher of anatomy, physics, and surgery who studied under Fabrici in Padua. Casserius's engraved anatomical plates were ornate and embellished with detailed landscapes. His first publication, *De vocis auditusque organis* (1601), about speech and hearing, had 34 plates, some comparing human vocal cords with those of the cat. His *Tabulae Anatomicae* (1627), an incomplete volume, was finished by others. He was succeeded by Spigelius.

Elizabeth Cellier

English midwife who established a corporation of midwives in London in 1687 and a foundling hospital. Cellier, who married Frenchman Peter Cellier and converted to Roman Catholicism, was falsely accused of treason in 1680, but defended herself and was acquitted. She wrote about the trial in a short but sensational publication entitled "Malice Defeated." She is believed to have been buried in Great Missenden Church, in Buckinghamshire, England.

Peter Chamberlen, the Elder
1560-1631

English midwife credited with introducing obstetric forceps to the midwifery profession. The Chamberlen family of male midwives, which included Peter Chamberlen the Younger and nephew Peter Chamberlen III, practiced in London and claimed to have a superior way to deliver babies with less pain to women. They allegedly kept Peter the Elder's forceps a "closely guarded secret," training select physicians to use them but for a price.

Giovanni Colle
1558-1630

Italian physician who, following the explanation of how blood circulation works in 1628 by William Harvey, was the first to give a brief account of how a blood transfusion is performed in the same year.

Valerius Cordus
1515-1544

German physician, botanist, pharmacologist, and chemist who won the admiration of his contemporary scientists even before his early death from malaria. Cordus synthesized ether; developed ways to categorize plants; invented phytography, the systematic science of describing plants; updated and augmented the authoritative first-century catalog of medicinal plants prepared by the Greek physician Dioscorides Pedanius Anazarbeus; and wrote the first pharmacopoeia, the Nuremberg *Dispensatorium* (1546). (A pharmacopoeia is a reference source detailing the dosages and administration of medicinal preparations and drugs.) Swiss naturalist Konrad von Gesner (1516-1565) edited and published several of Cordus's posthumous works, including the revision of Dioscorides (1561).

Jean-Baptiste Denis
1640-1704

French physician who was one of the first to attempt therapeutic transfusions of blood from animals to humans. In 1667, with the help of Paul Emmerez, a surgeon and anatomist, Denis injected blood from a lamb into a fifteen-year-old boy who had suffered from a fever. After several apparently successful experiments, a patient died after two injections of calf's blood. Although Denis was not found guilty of malpractice, Denis and Emmerez discontinued their experiments.

Pierre Dionis
1643-1718

French surgeon who, after studying at Confraternity of S. Come, was the first surgeon to Queen Maria Theresa. He was also a professor of surgery and anatomy and he wrote extensively on medicine.

Johannes Dryander
1500-1560

German anatomist and astronomer who is known as one of the first to dissect human cadavers in Germany. He also wrote anatomy books which he illustrated. He graduated from the University of Erfurt in Germany and lectured on astronomy and mathematics in Paris from 1528 to 1533.

Joseph Guichard Duverney
1648-1730

French anatomist who graduated from the University of Avignon and became a professor of anatomy in Paris. He named the nerves of the brachial plexus, which are spinal nerves supplying impulses to the arm and hand and was the first to produce a book regarding the functioning of the ear in 1683. Duverney also discovered a ganglion, or bundle of nerves, behind the eyes and was the first to note the tensor tarsi muscle of the eye.

Johann Sigmund Elsholtz
1623-1688

German doctor who was an early pioneer in the use of opium, whose main active ingredient is morphine, as an analgesic. Together with Johann Major, Elsholtz gave an opium solution intravenously to a dog and noted afterwards that the dog did not react to the prick of a pin. In 1676 Elsholtz observed the shimmering thermoluminescence of heated fluorspar. He died in Berlin in 1688.

Charles-François Felix
1635?-1703

French surgeon who single-handedly raised the status of surgeons above that of barbers and closer to that of physicians with his operation of an anal fistula on Louis XIV. At this time, surgeons did not have the social or educational status of physicians and competed with barbers for surgical work. This single surgery on the king of France gained royal support and did much to further the respect of the profession.

Jean François Fernel
1497-1558

French physician and astronomer who was the first to describe appendicitis, or the inflammation of the appendix, and the first to describe peristalsis which is the wavelike motion of muscle in the throat to push food down to the stomach. Fernel was the first to use the terms *physiology* and *pathology* and was the first to precisely record observations of endocarditis, or the inflammation of the lining of the heart.

Hans von Gersdorff
1454-1517

German surgeon who advanced the emerging profession of surgery by stopping the bleeding associated with amputation of limbs by applying pressure to the limbs and blood vessels. Gersdorff also published a book on surgery in which he first pictured an amputation of a leg. At that time the surgeon would wrap the wound on either side of the limb to be amputated and then cover the stump with the bladder of a bull or hog.

Valentine Greatrakes
1629-1683

Irish man who, with what would today be called faith healing, cured many people suffering from psychosomatic illness—illness that originates or is made worse by the patients' belief that they are ill. His ability to connect with the patients psychologically foreshadowed the concept of healing the whole man.

Nehemiah Grew
1641-1712

English botanist and microscopist who was one of the founders of plant anatomy. His pioneering text, *The Anatomy of Vegetables Begun* (1672) described the structure of bean seeds, the existence of cells, and introduced many of the terms used to describe the parts of plants. In *The Anatomy of Plants* (1682), he discussed the anatomy of flowers, the microscopic structure of plant tissues, and suggested that the stamen functioned as the male sex organ of plants and the pistil as the female organ.

Clopton Havers
1655?-1702

English physician who is known for his study of bone structure and growth, as well as his book *Osteologia Nova,* which was the first publication to describe bone structure in great detail. The intricate vessel system consisting of canals and glands that supplies nutrients in the bones, called the Haversian system, is named after him. Havers graduated with his medical degree from Cambridge University; he practiced medicine in London and was a Fellow of the Royal Society.

Nathaniel Highmore
1613-1685

English physician who precisely described the cavity of the superior maxillary bone, a bone below the nose and holding the upper row of teeth. This cavity, or sinus, is in the upper part of the maxillary bone and opens into the middle nose; it is called the antrum of Highmore. A mass of tissue on the back of the testis, called Highmore's body, is named for him as well. Highmore graduated from Oxford University in 1642 with his medical degree.

Juan Huarte de San Juan
1530?-1592

Spanish physician and philosopher who, at a time when Jews were being expelled from Spain, wrote of the superiority of Jewish physicians. Using geographic determinism as his rationale, he advocated study in Egypt where Jewish physicians taught because it was an excellent ground to develop the imagination. The assumption was that mental faculties necessary for science required imagination, and Spain's climate was not conducive to studying medicine. Huarte may have been a converso, a Spanish convert to Christianity.

Edward Jorden
1569-1632

English physician who wrote "A Briefe Discourse of a Disease Called the Suffocation of the Mother," in which he valiantly defended 14-year-old Mary Glover, accused of witchcraft. Jorden wrote that her fits, actions, and "passions of the body" were not because of the devil but because of her uterus and the problems that women had because of "womanhood." Unfortunately the judges did not accept his explanations.

Johannes de Ketham
?-1490?

German physician, also known as Johannes von Kirchheim, who was professor of medicine at the University of Vienna around 1460. His *Fasciculus medicinae* (Compilation of Medicine, 1491) contains the first printed anatomical illustrations, including the famous dissection scene. The book consists of many short, anonymous medical essays, some of which existed in manuscript 200 years prior. The Italian translation, *Fasciculo de medicina*, with more and better illustrations and text, was popular for about 30 years after its publication in 1493.

Daniel Le Clerc
1652-1728

Swiss physician who practiced medicine in Geneva and wrote extensively on the subject of medicine. His published works include *Surgery* (1695), *History of Medicine* (1696), and *Historia naturalis . . . nascentium* (1715). Le Clerc became a member of the French Academy of Sciences in 1699; he gave up the practice of medicine in 1704 for the public life.

Charles L'Ecluse
1526-1609

French botanist, also known as Carolus Clusius, who helped found the science of botany by cultivating exotic and fungal plants, including the potato (1588). L'Ecluse also wrote extensively on the subject of botany. He was appointed director of the emperor's garden in Vienna, Austria, from 1573-87 and in 1593 was made professor of botany at Leyden in the Netherlands. L'Ecluse first studied law in Belgium, then medicine in Germany; he received a medical degree in 1555.

Leonardo da Vinci
1452-1519

Italian artist and scientist whose insatiable curiosity and artistic and scientific imagination exemplify the genius of Renaissance humanism. Although primarily remembered for paintings such as *The Last Supper* and the *Mona Lisa,* Leonardo was skilled as an engineer, inventor, draftsman, mathematician, architect, sculptor, and anatomist. Leonardo believed that art required accurate anatomical knowledge that could only be obtained through dissection, comparative anatomy, and physiological experiments. He may have dissected as many as thirty human bodies, but he did not complete many of his ambitious projects, including his plan for a great treatise on human anatomy.

Andreas Libavius
1540?-1616

German physician who investigated the practical application of chemistry to the discipline of medicine. Libavius wrote a number of books on medicine and chemistry, the best known is probably *Alchymia,* published first in 1597 then in 1606. He described in this book the methods of preparation for various chemicals; it is regarded as the publication that set the standard for seventeenth-century chemistry texts in France.

Johann Daniel Major
1634-1693

German physician who is thought to be the first to successfully inject a medicinal compound into the vein of a human subject (1662). His book *Chirurgia infusoria* was concerned with the subject of infusory surgery, which are procedures in which a solution is directly injected into the veins. Major practiced medicine in Wittenberg as well as Hamburg, Germany; he was appointed professor of medicine at Kiel, Germany, in 1665, where he also planted a botanical garden.

Edme Mariotte
1620-1684

French physicist and ordained Roman Catholic priest who performed a number of experiments on light and sight, among other things. He discovered the punctum caecum, which is where the optic nerve enters the back of the eyeball from the brain. He experimented with rainbows and light diffraction, and discovered the macula lutea, which is in the center of the retina where the most acute vision is. Mariotte was one of the first members of the French Academy of Sciences.

François Mauriceau
1637-1709

French surgeon who contributed to the practice of obstetrics, the branch of medicine concerned with the care of pregnant women. He was the first to use the term pudendal, which refers to the external genitals of a woman, and the term fossa navicularis, which is a deep depression of the skin in the vagina. Mauriceau described a number of conditions during pregnancy and birthing, such as tubal pregnancy.

John Mayow
1640-1679

English chemist and physiologist who first suggested the existence of oxygen, which he called *spiritus nitroaereus*. According to Mayow, the purpose of breathing was to acquire this special life-giving substance from the air. Mayow investigated the anatomical and physiological basis of respiration and called attention to the analogy between combustion and respiration. He was the first to argue that the seat of animal heat was in the muscles. In 1674 he published the first description of a case of mitral stenosis.

Maria Sibylla Merian
1647-1717

German naturalist who devoted her life to the study and painting of insects and flowers. As a child she was fascinated by caterpillars, moths, and butterflies, and made numerous lifelike sketches. In 1679 she published "Wonderful Transformation and Singular Flower-Food of Caterpillars," which won her scientific recognition for its catalog and drawings of almost 200 European butterflies and moths in various stages of metamorphosis. Merian died in Amsterdam in 1717.

Nicolas Monardes
1507-1578

Spanish botanist and physician who contributed to the sciences of botany and zoology; he classified plants and animals of the East and West Indies. He is noted for having transplanted several plants from America to Spain, and his name is used in the genus classification of plants *Monarda*. Monardes wrote books on plant life, the medicinal use of plants from the Indies, and a history of medicine.

Jakob Nufer
fl. 1500

Swiss sow gelder who performed the first successful cesarean section on a living woman. When Nufer's wife went into labor, he sought the aid of a village midwife, then another, until the entire 13 midwives in Sigershaufen had been consulted. None could deliver the baby. In desperation, he decided to try. Experienced in genital surgery on pigs, he took a knife and opened her uterus, extracting the baby. His wife is reported to have recovered, bore additional children, and lived to age 77.

Guy Patin
1601-1672

French physician who first documented FOP, or fibrodysplasia ossificans progressiva, a disease wherein bones grow in connective tissue. In 1692, a patient with what is now known as FOP went to Patin, who recorded the encounter in his writings. Leader of the Paris medical community, Patin was outspoken in his opposition to the social-service programs administered by Théophraste Renaudot (1586?-1653), who advocated free medical treatment for the poor.

Jean Pecquet
1622-1674

French physician and anatomist credited with a number of discoveries. Among the aspects of human physiology Pecquet was first to observe are the course of the lacteal vessels, the cistern chyli (sometimes called the reservoir of Pecquet in his honor), and the termination of the thoracic duct at the place where it opens into the left subclavian vein. In 1651, he published a book in which he helped to popularize experiments in air pressure conducted earlier by Gille Personne de Roberval (1602-1675). The book introduced the term "elater" to describe the tendency of air to expand.

Heinrich von Pfolspeundt
fl. 1460

German physician and member of the Teutonic Order of Knights. Pfolspeundt was one of the first doctors in the late medieval, early Renaissance period to take medical practices beyond the crude conditions that had prevailed throughout much of the Middle Ages. During his time, a number of other German physicians, particularly those in Strasbourg, served to advance the study of medicine.

Felix Plater
1536-1614

French physician who studied in Montpelier at the time of an outbreak of plague in his hometown. Although his diary narrates the adventures in procuring corpses and observations of executions, most secondary sources credit him with psychiatric insights. His *De Corporis Humani Strucura et Usu* is illustrated with 50 engraved plates that draw on the previous images of Pare and Vesalius.

François Rabelais
1483?-1553

French physician and writer whose most scathing works, *Pantagruel* (1532) and *Gargantua* (1534), satires on the human condition and theology, were vehemently condemned by the Church. His name is a created anagram of Alcofribas Nasier. Originally a Franciscan priest, he engaged in literary exchanges with humanist scholars. Later, he changed to the Benedictines because his Greek language books were confiscated. In 1530 he went to Montpellier and became a doctor within a year. He then moved to Lyons where he lectured on human anatomy and practiced medicine.

Thomas Raynalde

English physician who published *The Byrthe of Mankynde* (1540, 1545). This was the first English translation of Rösslin's *Der Rosengarten,* a German manual for midwives, itself a synopsis from medieval writers. It was the only book that dealt with obstetrics apart from medicine and surgery. The copperplates depict subjects that had been exclusive women's knowledge: the birth chair, birthing room, development of the fetus, and its relation to the uterus. In its introduction, Raynalde wrote that many might not think it proper to have such matters written in "our mother and vulgar language for the detection and discovery by men."

Francesco Redi
1626-1697

Italian physician and scientist who demonstrated that maggots and flies do not arise by spontaneous generation in putrefying meat, but from eggs laid by adult flies. Redi designed a series of carefully controlled experiments using different kinds of meats in flasks that were open, or sealed to keep out air and flies, or covered with gauze to admit air, but keep out flies. Despite the putrefaction that occurred in all cases, maggots only appeared in the open and uncovered flasks.

Théophraste Renaudot
1586?-1653

French physician who organized an early state-supported medical care program for the poor. A longtime protégé of Cardinal de Richelieu, Renaudot served as court physician to King Louis XIII, who commissioned him to organize a system of public assistance. The result, in 1630, was the bureau d'adresse, an organization whose many functions included a free dispensary and a service to direct poor patients to doctors offering free medical care. Renaudot ran into opposition from the Paris medical establishment, led by Guy Patin (1601-1672), and in 1642, after the deaths of Richelieu and Louis, they denied him the right to practice medicine in the capital. In addition to his other activities, Renaudot is remembered as the father of French journalism.

Eucharius Rösslin
1490?-1526

German physician who published the first printed textbook for midwives. Rösslin's *Die swangern frawen und hebammen roszgarten* (A Garden of Roses for Pregnant Women and Midwives) was published in 1513 and went through forty editions. It was still in use in the 1730s. An English translation of the text by Richard Jonas, entitled *The Byrth of Mankynde*, published in 1540, was the first book of its kind to be printed in English. A revised, illustrated Latin translation of Rösslin's book was published by Jacob Fueff in 1554 as *De conceptu et geratione hominis.*

François Rousset
1535-1590

French physician who was the personal physician to the Duke of Savoy and who is known for publishing in 1581 a record of 15 cesarean sections. Rousset's account confirms that the dangerous procedure was indeed practiced at that time; these particular procedures were performed over a span of 80 years. As a point of interest, one woman was recorded to have undergone six of these procedures.

Olof Rudbeck
1630-1702

Swedish naturalist who in 1651 discovered the lymphatic system, notably that of the digestive system and the thoracic duct. The lymph system of vessels moves lymph from the tissues to the bloodstream. Rudbeck was a student of medicine at the University of Uppsala, Sweden, where he later became a professor of botany and anatomy and, at age 31, chancellor. While at Uppsala he also built a botanical garden.

Johannes Scultetus
1595-1645

German surgeon, also known as Johann Schultes or Shultes who wrote *Armamentarium chirurgicum* (Surgical Armamentarium), the late seventeenth and early eighteenth century's most popular guidebook of surgical instruments, techniques, and procedures. Edited by his nephew, also named Johannes Scultetus (d. 1663), the work was first published posthumously in 1655, then translated into most modern European languages, appearing in its final new Latin edition in 1741. Its contents include mastectomy, splinting, lithotomy (the removal of stones from the bladder), trephination, wound and injury repair, cesarean section, and amputation.

Marco Aurelio Severino
1580-1656

Italian anatomist, also known as Severinus, whose contributions to medicine include the description of abnormal formations of tissue and growths such as tumors, descriptions of abscesses, and the investigation of inflamed or swollen lymph nodes. Severino showed the similarity of function in the anatomy of different animals. He also studied how the venom of snakes can kill and investigated antidotes for such poisons. Severino received his medical degree from the University of Naples, Italy.

Jane Sharp
fl. 1650

English midwife who was the first to author a text for the instruction of other English midwives, entitled *The Midwives Book, or the Whole Art of Midwifery Discovered; Directing Child-bearing Women How to Behave Themselves* (London, 1671). The book consisted of six sections: anatomy, signs of pregnancy, sterility, the conduct of labor, diseases of pregnancy, and postpartum diseases. Sharp's book, which went through several editions, was an influential and practical manual that reflected the state of the art in midwifery, and also recognized restrictions on the contemporary education of women.

Adriaan van den Spieghel
1578-1625

Belgian physician who studied in Padua, taught anatomy and surgery, and like Vesalius came from Brussels. Spieghel wrote on fevers and botanical subjects. His anatomical texts, *De Humani Corporis Fabrici* (1627), and on the development of the fetus, *De Formato Foetu* (1626), were neither illustrated nor published during his lifetime. In his will, he asked that Bucretius perform the necessary tasks to do both. The plates in Spieghel's books were acquired from Giulio Casserio's relatives.

Nicolaus Steno
1638-1686

Danish anatomist and geologist, also known as Niels Steensen (or Stensen), whose landmark treatise *The Prodromus of Nicolaus Steno's Dissertation Concerning a Solid Body Enclosed by Process of Nature Within a Solid* (1669) advanced the study of geology and helped establish the science of crystallography. Steno described the structure of quartz crystals, suggested that fossils were the remains of ancient organisms, and that the study of the earth's strata could provide a history of the planet. His most famous anatomical contribution was the discovery of the parotid salivary duct, which is also known as Stensen's duct.

Gasparo Tagliacozzi
1546-1599

Italian surgeon known as the "father of plastic surgery." He became famous throughout Europe for his inventive operation, called Italian or tagliacotian rhinoplasty, which he performed on patients who lacked a complete nose either because of trauma or birth defect. The patient's nose was attached to a skin pedicle on the upper arm and held in place with a brace until it healed. Although he became quite famous, certain Church officials felt that such emphasis on physical appearance was blasphemy and persecuted him.

Edward Topsell
1572-1638

English clergyman and author who is best known for two works of natural history, *The Historie of Four Footed Beastes* (1607) and *The Historie of Serpents, Or the Second Book of Living Creatures* (1608). He completed his studies at Cambridge University in 1592. He served as curate or vicar at a number of Anglican churches while authoring at least three volumes of religious writings and sermons. Basing his later material in large part on the earlier studies of the Swiss naturalist Conrad Gesner, Topsell also included many of the prevailing zoological ideas of his time and place. Each of his animals, some of them fantastical (such as flying dragons), were given personal characteristics, reflecting moral lessons that Topsell wanted to impart. Most of his illustrations were taken from Gesner's woodcuts.

Jeremiah Trautman

German physician who performed the first completely documented cesarean section, in Wittenberg, Germany, on April 26, 1610. The mother is reported to have lived 25 days and then died, most likely from an infection of the genitals following the procedure. The child lived for nine years.

Nicolas Tulp
1593-1674

Dutch physician who is remembered for his description of the disease beriberi, which results from a deficiency in nutrition and is marked by inflammation of the nerves and swelling. Tulp is also known for his description of ileocecal valves, which are muscles that open from the small intestine into the large intestine. Tulp was the personal physician and friend of Dutch painter Rembrandt.

Giovanni Vigo
1460-1525

Italian surgeon who collected all that was known to medical science during his time and compiled the most comprehensive book on surgery as well. Vigo practiced medicine in Rome, where he was the personal physician to Pope Julius II.

St. Vincent de Paul
1581-1660

French priest who established hospitals for the poor in Paris. In 1625, Vincent founded the Congregation of the Mission, an organization of priests dedicated to helping the poor, and he later established the Confraternities of Charity. These were associations composed primarily of women from noble families, who not only assisted Vincent in caring for destitute patients, but also financially supported him in setting up hospitals for foundlings and others. He cofounded the Daughters of Charity (or Sisters of Charity of St. Vincent de Paul), a similar organization for nuns, with St. Louise de Marillac (1591-1660). Vincent was canonized in 1737.

Johann Conrad Wepfer
1657-1711

German physician who advanced the emerging science of pathology, the study of the cause and nature of diseases. Wepfer perfected the observation of a patient's case history and studied the progression of each disease; he related these observations with findings from the autopsies.

Richard Weston
1591-1652

English agriculturist who advanced the field of agriculture by developing the concepts of crop rotation and irrigation of hay fields to increase the volume of crops. Weston also brought forward the idea of canal locks to regulate the level of water in the canals. He was involved in making the Wey River navigable to Guilford, England.

Thomas Wharton
1610?-1673

British physician and Oxford professor who was one of the first physicians to study glands. In 1656 Wharton discovered the duct in the submaxillary gland (the submaxillary gland is either of the two salivary glands located inside the mouth). The duct has been subsequently named Wharton's duct. Wharton then suggested that all glands had ducts for the secretion of fluid; however, Wharton also believed that the function of the thyroid gland was to "round out the fullness of the neck."

Daniel Whistler
1619-1684

English physician who in 1645 provided the first scientific description of a vitamin D deficiency, or rickets. Whistler described the condition, wherein "the whole bony structure is as flexible as softened wax," in his Oxford doctoral dissertation, *Inaugural Medical Disputation on the Children's Disease of the English Which the Inhabitants Idiomatically Call the Rickets*. Five years later, in 1650, Francis Glisson (1597-1677) observed the same condition. Not until the early twentieth century would physicians be able to devise preventive measures against rickets.

Richard Wiseman
1622?-1676

British surgeon who served as personal surgeon to King Charles II. Wiseman's greatest contributions were collected in *Severall Chirurgical Treatises* (1676). The work includes classic descriptions of tuberculosis of the joints (tumor albus) and scrofula ("the king's evil," or tuberculosis of the lymph glands), instructions for amputating above the injury, and accounts of successful urethrotomies and other operations. He increased the knowledge of gunshot wounds, cancer, ulcers, rectal disorders, urological diseases, fractures, dislocations, and sexually transmitted diseases.

John Woodall
1556-1643

English physician for the East India Company whose handbook *The Surgeon's Mate* (1617) provided an extensive overview of the medicines available to a physician at the time. The book, which proved highly popular, also covered a number of conditions the surgeon might encounter, most notably "the scurvy called in Latine Scorbutum." The book noted a number of natural remedies for scurvy, in particular "the Lemmons, Limes, Tamarinds, Oranges, and other choices of good helps from the Indies...."

Gabriele de Zerbis
1445-1505

Italian physician, anatomist, and medical philosopher, also known as Gabriele Zerbi or Gabriello Zerbus. He wrote the first printed book on geriatrics (*Gerontocomia*, 1489), the first on medical ethics (*De cautelis medicorum*, c. 1495), and one of the most accurate anatomical texts before Belgian anatomist Andreas Vesalius (1514-1564) (*Liber anathomie corporis humani et singulorum membrorum illius*, 1502). Zerbis also wrote *Quaestionum metaphysicarum libri XII* (1482) and the posthumously published *Anatomia infantis* (1537), edited by Johann Dryander.

Theodor Zwinger

Swiss physician known for his work *Theatrum vitae humanae,* a type of universal encyclopedia in which he presented various texts. Born in Basel, Zwinger was educated in philology, languages, and medicine. He taught Greek and was a professor of theoretical medicine.

Bibliography of Primary Sources

Arcolani, Giovanni. *Practica.* 1483. Contains Arcolani's writings on surgery and, in particular, his specialization—dentistry.

Bacon, Francis. *Advancement of Learning.* 1605. Contains Bacon's proposal for a new science of observation and experiment to replace traditional Aristotelian science. The Baconian method, also known as the inductive method, involves the exhaustive collection of particular instances or facts and the elimination of factors that do not accompany the phenomenon under investigation. Generally suspicious of mathematics, deductive logic, and intuitive thinking, Bacon believed that valid hypotheses should be derived from the assembly and analysis of "Tables and Arrangements of Instances," and that the scientist must be actively involved in putting questions to nature, rather than passively collecting facts.

Bartholin, Caspar. *De studio medico.* 1628. Early medical textbook; Bartholin was the first to describe the olfactory nerves and the suprarenal glands.

Bartholin, Thomas. *Vasa lyphatica et hepatic exsequiae* (The Lymphatic Vessels and the Secretion of the Liver). 1653. Introduced the use of color plates in anatomical studies.

Bauhin, Gaspard. *Pinax theatri botani.* 1627. A comprehensive botanical work.

Belon, Pierre. *Les Observations de Plusieurs Singularitez et Choses Mémorables* (Observations of Several Curiosities and Memorable Things). 1553. Includes Belon's accounts of places, objects, animals, and plants mentioned by ancient writers, which he encountered while on a series of diplomatic missions to Constantinople and the Middle East. Belon's detailed record of scientific curiosities subsequently served as an itinerary for travelers interested in scientific discoveries.

Belon, Pierre. *L'Histoire de la Nature des Oyseaux* (The Natural History of Birds). 1555. Contains important comparative studies of bird and human skeletal structures, marking the beginning of comparative anatomy and becoming an early contribution to the future development of evolutionary theory.

Belon, Pierre. *L'Histoire Naturelle des Etranges Poissons Marins* (The Natural History of Strange Marine Fish). 1551. Contains descriptions and illustrations of fish and cetaceans, such as porpoises and whales, which Belon had dissected. This work also established a classification system for marine fish, a system based largely on Aristotle's animal classifications, which included both cetaceans and hippopotami under "fishes." However, Belon recognized that the milk glands of cetaceans were mammalian in type, and noted that these creatures were air-breathing mammals, even though they lived underwater.

Benivieni, Antonio. *The Hidden Causes of Disease.* 1507. Presented a new method of thinking in medical science—that of the discovery of cause of death by autopsy—by which the science of pathology was established.

Borelli, Giovanni. *De motu animalium* (On Motion in Animals). 1680-81. Contains a sustained analysis of the mechanics of muscle contraction, with the movements of individual muscles and groups of muscles treated geometrically in terms of mechanical principles.

Bourgeois, Louise. *Observations Diverse sur la Stérilité, Perté de Fruict, Foecundité, Accouchements, et Maladies des Femmes, et Enfants Nouveaux Naiz* (Diverse Observations on Sterility, Loss of Fruit, Fecundity, Childbirth, and Diseases of Women, and Newborn Infants). 1609. The first textbook on midwifery written by a woman. Bourgeois's book, based on the teachings of French surgeon Ambroise Paré and written in the vernacular, was widely read in her time and established her reputation as a pioneer of scientific midwifery.

Brunfels, Otto. *Herbarum vivae eicones* (Living Images of Plants). 1530-36. A three-volume work containing illustrated descriptions of plants, most of which with medical uses. This was the first printed botanical text to include accurate and realistic images.

Brunschwig, Hieronymus. *Buch der cirurgia* (Book of Surgery). 1497. The first printed, illustrated surgical textbook, dealing mainly with wounds and injuries, especially fractures, cuts, and gunshot wounds.

Brunschwig, Hieronymus. *Buch der Wund Artzney* (Book of Wound Dressing). 1497. Contains descriptions of how to treat soldiers wounded by newly introduced guns, cannons, and lead shot, which pierced the flesh, leaving gaping wounds and major infection. Also includes the earliest printed illustration of surgical instruments.

Brunschwig, Hieronymus. *Liber de arte distillandi* (Book on the Art of Distilling). 1512. A classic of botanical medicine, including demonstrations of how to obtain essential oils and other chemicals from plants.

Burton, Robert. *The Anatomy of Melancholy*. 1621. This treatise, which influenced English writing, includes astute descriptions of the causes, symptoms, and cures of depression disorders, though some of the etiologies were attributed to myth rather than fact. Similar in tone to St. Augustine's *Confessions,* Burton's self-reflective point of view provides insight into his own melancholia.

Caesalpinus, Andreas. *De plantis libri XVI.* 1583. The first textbook of botany, including description and classification of more than 1,500 plants.

Canano, Giovanni Battista. *Musculorum humani corporis picturata dissectio* (Picture of the Dissection of the Muscles of the Human Body). 1541. Important anatomical study based directly on structures of the human body and living animals, rather than the dissection of the ape as performed by Galen. Includes the use of copperplate illustrations, which allowed finer details than woodcuts, and featured the fine muscles of the hand and the valves of the deep veins, describing their function in controlling blood flow. His illustrations also showed a novel approach to the myology (study of muscles), as they depicted only a few muscles and their movement of the fingers.

Carpi, Jacopo Berengario da. *Isagogae breves.* 1523. An anatomical work containing illustrations of the human body with the skin removed, and the figures positioned to show their dissected muscles while standing and observing a landscape. Another convention of da Carpi's unique medical illustrations is the use of multiple views of a single part of a muscle, a technique that likely influenced Leonardo da Vinci.

Casserius, Julius. *De vocis auditusque organis.* 1601. An anatomical study dealing with speech and hearing, including 34 plates, some comparing human vocal cords with those of the cat.

Colombo, Realdo. *De re anatomica* (On Anatomy). 1559. A posthumously published anatomical treatise that included several important original observations. A clearly written and organized work, Colombo carefully described the organs within the thoracic cavity, the membrane surrounding the lungs, the membrane surrounding the abdominal organs, and, most importantly, general heart action and the minor, or pulmonary, circulation of the blood.

Cordus, Valerius. *Dispensatorium.* 1546. The first pharmacopoeia, or reference source detailing the dosages and administration of medicinal preparations and drugs.

Descartes, René. *L'Homme, et un traité de la formation du foetus* (Man, and A Treatise on the Formation of the Foetus) 1664. Includes a summation of Descartes's ideas about the human body based on dissections and physiological studies. Although Descartes argued that animals were purely mechanical and had no soul, he described human beings as a union of mind and body. The interaction between mind and body, according to Descartes, took place in the pineal gland, the only unpaired organ in the brain.

Díaz de Isla, Ruy. *Tractado contra el mal serpentino* (Treatise on the Serpentine Malady). 1539. An important work lending support for the belief that syphilis was a new disease. Written as early as 1505, Díaz suggested that the disease was brought back by Columbus and his men, and that he had treated one of them in 1493. As some of these men were known to be fighting at the siege of Naples, the spread of syphilis was blamed on them.

Dürer, Albrecht. *The Great Piece of Turf.* 1503. A life-sized representation of grasses and dandelions in a clod of earth. No one as famous as Dürer had ever painted anything so apparently insignificant as a piece of turf. Yet this single painting had a tremendous artistic and scientific impact in Europe. Like Leonardo da Vinci, Dürer took great pride in the accuracy of his plant drawings.

Dürer, Albrecht. *The Women's Bath.* 1496. An ink drawing depicting a group of nude women of all ages in a public bath, representing an attempt to show what the aging process does to the human body. Dürer's nudes perhaps most clearly reflect his devotion to capturing reality.

Eustachio, Bartolommeo. *Opuscula Anatomica.* 1563. A collectively titled work containing a remarkable series of treatises on, among other structures, the kidney (*De Renum Structura*), the auditory organ (*De Auditus Organis*), and the teeth (*De Dentibus*). The work on the kidney, the first to be dedicated to that organ, contains the first account of the suprarenal gland and the first detailed analysis of the concept of anatomical variation. *De Dentibus* was the first detailed study of the development of the teeth, describing the first and second dentitions, the structure of both the hard and soft tissues, and possible reasons for the sensitivity of the hard tissues of the tooth.

Fabrici, Girolamo. *De venarum ostiolis* (On the Valves of the Veins). 1603. Contains the first description of valves in veins.

Falloppio, Gabriele. *Observations Anatomicae.* 1561. An important anatomical work containing original descriptions of the eye, ear, teeth, and female reproductive system, and introducing many anatomical terms including vagina, placenta, cochlea, labyrinth, and palate. In this series of commentaries on the *De humani corporis fabrica* of Vesalius, Falloppio was the first to describe the uterine fallopian tubes as small trumpets, and to distinguish the vagina as a separate structure from the uterus. He also described structures that were filled with a watery fluid and others that have a yellow humor or fluid; these were possibly the ovaries with the egg encased in the follicle and fluid.

Fernández de Oviedo, Gonzalo. *Summary of the Natural History of the Indies.* 1526. Contains the earliest reference to a New World origin for syphilis. The belief that syphilis was a new disease was common from the fifteenth century on, though there were some disagreements in the medical community over its origins.

Fuchs, Leonhard. *De historia stirpium*. 1542. Botanical work including over 500 illustrations that accurately portray a wide variety of plants, most of which were useful in medicine. The book is particularly noteworthy because Fuchs chose superior artists to illustrate his text, which was both scholarly and accurate.

Gerard, John. *The Herball, or Generall Historie of Plantes*. 1597. A botanical work that combined medical knowledge of plants with poetic prose, personal observations, and elaborate illustrations. It is significant for its content, but also because, due to its popularity, it demonstrates the extent to which publishing transformed Elizabethan English society. It became a landmark in botanical publishing, and is considered one of the monuments of the English language.

Gersdorff, Hans von. *Feldbuch der Wundartzney* (Fieldbook of Wound Dressing). 1517. Includes descriptions of how to extract bullets with special instruments and how to dress wounds with hot oil. Gersdorff also showed how to enclose amputated stumps with animal bladders.

Gesner, Konrad von. *Bibliotheca universalis* (Universal Library). 1545-55. A comprehensive bibliography of European authors, for which Gesner is regarded as the "father of bibliography." Though without a specific medical section, this work inspired William Osler to create his monumental *Bibliotheca Osleriana* (1929).

Gesner, Konrad von. *De chirurgia scriptores* (Authors on Surgery). 1555. Contains the first useful bibliography of surgery.

Gesner, Konrad von. *Historia animalium* (History of Animals). 1551-87. A five-volume work that is commonly regarded as the basis of modern zoology. Unlike many of his contemporaries, Gesner supplemented knowledge of the natural world taken from antiquity with his own biological research. This 4,500-page work consists of a comprehensive catalog of all animals known up to that time and over a thousand beautiful woodcut illustrations, including the famous, but inaccurate, depiction of the rhinoceros by Albrecht Dürer.

Gesner, Konrad von. *Opera Botanica* (Botanical Works). 1551, 1571. Gesner was the first botanist to grasp the importance of floral structures to establish a systematic key for plant classification. This work features close to 1,500 plates of his own composition.

Glisson, Francis. *Anatomie hepatitis*. 1654. A medical text concerning the anatomy of the liver, including the first description of the layer of connective tissue that covers the liver, now known as Glisson's capsule. This book was chiefly the result of lectures Glisson made on the subject in 1640.

Glisson, Francis. *Tractatus de rachitide*. 1650. Contains one of the earliest clinical descriptions of rickets, a disease characterized by softening of the bones; Glisson attributed the condition primarily to nutritional deficiency. Glisson laid out his empirical findings in a academic manner and defended against any anticipated disagreements with his position. This manner of

conveying scientific information in published form was to be a model for years to come.

Grew, Nehemiah. *The Anatomy of Plants*. 1682. Includes discussion of the anatomy of flowers, the microscopic structure of plant tissues, and the suggestion that the stamen functions as the male sex organ of plants and the pistil as the female organ.

Grew, Nehemiah. *The Anatomy of Vegetables Begun*. 1672. A pioneering text that described the structure of bean seeds, the existence of cells, and introduced many of the terms used to describe the parts of plants.

Harvey, William. *De generatione animale*. 1651. Contains the famous statement, "all living things come from an egg," and the assertion that the egg is a composite of both parents. Harvey is credited with coining the term *epigenesis* to denote the developmental process of the embryo as it is gradually differentiated and emerges from the formless egg mass. Eventually a theory of epigenesis modified from that of Harvey was adopted and is currently accepted.

Harvey, William. *Exercitatio anatomica de motu cordis sanguinis in animalius* (On the movement of the heart and blood in animals). 1628. Landmark medical work in which Harvey proved that human blood flows from the left ventricle of the heart, through the arteries of the body, and then into the veins, which eventually returned the blood to the right side of the heart. Harvey confirmed that blood in the right ventricle went to the lungs and returned to the left side of the heart, as part of the pulmonary circulatory system. He also illustrated the functioning of the valves found in the heart and veins, and assumed the presence of capillaries connecting arteries with veins, though final confirmation would come later in the century with the improvement of microscopic study. Harvey's methodical research exposed the fallacies of the Galenic concept of blood circulation, proving that the heart is a hollow muscle that contracts regularly to provide the single motive force of the blood's movement.

Havers, Clopton. *Osteologia Nova*. 1691. The first publication to describe bone structure in great detail. The intricate vessel system consisting of canals and glands that supplies nutrients in the bones, called the Haversian system, is named after the author.

Hildanus, Fabricius. *De combustionibus* (On burns). 1607. One of the earliest books to deal exclusively with injuries caused by heat and fire. Hildanus categorized burns in a practical, systematic way.

Hildanus, Fabricius. *De gangraena et sphacelo* (On gangrene and sloughing). 1593. Contains the recommendation that amputation for gangrene, a potentially lethal result of major infection, should be made high above the infected part.

Hooke, Robert. *Micrographia*. 1665. A tremendously important review of observations concerning the development and improvement of the microscope. Hooke described in detail the structure of feathers, the stinger of a bee, and the foot of the fly. He also noted structures similar to cells in the tissue of trees and plants, and discerned that in some tissues the cells were filled with a liquid, while in others they were empty; he therefore supposed that the function of the cells was to transport substances throughout the plant.

Jorden, Edward. "A Briefe Discourse of a Disease Called the Suffocation of the Mother." 1602. Contains Jorden's valiant defense of 14-year-old Mary Glover, who was accused of witchcraft. Jorden wrote that her fits, actions, and "passions of the body" were not caused by the devil, but by her uterus and problems of "womanhood." Unfortunately the judges did not accept his explanations.

Ketham, Johannes de. *Fasciculus medicinae* (Compilation of Medicine). 1491. The first printed medical work with anatomical illustrations. The book consists of many short, anonymous medical essays, some of which existed in manuscript 200 years prior. The Latin version had six illustrations—a uroscopy chart in red and black, phlebotomy figure, zodiac man, the female viscera, wound man, and disease man. The 1493 Italian translation, *Fasciculo de medicina*, contains 10 illustrations, with the most notable addition being a dissection scene in which a professor presides over a group of barber-surgeons cutting open a human cadaver.

Leonardo da Vinci. Unpublished notebooks. c. 1470-1519. Contains, among many other things, meticulous anatomical sketches based on dissections of both human and animal corpses. Da Vinci created about 750 anatomical drawings, and planned an anatomical atlas of the stages of man from the womb to the tomb, though this was never completed.

Leoniceno, Niccolò. *De epidemia quan vulgo morbum gallicum vocant* (On the Epidemic Vulgarly Called the French Disease). 1497. Contains the assertion that ancient authors had indeed discussed syphilis; thus syphilis was not a new disease, but poor translations hindered modern understanding.

Lister, Martin. *Tractatus de Araneis.*The world's first scientific work on spiders, published in English as *Martin Lister's English Spiders* (1678). Modern entomologists study Lister's book and value both his astute observations about spiders and the precise drawings by Michael Roberts.

Lower, Richard. *Diatribae Thomae Willisii de febribus vindicatio* (Vindication of the discourse of Thomas Willis on fevers). 1665. A vigorous polemic, in defense of Thomas Willis, that refuted Edmund Meara's Galenic view of blood circulation and defended William Harvey's conclusions about the physiology of the blood. Lower's work contained some of the earliest correct observations about lung function, the interaction of the heart and lungs, the differences between arterial and venous blood, and the relation of respiration to blood color.

Lower, Richard. *Tractatus de corde* (Treatise on the Heart). 1669. Contains many noteworthy advances in cardiology, including the first accurate anatomical description of the structure of heart muscle. Lower improved Harvey's theory of blood circulation and speculated about why dark blood from a vein turns bright red when exposed to air. One particularly famous section of this book, "Dissertatio de origine catarrhi" (Dissertation on the origin of catarrh), disproved the traditional belief that nasal congestion was caused by mucus dripping down from the brain or the pituitary.

Major, Johann Daniel. *Chirurgia infusoria.* 17th century. Contains discussion of infusory surgery, which are procedures in which a solution is directly injected into the veins. Major is thought to be the first to successfully inject a medicinal compound into the vein of a human subject.

Merian, Maria Sibylla. "Wonderful Transformation and Singular Flower-Food of Caterpillars." 1679. A catalog of caterpillars, moths, and butterflies, illustrated with her own drawings of almost 200 European butterflies and moths in various stages of metamorphosis.

Paracelsus. *De generatione stultorum.* 1603. Revealed the association of cretinism with endemic goiter.

Paracelsus. *Grosse Wund Artzney* (Great Surgery). 1536. Offers a detailed analysis of gunshot wounds and argues against treating them with hot oil, which was common among military surgeons before the work of the French surgeon Ambroise Paré. Paracelsus's work was translated into Latin as *Chirurgia magna.*

Paracelsus. *Von der Bergsucht oder Bergkrankheiten.* 1567. One of the first books on occupational health hazards, focusing on the diseases of miners.

Paracelsus. *Von der frantzösischen Kranckheit.* 1553. Contains Paracelsus's studies of syphilis, which he called "French disease" or "French gonorrhea." He advocated mercury for its cure.

Paracelsus. *Von den Kranckheyten so die Vernunfft berauben als da sein.* 1567. Contains Paracelsus's rejection of the popular notion that unwelcome mental states are caused by demons. He instead described psychiatric disorders in terms of purely physical occurrences.

Paré, Ambroise. *Briefve collection de l'administration anatomique* (A short collection on governing the body). 1549. Includes description of Paré's improved podalic version, an obstetrical technique employed as a means of extraction in cases of difficult birth.

Paré, Ambroise. *Deux livres de chirurgie* (Two books on surgery). 1573. Contains both scientific and fanciful accounts of birth defects.

Paré, Ambroise. *Dix livres de la chirurgie* (Ten books on surgery). 1564. Includes discussion of dental and oral surgery, amputations, and several other topics.

Paré, Ambroise. *La méthode curative des playes et fractures de la teste humaine* (The method of curing wounds and fractures of the human head). 1561. Contains Paré's research on the treatment of head wounds, undertaken after the 1559 death of Henri II, who received a lance wound to the eye while competing in a tournament and died under Paré's care.

Paré, Ambroise. *La méthode de traicter les playes faictes par hacquebutes et aultres bastons à feu, et de celles qui sont faictes par fléches, dardz, et semblables* (The method of treating wounds made by harquebuses and other firearms, and those made by arrows, darts, and the like). 1545. Describes Paré's experiments with various substances in which he soaked bandages for the treatment of military wounds.

Pollaiuolo, Antonio. *The Battle of Ten Naked Men.* 1460s. An illustration depicting male nudes in violent action as they battled in mortal combat. Concentration on the exterior muscles of the body, to show the body in motion, was adopted by artists and reached its culmination with Pollaiuolo's large engraving. However, he crammed so many contracted muscles and sinews into each figure that the drawings are not realistic.

Rabelais, François. *Pantagruel* and *Gargantua*. 1532; 1534. Two scathing literary satires on the human condition and theology. Originally published under the pseudonym Alcofribas Nasier, both books were vehemently condemned by the Church.

Ray, John. *Historia Plantarum*. 1686-70. A three-volume botanical work in which Ray developed a classification scheme for plants, introduced the concept of species, and established the basic taxonomic principles later adopted by Linnaeus for his classification system.

Raynalde, Thomas. *The Byrthe of Mankynde*. 1540. An important obstetrical work in English, based on Eucharius Rösslin's *Die swangern frawen und hebammen roszgarten*. Raynalde's was the only book that dealt with obstetrics apart from medicine and surgery. The copperplates depict subjects that had been exclusive women's knowledge: the birth chair, birthing room, development of the fetus, and its relation to the uterus.

Rondelet, Guillaume. *Libri de Piscibus Marinis* (Book of Marine Fish). 1554. Contains detailed descriptions of 250 types of marine animals, including whales, marine invertebrates, and seals, each accompanied by illustrations.

Rösslin, Eucharius. *Die swangern frawen und hebammen roszgarten* (A Garden of Roses for Pregnant Women and Midwives). 1513. First printed textbook for midwives, indicating the beginning of gynecology as a medical specialty. The work, illustrated with woodcuts, went through 40 editions and was still in use in the 1730s.

Santorio, Santorio. *On Medical Measurement*. 1614. A landmark text that has been called the first systematic study of basal metabolism. The book was widely read and highly respected by his contemporaries.

Scultetus, Johannes. *Armamentarium chirurgicum* (Surgical Armamentarium). 1655. The most popular guidebook of surgical instruments, techniques, and procedures during the late seventeenth and early eighteenth centuries. Edited by the author's nephew and posthumously published, its contents include mastectomy, splinting, lithotomy (the removal of stones from the bladder), trephination, wound and injury repair, cesarean section, and amputation.

Serres, Olivier de. *Theatre d'agriculture*. 1600. An immensely popular work that provided a complete guide to agricultural practices and helped to outline the ideals of French Protestant culture. Divided into sections that explain specific agronomic and horticultural practices, the work covers subjects as diverse as methods of tilling, dressing, and sowing, the nature of soils, the art of grafting, the domestication of animals, the maintenance of kitchen gardens, the planting of trees, and the comportment of the country gentleman. In essence, this book operated as a guidebook for the French landowner.

Servetus, Michael. *The Restoration of Christianity*. 1553. Discussed the pulmonary transit of the lungs within the framework of how the Holy Spirit entered man. Since, according to the Bible, God breathed into man the breath of life or soul, Servetus reasoned that there must be a point of contact between the air and blood. He concluded that blood was pumped from the right side of the heart to the lungs through an artery and picked up the vital spirit or air. The "vital spirit" was then received in the left side of the heart, which then pumped blood into the arteries of the entire body.

Sharp, Jane. *The Midwives Book, or the Whole Art of Midwifery Discovered; Directing Child-bearing Women How to Behave Themselves*. 1671. First textbook by an English woman for the instruction of midwives. The book consisted of six sections: anatomy, signs of pregnancy, sterility, the conduct of labor, diseases of pregnancy, and postpartum diseases. Sharp's book, which went through several editions, was an influential and practical manual that reflected the state of the art in midwifery, and also recognized restrictions on the contemporary education of women.

Sydenham, Thomas. *Observationes medicae* (Medical observations). 1676. Contains the best descriptions of measles and scarlet fever up to that time, and the first clear distinction between these two diseases. The fourth edition (1685) reported significant progress in smallpox research.

Sydenham, Thomas. *Schedula monitoria de novae febris ingressu* (A warning essay on the emergence of a new fever). 1686. Includes description of chorea minor, or St. Vitus's dance, now called Sydenham's chorea or rheumatic chorea.

Sydenham, Thomas. *Tractatus de podagra et hydrope* (Treatise on gout and dropsy). 1683. Includes discussion of the difference between gout and rheumatism.

Vesalius, Andreas. *De humani corporis fabrica*. 1543. A landmark anatomical work, with more than 200 woodcut illustrations, that helped usher in the renaissance in medicine. Vesalius insisted on the importance of conducting human dissections (performed on executed criminals) and showed that the human body was the key to medical knowledge, not the received wisdom of ancient scholars such as Galen, whom he disproved on many points. The illustrations in Vesalius's work were once attributed to the famous Italian painter Titian, but were later credited to one of his assistants, Jan Stephen van Calcar. The poses are astounding, showing a body without skin standing or posing in front of landscapes of various kinds.

Vieussens, Raymond. *Neurographia universalis* (General Neurography). 1684. Contains the first precise description of the centrum ovale (sometimes called S's centrum). The beautifully executed copperplate illustrations make *Neurographia universalis* second in importance only to Willis's *Cerebri anatome* among seventeenth-century books on neuroanatomy.

Vieussens, Raymond. *Novum vasorum corporis humani systema* (New System of the Vessels of the Human Body). 1705. A classic of cardiology, including the earliest accurate descriptions of mitral stenosis, aortic insufficiency, aortic regurgitation, and several other heart diseases and circulatory disorders.

Vieussens, Raymond. *Tractatus duo* (Treatise on Two Subjects). 1688. A two-part study dealing with human anatomy and fermentation.

Whistler, Daniel. *Inaugural Medical Disputation on the Children's Disease of the English Which the Inhabitants Idiomatically Call the Rickets*. 1645. Contains the first description of the disease rickets, addressed by Whistler in his thesis at the University of Leyden.

Willis, Thomas. *Cerebri Anatome, cui accessit Nervorum descriptio et usus.* 1664. A comprehensive work presenting Willis's careful studies of the nervous system and various diseases. His work was the most accurate account of the nervous system to date and was the first to clearly identify the distinct sub-cortical structures. Willis's also detailed the concept of circulation of the blood and introduced the world to the term "reflex action." He described the circle of arteries located at the base of the brain (still called "The Circle of Willis) and explained its function.

Wiseman, Richard. *Severall Chirurgical Treatises.* 1676. Includes classic descriptions of tuberculosis of the joints (tumor albus) and scrofula ("the king's evil," or tuberculosis of the lymph glands), instructions for amputating above the injury, and accounts of successful urethrotomies and other operations.

Woodall, John. *The Surgeon's Mate.* 1617. A highly popular handbook that provided an extensive overview of the medicines available to physicians at that time. The book covered a number of conditions, most notably scurvy, along with its natural citrus remedies.

Zerbis. Gabriele de. *De cautelis medicorum.* c. 1495. The first printed work on medical ethics.

Zerbis, Gabriele de. *Gerontocomia.* 1489. The first printed book on geriatrics.

Zerbis, Gabriele de. *Liber anathomie corporis humani et singulorum membrorum illius.* 1520. One of the most accurate anatomical texts before that of Vesalius.

Zwinger, Theodor. *Theatrum vitae humanae.* 1565. A type of universal encyclopedia in which Zwinger brought together various texts.

JOSH LAUER

Mathematics

Chronology

1500 Hindu-Arabic numerals come into general use in Europe, replacing Roman numerals.

1525 Albrecht Dürer, a German artist, publishes *Underweysung der Messung,* in which he teaches the theory and method of perspective, a forerunner of projective geometry.

1545 *De Ludo Aleae* by Italian mathematician Girolamo Cardano marks the beginnings of probability theory.

1594 Scottish mathematician John Napier first conceives of the notion of obtaining exponential expressions for various numbers, which he eventually dubs "logarithms."

1637 Amateur mathematician Pierre de Fermat, who founded modern number theory and pioneered analytic geometry and calculus, writes his famous last (or great) theorem, the proof of which remains the greatest unsolved problem in mathematics for nearly 400 years.

1637 With his *La Géometrie,* René Descartes founds analytic geometry by showing how geometric forms may be systematically studied by analytic or algebraic means.

1638 Galileo Galilei publishes *Discorsi e Dimonstrazioni Matematiche Intorno a Due Nuove Scienze,* in which he points out fundamental differences between finite and infinite classes of numbers.

1640 French mathematician Blaise Pascal publishes a one-page *Essai sur les Coniques,* in which he makes important contributions to the study of conic sections by stating his theorem that the opposite sides of a hexagon inscribed in a conic intersect in three collinear points.

1662 English statistician John Graunt, in his *Natural and Political Observations... Upon the Bills of Mortality,* is the first to apply mathematics to the integration of vital statistics.

1666 Gottfried Wilhelm Leibniz, German mathematician and philosopher, inaugurates the study of symbolic logic by calling for a "calculus of reasoning" in his essay "De Arte Combinatoria."

1669 Isaac Newton circulates a paper, "De Analysi per Aequationes Numero Terminorum Infinitas," in which he lays the foundations for differential and integral calculus; four years later, and completely independent of Newton, G. W. Leibniz in Germany also develops calculus.

1679 Leibniz publishes *Characteristica Geometrica,* which marks the beginnings of topology as a mathematical discipline.

1687 Newton's *Philosophiae Naturalis Principia Mathematica,* one of the greatest scientific works ever written, inaugurates the practice of applied mathematics.

Overview:
Mathematics 1450-1699

Background

In the years after 1450 European mathematics flourished, yet during the previous 850 years almost all mathematical developments occurred in other parts of the world. In Medieval Europe the religious and philosophical works of the ancient Greeks and Romans were still preserved, but many mathematical texts had been lost. Yet while European mathematics suffered through the Dark Ages, elsewhere in the world mathematical innovation continued. In India mathematics was well advanced by A.D. 700, especially in trigonometry. The use of the zero helped improve the Hindu numerical system, which was the basis for the numbers we use today. Chinese mathematicians developed a decimal system and discovered solutions for a number of cubic and quadratic equations. The Arab regions of the Middle East were geographically fortunate, having easy access to Babylonian, Greek, and Indian mathematics. Arab mathematicians assembled, and then developed, a wide range of knowledge. However, while these cultures made fundamental discoveries in mathematics, they all reached a peak of activity, then faded. For various reasons the continuation of mathematical research was not given support or encouragement in these societies.

Europeans rediscovered many of the ancient Greek texts through contact with Arab mathematics. However, these works suffered somewhat from the multiple translation process, from Greek to Arabic to Latin, in which mistakes were often compounded. Innovations were often resisted, such as Gerbert's (c. 945-1003) and Leonardo of Pisa's (c. 1170-1240) separate attempts to popularize Arabic-Hindu numerals. Not until the sixteenth century were they widely accepted.

The Printing Press Aids the Development of Mathematics

By the middle of the fifteenth century translators such as Regiomontanus (1436-1476) had recovered most of the ancient Greek sources from Arabic copies. However, the books were few and extremely expensive, as they had to be painstakingly handwritten. The development of movable type and papermaking after 1450 resulted in an information revolution in Europe. Books became

easy and cheap to mass-produce, and language and type were slowly standardized.

One of the earliest mathematical texts printed was Luca Pacioli's (1445-1517) *Summa de Arithmetica* (1494). It was difficult to read, and many commentators have noted that there were superior alternatives, such as Nicolas Chuquet's (c. 1450-1500) *Triparty* (1484), which contained practical exercises in geometry and commerce and was clearly written. However, it remained in manuscript form, and so Pacioli's printed book reached a much wider audience.

Most texts early in the period relied heavily on ancient sources, such as Rafael Bombelli's (1526-1572) influential treatise on algebra (1572). This fascination with ancient texts also led to the revival of ancient mystical themes within mathematics. Scholars like John Dee (1527-1608) became interested in the occult power of numbers and symbols, assigning numerical values to letters in words to expose hidden meanings. These magical elements led some scientists, such as Francis Bacon (1561-1626) and Giordano Bruno (1548-1600), to ridicule or downplay the role of mathematics. Yet the popularity of numerology and astrology gave rise to many mathematical and scientific innovations, such as precise computations of the position and motion of the stars and planets.

Slowly European mathematics caught up with Indian, Chinese, and Arabic mathematical knowledge, but often information was not shared. Scipione del Ferro (1465-1525) discovered a method of solving some cubic equations, which his pupil, Antonio Maria Fior, kept secret in order to win mathematical contests. However, Fior was utterly defeated in 1535 by Niccolò Fontana, better known as Tartaglia (1500-1557), who figured out the method. Girolamo Cardano (1501-1576) heard of Tartaglia's victory and persuaded him to allow one of the cubic solutions to be printed. However, Cardano also worked out several more cases, and Tartaglia became furious when those too were published. Secretive findings in mathematics and science were common, but many thinkers eventually began to see the benefits of sharing information. A number of important figures, such as Fr. Marin Mersenne (1588-1648), acted as intermediaries between many mathematicians, corresponding widely and spreading new ideas.

Mathematical innovations came from many sources. The symbols + and − were originally used by German bookkeepers, as they allowed easy cocomputation of deficiencies and excesses in stock. A Welsh doctor, Robert Recorde (1510-1558), created the = sign, and François Viète (1540-1603) introduced the use of letters to represent numbers in algebraic equations. Such developments combined to give mathematics its modern look and allowed easy communication across language barriers.

Albert Girard (1595-1632) was one of a number of mathematicians who expanded the field of algebra, writing an influential text, *Invention nouvelle en l'algebre* (1629). Girard was a military engineer, and like others of this profession he found mathematics increasingly useful. The development of cannons led to the study of the flight of cannonballs. Cannons also made older fortifications obsolete, as tall, thin, straight walls were easily toppled, and so new, geometrically elaborate fortifications were constructed to withstand the impact of artillery.

John Napier Develops Logarithms

The development of logarithms by John Napier (1550-1617) had practical applications in many areas, from astronomy to bookkeeping. Starting from ideas suggested by Chuquet, Napier produced many tables of computed logarithms, numbers that reduced multiplication and division problems to simple addition and subtraction. Henry Briggs (1561-1630), a keen astronomer, expanded and helped popularize Napier's works. Napier is also remembered for his system of rods to make calculations easier, and many versions of such devices were constructed, eventually producing the slide rule, a device that remained popular until the invention of the electronic calculator.

The development of European painting produced a new interest in geometry. To create an illusion of depth in a two-dimensional painting, the art of perspective was revived. Gerard Desargues (1591-1661), an architect and engineer, studied the figures created when a plane intersects a cone, and introduced the concept of projective geometry. Geometrical studies also helped improve mapmaking and navigation.

The popularity of gambling led Blaise Pascal (1623-1662) and Pierre de Fermat (1601-1675) to calculate probabilities of winning at dice games. Both men also contributed to a wide range of other mathematical topics. Fermat left behind one of the most puzzling mathematical challenges, his Last Theorem, which remained unsolved until the final decade of the twentieth century.

The Invention of Calculus

Many fields of mathematics were developed in this period. Number theory, induction, new methods for calculating π, decimal fractions, and a host of other important work helped make the years 1450-1699 one of the most productive periods of mathematical research. However, it was the invention of calculus that was the crowning achievement of the time. A number of seemingly independent developments combined to give rise to this new field. René Descartes (1596-1650) introduced algebra into geometry, and along with others, such as Fermat, developed the field of analytic geometry, which describes curves in the form of equations. The study of curves led to a number of new techniques being developed to analyze their slope and rate of change. Descartes developed a method for finding the perpendicular to the tangent of a curve, while Fermat worked on the problem of finding the maxima and minima of variables. Bonaventura Cavalieri (1598-1647), a Jesuit priest, developed a method for calculating the area or volume of a geometric figure, the method of indivisibles. Then Isaac Newton (1642-1727) and Gottfried Leibniz (1646-1716) independently put all the pieces together and developed methods for differentiation and integration, the beginnings of calculus.

Newton is also important for the way he viewed mathematics. His writings popularized the use of mathematical analysis in scientific subjects, such as physics and optics. He helped make mathematics the new language of science. Mathematics had also been readily adopted in areas as diverse as commerce, mapmaking, navigation, astronomy, astrology, and even gambling. The continuation of mathematical development in Europe owed much to the practical applications the field offered. Financial support and patronage from wealthy individuals and governments enabled many mathematicians to dedicate their lives to their studies.

Looking ahead to the 1700s

Mathematics in the following century would be dominated by calculus. Newton and Leibniz's creation was to occupy the minds of the best and brightest mathematical minds of the eighteenth century, Leonhard Euler (1707-1783), Joseph Lagrange (1736-1813), and Pierre Simon

Laplace (1749-1827) to name a few. Yet, despite the work of these great mathematicians, by the end of the eighteenth century calculus was a field in trouble. While calculus worked, it seemed to have no logical foundation. It would be the mathematicians of the nineteenth century who had to deal with such objections. The eigh-teenth century also saw important work in other mathematical fields. Advances in analytic geom-etry, differential geometry, and algebra would all play important roles in the development of mod-ern mathematics.

DAVID TULLOCH

Advancements in Notation Enhance the Translation and Precision of Mathematics

Overview

During the Renaissance in Western Europe, a re-discovery and advancement of classical mathe-matics laid the foundation for the empiricism of the Scientific Revolution. One of the pillars of this intellectual reawakening in mathematics was the increased use of mathematical symbols that enabled scholars to communicate with each other more easily and accurately across geo-graphical, national, and linguistic boarders.

Background

Although the use of symbols in mathematical operations dates to antiquity, scholars often dif-fer in the awarding of credit for the first use of mathematical symbols. For example, according to some interpretations based on extant archaeo-logical evidence, some scholars credit ancient Egyptian civilizations with the use of specific symbols for the operation of addition and with a symbol that denoted equality. Other scholars cite the first use of such symbols by Greek, Hindu, or Arabic cultures. Regardless, although symbol-ic use was certainly present in indigenous math-ematical systems scattered throughout the world, there is no evidence of any concerted at-tempt at a universification of symbolism until after the advent of the printing press at the start of the European Renaissance, around 1450.

Beginning at this time, expansion in trade and foreign wars by European nations fueled the need for an easily translatable mathematics, espe-cially arithmetic and geometry that were exten-sively used in commercial trades and in the devel-opment of weapons. In addition, the growing availability and capability of printing technology made it possible to quickly and accurately ex-change scholarly ideas. Although these exchanges were not always altogether altruistic, with the goal of scholars often limited to the rapid incor-poration of ideas that might benefit political or commercial interests, in general the introduction of mathematical symbols can be viewed as an at-tempt not only to streamline mathematical opera-tions so that they could be more easily rendered by printed type but also to make such operations more precisely translatable.

Although the conventional view of histori-ans for centuries held that Arab scholars merely preserved the classical mathematics of the Greeks until Europe emerged from its Medieval Age, recent reevaluations of the contributions of Arab mathematicians have cast light on many important mathematical innovations advanced by Arabic scholars. With regard to the develop-ment of mathematical symbolism, the writings of Arab (Moroccan) mathematician al-Mar-rakushi ibn Al-Banna (1256-1321) and Arab (Moorish and Tunisian) mathematician Abu'l Hasan ibn Ali al Qalasadi (1412-1486) exhibit a use of mathematical symbols to describe alge-braic operations that scholars argue provides ev-idence of still earlier origins of algebraic notation in Arabic mathematics.

Regardless of intent or origin, an examina-tion of mathematical works produced during the European Renaissance reveals an increasing re-liance on mathematical symbols and the concur-rent steady reduction in text to describe mathe-matical processes. Of particular benefit to those attempting to duplicate mathematical operations was a reduction in the extensive use of often confusing word abbreviations in the description of mathematical methods.

Impact

As a direct consequence of increased commercial trade during the Renaissance, there arose a form of commercial arithmetic made more easily understandable by the introduction of various symbols. Prior to the widespread use of symbols to denote numbers and operations, reading mathematics was a cumbersome and tedious exercise. Moreover, differences in language—and different usage of words within individual languages—increased the errors in translation that usually manifested themselves in errant mathematical calculations. In particular, prior to the introduction of algebraic symbols, algebraic expressions were laboriously scripted. As a result of the sensitivities of language, particularly when translating from Arabic, there was often great confusion and difficulty in reproducing and advancing algebraic concepts.

German mathematician Johann Widman's 1489 work, *Behede und hubsche Rechnung* was the first printed book to utilize and set out the plus (+) and minus (-) signs for the mathematical operations of addition and subtraction. Although the introduction of symbols for such elementary operations may seem a modest accomplishment by modern mathematical standards, it is important to remember the state of mathematics during the fifteenth and sixteenth centuries. As late as 1550 most learned scholars did not know how to multiply or divide. Accordingly, the advancement of mathematical symbols, even for elementary operations, contributed to the slow evolution of standardized notational systems that could be incorporated more easily into texts that, in turn, promoted a wider growth of scientific literacy upon which mathematics itself could flourish. In 1557 English mathematician Robert Recorde (1510-1558) introduced the equals sign (=) in a work titled *Whetstone of Witte* and made popular the plus and minus symbols published by Widman.

The rediscovery of the work of Pisan (later a part of Italy) mathematician Leonardo Pisano Fibonacci (1170-1250)—who borrowed extensively from other cultures during his travels—fueled the development of a notational system equipped to translate Fibonacci's mathematical puzzles and word problems.

The increased use of mathematical symbolism was given an important boost in French mathematician François Viète's (1540-1603) *Canon Mathematicus*, which covered many aspects of trigonometry and included the use of decimal fractions (printed in differential smaller type). Moreover, Viète made a systematic compilation of notation related to algebraic methods. In subsequent works, Viète used letter symbols

to represent both known and unknown quantities (e.g., using vowels to represent unknowns and consonants to represent known quantities).

In his work *De Thiende* (Art of Tenths), Flemish-born Dutch mathematician Simon Stevin (1548-1620) demonstrated computation without the use of common fractions. Stevin formalized and synthesized prior concepts of decimal numbers into a notation system capable of representing decimal numbers, and he described the methods by which to add, subtract, multiply, and divide decimal numbers. Stevin's work was critical for the advancement of the empiricism and quantification of physics.

In his 1637 work *Discours de la méthode* (Discourse on Method) French philosopher René Descartes (1596-1650) championed the logic of mathematics as a paradigm for reasoning and, in so doing, elevated the use of mathematical symbolism into philosophical reflections on the nature and value of mathematics. Descartes incorporated the notation of English mathematician Thomas Harriot (1560-1621) with regard to the use of symbols for inequalities and made important modifications to the symbolism for exponential notation first advanced by Italian mathematician Rafael Bombelli (1526-1572).

In a later work titled *La Géometrie*, Descartes advanced the still-used convention wherein letters near the beginning of the alphabet are used to represent known quantities in equations and letters near the end of the alphabet are used to represent unknown quantities This innovation by Descartes allowed the rapid formulation and evaluation of equations with regard to the determination of unknown variables (e.g., in the equation $ax = b$ it is apparent upon inspection that the quantity designated by x is an unknown variable).

The development of symbolism in coordinate geometry in the works of Descartes and French mathematician Pierre de Fermat (1601-1665) allowed the easier quantification of geometry and the application of algebraic methods to geometric problems in navigation and engineering. Moreover, the advancements by Descartes laid a solid foundation for English physicist and mathematician Sir Isaac Newton's (1642-1727) profoundly important *Philosophiae Naturalis Principia Mathematica* (Mathematical Principles of Natural Philosophy).

Curiously, not everyone readily accepted the increased use of symbolism in mathematics. The English philosopher Thomas Hobbes (1588-

1679) argued that symbols only made the description of mathematics shorter—that symbols were nothing more than a convenience for the printer—and that they did nothing for the ultimate understandability of mathematical operations. Hobbs decried excessive use of symbolism as a "double labor" of the mind.

Other notable advances in seventeenth-century mathematical symbolism also include English mathematician William Oughtred's (1574-1660) advancement of a symbol for π as the ratio of the diameter of a circle to its circumference. In addition, in his work *Clavis Mathematica*, Oughtred advanced the use of "x" as a sign for mathematical multiplication. It was not, however, until the 1660s that the symbol ÷ was widely used to denote division. In his 1685 *Treatise on Algebra* English mathematician Rev. John Wallis (1616-1703) introduced a symbol used to represent numerical infinity in operations.

In the later decades of the seventeenth century German mathematician Gottfried Wilhelm von Leibniz (1646-1716) invented symbols of such operational utility that they quickly became the standard notation for the new calculus. The symbols still used for differentials (dx and dy) and the sign for the integral appear in papers that Leibniz published on the calculus.

This normalization of mathematics—occurring despite increased nationalism and provincialism in language—made possible the mathematical advances necessary to allow the quantification of accurate descriptions of uniform and accelerated motion (e.g. along parabolic curves) and the development of analytic geometry and calculus.

K. LEE LERNER

Further Reading

Abbott, D., gen. ed. *The Biographical Dictionary of Scientists: Mathematicians*. New York: Peter Bedrick Books, 1985.

Ball, W. W. Rouse. *A Short Account of the History of Mathematics*. New York: Dover Publications, 1960.

Cajori, F. *A History of Mathematical Notations*. Chicago: Open Court Publishing Co., 1928.

The Reappearance of Analysis in Mathematics

Overview

Important mathematical developments occurred during the sixteenth century, but it was not clear how to fit them into the framework of classical mathematics, which was still used as the center of the curriculum. In particular, the work in algebra did not look like part of the system of axioms and theorems used to set out the discipline of geometry. The French mathematician François Viète (1540-1603) presented the new mathematics in a way that could make it acceptable to those who insisted on having a classical background for mathematics. Thanks to his work, algebra could thereafter be presented in a way that was both easier to explain and to extend to further new results.

Background

In the sixteenth century a number of Italian mathematicians made advances in the branch of mathematics devoted to solving equations. Perhaps the most celebrated accomplishment was that of Niccolo Fontana (1500?-1557), better known as Tartaglia, who managed to solve cubic equations, in which the variable appeared to the third power. This was a notable advance over previous techniques, and in the course of the century equations with variables to the fourth power (called quartic equations) were also solved. There was argument at the time about who had first accomplished which stages of these advances, but there was no doubt that methods for solving of equations had been immensely improved.

These new discoveries, however, did not conform to those of classical Greek mathematics represented by the *Elements* of Euclid (330?-260? B.C.). Even within the world of Greek mathematics some of Euclid's successors had come up with additional approaches to mathematics beyond those of the *Elements*. Diophantus (fl. A.D. 250) was the first mathematician to approach the business of solving equations in more than one vari-

François Viète. *(Library of Congress. Reproduced with permission.)*

able systematically, but the system he used was not that of Euclid. In an effort to make the work of Diophantus at home in the world of Greek mathematics, Pappas (300?-350?) came up with the categories of zetetics and poristics as part of the machinery used by Diophantus. Zetetics involved setting up equations and poristics involved checking the truth of earlier results by use of equations.

Since little progress was made for more than a millenium in the areas of algebra in which Diophantus had worked, the terminology and divisions that Pappus used seemed to be sufficient. With the developments in the sixteenth century, however, change was called for, and the individual who made those changes to mathematics was François Viète, a French mathematician who wrote in Latin (the scholarly language of the time) under the name of Vieta. As a humanist he felt obliged to find room for the new mathematics under the heading of the Greek learning that humanism had reintroduced into Europe. Since the new mathematics did not quite fit, he had to expand the categories of Pappus to find room for the Italian advances in algebra already made, and for those likely to come in the future.

Viète introduced the new area of exegetics into the classification Pappus had created. This had to do with determining the value of the unknown in a given equation. What the techniques

of Tartaglia and others had required was the substitution of one expression into another in an effort to reduce an equation to a form that could be handled. Euclid's arithmetic (stated in the *Elements* in geometric form) allowed for the substitution of equals for equals, but since the algebraic expressions involved unknowns, different expressions could not be recognized as being equal. Viète argued that this process of replacing one expression by another was defensible if the resulting solution could be substituted back into the original equation and tested, according to Pappus. As a result, the algebraic techniques of the Italian equation solvers could be worked into a Greek scheme, although that was not part of what the Renaissance had inherited from classical mathematics.

Viète called his new approach the *ars analytice,* (analytic art). This referred to the technique of solving an equation by breaking it down (from which the word "analysis" comes). The presentation of mathematics in Euclid involved the method of synthesis, building up the subject from its elements. Algebra had never followed that mode of presentation, but this was partly connected with its looking like a bag of tricks rather than a connected discipline. Viète was the first to recognize that algebra could be presented in a more coherent fashion by dividing equations into classes which could be treated as a whole. This enabled mathematicians to proceed by solving a certain class of equations and then trying to reduce other equations to examples within that class. His method of classification and discovery brought the haphazard techniques for solving cubic and quartic equations into the framework of the heritage of classical mathematics.

Impact

The presentation of algebraic results under the heading of analysis accomplished a good deal to make the solving of equations more academically respectable. The mathematical curriculum in the sixteenth century remained centered on the *Elements,* and, in the next century, even the work of a mathematician of the stature of Sir Isaac Newton (1642-1727) was still presented in geometrical form. Those who had accomplished the feats of solving cubic and quartic equations could be dismissed as not being mathematicians (or philosophers, in the sense of natural philosophy) if their work did not adhere to the canons of Greek mathematics. Viète's stretching of the notions of Greek mathematics put algebraic progress on the map of the academic world as well.

Much of Viète's own nomenclature for algebra was supplanted in the seventeenth century by the work of René Descartes (1596-1650). Descartes's algebra and the simultaneous work of his colleague Pierre de Fermat (1601-1665) would have been impossible without Viète taking the isolated results of his Italian predecessors and systematizing them. The influence of Descartes and Fermat has continued throughout the subsequent development of mathematics, while that of Viète has ceased to be nearly so visible. Nevertheless, both Descartes and Fermat were well aware of what they had inherited in the form of the analytic art.

There was a good deal of discussion in the nineteenth century about whether algebra was about numbers or about symbols. The issue reflects Viète's ability to transform the discussion about specific numerical examples, which he received from earlier algebraists, into classes of equations that looked as though they could be solved regardless of the kinds of numbers that the variables represented. In the course of the nineteenth century, different rules for dealing with equations had to be developed according to the nature of the numbers being considered, but the possibility for extending algebra in this way depended on having a language for the communication of algebraic results. Viète supplied that language by merging Greek and algebraic traditions into the analytic art.

THOMAS DRUCKER

Further Reading

Klein, Jacob. *Greek Mathematical Thought and the Origin of Algebra,* translated by Eva Brann. Cambridge, MA: Massachusetts Institute of Technology Press, 1968.

Mahoney, Michael S. *The Mathematical Career of Pierre de Fermat.* Princeton, NJ: Princeton University Press, 1973.

Otte, Michael, and Marco Panza, eds. *Analysis and Synthesis in Mathematics: History and Philosophy.* Dordrecht: Kluwer Academic Publishers, 1997.

John Napier Discovers Logarithms

Overview

Logarithms are of fundamental importance to an incredibly wide array of fields, including much of mathematics, physics, engineering, statistics, chemistry, and any areas using these disciplines. However, until the early seventeenth century, they were unknown. Invented by a Scottish amateur mathematician named John Napier (1550-1617) after 20 years of work, they were met with almost immediate acceptance by mathematicians and scientists alike. In the intervening centuries, logarithms and their converse, exponents, have proven to be among the most useful mathematical tools of all time.

Background

Arithmetic (addition, subtraction, multiplication, and division) dates back to human prehistory. Of these most basic operations, addition and subtraction are relatively easy while multiplication and division are much more difficult to master. Until the Renaissance, however, mathematics and the sciences were not very dependent on mathematical calculation, and these difficulties, while vexing, were not insurmountable.

However, the flurry of discoveries in physics and mathematics that began in the fifteenth century gave rise to more and more need for calculation, and what had been inconvenient now became drudgery. Unfortunately, in those days before computers (or even slide rules), no other options existed, so scientists and their assistants plugged away, laboriously multiplying and dividing by hand.

Towards the end of the seventeenth century, John Napier, a Scottish laird, began to look for an easier way to undertake these operations. One observation he made dealt with multiplying powers of numbers. For example, $4 \times 8 = 32$. If we write these as powers of two, we have $2^2 \times 2^3 = 2^5$. Napier observed that the exponents on the left side of the equation add up to equal the exponent in the answer (that is, $2 + 3 = 5$). This meant that, instead of multiplying two numbers together, one could simply add the exponents together, and then calculate the final answer using exponents. Or, in terms of our example,

$$4 \times 8 = 2^2 \times 2^3 = 2^{(2+3)} = 2^5 = 32.$$

This series of operations has a few drawbacks. First, it is far too much work for such a

John Napier. *(Library of Congress. Reproduced with permission.)*

simple problem. Second, in this form, it will only work for integer multiples of integers—it is just too limited in scope to be of general use. Napier's breakthrough came in realizing that this method could be extended in such a way as to make it more generally useful. After this breakthrough, he spent the next 20 years calculating the first table of logarithms, publishing his results in 1614. It is worth noting that a Swiss mathematician, Joost Bürgi (1552-1632) apparently invented logarithms some time before Napier, but for some inexplicable reason, neglected to publish his results and their utility until 1620. Because of this, he lost his claim to scientific priority and is not given credit for his work.

There are, of course, some differences between what Napier called logarithms and our current definition. These differences are, however, not fundamental, and all of Napier's work easily translates to current usage with only minor adjustments. In general, it is safe to say that the fundamental concepts of Napier's system remain intact, even after nearly 400 years—a remarkable achievement. What is even more remarkable is that Napier performed this work in intellectual near-isolation. Or, as Lord Moulton said in a 1914 tribute marking the 300th anniversary of Napier's paper,

The invention of logarithms came on the world as a bolt from the blue. No previous work had led up to it, foreshadowed it, or heralded its arrival. It stands isolated, breaking in upon human thought abruptly without borrowing from the work of other intellects or following known lines of mathematical thought.

Impact

As mentioned above, Napier's work was greeted with instant enthusiasm by virtually all mathematicians who read it. The primary reason for this is because his tables of logarithms vastly simplified computation. Along these same lines, the principles upon which the slide rule works are dependent on the addition of logarithms, a fact that helped speed up computation in precalculator days. Mathematicians also quickly found other uses for logarithms, and invented other related concepts such as fractional exponents, the number *e,* and similar mathematical tools. Although these were not of great importance to the average person of the seventeenth century, Napier's invention was of such a tremendous boon that, directly or indirectly, it has affected virtually everyone in some way.

Simplifying computation

As mentioned above, the invention of logarithms greatly simplified mathematical operations. While this sounds relatively straightforward, its importance may not be obvious. Consider, however, the fate of an astronomer or physicist before and after 1614:

To determine a planetary orbit, an astronomer needed to make a number of relatively sophisticated calculations, many of which used rather large numbers and several operations of multiplication and division. To multiply, for example, two five-digit numbers required several minutes, five sets of multiplication, and adding the results of each of these sets. At any point of this laborious process a mistake could be made, so each calculation had to be checked for accuracy. This could take many minutes for each multiplication, and division was much more difficult. With Napier's system, on the other hand, this operation took just a few minutes. First, the astronomer would look up the logarithms of each factor. Next, he would add these logarithms together, and then would find in the tables the number for which this sum was the logarithm (called the antilogarithm).

Saving five minutes of calculation time is not significant—but saving five minutes of calculation time for each of several thousand calculations makes a dramatic difference. This single mathematical invention was responsible for an incredible decrease in mathematical drudgery while simultaneously increasing computational accuracy and scientific output. In fact, Pierre-Simon Laplace (1749-1827) commented that Napier had "by shortening labors doubled the life of the astronomer." The next such dramatic increase in computational efficiency would be the invention of the slide rule in the 1620s, very shortly after publication of Napier's original paper. By removing the necessity to look up values of logarithms and antilogarithms, the slide rule sped up computation even more, while further reducing the chance of error. In fact, the slide rule would remain virtually unchanged for nearly four hundred years, an indispensable tool for anyone performing computations. The only device to simplify calculations even further has been the electronic calculator, which has replaced the slide rule in virtually every office in the world.

Mathematical spin-offs

Logarithms not only led directly to the invention of slide rules, but also produced other intellectual spin-offs. Exponents, for example, were already well known in Napier's time, but were of limited utility because of the contemporary insistence of using only integers as exponents for other integers. For example, one could write the term 2^3, but not $2^{2.5}$. This limitation placed an unnecessary constraint on the use of exponents. However, shortly after publication of Napier's paper, mathematicians realized that logarithms were simply exponents. Since logarithms were also written in decimal notation, this opened the door to a wider use of fractions and decimals as exponents, again simplifying mathematical computation. Similarly, mathematicians realized that they could use exponents with fractions and decimal numbers as the base. Today, neither of these seems a revelation, but in the early 1600s, this was a major breakthrough.

Another spin-off of Napier's work was the realization that, when examined mathematically, his work produced the number e, the base for the natural logarithms. Like π, e is a transcendental number that will never terminate or repeat; it has also, like π, proven itself to be an incredibly versatile number that pops up in calculations performed in just about every field that uses mathematics. Compound interest, radioactive decay, the growth and decline of bac-

terial populations, astronomical calculations, and any number of engineering problems all make use of e, and solutions to many of these problems also require the use of logarithms. In the eighteenth century, the brilliant mathematician, Leonhard Euler (1707-1783) would help give logarithms and exponential functions an important place in higher mathematics and the calculus.

The effects of logarithms on society

While logarithms and exponents are basic mathematical concepts, they are rather esoteric for those who do not work with them regularly. This was especially true for the average person in the seventeenth century, who was certainly had no appreciation for any mathematics beyond simple arithmetic. Even today the average person is likely to have only a passing familiarity with these concepts. Logarithms, however, are so fundamental to the mathematics of many disciplines and are so ubiquitous in mathematical problem solving that most people's lives have been affected in some way by them. For example, the mathematics by which aircraft, internal combustion engines, electrical generators, and petroleum refineries are designed depend intimately on the use of these concepts. Compound interest calculations, important to anyone with either a bank account or a loan, are made simpler and more accurate by use of e. The mathematics upon which television and radio broadcast and reception is based also depend on these mathematical tools. Except for very isolated, primitive tribes, it is likely that everyone has been exposed to at least one of these categories of objects, or to the fruits of one.

Put simply, John Napier, laboring on his own, created a mathematical concept that has proven extraordinarily useful to mathematicians, scientists, financiers, and many others. Because of the incredible utility of logarithms and exponents, technical and statistical calculations were made much simpler and more reliable, and all of society has benefited as a result.

P. ANDREW KARAM

Further Reading

Maor, Eli. *e: The Story of a Number.* Princeton, NJ: Princeton University Press, 1994.

Boyer, Carl and Merzbach, Uta. *A History of Mathematics.* New York: John Wiley and Sons, 1991.

Militarizing Mathematics

Overview

Humanity has an uncanny knack of finding a military application for almost any discovery made by science. Mathematics is no exception and, in fact, because of its utility in describing many physical phenomena, has been extraordinarily useful to the military. This process began during the Renaissance, continues today, and is likely to be ever more important as we develop heavily computerized weapons and armies for the future.

Background

Early warfare was a rather hit-or-miss affair. Armies were generally composed of a large number of poor people on foot with sharp weapons, and a small number of wealthy people on horses with sharp weapons. They would rush into battle, shooting arrows and hacking at each other until one side realized it could not win and either withdrew or was slaughtered. Sometimes, one army would ensconce itself inside a castle or some other fortification, where they would be relatively immune from the ravages of the other side for as long as their food and water held out.

Sometime around the fourteenth century, with the introduction of gunpowder from China, all this began to change. Almost simultaneously, firearms and artillery pieces gave foot soldiers a weapon to use at longer range and against those in armor, and gave besieging armies a weapon that could bring down castle walls. Because of this, gunpowder is possibly the most significant advance in the history of warfare, and one of the most important inventions in human history.

With the advent of gunpowder and artillery, battlefield strategies changed. Enemy troops could now be attacked from a greater distance, giving the side with superior artillery a tactical advantage. Since city walls were no longer impenetrable, they began to be replaced by designs that were more effective against cannons, while making best use of the cities' own cannons. In addition, the fledgling science of ballistics, based almost entirely on mathematics, helped make the flight of shot from a cannon more predictable; another significant advance.

Much of the these new weapons' effectiveness was due to the use of mathematical analysis to perfect both their design and use. Of course, battles remained anything but predictable, but the characteristics of an army's main weapons were better known, making their use in battle more certain.

At sea, mathematics found increasing use, too. As navies and merchant fleets ranged further from shore, becoming true blue water fleets, their dependence on accurate navigation grew. This, in turn, was dependent on the accurate use of trigonometry to determine a ship's location from the position of the sun or various stars. Some sea captains, especially those with more scientific knowledge, began to try to balance the forces acting on their ships to maximize their performance. Finally, waging naval warfare on an intercontinental scale put a heavy emphasis on logistics to make sure that a ship could be supplied with food, water, gunpowder, shot, and other supplies, travel all over the world and still fight effectively. This, too, benefited from the increased use of mathematics rather than operating solely by rule of thumb.

Impact

The increased use of mathematics in warfare directly affected military technology and strategy, making warfare more efficient and more predictable. These are significant developments, and are in some ways intimately related because the process of making weapons more predictable allows them to be used to their maximum advantage. This, in turn, makes them more efficient in battle. The reason for this is not obvious, but is relatively simple, as the following example will demonstrate.

Land warfare

Take two cannons, both of which are the same size and weight and hold the same amount of gunpowder. One cannon was cast and bored by eye, based on the judgment of the person overseeing the process while the other was cast and bored according to a master set of plans drawn up by an engineer. The first is loaded with a batch of gunpowder mixed according to an old formula that has been handed down over the decades, and poured into the barrel to a level that "looks" right. The other receives a standardized amount of a standardized batch of powder. They are loaded with cannonballs, aimed and fired.

The second gun will fire more consistently every time. Because the barrel is a standard size, it can fire a shot that has been designed to fit as closely as possible without jamming, so the shot will travel farther. Because the powder is uniform and the same amount is used every time, the crew will know exactly how far the shot will travel for a given cannon elevation. In addition, the new science of ballistics will have provided the gun crew with a table of different trajectories and their effects. In short, the performance of the second gun is more predictable than that of the first. Because of this, a general will know exactly where to place his artillery pieces to get the best use out of them. Of course, this example is a great simplification, but the fact remains that the mathematical control of ammunition and weapons use could give a great advantage to the gun crews using those weapons.

Naturally, both attackers and defenders used artillery, and defenders gained the same advantages in their use of guns. In addition, defenders soon learned that elevating their weapons gave even greater range, while replacing their high walls with lower, sloping embankments helped mitigate the worst effects of cannonballs. In this, too, mathematical analysis of the variety of angles from which attacks could come helped prepare for the most likely scenarios.

Naval warfare

Naval gunnery benefited from mathematics in the same way that land-based artillery did, although firing a gun from a rolling ship at a moving target was much more difficult than hitting a stationary target with stationary weaponry. Mathematics was also useful in such seemingly mundane tasks as determining the optimum amount of food and water with which to provision a ship. (Even these simple tasks can help a ship to fight more efficiently: carrying too much food left less room for powder and shot, while scrimping on supplies adversely affected a crew's efficiency and effectiveness over long periods of time.) In addition, improved navigational techniques facilitated coordinating with other ships, and a better understanding of the physical stresses on naval vessels helped captains to sail them better and more quickly. All of these im-

provements, in turn, helped navies defend their national interests and project national power to newly discovered colonies and trading partners.

Effects on society

The same technology that produced cannon barrels of precisely the same dimensions helped make the first internal combustion engines efficient enough for use in automobiles and aircraft. It also helped create piston cylinders on early steam engines, an invention that powered both the Industrial Revolution and our current automotive age. Making gunpowder as consistent and effective as possible helped produce what is now the science of chemistry and all that it has given humanity. The navigational improvements that made navies so much more efficient also made commercial travel more efficient, opening up the world for exploration, exploitation, and colonization by the European powers.

These spin-offs weren't always positive. The military changes that accompanied the use of gunpowder made noncombatants more vulnerable during wartime because they had no place to retreat for protection. The growing efficiency and complexity of warfare also helped spur the formation of professional armies, specialists in warfare who had mastered the weapons and techniques of their trade. The growing effectiveness of weapons on killing noncombatants, helped lead to rules of warfare, which were designed to protect civilian populations while allowing the wholesale slaughter of soldiers. So warfare changed, from the standpoint of the military and civilians both.

This is not to suggest that mathematics alone is to blame for these trends. Most of these "improvements" would have happened regardless of the use of mathematics. Just about everything mentioned above would have been determined empirically at some point. The use of mathematics simply made the process more efficient, a trend that has continued to this day.

P. ANDREW KARAM

Further Reading

Dyer, Gwynne. *War.* New York: Crown Publishers, 1985.

Algebraic Solution of Cubic and Quartic Equations

Overview

The solution of the cubic and quartic equations was one of the major achievements of Renaissance algebra. The publication of the results in Girolamo Cardano's book *The Great Art* brought charges that Cardano had broken his promise to Tartaglia, who claimed he had made the major discovery in the cubic case. Attempts to identify all solutions of the cubic, quartic, and higher order equations would require the invention of complex numbers and would lead to the discovery of the theory of groups, one of the most important ideas in modern abstract algebra.

Background

Our word "algebra" is derived from the Arabic *Kitab al-jabr w'almuqabala,* a book by the Arabic Mathematician Muhammad ibn-Musa al-Khwarizmi (c. 780-c. 850) which described an art of "restoration and reduction," that is, finding the value of unknown quantities in an equality by rearranging terms. The book was translated into Latin in 1145 as the *Liber algebrae et almucabala* by Robert of Chester (fl. 1145), an English scholar living in Muslim Spain. In Christian Western Europe mathematics was in a far less advanced state than in countries under Muslim control. The Greek scholarly tradition had continued in the Byzantine Empire, which, however, was under frequent attack by its neighbors. In 1543 Turkish forces overran its capital, Constantinople. Byzantine scholars found refuge in Italy, where rich and powerful families like the Medici added scholars to their entourage and manuscripts to their libraries. The appearance of Gutenberg's printing press made mathematical ideas far more widely available. Over 200 new books on mathematics appeared in Italy before 1500.

In 1545 a book entitled *Ars Magna,* or *The Great Art,* by the Italian mathematician Girolamo Cardano (1501-1576), appeared. This work incorporated significant new results—the solution of the cubic and quartic equations. In modern treatments of algebra, a quadratic equation is any equation of the form

$$Ax^2 + Bx + C = 0$$

where A is a number other than zero, and B and C are constants that can be positive, negative, or zero. The letter x, of course, is the unknown to be found. Until quite recently, however, mathematicians did not have the tools available to deal with the case in which A, B, and C are all positive numbers. Further, there was a tendency to avoid the appearance of negative numbers. A method of solution for some quadratic equations had been developed by the ancient Babylonians based on a process now taught as "completing the square." The *al-jabr* discusses the six possible variations of the quadratic equation that can be written without negative numbers or a zero, for example, $Ax^2 = Bx$, or $Ax^2 + Bx = C$. In the latter case there are two possible solutions given by the quadratic formula, which involves the common operations of arithmetic, addition, subtraction, multiplication, and division as well as taking a square root.

A cubic equation has the modern form

$$Ax^3 + Bx^2 + Cx + D = 0.$$

With B set to zero, this is known as the reduced cubic equation. The quartic or biquadratic equation has the form

$$Ax^4 + Bx^3 + Cx^2 + Dx + E = 0.$$

As with the earlier solution of the quadratic, in treating these equations Cardano had to consider many special cases to avoid negative quantities and the need to take the square root of a negative number. As with the quadratic, Cardano's solution to the cubic and quartic involved adding subtracting, multiplying, dividing, and taking roots, in this case including cube roots as well as square roots.

Cardano freely admitted that the solutions he presented were not his original discovery. The solution to the reduced cubic equation $x^3 + Ax = B$ had been found in 1515 by Scipio del Ferro (1465-1526), a professor at the University of Bologna. He communicated this solution to his student Antonio Fior some 20 years later. The Italian mathematician Niccolò Fontana, better known as Tartaglia (1500?-1557), then announced that he had discovered the solution to the cubic equation lacking a first order term, $x^3 + Ax^2 = B$, as well as the solution to del Ferro's case. Fior doubted that Tartaglia could have found such a solution and arranged a contest in which he and Tartaglia exchanged sets of 30 problems. At the end of the agreed-upon time

Tartaglia had solved all of Fior's problems while Fior could solve none of Tartaglia's. Cardano invited Tartaglia to his home, hinting that he might be able to introduce him to a possible patron. There, Tartaglia disclosed his method to Cardano in return for assurances that it would not be published.

The solution to the quartic was obtained by the Italian Ludovico Ferrari (1522-1565), who had been Cardano's secretary and would become his son-in-law. Cardano wrote that Ferrari had developed it at his request. The essence of the solution was to define a new variable, related to the unknown, in such a way that the quartic could be written as a cubic, which could then be solved.

Tartaglia responded to Cardano's book by publishing one of his own, describing his own research on the cubic equation and attacking Cardano's integrity for breaking his promise. The meeting between Cardano and Tartaglia had taken place in 1539. Cardano learned of del Ferro's solution in 1542, however, and felt he was no longer bound by his promise, as Tartaglia's results had been in good measure anticipated by del Ferro. Ferrari rose to the defense of Cardano, and issued a public mathematical challenge to Tartaglia, to which Tartaglia responded. After six such exchanges Ferraro and Tartaglia engaged in a public oral debate in a church in Milan in 1548.

Despite the argument over whether Cardano had acted properly, there are important discoveries that are undoubtedly Cardano's. It was Cardano who discovered a systematic method to get the general cubic equation into reduced form so that del Ferro's solution could be used. Cardano was also the first to show that the cubic equation could have three real solutions. He was also among the first mathematicians to use imaginary numbers in expressing the solutions of algebraic equations, although a full understanding of their properties would not come for nearly three centuries.

Impact

The extreme competitiveness of the mathematicians involved in solving the cubic and quartic equations is consistent with the aggressive individualism of the Renaissance. Cardano was more flamboyant than most. The illegitimate son of a lawyer, he played dice and chess to gain income. He received a medical degree from the University of Padua in 1526, but was prevented until 1534 from practicing in his native Milan because

of his illegitimacy. Although his patients once included the Pope, he went to jail for heresy in 1570 for having cast a horoscope of Jesus, only to become the papal astrologer a year later. Tartaglia too had engaged in questionable practices. He published a translation of Archimedes by the Belgian scholar William of Moerbeke (c. 1220 -1286) in a manner suggesting that it was his own work. Ferrari was probably involved in intrigues as well. It is reported that he was poisoned by a relative.

A major step forward in algebra would occur with the work of the French lawyer and writer François Viète (1540-1620). While the above discussion has followed modern practice in using letters to represent both known and known numbers, it is only in the work of Viète that this is accomplished. The *al-jabr* used no mathematical symbols at all, while Cardano's work used letters for known quantities but not the unknown. With Viète's new notation it became easier to think of solving an algebraic equation as finding the values of x for which a definite function of the variable x would equal zero. This set the stage for the study of functions themselves and the study of transformations of functions caused by introducing new variables, ideas important in modern algebra, trigonometry, and calculus.

The solution of equations involving powers of unknown quantities has repeatedly served to inspire new developments in mathematics. The Greek mathematician Diophantus (fl. third century A.D.) raised the question of the existence of whole number solutions for equations involving whole number powers of different unknowns. The French mathematician Pierre de Fermat (1601-1665) conjectured that there would be no whole number solutions, A, B, C, to equations of the form

$$A^n + B^n = C^n$$

When n is greater than two. The search for a proof of Fermat's conjecture would occupy mathematicians for centuries.

The introduction of complex numbers, that is numbers of the form, $a + ib$, where the "imaginary" unit i has the property that $i^2 = -1$, was motivated in part by the study of algebraic equations. If complex numbers are allowed as solutions, then every quadratic equation has two solutions, every cubic three, and every quartic four. Understanding the nature of complex numbers was a source of many new ideas in nineteenth-century mathematics.

With the solutions of the cubic and quartic equations known, it might seem that the solution of the quintic equation, which included the fifth power of the unknown, and even higher order equations would be achieved eventually. Despite several centuries of effort, these solutions were not found. That such exact solutions, involving roots and arithmetic operations only, are not possible was demonstrated in 1824 by the Norwegian mathematician Niels Henrik Abel (1802-1829). Abel's results were generalized to all equations involving powers of the unknown higher than fourth by the French mathematician Evariste Galois (1811-1832). The study of algebraic transformations by these two mathematicians would lead to a general theory of transformations now known as group theory, which is considered one of the major areas of both abstract and useful mathematics by modern mathematicians and scientists.

DONALD R. FRANCESCHETTI

Further Reading

Bell, Eric Temple. *Development of Mathematics*. New York: McGraw-Hill, 1945.

Boyer, Carl B. *A History of Mathematics*. New York: Wiley, 1968.

Kline, Morris. *Mathematical Thought from Ancient to Modern Times*. New York: Oxford University Press, 1972.

The Development of Analytic Geometry

Overview

The fundamental idea of analytic geometry, the representation of curved lines by algebraic equations relating two variables, was developed in the seventeenth century by two French scholars, Pierre de Fermat and René Descartes. Their invention followed the modernization of algebra and algebraic notation by François Viète and provided the essential framework for the calculus of Isaac Newton and Gottfried Leibniz. The calculus, in turn, would become an indispensable mathematical tool in the development of physics, astronomy, and engineering over the next two centuries.

Background

The relationship between geometry and algebra has evolved over the history of mathematics. Geometry reached the greater degree of maturity sooner. The Greek mathematician Euclid (335-270 B.C.) was able to organize a great many results in his classic book, *the Elements*. Algebra was a far less organized body of ideas, drawing on Babylonian, Egyptian, Greek, and Hindu sources, and dealing with problems ranging from commerce to geometry. Until the Renaissance, geometry might be used to justify the solutions to algebraic problems, but there was little thought that algebra would shed light on geometry. This situation would change with the adoption of a convenient notation for algebraic relationships and the development of the concept of mathematical function that it permitted.

To illustrate both the importance of notation and the function concept, we could consider one of the classic problems in algebra, solution of the quadratic equation. In modern notation such an equation would be written $Ax^2 + Bx + C = 0$. Here it is understood that A, B, and C represent numbers, x represents the unknown quantity to be found, and the small 2 appearing in the first term means that the unknown x is to be squared or multiplied by itself. While the solutions to some forms of this equation were known to the ancient Babylonians, the notation was not fully developed until the work of the French mathematician François Viète (1540-1602), who standardized the use of letters to represent both constants and variable quantities. Given this notation, it is then an easy thing to think about the equation as having the form $f(x) = 0$. Where the function $f(x) = Ax^2 + Bx + C$. One can then think of a second variable, say y, being defined by the function, $y = f(x) = Ax^2 + Bx + C$, so that we have a relation between the two variables x and y that can be studied in itself.

The essential idea behind analytic geometry is that a relation between two variables, such that one is a function of the other, defines a curve. This idea appears to have been first developed by the French lawyer and amateur mathematician Pierre de Fermat (1601-1665). In his book *Introduction to Plane and Solid Loci,* written in 1629 and circulated among his friends but not published until 1679, he introduced the idea that any equation relating two unknowns defines a locus or curve. Fermat allowed one of the variables to represent a distance along a straight line from a reference point. The second variable then denoted the distance from the line. Fermat went on to derive equations for a number of simple curves including the straight line, the ellipse, the hyperbola, and the circle. Since Fermat did not consider negative distances, he could not display the full curves, but other mathematicians would soon overcome this problem.

The French philosopher René Descartes (1596-1667) also discovered an algebraic approach to geometry, apparently independently. Descartes was one of the dominant intellectual figures of the seventeenth century, best known as a philosopher, the author of several important physical theories, and a major contributor to mathematics. Descartes's work on geometry appears as one of the three appendices to his famous book *Discourse on the Method of Rightly Conducting Reason and Reaching the Truth in the Sciences*. The other two appendices are on optics and meteorology. As the title suggests, Descartes saw mathematics primarily as a route to sure knowledge in the sciences.

In his appendix on geometry, Descartes began by pointing out that the compass and straight-edge constructions of geometry involve adding, subtracting, multiplying, dividing, and taking square roots. He proposed assigning a letter to represent the length of each of the lines appearing in a construction, and then writing equations relating the lengths of the lines, obtaining as many equations as there are unknown lines. Finding the unknown lengths then becomes a matter of solving the set of equations thus obtained.

After thus showing that algebra can be applied in solving classic geometric problems, Descartes then discussed solving problems that have curves as their solution. In this type of problem there are not enough equations to determine all the unknown quantities, and one ends up with a relation between two unknowns. It is at this point that Descartes suggested using the length away from a fixed point on a given line to represent x, and the distance from x on a line drawn in a fixed direction to represent y. If the fixed direction is chosen at a right angle to the first line, we obtain the modern system of rectangular or Cartesian coordinates, named after Descartes.

Descartes then proposed that any equation involving powers of x and y describes an accept-

able geometrical curve and showed that those special curves known as conic sections—the circle, ellipse, hyperbola, and parabola—are all described by algebraic equations in which the highest power of x or y is two. The study of such curves was gaining in importance as a result of discoveries in physics and astronomy, particularly the discovery by the German scientist Johannes Kepler (1571-1630) that the planets move not in perfect circles or combinations of perfect circles but in ellipses. Further, Kepler had shown the planets do not move at constant speed but at a speed that varies with their distance from the Sun. Analytic geometry provided a useful description of the shape of such orbits. An explanation of the actual motion would soon follow as Isaac Newton (1642-1727) proposed his laws of motion and universal gravitation and developed the techniques of the calculus to apply them.

The two basic problems of the calculus are easily expressed in terms of analytic geometry. The first is finding the tangent line to the curve described by $y = f(x)$ at any point, and the second is finding the area between a segment of the curve and the line $y = 0$. Solving these problems leads directly to the solution of two others: finding the values of x for which $y = f(x)$ is a minimum or maximum, and finding the length of a segment of a curve. Fermat had solved the problem of finding tangents and the associated problem of finding maxima and minima by 1629. When Descartes' *Geometry* appeared in 1637, Fermat criticized it for not including a discussion of tangents or maxima and minima. Descartes replied that these results could be readily obtained by anyone who understood his work, and that Fermat's work showed far less understanding of geometry than his own. The dispute over the importance and priority of Fermat's and Descartes' contributions would eventually subside, with each man acknowledging the other's contributions.

The full development of the calculus would be achieved by Newton and German scientist Gottfried Wilhelm Leibniz (1646-1716), working independently of each other. As in the case of Fermat and Descartes, a dispute as to priority broke out, but in this case a more bitter and long lasting one.

In 1696 the Swiss mathematician Johann Bernoulli (1667-1748) published a problem in applied calculus as a challenge to other mathematicians. The problem, known as the *brachistochrone*, is to find the curve along which a sliding bead will move from one point to another in the least time with gravity as the only external force. The answer is an upside-down version of the cycloid, the curve generated by a point on the circumference of a wheel as it rolls along a level surface. In 1697 Bernoulli was able to publish his own solution along with those obtained from four other mathematicians, Newton and Leibniz among them. Newton's solution had been submitted anonymously, but this did not fool Bernoulli, who said, "I recognize the lion by his claw."

Impact

Analytic geometry represents the joining of two important traditions in mathematics, that of geometry as the study of shape or form and that of arithmetic and algebra, which deal with quantity or number. This combination was needed if the physical sciences were to progress beyond the notions of Aristotelian philosophy about perfect and imperfect motions to a natural philosophy based on observation and experiment. It is not surprising then that both Fermat and Descartes were concerned with the scientific issues of their day, both with optics in particular, and Descartes more generally with all areas of physics and astronomy.

The techniques of calculus, built on the insights of analytic geometry, have become the fundamental mathematics of the physical sciences and engineering. With the addition of differential equations, which represent a further development of the basic ideas of calculus and analytic geometry, the mathematical framework has proven robust enough to incorporate the new areas of thermal physics and electromagnetism in the nineteenth century and quantum theory in the twentieth. Thus the modern university curriculum for future scientists and engineers always includes several semesters devoted to analytic geometry and calculus.

DONALD R. FRANCESCHETTI

Further Reading

Bell, Eric Temple. *Development of Mathematics*. New York: McGraw-Hill, 1945.

Boyer, Carl B. *A History of Mathematics*. New York: Wiley, 1968.

Kline, Morris. *Mathematical Thought from Ancient to Modern Times*. New York: Oxford University Press, 1972.

The Printing of Important Mathematics Texts Leads the Way to the Scientific Revolution

Overview

By the late fifteenth century the scholars of Europe were poised to fully reclaim the classical mathematical heritage that was nurtured and expanded by Islamic scholars and reintroduced into the West in the thirteenth century. Crucial in the growth of the mathematics of the period was the publication of mathematical texts. These texts helped pave the way for the growth of commerce and the onset of the Scientific Revolution.

Background

Many changes in economics, culture, and education were taking place in Europe in the fourteenth and fifteenth centuries. These changes had an effect on the mathematics of that time. These changes can be traced back even further to the Middle Ages. National boundaries became stabilized. In order for these new nations to expand economically and culturally, both secular and religious leaders initiated the founding of universities in major cities. The Crusades, while largely an exercise in futility, did manage to open trade routes to the Byzantine and Islamic peoples. Both of these events led to the recovery of classical learning. At first, scholars set to work on translating Greek and Arabic texts into Latin. Next, some scholars, notably Leonardo of Pisa (a.k.a. Fibonacci, 1170-1240) and Jordanus de Nemore (fl. 1200s), contributed original mathematics. New mathematical techniques were necessary to accommodate the new business of the trading companies of Italy. In order to succeed, these new kinds of merchants needed to learn this new mathematics, so some mathematicians soon found themselves in demand as teachers and textbook authors. All of this work was more widely distributed with the invention of the printing press by Johannes Gutenberg (c. 1398-1468) around 1450, and the subsequent rapid growth of printing with moveable type.

What are some of the influential books printed during this period? The first printed arithmetic text was published in 1478. This work had no title and was by an unknown author. It is now commonly referred to as the *Treviso Arithmetic* because of its place of publication—Treviso, north of Venice in Italy. The text is meant to be a training manual for young students seeking to enter the new trading businesses in Italy. In it one finds how to write and compute with numbers (only the four basic operations) and how to apply these techniques to questions of partnerships and trading. For example, one problem tells of three partners who invest unequal amounts in a partnership. If the partnership earns a profit, one must solve the problem of how the profit should be fairly divided. The *Treviso Arithmetic* itself was not very influential, except for the fact that its publication paved the way for an incredible amount of printed mathematical texts in the years following.

The first significant work in mathematics to be printed was the *Elements* of Euclid (c. 330-c. 260 B.C.), printed in Venice in 1482. All mathematics of the time relied heavily on the *Elements*. Indeed the *Elements* was a foundational work for all mathematics well into the eighteenth and nineteenth centuries. It is natural, then, that this work should be among the first printed. Euclid's work had been translated before by Islamic mathematicians and in turn these Arabic versions were translated into Latin, notably by the English monk Adelard of Bath (1075-1164) and Gerard of Cremona (1114-1187). Adelard translated the *Elements* from the Arabic version of al-Hajjai (c. 786-833). To Adelard's translation the Italian Johannes Campanus of Novara (1220-1296) added proofs for the propositions and other supplementary material. This became the standard translation of the *Elements* and was the one published in 1482. The fact that this version was the one that was printed illustrates its importance during this period.

Printed versions of other translations of the *Elements* later appeared on the scene. Interestingly, one of the defenders of the Campanus version was the Italian friar Luca Pacioli (1445-1517), who published corrections and commentary to the Campanus version. However, it is Pacioli's work *Summa de Arithmetica, geometria, proportioni et proportionalita*, finished in 1487 and first printed in 1494, that is the next significant publication under consideration.

The *Summa* is regarded as the first printed work on algebra. Again the honor does not go to an original piece of work. Concerned with the lack of teaching materials, Pacioli gathered mathematical materials from various sources and published them in a large comprehensive text. Very little in the book is the original work of Pacioli; in fact, much of the material can be found in the *Liber abbaci* of Leonardo of Pisa, a very influential work of its time. However, given the book's scope and the fact that is was one of the first texts printed, it became widely circulated and hence very influential. Its circulation and influence was also extended due to the fact that the book was printed in the vernacular Italian and not Latin.

The arithmetic in the *Summa* is standard for the time, including many techniques for the four arithmetical operations and the extraction of square roots. There are also problems in commercial arithmetic and an important study of double entry bookkeeping. The *Summa* also deals with algebra up to and including the solution of quadratic equations. The algebra here is verbal with no symbols used. Pacioli is one of the first to write on the so-called problem of the points for determining the division of the stakes in a game of chance if the game is stopped before it is finished. The study of this problem later led Blaise Pascal (1623-1662) and Pierre de Fermat (1601-1665) to the invention of the science of probability. The geometry in the *Summa* is basic, being a rehash of portions of Euclid along with some work on using algebra to solve geometric problems, which was first done by Regiomontanus (born Johann Müller, 1436-1476).

Regiomontanus' greatest work is *De Triangulis Omnimodis* (On Triangles of Every Kind) finished in 1464 (so Pacioli certainly had access to it) but not printed until 1533. Regiomontanus was one of the scholars involved in translating classical works, having translated Ptolemy's (100-170) *Almagest* from Greek to Latin. Regiomontanus was also a leader in using the new printing technology to publish mathematical works.

De Triangulis Omnimodis is the first European trigonometry text, even though it does rely heavily on earlier Islamic texts. Trigonometry was important to kings and merchants alike—for astrological predictions and making calendars and for astronomy and navigation. Trigonometry at the time was primarily concerned with the sine of an arc of a circle, which is defined to be half of the chord of double the arc on a circle. Regiomontanus' work discussed this function and also made use of the sine of the complement of the arc, the precursor of the modern cosine function.

These are the only two trigonometric functions used in *De Triangulis Omnimodis*.

Regiomontanus's work is made up of five books and is a compilation of the trigonometry known in Europe at the time. The first book is concerned with lengths and ratios and is based on Euclid's *Elements*. In fact the entire work is written in the style of the *Elements*, with theorems proven based on axioms and results from Euclid's work. However, Regiomontanus also includes many examples. In this book Regiomontanus shows how to solve triangles; that is, given various combinations of sides and/or angles, to find the other sides and angles. For example, Theorem 28 of Book I states, "When the ratio of two sides of a triangle is given, its angles can be ascertained." Theorem 1 of Book II of *De Triangulis Omnimodis* is a perfectly modern statement of the law of sines for plane triangles. The remainder of Book II uses the law of sines to solve other triangles. This is standard material in a modern course on triangle trigonometry. To solve these triangles, Regiomontanus used his sine table, which was based on a circle of radius 60,000. The sines for other radii could be determined from this using the properties of similar triangles.

In *De Triangulis Omnimodis* Regiomontanus was also the first European to make use of algebra when solving triangles, solving a quadratic equation to find the remaining sides of a triangle given the lengths of the base, the altitude to the base, and the ratio of the unknown sides. This can be accomplished using the Pythagorean theorem and the given ratio. Book III of *De Triangulis Omnimodis* deals with spherical geometry, and correspondingly Book IV deals with spherical trigonometry. Naturally this material is of great use in astronomy. In these books Regiomontanus shows how to solve spherical triangles, whose sides are arcs of great circles on a sphere. Included is the spherical law of sines and the first use of the law of cosines for spherical triangles.

Because of their trading companies and banks, the Italians led the way in arithmetic and algebra related to commerce, followed in turn by the Germans with their trading towns on the Baltic Sea and the support of German banking tycoons. It is German texts that greatly influenced the important texts written in English.

The first English text author of any importance is Robert Recorde (1510-1558). Recorde is responsible for the first series of mathematics texts in England written in English, the first of which is *The Ground of Arts* (1543), an arithmetic text with commercial applications along the lines of Pacioli's *Summa*. Recorde's works are written in dialogue form, with an engaging style in which all steps are carefully explained. This might explain the fact that *The Ground of Arts* remained in print for over 24 editions over a period of 150 years. Recorde also wrote less successful companion texts on geometry (*The Pathway to Knowledge*, 1551), algebra (*The Whetstone of Witte*, 1557), and astronomy (*The Castle of Knowledge*, 1556).

The algebra texts of the time concentrated on solving equations no higher than second degree. Indeed, Pacioli deemed that the solution of the general cubic equation by a formula analogous to the quadratic formula could not be found. Pacioli was proven wrong by Scipione del Ferro (1465-1526) and Niccolò Fontana, better known as Tartaglia (1499-1557). The publication of the solution of the general cubic equation occurred in 1545 with the publication of *Ars Magna, sive de Regulis Algebraicis* (The Great Art, or On the Rules of Algebra) by Girolamo Cardano (1501-1576).

The *Ars Magna* is a book on solving algebraic equations, including cubic and quartic equations. The major breakthrough in the work is the first publication of the solution of cubic and quartic equations by radicals, that is, with a formula similar to the quadratic formula. Cardano also shows his awareness of imaginary numbers in *Ars Magna* because they appear in the formula for the solution of the cubic equation. Our modern algebraic symbolism is not present in the work, nor is the use of general coefficients. Cardano lists many different cases of cubic equations corresponding to the various ways necessary to write the equation with all positive coefficients. Cardano also supplies geometric proofs of the validity of his results. While this renders the work quite difficult for a modern reader, it was a revelation to the readers of the time. Until the invention of a suitable algebraic notation, the *Ars Magna* was the best that any mathematician could do on the subject.

By the time of René Descartes (1596-1650), suitable algebraic notation and the use of general coefficients had both been developed. Descartes uses these to good effect in *La Géométrie*, an appendix to his 1637 work *Discours de la méthode pour bien conduire sa raison et chercher la vérité dans les sciences* (A Discourse on the Method of Rightly Conducting the Reason and Seeking Truth in the Sciences). *La Géométrie* is a groundbreaking work, one of the greatest scientific works of the Renaissance. However, its influence was not initially felt. Descartes wrote in French rather than Latin, and while writing in the vernacular was essential for works that wished to popularize mathematics (see Pacioli and Recorde), for works that were on the cutting edge, Latin was the language of the learned. Also, Descartes was not very forthcoming with details and explanations in his work. The profound influence that *La Géométrie* exerted on mathematics is due to Franz van Schooten (1615-1660) and his publication of *Geometria a Renato Des Cartes* in 1649. Analytic geometry is truly born in this work, and from there it is a short step to the invention of the calculus and a whole new era of mathematics. In his Latin edition van Schooten added commentary, figures, alternate techniques, and a logical ordering of the material. He also added the work of other mathematicians, notably Sluse, van Heuraet, and de Witt. It is important to realize that Descartes's work accounts for approximately 100 pages, while the additional work of van Schooten and the others accounts for nearly 900 pages. This and later editions were the primary source of analytic geometry for mathematicians until the middle of the eighteenth century.

Impact

All of these works exerted significant influence. They became the standard textbooks of the time in the universities and training schools. Important pioneers in mathematics and the sciences studied from these works. Isaac Newton (1642-1727) mastered Euclid's *Elements*, although it is unclear what version he used. However it is known that Newton studied from van Schooten's *Geometria a Renato Des Cartes*, as did Christiaan Huygens (1629-1665), Gottfried Leibniz (1646-1716), John Wallis (1616-1703), Jakob and Johann Bernoulli (1654-1705 and 1667-1758, respectively), and Leonhard Euler (1707-1783). Pacioli was well versed in Campanus' version of Euclid, as was every other mathematician of the time, including John Dee (1527-1608), the editor of the first English version of the *Elements*.

Pacioli's *Summa* was studied by the sixteenth-century Italian mathematicians; it was the common starting point from which they advanced the study of algebra. Scipione del Ferro and Cardano were known to be familiar with the *Summa*. It also helped popularize the use of abbreviations for mathematical terms, which led to the invention of algebraic symbolism, which in turn was crucial to the growth of mathematics.

The *Summa* also popularized using the methods of algebra to solve geometric problems, which were initiated in the West in Regiomontanus' *De Triangulis Omnimodis*. This started mathematicians down the road towards analytic geometry, a road that ended with van Schooten's *Geometria a Renato Des Cartes*.

After Regiomontanus' work, many other trigonometry works were published, and European trigonometry began to surpass the trigonometry of the Islamic mathematicians. In addition, trigonometry supplied the tools necessary for the advances in astronomy soon to follow, culminating with the overthrow of the Ptolemaic system of the world in favor of the Copernican system.

Research in algebra was stimulated by the *Ars Magna*. For the next 200 years mathematicians, inspired by the discovery of solutions in radicals to the cubic and quartic equations, sought to generalize those results to higher degree equations. Although we now know that they would search in vain (general equations of degree five and higher cannot be solved using radicals only), a great deal of important mathematics in the fields of higher algebra, which includes the theory of groups, rings, and fields, was discovered along the way. In addition, a new kind of number was born of the solution of the cubic. Mathematicians turned to the study of imaginary numbers to better understand the solutions given by Cardano's cubic formula.

Cardano's book also showed that analytic methods could be quite powerful. Problems could be solved without having to resort to geometry, which up to this time was the foundation for all of mathematics. The *Ars Magna* exerted such influence that it could be said that too much emphasis was placed on algebraic methods, perhaps postponing the marriage of algebra and geometry later found in van Schooten's Latin *Géométrie* and in the development of the calculus.

GARY S. STOUDT

Further Reading

Boyer, Carl. *History of Analytic Geometry*. New York: Scripta Mathematica, 1956.

Ore, Oystein. *Caradano: The Gambling Scholar*. Princeton, NJ: Princeton University Press, 1953.

Swetz, Frank J. *Capitalism and Arithmetic: The New Math of the 15th Century*. LaSalle, IL: Open Court, 1987.

Marin Mersenne Leads an International Effort to Understand Cycloids

Overview

French mathematician, theologian, and educator Marin Mersenne (1588-1648) made numerous contributions to mathematics, including the prompting of a greater understanding of cycloids, which in turn directly affected the development of the pendulum clock. By opposing irrational or superstitious interpretations of phenomena, including numbers, Mersenne helped elevate the level of mathematics and mathematical research. Indeed, it was his insistence on empirical evidence, as well as his curiosity about cycloids, that affected Christiaan Huygens's (1629-1695) work, which in turn resulted in Huygens patenting a pendulum clock. Without a doubt, though, Mersenne's great contribution to his time was his indefatigable devotion to collecting, sharing, and distributing scientific and mathematical information among a wide community of correspondents and scholars. In this way, he was a sort of one-man repository or clearinghouse, very much aware of the vital importance of feedback and commentary to scientific progress.

Background

Marin Mersenne was the child of a laborer who escaped the poverty of his birth by education—which may have been subsidized by the Jesuits. He studied at the College of Mans, and from 1604 to 1609 studied at Le Fleche, a Jesuit institution, followed by two years of theological studies at the Sorbonne in Paris. In 1611 Mersenne joined the Minim Friars, an order that took its name from its commitment to reducing worldly involvement and focusing on prayer and contemplation, study and reflection. In 1619 Mersenne took up residence in a Minim monastery in Paris, where he lived for the rest of his life.

Minimizing worldly concerns may have been a major concern of Mersenne's religious order, but it hardly kept him from contact with the larger world. His correspondence with the leading scientific and philosophical thinkers of his time quickly became legendary. Mersenne's letters spanned much of the known world, and consisted not only of his own insights, but also of exhortations to the recipients of his letters that they pursue certain fields of inquiry or investigation. Furthermore, Mersenne invariably

introduced his own correspondents to others with whom they might pursue independent correspondence. Thus he established a network of communications that reinforced, enriched, and extended the body of knowledge.

Mersenne also devoted much effort to theories of the nature of prime numbers, particularly large prime numbers. Although some of his theories proved to be flawed, today a number of computer programs are devoted to searching for what are still called Mersenne Primes.

Among the many other specific problems that Mersenne pursued was the nature of a particular type of curve known as a cycloid. Cycloids are achieved by tracing the path of a fixed point on the edge of a cylinder rolling smoothly along a straight lane. Cycloids provided a means for calculating the "curve of least time"—the curve by which a fixed point takes the least amount of time to make the transit between two points, A and B. The origins of research into cycloids remain in some dispute. While various sources for its initiation are invoked, there is general agreement that French mathematician Charles Bouvelle (c. 1471-c. 1553) was in 1501 the first to accurately describe the curve's nature, but he did not produce the equations necessary for deriving the specific properties of cycloid curves.

Because those properties promised great practical as well as theoretical and mathematical benefits—particularly in determining the mathematical mechanics of moving objects, as well as in calculating the nature of curves for arches, and the curves described by a pendulum—the search for those properties, including the area of the space contained within the curve, grew heated and acrimonious, with arguments over the accuracy of solutions, and even the authorship of solutions generating anger and accusations.

As a result of these conflicts, and the near-universal attraction of mathematicians to the cycloid question, the cycloid itself came to be known as either the "apple of discord" or the "Helen of Geometers"—a source of irresistible appeal, but over whose appeal battles would be waged.

Galileo (1564-1642) in the late 1500s made an empirical approach to solving the cycloid. As a result of experiments he determined that the

Marin Mersenne. *(The Granger Collection, Ltd. Reproduced with permission.)*

area of the cycloid is approximately three times that of the circle that generates the curve, a property he believed accurately would be of use to the building of arches in bridges and other structures. Galileo is also believed to have given the cycloid its name.

Although Galileo's experimental conclusions were useful, the search for a purely mathematical proof continued. It was Marin Mersenne who put into motion the research that resulted in that proof. Along with every other mathematician of the time, Mersenne felt the attraction of the cycloid. His network of correspondents was crucial here, for he used the network to make the cycloidal challenge more widely known. One of the recipients of Mersenne's letters was French mathematician Gilles Personne de Roberval (1602-1675), who used purely mathematical techniques in 1634 to prove that the area of the cycloidal arch is three times the area of the circle. Unfortunately, Roberval neglected to publish his proof, with the result that his achievement was considered plagiarism by those who proved the area later but published sooner.

Other mathematicians applied themselves to determining other properties of the cycloid. Blaise Pascal (1623-1662) had enjoyed a brilliant career as a mathematician, making large contributions to pure mathematics, but had abandoned mathematics in favor of devotion to

religious philosophy. Beset by a toothache in 1658, however, Pascal claimed to have experienced a mathematical vision, the result of which was the development of mathematical functions that allowed the determination of the areas of sections of cycloids, the volumes of areas generated by the rolling sphere, as well as the centers of gravity for these sections. According to legend, Pascal derived all of these functions within eight days of his toothache.

What makes the achievement of these mathematicians the more remarkable in regard to cycloidal properties is that these properties and functions were determined without benefit of the most appropriate mathematical tool for tackling problems of motion over space: the calculus. The calculus would have simplified much of the investigation into cycloids, but it had not been invented at the time Galileo, Mersenne, Roberval, Pascal, and others were performing their cycloidal research.

The cycloid's mechanical properties were most determinedly attacked by Christiaan Huygens, who perceived that a pendulum described a cycloidal curve as it moved rhythmically back and forth. Combining his understanding of cycloids with the emergent knowledge of gravity, Huygens found that a body drawn by gravity along a cycloidal arch requires the same amount of time to reach the bottom of the arch.

Huygens's application of his finding to the pendulum, using devices that insured that the pendulum swung in a cycloid, enabled the attachment of a pendulum to the workings of a clock. Huygens built his first pendulum clock in 1657. This represented the largest advance in the clockmaker's art since the development of the waterclock in Greece of the second century B.C.

Huygens would spend much of the rest of his life refining the relationship of the pendulum to the cycloid, moving ever closer to the accurate tracking of the passage of time. As important—and perhaps more so—as the availability of accurate, pendulum-kept time on land, was the applicability of the pendulum clock to keeping accurate time at sea or in transit over large distances, allowing for a far more accurate and efficient means of determining a traveler's precise location.

Above all, it was Huygens's application of his understanding of the cycloid to specific questions of motion, velocity, and the effects of gravity on bodies in motion that extended the realm of the cycloid from pure mathematics and applied mathematics to the world of physics, and the revolution in the understanding of physics that would take place in the century after Huygens's death.

Impact

The widespread—if often angry and accusatory—nature of investigations into the properties of cycloids is perhaps the best early example of the growth of *scientific community*. Thanks in no small part to Mersenne's energy as a correspondent and instigator, the full and diverse power and ability of a group of mathematicians and scientists was brought to bear on a single problem. While competition certainly figured in the search for cycloidal solutions, there was also an underlying current of collaboration, however unofficial or even unwanted. This sort of collaboration and review by fellow scientists would become far more common in the century ahead, and would ultimately become the very essence of the worldwide scientific community. Of no small consequence to all of the various approaches to the cycloid was the fact that the knowledge gained, when coupled with Huygens's mechanical insights, established accurate timekeeping as an achievable goal, radically altering the nature of a world whose rhythms had previously been governed by only the loosest sense of time.

KEITH FERRELL

Further Reading

Dunham, William. *Journey through Genius: The Great Theorems of Mathematics.* New York: John Wiley & Sons, 1990.

Katz, Victor. *A History of Mathematics: An Introduction.* 2nd ed. New York: Addison-Wesley, 1998.

Swetz, Frank J., ed. *From Five Fingers to Infinity: A Journey through the History of Mathematics.* Chicago and Lasalle: Open Court, 1994.

Mathematicians Revolutionize the Understanding of Equations

Overview

The progress of mathematics from its origins in simple counting to its ability to handle and manipulate variables, unknowns, and changing properties accelerated during the 1600s. Mathematics had from earliest times been a supremely practical science, either for enumerating items (counting) or for establishing the relationships among shapes (geometry). Most ancient mathematics took the form either of simple counting or of geometric measurement. The evolution of mathematics in the 1600s was nothing short of explosive, and the evolution of the equation—and the power of the equation as a tool for solving complex problems of many sorts—lay at the heart of the growth of mathematical capability. Equations are mathematical formulae whose purpose is, at least in principle, simple: equations consist of factors that must be balanced or made equal. By introducing systems for accommodating variables and unknowns, and the tools for solving those items, the equation became a powerful mathematical tool, with applications that reached far beyond the realm of numbers, affecting everything from construction (geometry) to theories of the motions of the planets (calculus) and embracing virtually every field of human endeavor. If in the centuries to come mathematics would prove to be the key to the universe, the equation, and the rules for equations that were refined, established, and developed during the 1600s, would prove to be the key to mathematics.

Background

The process of establishing equivalencies, of making both sides of an equation balance out, is the essence of much of mathematics. Put simply, an equation challenges the mathematician to prove that something is equal to something else, that the propositions on both sides of the equal sign can indeed be made to equal each other. From that simple proposition, vastly complex mathematical structures can be constructed and solved.

In order for equations to become effective, however, a large leap of process had to be made. In short, the mathematics of equations had to become formalized—a set of rules had to be established and agreed to. While much of that standardization took place during the 1500s and 1600s (and was aided immeasurably by the fact that printing was by that time well established), equations themselves are at least as old as the ancient Greeks.

Diophantus of Alexandria (c. 210-c. 290) was among the first to extend complex mathematics beyond the more common geometries of his time and into what is today known as algebra. Among other things, Diophantus introduced the use of symbols into equations; previously equations had been written in words. Diophantus's employment of Greek symbols to represent frequently used quantities and other factors played an important role in simplifying the construction (and solution) of complex equations.

It was the Arabian mathematician Muhammad ibn Musa Al-Khwarizmi (c. 780-c. 850) who translated and expanded Diophantus's work, in a book that gave the science of equations its modern name. Al-Khwarizmi's book was called *ilm al-jabr wa'l muqubalah* (The Study of Transposition and Cancellation), although al-jabr (or al-jebr) can also be translated as "the reunion of broken parts." That reunion meant the balancing of both sides of an equation, and al-jabr has, of course, been transformed over the centuries into the word *algebra*, which is the branch of mathematics most closely associated with equations. (In addition to naming algebra, Al-Khwarizmi imported certain concepts, including the zero and numerals, from Hindu sources. When Al-Khwarizmi's work was translated into Latin, those numerals became known as Arabic numerals.)

The introduction of standardized symbology laid the groundwork for modern equations. Another large forward step was taken by French mathematician François Viète (1540-1603), who first used letters of the alphabet to represent unknowns and constants in equations. In Viète's system vowels were used to represent unknowns, and consonants represented constants. Although he is known to this day as the "father of algebra," Viète disliked the term, preferring "analysis" to refer to the process of using equations to solve propositions.

By the turn of the seventeenth century, equations and algebra were in flux with new discoveries and approaches. Among the most dramatic was the work of Albert Girard (1595-1632). A French mathematician, Girard was an engineer as well as a mathematician, and he applied many of his mathematical insights to his engineering work, particularly the development of fortifications. But it was his 1629 book *L'invention en Aalgèbre* (The Invention of Algebra) that solidified his reputation as a major contributor to the development of equations.

In that book, Girard made a formal approach to what would later become known as the Fundamental Theorem of Algebra. In essence, the Fundamental Theorem states that for any polynomial equation (an equation with at least one algebraic term multiplied by at least one positive variable raised to an integral power) there is a root in the complex numbers. In mathematical terms, this root is expressed as $a + bi$, with a and b representing real numbers and i representing the square root of -1. Numbers expressed as $a + bi$ are called complex numbers.

Refined and extended by mathematicians more gifted than Girard, and proved in 1799 by German mathematician Johann Karl Friedrich Gauss (1777-1855), the Fundamental Theorem remains central to algebra today.

The greatest of all contributors to the evolution of algebra and equations (up to his lifetime at least) was the French philosopher René Descartes (1596-1650). In the course of his life, Descartes so revolutionized mathematics and made so many large contributions to philosophical thought that aspects of both disciplines are still referred to as Cartesian in his honor. (And so influential was his writing that his theory of the working of the universe, although false, was accepted as accurate until disproved by Isaac Newton [1642-1727].)

Descartes's philosophy—and his approach to mathematics—rested upon his belief in absolute fact, in the "mechanical" nature of the universe. By beginning with an absolute fact, he thought, one could progress to an understanding of the whole of the universe. This search for absolutes, and absolute accuracy, guided his great mathematical works. (He even found a way to overcome any doubts: the very act of doubting proved the existence of the doubter. From this he derived his famous maxim, *Cogito, ergo sum* (I think, therefore I am).

In applying his search for absolutes to mathematics, Descartes began by further formalizing the nature of equations. Adapting Viète's work with alphabetical symbols for numbers, and further focusing it, Descartes used the early letters of the alphabet to represent constants, and letters at the end of the alphabet to represent variables. It is to Descartes that we owe the familiar use of x and y variables in algebra. In addition he devised a system for displaying exponents, and he was the first to use the square root symbol.

This systematic approach served Descartes well as he undertook the great mathematical work of his life, the unification of algebra and geometry. According to legend, Descartes was restricted to bed rest as a result of ill health, and thus confined amused himself by watching the movements of a housefly around his sickroom. Insight struck as he watched the fly flit about—every position of the fly's constantly varying motion could be expressed as a point that could be located in three dimensions by determining the coordinates of three intersecting lines, representing east/west, north/south, and up/down.

From that insight it was another step for Descartes to develop an equation that could translate any point into an equation, and conversely any equation composed of representations of points into a geometrical curve. Descartes then separated the world of curves into two types: geometric curves were those that were mathematically pure and could be expressed as equations; mechanical curves cannot be expressed in equations.

Descartes's fusion of algebra with geometry transformed both, and set in motion a wave of mathematical progress that led directly to Newton's development of calculus, which further extended Descartes's insight by applying algebraic equations to variables that are constantly changing, such as objects in motion.

Because of the great power and usefulness of Descartes's application of equations to analyzing geometric functions, the resultant discipline became known, and is still known, as analytic geometry, and the coordinates represented in its equations are known as Cartesian coordinates. His insistence that virtually all natural phenomena could be expressed as equations was a vital contribution to the development of modern science.

Impact

Without a formalized system for expressing mathematical variables, equations would have remained cumbersome and unwieldy, expressed in different ways by different mathematicians. The accretion of standardized approaches to equations, though, played an enormous role in transforming mathematics into a collaborative effort, one in which researchers and thinkers throughout the world shared a common means of expression, however different their approach to similar problems might be. Mathematics began to develop its now-familiar set of rules expressed in a common language. (All of this, it should be repeated, was aided immeasurably by the printing press and mass distribution of ideas and treatises.) Thus one generation could build upon the work of previous generations without having to re-discover or re-invent basic principles, approaches, and proofs. Descartes's combination of algebraic equations with geometric points, quite simply, provided the basis for Isaac Newton to develop calculus, which in turn altered and refined our understanding of how the universe works.

KEITH FERRELL

Further Reading

Dunham, William. *Journey through Genius: The Great Theorems of Mathematics*. New York: John Wiley & Sons, 1990.

Katz, Victor. *A History of Mathematics: An Introduction*. 2nd ed. New York: Addison-Wesley, 1998.

Swetz, Frank J., ed. *From Five Fingers to Infinity: A Journey through the History of Mathematics*. Chicago and Lasalle: Open Court, 1994.

Girard Desargues and Projective Geometry

Overview

In 1639 Girard Desargues (1591-1661), a military engineer and architect, published the *Brouillon project d'une atteinte aux événements des rencontres d'un cône avec un plan* (Proposed Draft of an Attempt to Deal with the Events of the Meeting of a Cone with a Plane). In this work Desargues established the principles of projective geometry, an alternative to traditional Euclidean geometry. Projective geometry was concerned with projections, or the extent to which the shape of a figure is changed by the perspective from which that figure is viewed.

Desargues's projective geometry, however, was overshadowed by the work of his contemporary, René Descartes (1596-1650). These two, along with other important scientists and mathematicians of the time, helped to move math and science towards more practical applications. Also, the frequent collaboration between these figures led to the creation of the powerful and important French Académie des Sciences.

Background

Mathematic techniques of the seventeenth century responded rapidly and radically to the conclusions that figures such as Johannes Kepler (1571-1630) and Galileo Galilei (1564-1642) derived from their studies of astronomy. A text such as Galileo's *The Assayer* (1623), for instance, succeeded in both ridiculing the explanation of comets proposed by Jesuit scholars and establishing new views on scientific reality and the scientific method. Likewise, Kepler's *The New Astronomy* (1609) overturned the traditional concept of the universe as a series of interlocking celestial spheres, and necessitated the revision of the laws of planetary motion.

This revision was achieved through a combination of mathematical and philosophical analysis that occurred in the work of Descartes, a French mathematical philosopher intent on establishing a new approach to the universe. Descartes's *Discourse on Method,* published in 1637, worked towards a new mathematics able to surpass the philosophical and analytical limitations of traditional geometry.

In "Geometry," the third appendix of the *Discourse on Method,* Descartes explained the fundamentals of algebraic geometry. However, this new geometry did not consist only of the application of algebra to geometry. Instead, it may be better characterized as the translation of algebraic operations into the language of geometry. Indeed, the overall theme of Descartes's "Geometry" is established by its opening sentence:

> *Any problem in geometry can easily be reduced to such terms that a knowledge of the lengths of certain lines is sufficient for its construction.*

Desargues developed a projective geometry, which may be seen as a counterpoint to Descartes's analytic geometry. In fact, Desargues spent many years in Paris with a group of mathematicians that included Descartes and Pascal as well as the Jesuit scientist Marin Mersenne (1588-1648) and Etienne Pascal (1588-1651). Desargues's work on projective geometry was printed principally for this limited readership of friends. Unfortunately, however, his views were very unorthodox and unpopular during his life—Blaise Pascal (1623-1662) was one of his

few admirers. Only 50 copies of the *Brouillon project* were printed, many of them later destroyed by the publisher. Desargues's projective geometry slipped into obscurity for nearly 200 years after the publication of his defining text on the subject.

The group of mathematicians and scientists with whom he associated, however, is notable for more than the academic achievements of its members. The group's informal meetings, which began in Mersenne's "cell" in the Convent of the Annunciation in Paris, later became what is still today the principle scientific organization of France, the Académie des Sciences. As the gatherings of these scientists grew larger, their influence on government increased as well. The author and administrator Charles Perrault suggested the establishment of a scientific academy to Jean-Baptiste Colbert, Minister of Finance to Louis XIV. The academy allowed for and encouraged practical applications of scientific discoveries while providing a forum for intellectual exchange on a previously unprecedented scale. After its establishment in 1666, the academy provided a royal pension and financial assistance for research to its members.

Impact

As an architect and military engineer, Desargues was interested in problems concerning the role of perspective in architecture and geometry. His principal work, *Brouillon project,* was immensely unsuccessful, even though it laid the foundations of projective geometry. The title was cumbersome (consider the simplicity of Apollonius's [245?-190? B.C.] *Conics*) and the prose was remarkably unwieldy. He introduced more than 70 new terms in this text, of which only one, *involution,* has survived. This term, which denotes quite literally the twisted state of young leaves, is used to designate the projective transformation of a line that coincides with its inverse. (Most of the terms that Desargues proposed were based on obscure botanical references.) The deliberate use of these confusing terms resulted in a decidedly negative reception. Indeed, Blaise Pascal was one of the few able to comprehend Desargues's deliberate obfuscation. The work was ignored for roughly two centuries, until the French mathematician Michel Chasles (1793-1880) completed his standard history of geometry.

In addition, Descartes's contemporaneous work further limited interest in Desargues's volume. The crystalline simplicity of Descartes's prose eased readers through difficult concepts. As a result, mathematicians were far more interested in developing the applications made possible by Descartes's powerful contribution to mathematics.

Despite its unwieldy explication, the thought behind Desargues's work is actually quite simple. Projective geometry was indebted to Leon Battista Alberti's (1404-1472) treatment of perspective and Kepler's principle of continuity. Alberti's account of perspective was of key importance for Renaissance painting and architecture. His *De pictura* seeks to link painting to mathematics and provides criteria for artists interested in creating the illusion of reality in their works. However, in Alberti's treatment of perspective, artists had to use an "eye point" that existed somewhere beyond the edge of the picture. This point was ordinarily positioned at a distance equal to the distance between the picture and the eye of the observer. This was one of the key problems with Alberti's perspective, and a popular problem for mathematicians to attempt to rectify. Desargues's construction allowed this function to be carried out by a point that always lies within the picture, thereby eliminating distortion.

This attention to visual perception characterizes projective geometry. It seeks to understand the extent to which different shapes, or appearances, share the same origin. For example, when a circle is viewed obliquely, it resembles an ellipse. Likewise, the outline of the shadow of a lampshade will appear as either a circle or a hyperbola, depending on whether it is projected on the ceiling or on the wall. In other words, the shape of a figure may change but, despite these changes, the figure maintains many of the same properties. The circle and hyperbola emitted by the lamp appear quantitatively and qualitatively distinct. However, the method of analysis proffered by projective geometry allows one to measure their similarities.

Desargues's projective geometry had little effect on the seventeenth century. It lay dormant until the nineteenth century, when great advances in the subject were made by figures such as Chasles, Charles Dupin (1784-1873), and Victor Poncelet (1788-1867). While Desargues's contemporary, Blaise Pascal, expressed appreciation for the work, Descartes could not stifle his dismay when he heard of the *Brouillon project.* For Descartes, the notion of treating conic sections without the use of algebra seemed both impossible and implausible. Descartes believed that new achievements in geometry could only be obtained through the use of algebra. Descartes's

views echoed those of most of his contemporaries, and this perception dominated mathematics for quite some time.

Prior to the rediscovery of projective geometry, Desargues was known instead for a proposition that does not even appear in the *Brouillon project.* This theorem, applicable to either two or three dimensions, states:

> If two triangles are so situated that lines joining pairs of corresponding vertices are concurrent, then the points of intersection of pairs of corresponding sides are collinear, and conversely.

This theorem was first published in 1648 by Abraham Bosse (1602-1676), an engraver, in *La Perspective de M. Desargues* (Desargues's Perspective). It became one of the most important propositions of projective geometry.

While Desargues's work was overlooked in the seventeenth century, his involvement with his contemporaries, indicated through his publications, correspondences, and frequent public speeches, was of vital importance. The development of organizations such as the Académie des Sciences in France and the Royal Society in England testify to the growing power of scientists at this time. These groups, which used mathematics as a vehicle for practical applications and philosophical speculations, formed at a time when nation-states were also beginning to develop into their modern incarnations. Mathematic developments of the seventeenth century are emblematic of the transition from the medieval, which focused on exterior causes and the cosmological, to the Age of Reason, which was concerned with interior perception and the individual.

DEAN SWINFORD

Further Reading

Boyer, Carl B., and Uta C. Merzbach. *A History of Mathematics.* 2nd ed. New York: John Wiley & Sons, 1989.

Coxeter, H.S.M. *Projective Geometry.* 2nd ed. New York: Springer-Verlag, 1987.

Eves, Howard. *An Introduction to the History of Mathematics.* 4th ed. New York: Holt, 1976.

Field, J. V., and J. J. Gray. *The Geometrical Work of Girard Desargues.* New York: Springer-Verlag, 1987.

Rosenfeld, B. A. *A History of Non-Euclidean Geometry: Evolution of the Concept of Geometric Space.* New York: Springer-Verlag, 1988.

Young, Laurence. *Mathematicians and Their Times.* Amsterdam: North-Holland Publishing Company, 1981.

Mathematical Induction Provides a Tool for Proving Large Problems by Proceeding through the Solution of Smaller Increments

Overview

The development of mathematical induction was one of the great forward steps in mathematics. An elegant principle that played a large part in the continuing evolution of mathematical logic, and affected the development of other mathematical disciplines, including algebra and analytic geometry, mathematical induction is related to the nonmathematical process called inductive reasoning. As opposed to deductive reasoning, by which a large general truth is taken as the starting point from which smaller, more specific truths are derived, induction provided a tool for moving from the specific to the general, from small individual truths to larger overall ones. By the process of induction, the truth of an entire mathematical proposition is proved one step at a time, with each step being used as a building block toward proving the next step, with the ultimate goal of proving the entire proposition. The nature of mathematical induction is such that it offers effective proofs of certain propositions while at the same time eliminating the need to prove every example of a proposition. While many feel that induction was perceived as long ago as ancient Greece, the method was not clearly expressed until 1575, when Sicilian mathematician Francisco Maurolico (1494-1575) used the method to prove a theorem. Maurolico's approach did not have a name: that waited for English mathematician John Wallis (1616-1703), who described the method as induction in his 1655 book *Arithmetica Infinitorum* (Infinitesimal Arithmetic). Nearly a decade later, with the posthumous publication of Blaise Pascal's (1623-1662) *Traite du Triangle Arithmetique* (Treatise On the Arithmetic of Triangles), mathematical induction became widely known as an effective and indispensable mathematical tool.

Background

Deductive reasoning was one of the great advances in human knowledge. By the process of deduction a large truth serves as a starting point from which smaller and more specific truths are logically derived. But deductive reasoning was not applicable to all areas of knowledge. In mathematics, particularly, large truths—or proofs—were often elusive, approachable only fitfully, in increments.

But as mathematics became a more and more important and precise tool, much thought and investigation was applied to the task of proving large theorems. That thought, research, and experimentation led to the development of mathematical induction, also known as the induction principle.

Some scholars see evidence of mathematical induction in the works of Greek mathematicians including Pappas (c. 260-?), whose works collected most of the Greek mathematical work that has survived to the present. Because Pappas was primarily a collector of mathematical ideas, rather than an originator of them, it is likely that his work with induction was derived from earlier thinkers.

Other evidence of early inductive reasoning can be found in works of both Islamic and Talmudic scholars, particularly Levi ben Gerson (1288?-1344?). In referring to his approach to solving complex problems, Levi ben Gerson wrote that he pursued a mathematical process of "rising step-by-step without end." That step-by-step approach is the very essence of induction, although that essence would not be formalized for another 250 years.

Part of the problem faced by Levi ben Gerson was his reliance on words rather than symbols when presenting his insights into algebra. By the 1500s mathematics was in a state a rapid evolution, with new insights and methodologies being developed at a steady pace, and the science itself becoming a more purely symbolic activity.

Italian (Sicilian) Francisco Maurolico, a Benedictine monk and also the head of the Sicilian mint, applied himself through most of the 1500s to collecting and translating the world's mathematical knowledge. He also wrote original treatises on mathematics, including the *Arithmeticorum Libri Duo* (Two Mathematics Books), in which he used the inductive principle to prove a theorem.

Put simply, mathematical induction reduces a mathematical proposition or theorem to simple statements that can be proved, each statement serving as a step toward the solution of the larger proposition. When searching, for example, for properties of whole numbers, mathematical induction solves for the simplest example of a whole number's property, which is of course the number 1. The next step is to select a random whole number, represented as k. If you can prove that the statement that was true for 1 is also true for the number k, you have also proved that is true for the number $k+1$.

By proving those two statements, you have *induced* that the statement is true for all whole numbers, or n.

Maurolico's principle attracted some attention but remained nameless until the work of English mathematician John Wallis. Considered by many to be the most important English mathematician before Isaac Newton (1642-1727), Wallis made many large contributions to both mathematics and science, including helping to found England's Royal Society (the most prestigious of all scientific bodies) and formulating for the first time the law of conservation of momentum. He also engaged in many bitter scientific and mathematical quarrels.

It is perhaps as a writer on mathematics that Wallis made the largest of his contributions. He was obsessed with the history of mathematics and devoted to preserving that history for the modern world. Among his many books was *Arithmetica Infinitorum*, published in 1656. In that book, among many examples of mathematical properties, including infinite series and anticipations of integral calculus, Wallis recapitulated mathematical induction, referring to it as *per modum inductonis* (by the method of induction) and gave the procedure the name by which it is still known.

For mathematical induction to become *well* known, however, it would require almost another decade and the 1665 publication another book, which, ironically had been written *before* Wallis's volume.

This was Blaise Pascal's *Traite du Triangle Arithmetique* (Treatise on the Triangle), one of the key mathematical treatises of its time. Pascal was the son of a mathematician, although his father initially opposed the child's early interest in mathematics. Pascal's genius quickly became obvious and his father relented; the boy immersed himself in mathematics. Pascal's interest went beyond the theoretical: by the time was 19 he had invented an early version of the mechanical calculator. Only the excessive cost of manufacturing his calculator kept the machines from becoming successful.

While Pascal's brilliance led him into many fields of research and reflection, including religious philosophy, it was in pure mathematics that he made his largest contributions, laying the groundwork (along with Pierre de Fermat [1601-1675]) for the modern science of probability, as well as insights into calculus and geometry.

His "Treatise on the Triangle" focused on the properties of a triangle composed of numbers, but also on other mathematical ideas and concepts. Among them was his approach to proving propositions for which there are infinitely many cases. He informed his readers that faced with such a proposition they should initially prove that the proposition is true for the first case. Following that, they should prove the proposition for a given (or random) case. With those two proofs, the proposition is solved for the next case, and for an infinite number of other cases.

Using induction, Pascal provided a method for solving the binomial theorem, which was essential for solving problems in which two quantities vary independently. The success of Pascal's book, along with the name Wallis gave to the process, insured that Maurolico's induction principle became widely known, and continues to serve mathematics to this day.

Impact

Mathematical induction, the ability to prove for *all* cases of a numerical property, was a major step forward in the transition of mathematics from the purely practical—counting—to the more theoretical. Rather than being restricted to concrete items, mathematics was able to deal with variables, with unknowns, with relationships among variables and unknowns, with infinite series of properties. These abilities vastly extended the reach of mathematics, both in relation to calculations and equations related to the real world, and to more purely theoretical pursuits. The ability to solve for *all* cases of a proposition yielded a mathematical tool that in turn helped mathematicians determine many of the properties of numbers in sequence. The induction principle also offered an important, and still-used, tool for both the devising and also the solving of complex codes.

KEITH FERRELL

Further Reading

Dunham, William. *Journey through Genius: The Great Theorems of Mathematics*. New York: John Wiley & Sons, 1990.

Mathematics

1450-1699

Katz, Victor. *A History of Mathematics: An Introduction.* 2nd ed. New York: Addison-Wesley, 1998.

Swetz, Frank J., ed. *From Five Fingers to Infinity: A Journey through the History of Mathematics.* Chicago and Lasalle: Open Court, 1994.

The Emergence of the Calculus

Overview

The calculus describes a set of powerful analytical techniques, including differentiation and integration, that utilize the concept of a limit in the mathematical description of the properties of functions, especially curves. The formal development of the calculus in the later half of the seventeenth century, primarily through the independent work of English physicist and mathematician Sir Isaac Newton (1642-1727) and German mathematician Gottfried Wilhelm Leibniz (1646-1716), was the crowning mathematical achievement of the Scientific Revolution. The subsequent advancement of the calculus profoundly influenced the course and scope of mathematical and scientific inquiry.

Background

Important mathematical developments that laid the foundation for the calculus of Newton and Leibniz can be traced back to mathematical techniques first advanced in ancient Greece and Rome. In addition to existing methods to determine the tangent to a circle, the Greek mathematician and inventor Archimedes (c. 290-c. 211 B.C.) developed a technique to determine the tangent to a spiral, an important component of his water screw.

The majority of other ancient fundamental advances ultimately related to the calculus were concerned with techniques that allowed the determination of areas under curves (principally the area and volume of curved shapes). In addition to their mathematical utility, these advancements both reflected and challenged prevailing philosophical notions regarding the concept of infinitely divisible time and space. Two centuries before the work of Archimedes, Greek philosopher and mathematician Zeno of Elea (c. 495-c. 430 B.C.) constructed a set of paradoxes that were fundamentally important in the development of mathematics, logic, and scientific thought. Zeno's paradoxes reflected the idea that space and time could be infinitely subdivided into smaller and smaller portions, and these

paradoxes remained mathematically unsolvable in practical terms until the concepts of continuity and limits were introduced.

Archimedes also built upon the work of Greek astronomer, philosopher, and mathematician Eudoxus of Cnidus (c. 400-c. 350 B.C.). Eudoxus developed a method of exhaustion that could be used to calculate the area and volume under curves and of solids (e.g. the cone and pyramid) that relied on the concept that time, space, and matter could be divided into infinitesimally small portions. Moreover, the method of exhaustion pointed the way toward a primitive geometric form of what in calculus terms would become known as integration.

Although other advances by classical mathematicians also set the intellectual stage for the ultimate development of calculus during the Scientific Revolution, it is apparent that ancient Greek mathematicians failed to find a common link between problems dedicated to finding the area under curves and to the problems requiring the determination of a tangent. That these process are actually the inverse of each other became the fundamental theorem of the calculus eventually developed by Newton and Leibniz.

During the Medieval Age philosophers and mathematicians continued to ponder questions relating to the movement of objects. These inquiries led to early efforts to plot functions relating such variables as time and velocity. In particular, the work of French Roman Catholic bishop Nicholas Oresme (c. 1325-1382) proved an important milestone in the development of kinematics (the study of motion) and geometry, especially Oresme's proof of the Merton theorem, which allowed for the calculation of the distance traveled by an object when uniformly accelerated (e.g. by acceleration due to gravity). Oresme's proof established that the sum of the distance traversed (i.e. the area under the velocity curve) by a body with variable velocity was the same as that traversed by a body with a uniform velocity equal to the middle instant of whatever period was measured. In other words, the area under the curve was a sum of all distances covered by a series of instantaneous velocities. This work was to prove indispensable to the quantification of parabolic motion by Italian astronomer and physicist Galileo Galilei (1564-1642) and later influenced Newton's development of differentiation techniques.

During the Renaissance in Western Europe, a rediscovery of ancient Greek and Roman mathematics spurred the increased use of mathematical symbols, especially to denote algebraic processes. The rise in symbolism also allowed the development and increased application of the techniques of analytical geometry principally advanced by French philosopher and mathematician René Descartes (1596-1650) and French mathematician Pierre de Fermat (1601-1665). Beyond the practical utility of establishing that algebraic equations corresponded to curves, the work of Descartes and Fermat laid the geometrical basis for calculus. In fact, Fermat's methodologies included concepts related to, and to the determination of, minimums and maximums for functions that are mirrored in modern mathematical methodology (e.g. setting the derivative of a function to zero). Both Newton and Leibniz were to rely heavily on the use of Cartesian algebra in the development of their respective calculus techniques.

Although many of the fundamental elements for the calculus were in place, the recognition of the fundamental theorem relating differentiation and integration as inverse processes continued to elude mathematicians and scientists. Part of the difficulty related to a lingering philosophical resistance toward the philosophical ramifications of the limit and the infinitesimal.

Accordingly, in one sense the genius of Newton and Leibniz lay in their ability to put aside the philosophical and theological ramifications of the utilization of the infinitesimal to develop a very practical branch of mathematics. Neither Newton or Leibniz gave serious address to the deeper philosophical issues regarding limits and infinitesimals in their publications on technique. In this regard Newton and Leibniz worked in the spirit of empiricism that grew during throughout the Scientific Revolution.

Impact

Although largely carried over into the eighteenth century, and affecting more the elaboration of the calculus rather than the initial development of the techniques, the acrimonious controversy sur-

Gottfried Leibniz. *(Library of Congress. Reproduced with permission.)*

rounding whether Newton or Leibniz deserved credit for the development of the calculus was grounded in the actions of both men during the late seventeenth century. There is clear and unambiguous historical documentation that establishes that Newton's unpublished formulations of the techniques of calculus came two decades prior to Leibniz's preemptory publications in 1684 and 1686. Although their correspondence (mostly through a third party) makes Leibniz's path to calculus less clear, scholars generally conclude that Leibniz independently developed his own set of the techniques. Although the mathematical outcomes were identical, the differences in symbolism and nomenclature used by Newton and Leibniz evidence independent development. The dispute regarding credit for the calculus quickly evolved into a feud that drew in supporters along blatantly nationalistic lines that subsequently divided English mathematicians who relied on Newton's "fluxions" from mathematicians in Europe who followed the conventions established by Leibniz. In particular, the publications and symbolism of Leibniz greatly influenced the mathematical work of two brothers, Swiss mathematicians Jakob Bernoulli (1654-1750) and Johann Bernoulli (1667-1748).

Working separately, the Bernoulli brothers widely applied the calculus. Johann Bernoulli was the first to apply the term *integral*, and he spread Leibniz-based methodologies and nomenclature among influential French mathematicians. Jakob Bernoulli incorporated the calculus into his work regarding probability and statistics. In addition to the greater utility and translatability of Leibniz's notational systems, Bernoulli's spread of Leibniz's notation is one of the major reasons modern calculus much more closely resembles the original notations set forth by Leibniz than those of Newton.

In modern mathematical texts, Newton is often cited as the inventor of the differential calculus, and Leibniz is given credit for the development of integration. Both men, however, developed techniques for differentiation and integration. Accordingly, any awarding of credit for the development of the respective techniques more properly recognizes the varying mathematical and philosophical emphasis exhibited by Newton and Leibniz.

Following Leibniz's publication, Newton published his own work in his *Philosophiae Naturalis Principia Mathematica* (Mathematical Principles of Natural Philosophy), *Opticks*, and in John Wallis's works. Newton's writing carefully reflected ancient Greek philosophical ideas. In fact, the concept of the limit in *Principia* is defined as a "ratio of evanescent quantities."

The publications of Newton and Leibniz emphasized the utilitarian aspects of calculus. Nevertheless, the respective development of nomenclature and techniques by Newton and Leibniz also mirrored their own philosophical leanings. Newton developed the calculus as a practical tool by which to attack problems regarding the effects of gravity and as an accurate calculator of planetary motion. Accordingly, Newton emphasized analysis, and his mathematical methods attempted to describe the effects of forces on motion in terms of infinitesimal changes with respect to time. Leibniz's calculus sought to derive integral methods by which discrete infinitesimal units could be summed to yield the area of a larger shape. Thus, he derived inspiration from the idea that incorporeal entities were the driving basis of existence and change in the larger world experienced by mankind.

Although philosophical debates regarding the underpinning of the calculus simmered, the first texts in calculus were able to appear before the end of the seventeenth century. Despite the fact that modern scholars now credit much of the content of the text to Johann Bernoulli, the first textbook in calculus was published by French mathematician Guillaume François Antoine l'Hospital (1661-1704). L'Hospital's *Analyse des Infiniment Petits four l'intelligence des lignes courbes* first appeared in 1696 and helped bring the calculus into wider use throughout continental Europe.

Although the philosophical debate on the logical consistency of the calculus would gain importance in the eighteenth century, it is a telling note of the intellectual climate of the time that within a few decades the calculus was quickly embraced and applied to a wide range of practical problems in physics, astronomy, and mathematics. Why the calculus worked, however, remained a vexing question that would eventually open calculus to attack on philosophical and theological grounds. This school of critics—eventually to be led in eighteenth-century England by the Anglican Bishop George Berkeley (1685-1753)—argued that the fundamental theorems of calculus derived from logical fallacies and that the great accuracy of calculus actually resulted from the mutual cancellation of fundamental reasoning errors.

Within a century the attacks upon calculus, because they resulted in an increased rigor in mathematical analysis, ultimately proved beneficial to the development of modern mathematics. The practical genius of Newton and Leibniz, grounded in their respective recognition of the fundamental theorem of calculus (that differentiation and integration are inverse processes), endured to provide a powerful analytical tool that fueled Enlightenment Age inquiries into the natural world and offered the mathematical basis for the development of modern science.

K. LEE LERNER

Further Reading

Boyer, Carl. *The History of the Calculus and Its Conceptual Development.* New York: Dover, 1959.

Boyer, Carl. *A History of Mathematics.* 2nd ed. New York: John Wiley and Sons, 1991.

Edwards, C. H. *The Historical Development of the Calculus.* New York: Springer Press, 1979.

Hall, Rupert. *Philosophers at War: The Quarrel Between Newton and Leibniz.* New York: Cambridge University Press, 1980.

Kline, M. *Mathematical Thought from Ancient to Modern Times.* New York: Oxford University Press, 1972.

Kuhn, Thomas S. *The Structure of Scientific Revolutions.* Chicago: University of Chicago Press, 1970.

The Enduring and Revolutionary Impact of Pierre de Fermat's Last Theorem

Overview

Pierre de Fermat (1601-1665) was a contemporary of the renowned philosopher and mathematician René Descartes (1596-1650). Fermat, like Descartes, was fascinated with numbers and their properties and relationships, and indeed corresponded with Descartes about his insights and conjectures. Unlike Descartes, however, Fermat was neither a professional mathematician nor a professional philosopher. Nevertheless, though he was considered an amateur mathematician, Fermat now is known as the "Prince of Amateurs."

Despite his amateur status, Fermat contributed much to mathematics, including providing the necessary groundwork for the fields of analytic geometry and infinitesimal analysis. What is more, Fermat is credited with founding number theory, the calculus, and, along with Blaise Pascal (1623-1662), inventing probability theory.

Notwithstanding these incredible accomplishments, Fermat perhaps is most famous for his Last Theorem, a theorem whose solution evaded the brightest minds of mathematics for over 350 years, but whose solution—and quest for the same—revolutionized number theory. According to one contemporary mathematician, the proof of Fermat's Last Theorem, which was finally completed in the fall of 1994, is the historical and intellectual equivalent of "splitting the atom or finding the structure of DNA."

Background

Fermat was born into a wealthy family in the town of Beaumont-de-Lomagne in southwest France. He was well educated both at a Franciscan monastery and the University of Toulouse, although there is no indication that he was particularly attracted to mathematics during this period.

Due in large part to familial influence, Fermat eventually became a lawyer and later entered the civil service. Upon entering the civil service, he served the local parliament as a lawyer, then as a councillor, acting mainly as a liaison between the local municipality and the king. Thus, it appears that politics was his profession. Although Fermat succeeded well in his professional life, his true love and devotion increasingly turned toward the study of numbers.

Fermat's fascination with numbers was rooted in the writings of the ancient Greek mathematician and philosopher Diophantus (fl. A.D. 250), author of the *Arithmetica*. The *Arithmetica* originally was comprised of 13 books, though only six are known to have survived the Dark Ages. These books comprised a collection of mathematical problems for which only whole number solutions are possible. As Diophantus was the author of these mathematical problems, they now are known, not surprisingly, as Diophantine problems. Pythagoras's famous theorem is an example.

Interestingly, in ancient Greece, *arithmetica*, meaning "arithmetic," did not mean mere computation, as it presently does. Rather, arithmetic meant "theory of numbers," and as such was more philosophical in its approach toward understanding numbers and their properties than straightforward mathematical theory.

So it was through Fermat's study of Diophantus's *Arithmetica* that he became interested in exploring further the mysteries of numbers and their properties, especially with regard to Diophantine problems. Indeed, one of Fermat's favorite pastimes was to write to other mathematicians, asking them whether they could prove a particular theorem, some of which he found in other texts, and some of which he created on his own. Fermat then would taunt these mathematicians, who struggled to provide proofs to Fermat's challenging theorems, by stating that he had the proofs although he would refuse to reveal them.

Needless to say, the mathematicians with whom Fermat corresponded grew increasingly frustrated. Indeed, such taunting moved Descartes to label Fermat a braggart, and the English mathematician John Wallis (1616-1703) to refer to Fermat as "that damned Frenchman." Such responses proved only to motivate Fermat even more as his challenges became ever more complicated and frustrating. These challenges, however, paled in comparison to his ultimate taunt, known as Fermat's Last Theorem, referred to as such because it was the last of Fermat's theorems to be proved.

Impact

Pythagoras's theorem states that the square of the hypotenuse of a right triangle is equal to the

Pierre de Fermat. *(Corbis-Bettman. Reproduced with permission.)*

sum of the square of its sides. Symbolically, the theorem may be expressed as $x^2 + y^2 = z^2$. If x is assigned the value 3; y, the value 4; and z, the value 5, the equation may be solved: $9 + 16 = 25$. Of course, the combination of 3, 4, and 5—known as a Pythagorean triple—is not the only solution to this theorem. Indeed, mathematicians have proved that there exists an infinity of Pythagorean triples that provide solutions to the equation.

Around 1637, when he was 36 years old, Fermat wondered whether there might also be an infinity of Pythagorean Triples to a slight variation of Pythagoras's theorem. Specifically, Fermat wondered whether a cubed version of Pythagoras's theorem—$x^3 + y^3 = z^3$—had any solutions. Surprisingly, Fermat could not find any Fermatian triples, that is, Fermat was unable to find any solutions to his cubed variation of Pythagoras's theorem. Indeed, Fermat could not find any triples for any Pythagorean variation where the exponents were integers greater than two.

As a result, Fermat noted in the margin of his prized copy of *Arithmetica* that "It is impossible for a cube to be written as a sum of two cubes or a fourth power to be written as the sum of two fourth powers or, in general, for any number which is a power greater than the second to be written as a sum of two like powers." This statement became known as Fermat's Last

Theorem. Another way of stating the theorem is that there are no integer solutions to the equation $x^n + y^n = z^n$, where n is an integer greater than two. (The equation contained within the Last Theorem was henceforth referred to as Fermat's Equation.)

However, a significant problem remained: Could Fermat actually prove this theorem? In a cryptic notation left in his copy of *Arithmetica* Fermat claimed to have found a proof for the theorem: "I have discovered a truly remarkable proof which this margin is too small to contain." Unfortunately, Fermat died on January 9, 1665, without ever revealing the proof of his Last Theorem, thereby leaving both an enduring mystery and an incredible taunt for the world to ponder.

For years afterward mathematicians tried in vain to prove the theorem (which more appropriately should have been labeled a conjecture, inasmuch as no proof existed). Partial proofs up to $n=7$ were achieved, but as there are an infinity of numbers, such proofs amounted to very little toward proving the theorem completely. Some were so intrigued by the possibility of a proof that they offered monetary rewards to those who discovered a proof for all n greater than two. In the early 1900s, for example, Dr. Paul Wolfskehl, a German industrialist and amateur mathematician, offered a prize of 100,000 marks (approximately equivalent to $1 million today) to the first person to solve Fermat's Last Theorem. Needless to say, no serious contenders for the prize money came forward. Indeed, many mathematicians grew skeptical of whether a proof even was possible.

Despite overwhelming skepticism and seemingly impossible odds, on June 27, 1997, approximately 360 years after Fermat conjured up his infamous theorem, and nearly 100 years after the establishment of the Wolfskehl Prize, the prize money was finally bestowed on an unassuming, yet profoundly brilliant mathematics professor from Princeton University. After struggling in secrecy and the isolation of his attic for eight years, Andrew Wiles (1953-) published a 100-page proof of Fermat's Last Theorem in the *Annals of Mathematics* (May 1995).

What was just as amazing as the fact that Fermat's Last Theorem was finally proved, was the process by which Wiles achieved the proof. To prove Fermat's Last Theorem, he had to unite two completely disparate branches of mathematics, thereby developing a completely novel approach to number theory in order to achieve the proof.

The two disparate areas of mathematics Wiles united concerned elliptic curves and modular forms. Elliptic curves are equations used to measure the perimeter of ellipses and, as a result, have been famously used to compute the elliptical trajectory of planetary orbits. Elliptic equations take the form of $y^2 = x^3 + ax^2 + bx + c$, where $a, b,$ and c are any whole numbers. Hence, elliptic equations, like the Fermat's Equation, are also Diophantine equations; indeed, Diophantus devoted a large portion of the *Arithmetica* to the study of elliptic equations.

In contrast, modular forms are highly abstract, complex mathematical objects that display an unusual amount of symmetry. Modular forms are symmetric in the sense that they can be moved about in mathematical space in any conceivable way, but remain unchanged. To take a highly simplified example, imagine a perfect square drawn on a sheet of paper. If one were to rotate the sheet of paper exactly one-quarter turn, the square would appear to remain completely unchanged. Indeed, the square would appear to remain unchanged if one were to turn the paper one-half turn, three-quarters, or even, of course, one full turn. This unchanging aspect of the square as it is moved about on the two-dimensional surface of the sheet of paper demonstrates the square's symmetry.

The study of elliptic equations and of modular forms have historically been completely unrelated endeavors: elliptic equations were used to study real world phenomena, whereas modular forms were studied for their interesting properties in imaginary, mathematical space. However, in September 1955 two young mathematicians at the University of Tokyo—Goro Shimura and Yutaka Taniyama—posed the following conjecture: for every elliptic equation there is an equivalent modular form. This conjecture, generally known as the Taniyama-Shimura conjecture, was so shocking that, initially, few paid much attention to it. Over time, however, the mathematical community discovered that if the Taniyama-Shimura conjecture was correct, many useful applications could be developed, including solving mathematical problems that remained resistant to resolution.

The problem with the Taniyama-Shimura conjecture, like the problem with Fermat's Last Theorem, was that nobody knew how to prove the conjecture. Twenty-nine years later, in the fall of 1984, and half a world away, a small group of number theorists met in Oberwolfach, Germany, to discuss current research in elliptic equations. Work was still being done toward proving the Taniyama-Shimura conjecture, but little progress had been made. Gerhard Frey, a mathematician from Saarbrücken, had discovered something interesting. Frey discovered that Fermat's Equation could be translated into the form of an elliptic equation, but this elliptic equation was quite odd in that it suggested a possible solution to Fermat's Equation, and furthermore, the elliptic equation had no modular form equivalent. Consequently, if Frey's elliptic version of Fermat's Equation was valid, then two conclusions followed: (1) it was possible for there to be a solution to Fermat's Equation, which would prove that the Last Theorem was false; and (2), the Taniyama-Shimura conjecture was false because there existed an elliptic equation without a modular form equivalent.

Stated conversely, what Frey essentially had discovered was that if the Taniyama-Shimura conjecture was correct, then the elliptic version of Fermat's Equation was invalid, which meant that, indeed, there were no solutions to Fermat's Last Theorem. So, in a nutshell, Frey had proved that if the Taniyama-Shimura conjecture was correct, then so was Fermat's Last Theorem.

Ultimately, proving the Taniyama-Shimura conjecture was precisely what Andrew Wiles did in order to finally prove Fermat's Last Theorem. As a result, Wiles opened up new avenues of research previously unavailable, indeed, unknown, to mathematicians. Thereafter problems in elliptic equations could be solved in the modular forms world, and vice versa. Furthermore, the impact of recognizing the underlying unity between these two branches of mathematics would allow mathematicians to better understand each branch.

In the history of knowledge, such unification of disparate branches of thought is not unknown. For example, prior to the nineteenth century physicists studied magnetism and electricity completely separately. But then it was discovered that these two areas of physics were inextricably linked, thereby creating the study of electromagnetism. From the study of electromagnetism came the further discovery that light was nothing more than electromagnetic radiation, which in turn allowed physicists to better understand the nature of the world. Likewise, the quest for the proof of Fermat's Last Theorem has greatly impacted the future of mathematics, and our knowledge of the world generally.

Despite the fact that the mystery of Fermat's Last Theorem has finally been solved, the mystery of whether Fermat actually solved it remains. Although it is highly improbable that Fermat did in fact have a proof, there nevertheless remains the possibility. Consequently, the mystery of whether there exists Fermat's proof to his Last Theorem persists. If in fact Fermat did prove his Last Theorem, the discovery of just what that proof is may further illuminate our understanding of mathematics.

In any event, there are many more mathematical mysteries challenging—indeed, taunting—the world's best mathematicians to this day. On May 24, 2000, the Clay Mathematics Institute published the Millenium Prize Problems, offering a $1 million prize for the solution to each of seven unsolved mathematical problems. As some of these problems are over 100 years old, the Institute hopes that the prize money, like that of the Wolfskehl Prize, will motivate and inspire current and future mathematicians to work on these seemingly intractable mysteries. Importantly, like the revolution sparked by the quest for the proof of Fermat's Last Theorem, solutions to these problems, should they be found, likely will provide us with a better understanding of the nature of numbers. As Pythagoras (580?-500? B.C.) understood so long ago, the nature of numbers is the language of nature.

MARK H. ALLENBAUGH

Further Reading

Bell, Eric T. *The Last Problem*. Washington, DC: Mathematical Association of America, 1990.

Boyer, Carl B. *A History of Mathematics*. 2nd ed. New York: John Wiley & Sons, 1991.

Mahoney, Michael. *The Mathematical Career of Pierre de Fermat*. Princeton, NJ: Princeton University Press, 1994.

Singh, Simon. *Fermat's Enigma*. New York: Anchor Books, 1997.

Mathematics, Communication, and Community

Overview

Perhaps more than any other scientific subject, mathematics seems to depend upon individual genius and moments of inspiration. Mathematical theorems and concepts are even named after their discoverers. But in fact, mathematical research did not begin to flourish until the sixteenth century, when reliable communication networks helped support an international community of like-minded scholars who stimulated each other's work through the exchange of ideas and the spirit of competition. A dynamic balance between individual discovery and communal validation characterizes mathematics to this day.

Background

The invention of the printing press in the fifteenth century gave an enormous boost to the field of mathematics. While the great mathematical works of antiquity had been preserved in Europe, they were often studied through incomplete manuscripts, and advances to algebra, trigonometry, and geometry made in Islamic countries during the Middle Ages were little known. There was little enthusiasm for innovative ideas, and few opportunities for collaboration or exchange. But with the advent of printing, mathematical communication became newly convenient. Regiomontanus (1436-1476), one of the most important early printers, was also probably the most important mathematician in Europe during the fifteenth century. Although he died prematurely, Regiomontanus did begin an important era in the publication of the major texts of classical science and mathematics, and of textbooks based upon them. The first printed edition of Euclid's (330?-260? B.C.) famous book on geometry appeared in 1482. During the next century, more than one hundred different editions of Euclid alone were published, along with scores of other mathematical treatises.

The proliferation of mathematical books helped to stimulate mathematical education, not only in universities but also in schools that helped to prepare men to work in areas such as surveying, which required the use of arithmetic and geometry. Mathematics began to be applied to new fields and to older subjects in new ways. Bookkeeping, mechanics, surveying, art, architecture, cartography, optics, and music were all transformed by the application of mathematical techniques, and mathematics was in turn influenced by the demands of these real-world activities.

The availability of somewhat standardized printed versions of texts by Euclid and Archimedes (287?-212? B.C.), the increase in interest in mathematical education, and the expansion of the application of mathematics to important new areas established promising conditions for mathematical activity. The printed texts provided a kind of uniform starting point for those interested in addressing problems posed by the classical authors, and the use of mathematics to solve practical problems provided another outlet for mathematical novelty. The desire to exchange mathematical ideas with others interested in the still-esoteric subject drove far-flung investigators to begin correspondence with each other.

Communication among scholars living in different countries throughout Europe was greatly facilitated by the widespread use of Latin in scholarly work. Well into the seventeenth century, most scientific and mathematical books were published in Latin, and virtually every educated person would read and write Latin fluently. While vernacular languages were used to write and publish some elementary texts, Galileo (1564-1642) was in the 1630s the first prominent scientist to publish important scientific texts in his native language. But throughout the sixteenth and seventeenth centuries, the use of Latin as a universal language for European scholars made correspondence between mathematicians much simpler.

But how did these scholars find one another? Prior to the existence of scientific societies or scholarly journals, identification of others who shared common research interests or mathematical talent depended upon individual encounters and connections. Some mathematicians were affiliated with universities or monasteries and so had institutional relationships that put them in touch with others at similar posts, but many active mathematicians in this era were more or less independent. It fell to enthusiastic individuals to bring other mathematicians into contact. One of the most important of the mathematical impresarios was a Jesuit priest by the name of Marin Mersenne (1588-1648).

Mersenne's duties to his Jesuit order were for the most part strictly intellectual. He published several books of his own, beginning in 1623, on a variety of topics including theology, ancient and modern science, and mathematics. These publications often included the first published accounts of the work of other mathematicians with whom Mersenne corresponded. This was one of many ways that Mersenne encouraged the exchange of mathematical ideas.

Mersenne became the coordinator of mathematical activity for scholars based in Paris, as well as for interested foreigners who visited the French capital. He held frequent *salons* at his convent that brought together leading thinkers in a number of other scholarly fields as well as mathematics. These evolved into something resembling formal scientific conferences by the mid-1630s. Another significant encounter that Mersenne brought about was the first meeting of the great French mathematicians Blaise Pascal (1623-1662) and René Descartes (1596-1650). In addition to these important personal meetings and exchanges, Mersenne built up a correspondence network that came to include most of the important mathematicians in Europe. It was through Mersenne that Galileo and his followers maintained contact with scientists elsewhere in Europe. Mersenne relayed correspondence between mathematicians interested in common problems, and he often smoothed relationships among various personalities, such as those between the notoriously difficult Descartes and France's other leading mathematicians Pierre de Fermat (1601-1665), Gilles Personne de Roberval (1602-1675), and Pascal.

But Mersenne was something more than a secretary. In addition to establishing and maintaining contact among Europe's mathematicians, Mersenne also directly influenced mathematical work. He often criticized or challenged the mathematical work of others, setting one mathematician against another to advance work in an area where problems seemed to fester. He also set out mathematical problems for others to solve, thus contributing to the era's spirit of competition among mathematicians eager to claim new solutions and techniques for themselves. Mersenne's extensive correspondence, covering more than 20 years, from the early 1620s up to his death in 1648, was finally itself published in the twentieth century and stands as a tangible monument to his great influence on the development of mathematics and the international mathematics community. While Mersenne was undoubtedly the most significant mathematical correspondent of his century, others such as John Collins of London performed a similar role in their own countries.

Impact

Many conditions came together to bring about the flowering of mathematics in the sixteenth century. The dissemination of published mathematical books, the expansion of education in mathematics, new areas of application, and the establishment of an international community of scholars connected via correspondence, transformed mathematics from a limited academic subject to an actively evolving scientific endeavor. Historians agree that by the seventeenth century, the practice of mathematics had come to fully resemble the modern mathematics of later centuries.

From the start they were given by Mersenne, European mathematicians established ever-closer ties among one another. Detailed exchanges of problems and proofs, counterexamples, and challenges took place across national boundaries and between (mostly) men of very different social and professional standing. New ideas, techniques, and applications began to develop quickly as enthusiasts stirred each other to work harder and more rigorously on problem after problem. While this often had the positive effect of stimulating fresh discoveries, it also set the stage for a number of bitter disputes over who first came upon an important idea. Most notorious among these "priority disputes" was the bitter battle waged between Isaac Newton (1642-1727) and Gottfried Leibniz (1646-1716) over who had discovered the calculus.

Communication among mathematicians grew faster and easier as technology improved. Letters moved more and more quickly thanks to new roads and later railroads and improved ships. The telephone, telegraph, and eventually computers sped communication further. At the same time the mathematical community fostered institutions to aid communication among researchers. Mathematicians joined the earliest scientific societies as they formed in the late seventeenth century, and starting in the nineteenth century began to form societies and publications that were specialized to their interests alone. As

mathematics became more and more specialized, so did its institutions. By the end of the twentieth century, hundreds of mathematical journals and dozens of mathematical societies reflected the diversity of the field and its profound degree of international cooperation. English is treated as a common, although not universal, language of exchange, and the highly symbolic nature of mathematics itself helps scholars from different nations to easily understand one another's work.

Mathematics developed a public face in the sixteenth century that has lasted to the present: a new mathematical concept can be accepted only after the mathematical community has been persuaded of its truth. A mathematician with a new discovery or proof must convince his peers of his accomplishment, otherwise it is no accomplishment at all. The importance of a community to modern mathematics, then, cannot be exaggerated—mathematics is at its very core a social activity, no matter how essential the contribution of individual work might be.

LOREN BUTLER FEFFER

Further Reading

Boyer, Carl. *A History of Mathematics.* Rev. by Uta Merzbach. New York: Wiley, 1989.

Cooke, Roger. *The History of Mathematics.* New York: Wiley, 1997.

Dear, Peter. *Discipline and Experience: The Mathematical Way in the Scientific Revolution.* Chicago: University of Chicago Press, 1995.

Hay, Cynthia, ed. *Mathematics from Manuscript to Print, 1300-1600.* Oxford: Clarendon Press, 1988.

Hollingdale, Stuart. *Makers of Mathematics.* London: Penguin, 1989.

Katz, Victor. *A History of Mathematics.* Reading, MA: Addison-Wesley, 1998.

Struik, Dirk, ed. *A Source Book in Mathematics, 1200-1800.* Princeton: Princeton University Press, 1986.

Mathematicians Develop New Ways to Calculate π

Overview

During the Renaissance, mathematicians continued their centuries-old fascination with π. As part of this, they calculated π to ever-greater precision

while developing new formulae to add digits more quickly. Part of this fascination with π was the equally old quest to "square the circle," and part was simply human curiosity. In spite of their lack of success in quadrature (the term for squaring the circle), these efforts did yield greater insights into the nature of π, some aspects of geometry, and other areas of mathematics.

Background

The ancient Egyptians and Babylonians may have been the first to notice that the ratio of a circle's circumference to its diameter was not an even number. By about 2000 B.C., in fact, the Egyptians had determined this ratio was about 31/8, while the Babylonians put it at about 4 x $(8/9)^2$. These come out to be equal to 3.125 and about 3.161, respectively; not far from the modern value of 3.1416.... These relationships remained largely unchanged until the time of the Greeks, nearly 1,600 years later.

Among the ancient Greeks, Archimedes stands out as a mathematician who made particularly important advances in determining the value of pi, among his many other accomplishments. Using a method that remained largely unchanged for nearly two millennia, Archimedes determined that π had a value of between 3.14084 and 3.14258; remarkably close to the actual value. In reaching this value, he also developed mathematical techniques that seemed to anticipate the development of some parts of differential calculus.

One of the driving forces behind calculating π to greater and greater numbers of digits was a fascination, in some cases an obsession, with the quadrature problem. This problem simply asked a person, using only an unmarked straight edge and a compass, to create a square with exactly the same area as a circle of a given diameter. The quadrature problem seems to have originated with the ancient Greeks, and occupied an inordinate amount of attention for nearly 2,000 years. Part of the reason for the attention is that success seemed, in many cases, tantalizingly close, while remaining just out of grasp. Somewhere along the way, mathematicians and other "circle-squarers" realized that, if only they could determine the exact value of π, they would have a much better chance of developing an algorithm that would let them achieve their goal. So work continued.

Archimedes' method for calculating π relied on the fact that a polygon drawn on the outside of a circle would always have a greater circumference than the circle itself, while a polygon drawn on the inside of a circle would always have a smaller circumference. By adding more sides to each of these polygons, they would gradually approach the same circumference, and if they had infinitely many sides, they would have exactly the same circumference as the circle itself. In effect, the circumferences of the two polygons would gradually approach that of the circle that was being "squeezed" between the interior and exterior shapes.

The next advance in calculating the value of π came in the 1500s when François Viète (1540-1603) developed an elaboration on Archimedes' method. While his basic method would have been both understood and appreciated by the Greek, Viète became the first to use an analytically derived infinite series to aid in the calculations. This, in and of itself, was a major accomplishment for not only π-calculators, but for all of mathematics.

The first attacks on Archimedes' method itself were not launched until 1621, when Dutch physicist and mathematician, Willebrord Snell (1580-1626) developed a superior method of calculation. This was verified by another Dutchman, Christiaan Huygens (1629-1695). His method would yield as many accurate digits for π in just a few iterations as could be garnered by Archimedes' method using a polygon of over 400,000 sides.

The next significant advance took only another 44 years, when English mathematician John Wallis (1616-1703), like Viète, used an infinite series to calculate the value of π. Unlike Viète, Wallis's formula did not rely on a series of square roots, which are always difficult to calculate by hand. Instead, Wallis developed a method that used ratios of whole numbers and trigonometric operations, giving a much faster and more accurate way to calculate π, and allowing many more decimal places to be added to its known value. Wallis's contribution is significant from another standpoint; it was the last important advance to be made that did not use calculus and the methods of analytical geometry, recently invented by Isaac Newton (1642-1727) and Gottfried Leibniz (1646-1716).

Impact

The impacts of these advances range from significant to almost trivial. On the trivial side, it became ever easier to calculate π to an ever-longer string of digits. More important than simply knowing a string of digits for π, however, was the mathematical understanding that came from this process. Finally, the analytical process that began with Viète led to a much deeper understanding of the nature of π, eventually proving that the exact value of π could never be known because it is a transcendental number; a number that neither repeats itself nor terminates. This discovery, in turn, was to forever quash attempts to square the circle, something that had remained in the public imagination for many centuries.

From one standpoint, adding calculated decimal places to the known value of π was akin to mounting the heads of dead animals on the wall; it accomplished little except to give some fleeting fame to a mathematician. However, from another standpoint, even so apparently mundane a task could still carry some degree of significance. For example, Viète was able to show his method generated digits much more quickly than did Archimedes'. By reproducing the digits of those who came before, each method was validated, as was the math that went into developing that method. And, by generating an increasingly long string of digits for π, mathematicians began to suspect that it was not so simple a number as it at first appeared. These suspicions gained ground, even as circle-squarers appeared, each convinced he or she had accomplished a millennia-old task. Finally, a very small subdiscipline began to take shape that looked for patterns in the digits of π. It was thought that, by studying the digits closely enough, one might be able to find a method to predict the next digit without having to go through the laborious calculations already developed. Unfortunately, such efforts were in vain, and many properties of π remain inscrutable to this day. However, it is also of interest to note that, even today, stories appear in the media on a regular basis noting that someone (or, increasingly, a new computer) has broken the record for calculating π by generating another few million digits. So, even today, the digit hunters continue to be noticed by the general population.

There was, however, a great deal of mathematical understanding that came from these attempts to determine the value of π once and for all. For example, in addition to being the best way to calculate π for nearly 2,000 years, Archimedes' method for calculating π was one stepping stone on the way to developing differential calculus. He, and many centuries of mathematicians who followed him, was limited by the lack of decimal notation (i.e. writing 3.25 instead of $3\frac{1}{4}$), but he was squarely on the track of methods that would later be used to develop the concepts of calculus. Further along in history, Viète's use of infinite series to help calculate π

was revolutionary, and was, in fact, the first known use of analytically derived infinite series in any part of mathematics. The improvements made by Snell and Wallis helped to greatly extend the utility of infinite series, which are today used in a very large number of scientific, mathematical, and engineering applications. These mathematical tools would likely have been developed regardless of their use in attacking this one problem, but it is very likely that continued frustration in attempts to solve the quadrature problem served to accelerate progress in these, and other related areas.

Finally, over time the quadrature problem had attained a notoriety approaching that of Fermat's Last Theorem, finally solved in 1993. Although the quadrature problem was not resolved during this period (indeed, it was later found to be impossible to do), it did capture the public attention. Or, more appropriately, it captured the attention of those who were literate, educated, and had some amount of free time in which to pursue such idle interests; a very small fraction of the total population. However, this relatively small number of people was very important in the society of the times because they were the ones upon whom much of society rested. The educated elite helped set public policy, wrote the works by which their age is known, advised the government, ran businesses, taught, and did the myriad other things that helped establish, preserve, and spread their culture and government throughout the world. Those nations with a relatively large middle and upper class, such as the Dutch and English, dominated world affairs more than their relatively small size would suggest; while larger nations, such as Russia, remained relative backwaters both culturally and politically for many centuries.

The nations that became dominant—especially France, Holland, and England—were the same nations in which the quadrature problem seemed to gain the most notoriety, because it was these nations that had relatively large groups of intellectuals who could appreciate and try to solve it. Their interest and near misses, in turn, encouraged their contemporaries to try their hands at the problem, or to at least encourage further attempts. This culminated in a decree by the French Academy that they would no longer accept for publication any papers claiming to have solved this problem. The final proof, in the nineteenth century, of π's status as a transcendental number, drove the last nail into this particular coffin, forever dashing mathematical hopes of a solution. However, it should be noted that no small number of mathematically uninformed continue to try to repeal rigorous mathematical proof, including the government of one state in the United States. This fact alone demonstrates the continuing hold of π on the imagination of the general public, and the continuing impact of this number on the general public, as well as on mathematicians and scientists.

P. ANDREW KARAM

Further Reading

Backmann, Petr. *A History of Pi*. New York: St. Martin's Press, 1971.

Boyer, Carl and Uta Merzbach. *A History of Mathematics.*New York: John Wiley and Sons, 1991.

Dunham, William. *Journey through Genius: The Great Theorems of Mathematics.*New York: Penguin Books, 1990.

Mastering the Seas: Advances in Trigonometry and Their Impact upon Astronomy, Cartography, and Maritime Navigation

Overview

Until the advent of modern navigational tools in the sixteenth century, mariners had since ancient times used similar methods of navigating, largely by instinct. Even as late as the earliest voyages to the New World by Spanish and Portuguese explorers, mariners who embarked on voyages across open waters, out of the sight of land, could primarily only navigate by keeping a daily record of the general distances and directions they traveled, surrounding currents, wind patterns, hazards, and sightings of land. These journals, or ship logs, were used to notice "landmarks" at sea and retrace one's path back to their port of origin. Though pin-point navigation from these journals was difficult, the body of information collected over numerous voyages was immensely useful and often later incorporated into more mathematically accurate charts.

Practical inventions of the sixteenth through the nineteenth centuries, and innovations of existing instruments, were largely responsible for the modernization of navigation on the high seas. With increased ability to accurately plan voyages, trade boomed, transforming forever the shape of Europe and the Americas. Though early European exploration sparked interest in the lands across the Atlantic, advancements in navigation made colonial settlement and international trade a perceptible reality. These marvels, however, would have been impossible without the significant developments in the field of mathematics—most especially in trigonometry—that underscored scientific and engineering advancements of the seventeenth and eighteenth centuries.

Background

Trigonometry, the subdivision of mathematics concerned with the unique functions of angles and their applications in geometry calculations, was primarily developed and advanced from a need to compute measurements in various scientific fields such as map making, navigation, astronomy, and surveying. Incorporating elements of geometry, algebra, and simple arithmetic, plane trigonometry covers problems involving angles and distances on one plane. Angle and distance problems in three-dimensional space, which occupy more than one plane, are the subject of spherical trigonometry. The latter branch of trigonometry was almost always applied to questions in astronomy and navigation and thus was mastered and developed, most especially by the Arab and Chinese civilizations, faster than plane trigonometry. In fact, trigonometry did not separate itself as a unique discipline from astronomy until the later thirteenth century.

The computation of trigonometric problems is reputed to have developed following the calculation of a table of chords by Hipparchus (fl. c. 100 B.C.) in the second-century B.C. However, to what extent he developed and applied the uses of the tables are unknown as the complete original work is lost. The oldest extant work on trigonometry, dating to the middle of the second-century, is contained in Ptolemy of Alexandria's (fl. A.D. 127-145) voluminous work on astronomy, *The Almagest*. The cornerstone equations of trigonometric problems, those involving right triangles (both plane and spherical), were most likely derived by Hindu astronomers and mathematicians, translated into Arabic around 750 A.D., and then slowly filtered into Western Europe through contact with Arab civilizations in the Middle East and in Spain. Moorish astronomers in Spain improved upon these basic formulas over the course of the next several hundred years, adding the laws of cosines, and the use of tangents and cotangents. In the mid-tenth century, Arab astronomer and mathematician Abu al-Wafa' (940-998) discovered a more accurate method for computing sines, one of the primary mathematical operations used in navigational calculations, and also introduced secant and cosecant functions (assigning them both the more familiar terms of *sine* and *cosine*.)

Western Europe, through repeated contact with the Arab world—mostly during the Crusades (1095-1228)—became more familiar with advanced mathematics, including trigonometry. In Europe, Prussian astronomer Johann Müller (1436-1476), known as Regiomontanus, was responsible for systematizing plane and spherical trigonometry and establishing it as a discipline separate of astronomy. Later developments in algebra allowed Regiomontanus's successors to unify some elements of plane and spherical trigonometry (substituting ratio for a trigonometric line, and the angle for the arc), thus simplifying the lengthy equations that the original author had previously scripted.

In the years following the initial explorations to the New World, the need for creating accurate maritime charts pushed for a deeper understanding of the relationships between spherical trigonometry (used in collecting data and navigating) and plane trigonometry (until the late seventeenth century, the predominate mathematical system used to create charts and maps). Sixteenth-century French mathematician François Viète worked to figure out properties of plane triangles similar to those that were known for arc-triangles, such as the cosine and tangent laws. Viète's success inspired others to find similar formulas and properties. Similar means of computing plane triangle properties, as well as the formula for half-angles, appeared in the work of Austrian mathematician Georg Rheticus (1514-1574) in 1568, and nearly 90 years later in a publication by English mathematician William Oughtred (1574-1660).

English mathematician John Napier (1550-1617) invented logarithms, which he called "analogies," in 1619. Logarithms, defined as the powers of a base such that a certain numerical result is obtained as a power of the base, challenged other mathematicians to develop formulas suitable to their use. The eventual outcome of this line of inquiry was the understanding of calculus, which in turn enabled English physicist Sir Isaac Newton (1642-1727) to set forth the first modern physical theories, transforming the predominant modes of scientific and theoretical inquiries in physics, mathematics, and astronomy.

Impact

Developments in trigonometry aided navigation not only through astronomy, but also in the development of more systematized and accurate methods of cartography, the science (and arguably, art) of graphically representing a certain geographic region on a map or chart. Many, but certainly not all, ancient and medieval maps in Europe represented more stylized depictions of land forms, rarely endeavored to identify distances, and often depicted mythological creatures or reflected prevailing religious thought. Contact with the New World provided a need for more scientifically sophisticated maps, which could represent landforms with a greater degree of accuracy and aid in the establishment of location at sea.

On longer journeys, as became standard during the Age of Exploration, the ancient method of dead reckoning did not work. Over great distances, the approximated rhumb lines of the Mediterranean chart could not be taken as straight, and the equations devised by astronomers and mathematicians failed to approximate location. In other words, longer voyages required some means of taking into account the curvature of Earth. To this end, the Mercator projection was developed in 1569 by Gerardus Mercator (1512-1594) to represent sections of the spherical Earth on flat charts. Instead of bearings and distances, location was defined by the larger and more defined latitude and longitude. Mercator's charts featured equally spaced representations of the lines of longitude, or meridians, and compensated for the distortions in distances that appear in attempting to illustrate a curved surface by representing the lines of latitude, or parallels, closer together at the Equator and further apart at the Poles. Mercator guarded the secret of calculating his maps, but mariners soon realized that east-west distances were slightly distorted at some latitudes. Not until 1599, when English mathematician Edward Wright provided an explanation of the trigonometry involved in Mercator's map projections, could the distorted distances be corrected.

Innovations in cartography continued to be made in the seventeenth and even into the eighteenth centuries, but practical applications of spherical trigonometry in astronomy and planar trigonometry in chart making did not yield the solution to perhaps the most problematic piece of the puzzle of navigation: determining longitude while at sea. Latitude, one's north-south position, could be determined by measuring the altitude of the Sun at noon or the altitude of any star, provided it was tabulated in one of the various astronomical almanacs of the time, when it crossed the local meridian. In higher latitudes, most marine navigators determined latitude by observing the altitude of the polestar, or the angle between its direction and the horizontal; as ships approached the equatorial latitudes, stars were often below or too close to the observable horizon and navigators had to rely on measurements of the position of the Sun. Knowledge of one's east-west position—longitude—was critical, especially on long transatlantic voyages on which water and foodstuffs were prone to ruin or were in short supply. Calculating longitude required determining the precise time as one element of the equation—a measurement that could not readily be determined at sea. The calculation of longitude using astronomical observations and trigonometry was a mathematical possibility first introduced in 1474 by Regiomontanus. Nevil Maskelyne (1732-1811), who was appointed British Astronomer Royal in 1765, also proposed that complex trigonometry could be used in conjunction with a voluminous catalog of observed and predicted star positions to calculate longitude. The mathematically intensive task was cumbersome and impractical, and the problem of determining longitude was eventually settled with the invention and acceptance of a portable timepiece by English inventor John Harrison (1693-1776) in 1773.

Besides the chronometer, other practical mechanical devices were employed in aiding navigation. The patent log, designed by English inventor Humphry Cole in 1688, aided in approximating the speed at which a ship was traveling. Innovations to the original design of the patent log made it more reliable, and an 1861 design by another Englishman, Thomas Walker, remains in use. Telescopes were employed in the eighteenth century to determine the length of a degree of longitude. The principle of the magnetic compass, long known in China and the Arab world, was introduced in Europe with the basic modern design emerging in the West in the seventeenth century. The advent of ironclad and steel ships required the addition of magnetic stabilizers to keep compass readings of magnetic north accurate. The determination of true directional north still required mathematical computation.

As the problems of navigating the open waters came to be solved, there was an increased need to refocus attention back to the careful navigation of those waters closer to port. The enormous increase of commercial vessel traffic, a result of the discovery of the New World and the establishment of large-scale maritime mercantile trade, made navigating ships into harbor precarious. A growing concern among ship pilots was the avoidance of other vessels. The largest of the merchant and military ships were easy to see, but were also more difficult to maneuver. Sail power and varying winds close to shore complicated matters further. Simple trigonometric formulas could be employed to predict the velocity and path of another ship, but a more practical solution, the designation and division of incoming and outgoing shipping lanes, was also needed to manage traffic.

Rising investments in merchant endeavors and bitter rivalries over colonial territories led some European nations to begin to build strong navies. Designing military vessels was a special challenge, as they needed to be sturdy, yet also quick and maneuverable. Simple trigonometry advanced sail designs and allowed for design and implementation of heavy weaponry. Relatively new advances in war weaponry, especially cannons and mortars, required a basic knowledge of the principles of trigonometry to aim the implements and determine the trajectory of their projectiles. Thus, new mathematical understanding facilitated nearly all aspects of the European maritime and colonial endeavors.

ADRIENNE WILMOTH LERNER

Further Reading

Sobel, Dava. *Longitude: The True Story of the Lone Genius Who Solved the Greatest Scientific Problem of His Time.* New York: Penguin, 1996.

Turner, Gerald L'estrange. *Scientific Instruments 1500-1900: An Introduction.* Berkeley: University of California Press, 1998.

Mathematics, Science, and the Society of Jesus

Overview

Founded in the year 1540 by St. Ignatius Loyola, the Society of Jesus quickly became one of the preeminent religious orders of Europe and the world. In addition to teaching, the Jesuits considered the acquisition of knowledge to be a source of spirituality because it could help humans to better understand God's universe. Following this philosophy, many Jesuits became mathematicians and scientists, conducting research and teaching at universities as they contributed to man's store of knowledge. This led the Society of Jesus to become perhaps the world's most scientifically prolific religious order as well as some of the world's best teachers, traditions that continue to this day.

Background

In the early 1530s Ignatius of Loyola, a Spanish soldier, was wounded in battle. He experienced a profound religious conversion during his convalescence and, along with six companions, vowed to follow a life of poverty and chastity and to make a pilgrimage to Jerusalem. In 1539, realizing they would not be able to make this journey, the seven men promised to accept any work assigned them by the Pope that would help the Church. In 1539, Ignatius presented the Pope with an organizational outline for a new religious order, the Society of Jesus. Unlike existing religious orders, the Society of Jesus swore an oath of personal loyalty to the Pope. In addition, there was great emphasis placed on flexibility, independent thought, and similar innovations. This resulted in a religious order that was to become important as the Church's emissaries abroad, as teachers, missionaries, diplomats, and the like.

Almost from the start, the Society of Jesus had its detractors. In fact, its familiar name, the Jesuits, was first given as a pejorative nickname, and was adopted by the Society soon after its founding. Because their training emphasized obedience to the Pope and to their hierarchy, education, and a rational approach to religious belief, the Jesuits quickly became Europe's teachers, opening colleges and universities throughout Europe. In fact, at one point Francis Bacon (1561-1626), a prominent Protestant and an op-

ponent of the Catholic Church commented, "They are so good that I wish they were on our side.'" In 1556, when Ignatius died, there were 1,000 Jesuits. Seventy years later, the order had over 15,000 members teaching at several hundred colleges and universities, most of which were Jesuit-run.

At that time only royalty, the aristocracy, the clergy, and the upper class of society received much in the way of formal education. The Jesuit's prominence in education in Catholic nations gave them unparalleled access to those in power. This, in turn, helped them to become respected and feared. The universities also gave the Jesuits the impetus to produce scientists of their own, which was encouraged by their admonition to find God in the laws of nature as well as in the Church.

Most Jesuit mathematicians of the sixteenth and seventeenth centuries are referred to as "geometers" and, because of this, it is often assumed that Jesuits concentrated on geometry and the mathematics of classical Greece to the exclusion of other methods. However, it is important to recall that, until Isaac Newton's (1642-1727) invention of the calculus, and the mathematical advances that accompanied this, most mathematical proofs were geometrical in nature. In fact, even Newton used geometry to prove many aspects of his calculus, and it was only in later decades that purely mathematical proofs, without geometry, became accepted. In particular, it is interesting to note that the work of Andre Tacquet (1612-1660) on infinitesimals helped provide the groundwork for the development of the calculus by exploring aspects of limits that were important to understand fully. In this, he ran counter to some religious arguments regarding the nature of infinity, since many felt that God's infinity should not have to accommodate a mathematical infinity. However, Tacquet's work remained important, and was among the first to describe many of the concepts later expanded on by Newton, Gottfried Leibniz (1646-1716), and Blaise Pascal (1623-1662) in their work.

Other noteworthy Jesuit mathematicians of this time included Ignace Pardies (1636-1673), Gregory St. Vincent (1584-1667), Honoré Fabri (1607-1688), and Christoph Clavius (1538-

1612). There were, of course, many other Jesuit mathematicians and geometers; in fact, one reference notes no fewer than 631 Jesuits involved in mathematical work in the first two centuries of the order's existence, whose work ranged from mediocre to superb, as well as noted Jesuit astronomers, physicists, biologists, and geologists. All in all, the Society of Jesus has likely contributed more to the sciences than any other religious order in the world. In the next section, we will examine the impacts that this devotion to learning has had on science and on society.

Impact

The impact of the Society of Jesus and its priest-scientists can be seen in three primary areas: advances in scientific and mathematical knowledge, the impact of this research on Jesuit-led education, and the effects on the Catholic Church and its followers.

The first and most obvious impact, of course, is that on science itself. Jesuit mathematicians and scientists made a number of significant discoveries that helped expand our knowledge of mathematics, physics, and the world around us. Their discoveries helped pave the way for many of Newton's discoveries, as well as helped to consolidate the intellectual territories he and his contemporaries opened for inquiry. Jesuits were also among the first to differentiate between science and pseudo-science, questioning some of the more dubious claims and "discoveries" made in an era that did not always clearly differentiate between the natural and the supernatural. On the other hand, Jesuits were also in the forefront as the Church attempted to combat the effects of the Reformation, and they often took the role of reactionaries trying to protect the Church from change. This also led, in some cases, to attempts to hew to the status quo, including an emphasis on geometrical proofs as noted above, but it did not preclude Tacquet's work on infinitesimals or Fabri's support of Galileo (which landed him in prison for 50 days).

In addition, Jesuit mathematicians had a significant impact on the mathematics of this era. According to one writer, "One cannot talk about mathematics in the sixteenth and seventeenth centuries without seeing a Jesuit at every corner." Fabri, for example, worked on the problems of tides, optics, heliocentrism (whether the earth or the sun was the center of the universe), and tried to unify all of physics in a manner similar to geometry. Gregory Saint Vincent (1584-1667) de-veloped polar coordinates, which are crucial to solving some types of math and physics problems. He was also important in developing analytic geometry, taught today alongside calculus in most colleges and universities. Pardies investigated problems in a manner that later inspired Newton, and even questioned some of Newton's findings in a way that forced Newton to go back to clarify his thinking on some crucial points. Other Jesuits, particularly those who were born in the sixteenth century, seemed more likely to concentrate on the classical teachings and wisdom of Aristotle, Euclid, and other ancient thinkers, rather than pursuing new avenues of inquiry. However, such teachers and thinkers became fewer and fewer over time, and their impact on society and on science lessened with the passing years.

As important as the Jesuit's research was, they clearly made their most significant contribution to European society through their teaching; and their teaching was, in turn, influenced by their research.

Because of the Jesuit's emerging tradition of intellectual independence and inquiry, their colleges and universities also tended to incorporate these traits into their teachings. There is no substitute for learning from those who are making important discoveries, and many of Europe's leaders were educated by priests who were active in adding to the sum of human knowledge. At the same time, these priests were celebrating the joy and the importance of learning, and teaching that physics, mathematics, and other branches of knowledge were not incompatible with religious beliefs. This combination may well have helped encourage the same spirit of rational inquiry that furthered the Renaissance and led to the Enlightenment, which led in turn to the French Revolution and the revolutions in North and South America.

Finally, and possibly most important, is the effect that all of the above had on European society of the sixteenth and seventeenth centuries. This was the time of the Renaissance, the rebirth of Western civilization after centuries of feudal rule. Although the Medieval period was not necessarily the long centuries of intellectual darkness so often pictured, it was also not the hotbed of new ideas that was to come. Man was beginning to realize that it might just be possible to understand nature and, to some extent, to predict and control it, and this was heady knowledge. It affected philosophy, politics, and religion, and all of these, to some extent, affected all of society.

The two biggest controversies in which Jesuit scientists became embroiled were those regarding the legitimate place of Earth in the universe and the nature of the infinite. In both cases, there were those who felt that science was intruding on God's prerogatives or that science was attempting to dethrone humans and God from their rightful places in the cosmos. The Church came down as squarely opposed to the Copernican solar system (in which the Earth and planets orbit the Sun), just as it was not in favor of mathematical infinities (or their converse, the infinitesimal). And, in both cases, scientists and mathematicians, including some Jesuits, showed the Church to be in error. These were but two steps in a process that continues to this day, in which scientific findings seem to detract from the authority of the Church, weakening it ever so slightly.

It is important, however, to recognize one important fact: most current scientific and religious leaders do not perceive that science and religion are mutually exclusive. While the scientific controversies of earlier years may have fostered this view, it should be obvious by the continuing presence of Jesuit (and other orders) priest-scientists through the centuries that this belief is overly simplistic and, in fact, is just wrong. And this may have been one of the more important societal contributions of Jesuit scientists and mathematicians at the time; their demonstration that a priest could also be a good scientist helped show Europe, recently emerged from superstition, that a good Christian could also believe in science and view the world in a rational way. This realization has served mankind well for over 400 years.

P. ANDREW KARAM

Further Reading

Books

MacDonnell, Joseph. *Jesuit Geometers: A Study of Fifty-Six Prominent Jesuit Geometers during the First Two Centuries of Jesuit History.* The Institute of Jesuit Sources, 1989.

Websites

"Jesuits and the Sciences." Loyola University of Chicago. http://www.luc.edu/science/jesuits/jessci.html.

"Jesuit Geometers." http://www.faculty.fairfield.edu/faculty/jmac/sj/sjgeom.html.

Mathematical Challenges and Contests

Overview

During the sixteenth century, mathematics was transformed from the traditional study of classical texts and problems to a dynamic science characterized by active research in problems both abstract and applied. Such research depended on the lively exchange of ideas and techniques, which fostered a spirit of competition among investigators. The practice of offering challenges and contests characterized sixteenth- and seventeenth-century mathematics, and left a permanent legacy of mathematical competitions.

Background

Sixteenth-and seventeenth-century European mathematicians were keenly aware of one another's work, thanks to efficient correspondence networks and the rise of printed mathematical books. This awareness helped speed the pace of research, but it also gave mathematical work a competitive spirit. Mathematicians became anxious to identify and solve important problems, and to be the first to do so. The rewards for such competition were primarily personal—a triumphant solution would bring the author prestige among his mathematical peers—but were sometimes more tangible. Mathematical teachers who lost competitions or challenges could find their jobs to be in jeopardy. On the other hand, improved posts or access to patronage could come to victorious mathematical debaters, and some mathematical contests later in the seventeenth and eighteenth centuries even offered prize money to the successful author, augmenting the spoils of intellectual victory.

Two of the most famous mathematical challenges of the sixteenth century involved the Italian mathematician Niccolo Tartaglia (1499?-1557). Tartaglia was a largely self-taught scholar who overcame poverty and a traumatic childhood disfigurement—he was stabbed in the head during a military assault on his hometown of Brescia, and formally adopted his nickname

Tartaglia, or "stammerer." He made his living as a mathematics teacher in Verona and Venice, and became well known in mathematical circles early in his career by his successful participation in a number of mathematical debates, and later by publications in pure mathematics and the application of mathematics to problems of warfare.

One of the most popular fields of mathematical study in the sixteenth century was algebra; of particular interest was the search for methods to solve third (and higher) order equations. A solution to the basic problem of solving cubic equations was found sometime in the first two decades of the sixteenth century by Scipione del Ferro (1465-1526), a mathematics lecturer at the University of Bologna. He did not publish his work, but he did share it with a few disciples. In 1535 one of these disciples, Antonio Fiore, sought to exploit his master's secret and issued a challenge to Tartaglia, by then known as a master debater. The two exchanged 30 mathematical problems for each other to solve; Tartaglia's challenge included a range of mathematical problems, but all of Fiore's problems for Tartaglia were cubic equations. Under the pressure of competition, Tartaglia figured out the method for a general solution of such problems (not knowing that Fiore was in possession of such a method), and was able to dispatch all 30 of Fiore's problems in less than two hours. Fiore made little progress solving any of Tartaglia's problems, and Tartaglia emerged from the encounter with his reputation greatly enhanced.

Tartaglia did not publish the method of solving cubic equations either, perhaps hoping to keep it secret for use in future challenges or debates. This decision came to cause him grief. In 1539 Tartaglia was approached by a wealthy and well-connected Milanese mathematician and physician named Girolamo Cardano (1501-1576). Cardano was anxious to have the method for solving cubic equations, and he tried several strategies to make Tartaglia share his secret. First, he asked Tartaglia to include his method in a book Cardano was publishing; when that failed, he challenged Tartaglia to a debate. Although Tartaglia did not agree to debate Cardano, he was finally persuaded by hints of patronage opportunities to come to Milan. While he was in Milan as Cardano's guest, Tartaglia did finally share the formula—disguised as a poem—after extracting a pledge of secrecy from his persistent host.

While Cardano initially honored his promise to keep Tartaglia's formula a secret, he used it as a basis for further discoveries of his own. When he

Niccolo Tartaglia. *(Library of Congress. Reproduced with permission.)*

found out some years later that Ferro had discovered a method for solving cubics prior to Tartaglia, Cardano felt justified in including Ferro's method in a book he published in 1545. This infuriated Tartaglia, and led to an angry feud and finally a debate between Tartaglia and Cardano's disciple Lodovico Ferrari (1522-1565); Cardano himself refused to enter into a challenge with Tartaglia. Tartaglia traveled to Milan again, hoping that he could vanquish Ferrari in a debate and thereby secure a superior teaching post. But Ferrari quickly showed a better grasp of the problems offered than Tartaglia had, and Tartaglia fled the city after the first day of their contest, thinking it better to leave the contest unresolved than to lose outright to Ferrari. But news of his poor performance in Milan dogged Tartaglia, and his career suffered accordingly.

Tartaglia's two debates are perhaps the most famous in mathematics history, but they were by no means unusual events for their time. Challenges were a vital part of mathematical practice. They helped to identify superior mathematicians, to spur work on the solution of particular problems, and to display for mathematical audiences the range of active problems at a given time. For example, the challenge between Tartaglia and Ferrari including problems not only associated with cubic and quartic equations, but also ranged through topics including astronomy, optics, architecture, and cartography. Prior to the establishment of mathematical societies and periodical journals, challenges and the correspondence and books they inspired were one of the primary means of identifying and disseminating mathematical progress.

Impact

Tartaglia's debates were like mathematical duels —one mathematician would issue a challenge to another, and the dispute would usually be resolved publicly in a face-to-face exchange. Mathematical challenges evolved into another form in the seventeenth century, when mathematicians set out important unsolved problems for general solution by their peers. Skilled investigators throughout Europe would take up such problems, and compete with one another to have the first, and the best, solution.

Some of the best-known mathematical challenges of the seventeenth century came from Pierre de Fermat (1601-1665). Fermat is remembered today for his extraordinary contributions to mathematics, especially number theory, but he was a lawyer by profession and mathematics was just one of his several avocations. These other interests may explain Fermat's lack of publications, and his practice of presenting his mathematical ideas rather casually in letters to friends. In 1657 Fermat issued a series of problems as challenges to other mathematicians. These problems were stated in the form of theorems to be proved. Because of Fermat's failure to publish or present formally his own work, historians are not sure how many of these theorems Fermat had solved himself, or what his methods might have been. The challenge problems reflected Fermat's interest in prime numbers and divisibility, and in producing general solutions based upon a single paradigmatic solution. The final one of these problems (to show there is no solution for $x^n + y^n = z^n$ for $n > 2$) resisted conquest until the late 1990s. The solution by Andrew Wile (1953-) of what had come to be known as Fermat's Last Theorem received enormous attention from even nonmathematicians, and is considered one of the greatest mathematical achievements of the twentieth century.

By the eighteenth century, the practice of mathematical contests had been well established. Scientific societies large and small issued challenges to draw attention to themselves and to particular mathematical problems, but perhaps most important for the advance of research have been challenges like those of Fermat issued by prominent mathematicians themselves. By singling out particular theorems or problems for attention, mathematicians such as David Hilbert (1862-1943), who issued 23 important problems for study in 1900, have helped shape research agendas for the entire mathematical community. Challenges may have lost the personal vitriol that colored Tartaglia's debate, but the spirit of competition continues to surge through research mathematics. Young mathematics students routinely enter mathematical competitions, ranging from local contests to international olympiads. Intense pressure to produce the first proof of an unsolved theorem, or to find a counterexample to discredit it, is an essential feature of mathematical practice. Careers are still made or lost on the basis of priority of solution, and the entire international mathematical community can be caught up in the evaluation of the reported solution of important unsolved problems such as Fermat's Last Theorem. In this way, research mathematics retains some of the spirit of an age when many kinds of disputes could be settled by a duel.

LOREN BUTLER FEFFER

Further Reading

Boyer, Carl. *A History of Mathematics*. Rev. by Uta Merzbach. New York: Wiley, 1989.

Cooke, Roger. *The History of Mathematics*. New York: Wiley, 1997.

Dear, Peter. *Discipline and Experience: The Mathematical Way in the Scientific Revolution*. Chicago: University of Chicago Press, 1995.

Hay, Cynthia, ed. *Mathematics from Manuscript to Print, 1300-1600*. Oxford: Clarendon Press, 1988.

Hollingdale, Stuart. *Makers of Mathematics*. London: Penguin, 1989.

Katz, Victor. *A History of Mathematics*. Reading, MA: Addison-Wesley, 1998.

Struik, Dirk, ed. *A Source Book in Mathematics, 1200-1800*. Princeton: Princeton University Press, 1986.

Isaac Barrow
1630-1677
English Mathematician

A geometer and theologian who also contributed to the field of optics, Isaac Barrow is best known for the influence he exerted over the career of the young Isaac Newton (1642-1727). Among his most important published works were *Euclidis elementorum libri XV* and *Lectiones geometricae,* in which he interpreted and synthesized the ideas of more well-known geometers for a popular audience.

Barrow's father, Thomas, was a merchant and linen draper for King Charles I, and his mother Anne died shortly after her son's birth in London in 1630. The boy proved an unruly student, and after a stint at the Charterhouse school, he was sent to Felsted School in Essex, where schoolmaster Martin Holbeach had a reputation as a stern disciplinarian. Barrow thrived in this environment, and became immersed in subjects that included Latin, Greek, Hebrew, French, logic, and the classics.

In the unrest leading up to the English Civil War (1642-48), Barrow's father suffered financial losses, and the son was forced to take a job as a tutor. In 1646, however, he obtained a scholarship to Trinity College, Cambridge, from whence he graduated two years later. In 1649, he was elected as a college fellow, and in 1652 earned his M.A., whereupon he went to work as a college lecturer and university examiner. He published his first work, *Euclidis elementorum libri XV,* in 1654. A translation of writings by Euclid (c. 325-c. 250 B.C.), the book became highly popular, and eventually appeared in a pocket-sized edition.

With his rising fortunes, Barrow seemed a natural choice to take a highly respected Regius professorship in Greek, but again political tensions intervened. The university chancellor responsible for choosing among the candidates was none other than Oliver Cromwell, leader of the Puritans who overthrew Charles in the Civil War; and given Barrow's ties to the monarchy, he had no chance of obtaining the position. Embittered, Barrow spent nearly five years away from England, travelling through France, Italy, and Turkey on a Trinity College fellowship.

Isaac Barrow. *(Library of Congress. Reproduced with permission.)*

By 1660, when Barrow returned to his homeland, the political tides were turning again, and with the restoration of King Charles II to the throne, Barrow was ordained in the Anglican Church and appointed to the Regius professorship. He had become increasingly interested in mathematics while abroad, and now began supplementing his income by teaching geometry and astronomy at Gresham College. This led to his 1663 appointment as Lucasian professor of mathematics at Cambridge, a position that carried with it enough funds that he could give up his other teaching jobs.

As a further hallmark of his restored fortunes, Barrow was appointed as the first fellow of the recently founded Royal Society of London in 1663. In the six years that followed, he developed a series of lectures on geometry, or *Lectiones geometricae,* in which he brought together ideas from René Descartes (1596-1650), John Wallis (1616-1703), and James Gregory (1638-1675) in a format comprehensible to the rising generation of scholars.

Among the latter was Isaac Newton, for whom Barrow served as scholarship examiner in

1664. Biographers have generally credited Barrow with providing the initial spark of inspiration that, under considerable development by Newton, would lead to the latter's epochal work in physics. Particularly notable was Barrow's work in optics, which has long been overshadowed by Newton's considerably more impressive achievements in that field. As the student became greater, the teacher's impact receded: in 1669, Barrow stepped down as Lucasian professor in favor of Newton.

Barrow went on to serve as royal chaplain, and in 1673 returned to Trinity College at the king's request. He became vice chancellor in 1675, but died two years later at age 47. Barrow had never married, and the evidence indicates that he died of a drug overdose.

JUDSON KNIGHT

Jakob Bernoulli
1645-1705

The first member of his distinguished family to attain international notoriety, Jakob Bernoulli contributed greatly to mathematicians' understanding of calculus and probability theory. He maintained a correspondence with Gottfried Wilhelm von Leibniz (1646-1716), and was one of the first scholars to fully grasp the latter's "infinitesimal calculus," as that branch of mathematics was then called.

The Bernoullis came from Holland, which at the time was controlled by Spain, and fled Spanish oppression, winding up in the Swiss town of Basel. Against his father's wishes, Jakob studied mathematics and astronomy, but it was in theology that he earned his degree in 1676, at age 31. He then went to work as a tutor in France, where he became acquainted with the writings of René Descartes (1596-1650), and in 1681 traveled to Holland and England, where he met physicist Robert Boyle (1627-1691).

Upon his return to Basel in 1682, Bernoulli established a school for science and mathematics, and published a number of articles in Europe's two leading scientific journals, *Journal des sçavans* and *Acta eruditorum*. In a 1684 article, he showed his grasp of calculus at a time when the discipline was new, and in subsequent pieces expanded on the base of understanding he had developed. It was Bernoulli who in 1690 coined the term "integral calculus" to describe the branch concerned with determining a function where a derivative is known.

Jakob Bernoulli. *(Corbis-Bettman. Reproduced with permission.)*

Bernoulli took a position as professor of mathematics at the University of Basel, where he would spend the remainder of his career, in 1687. Two years later, he published his famous Bernoulli inequality, a theorem already derived—unbeknownst to Bernoulli—by Isaac Barrow (1630-1677) in the latter's *Lectiones geometricae.* He also solved a problem of long standing concerning a catenary, a shape that results when a flexible, nonelastic cable is suspended between two fixed points. Mathematicians had traditionally maintained that a catenary is a parabola, but Bernoulli showed that it was not.

Another shape that intrigued Bernoulli was the brachistochrone, a curve of quickest descent between two points A and B, where B does not lie directly beneath A. The latter problem received the attention of both Leibniz and Bernoulli's younger brother Johann (1667-1748). So intense was the rivalry between the two brothers that Johann claimed Jakob's brachistochrone solution as his own. This set a pattern for intrafamily competition, much of it centering around Johann, that would continue into the next generation, locking the latter in a struggle with his son Daniel (1700-1782), most famous member of the clan.

As for Jakob and the shapes that interested him, one of the foremost was the logarithmic spiral, the shape made by the cross-section of a chambered nautilus. Among its most interesting properties is self-similarity: any section, if properly scaled, is congruent to other parts of the curve. When he died, having never married or fathered children, he had this spiral engraved on his headstone, along with the motto *Eadem mutato resurgo* (Though changed, I arise again the same.) His *Ars conjectandi* (The Art of Conjecture), published posthumously in 1713, is considered one of the foundational texts on probability.

JUDSON KNIGHT

Johann Bernoulli
1667-1748
Swiss Mathematician

So great was the reputation of the Bernoulli family, and so numerous its members—a condition not unlike that of their German contemporaries the Bachs—that historians often have a hard time sorting out the various personalities. Thus Johann Bernoulli was also known as Jean, while his equally famous older brother Jakob (1654-1705) has been variously identified as Jacques, James, or Jakob I to distinguish him from other Bernoullis of that name. Another abiding characteristic of the Bernoulli family was a tendency toward rivalries, conflicts that always seemed to center around Johann—who struggled first with his brother, and later with his son Daniel (1700-1782), most notable among a prominent line.

Johann originally studied medicine, but from the beginning of his academic career, his passion was mathematics. Thus his doctoral dissertation was a mathematical treatise presented as a study of muscle contractions. By 1691, when he was 24, Bernoulli had travelled from the family's home in Switzerland to Paris, where he became associated with some of the leading scientists and philosophers of the day. Of these, none was more significant than Gottfried Wilhelm von Leibniz (1646-1716), whose friend, correspondent, and loyalist Bernoulli would remain for the rest of his life.

Hired by the young Marquis de L'Hospital (1661-1704) as the latter's tutor in 1692, Bernoulli eventually "sold" a number of his mathematical discoveries to the wealthy young man. Thus L'Hospital's Rule, which enabled mathematicians to determine the limiting value of a fraction in which both numerator and denominator tend toward zero, was actually a dis-

Johann Bernoulli. *(Corbis-Bettman. Reproduced with permission.)*

covery of Bernoulli. L'Hospital was, however, a talented thinker in his own right as well. Thus some years later, when Bernoulli presented his famous brachistochrone problem, Leibniz correctly posited that only five people would solve it: Leibniz himself, the two Bernoulli brothers, Sir Isaac Newton (1642-1727), and L'Hospital.

The brachistochrone problem, which concerned the curve marking the quickest descent of an object between two points at different altitudes (but not along a vertical line), was but one of several points of contention for the Bernoullis. Jakob claimed to have found the solution first, and in the case of the isoperimetric problem—involving the comparison of different polygons with equal perimeters—there is no doubt that Jakob was first, though Johann did improve on his brother's findings.

This rivalry was a driving force in Johann's career: in 1695, he took a position at the University of Gröningen in the Netherlands, knowing that Jakob would keep him off the faculty at his own University of Basel. Following Jakob's death in 1705, however, Johann took his place as professor of mathematics at Basel, where his stature as a leading figure in the European intellectual world would grow during the four subsequent decades. Soon he had a new rival within the family, however: his brilliant son Daniel, whose famous law concerning fluid pressure (Bernoulli's principle) he attempted to present as his own.

Despite his apparent disloyalty to blood kin, Johann proved a loyal supporter of Leibniz throughout his career. In early years, he defended the German mathematician as the founder of calculus against supporters of Newton, who had simultaneously developed his own calculus. Late in life, Bernoulli made a name for himself in the realm of mechanics, supporting Leibniz's views on what became known as the conservation of energy. Thanks in part to Bernoulli's efforts, the Leibnizian concept of "living force" would gain acceptance over countervailing (but not entirely inaccurate) ideas presented by René Descartes (1596-1650). Today living force is known as kinetic energy, and Bernoulli as credited as the one of the first thinkers to recognize the principle of conservation.

JUDSON KNIGHT

Rafaello Bombelli
1526-1573
Italian Mathematician

The career of Italian algebraist Rafaello Bombelli helped bridge the late Renaissance and the early period of the Enlightenment. The last among many Italian mathematicians who contributed to a developing theory of equations, Bombelli became the first to conceive a consistent theory of imaginary numbers including rules for operations on complex numbers. His work, whose implications for complex numbers Bombelli never fully appreciated, won him admiration among future mathematical giants such as Gottfried Wilhelm von Leibniz (1646-1716) and Leonhard Euler (1707-1783).

Born in Bologna, Bombelli opted not to follow his father, Antonio, in the latter's profession as wool merchant. Instead, he became an engineer, and rather than attend university, received his training as apprentice to engineer-architect Pier Francesco Clementi of Corinaldo. He spent much of his career under the patronage of the Bishop of Melfi, Monsignor Alessandro Rufini, and was responsible for the draining of the Val di Chiana marshes (1551-60), as well as the unsuccessful 1561 attempt to repair the Ponte Santa Maria, a bridge in Rome.

Bombelli's work in mathematics began during the 1560s, first with the writing of *Algebra.* The latter is particularly significant because it reintroduced scholars to the work of the ancient

Greek mathematician Diophantus of Alexandria (3rd century A.D.) In addition to reproducing 143 problems of Diophantus, whose writings Bombelli had discovered in the library of the Vatican, the book also introduced a number of symbols in algebraic notation.

Attempting to improve on the already established Cardano-Tartaglia formula, Bombelli set out to develop his own highly precise theory of imaginary numbers (e.g., the square root of a negative number). This led to the conclusion that real numbers can result from the operations of complex numbers (i.e., numbers that are a mixture of real and imaginary). His findings on complex numbers would have a larger impact than Bombelli, who considered such numbers irrelevant, might have guessed. Euler would quote him in his own *Algebra,* and Leibniz called him an "outstanding master of the analytical art."

JUDSON KNIGHT

Charles de Bouelles
c. 1470-c. 1553
French Mathematician

French priest Charles de Bouelles, whose name is variously rendered as de Boville, Bovillus, Bovelles, and Bouvelles, was responsible for a number of contributions to mathematics. Most notable among these were his work on the quadrature, or squaring, of the circle, and his writings on perfect numbers. He also published the first book on geometry written in French, and conducted an early study of the cycloid, the shape generated by following a fixed point on the circumference of a circle that rolls along a straight line.

Little is known about Bouelles's early life, except that he came from an aristocratic family in the town of Saucourt, located in the Picardy region of France. He studied in Paris until about the age of 25, part of this time under the noted educator Jacques Lefèvre d'Etaples (c. 1455-1536), but left in 1495 after a new outbreak of the Plague. For more than a decade, he traveled throughout Switzerland, Germany, Italy, Spain, and other parts of Europe before returning to France and the priesthood in 1507. His first appointment was as a canon at the cathedral of Saint Quentin, followed by a stint in the town of Noyon, where he also served as a theology professor.

Thanks in large part to the patronage of Charles de Hangest, an official at Noyon, Bouelles was free to engage in mathematical investigations. His first mathematical text appeared as *Geometricae introductionis* in 1503, though it proved so popular that translations in French and Dutch eventually made their appearance as well—a remarkable feat at a time when virtually all scientific material was published exclusively in Latin.

In *Geometricae,* Bouelles addressed the age-old problem of squaring the circle, which concerned the attempt to map the area of a circle onto a square of equal size. Thanks to the highly accurate determination of pi in circulation today, the quadrature of the circle is no longer a challenge, but in Bouelles's day and afterward, it continued to bedevil mathematical scholars.

Bouelles in 1510 published *Liver de XII numbers.* The latter addressed the subject of perfect numbers, or those integers which are a sum of all their factors excluding the number itself: for instance, 6 = 1 + 2 + 3, and 28 = 1 + 2 + 4 + 7 + 14. During the following year, he published *Le livre de l'art et science de géométrie,* the first known work on geometry in French. Later years saw the publication of *Prover biorum vulgarium libri tres* (1531) and *Liber de differentia vulgarium linguarum et gallici sermonis varietate* (1533). Bouelles died in Noyon at about the age of 83.

JUDSON KNIGHT

Viscount William Brouncker
1620-1684
English Mathematician

In 1662, Viscount William Brouncker proposed to the newly restored English monarch Charles II that an institution be established to advance scientific discussion and learning. The result was the Royal Society of London, of which Brouncker served as first president. In a career that put him into contact with such preeminent figures as Pierre de Fermat (1601-1665) and John Wallis (1616-1703), Brouncker examined a number of problems, particularly the use of continued fractions to express π.

Other than the fact that he was born in Castle Lyons, Ireland, to Sir William and Lady Winefrid Brouncker in 1620, the details of Brouncker's childhood are sketchy. He and his younger brother Henry probably studied with tutors; then at age 16, Brouncker entered Oxford, where he excelled at a range of subjects that included mathematics, music, languages, and medicine. While he was still in school, in 1645, his father was named a viscount by King Charles I. The

William Brounker. *(Archive Photos, Inc. Reproduced with permission.)*

elder William died a year later, and thus at age 26, the son became a peer of the realm.

Brouncker received the degree of Doctor of Physick in 1647, a year that would prove monumental in more than one respect. It was then that the forces led by Oliver Cromwell deposed Charles I, establishing a theocratic and egalitarian dictatorship under which all persons with ties to the royalty or nobility were in danger of imprisonment or even death. Therefore with apparent deliberation, Brouncker began a period of virtual invisibility that would last until the restoration of the monarchy 13 years later. During this time of self-imposed internal exile, his only notable work was a translation of *Musicae compendium* by René Descartes (1596-1650). Indeed, this was the only book of Brouncker's entire career, his other writings being confined to correspondence, manuscripts, and contributions published in other mathematicians' works.

In 1660, with the restoration of the monarchy under Charles II, son of Charles I, Brouncker suddenly reappeared at center-stage of English public life. He won election to parliament in that year, and in 1662, when Charles established the Royal Society as a result of a proposal put forth by Brouncker, the latter was elected its first president with no opposition. Brouncker would be re-elected annually until 1677, when he decided to step down from leadership of the Royal Society.

Brouncker enjoyed a lively interaction with other mathematicians of the day. When Wallis requested help in developing an expression of π other than as an endless decimal, Brouncker applied the use of continued fractions to the problem. He also used continuous fractions for the quadrature, or squaring, of a rectangular hyperbola. He and Fermat both worked on the Pell Equation, and he interacted with James Gregory (1638-1675) on the subject of binomial series.

The fact that he never married made it easier for Brouncker to devote his time to a number of institutions and offices. In 1664, he became president of Gresham College, and entered government service as commissioner of the navy. He held the latter position until 1668, at which time he became comptroller of the treasurer's accounts. He also served as master of St. Catherine's Hospital in London from 1681 to 1684, the year he died. Since he had no heirs, his title passed to his brother Henry, who was also unmarried; therefore upon Henry's death in 1687, the family line and its title came to an end.

JUDSON KNIGHT

Joost Bürgi
1552-1632
German Mathematician

In his work as a clockmaker and astronomer, Joost Bürgi needed accurate mathematical information, and for this reason developed the concept of logarithms into a practical method of computation. He did this as much as a decade before John Napier (1550-1617), the Scottish mathematician generally credited with the foundational work in logarithms; but because Bürgi did not consider himself a mathematician, he waited to publish his findings in 1620, long after Napier. Thus Bürgi was destined to become much less famous than either Napier or his young assistant, Johannes Kepler (1571-1630).

Bürgi has the distinction of being one of the few people in world history from Liechtenstein, a principality smaller in area than Washington, D.C. At the time of his birth in 1552, however, the region was still part of the German-controlled Holy Roman Empire. He probably received little in the way of a formal education, since he was unable to read Latin, the language of educated men during that era, but from an early age he excelled in the practical realm of clockmaking. In 1579, 27-year-old Bürgi received an appointment as court watchmaker for Duke Wilhelm IV of Kassel, a city in what is now western Germany.

The art and science of making and maintaining clocks was very much in its infancy then, but rulers and merchants quickly recognized the importance of these skills; thus Bürgi's position was analogous to that of a youthful computer whiz in the 21st century. Just as the latter might go to work for a major corporation to avail him- or herself of the company's resources and equipment, Bürgi made use of the duke's observatory for astronomical work. Not only did he build clocks, but he developed instruments for making astronomical measurements, among them a proportional compass that in some respects improved on a similar instrument constructed by Galileo (1564-1642).

His most important work while at Duke Wilhelm's court, however, was his computation of logarithmic tables. Bürgi did not approach this as a theoretical problem, but as a practical one: in order to process astronomical data, he needed to make computations quickly. Thus as early as 1584, he began making improvements to the existing system of *prosthaphairesis,* a method of applying trigonometric formulae to problems as a means of converting multiplication to addition. Eventually he chanced on the idea of logarithms, exponents that indicate the power to which a number is raised to produce a given number. These would greatly facilitate quick multiplication, and to this end, Bürgi compiled an extensive logarithmic table.

Eventually Rudolf II, a Holy Roman emperor noted as much for his mental instability as for his interest in science, took an interest in Bürgi. Upon the death of Wilhelm, Rudolf summoned Bürgi to his court, and gave him Kepler as an assistant. Arriving in 1603, Bürgi again set to work in the observatory, working on astronomical calculations and the development of better instruments for measuring. Among Kepler's unpublished notes are papers containing evidence that Bürgi used the decimal point, and improved on a method for calculating the roots of algebraic functions.

Bürgi remained at the court of the Holy Roman emperor after the unstable Rudolf ceded power to his brother, Matthias, in 1606, and again after Ferdinand II succeeded Matthias in 1619. His stint at the court lasted until 1631, at which point he returned to Kassel, where he died a year later. By then Napier, who had published two works on logarithms and established himself as the mathematician who developed them, was long dead. The sole complete copy of Bürgi's logarithmic table is in a library in Gdansk, Poland.

JUDSON KNIGHT

Girolamo Cardano
1501-1576
Italian Mathematician

A mathematician and physician, Girolamo or Geronimo Cardano lived a turbulent personal and professional life, and became embroiled in a conflict over cubic equations so full of drama and surprises that it sounds more like a movie script than an incident from the history of mathematics. He was also one of the first mathematicians to conceive the idea that negative numbers have square roots, but lacked the conceptual framework for understanding these imaginary numbers.

Born in Pavia, Italy, on September 24, 1501, Cardano was the illegitimate son of Fazio Cardano and Chiara Micheri. The father was a successful lawyer and friend of Leonard da Vinci (1452-1519), but the fact that his parents did not marry until after his birth—and the father did not begin living with the family until Cardano was seven years old—provided a constant source of stigma for the young Cardano.

Cardano studied mathematics, astrology, and the classics at the University of Pavia, and went on to earn his doctorate in medicine there in 1526. He then began his medical practice in Saccolongo, a town near Padua, and in 1531 married Lucia Bandareni, with whom he had two sons and a daughter. Cardano supplemented his income by teaching mathematics at a school in Milan from 1534 to 1536, but soon his medical practice had become so successful that he was able to devote himself fully to it. His interest in mathematics continued for the rest of his life, however, and in his first mathematical work, *Practica arithmetice et mensurandi singularis* (1539), he proved himself adept at solving cubic equations.

His interest in the latter led Cardano into a fascinating entanglement with Nicolò Fontana, a.k.a. Tartaglia (1499-1557). Tartaglia had solved even more difficult cubic equations than those in *Practica arithmetice*, but refused to explain how he had done so, yet after much pressure from Cardano, he agreed to share his secret—provided that Cardano would not pass it on to anyone until Tartaglia had published it. Soon afterward, however, Cardano took as his servant a promis-

Girolamo Cardano. *(Library of Congress. Reproduced with permission.)*

ing young mathematician, Ludovico Ferrari (1522-1565), and shared Tartaglia's methods with him. Ferrari had learned how to solve a type of cubic equation called a "depressed cubic," which lacks a second-power term, and the two men discovered a method for reducing any generalized cubic equation to a depressed cubic. With the use of Tartaglia's methods, they could solve the equation.

They could not share this information with the mathematical community, however, because Tartaglia had yet to publish his method, and the prospects of him doing so any time soon appeared dim. It so happened, however, that in 1534 Cardano and Ferrari were examining the papers of the late Sciopione dal Ferro (1465-1526) when they discovered that he had solved the depressed cubic equation two decades earlier. Practically on his deathbed, dal Ferro had explained the secret to his student, Antonio Fior, who later foolishly challenged Tartaglia to a mathematical contest. As a result, Tartaglia had been able to discover the method that he later claimed as his own.

In 1545, Cardano—who reasoned that he was no longer under any obligation to Tartaglia—published his findings in his *Ars magna*. Tartaglia was incensed, and began a letter-writing campaign against Cardano, which coincided with a series of other misfortunes spanning nearly two decades in the latter's life: in 1546, his wife died; in 1560, his son Giambattista was executed for murdering his own wife; in 1565, Ferrari died of poisoning; and in 1570, Cardano was accused of heresy by the Inquisition. The charge, that he had claimed astrological causes and not divine intervention as the force behind events in the life of Jesus Christ, was enough to land him in jail. He was released upon agreeing to stop teaching, and only in 1573, when Pope Gregory XIII granted him a lifetime pension, did his world return to some semblance of normalcy.

Cardano had only a few years left, but even in the midst of the turmoil resulting from the heresy charge, he had revised his *Ars magna* by adding a section on what are now known as imaginary numbers. He did not recognize them as such, however, and regarded them as a mere mathematical novelty. He died in Rome on September 21, 1576.

JUDSON KNIGHT

Pietro Antonio Cataldi
1548-1626
Italian Mathematician

In the course of a career that spanned more than six decades, Pietro Antonio Cataldi made contributions in a number of areas. Not only was he one of the first mathematicians to work on continued fractions or infinite algorithms, providing definitions, common forms, and symbolism, but his research in algebra and perfect numbers, as well as his extensive writing and editing of texts, added greatly to the sum of mathematical knowledge. In his will, the unmarried Cataldi requested that a school of math and sciences be established in his home, a wish that went unfulfilled.

Cataldi, who was born in Bologna on April 15, 1548, studied at the Academy of Design in Florence before going to work as a math instructor at age 17. He continued in this position, lecturing in Italian rather than Latin (the accepted language of scholarly discourse at that time), until his early to mid-twenties. From 1570 or 1572 until 1584, he taught at the University of Perugia and the Perugia Academy; then he returned to Bologna at age 36, at which point he received his diploma. It was around this time that Cataldi—who would publish more than 30 books during his career—began his most important writing.

Practica aritmetica, published in four parts between 1606 and 1617, was Cataldi's first significant published work. In a generous if perhaps eccentric gesture, he arranged for Franciscan monks to distribute copies of the book to poor children. Perhaps his most important work was *Trattato del modo brevissimo di trovar la radice quadra delli numeri,* which though he seems to have completed it in 1597, did not see publication until 1613. In *Trattato,* Cataldi employed infinite series and unlimited continued fractions to find a number's square root. This was perhaps the first serious examination of continued fractions on record.

In the area of applied mathematics, Cataldi—both in the *Trattato* and in other works—addressed the topic of artillery range. He also edited a 1620 edition of the first six books in Euclid's (c. 325-c. 250 B.C.) *Elements*. In latter years, he attempted to establish a Bologna mathematics academy, with uncertain results. He died in his home town on February 11, 1626.

JUDSON KNIGHT

Bonaventura Cavalieri
1598-1647
Italian Mathematician

Praised by thinkers ranging from his friend Galileo (1564-1642) to twentieth-century writer Isaac Asimov, Bonaventura Cavalieri is best known for his work on the concept of indivisibles. This laid the foundation for the development of the infinitesimal, and with it calculus as conceived by Sir Isaac Newton (1642-1727).

Cavalieri's true first name is not known: Bonaventura was a religious name adopted when he joined a monastery at age 17. As for the circumstances of his birth, it is known only that he was born in Milan in 1598. In his mid-teens, Cavalieri joined the Jesuatis, an order under the Augustinian rule not to be confused with the Jesuits, and in 1615 took minor orders at a Milan monastery.

The following year found him at another monastery in Pisa, where he came under the influence of Benedetto Castelli, a former student of Galileo. Castelli inspired in his young friend an interest in geometry, and eventually introduced him to Galileo himself. Cavalieri became a devoted protege of the latter, and sent him more than a hundred letters over the course of the years that followed.

Cavalieri was still only 23 years old when he received ordination as a deacon under Cardinal Federigo Borromeo (1564-1631). Again, he had the good fortune of positive association: Borromeo encouraged Cavalieri's scholarship, and helped him obtain a position as teacher of theology at the monastery of San Girolamo in Milan. Cavalieri first began work on his method of indivisibles at San Girolamo, and continued this after receiving an appointment as prior of St. Peter's at Lodi.

Despite his youth, Cavalieri was soon struck down with an attack of gout, which occurred while on a visit back to Milan. As a result, he was confined to his bed for several months, and during this time wrote much of the book that would appear in 1635 as *Geometrica*.

Meanwhile, in 1628, Cavalieri took an interest in a teaching position at the University of Bologna—Europe's first university, established 470 years before. He sought and received a glowing letter of recommendation from Galileo, who informed the patron of the university that "few, if any, since Archimedes have delved as far and as deep into the science of geometry" as Cavalieri. Not only did Cavalieri receive the appointment, which he held until his death, but his order appointed him prior of Bologna's Church of Santa Maria della Mascarella. During the nearly two decades that remained for him, he published 11 books.

Drawing on ideas first explored by Archimedes (c. 287-212 B.C.), but barely considered in the intervening centuries, Cavalieri proposed that indivisibles could be used for the determination of area, volume, and center of gravity. Later, Evangelista Torricelli (1608-1647), who improved on the idea, wrote that "the geometry of the invisible was, indeed, the mathematical briar brush, the so-called royal road, and one that Cavalieri first opened and laid out for the public as a device of marvelous invention." It would later prove, as Asimov observed more than four centuries later, "a stepping-stone toward . . . the development of the calculus by Newton, which is the dividing line between classical and modern mathematics."

Cavalieri also put forth what came to be known as Cavalieri's Theorem. The latter states that for two solids of equal altitude, if sections made by planes parallel to and at equal distance from the bases always have a given ratio, then the volumes of the two solids will have the same ratio. Furthermore, Cavalieri developed a general proof of Guldin's theorem, which concerns the area of a surface and the volume of rotating solids. He died in Bologna on November 30, 1547.

JUDSON KNIGHT

Giovanni Ceva
1647?-1734
Italian Mathematician

As an engineer, Giovanni Ceva concerned himself with applied mathematics; but his other career, as a geometer, took him deep into the realm of pure math. He became the era's foremost authority on geometric problems, working particularly with transversals, and is also credited with Ceva's theorem, on a triangle's center of gravity. Other writings addressed areas of mathematical application ranging from mechanics to economics.

The son of a wealthy and influential family—his younger brother Tomasso (1648-1737), was destined to become a famous mathematician as well—Ceva studied at Pisa. Some of his most significant writing appeared during the late 1670s and early 1680s, when he was in his early to mid-thirties. Perhaps his most notable work was *De lineis rectis* (Concerning Straight Lines), published in 1678. This work contained Ceva's theorem, on the geometry of triangles, which in turn related to an area of interest throughout much of Ceva's career: center of gravity. His theorem found the center of gravity for an equilateral triangle at the place where lines drawn from the vertices to opposite sides intersected.

Ceva followed *De lineis* with *Opuscula mathematica* (A Short Mathematical Work) in 1682. Here he again returned to centers of gravity, this time expanding his studies to other shapes. The book also contained an infamous error on Ceva's part, his statement that the periods of oscillation pendulums are on the same ratio as their lengths.

It is easy to understand how he could have mistakenly deduced such a correlation, and in any case Ceva corrected his mistake in a later book, *Geometrica motus* (The Geometry of Motion, 1692). The latter concerned the geometry of motion, and contained elements that foreshadowed the infinitesimal calculus soon to be developed independently by Gottfried Wilhelm von Leibniz (1646-1716) and Sir Isaac Newton (1642-1727).

Ceva married in the 1670s and fathered a daughter in 1679. When he was a little more than 40 years old, in 1686, he went to work for the Duke of Mantua, a region where he would spend much of his career. In addition to holding a professorship at the local university, Ceva served the duke in a variety of official capacities. He continued his mathematical studies, however, and in 1711 published a study of economics, *De re numeraria* (Concerning Money Matters), in which he applied his knowledge.

Indeed, applied mathematics was at the center of Ceva's later career. As a hydraulic engineer in Mantua, he was responsible for a number of projects, but—in an act that suggests he may have been an early conservationist—he successfully opposed a scheme to divert the River Reno into the Po. As time went on, Ceva found that professional and family pressures kept him away from his mathematical studies, however. He died on February 3, 1737, and was buried at St. Teresa de Carmelitani Scalzi, a church in Mantua.

JUDSON KNIGHT

Nicolas Chuquet
1445-1488
French Mathematician

The origins of modern exponential notation—the 2 in x^2, for instance—can be traced to Nicolas Chuquet, a mathematician who at the dawn of the modern era struggled to find symbols corresponding to the ideas with which he grappled. He was one of the first to treat zero and negative integers as exponents, and appears also to have been a pioneer in his isolation of a negative number within an algebraic equation. Chuquet also approached the subject of logarithms, and even touched on imaginary numbers, concepts far beyond his time.

Chuquet was born in Paris in 1445, and from about 1480 worked in Lyon as a medical doctor and copyist or master of writing. In 1484, he published his principal work, *Triparty en la science des nombres*. At this time, arithmeticians lacked even the most basic notational symbols such as those for addition, subtraction, multiplication, and division. Chuquet became one of the first to offer symbols—though these bore little resemblance to the ones in use today—for what he called, respectively, *plus, moins, multiplier par,* and *partyr par.* He also provided notation for the previously inexpressible concept of square root: rather than $\sqrt{4}$, he would have written R)²4.

Triparty also contains a rule for average numbers: if $a, b, c,$ and d are all positive integers, then $(a + c)/(b + d)$ is greater than a/b, but less than c/d. The book also offered new names for

variables and exponents, names which failed to catch on because mathematicians already had accepted terms for these, such as the Latin *census* for the second power.

More significant was Chuquet's use of exponential notation. Lacking symbols for multiplication or variables, his version of $3x^2$, for instance, would have looked like this: .3.². He also accepted the use of 0 as an exponent, which always yields a result of 1; and of negative numbers, which when used as exponents yield a decimal fraction.

Equally interesting—though perhaps not as significant, since Chuquet was not in a position historically to carry it forward—was his work on what would become known as logarithms. He recognized that the sum of two indices and the product of their powers is the same. For example, $3^2 \times 3^3 = 3^{2+3}$; or, to put it another way, $9 \times 27 = 243 = 3^5$. Using this knowledge, Chuquet created a rudimentary logarithmic table for all the powers of 2 from 0 to 20; as a result, he discerned what would be shown in modern notation as $1 = \log_2 0, 2 = \log_2 1, 4 = \log_2 2$, and so on.

Finally, Chuquet was able to provide the solution to a problem that in modern terms would be written as $4x = -2$, perhaps the first notable use of a negative number in an algebraic equation. Chuquet even came close to the idea of an imaginary number—e.g., the square root of a negative integer—but failed to recognize how this could be of practical value. (The book *One Two Three... Infinity* by George Gamow [1904-1968] provides a good example of imaginary numbers applied to the solution of a practical problem.)

Chuquet died in Lyon in 1488. Today a street in the 17th arrondisement of Paris, rue Nicolas Chuquet, is named after him.

JUDSON KNIGHT

Scipione dal Ferro
1465-1526
Italian Mathematician

Scipione dal Ferro left behind no published writings, and were it not for papers found after his death, his role in unlocking one of the key mathematical challenges of his day might never be known. At the time, mathematicians were struggling with the solution to third-power equations, and the frustration associated with the quest had prompted Luca Pacioli (1445-1517) to suggest that such a solution was impossible. Unbeknownst to his colleagues, however, dal Ferro had found a way to solve such problems, but he kept this knowledge to himself.

The son of Floriano, a papermaker, and Filippa dal Ferro—the family name is sometimes rendered as Ferreo, Ferro, and del Ferro—dal Ferro was born in Bologna on February 6, 1465. He probably attended the University of Bologna, Europe's oldest institution of higher education; but other than this supposition, virtually nothing is known about his early life. It was at Bologna in 1496 that dal Ferro obtained what turned out to be lifetime employment as a lecturer in arithmetic and geometry. At some point he must have married, though there is no record of such, and fathered a daughter, Filippa, who grew up to marry one of his students, Hannibal Nave or Annibale dalla Nave.

Dal Ferro made his most important discovery some time between 1505 and 1515, but the fact that he was so secretive about it makes it difficult to pinpoint the date. Since the time of the Babylonians, mathematicians had known how to solve quadratic equations, ones in which (despite the somewhat deceptive name) the highest power is 2; but the solution to cubic equations, or those in which the highest order is 3—e.g., $ax^3 + bx^2 + cx + d = 0$—had eluded them.

After Pacioli, with whom dal Ferro was acquainted, made his statement in 1494, there followed many years of competition between the leading mathematicians of the day, each of them eager to find a solution to the cubic equation. Meanwhile, dal Ferro worked out the solution to what was called the depressed cubic, or one in which the second-power term was missing. Though this did not completely solve the larger problem, it would make such a solution possible—yet dal Ferro kept his discovery under his hat.

His reason for this may have been the then-common practice of public challenges, in which two prominent mathematicians engaged in a sort of intellectual boxing match. Often a mathematician's patronage, or his continued economic support by a wealthy benefactor, was at stake; and dal Ferro, who greatly feared the threat of a challenge, may have held out his secret as a sort of trump card to ensure victory.

But dal Ferro, who died on November 5, 1526, did not take his secret to the grave. He had written it in a notebook, which his son-in-law Hannibal kept; and he had shared it with his lackluster assistant, Antonio Fior. Fior was even more paranoid about a challenge than his former mentor, and he recklessly engaged the superior mind of Nicolò Fontana, a.k.a. Tartaglia (1499-1557). As a result of the competition, which he won, Tartaglia discovered the depressed cubic method and appropriated it as his own.

Only in 1543, when Girolamo Cardano (1501-1576) and his assistant Ludovico Ferrari (1522-1565) paid a visit to Hannibal and saw the notebook, did the truth come out. Tartaglia had deceived Cardano into believing that *he* was the original discoverer of the depressed cubic solution, but now Cardano published the new information. Unlike Tartaglia, Cardano was eager to give credit where credit was due, referring to dal Ferro's work as "a really beautiful and admirable accomplishment." Ironically, however, neither dal Ferro nor Tartaglia ultimately received credit; rather, the depressed cubic is known today as Cardano's formula.

JUDSON KNIGHT

Girard Desargues
1591-1661
French Mathematician

The significance of Girard Desargues's work did not become apparent until long after his death; indeed, at the time of his passing in 1661, it would have been hard to imagine that anyone would remember him a generation later. His ideas about geometry broke with Euclidean tradition, and thereby aroused the criticism of no less a figure than his acquaintance René Descartes (1596-1650); yet Desargues is today recognized as a prophet of projective geometry nearly two centuries ahead of his time.

Desargues's father Girard, who with his wife Jeanne Croppet had nine children, served as a tithe collector in Lyon. In an era when the church still dominated political affairs, this was the equivalent of a tax collector; and as in the time of Christ, it appears that collectors of the church's tax fared well: the Desargues family owned several houses, a chateau, and a vineyard. Little is known about the early life of the younger Girard, born on February 21, 1591, but it appears that by 1626, when he was 35, he had gone to work in Paris as an engineer.

His career was an intriguing one. At one point, Desargues proposed a scheme to pump waters from the Seine throughout Paris—an idea which, like his geometry, was ahead of its time. Later, as an architect during the 1640s and 1650s, he designed a number of houses in Paris, applying his concepts on perspective to the design of staircases. Employed by the French government under Cardinal Richelieu, he also assisted in the 1627-28 siege of the Huguenot stronghold at La Rochelle, where members of the Protestant sect were forced through starvation to capitulate. It appears that during the siege, he met Descartes, and by 1630 had become part of a Parisian intellectual circle that also included Marin Mersenne (1588-1648) and Etienne Pascal (1588-1651). In time, Desargues would exert a powerful influence on Pascal's son Blaise (1623-1662), who was destined to shine brighter than any of the group other than Descartes.

During the 1630s, Desargues published several works, one of which was particularly notable. *Traité de la section perspective* (Treatise on the Perspective Section, 1636), challenged existing ideas of perspective, and presented what came to be known as Desargues's theorem. According to the latter, two triangles may be placed such that the three lines joining corresponding vertices meet in a point, if and only if the three lines containing the three corresponding sides intersect in three collinear points. This theorem, which holds true in either two or three dimensions, would provide an early foundation for what came to be known as projective geometry.

In 1639, Desargues published *Brouillon project d'une atteinte aux événements des recontres d'une cône avec un plan* (Proposed Draft of an Attempt to Deal with the Events of the Meeting of a Cone with a Plane). As its title suggests, the essay dealt with conic sections, and presented what amounted to a unified theory of conics. As with Desargues's other groundbreaking ideas, however, those in *Brouillon* were slow to find acceptance, in part because of the highly arcane terminology the author used, substituting tree, stump, and other botanical terms for various geometrical constructs.

Desargues meanwhile came under attack from Descartes for what the latter perceived as an attack on geometry—which indeed Desargues's work was, though what was really threatened was the limited Euclidean worldview that then dominated. Jean Beaugrand (1595-1640) charged that Desargues had lifted his ideas from the ancient Greek mathematician Apollonius of Perga (262-190 B.C.), and the two exchanged attacks in print for many years. His ideas on perspective also won Desargues a number of detractors outside the world of mathematics, including artists, other architects, and even the stonecut-

ters' guilds, whose established method of performing their job was under challenge.

Desargues, who also pioneered a method of using cycloidal or epicycloidal teeth for gear wheels, died a nearly forgotten man in September 1661. Had it not been for the efforts of his friend, the engraver Abraham Bosse, his writings might have been lost. Only with the rediscovery of Desargues's ideas by Jean-Victor Poncelet (1788-1867) in the nineteenth century did the significance of his work for projective geometry become known.

JUDSON KNIGHT

Pierre de Fermat
1601-1665
French Mathematician

One of the most intriguing figures in the history of mathematics, Pierre de Fermat was the classic talented amateur. A lawyer and government official, he spent much of his time too busy with other affairs to devote any attention to his mathematical studies. But in the time that he did have, he conceived the principles of analytic geometry, independent of cofounder and acquaintance René Descartes (1596-1650); established number theory and, with friend Blaise Pascal (1623-1662), the theory of probability; laid down the fundamentals of differential calculus; and left behind a problem that bedeviled mathematicians for 325 years.

Fermat, who added the aristocratic "de" to his name in his early 1630s, was the son of Dominique Fermat, a successful leather merchant, and Claire de Long, who came from a highly respected family of lawyers. In 1631, Pierre Fermat married his fourth cousin, Louise de Long, with whom he had five children. By then he had studied at a number of institutions, including the universities of Toulouse, Bourdeaux, and Orleans. Having earned his degree in civil law from the latter, he began his law practice, and with the purchase of several key posts started a climb to the upper echelons of French jurisprudence. By 1648 he had received an appointment as king's councilor.

With no mathematical training, Fermat in the 1630s he became involved with the Paris mathematical circle of Marin Mersenne (1588-1648), Gilles Personne de Roberval (1602-1675), and Etienne Pascal (1588-1651), father of Blaise. He quickly earned a reputation as a brash upstart who in his first communication with the group claimed to have found fault with a statement of Galileo (1564-1642) regarding the path of a freely falling cannonball. As time went on, Roberval and Mersenne became increasingly irritated with Fermat, who had a habit of presenting them with incredibly difficult problems. In time they began to suspect that he did not know the solution to such problems himself, and started requesting that he provide full explanations regarding how such solutions were derived.

Fermat also ran afoul of Descartes, an almost inevitable result of the fact that both men discovered analytic geometry. Descartes actually made the discovery earlier, but Fermat, in his 1637 *Introduction to Plane and Solid Loci,* was first to present his findings to the Paris group. The latter continued to regard Fermat as an outsider, and his brash ways did little to win friends during the ensuing dispute with Descartes.

With his restoration of writings by the Greek mathematician Apollonius of Perga (262-190 B.C.), as well as his *Method for Determining Maxima and Minima and Tangents to Curve Lines* (1636), Fermat laid the groundwork for differential calculus. Yet from 1643 to 1654, he remained so occupied with professional and political concerns—and, thanks to an outbreak of the plague in 1651 that very nearly killed him, personal ones—that he had little communication with other mathematicians. During this time, however, he did manage to develop Fermat's Theorem for determining whether or not a number is prime: if n is any whole number and p any prime, then $n^p - n$ is divisible by p.

A 1654 letter from Pascal, in which the latter requested Fermat's help with a problem involving consecutive throws of a die, led to a series of communications in which the two men set down the elements of probability theory. Pascal took less interest in another growing area of interest for Fermat, number theory. Fermat was, however, able to correspond with Dutch physicist and astronomer Christiaan Huygens (1629-1695) about that subject until Huygens, too, concluded that number theory—now an important branch of mathematics—was useless.

Fermat had been weakened by the plague, and his latter years saw him in increasingly poorer health. He died on January 12, 1665, and was buried in the Chapel of St. Dominique in Castres. During the last decade of his life, he conducted experiments with optics, and made a discovery known as Fermat's Principle, which states that light travels by the path of least duration. It was

perhaps also during this time that, while reading a Latin translation of *Arithmetic* by Diophantus of Alexandria (3rd century A.D.), Fermat jotted down a theorem in the margin: for the equation $x^n + y^n = z^n$, where n is greater than 2, there are no positive integer solutions for $x, y,$ and z.

Though the theorem was apparently true, its proof remained elusive to mathematicians, who would grapple with the problem for more than three centuries. Only in 1994 did English mathematician Andrew Wiles (1953-), who had devoted much of his career to the quest, present a correct proof of what had long since become known as Fermat's Last Theorem. Given the complexity of Wiles's proof, some mathematicians have questioned whether Fermat himself was able to prove the theorem. As for Fermat himself, his only answer was a note in the space beside his theorem: "I have discovered a truly remarkable proof which this margin is too small to contain."

JUDSON KNIGHT

Albert Girard
1595-1632
French Mathematician

A mathematician who contributed to a number of areas ranging from arithmetic to algebra, Albert Girard enjoyed little recognition during his lifetime. A 1626 treatise on trigonometry by him contains the first use of the abbreviations *sin, cos,* and *tan,* and he was also perhaps the first mathematician to offer a formula for the area of a triangle inscribed on a sphere.

Girard was born in 1595 in St. Mihiel, France, but he was destined to spend most of his life away from his homeland. The reason for this was that as a member of the Reformed Church, he found himself ostracized in stridently Catholic France. He therefore moved to the Netherlands, where he studied under Willebrord Snell (1580-1626) at the University of Leiden, and later he went to work for Prince Frederick Henry of Nassau. Yet just as the French did not consider him fully one of their own because of his religion, he was not Dutch either, and he never succeeded in securing the patronage so necessary to the career of a mathematician in the seventeenth century. On the other hand, he was able to supplement his income in an unusual way, as a professional lute-player.

In his writings on geometry, Girard identified multiple types of quadrilaterals and pentagons, defined 69 of 70 types of hexagons, and became the first mathematician to state that the area of a spherical triangle is proportional to its spherical excess. He became the first to establish the definition $f_{n+2} = f_{n+1} + f_n$ for the Fibonacci sequence, and improved on the work of Rafaello Bombelli (1526-1573) for extracting the cube roots of binomials. In addition, Girard developed a simplified means for demarking the cube root still in use today.

Also a widely published translator, Girard was responsible for translating a number of works from French into Flemish, the language of Holland, and from Flemish into French. Many of these concerned military applications of mathematics, or specifically fortifications, and it is likely that he served in the Dutch army as an engineer—yet another way he managed to support himself. Poor, obscure, and just 37 years of age, Girard died in Leiden on December 6, 1632. Were it not for an explanation of his circumstances included in an edition of works by Simon Stevin (1548-1620) which he translated, historians would know almost nothing about his life.

JUDSON KNIGHT

James Gregory
1638-1675
Scottish Mathematician

James Gregory published papers on a number of mathematical and scientific subjects, and a look at his unpublished papers suggests an even more wide-ranging talent. Among his most notable contributions were his laying of the groundwork for the development of calculus, as well as his experiments in optics, which greatly influenced the later work of Isaac Newton (1642-1727). Yet Gregory had the misfortune to be caught up in political struggles that pitted his new ideas against a stodgy and powerful academic establishment, and this greatly limited the influence and perhaps even the length of his career.

Gregory was born in Drumoak, Scotland, the son of John, a minister, and Janet Anderson Gregory. Because her son was sickly, Janet Gregory proceeded to teach him at home, including in the curriculum subjects—most notably, geometry—barely known to most men at the time, let alone most women. From 1651 to 1662, Gregory studied in Aberdeen, first at grammar school, then at the city's Marischal College. He then traveled to London, where he published his *Optica promota* (1663). The latter suggested that telescopes should use concave mir-

James Gregory. *(Library of Congress. Reproduced with permission.)*

rors, an idea upon which Newton later drew without giving Gregory credit.

Frustrated in his efforts to secure a position in London, Gregory in 1664 travelled to Italy, where he spent four years in study and research. His *Vera circuli et hyperbolae quadratura* (1667) discussed the means of finding the area for a circle or a hyperbola, and *Geometriae pars universalis* the following year examined convergent and divergent series. The latter work also contained Gregory's foundational ideas for what became calculus.

By 1668, Gregory was back in London, where he won election to the Royal Society and an appointment as chairman of mathematics at St. Andrew's College in Scotland. It seemed that at 30, he had found lasting career success, but this was not to be: at St. Andrew's, Gregory was confronted with an extremely conservative college governing board, who greeted all his efforts to update the curriculum with hostility.

He was not the only one dissatisfied with the current state of academic affairs: in an incident more like something from the twentieth than the seventeenth century, a group of students revolted against the university establishment, demanding change. Gregory was away in London at the time, hoping to secure support for his plan to establish the first public observatory in Britain at St. Andrew's; nonetheless, the administration found him a convenient scapegoat for the uprising, and punished him by withholding his salary.

When Edinburgh University offered him a position as chairman of its mathematics department, it seemed that once again Gregory's fortunes had taken a turn for the better. But within a year he was dead at age 37, having suffered a massive stroke. His papers lay dormant for many years, and only began receiving attention after some of them were published in 1939. Based on the many subjects covered in his unpublished writings, it seems clear that if Gregory had been allowed greater intellectual freedom, he might well have emerged as one of the preeminent mathematical thinkers of his time. Ironically, St. Andrew's—now St. Andrew's University—today operates one of the World Wide Web's best sites on the history of mathematics (http://www-groups.dcs.st-andrews.ac.uk/%7Ehistory).

JUDSON KNIGHT

Thomas Harriot
1560?-1621
English Mathematician

Thomas Harriot invented the signs for "greater than" (>) and "less than" (<) in use today, and was one of the first mathematicians to use a number of now-commonplace symbols. Much of his work involved astronomy, navigation, and geometry: an employee and associate of Sir Walter Raleigh (1554-1618), he was at the center of English efforts to conquer the seas and the New World.

Harriot was educated at St. Mary's Hall, Oxford University, from whence he received his B.A. in 1580. For a time, he appears to have worked as a mathematics tutor in London before securing employment with Raleigh in 1584. The famous gentleman-explorer needed someone to teach navigation to his sailors, and for this purpose Harriot composed a manuscript—long since lost—called the *Articon*.

During the following year, Raleigh sent Harriot with a group of colonists to Roanoke Island off the coast of what is now North Carolina. (These were not the inhabitants of the famous "Lost Colony": the first settlement lasted only 10 months before being disbanded, and the doomed Lost Colony settlers arrived in 1587.) Working with artist John White (d. 1593?), Harriot was responsible for studying the indigenous peoples, as well as the local vegetation, animal life, and other

natural resources. He published *A Briefe and True Report,* an account of his findings, in 1588.

During the three decades that followed, Harriot's patrons—first Raleigh and then Henry Percy, Earl of Northumberland (1564-1632), for whom he went to work in 1595—ran afoul of the English royal house. Harriot himself, though occasionally caught up in the turmoil even to the extent of being accused of atheism in 1603, in general passed the time unscathed, and continued his scientific observations. He studied the parabolic path of projectiles; determined the specific weights of materials; calculated the areas of spherical triangles, and thus confirmed that the Mercator projection preserves angles; independently discovered the sine law of refraction associated with Willebrord Snell (1580-1626); built telescopes; and in 1607, long before the birth of Edmund Halley (1656-1742), observed what came to be known as Halley's Comet.

As an astronomer, Harriot also made a map of the Moon (1609), calculated the orbits of Jupiter's moons (1610-12), studied sunspots and the Sun's rotation speed (1610-13), and observed another comet (1618). In his writings as a mathematician, not only did he become the first to use > and <, he was one of the first to adopt the plus sign and minus sign, lowercase letters for variables, and the equal sign of Robert Recorde (1510-1558). He was also among the first to write an equation with the sum of all terms equal to zero, as is common today.

For a decade, from the mid-1580s to the mid-1590s, Harriot had lived on an Irish property granted him by Raleigh. From 1595, however, he had resided at Northumberland's estate at Syon, and continued there even after Northumberland was arrested by King James in 1605. By the time his patron was released from the Tower of London in 1622, Harriot was dead, having succumbed to a cancer of the nostril on July 2, 1621. (Raleigh was dead, too, executed by James in 1618.) Harriot left behind a vast array of papers, many of them published as *Artis analyticae praxis,* a significant algebra text, in 1631.

JUDSON KNIGHT

Gottfried Wilhelm von Leibniz
1646-1716
German Mathematician and Philosopher

His invention of differential and integral calculus, which he developed independent of

Sir Isaac Newton (1642-1727), was but the most visible of philosopher and mathematician Gottfried Wilhelm von Leibniz's contributions. In his writings, he commented on areas ranging from physics to religion, and developed a philosophical system that placed him on a level with the two other most prominent Continental rationalists, René Descartes (1596-1650) and Benedict Spinoza (1632-1677). Many of his ideas were ahead of their time; yet he was ridiculed by Voltaire (1694-1778) and others, and at his death, his passing hardly attracted any notice.

Born in Leipzig, Saxony, on July 1, 1646, Leibniz was the son of Friedrich and Catherina Schmuck Leibnütz, whose son later altered his last name and added the aristocratic "von." Catherina was the third wife of Friedrich, a lawyer and professor who had one other child with her, as well as two from previous marriages. The father died when Leibniz was six, and already by that point the boy had shown himself an intellectual prodigy. At the age of 12, he could read Latin, and had already begun conceiving such grandiose ideas as the composition of a universal encyclopedia.

Leibniz entered the University of Leipzig when he was 15 years old, and earned his bachelor's degree by the age of 17. By 1664, the 18-year-old had received a master's degree and written a dissertation for his doctorate in law, which the university refused to award because of his youth. He then moved to the University of Altdorf in Nuremberg, which agreed to award him his doctorate at the grand old age of 20.

After holding several jobs, Leibniz went to work for the Elector of Mainz, a powerful nobleman of the Holy Roman Empire. Leibniz would produce vast quantities of writings on a variety of subjects while in the Elector's employ, and still managed to write more than 15,000 letters, many of them to other leading intellectuals of the time.

Tensions between the German states and France led to his appointment to a diplomatic mission in Paris in 1672. There Leibniz met the astronomer Christiaan Huygens (1629-1695), destined to become a lifelong friend, as well as the mathematicians Descartes and Blaise Pascal (1623-1662). In the following year he traveled to London, where he became acquainted with such luminaries as the chemist Robert Boyle (1627-1691) and the microscopist Robert Hooke (1635-1703). The elector died around this time, and by 1676 Leibniz was en route from Paris to Hanover, where he would serve the Duke of Brunswick for the rest of his life. On his way, however, he stopped in Holland, where he met with Spinoza, microscopists Jan Swammerdam (1637-1680) and Anton van Leeuwenhoek (1632-1723), and others.

Leibniz introduced his differential calculus in a 1684 paper, and unveiled integral calculus two years later. In 1689, Newton published his own work, and eventually it became apparent that the two men had developed more or less the same method from quite different approaches—geometric in Newton's case, algebraic in Leibniz's. There would follow a heated debate as to whose calculus was better, and even the two great men became drawn into the partisan squabbling. In time Leibniz's methods and notation, which were simply more efficient, prevailed; but for a century, English mathematicians fell behind their Continental counterparts due to a nationalistic insistence on Newtonian methods.

In addition to calculus, Leibniz was the first mathematician to use the integral sign. He introduced the term "function"; put forth a theory of special curves; developed a general theory of tangents; and made a number of other contributions to mathematics. His calculating machine, which could perform arithmetic functions, was a precursor to the computer, and won him election to the Royal Society in 1673. In his philosophical writings, he posited the existence of "monads," or thinking entities that occupied no space, and whose every part comprised its totality. The concept seemed preposterous until the late 20th century, when it became apparent that the structure of human brain cells, as well as that of holographic images, resembles Leibniz's monads.

But Leibniz was also cursed by his statement, again in his philosophical writings, that the present life is "the best of all possible worlds"—an idea for which Voltaire, perhaps misunderstanding Leibniz's point, skewered him in *Candide,* whose Dr. Pangloss was based on Leibniz. Though he enjoyed some honors during his lifetime, including his election as first president of the Berlin Academy when it was founded in 1700, as well as election to the Académie Royale des Sciences in Paris, Leibniz was largely forgotten by the time of his death on November 14, 1716. He was buried in an unmarked grave, and neither the Royal Society nor even the Berlin Academy saw fit to publish an obituary.

JUDSON KNIGHT

Marin Mersenne
1588-1648
French Mathematician

As a mathematician and scientist, Marin Mersenne was far from the equal of his more well-known friends and acquaintances, including Galileo (1564-1642), René Descartes (1596-1650), Pierre de Fermat (1601-1665), Blaise Pascal (1623-1662), and Christiaan Huygens (1629-1695). Yet without Mersenne, the world would know far less about these giants: in a time when there were no scientific journals, Mersenne served as a disseminator of knowledge. An avid correspondent, he also conducted weekly scientific discussions in Paris that became the basis for the French Académie Royale des Sciences. In addition, he inspired investigations into number theory and prime numbers, as well as Huygens's invention of the pendulum clock.

Mersenne was born on September 8, 1588, near the town of Oize in France, and at the age of 23 became a Catholic friar. While in school, he met Descartes, destined to become a lifelong friend. He also became acquainted with Galileo, who he defended against attacks from the church. Eventually Mersenne would be responsible for a "meeting of the minds" between his two distinguished friends. He passed on to Descartes a question from Galileo regarding the path of falling objects on a rotating Earth, and this led to Descartes's suggestion of the logarithmic spiral as the likely path.

Again and again, Mersenne performed this role of bringing great minds together, and both with his correspondence and his weekly meetings, he began creating what a modern person might call a vast database of knowledge. At the meetings, figures such as Descartes, Fermat, and Pascal assembled under one roof to share ideas and argue for opposing theories. Mersenne also had an opportunity to read manuscripts by Descartes and others, in some cases long before they were published.

Though Mersenne wrote widely on mathematical and scientific subjects, he would always be more well-known as a disseminator of works by other thinkers—including ancient Greeks such as Euclid (c. 325-c. 250 B.C.), Archimedes (c. 287-212 B.C.), Apollonius of Perga (262-190 B.C.), and others, whose works he edited. Of his own writings, the most significant was *Cogitata physico-mathematica* (1644), in which he put forth several theorems regarding prime num-

bers. The theorems turned out to be incorrect, but inspired later investigations into number theories and large primes—sometimes called Mersenne primes.

Mersenne was also the one who suggested that Huygens use a pendulum as a timekeeping device, and this led to the latter's introduction of the pendulum clock in 1656. By then, however, Mersenne was long gone, having died in Paris on September 1, 1648, at the age of 60.

JUDSON KNIGHT

John Napier
1550-1617
Scottish Mathematician

In 1914, on the brink of World War I, the Royal Society of Edinburgh took time to commemorate the 300th anniversary of *Mirifici logarithmorum canonis descriptio,* in which John Napier first presented his system of logarithms. Fifty years later, on the verge of the computer revolution, Napier University of Edinburgh was named in honor of Scotland's great mathematician—a man who, as a pioneer of logarithms and an inventor of an early calculator, helped make that revolution possible.

Members of the Scottish nobility, both Napier's father, Sir Archibald Napier, and his mother, Janet Bothwell, came from old and highly distinguished families. At the age of 13, Napier entered the University of St. Andrews, but did not complete his education; rather, he traveled in Europe for a time before returning to Scotland at age 21. In the following year, he married Elizabeth Stirling, with whom he had two children. Elizabeth died in 1579, and Napier married Agnes Chisholm, with whom he fathered 10 more children. His father died in 1608, at which point Napier inherited Merchiston Castle and the title eighth laird of Merchiston; hence his later nickname, "the Marvelous Merchiston."

Much of Napier's most important mathematical work, including the *Descriptio* and its companion the *Constructio* (1619), dates from the final decade of his life, the 1610s, when he was in his sixties. In these and other writings, he became one of the very first mathematicians to use the decimal point. Among the concepts he introduced were "Napier's analogies," formulae for spherical triangles; "Napier's rules of circular parts," tables for the right spherical triangle; and of course logarithms.

He arrived at the latter not through pure mathematics, but through his interests as an astronomer, which forced him to make large and detailed calculations. Eventually Napier hit upon the idea of using a logarithm, or the power to which a given number must be raised to yield a given product, as a means of simplifying such computations. The sum of two indices and the product of their powers is the same: thus $2^2 \times 2^3 = 2^{2+3}$; or, to put it another way, $4 \times 8 = 32 = 2^5$. By putting together logarithmic tables, as he did with the help of mathematician Henry Briggs (1561-1630), Napier provided a means of relatively easy computation that remained in use until the advent of the calculator and computer in the twentieth century.

Napier also invented several rudimentary calculators, the most famous of which was "Napier's bones," a set of rods marked with numbers which he introduced in *Rabdologia* (1617). So named either because they were often made of bone or ivory, or because of the bone-like shape of the rods, the bones were but the most famous of many practical (and some impractical) creations from "the Marvelous Merchiston." In his role as landowner, Napier experimented with fertilizers, invented a hydraulic screw, and developed a revolving axle for pumping water. He also designed an early submarine and tank, as well as a mirror for using the Sun to set enemy ships on fire, and claimed to have created a prototype for what would later become known as the machine gun.

Napier was not without his eccentricities: an ardent Protestant, he wrote a highly popular tract identifying the pope as the Antichrist. Detractors, no doubt awed by his wide-ranging genius, claimed that he practiced black magic, and kept a black rooster as a spiritual familiar.

The great mathematician and inventor was only 67 years old when he succumbed to gout, no doubt exacerbated by overwork, on April 4, 1617. His burial place is unknown, though it may be the church of St. Cuthbert's Parish in Edinburgh.

JUDSON KNIGHT

Nicholas of Cusa
1401-1464
German Mathematician and Philosopher

Nicholas of Cusa is a figure difficult to assess within the context of mathematics. Certainly he wrote extensively about the subject, in particular on the properties of circles, but it is pri-

marily as a philosopher and mystic that he is remembered. On the one hand, he held firmly to mindsets associated with the medieval world, in particular with his belief that all knowledge has its roots in theology. On the other hand, he displayed an openness to new ideas more characteristic of the Renaissance and the modern age that lay beyond it.

Born Nicholas Krebs in the German town of Kues in 1401, Cusa studied law and mathematics at the University of Padua in Italy. He later received his doctorate in canon law before moving to Cologne in the 1420s. There he took an interest in the philosophical writings of the ancients, particularly Plato (427-347 B.C.), and began forming the foundations of his own mystical philosophical system.

In 1431, a year after he entered the priesthood, Nicholas took part in the Council of Basel, convened in an attempt to shore up the church against the rising tides of dissent that would culminate in the Reformation. Six years later, he took part in a failed mission of reconciliation between the Western and Eastern churches, travelling to Constantinople—which, unbeknownst to anyone at that time, would fall permanently into Muslim hands in less than two decades' time.

Despite the mission's lack of success, Nicholas won recognition for his diplomatic work, and was ultimately granted the position of cardinal. He gained even greater favor with the papacy when an old friend, Italian humanist Enea Silvio Piccolomini, assumed St. Peter's throne as Pius II in 1458. Two years later, Nicholas settled permanently in Rome.

By then he was just six years away from the end of his life, and had long since formed his rather idiosyncratic philosophical system. Though he had an interest in mathematics far beyond that of a typical medieval, mathematical knowledge in Nicholas's mind served to increase the mystery in the world rather than to unravel it. Certainly he was not the first to see the discipline in those terms: Pythagoras (c. 580-c. 500 B.C.), who he greatly admired, was perhaps the most notable of all mathematical mystics.

Among Cusa's mathematical interests were the ideas of infinity and of squaring the circle—that is, mapping the area of a circle onto an equally large square, using only a ruler. In his mind, these concepts were linked, because an infinitely large circle would be the same as a square—and, as he noted, would have neither a center, a radius, nor a diameter.

Nicholas of Cusa. *(Corbis Corporation. Reproduced with permission.)*

In the realm of astronomy, Nicholas anticipated Copernicus (1473-1543) by many years in saying that the center of the universe was the Sun rather than the Earth. He also suggested that many stars in the universe had their own worlds revolving around them; however, even these scientific statements were heavily laced with Nicholas's mysticism, and his belief that the center of the universe is God. He died on August 11, 1464, in the town of Todi—then part of the Papal States—near Rome.

JUDSON KNIGHT

William Oughtred
1574-1660
English Mathematician

An Anglican priest who tutored students in mathematics, William Oughtred invented both the linear and the circular slide rules. He also introduced the symbol x for multiplication, and :: for proportion.

Oughtred was born in Eton on March 5, 1574. His father, Benjamin, taught writing at Eton School, which Oughtred attended as a king's scholar. At age 15, he entered King's College at Cambridge, where he earned his B.A. degree seven years later, as well as an M.A. in 1600. Though mathematics was not his area of study, it was during this time that Oughtred developed a keen interest in the subject such that even after he was ordained as a priest, he tutored math students without receiving any pay. His salary as a clergyman, he said, was sufficient to cover his needs.

Following his ordination in 1603, Oughtred received an assignment to serve as vicar of Shalford in 1604. Six years later, he was appointed rector of Albury, where he would remain until his death 50 years later. During the 1620s, he began tutoring, and in 1628 the Earl of Arundel offered to become Oughtred's patron in return for his teaching the earl's son, Lord William Howard.

Oughtred's *Clavis mathematicae,* published in 1631, was probably intended as a textbook for young Lord William, but with its summation of all significant arithmetic and algebraic knowledge up to that time, it proved to have much wider appeal. The author used the book to introduce a number of proposed mathematical symbols, but most were cumbersome and failed to catch on; yet x and :: became permanent fixtures of the landscape.

Ultimately Oughtred, encouraged by the warm reception *Clavis mathematicae* received, would publish numerous other books. In *The Circles of Proportion and the Horizontal Instrument* (1632), he discussed the idea of a slide rule as an instrument to aid in navigation. Though one of his students later claimed that *he* had invented it, Oughtred is generally credited with both the circular and the linear slide rule, which he may have invented as early as 1621. The slide rule would remain an important tool of computation for more than 350 years, until the advent of handheld calculators.

In the English Civil War (1642-60), Oughtred supported the Crown against Oliver Cromwell's radical Puritans, and with the triumph of Cromwell was forced to keep a low profile. By that point he was already up in years, and he died on June 30, 1660, not long after King Charles II had been restored to the throne.

JUDSON KNIGHT

Blaise Pascal
1623-1662
French Mathematician and Philosopher

A mathematical prodigy who first made a name for himself at age 16, Blaise Pascal had a meteoric career that concluded before he was 40. His efforts were further curtailed by his growing interest in a religious sect during the latter half of his life. Yet during his brief years of fruitful work, he helped develop the foundations of projective geometry with Girard Desargues (1591-1661); established probability theory with Pierre de Fermat (1601-1665); made possible new forms of calculus; and created a number of inventions, including the syringe, the hydraulic press, and the world's first mechanical calculator.

Today the French town of Clermont, or Clermont-Ferrand as it is now known, is famous as the birthplace of three things: Michelin tires, the Crusades (Pope Urban II preached the sermon beginning the First Crusade here in 1095), and Blaise Pascal. Son of mathematician and civil servant Etienne Pascal (1588-1651) and his wife Antoinette Bégon, Pascal came from a tightly knit family. His mother died when he was three, and this only drew him closer to sisters Gilberte and Jacqueline, as well as his father.

When Pascal was eight, the family moved to Paris, and there he began to excel as a student of mathematics and ancient languages. He and his father also became associated with the discussion group that centered around Marin Mersenne (1588-1648), and included such luminaries as René Descartes (1596-1650) and Fermat. The family moved to Rouen in 1639, but Pascal and his father continued to visit Paris, and in the following year the 16-year-old boy presented a pamphlet that impressed no less a figure than Descartes.

The title was *Essai sur les coniques,* and Pascal's purpose in writing it was to clarify ideas Desargues had presented, using highly complex and confusing terminology, in a 1639 publication. As he continued work on it, however, the youth went far beyond Desargues's original point, developing a theorem concerning what came to be known as "Pascal's mystic hexagram." According to the theorem, from which he deduced some 400 corollary propositions, the three points of intersection of the pairs of opposite sides of a hexagon inscribed in a conic are collinear. Along with Desargues's ideas, these helped form the basis for projective geometry.

Despite health problems in the 1640s, Pascal developed his calculator, which used cogged wheels to perform its computations, and in 1649 received from the French crown a monopoly for its manufacture. In fact production of the calculator turned out to be prohibitively expensive in that preindustrial era, but the mechanical calculators that did eventually appear—and

Blaise Pascal. *(Library of Congress. Reproduced with permission.)*

which proliferated in the era before electronic devices—were modeled on Pascal's design.

From the age of 23, Pascal became increasingly involved with the Jansenists, a sect that believed in predestination and divine grace as the sole means of salvation. (Though they called themselves Catholics, the Jansenists most closely resembled Calvinists, and in fact the sect was later declared heretical by the Vatican.) Despite his growing preoccupation with spiritual matters, Pascal during this period conducted a number of experiments to measure atmospheric and barometric pressure, and used the information he gathered to invent the syringe and the hydraulic press.

In 1647, the family returned to Paris, and the father died three years later. Pascal in 1654 had a riding accident that nearly took his life, and as a result decided to join his sister Jacqueline at the Jansenist convent of Port-Royal. Thereafter his scientific work tapered off, while his writings in religious philosophy increased; however, just before this happened, an exchange of letters with Fermat regarding a game of dice led to the development of probability theory.

Also during the late 1650s, Pascal resumed his interest in geometry. Among the figures that attracted his attention were the arithmetic triangle—his work ultimately influenced the general binomial theorem later put forward by Sir Isaac

Newton (1642-1727)—and the cycloid. The latter is a curve traced by the motion of a fixed point on the circumference of a circle rolling along a straight line. Though mathematicians had been investigating cycloids for many years, Pascal was able to solve most of the remaining problems involving them in just eight short days.

As it turned out, time was running short for Pascal, who had always been sickly. In 1662, he devoted himself to designing a public transportation system of carriages for Paris, but before the system became operational, he died on August 19 of a malignant stomach ulcer at his sister Gilberte's home.

JUDSON KNIGHT

Bartholomeo Pitiscus
1561-1613
German Mathematician

Bartholomeo Pitiscus is known primarily for his coining of the term "trigonometry," which appeared in the title of his *Trigonometria: sive de solutione triangulorum tractatus brevis et perspicuus* (1595). The latter, which used all six trigonometric functions, became a highly respected mathematical text, and was translated into several languages.

Pitiscus was born on August 24, 1561, in Grünberg, Silesia, which is now the town of Zielona Góra, Poland. At Zerbst and later Heidelberg, he studied theology under Calvinist teachers, and throughout his life remained a committed proponent of Calvinism, an early Protestant sect that placed an emphasis on predestination and the work ethic.

In 1584, Pitiscus received an appointment to tutor 10-year-old Friedrich der Aufichtige, or Frederick IV, elector Palatine of the Rhine. The boy was destined to hold an important position within the nobility of the Holy Roman Empire; thus for Pitiscus, who later became his court chaplain, the position was a secure one.

Trigonometria, first published in 1595, consisted of three sections. The first addressed plane and spherical geometry, while the second contained tables for the six trigonometric functions. (Pitiscus carried these to five or even six decimal places.) The third section consisted of assorted problems in areas ranging from astronomy to geodesy. Revised editions followed in 1600, 1609, and 1612, along with translations into English in 1614 and French in 1619.

Pitiscus remained a staunch defender of Calvinism, and his influence expanded when Frederick took power after his uncle, John Casimir, died in 1592. Frederick ruled until his death in 1610, and Pitiscus followed his patron by three years, dying on July 2, 1613, in Heidelberg.

JUDSON KNIGHT

Robert Recorde
1510-1558
Welsh-English Mathematician

Robert Recorde introduced the "equals" symbol (=) to mathematical notation, and greatly advanced mathematical education in the British Isles. Not only was he the first to write on arithmetic, geometry, and astronomy in English rather than Latin, he introduced the study of algebra to England. Unfortunately, political intrigue cut short his career.

Born in 1510 in Tenby, Wales, Recorde was the son of Thomas and Rose Johns Recorde. His paternal great-grandparents had been English, and he spent his career in England, beginning with his studies at Oxford. He earned his B.A. from the latter in 1531, and after a stint at All Soul's College, moved to Cambridge, where he received his M.D. degree in 1545. Soon afterward, he gained a prestigious appointment at the court of King Edward VI. Recorde later married, fathering nine children.

Undoubtedly, Recorde's position at court, though it would eventually pose a liability that proved fatal, in the short run provided him with the means and the freedom to embark on his career as a mathematical writer. The first of his known publications was *Grounde of Arts* (1541), which in addition to its scholarly overview of contemporary mathematics provided practical knowledge concerning commercial math. *The Pathway to Knowledge* (1551) was a translation of the first four volumes in Euclid's (c. 325-c. 250 B.C.) *Elements,* and is the only one of Recorde's books not written in the form of a dialogue between master and student.

In *The Whetstone of Witte* (1557), Recorde presented the equals sign, using what he considered a hallmark of equality: two parallel lines of equal length. Other mathematical works included *Castle of Knowledge* (1556), on the properties of spheres, and *The Gate of Knowledge,* a text on measurement and the quadrant which has been lost. He also wrote an early urological treatise, *The Urinal of Physick* (1547).

Most of Recorde's books appeared in verse form, to make memorization easier, and provided students with detailed knowledge regarding how solutions were derived. His writing was highly readable by the standards of his time, and thus made his work popular in England if not on the Continent, where there were few readers of English.

In his political career, Recorde did not prove as successful as he had been in the world of mathematics. While serving as comptroller of the Bristol Mint in 1549, he came into conflict with Sir William Herbert, later Earl of Pembroke, and this led to his ostracism from court and his imprisonment for 60 days. Nor did an appointment as surveyor of mines in Ireland serve to recover his good standing. He again found himself at loggerheads with Pembroke, who he charged with malfeasance, and when the mines proved unprofitable, Recorde was removed from his position.

By then it was 1553, the same year Queen Mary I (Bloody Mary) took the throne following the death of her half-brother Edward. Three years later, when Recorde tried to gain reinstatement at court, Pembroke responded to the malfeasance charge by suing him for libel. Mary and her husband, King Philip of Spain, sided with Pembroke, and Recorde was sent to King's Bench Prison, where he was executed in 1558.

JUDSON KNIGHT

Michel Rolle
1652-1719
French Mathematician

The name of Michel Rolle is primarily associated with Rolle's Theorem, which concerns the position of roots in an equation. He also developed the modern expression for the term nth root of x, and presented what he called his "cascade" method for separating the roots in an algebraic equation. Rolle, whose most famous work was *Traité d'algèbre* (1690), is also remembered for his opposition to techniques pioneered by René Descartes (1596-1650).

The son of a shopkeeper, Rolle was born in the French town of Ambert on April 21, 1652. His origins prevented him from obtaining a formal education, and instead he went to work in his teens as a scribe. By the age of 24 he was in Paris, earning his living as a secretary and an accountant. He married and had children, and was

sufficiently successful in his profession that from the late 1670s (when he was in his late twenties) onward, he was able to devote considerable time to his avocation of mathematics.

A prominent self-taught mathematician of the time—and no doubt something of a model for Rolle—was Jacques Ozanam (1640-1717). Recreational mathematics, such as puzzles and tricks, were popular among educated Frenchmen of that era, and Ozanam was a master, putting forth problems such as the following: find four numbers such that the difference between any two is a perfect square, and is also the sum of the first three. It was Rolle's solution to this problem in 1682 that first brought him to the attention of Ozanam and to the larger mathematical public. The publication of his solution in *Journal des scavans,* the leading scientific journal, led to his earning an honorary pension, as well as his appointment to tutor the son of an influential government official.

In 1690, Rolle published *Traité d'algèbre.* The work contained what was then the novel use of notation for the nth root of a number, as well as Rolle's method of cascades. The latter used principles first put forth by Dutch mathematician Johann van Waveren Hudde (1628-1704) for finding the highest common factor of a polynomial, a method Rolle used to separate the roots of an algebraic equation. When other mathematicians complained that the book contained inadequate proofs, Rolle published *Demonstration d'une methode pour resoudre les egalitez de tous les degrez* (1691). This book included Rolle's Theorem, a special case of the mean-value theorem in calculus.

Beginning in 1691, Rolle began speaking out against errors in Cartesian methodology—first with regard to Descartes's ordering of negative integers on the same path as positives, such that -2 was smaller than -5. In 1699, the same year he was awarded a geometry pension by the Académie Royale des Sciences, he published a significant paper on indeterminate equations. He also weighed in against the validity of infinitesimal analysis, and though he was eventually forced to accept the discipline, his critique helped its supporters work out difficulties in their methodology.

Rolle suffered a stroke in 1708, and was never as strong thereafter. A second attack in 1719 proved too much, and he succumbed on November 8 of that year.

JUDSON KNIGHT

Christoff Rudolff
1499?-1545?
Polish-Austrian Mathematician

Christoff Rudolff wrote *Coss* (1525), the first book of algebra to appear in German. This was significant in that German was the vernacular in much of northern and central Europe, and few among the rising bourgeois class could read Latin. As for the title, it referred to the word *cosa* or "thing," which was used to refer to anything unknown or indeterminate; since algebra dealt with such perplexities, it was known as the cossic art.

Rudolff was born, probably in 1499, in the village of Jauer in Silesia, which is now Jawor, Poland. In fact the area had been culturally Polish for centuries prior to his birth, but it had been dominated by Bohemians for some time, and would fall into Austrian hands in 1526. By that time Rudolff, who was probably brought up speaking German, had long since graduated from the University of Vienna.

Following his education at the university (1517-21), where he studied algebra, Rudolff continued living in the Austrian capital, where at the age of 26 he produced *Coss*. The book consisted of two parts, the first covering a number of topics—such as square and cube roots—necessary to the study of algebra in the second half. The latter was in turn divided into three sections, respectively covering first- and second-degree equations, rules for solving equations, and a series of algebraic problems.

No doubt in part because of its author's youth and his shocking use of German rather than Latin, the book attracted the opprobrium of other mathematicians, who claimed that Rudolff had lifted many of his problems and examples from existing works in the university library. On the other hand, German mathematician Michael Stifel (1487-1567) defended Rudolff's work, and even wrote a preface to a second edition.

Rudolff published *Künstliche Rechnung mit der Ziffer und mit der Zahlpfennigen* (1526), which addressed questions of computing and offered problems applicable to the rising commercial and industrial culture of Renaissance Europe. He followed this in 1530 with *Exempelbüchlin,* which contained nearly 300 more problems. Rudolff died in Vienna in 1545.

JUDSON KNIGHT

Takakazu Seki Kowa
1642-1708
Japanese Mathematician

Takakazu Seki Kowa was one of the few notable Japanese mathematicians prior to the late modern era. At a time when his country was shut off from most of the world and controlled by military dictators called shoguns, Seki Kowa—himself a samurai under the shogunate—sparked Japanese scholars' interest in mathematics. In the course of his career, he developed his own system of notation, as well as an early form of calculus.

The son of a samurai named Nagaakira Utiyama, Seki was born in the town of Fujioka in March 1642. It is not clear whether his father died, or if the father himself sent the boy to live with the family of accountant Seki Gorozayemon; but in any case this occurred when the boy was very small, and he took on the name of his adoptive father. From an early age, Seki Kowa showed himself a mathematical prodigy, earning the nickname "divine child" because of his remarkable abilities.

At some point Seki Kowa married, though he never had any children, and went to work as an examiner of accounts for the Lord of Koshu. The latter eventually became shogun, and thus Seki Kowa—a member of the samurai class by birth—was named a shogunate samurai. His work as an accountant, a trade he must have learned from his adoptive father, naturally dovetailed with his interests as a mathematician, and during his late twenties, he began to teach and write on the subject.

Seki Kowa's one notable publication was *Hatubi sanpo,* which appeared in 1674. The book was written in response to the announcement of some 15 supposedly unsolvable problems that had been put forth four years before; in *Hatubi sanpo,* Seki Kowa solved them all. However, it was not the Japanese custom to show how one had arrived at one's solutions, and it appears that even Seki Kowa's students remained unaware of his methodology.

Much of what the Japanese knew about mathematics had been derived (as had many other aspects of their civilization) from older Chinese models. Chinese math at the time made it possible to solve equations with a single variable, but Seki Kowa took this several steps further. In the course of his work, he implemented notation he had created to express these unknown quantities.

His work was all the more remarkable in light of the fact that Japanese mathematicians were unaware of algebra. It was Seki Kowa's achievement, however, to provide a number of advancements without the benefit of input from other scholars, a situation quite unlike that of his counterparts in Europe. Like Sir Isaac Newton (1642-1727), who was born the same year as he, he developed a method for approximating the root of a numerical equation, and he created his own table of determinants at about the same time this idea made its debut in Europe. In addition, he calculated the value of π to some 20 places.

In his latter years, Seki Kowa was granted the title of master of ceremonies for the shogun's household, an esteemed position. He died on October 24, 1708, in the village of Edo, which is now Tokyo.

JUDSON KNIGHT

Michael Stifel
1487-1567
German Mathematician

Known for his advancement of mathematics in general, and of German mathematical education in particular, Michael Stifel was a fervent Lutheran given to sometimes bizarre numerological theories. As a mathematician, he was one of the first to use plus and minus signs; developed a system of logarithms independent of John Napier (1550-1617); and helped make algebra more comprehensible to Germans by writing about it in their own language.

Born in the German town of Esslingen in 1487, Stifel became a monk in his twenties, and in 1511 was ordained to the priesthood. Almost immediately, however, he found himself disillusioned with aspects of Catholicism, in particular the church's habit of paying its officials out of alms collected from the poor. Similar issues had begun to enrage another Catholic monk, Martin Luther, who in Wittenberg in 1517 would post his 95 theses challenging the established teachings of Catholicism.

By 1520, Stifel's interests in numbers and his growing antipathy toward Catholicism had led him to a complicated numerological interpretation of the prophetic biblical books of Daniel and Revelation. The upshot was that by analyzing figures such as Revelation's seven-headed beast with 10 horns, he concluded that Pope Leo X was the Antichrist. Within two years, he was forced to leave the monastery.

The next 13 years of Stifel's life were a hodgepodge, a crazy-quilt pattern in which he traveled to one town, secured a pastorate, got into trouble (sometimes through his own fault, sometimes not), fled, and repeated the cycle. Only through the intervention of influential friends such as Luther, who he befriended in 1523, and the religious reformer Philip Melanchthon, was he able to keep obtaining new jobs in new towns.

Luther helped Stifel become pastor for the Count of Mansfield (1523-24), a job he soon lost due to anti-Lutheran sentiment. Then he went to work for a powerful noblewoman, Dorothea Jorger, before leaving again to join Luther. The latter helped him find both a pastorate and a wife in Lochau (now Annaberg) in 1528, but by 1533 Stifel had been forced out of the town for preaching that—according to more numerological calculations—the world would end on October 18 of that year.

Luther and Melanchthon helped him secure a position at Holzdorf in 1535, and for a time Stifel's life became more settled. During the 12 years that followed, he earned his master's degree from the University of Wittenberg, tutored a number of students, and wrote all of his known mathematical works. In 1544, he published *Arithmetica integra,* a summation of mathematical knowledge up to that point. He followed this a year later with *Deustche arithmetica,* and in 1546 with *Welsche Practick.* In his works, Stifel broke with the cossists, as algebraists in Germany at that time were called, by presenting a general method for solving equations to replace the cossists' 24 rules.

Stifel's time at Holzdorf was brought to an end by the religious Schmalkaldic War of 1547. He fled the town, and by 1551 had secured a parish at Haberstroh in Prussia. Eventually, however, more conflicts forced him to give up the ministry altogether, and in 1559 he went to work as a lecturer in arithmetic and geometry at the University of Jena. There he died on April 19, 1567.

JUDSON KNIGHT

Tartaglia
1499-1557
Italian Mathematician

Tartaglia, whose given name was Nicolò Fontana, is remembered for a number of achievements in applied mathematics, as well as for his translations of Euclid (c. 325-c. 250 B.C.) and Archimedes (c. 287-212 B.C.) His most memorable achievement, however, was his work in algebra leading to a generalized solution of cubic equations. The latter placed him at the center of a heated conflict involving fellow mathematicians Girolamo Cardano (1501-1576) and Cardano's assistant Ludovico Ferrari (1522-1565).

Born in Brescia, Italy, in 1499, Nicolò Fontana was the son of a humble postal courier who died when he was seven. The family was rendered destitute by the father's death, and as if this were not enough, the French army attacked the town five years later. A French soldier disfigured young Nicolò's face, cutting his mouth so badly it was difficult for him to talk thereafter. Therefore he acquired the nickname Tartaglia, drawn from the Italian word *tartagliare,* "to stammer." Rather than be ashamed, however, Fontana took on the epithet as his name.

Tartaglia was almost entirely a self-made man, and by sheer force of will taught himself enough that by about the age of 18 he had obtained a position as a teacher of practical mathematics in Verona. He remained in Verona for more than 16 years, during which time he rose to the position of headmaster and may have married and had children. In 1534, he moved to Venice, where he would spend virtually the remainder of his life.

During the following year, Tartaglia inadvertently became involved in the cubic-equation problem. It so happened that Scipione dal Ferro (1465-1526) had developed a method for solving a cubic equation lacking a second-power term—a so-called "depressed cubic"—and before his death had shared it with his assistant, Antonio Fior. The latter was a man of no great genius; moreover, Fior quaked with fright at the prospect of a public challenge from a competing mathematician (a not-uncommon occupational hazard for mathematical scholars of the day), which would reveal his ignorance for all to see. Therefore he took what he thought was a preemptive strike, and challenged Tartaglia to a contest of wits.

Whereas Tartaglia presented his opponent with 30 problems involving a variety of mathematical topics, Fior hit Tartaglia with 30 problems that required the use of the depressed cubic for their solution. Thus Tartaglia was forced to figure out the depressed cubic ace public disgrace and probable loss of his position—

so he did, soundly defeating Fior, who could only answer a few of his questions.

Later Cardano begged Tartaglia to teach him his secret, and Tartaglia agreed on condition that Cardano would not reveal it to anyone else until Tartaglia published it himself. Of course Tartaglia had no intention of publishing information that gave him an advantage over other mathematicians, and in time Cardano revealed what he knew to his assistant Ferrari. As a result, the two men developed a method for reducing any generalized cubic equation to a depressed cubic—and thus they unlocked the secret of solving cubic equations.

Later, when they discovered that dal Ferro had developed the depressed cubic first, Cardano no longer considered himself under obligation to Tartaglia, and published his findings, giving credit both to dal Ferro and Tartaglia. Tartaglia was furious, and began conducting a fierce letter-writing campaign against Cardano. Ferrari defended his mentor, and the conflict came to a head in a public debate in Milan in 1548. It so happened that Ferrari was from Milan, and thus possessing the "home-field advantage," he and his supporters forced Tartaglia to back down.

Though he is known primarily for his involvement in the fascinating cubic-equation imbroglio, Tartaglia also contributed to areas including ballistics and surveying. His translations of Euclid and Archimedes into Italian marked the first time that many works by these ancients had appeared in a modern language. Despite his many achievements, however, he lived most of his life in poverty, and died poor in Venice on December 13, 1557.

JUDSON KNIGHT

François Viète
1540-1603
French Mathematician

In his *In artem analyticum isagoge* (Introduction to the Analytical Arts," 1591), François Viète established the letter notation still used in algebra: vowels for unknown quantities or variables, consonants for known quantities or parameters. He wrote a number of other mathematical texts; promoted the use of trigonometry for solving cubic equations; introduced a number of terms, including "coefficient"; and contributed to a variety of other mathematical areas.

The son of Etienne and Marguerite Dupont Viète was born in the French town of Fontenay-la-Comte in 1540. By the age of 16, he was studying law at the University of Poitiers, and following his graduation obtained a position as a lawyer in Fontenay. His contacts with royalty and nobility, which would characterize his career, began early: even at this point, Viète could count among his clients Queen Eleanor of Austria and Mary Stuart of Scotland.

From 1564 to 1570, Viète worked for the Soubise family in La Rochelle, eventually leaving his law practice to serve first as private sectary and later as tutor to one of the aristocratic family's daughters. It was probably during this period that he married his first wife, Barbe Cotherau. (After her death, he married Juliette Leclerc; he also had one child.) Also while working with the Soubise family, Viète embraced Protestantism as a member of the Huguenot sect—an extremely risky step in France at that time.

Viète served the French court, and worked in a variety of official capacities, from 1570 to 1584. During this time, he published his first significant mathematical treatise, *Canon mathematicus seu ad triangula cum appendibus* ("Mathematical Laws Applied to Triangles," 1579.) The text promoted trigonometry, then an underutilized discipline, and made use of all six trigonometric functions.

For five years beginning in 1584, Viète found himself out of favor with King Henry III for his Huguenot sympathies. Given the level of hatred and tension over this issue at the time, it is amazing that he suffered no worse punishment from the Catholic monarchy—and that Henry IV reinstated him on assuming the throne in 1589. Perhaps just to stay on the safe side, Viète rejoined the Catholic Church in 1594.

Several important publications occurred during the years following his reinstatement, beginning with *Isagoge,* considered by some scholars to be the first algebra textbook of the modern era. In 1593, Viète published *Supplementum geometriae,* which addressed topics such as the trisection of an angle; the doubling of a cube (a problem that had bedeviled many ancient Greek mathematicians); and the first known explicit statement of π as an infinite product.

With *De numerosa* in 1600, Viète presented a method for approximating roots of numerical equations. He retired two years later, and died in Paris on December 13, 1603. The posthumous *De aquationem recognitione et emedatione libri duo* ("Concerning the Recognition and Emendation of Equations," 1615) offered methodology for solv-

ing second, third, and fourth degree equations, and contained the first use of the term coefficient.

JUDSON KNIGHT

John Wallis
1616-1703
English Mathematician

John Wallis coined the mathematical use of the word "interpolation," and was the first to use the infinity symbol (∞). He introduced a number of other terms and varieties of notation, and made the first efforts at writing a comprehensive history of British mathematics. A founding member of the Royal Society, Wallis was also involved in a number of other scientific endeavors: for instance, he was the first hearing person to develop a means of teaching deaf-mutes.

Wallis was born on November 23, 1616, to John, a rector, and Joanna Chapman Wallis in the Kentish town of Ashford. His father died when he was six, and at age nine an outbreak of the plague forced Wallis to leave Ashford. He attended several boarding schools, and while on Christmas break one year during the equivalent of grammar school, he first displayed his prodigious mathematical talents. After asking his brother to teach him arithmetic, he mastered the subject in two weeks, and soon proved himself able to perform extremely difficult computations—for instance, calculating the square root of a 53-digit number—in his head.

During his studies in medicine at Emmanuel College, Cambridge, Wallis wrote one of the first papers on the circulatory system. He graduated in 1637, and earned a fellowship to Queen's College. By 1640 he had been ordained in the Church of England, and five years later married Susanna Glyde, with whom he had three children. The English Civil War (1642-60) had broken out in the meantime, and Wallis was recruited to Oliver Cromwell's cause, using his intellectual skills as a cryptographer. For his help in deciphering Royalist communiqués, he was granted the Savilian professorship in geometry at Oxford, which he would hold for the rest of his life.

Arithmetica infinitorum, published in 1655, was Wallis's first important work, and was destined to become a standard information source for mathematicians during the coming decades. He followed this with a treatise on conics in 1658, and produced a number of other writings on a wide variety of mathematical subjects. His *Algebra: History and Practice* (1685) not only discussed the

John Wallis. *(Library of Congress. Reproduced with permission.)*

whole history of the discipline, but marked the first recorded attempt at graphically displaying the complex roots of a real quadratic equation. A second edition contained the first systematic use of algebraic formulae, including numerical ratios.

In addition to his many accomplishments—including a celebrated public disagreement with noted philosopher Thomas Hobbes (1588-1679)—Wallis seems to have been adept at steering the ever-changing political tides of seventeenth century England. Thus despite his serving the Puritan cause, he went to work as chaplain for King Charles II following the restoration of the monarchy. After the Glorious Revolution of 1688, a bloodless change of governments, the newly installed William III put Wallis to work at a familiar task, deciphering enemy communications. Wallis died on October 28, 1703, in Oxford.

JUDSON KNIGHT

Biographical Mentions

Adriaen Anthoniszoon
c. 1543-1643

Dutch mathematician who made the most accurate calculation of π up to his time. In 1600, Anthonis-

zoon and his son calculated the number to seven decimal places, or 3.1415929. Today mathematicians know π to be an irrational number, or an infinite decimal, and have corrected Anthoniszoon's seventh decimal place. Thus it is more accurately represented as 3.14159265358979323846 2644....

Florimond de Beaune
1601-1652

French mathematician and astronomer. Beaune studied law at Paris, but his status and wealth enabled him to build an extensive library and observatory to indulge his scientific interests. He became friendly with many leading scientific figures, and is best remembered for writing an important Latin summary of René Descartes' geometric ideas. He also published several papers on algebra and left a number of unpublished writings in mathematics, mechanics, and optics that have unfortunately been lost.

Bernard Frenicle de Bessy
1605-1675

French mathematician. Bessy held an official position at the Court of Monnais, and was an amateur mathematician with a particular interest in number theory. He corresponded with many mathematicians, such as René Descartes, Pierre de Fermat, Christiaan Huygens, and Marin Mersenne. He solved many of the problems posed by Fermat, and introduced new ideas and further problems. He also did early work on magic squares, publishing *Des quassez ou tables magiques*. He was elected to the Académie Royale des Sciences in 1666.

Giovanni Alfonso Borelli
1608-1679

Italian mathematician and physiologist who developed a three-forces theory to explain the elliptical orbits of Jupiter's satellites (1666), which influenced Isaac Newton's work on universal gravitation. Borelli's observations of comets helped undermine the Aristotelian concept of the heavens by showing their absolute distance from Earth changed and that they were above the Moon. He also correctly explained muscular action and bone movements in terms of levers, but incorrectly attempted to extend this analysis to the internal organs.

Henry Briggs
1561-1630

English mathematician best known for his work in refining and popularizing John Napier's (1550-1617) logarithms. Briggs's wide-ranging talents led to him teaching medicine and mathematics. He was a keen astronomer, and became interested in logarithms as a means of making astronomical calculations easier. His joint work with Napier on the development of logarithms resulted in the form we still use today. He became the first professor of geometry at Gresham College, later going on to teach at Oxford.

Edward Cocker
c. 1631-1675

English mathematician and engraver. A skillful and exuberant calligrapher, Cocker also taught writing and arithmetic. He wrote a number of works that bore his name, such as *Cocker's Urania* and *Cocker's Morals*. His most famous work was *Cocker's Arithmetic*, which ran to more than 100 editions over a period of 100 years. However, some sources suggest that Cocker's editor and publisher forged the book. The phrase "according to Cocker," meaning absolutely correct, was common usage for many years.

John Collins
1625?-1683

English mathematician who wrote about sundials, cartography, accounting, navigational trigonometry, and the quadrant. An apprentice bookseller in Oxford, Collins became a clerk at Court and studied mathematics. To avoid the English Civil War he went to sea, where he continued his studies. Collins had a number of jobs, from mathematics teacher to accountant. He corresponded with many mathematicians, including Isaac Newton and Gottfried Leibniz, and helped publish important mathematical texts. Collins also became a member and librarian of the Royal Society.

Johan de Witt
1625-1672

Dutch mathematician and statesman who applied his mathematical talents to the financial problems of Holland during his career as grand pensionary, by arguing probabilistically that life annuities were offered at too high a rate of interest in comparison with fixed annuities. De Witt applied the concept of expectation to form equal contracts developed in 1657 by Christiaan Huygens in his *De ratiociniis in aleae ludo* (On Calculation in Games of Chance), which was important in the development of probability theory. In his work in pure mathematics de Witt gave one of the first systematic treatments of the analytic geometry of the straight line and conic in the Cartesian algebraic tradition.

John Dee
1527-1608

English alchemist, geographer and mathematician. Dee was educated at Cambridge, and also studied widely throughout Europe, returning to England with new astronomical instruments. Dee became astrologer to Queen Mary, but was imprisoned for sorcery. He later found favor with Queen Elizabeth, casting horoscopes for her. He wrote on a variety of topics, such as astronomy, astrology, alchemy, navigation, music, calendar reform, and geography. Many English explorers, particularly those in search of the fabled Northwest Passage, consulted Dee.

Albrecht Dürer
1471-1528

German painter, printmaker, and engraver who is considered the foremost artist of the Renaissance. In attempting to represent nature accurately in his work, he tried to develop mathematical formulations for ideal beauty, including the human body. He studied space, perspective, and proportion, and used both arithmetic and geometric methods in composing his subject matter on the canvas. His work, published in 1528, after his death, has had a major influence on subsequent developments in art.

Lodovico Ferrari
1522-1565

Italian mathematician who came to early prominence as assistant to Girolamo Cardano (1501-1576). Ferrari played a key role in the imbroglio over the solution to cubic equations, which pitted his mentor against Tartaglia (1499-1557). With Cardano, Ferrari discovered that Scipione dal Ferro (1465-1526), and not Tartaglia, had been the first to find the solution to the depressed cubic, itself a principal element in solving cubic equations. Later, Ferrari brought an end to the heated disputes between the two older mathematicians by soundly defeating Tartaglia in a 1548 public competition in Milan. As a result of his victory, Ferrari received a series of prominent appointments, but was poisoned in 1565, possibly by his own sister.

Antonio Fior

Assistant to Scipione dal Ferro (1465-1526), who unwittingly assisted Tartaglia (1499-1557) in learning his master's solution to depressed cubic equations. Before his death, dal Ferro had shown Fior his method for solving the depressed cubic, a cubic equation lacking a second-power term. Up to that point, dal Ferro alone had known the solution, a powerful trump card if he were forced into one of the public mathematical competitions common at the time. As for Fior, he so feared a public challenge that he preemptively approached Tartaglia, giving him 30 problems involving the depressed cubic. As a result, Tartaglia was forced to find the solution, which he did. Meanwhile Fior, who could only answer a handful of the questions put to him by Tartaglia, suffered a humiliating defeat.

John Graunt
1620-1674

English statistician generally credited as the founder of scientific demography. A founding member of the Royal Society, Graunt began studying London death records dating back to 1532. He noticed a number of patterns, which he discussed in *Natural and Political Observations... made upon the Bills of Mortality* (1662), classifying death rates according to cause, and identifying overpopulated conditions as a mortality-increasing factor. He also developed one of the earliest life-expectancy charts, which was based on his studies of survivorship. Graunt's ideas had a profound effect on the demographic efforts of Sir William Petty (1623-1687), and on the mathematical studies of Sir Edmund Halley (1656-1742).

Gregory of St. Vincent
1584-1667

Italian mathematician who, in his efforts at "squaring the circle," developed the rudiments of the method now known as integration. A Jesuit, Gregory taught in Rome, Prague, and Spain, where he served as tutor in the court of King Philip IV. He also helped write the curriculum for a Jesuit mathematical school in Antwerp, where he taught from 1617 to 1621. Gregory became intrigued by the idea of constructing a square equal in area to a circle, using as his only tools a straight edge and compass—an operation mathematicians now know to be impossible. In the course of his efforts, however, Gregory discovered the expansion for $\log(1+x)$ for ascending powers of x, and integrated x^{-1} as a geometric form equivalent to the natural logarithmic function.

Paul (Habakkuk) Guldin
1577-1643

Swiss/Italian goldsmith and mathematician. Of Jewish descent, Guldin was trained as a goldsmith and worked in several German towns until his conversion to Catholicism at the age of twenty. He joined the Jesuit Order and changed his name to Paul. He studied mathematics and later taught at Jesuit Colleges in Rome and Graz. His most important published work was a four-volume treatise on various mathematical topics, including solid geometry, mechanics, and centers of gravity, including that of the Earth.

Edmund Gunter
1581-1626

English mathematician and astronomer. Gunter received a divinity degree from Oxford, and became a Rector in Southwark. In addition he became professor of astronomy at Gresham College, London. He published some navigation works, tables of logarithms, sines, and tangents, and coined the terms cosine and cotangent. Gunter made a number of measuring instruments that bore his name: Gunter's scale, Gunter's chain, Gunter's line, and Gunter's quadrant. He was the first to observe the secular variation of the magnetic compass.

Guillaume-François-Antoine de l'Hospital
1661-1704

French mathematician who was instrumental in introducing calculus into France. His *Analyse des infiniment petitis* (1696) was the first textbook on differential calculus and dominated eighteenth-century thinking on the subject. Jean Bernoulli taught him calculus and agreed to turn his mathematical discoveries over to l'Hospital for a salary. Consequently, Bernoulli's result on indeterminate forms is known as L'Hospital's rule. The pedagogical qualities of l'Hospital's *Traité analytique des sections contiques* (1707) made it the standard eighteenth-century analytic geometry text.

Francesco Maurolico
1494-1575

Italian mathematician also known as Marol Marul. Ordained as a priest, Maurolico served as head of a mint, supervised fortifications at Messina, was appointed to write a history of Sicily, and wrote about Sicilian fish. He also wrote a number of works on Greek mathematics, translating many ancient writings, and restoring some damaged texts. He worked on geometry, the theory of numbers, optics, conics, and mechanics. Maurolico published a method for measuring the Earth, and made detailed studies of the 1572 supernova.

Pietro Mengoli
1625?-1686

Italian mathematician. Mengoli studied at Bologna, receiving doctorates in philosophy and civil and canon law. He was professor of arithmetic, then professor of mechanics, and finally professor of mathematics at Bologna, as well as becoming a parish priest. Mengoli studied infinite series, especially the harmonic series. He wrote on the theory of limits and infinite series, and found an infinite product expansion for pi divided by two. He also published works on astronomy, refraction in the atmosphere, and music theory.

Georg Mohr
1640-1697

Danish mathematician. Initially taught mathematics by his parents, Mohr later studied in Holland, France, and England. He fought in the Dutch-French wars, and was briefly a prisoner of war. Mohr was little known in mathematics until the rediscovery of his lost book *Euclides danicus* (1672) in a bookstore in 1928. The book contains the theorem and its proof that all Euclidean constructions can be carried out with compasses alone, a result not found again until 125 years later.

William Neile
1637-1670

English mathematician who studied mathematics at Oxford, and then law at the Middle Temple in London. He went on to become a member of Charles II's privy council. Neile was the first person to find the arc length of an algebraic curve. He studied the theory of motion, and made many astronomical observations from the roof of his father's house. Neile was one of the first members of the British Royal Society.

Isaac Newton
1642-1727

English physicist and mathematician who invented differential and integral calculus. He used this tool in developing the concept of universal gravitation and his three laws of motion, which appeared in 1687 in *Philosophiae Naturalis Principia Mathematica*—considered by many the greatest work of science ever written. Newton also demonstrated that light is composed of colors and invented a reflecting telescope. He discovered the general binomial theorem in 1665, as well as working on analysis, algebra, number theory, analytic geometry, and probability.

Pedro Nunes
1502-1578

Portuguese mathematician, astronomer, and cosmographer whose work helped maintain the Portuguese Empire. As mathematics professor, Nunes pioneered mathematical education for low-status navigators and ship pilots, while also serving as tutor for the brothers of King John III. From his astronomical research, Nunes invented the nonius, enabling mariners to measure fractions of degrees in finding their latitude. Cosmographically, Nunes' distinction between rhumb lines and great circles predates Gerardus Mercator's more famous incorporation of this distinction into map projections.

Luca Pacioli
c. 1445-1517

Italian mathematician also known as Lucus Paciuolo. A Franciscan friar and something of a traveler, Pacioli became professor of mathematics at Perugia, Rome, Naples, Pisa, and Venice. He produced a number of large volumes of material on many topics. While these books contained little original material, they offered masterful summaries of the mathematics known at the time. His work provided a basis for later developments to build on. He published an important edition of Euclid's work and co-wrote a book with Leonardo da Vinci (1452-1519).

Francisco Pellos
fl. 1450-1500

Italian mathematician who anticipated the use of the decimal point. In 1492, Pellos published *Compendio de lo abaco,* a book on commercial arithmetic that used the decimal dot (.) to indicate division by 10.

Jean Pena
?-1558

French mathematician who provided a unique theory to the celestial rather than conventional terrestrial origin of comets. In a period of critical rethinking on the legitimacy of the four elements and placement of atmospheric and celestial phenomena, Pena, royal mathematician at Paris, followed other unconventional thinkers in these matters, accepting the logic of Copernicanism and the rejection of wholesale Aristotelianism. In his treatise on geometrical optics of lens and mirrors (*Euclidis Optica et Catoptrica*, 1557), Pena noted that a comet's tail pointed away from the Sun, prompting him to theorize that comet's were made of some celestially transparent substance that refracts light and causes combustion and thus the tail.

William Petty
1623-1687

English demographer and economist who was the first economic theorist to make an enduring attempt to base economic policy upon statistical data. He called this line of reasoning "political arithmetic." Because this was a time when reliable census or sampling data were not yet available, he relied on indirect indices of population, such as the number of chimneys. Petty's talents were wide-ranging. For instance, he designed and built twin-hulled ships, a predecessor of the modern catamaran.

Georg Peurbach
1423-1461

Austrian mathematician and astronomer who studied at the University of Vienna, then traveled through Europe lecturing on astronomy in Germany, France, and Italy. He became the court astronomer of Hungary in 1454, and professor of astronomy at Vienna. He published detailed observations of the 1456 visit of what was later named Halley's comet. He also wrote on the computation of sines and chords, and produced tables of eclipse calculations. He taught Regiomontanus (1436-1476), and they collaborated on a number of texts.

(Johann Müller) Regiomontanus
1436-1476

German mathematician and astronomer who introduced Arabic algebraic and trigonometric methods to Europe, thus providing a systematic basis for their further development. In *De triangulis omnimodis* (1533) he developed the earliest statement of the cosine law for spherical triangles, and *Tabulae directionum* (1475) contains a valuable table of tangents. Regiomontanus also played a key role in reforming astronomical studies in fifteenth-century Europe by emphasizing and acting on the need for new and improved observations.

Hudalrichus Regius
fl. 1530s

Mathematician who showed that not all numbers of the form 2^n-1 for all primes n are themselves prime. Up to Regius's time, mathematicians had assumed that this was so, since $2^2-1=3$, $2^3-1=7$, $2^5-1=31$, and so on. However, Regius showed that the result of $2^{11}-1$ is not a prime number: its result, 2047, is equal to 23 multiplied by 89.

Rheticus (Georg Joachim von Lauchen)
1514-1574

German mathematician whose *Narratio prima* (1540) was the first published account of Nicolaus Copernicus's work. A disciple of Copernicus, Rheticus was instrumental in convincing him to publish *De revolutionibus* (1542). Rheti-

cus also was first to relate trigonometric functions to angles rather than arcs of circles; prepared the best trigonometric tables of his time, which contained values for sines, tangents, secants, and their complementary functions to ten decimal places; and wrote a biography of Copernicus which has since been lost.

Michelangelo Ricci
1619-1682

Italian mathematician whose work represents an early example of induction. A cardinal, Ricci was known in his lifetime for his many letters on mathematical topics, and for his correspondence with mathematicians and scientists including Evangelista Torricelli (1608-1647). In later years, however, his fame rested on a 19-page pamphlet called *Exercitatio geometrica, De maximis et minimis* (1666), in which he found the maximum of $x^m(a-x)^n$ and the tangents to $y^m=kx^n$. Among the other objects of Ricci's study were spirals and generalized cycloids.

Gilles Personne de Roberval
1602-1675

French mathematician who became a professor of mathematics at the Collège Royale in Paris, despite his peasant background. He traveled widely through France teaching mathematics and meeting many important mathematicians. He developed new methods of integration, and did foundational work on kinematic geometry. A founding member of the Académie Royal des Sciences, he invented the Roberval balance, which is still used today. He also studied cartography and did experimental work on vacuums.

Wilhelm Schickard
1592-1635

German astronomer and mathematician. Educated at the University of Tübingen, Schickard studied theology and oriental languages, and became a Lutheran minister in 1613. He was appointed professor of Hebrew at the University of Tübingen, and then in 1631 changed subjects to become professor of astronomy. He studied astronomy, mathematics, and surveying, and invented many calculating machines. He made significant advances in mapmaking, and corresponded with many European scientists. Schickard was also renowned as an engraver both in wood and in copperplate.

Frans van Schooten
1615-1660

Dutch mathematician who translated and published René Descartes' *Géométrie* into Latin and trained a large number of students to give an algebraic treatment of geometry in the Cartesian style, including Jan de Witt and Christiaan Huygens. Van Schooten disseminated the work of these students as appendices to his own publications, most famously Christiaan Huygens' *De ratiociniis in aleae ludo* (On Calculation in Games of Chance), which was important in the development of probability theory.

Willebrord Snell
1580-1626

Dutch mathematician and physicist who discovered the law of refraction of light rays, which states that the ratio sin i : sin r is a constant, dependent on the medium, where i stands for the angle of the incident ray and r for the angle of the refracted ray. Snell's most sustained work was his determination of the length of the meridian, for which he improved the method of triangulation of Gemma Frisius and measured the distance between Alkmaar and Bergen-op-Zoom (around 80 miles or 129 km) and later extended the network of triangles all the way to Mechelen, Belgium.

Simon Stevin
1549-1620

Dutch mathematician and military engineer who founded the science of hydrostatics by showing that the pressure exerted by a liquid upon a given surface depends on the height of the liquid and the area of the surface. While a quartermaster in the Dutch army, Stevin invented a way of flooding the polders in the path of an invading army by opening selected gates in the dike. He advised the Prince Maurice of Nassau on building fortifications for the war against Spain. Stevin in 1590 showed that Aristotelian physics was mistaken by showing that two lead balls of unequal weight hit the ground simultaneously when dropped from the tower of Delft. For a long time credit for this demonstration was given to Galileo.

Evangelista Torricelli
1608-1647

Italian mathematician and physicist best known for inventing the mercury barometer (1644). Torricelli made significant contributions to the development of calculus—a subject he possibly would have invented if he had lived longer. Using infinitesimal methods he produced the first modern rectification of a curve (1645), independently discovered the quadrature and center of gravity of the cycloid, and produced what is perhaps the first graph of a logarithmic function (1647).

Ehrenfried Walther von Tschirnhaus
1651-1708

German mathematician, physicist, and philosopher who traveled extensively between the various intellectual centers in his time, and became a veritable clearinghouse of new ideas in science and technology. In his book *Medicina Mentis* ("Mental Medicine," 1687), Tschirnhaus laid out a method of discovering rational truths as a basis of a happy life derived from Cartesianism, Spinoza, the English empiricists, and Leibniz, whom he greatly admired. Only true knowledge can tame the emotions, he believed, which are the sources of error and therefore of unhappiness. Later in life he rediscovered how to make hardpaste porcelain.

Cuthbert Tunstall
1474-1559

English mathematician. The Bishop of London, and later of Durham, Tunstall was a key Catholic figure in the troubled years of the English Reformation. An outstanding classical scholar, he studied theology and law at Oxford, Cambridge, and Padua. Tunstall wrote *De arte supputandi libri quattuor* (1522), one of the first printed works published in England devoted exclusively to mathematics. It contained no original material, but was a very practical arithmetic text. He also composed many religious works.

Johannes Widman
1462-1498

German mathematician who studied at the University of Leipzig, and became a teacher there, lecturing on arithmetic and algebra. He was the first to teach the subject of algebra in Germany. Widman is best remembered for an early German arithmetic book in 1489 that has the first use of plus and minus signs. It was very successful, containing a wider range of examples than previous texts; and it remained in print until 1526. Little is known of Widman's life after 1489.

John Wilkins
1614-1672

English mathematician and inventor who studied at Oxford and became an Anglican clergyman. During the English Civil War he sided with parliament, which helped him obtain a Wardenship at Oxford. He married Robina, sister of Oliver Cromwell (1599-1658). He promoted group discussion among scientists, and was a founder and first secretary of the Royal Society. Wilkins published works on astronomy, mechanical devices, linguistics, and codes and ciphers. He developed a new plough, a transparent beehive, an improved carriage, and many other machines.

Bibliography of Primary Sources

Barrow, Isaac. *Euclidis elementorum libri XV* (1654). A translation of writings by Euclid, this book became highly popular and eventually appeared in a pocket-sized edition.

Bernoulli, Jakob. *Ars conjectandi* (The art of conjecture, 1713). Published posthumously, this book is considered one of the foundational texts on probability.

Bouelles, Charles de. *Geometricae introductionis* (1503). Here Bouelles addressed the age-old problem of squaring the circle, which concerned the attempt to map the area of a circle onto a square of equal size.

Cardano, Girolamo. *Practica arithmetice et mensurandi singularis* (1539). In his first mathematical work, Cardano proved himself adept at solving cubic equations.

Cardano, Girolamo. *Ars Magna, sive de Regulis Algebraicis* (The great art, or on the rules of algebra, 1545). A book on solving algebraic equations, including cubic and quartic equations. The major breakthrough in the work is the first publication of the solution of cubic and quartic equations by radicals, that is, with a formula similar to the quadratic formula. Cardano also shows his awareness of imaginary numbers in *Ars Magna* because they appear in the formula for the solution of the cubic equation. Our modern algebraic symbolism is not present in the work, nor is the use of general coefficients.

Cataldi, Pietro. *Trattato del modo brevissimo di trovar la radice quadra delli numeri* (1613). Here Cataldi employed infinite series and unlimited continued fractions to find a number's square root. This was perhaps the first serious examination of continued fractions on record.

Ceva, Giovanni. *De lineis rectis* (Concerning straight lines, 1678). This work contained Ceva's theorem, on the geometry of triangles, which in turn related to an area of interest throughout much of Ceva's career: center of gravity. His theorem found the center of gravity for an equilateral triangle at the place where lines drawn from the vertices to opposite sides intersected.

Ceva, Giovanni. *Opuscula mathematica* (A short mathematical work, 1682). Here Ceva again returned to centers of gravity, this time expanding his studies to other shapes. The book also contained an infamous error on Ceva's part, his statement that the periods of oscillation pendulums are on the same ratio as their lengths.

Ceva, Giovanni. *Geometricamotus* (The geometry of motion, 1692). This work concerned the geometry of motion and contained elements that foreshadowed the infinitesimal calculus soon to be developed independently by Gottfried Wilhelm von Leibniz and Sir Isaac Newton.

Chuquet, Nicolas. *Triparty en la science des nombres* (1484). At the time of this work's publication, arithmeticians lacked even the most basic notational symbols such as those for addition, subtraction, multiplication, and division. Chuquet became one of the first to offer symbols—though these bore little resemblance to the ones in use today—for what he called, respectively, *plus, moins, multiplier par,* and *partyr par.* He also provided notation for the previously inexpressible concept of square root: rather than $\sqrt{4}$, he would have written R)^{2}4.

Desargues, Girard. *Traite de la section perspective* (Treatise on the perspective section, 1636). This work challenged existing ideas of perspective and presented what came to be known as Desargues's theorem. According to the latter, two triangles may be placed such that the three lines joining corresponding vertices meet in a point, if and only if the three lines containing the three corresponding sides intersect in three collinear points. This theorem, which holds true in either two or three dimensions, would provide an early foundation for what came to be known as projective geometry.

Desargues, Girard. *Brouillon project d'une atteinte aux evenements des recontres d'une cone avec un plan* (Proposed draft of an attempt to deal with the events of the meeting of a cone with a plane, 1639). As its title suggests, this essay dealt with conic sections and presented what amounted to a unified theory of conics. As with Desargues's other groundbreaking ideas, however, those in *Brouillon* were slow to find acceptance, in part because of the highly arcane terminology the author used, substituting tree, stump, and other botanical terms for various geometrical constructs.

Descartes, René. *La Géométrie* (1639). An appendix to his 1637 work *Discours de la méthode pour bien conduire so raison et chercher la vérité dans les sciences* (A Discourse on the Method of Rightly Conducting the Reason and Seeking Truth in the Sciences), *La Géométrie* is a groundbreaking work, one of the greatest scientific works of the Renaissance. Analytic geometry is truly born in this work, and from there it is a short step to the invention of the calculus and a whole new era of mathematics.

Fermat, Pierre de. *Method for Determining Maxima and Minima and Tangents to Curve Lines* (1636). With this work Fermat laid the groundwork for differential calculus.

Graunt, John. *Natural and Political Observations ... Made upon the Bills of Mortality* (1662). A founding member of the Royal Society, Graunt began studying London death records dating back to 1532. He noticed a number of patterns, which he discussed in this book. He classifyed death rates according to cause, and identified overpopulated conditions as a mortality-increasing factor.

Gregory, James. *Vera circuli et hyperbola quadratura* (1667). This work discussed the means of finding the area for a circle or a hyperbola.

Gregory, James. *Geometriae pars universalis* (1668). Examined convergent and divergent series and contained Gregory's foundational ideas for what became calculus.

L'Hospital, Guillaume-François-Antoine de. *Analyse des infiniment petitis* (1696). The first textbook on differential calculus, and the leading book on the subject during the eighteenth century.

Mersenne, Marin. *Cogitata physico-mathematica* (1644). In this work Mersenne put forth several theorems regarding prime numbers. The theorems turned out to be incorrect but inspired later investigations into number theories and large primes—sometimes called Mersenne primes.

Napier, John. *Mirifici logarithmorum canonis descriptio* (1614). Here Napier first presented his system of logarithms.

Napier, John. *Rabdologia* (1617). Napier invented several rudimentary calculators, the most famous of which was "Napier's bones," a set of rods marked with numbers, which he introduced in this work. So named either because they were often made of bone or ivory, or because of the bone-like shape of the rods, the bones were but the most famous of many practical (and some impractical) creations from "the Marvelous Merchiston."

Oughtred, William. *Clavis mathematicae* (1631).This work was probably intended as a textbook for young Lord William, but with its summation of all significant arithmetic and algebraic knowledge up to that time, it proved to have much wider appeal. The author used the book to introduce a number of proposed mathematical symbols, but most were cumbersome and failed to catch on; yet x (for multiplication) and :: (for proportion) became permanent fixtures of the landscape.

Oughtred, William. *The Circles of Proportion and the Horizontal Instrument* (1632). Here Oughtred discussed the idea of a slide rule as an instrument to aid in navigation. Though one of his students later claimed that *he* had invented it, Oughtred is generally credited with both the circular and the linear slide rule, which he may have invented as early as 1621. The slide rule would remain an important tool of computation for more than 350 years, until the advent of handheld calculators.

Pacioli, Luca. *Summa de Arithmetica, geometria, proportioni et proportionalita* (1494). Regarded as the first printed work on algebra, even though it was not an original work. Concerned with the lack of teaching materials, Pacioli gathered mathematical materials from various sources and published them in a large comprehensive text. Much of the material can be found in the *Liber abbaci* of Leonardo of Pisa, a very influential work of its time. However, given the scope of Pacioli's book and the fact that is was one of the first texts printed, it became widely circulated and hence very influential. Its circulation and influence was also extended due to the fact that the book was printed in the vernacular Italian and not Latin.

Pascal, Blaise. *Essai sur les coniques* (1640). Written when Pascal was 16, this pamphlet clarified ideas that Girard Desargues had presented, using highly complex and confusing terminology, in a 1639 publication. As he continued work on it, however, Pascal went far beyond Desargues' original point, developing a theorem concerning what came to be known as "Pascal's mystic hexagram." According to the theorem, from which he deduced some 400 corollary propositions, the three points of intersection of the pairs of opposite sides of a hexagon inscribed in a conic are collinear.

Along with Desargues' ideas, these helped form the basis for projective geometry.

Pascal, Blaise. *Traité du Triangle Arithimétique* ("Treatise on the Triangle," 1665). One of the key mathematical treatises of its time. It focused on the properties of a triangle composed of numbers, but also on other mathematical ideas and concepts. Among them was Pascal's approach to proving propositions for which there are infinitely many cases. He informed his readers that faced with such a proposition they should initially prove that the proposition is true for the first case. Following that, they should prove the proposition for a given (or random) case. With those two proofs, the proposition is solved for the next case, and for an infinite number of other cases. In other words, he described the principle of induction.

Pellos, Francisco. *Compendio de lo abaco* (1492). A book on commercial arithmetic that used the decimal dot (.) to indicate division by 10.

Pitiscus, Bartholomeo. *Trigonometria: sive de solutione triangulorum tractatus brevis et perspicuus* (1595). Pitiscus is known primarily for his coining of the term "trigonometry," which appeared in the title of this work. The book, which used all six trigonometric functions, became a highly respected mathematical text and was translated into several languages.

Recorde, Robert. *Grounde of Arts* (1541). In addition to its scholarly overview of contemporary mathematics, this book provided practical knowledge concerning commercial math.

Recorde, Robert. *The Whetstone of Witte* (1557). Here Recorde presented the equals sign, using what he considered a hallmark of equality: two parallel lines of equal length.

Recorde, Robert. *Castle of Knowledge* (1556). A work on the properties of spheres.

Regiomontanus. *Tabulae directionum* (1475). Contains a valuable table of tangents.

Regiomontanus. *De triangulis omnimodis* (1533). Contains the earliest statement of the cosine law for spherical triangles.

Ricci, Michelangelo. *Exercitatio geometrica, De maximis et minimis* (1666). In this 19-page pamphlet, Ricci found the maximum of $x^m(a-x)^n$ and the tangents to $y^m=kx^n$.

Rolle, Michel. *Traité d'algebre* (1690). The work contained what was then the novel use of notation for the nth root of a number, as well as Rolle's method of "cascades." The latter used principles first put forth by Dutch mathematician Johann van Waveren Hudde (1628-1704) for finding the highest common factor of a polynomial, a method Rolle used to separate the roots of an algebraic equation.

Rolle, Michel. *Demonstration d'une méthode pour resoudre les egalitez de tous les degrez* (1691). This book included Rolle's theorem, a special case of the mean-value theorem in calculus.

Rudolff, Christoff. *Coss* (1525). The first book of algebra to appear in German. This wassignificant in that German was the vernacular in much of northern and central Europe, and few among the rising bourgeois class could read Latin. As for the title, it referred to the word *cosa* or "thing," which was used to refer to anything unknown or indeterminate; since algebra dealt with such perplexities, it was known as the cossic art.

Rudolff, Christoff. *Künstliche Rechnungmit der Ziffer und mit der Zahlpfennigen* (1526). This book addressed questions of computing and offered problems applicable to the rising commercial and industrial culture of Renaissance Europe.

Seki Kowa, Takakazu. *Hatubi sanpo* (1674). This book was written in response to the announcement of some 15 supposedly unsolvable problems that had been put forth four years before; in *Hatubi sanpo*, Seki Kowa solved them all. However, it was not the Japanese custom to show how one had arrived at one's solutions, and it appears that even Seki Kowa's students remained unaware of his methodology.

Stevin, Simon. *De Beghinselen der Weeghconst* (1586). Here Stevin introduced what is perhaps his most famous discovery, the law of the inclined plane. He showed geometrically that a linked chain of spheres must remain motionless when hung-over two inclined planes joined to form a triangle, in effect demonstrating that the gravitational force is inversely proportional to the length of the inclined plane. His geometric proof is the basis for the parallelogram method for analyzing forces.

Stifel, Michael. *Arithmeticaintegra* (1544). A summation of mathematical knowledge up to 1544.

Treviso Arithmetic (1478). The first printed arithmetic text. This work had no title and was by an unknown author. It is now commonly referred to as the *Treviso Arithmetic* because of its place of publication—Treviso, north of Venice in Italy. The text is meant to be a training manual for young students seeking to enter the new trading businesses in Italy. In it one finds how to write and compute with numbers (only the four basic operations) and how to apply these techniques to questions of partnerships and trading.

Tschirnhaus, Ehrenfried. *Medicina Mentis* (Mental medicine, 1687). Here Tschirnhaus laid out a method of discovering rational truths as a basis of a happy life derived from Cartesianism, Spinoza, the English empiricists, and Leibniz, whom he greatly admired. Only true knowledge can tame the emotions, he believed, which are the sources of error and therefore of unhappiness.

Tunstall, Cuthbert. *De arte supputandi libri quattuor* (1522). One of the first printed works published in England devoted exclusively to mathematics. It contained no original material, but was a very practical arithmetic text.

Viète, François. *Canon mathematicus seu ad triangula cum appendibus* (Mathematical laws applied to triangles, 1579). The text promoted trigonometry, then an underutilized discipline, and made use of all six trigonometric functions.

Viète, François. *In artem analyticum Isagoge* (Introduction to the analytical arts, 1591). Here Viète established the letter notation still used in algebra: vowels for unknown quantities or variables, consonants for known quantities or parameters.

Viète, François. *Supplementum geometriae* (1593). Addressed topics such as the trisection of an angle; the doubling of a cube (a problem that had bedeviled many ancient Greek mathematicians); and the first known explicit statement of π as an infinite product.

Viète, François. *De numerosa* (1600). With this work Viète presented a method for approximating roots of numerical equations.

Viète, François. *Deaquationem recognitione et emedatione libri duo* (Concerning the recognition and emendation of equations, 1615). Posthumously published, this work offered methodology for solving second, third, and fourth degree equations, and contained the first use of the term "coefficient."

Wallis, John. *Arithmetica infinitorum* (1655). This was Wallis's first important work, and was destined to become a standard information source for mathematicians during the coming decades. In it, he became the first person to use the term "induction" as a mathematical principle.

Wallis, John. *Algebra: History and Practice* (1685). This book not only discussed the whole history of the discipline of algebra but marked the first recorded attempt at graphically displaying the complex roots of a real quadratic equation. A second edition contained the first systematic use of algebraic formulae, including numerical ratios.

NEIL SCHLAGER

Physical Sciences

Chronology

1543 Nicolaus Copernicus's publication of *De Revolutionibus Orbium*, in which he proposes a heliocentric or Sun-centered universe, sparks the beginnings of the Scientific Revolution.

1546 German mineralogist Georgius Agricola publishes *De Natura Fossilium*, the first scientific classification of minerals and the first handbook of mineralogy.

1572 Danish astronomer Tycho Brahe observes a galactic supernova, or exploding star, an event that puts to rest the long-held Aristotelian notion that the heavens are perfect and unchanging.

1584 Inspired by the ideas of Copernicus, Italian philosopher Giordano Bruno begins expounding a new idea of the universe, including concepts such as the infinity of space and the potential habitability of other worlds; for this act of "heresy" he is burned at the stake in 1600.

1587 Galileo begins experiments that lead to his law of falling bodies, showing that the rate of fall of a body is independent of its weight, and that all objects will fall at the same rate in a vacuum.

1609 Johannes Kepler's *Astronomia Nova* states his two laws of planetary motion: that the orbits of the planets can be drawn as ellipses, with the Sun always at one of their foci; and that a planet will move faster the closer it is to the Sun.

1619 *De Chymicorum*, by German physician Daniel Sennert—who writes of atoms and "second-level atoms" or molecules—is the first application of Greek atomic theory to chemistry.

1660 The Royal Society is founded in London to promote scientific inquiry.

1661 English physicist and chemist Robert Boyle publishes *The Sceptical Chymist*, a work regarded by many as the beginnings of scientific chemistry.

1667 Danish anatomist and geologist Nicolaus Steno first describes rock stratification.

1673 Building on Galileo's principle of isochronicity, Dutch physicist and astronomer Christiaan Huygens details his invention of the pendulum or "grandfather clock" in *Horologium Oscillatorium,* which begins the era of accurate timekeeping that is essential to the advancement of physics.

1687 Isaac Newton publishes *Philosophiae Naturalis Principia Mathematica*, generally considered the greatest scientific work ever written, in which he outlines his three laws of motion and offers an equation that becomes the law of universal gravitation.

Overview: Physical Sciences 1450-1699

The Medieval Foundation

Medieval science and intellectual thought were based not on direct observation and experience, but were heavily influenced by the Aristotelian view of nature (such as the four elements), and were further formalized by church teachings. Yet by the mid-thirteenth century Franciscan thinkers plied an observational/empirical logic to question wholesale acceptance of ancient scientific ideas, as in the impressive study of optics and the rainbow by Robert Grosseteste (c. 1175-1253) and others, with important contemporary efforts by Muslim thinkers. The heart of this critical view formalized into the new logic of nominalism, most familiarly recognized in William of Occam (c. 1285-1349) and overall as a late medieval disagreement with ancient, particularly Aristotelian, rationalization of abstractions and universals. More refined methodology resulted most effectively in the Parisian School of physical theorists, headed by Jean Buridan (c. 1297-c. 1358), who developed early theories of impetus as the causal agent of motion. His follower Nicole Oresme (c. 1320-c. 1382) criticized Aristotle's celestial ideas by hypothesizing the realistic logic of Earth rather than the universe rotating, one of Nicolaus Copernicus's later heliocentric theory arguments. These steps led to the physical science of the next 250 years, to the dawn of the eighteenth century, a time of profound transition to and foundation of modern physical science.

Renaissance Science

By the late fourteenth century European thinkers began a process of turning to original Greek thought as a new foundation of critical reappraisal of the ancient legacy. This was the so-called Renaissance, roughly continuing in spirit until 1600. The Renaissance was a period of European transitions, one much more complicated than simply the passing of medieval thought and the beginning of modern thought. It was a time of economic and social upheaval, emphasized by the age of exploration and discovery. In physical science the transition was outwardly noticeable, for without a new systematic base to replace that of ancient Greece, conservative thought mingled with changing views, all laced with a persistent traditional intuitive conception of knowing nature by occult process-

es, particularly astrology and alchemy. Interestingly, the Renaissance started roughly with one of the greatest gifts to intellectual stimulus, the printing press, which provided a dissemination of knowledge of phenomenal breadth via the printed word.

By the mid-fifteenth century physical science was also finding vision for this new Renaissance. The term "Renaissance Man" was first given to Leonardo da Vinci (1452-1519), an artist, inventor, and scientific polymath who delved at understanding nature with a stubborn brilliance for the thought process itself, rather oblivious of formal learning. Of the more formal variety of thought, the Renaissance spirit was effective in astronomy, initially through the efforts of Johann Müller (a.k.a. Regiomontanus, 1436-1476) and select others emphasizing a base of original ancient astronomical thought along with accurate instruments and observational technique. By the early sixteenth century investigators were poring over other areas of the physical sciences so comprehensively demarcated by Aristotle in physics, the earth sciences, and chemistry. And thinkers such as mathematician Girolamo Cardano (1501-1576) were challenging such ancient tenets as the legitimacy of the so-called four elemental building blocks of the terrestrial world; the delineation of terrestrial and celestial boundaries and the phenomena of each; and the immutability and perfection of celestial space with its curious essence, the "ether."

Nicolaus Copernicus (1473-1543) provided what has popularly become known as the revolutionary heliocentric theory of the universe, with Earth not only revolving around the Sun but rotating on its axis. The theory's relevance for the time was not so definitive, affecting only a modest few thinkers and in more practical aspects of observational astronomy. Still, there were a few scientists who took the heliocentric theory more profoundly. A school of heliocentric thought developed in England under Thomas Digges (c. 1543-1595), who drew from it a infinite cosmos instead of that fixed by the ancients. In Germany a large community of astronomers was greatly influenced by its implications in pro and con arguments. In Italy an unconventional philosopher/priest named Giordano Bruno (1548-1600) used it as part of his personal rebellion to church authority and was burned at the stake for it. This theory provided an impressive backdrop to a century of exploratory physical thought, groping toward systematic knowledge.

A characteristic part of that search was observational conscientiousness, stressing collecting and cataloguing not only physical specimens in natural history, mineralogy, and geology but also recording hard data of everything from comets and appearance of the Milky Way (not considered celestial) to rainbows and the odd shapes of hailstones. Among notable advances in physical science were: Georgius Agricola's (a.k.a. Georg Bauer, 1494-1555) systematic geological thought, William Gilbert's (1544-1603) landmark magnetic and electrical studies, and Tycho Brahe's (1546-1601) accurate astronomical measurements and their implications, one of which was application to the new Gregorian Calendar (1582).

The Seventeenth Century: Fundamental Base of Physical Science

Though Brahe and some astronomers and other thinkers into the seventeenth century were still tied to astrological sympathies, that century's astronomers were fully occupied with accurate planetary and stellar observations. These observations were made possible with new instruments of unparalleled sophistication. Nonetheless, Johann Kepler (1571-1630) conceived his monumental three laws of planetary movement (1609, 1619) partially out of his belief in a mystical geometry of a harmonious cosmos. But occult intuitiveness faded with its failure to compete with empirical and mathematical innovation in explaining nature as the seventeenth century wound toward its end.

As in the previous century, the appearance of comets and a host of variation in the supposed fixed star field continued to cast theoretical doubts about traditional ancient beliefs about the heavens. The introduction of the refracting lens telescope early in the century also opened up a closer look and new perspective on the heavens. Galileo Galilei (1564-1642) was the first to turn it toward the planets and discover the satellites of Jupiter, before he turned to his landmark studies in the physics of mechanics and his later crisis in the controversy over heliocentricity. Telescopic study of sunspots and comets with a clearer vision of the field of stars, resulting in significant advances in star catalogues and atlases through the century, paralleled advances in telescope technology.

The century's experimental science developed a growing methodology of laboratory investigation and instrumentation. At the forefront of this movement was René Descartes (1596-1650), grounding his mechanistic logic in physical and chemical phenomena. The building and testing of various thermometers, barometers, and hygrometers focused on the study of air and its properties. Jan Baptist van Helmont (1577-1644) first defined the term "gas." Galileo devised early forms of thermometers, while several investigators worked on measuring scales, thereby setting the stage for later studies in heat phenomena. Galileo's student Evangelista Torricelli (1608-1647) constructed the first mercury barometer. Robert Boyle (1627-1691) coined the term "barometer" and constructed many instruments, including hygrometers for studying weather changes. He rejected the traditional base of alchemy and promoted analysis of matter and substance by its composition. His experiments with gases culminated in the pressure/volume gas law. His colleague Robert Hooke (1635-1703), adding to a period of conjecture about the structure of Earth, pondered Earth's great age, developed atmospheric instruments, and—in experiments on mechanical elasticity—discovered the law of elastic force (1678).

The scientific thought of the seventeenth century was an arrangement of stepping stones toward a synthesis of ideas, as more outstanding minds provided plausible theories to solve physical problems. After mid-century that importance was made known in the transition of private and limited patronage of science to the government level. The great scientific societies were beginning to appear: the Academia del Cimento (Florence, 1657), the English Royal Society (London, 1662), and the French Académie Royal des Sciences (Paris, 1666). Institutional astronomy followed suit with, for example, the Observatoire de Paris (1667) and the Greenwich Royal Observatory (1675). Paralleling these outward signs of the credence in science were the ideas of some of the best thinkers of the latter part of the century. Christiaan Huygens (1629-1695), in addition to telescopic discovery of Saturn's rings and moon Titan, made important studies and applications in optics. He invented the pendulum clock and presented the mechanistic theory of light (1690) as an impulse (incipient wave theory), explaining light's optical properties. Edmond Halley (1656-1742) concluded that comets moved in closed orbits and appeared to be periodic. His theoretical base in this was a mathematical assurance in physical motion courtesy of Isaac Newton's (1643-1727) method of calculating a comet's apparent orbit.

Newton, who invented the reflecting telescope and derived calculus (1671) along with theorizing light behavior based on the corpuscular (particle) theory, had been developing a unifying mechanical theory for both celestial and terrestrial motion based on mathematical principles. This theory was a monumental culmination of the century's cumulative scientific experience, a long-awaited systematic plan for delving into nature. Newton's mathematical concept was really a synthesis of the two basic scientific trends of the seventeenth century. The one was a mathematical rationalism, basically a deductive method seen in Descartes or Galileo, while the other was a mathematical empiricism, the inductive method of experimentation, declared by Francis Bacon (1561-1626) and carried out by this ex-

perimental century, a significant representation being the body of the Royal Society's work.

"Newtonian" science was a secularized freedom from centuries of religious stricture on scientific thought, for it meant the discovery of independent and fundamental laws to the universe that were mathematically derived, one of its most essential being his universal law of gravitation between bodies. Newton's principles of scientific method were the immediate legacy of what would become physical science, what Newton himself termed natural philosophy "discovering the frame and operations of nature" in established rules and laws based on observation and experiment. These rules and laws became the now-defined classical physics thread that bound the fabric of physical science for the next two hundred years.

WILLIAM MCPEAK

Science and Christianity During the Sixteenth and Seventeenth Centuries

Overview

In the late 1800s, John W. Draper and Andrew D. White established a widespread belief in an irreconcilable "warfare" of science with "dogmatic theology." Both depicted a historical battle of enlightened, progressive, objective reason (science) continually advancing against blind ignorance, superstition, and prejudice (religion), with Galileo's trial and condemnation as the central illustration. Although historians have long known that this portrait relied on highly biased selections and interpretations of the historical evidence, it remains fixed in the popular imagination. In fact, relations between Christianity and science in the sixteenth and seventeenth centuries were extremely complex—by turns mutually antagonistic, indifferent, or supportive—as both underwent profound changes during the parallel courses of the Reformation and the Scientific Revolution.

Background

Prior to the sixteenth century, relations between natural philosophy (as science was called until the nineteenth century) and Christianity were generally harmonious, if not entirely free from tension.

Early Christian theology adopted a somewhat equivocal attitude toward science. On the one hand, the Scriptures viewed the universe as an orderly and purposive realm, originally created by God as "good" and reflecting His nature, above all in man as a rational being made "in the image and likeness" of God. The study of nature was thus not only possible but desirable, offering a means to the greater knowledge and glorification of God. The Genesis creation account of God's granting stewardship over Earth to Adam and his descendants likewise encouraged investigation of nature. On the other hand, the Scriptures also depicted the Creation as fallen and distorted by sin, with the New Testament emphasizing the renunciation of worldly concerns for heavenly ones as vital to personal redemption. This implied that curiosity about nature should be a matter of relative indifference to the believer, lest it become a snare to sin by distracting attention away from spiritual devotion to God and the pursuit of salvation.

In adapting ideas about nature from pagan philosophies (Aristotelianism and Platonism) it found compatible, Christian thought strove for a balance between two extremes. The natural order was neither to be worshiped as divine (Stoicism), nor rejected as without purpose (Epicure-

Galileo before the Inquisition. *(Bettmann/Corbis. Reproduced with permission.)*

anism) or evil (Gnosticism), but contemplated as an originally good creation, corrupted by sin but destined with man to divine redemption and restoration. Faith in God and scriptural revelation was not a restriction on the exercise of human thought, but rather the initial foundation which made right reasoning possible. Philosophy, including natural philosophy, assumed its proper role as the "handmaiden of theology"; along with prayer and the sacraments, it was instrumental to salvation, providing rational methods for understanding God's revelation. As the primary literate class in medieval Europe, priests and monks were the chief practitioners of natural philosophy, focusing upon mathematics, physics, astronomy, zoology, and botany.

The recovery of ancient Greek scientific and medical texts by Aristotle (384-322 B.C.), Ptolemy (c. 140) and Galen (130-200) during the thirteenth century dramatically expanded Western scientific knowledge and altered relations between natural philosophy and theology. Scholastic theologians distinguished between the created "book of Nature" and the written "book of Scripture" as two complementary avenues of divine revelation. Each book had its own integrity and proper principles of interpretation, and knowledge gained from either one could be used to correct understanding of the other. Natural philosophy thus implicitly assumed a position of autonomy rather than sub-

servience with respect to theology. When Aristotle's metaphysical notions contradicted basic Christian doctrines, such as in asserting the eternity of the universe instead of its creation in time by God, they were modified or rejected. The famous condemnation in 1277 by the Bishop of Paris of 219 propositions drawn from Aristotle may actually have advanced rather than impeded medieval natural philosophy, by opening avenues to alternative approaches.

Impact

Relations between natural philosophy and Christian theology became increasingly unsettled beginning in the sixteenth century. The recovery during the Renaissance of additional ancient Greek texts gave rise to a Christian humanist movement that challenged both Aristotle's natural philosophy and existing scholastic methods of textual interpretation, especially as applied to Scripture. Advances in astronomical, mathematical and medical techniques generated new observations that existing theories could not accommodate. Thousands of novel plants and animals observed in the newly discovered Americas rendered the ancient zoological and botanical systems of Aristotle and Theophrastus (372-287 B.C.) obsolete. The heliocentric theory of Copernicus (1473-1543) confronted the geocentric theory of Ptolemy in astronomy; Paracelsus (1493-1541) with his system of immaterial chemical principles rejected both

Aristotle's theory of material elements and Galen's medical theory of bodily humors; and the observations of Andreas Vesalius (1514-1564) overthrew Galen's authority in anatomy.

At the same time, deepening theological divisions between Roman Catholicism, Lutheranism, and Calvinism resulted in significantly revised approaches to biblical interpretation. Patristic and medieval theologians had exercised considerable latitude in expounding scripture, understanding many passages figuratively or allegorically. Against this, however, most Protestant Reformers stressed a more strictly literal interpretation of biblical texts as the only one assuring fidelity to their true meanings. Similarly the Catholic Church, seeking to counteract both Protestantism and extreme heretical sects, narrowed the range of acceptable interpretive approaches, by systematizing church doctrine in authoritative pronouncements such as those of the Council of Trent. One consequence of these trends was that scientific theories which previously were of little concern now came under scrutiny for their possible theological implications.

The most famous adversarial encounter between science and Christianity arising from these myriad cross-currents was the case of Galileo Galilei (1564-1642). For some theologians, the displacement of Earth by the Sun from the center of the universe in the Copernican theory also implied a heretical displacement from the focus of the Creation of man's redemption by Christ. In 1616, the Catholic Church briefly placed Copernicus' *De Revolutionibus* on the Index of Prohibited Books until a few objectionable passages could be excised. Galileo, the leading advocate of the Copernican theory, was summoned to Rome, where he accepted a certificate from Cardinal Robert Bellarmine pledging that he would not present the theory as anything other than a hypothesis. Bellarmine died in 1621, and in 1623 Galileo's friend Cardinal Maffeo Barberini ascended to the papal throne as Urban VIII. After six private audiences with Urban, Galileo rashly believed he had permission to publish a defense of Copernicanism, and in 1632 his *Dialogue on the Two Chief World Systems* appeared. Summoned to Rome and interrogated by the Inquisition, the aged Galileo was compelled in 1633 to renounce Copernicanism as a heresy contrary to the Catholic faith, and his book was placed on the Index. A sentence of life imprisonment was commuted to house arrest at his estate in Florence.

The Galileo affair involved far more than a supposed conflict between new science and old dogma, however. Attacks on and defenses of Copernicanism often reflected not just opposing theological views, but also power struggles between different priestly orders within the Catholic Church, and the theory was also an issue in Jesuit efforts to reconvert Protestant areas of Germany back to Catholicism. As a recipient of royal patronage from the Venetian, Pisan, and Florentine courts, Galileo was enmeshed in a complex web of political intrigue and envious competition for prestige between rival Italian city-states. In an age of violent personal invective unrestrained by libel laws, Galileo also excelled in penning scorching insults of his Jesuit opponents, earning several bitter and influential enemies. With a singular lack of humility and tact, he even presumed to instruct cardinals on points of biblical exegesis and doctrinal interpretation, and offended Urban VIII by inserting one of his arguments into the mouth of Simplicio, the Aristotelian dupe of the *Dialogue*. Galileo's support for atomism may have been yet another complicating factor. Far from being simply an instance of persecution of enlightened science by religious obscurantism, Galileo's condemnation percolated in a seething cauldron of potent political and personal rivalries, containing a poisonous stew of hostile theological and philosophical agendas, spiced by his own arrogance.

Unfortunately, the notoriety of the Galileo affair has obscured its exceptional nature. Historians have discredited apocryphal criticisms of Copernicus attributed to Martin Luther (1483-1546) and John Calvin (1509-1564), and the Copernican theory faced little formal theological opposition in Protestant areas. Even in Roman Catholic dominions, little effort was made outside of Italy and Spain to hinder its study and use. Although the French philosopher René Descartes (1598-1650) withheld his recently completed *Le Monde* from publication for several years because it endorsed Copernicanism, his colleagues Pierre Gassendi (1592-1655) and Marin Mersenne (1588-1648), both priests, continued to discuss the theory in public correspondence without interference. By 1700, even Jesuit astronomers in the many church-based observatories throughout Catholic parts of Europe used the Copernican theory to make astronomic calculations. The belated removal of Galileo's *Dialogue* from the Index in 1832 was merely an acknowledgment of what had been a dead letter for over a century.

A second area of tension between scientific theories and religious doctrines concerned seventeenth century corpuscular matter theories.

Because ancient atomic theory had posited that atoms were uncreated, eternal, indivisible, infinite in number, and in unceasing motion according to immutable laws of impersonal fate, with occasional completely random swerves, they were long associated with atheism, as denying God's creation of matter and providential governance of events. Atomism also implicitly contradicted the key Roman Catholic doctrine of transubstantiation—that the underlying substance of consecrated bread and wine in the Eucharist changes into the Body and Blood of Christ, while the equally real external accidents remain the same—by denying the Aristotelian doctrine of substantial essence and making the qualities merely phenomenal appearances. In response, advocates of corpuscularian theories such as Gassendi adapted them to conform to Christian doctrine, with God first creating a finite number of atoms in time out of an infinitely divisible matter and then setting and maintaining them in motion by His power according to His purposes. While this did not allay all suspicions of heresy, corpuscular theories nonetheless were widely accepted by the end of the century.

A third and very different challenge to Christian orthodoxy came from neo-Platonic advocates of "natural magic." Along with many Paracelsians, they championed a science based upon mastery of immaterial cosmic powers, according to an occult or esoteric knowledge of divinely created "signatures" embedded in every natural object. The close affiliation of such notions to Gnosticism and sorcery, and belief by many of their partisans in anti-Christian heresies such as denial of the Trinity and the Incarnation, led both Roman Catholic and Protestant authorities to oppose these ideas strenuously, a famous instance being the execution of Giordano Bruno (1548-1600) by the Inquisition. Also, advocacy of atomism and Copernicanism by some neo-Platonists imparted a suspicious taint to those theories that was not easily removed.

Aside from these occasional points of friction, however, most theologians and natural philosophers alike viewed the study of nature as cultivating rather than undermining real faith and devotion. The regularities, harmony, and hierarchical structure of the Creation, discovered by science and embodied in natural laws, revealed for many the marvels of God's creative omnipotence and goodness. Both the published books and private letters of virtually every scientist of the era are studded with innumerable religious references, expressing unquestionably genuine piety rather than mere lip service, and many scientists were deeply involved in church matters. Two founders of the Royal Society of London, John Wilkins (1614-1672) and Thomas Sprat (1635-1713), became Anglican bishops; the anatomist and geologist Nicolaus Steno [Niels Stensen] (1638-1686) became a Roman Catholic bishop and was canonized as a saint in 1988; the physicist Blaise Pascal (1623-1662) is more famous today for his devotional *Pensées* than his scientific research; and Sir Isaac Newton (1642-1727) left thousands of pages of unpublished theological notes that far exceeded his published scientific output.

Consequently, most historians now ask whether and how Protestant or Catholic theology may have actively encouraged the development of science, and whether and how the religious convictions of individual scientists influenced their scientific theories. Much attention has been given to the "Merton thesis," which argues that English Puritans constituted an usually high percentage of English scientists, due to a strong emphasis in Calvinist theology on the study of nature as a chief means to the glorification of God, a sense of divine calling to a specific earthly vocation, and the fruitful production of good works as a sign of predestination to eternal salvation. While a connection clearly existed in England between Protestant nonconformity and science, it is debatable whether this was due to a special Puritan theological affinity for science, or because dissenters were legally excluded from preferment in government, university, and established Church positions, and therefore sought other avenues for social advancement. The same applies to the large number of Protestants among scientists in Catholic France. At present, most scholars believe that while Catholicism as such was no less favorable to science than Protestantism, the greater institutional centralization and uniformity of the Catholic Church reinforced a more conservative outlook among Catholic scientists, making them less receptive to novel theories at variance from Aristotelian natural philosophy.

New tensions between science and religion emerged, however. The development of the scientific concept of natural law, and its successful application to the explanation of many hitherto puzzling phenomena, reduced the scope previously allowed for divine miracles. The tremendous increase in scientific knowledge mitigated the loss of ancient knowledge once thought to form part of original sin. The abandonment of teleological (purposive) explanations of natural phenomena for merely descriptive ones—of questions asking why for ones merely asking how—appeared to neglect or even deny the role of Divine providence. Reliance on reason and sensory observation eroded trust in divine revelation, as rationally demonstrable "natural religion" and Deism rejected any mystical aspect to faith. While many scientists successfully resolved these tensions, the very fact that some felt it necessary to marshal science in defense of Christianity against scepticism—the chemist Robert Boyle (1629-1698) endowed a lectureship for this purpose—indicated a widespread uneasiness over an increasing estrangement of science from religion.

Early modern science and Christianity thus were neither necessarily allies nor enemies. Instead, due to profound changes in each, a growing divergence between them gradually surfaced; science ceased to be the handmaid of theology and assumed an independent status. Because of its spectacular and continuing success in adding to man's knowledge and control of the natural world, science has gained such power and prestige that for many people it has displaced rather than supported religion as the foundation for their most fundamental beliefs. As scientific and religious thought both have continued to deepen and change, their mutual relations have also continued to converge and diverge in complex, unpredictable and fascinating ways.

JAMES A. ALTENA

Further Reading

Books

Brooke, John H. *Science and Religion: Some Historical Perspectives.* Cambridge: Cambridge University Press, 1991.

Draper, John W. *History of the Conflict Between Religion and Science.* New York: D. Appleton and Co., 1875. Reprint ed. New York: De Young, 1997.

Harrison, Peter. *The Bible, Protestantism, and the Rise of Natural Science.* Cambridge: Cambridge University Press, 1998.

Hooykaas, Reijer. *Religion and the Rise of Modern Science.* Edinburgh: Scottish Academic Press, 1972.

Langford, Jerome L. *Galileo, Science and the Church.* 3rd rev. ed. Ann Arbor: University of Michigan Press, 1992.

Merton, Robert K. *Science, Technology, and Society in Seventeenth-Century England.* New York: Harper and Row, 1970. Originally published in *Osiris* 4 (1938): 360-632.

Numbers, Ronald L., and David C. Lindberg, eds. *God and Nature: Historical Essays on the Encounter Between Christianity and Science.* Berkeley: University of California Press, 1986.

Westfall, Richard. *Science and Religion in Seventeenth-Century England.* New Haven, CT: Yale University Press, 1958.

White, Andrew D. *History of the Warfare of Science with Theology in Christendom.* 2 vols. New York: D. Appleton and Co., 1896. Reprint ed. New York: Dover, 1960.

Articles in Books

Dear, Peter. "The Church and the New Philosophy." In *Science, Culture and Popular Belief in Renaissance Europe*, ed. by Stephen Pumfrey, Paolo L. Rossi and Maurice Slawinski. Manchester: Manchester University Press, 1991: 119-139.

Pedersen, Olaf. "Science and the Reformation." In *University and Reform: Lectures from the University of Copenhagen Symposium*, ed. by Leif Grane. Leiden: E. J. Brill, 1981: 35-62.

Russo, François. "Catholicism, Protestantism, and the Development of Science in the 16th and 17th Centuries." In *The Evolution of Science*, ed. by Guy S. Métraux and François Crouzet. New York: New American Library, 1963: 291-320.

Nicolaus Copernicus Begins a Revolution in Astronomy with His Heliocentric Model of the Solar System

Overview

The publication of Nicolaus Copernicus's (1473-1543) *De Revolutionibus Orbium Celestium* in 1543 was attended by no official opposition. The heliocentric system Copernicus presented was initially viewed as a hypothetical model devised merely to facilitate computation. For many, the most attractive feature of the new system was

Copernicus's abolition of the equant, which restored uniform circular motion as the basic axiom of astronomy. Most early supporters passed over in silence the question of the system's physical reality. Theoretical improvements made possible by Copernican theory and new observations helped undermine Aristotelian physics and with it geocentrism—the idea that the Sun and all other planets in the Solar System revolved around Earth. By the mid-seventeenth century the heliocentric view reigned supreme, though Copernicus's circular orbits had by then been replaced by Johannes Kepler's (1571-1630) elliptical orbits.

Background

The theoretical framework of pre-Copernican astronomy was established in the *Almagest* of Ptolemy (c. 100-c. 170). Drawing heavily on the work of previous Greek astronomers, especially Hipparchus (c. 170-c. 120 B.C.), this work developed a theory of the universe employing geocentric models to predict planetary motions. Ptolemy appealed to Aristotelian physics to show Earth was at rest. He then geometrically demonstrated Earth is the center of the universe with the fixed stars moving together as a sphere. He further assumed, in accordance with Aristotelian teaching, that all celestial bodies move about Earth in perfect circles. To obtain agreement between observations and predictions based on circular motions, Ptolemy used epicycles, deferents, eccentrics, and other noncircular motions. He also found it necessary to introduce the equant—an imaginary point around which celestial objects moved—to approximate uniform motion about an off-center point.

Aristotle (384-322 B.C.) also regarded the planetary orbs as solid celestial spheres that formed a unified mechanism explaining the movements of celestial bodies. Though relying on Aristotelian physics, the *Almagest* makes no attempt to interpret epicycles and deferents as physically real, nested spheres. Regardless of whether or not Ptolemy actually held this view, it was adopted by Medieval astronomers who considered it an essential component of the Ptolemaic system.

Models resembling those of the *Almagest* were used in the Middle Ages to calculate tables of planetary motions, an example being the *Alfonsine Tables* (1252). However, most medieval European astronomical work amounted to little more than the collection and reorganization of Arabic and ancient Greek material. No new observations of importance were undertaken, and by the mid-fifteenth century the *Alfonsine Tables* (1252) were sorely in need of revision.

Georg von Purbach (1423-1461) pointed out to Johannes Regiomontanus (1436-1476) the inaccuracies of existing tables as well as the need for better translations of Greek texts. Pur-

Nicolaus Copernicus. *(Library of Congress. Reproduced with permission.)*

bach attempted to produce a revised and corrected version of the *Almagest* but died before finishing it. Regiomontanus completed Purbach's *Epitome of Astronomy*. This work contained, in addition to the Purbach-Regiomontanus translation of the *Almagest*, critical commentary and revised computations. Published in 1496, the work was a great success and attracted the attention of Copernicus, who was particularly struck by the errors inherent in Ptolemaic lunar theory.

Copernicus was not so much exercised by inaccurate predictions as he was by the lack of "perfection" exhibited by the Ptolemaic system, especially with respect to uniform circular motion. His solution was to give Earth a simple circular orbit about a static, off-center Sun. Furthermore, in this heliocentric model (more accurately *heliostatic*, since the Sun is not really the center of the system) the diurnal rotation of the heavens is accounted for by the rotation of Earth. A canonical wobble about Earth's axis of rotation was also proposed to explain precession.

As radical as Copernicus's system was, his firm commitment to maintaining uniform circular motion precluded any real simplification over the Ptolemaic system. He continued using epicycles, deferents, and eccentrics. By replacing Earth with the Sun, Copernicus immediately eliminated five large planetary epicycles. However, his rejection of the equant, because it failed to preserve uniform circular motion, required the introduction of secondary epicycles. The real power of Copernicus's system was to be found in the trigonometric methods he used and his improved lunar theory. It should also be mentioned that Copernicus believed his heliocentric model was physically true.

Copernicus first described his heliocentric system in the brief essay "Commentariolus." Composed sometime before 1514, it was privately circulated. *De Revolutionibus* was completed in the early 1530s but Copernicus delayed publication. Hearing of this work, Georg Joachim von Lauchen, self-named Rheticus (1514-1574), traveled to Frauenburg in 1539 to examine the manuscript. Rheticus realized the full revolutionary impact of the work and urged Copernicus to publish immediately. Copernicus refused but did allow Rheticus to publish a summary account that appeared under the title *Narratio prima* (1540). Copernicus finally relented and agreed to let Rheticus prepare *De Revolutionibus* for publication. Andreas Osiander (1498-1552), a Lutheran minister who saw the book through press, penned an anonymous and unauthorized preface stating the heliocentric hypothesis was merely a mathematical model and not intended as a true description of the universe.

Impact

De Revolutionibus appeared in March 1543. Though Osiander's preface may have forestalled official opposition to the work, many theologians denounced Copernican heliocentrism because of its conflict with the Bible, and Aristotelians everywhere objected to the very idea that Earth could be in motion. Despite such objections, *De Revolutionibus* acquired a small following.

Erasmus Reinhold (1511-1553) was one of the earliest supporters of the Copernican system. He made extensive corrections to *De Revolutionibus* and calculated the *Tabulae Prutenicae* (1551). This was the first set of practical planetary tables based on Copernicus's theory. More accurate than the *Alfonsine Tables*, Reinhold's tables were widely adopted and provided a strong argument in favor of Copernicanism. However, Reinhold's focus on Copernicus's mathematical modeling and silence regarding the physical reality of heliocentrism encouraged a similar attitude in German astronomers that persisted well into the 1570s.

Observations of the nova of 1572 by Tycho Brahe (1546-1601) and Michael Mästlin (1550-

1631) indicated this phenomenon was located among the fixed stars. This undermined the Aristotelian notion that the heavens were perfect and unchanging. Brahe and Mästlin's observations of the comet of 1577 dealt another blow to Aristotelian cosmology. Because the erratic behavior of comets was incompatible with the immutability of the heavens, Aristotle maintained they were atmospheric exhalations. Brahe and Mästlin's parallax measurements for the 1577 comet indicated it was more distant than the Moon. Furthermore, Brahe concluded that its orbit was elongated, suggesting it had passed through several planetary spheres—impossible if planetary spheres were solid.

These celestial events led Mästlin to completely reject Aristotelian cosmology and adopt Copernican heliocentrism. After attending Mästlin's lectures on the superiority of Copernicus's cosmology, Kepler embraced Copernicanism as well. Meanwhile, Thomas Digges (c. 1546-1595) had already taken up the Copernican cause, becoming its foremost exponent in England. Digges translated portions of Copernicus's *De Revolutionibus* and appended his own views on an infinite universe with fixed stars at varying distances from Earth (1576).

Brahe also rejected the Ptolemaic system as incompatible with his observations, but belief in the Bible and a lack of stellar parallax prevented him from accepting Copernican heliocentrism. As a compromise he advanced his Tychonic theory, in which all of the planets but Earth orbit the Sun, with the Sun and its train of planets revolving about a stationary Earth (1588). This system avoided many of the pitfalls of the Ptolemaic system while preserving a stationary Earth. Brahe's theory gained acceptance in many quarters over the next 50 years.

Galileo (1564-1642) announced his support of Copernicanism with the publication of his telescopic discoveries in *Sidereus nuncius* in 1610. Galileo's discovery of four satellites orbiting Jupiter contradicted the widely held belief that Earth was the center of rotation for all celestial bodies, and his observation of mountains and depressions on the lunar surface refuted the Aristotelian notion that the Moon was a perfect sphere. Galileo also discovered sunspots and the phases of Venus. The latter discovery removed a serious objection to Copernican heliocentrism. The primary effect of Galileo's findings was to further undermine Aristotelian cosmology, which served as the foundation of the Ptolemaic system. Galileo furnished further support for

Copernicanism in his *Dialogue Concerning the Two Chief World Systems* (1632). Having introduced the principle of inertia, he demonstrated that objects on a rotating Earth would behave no differently than on a stationary Earth.

Johannes Kepler provided the crucial advance in the ascendancy of heliocentrism. He joined Brahe in Prague in 1600. After Brahe died the next year, Kepler secured control of his incomparable data set and spent the next eight years devising various geometrical schemes to account for the observations of Mars. Kepler finally determined that the orbit of Mars, as well as those of the other planets, describes an ellipse with the Sun occupying one foci. This is known as Kepler's first law. At once, the Ptolemaic and Copernican epicycles and eccentrics were completely eliminated. Furthermore, since one of the foci of each planetary ellipse was anchored on the Sun, the Sun truly occupied a central position within the solar system. Kepler also showed that planets sweep out equal areas in equal times as they move about in their orbits. Known as Kepler's second law, this implies that planets move faster when closer to the Sun. Published in *Astronomia nova* (1610), Kepler's first two laws directly challenged the traditional canon of uniform circular motion. Kepler's third law appeared in *Harmonices mundi* (1619).

Kepler completed the *Tabulae Rudolphinae* (Rudolphine Tables) in 1627, having developed their theory in accordance with his new planetary laws. The predictive accuracy of these tables was two orders of magnitude better than anything previously achieved and did much to recommend Kepler's heliocentric theory to right-thinking minds. The success of heliocentrism was sealed when Pierre Gassendi (1592-1655) in 1631 observed the transit of Mercury across the disk of the Sun—just as Kepler predicted.

STEPHEN D. NORTON

Further Reading

Books

Cohen, I. Bernard. *The Birth of the New Physics.* New York: W. W. Norton and Company, 1985.

Copernicus, Nicolaus. *Nicolaus Copernicus On the Revolutions.* London: Macmillan, 1978.

Gingerich, Owen. *The Eye of Heaven: Ptolemy, Copernicus, Kepler.* New York: American Institute of Physics, 1993.

Grant, Edward. *Planets, Stars, and Orbs: The Medieval Cosmos, 1200-1687.* Cambridge: Cambridge University Press, 1994.

Hall, A. Rupert. *The Scientific Revolution: 1500-1800.* London: Longmans, Green and Co., 1954.

Kuhn, Thomas. *The Copernican Revolution.* Cambridge, MA: Harvard University Press, 1957.

Articles

Jardine, Nicholas. "The Significance of the Copernican Orbs." *Journal for the History of Astronomy* 13 (1982): 168-94.

Kraft, F. "The New Celestial Physics of Johannes Kepler." In S. Unguru, ed., *Physics, Cosmology, and Astronomy, 1300-1700: Tension and Accommodation.* Dordrecht: Kluwer, (1991): 185-227.

The Gregorian Reform of the Calendar

Overview

Few of us have cause to question whether or not calendars are correct. We hang them on the wall to remind us of what day and month it is, or when holidays fall. In the Middle Ages, one of the most important functions of the calendar was to set the dates for important religious festivals, such as Easter. By the thirteenth century, astronomers began to notice that the calendar currently in use did not correspond to observations; most significantly, the equinoxes did not fall on the days on which they were supposed to fall. In 1582, after centuries of calls for calendar reform, the Gregorian reform was finally established.

Background

Calendars are human constructs that allow us to track the passing of time. Certain aspects of the calendar have an astronomical basis. A day is the time it takes for the Earth to rotate one time, a month is related to the time it takes for the Moon to complete its cycle of phases, and a year is the time is takes for the Earth to revolve one time around the Sun. The difficulty with using these particular events to mark the passing of time is that they are not commensurate; that is, they do not divide evenly one into the other. For example, there are 365.2422 days within a year. If we counted only 365 days within a single year, we would soon find that the seasons were occurring at the wrong time of the year.

The problem of commensurability was recognized in antiquity. The Julian calendar, named for Julius Caesar, was established in the year 46 B.C., and was used subsequently throughout Europe. This calendar corrected the errors that had accumulated, used 365 days per year, and added one extra day every four years (the leap year). Under this scheme, the year was assumed to have 365.25 days. Although this is very close to the true number of 365.2422, even this small difference would amount to an appreciable error over a few centuries.

This error was eventually recognized during the Middle Ages. The most recognizable problem with the calendar was that the vernal (spring) equinox, a phenomenon that could be observed by a trained astronomer, was falling a few days before the traditional date of March 21. This might have been a trivial problem had the peoples of Europe not been Christian. In the fourth century, a church council had decided that Easter would fall on the first Sunday after the first full moon after the vernal equinox. Because calendars could not be mass-produced and distributed over all of Christendom every year, church officials created tables that allowed local priests to figure out when Easter should fall on any given year. The tables, however, assumed that the equinox fell on March 21, and calculated Easter based on that assumption. Because the equinox was clearly falling some days earlier than that, it was feared that Easter was being celebrated on the "wrong" date, and thus calendar reform was needed. While a concern for astronomical accuracy might have been significant to some persons, the overwhelming concern for those who proposed reforming the calendar was a religious one: ensuring that the most important religious festival of the year was celebrated at the proper time.

Impact

Despite the importance that celebrating Easter held for European society, and despite recurring proposals from astronomers, it was nearly three centuries before the calendar was finally reformed. In the thirteenth century, when reform was first seriously pursued, the Pope had been the natural authority to take on such a task. Calendar reform was, after all, primarily a religious concern, and at this time, the Pope was acknowledged throughout Europe as the head of the Christian religion. Various problems, including wars, schisms, and the Protestant Reformation, prevented the papacy from actually accomplishing the reform but in the last quarter of the sixteenth century, Pope Gregory XIII brought together a commission of clergy, mathematicians, and astronomers to reform the calendar. That reform would bear his name: the Gregorian reform.

The basic scheme of the reform came from the work of Luigi Lilio, also commonly referred to by modern authors as Luigi Giglio or Aloisius Lilius. In 1582, when the reform was to take place, the vernal equinox fell on March 11. To bring the equinox back to the traditional date of March 21, 10 days would be removed from the calendar; the dates October 5 through October 14 were dropped from that year (Friday, October 15 came right after Thursday, October 4). In addition, there would be fewer leap years. Leap years would still occur every four years, except that years ending in double zeroes that were not divisible by 400 (i.e. 1700, 1800, 1900, 2100, etc.) would not be leap years. The more complex part of the Gregorian reform was the calculation of Easter by using *epacts*. An epact is the age of the Moon on January 1. Using a table of epacts in combination with astronomical tables listing the dates for new moons, the date of Easter could be calculated perpetually.

The reform was set into effect by a papal bull—an official document of the office of the Pope—entitled *Inter gravissimas*. In addition, a document summarizing the reform, the *Compendium of the New Plan for Restoring the Calendar*, had been produced and sent to various universities to inform scholars of the changes that would take place. The reform, however, was not adopted uniformly throughout Europe. Catholic countries such as Italy, Spain, Portugal, and Poland adopted the reform immediately in 1582. Catholic regions of France, Germany, Belgium, Switzerland, and the Netherlands, adopted the reform within the next two years (dropping a different ten-day period, depending on when the local authorities decided to change their calendars). Other parts of Europe would not change their calendars until the eighteenth century, while nations in other parts of the world did not adopt the calendar until the twentieth century. One famous result of the delay in adopting the calendar was that the 1908 Imperial Russian Olympic team missed the competition because they did not arrange their travel plans in accordance with the Gregorian dates. Today, the calendar is nearly universal, though some religious calendars (such as the Jewish and Islamic) are kept concurrently.

Why was it that the reform was not universally embraced when it was originally mandated in 1582? Influenced by the Protestant Reformation, which had been underway for a number of decades, numerous Protestant rulers throughout Europe did not wish to adopt a reform dictated by the Roman Catholic Church. There was little scientific objection to the reform, as the modification of leap years accomplished the basic task of reconciling the incommensurability of the length of the day and year, while the scheme of epacts was an effective way to establish the date of Easter. Objections were most often framed in political terms stating that the Pope had no authority to enact a reform. This would remain a sticking point for a number of decades.

Some prominent scientists, such as J. Scaliger (1484-1558) and Michael Mästlin (1550-1631), did pose idiosyncratic objections to the reform, usually based on simplifications of astronomical calculations that the reform utilized. None of these objections was especially significant in the long run. The primary Catholic defender of the reform was Cristoph Clavius (1537-1612), a Jesuit mathematician and astronomer, who had been on the papal commission that had approved the reform; he argued that the simplifications were acceptable and indeed necessary for those not trained in astronomy to be able to use the calendar effectively. Two preeminent Protestant astronomers, both imperial mathematicians, also endorsed the reform: Tycho Brahe (1546-1601) immediately began to use the new calendar, while Johannes Kepler (1571-1630) argued that the Gregorian reform was the most effective one available.

In the end, it was recognized that both diplomacy and commerce could be more effectively pursued if all parties operated under the same calendar. In Germany, Gottfried W. Leibniz (1646-1716) advocated a reform based on the proposal of the astronomer Erhard Weigel. This reform, established in 1700, made the same changes as the Gregorian reform with regards to dropping days from the calendar and adjusting the leap years, but used Kepler's Rudolphine Tables, rather than the epacts, to calculate Easter (thus avoiding using a Roman Catholic method for determining the date of Easter, and thereby denying the authority of the Pope in such matters). This brought the calendar into step with countries that had adopted the Gregorian calendar, with the exception that Easter occasionally fell on different dates under the two calendars (this problem was corrected later in the century). In England, the change would not come until the middle of the eighteenth century, when the British government passed a bill in 1752. The days of September 3 to September 13 of 1752 were dropped from the calendar in England and its colonies (eleven days had to be dropped because 1700 had been a leap year under the Julian calendar in England, but not under the Gregorian calendar).

Though the calendar is in large part based on astronomical phenomena, it is ultimately an artificial construct designed to meet the needs of human society. The Gregorian reform of the calendar in the sixteenth century was enacted for religious reasons, and, in some places, was rejected on religious and political grounds. Where it had been rejected on such grounds, it was eventually adopted for practical purposes, as the religious objections no longer carried the political import that they had originally.

MATTHEW F. DOWD

Further Reading

Books

Coyne, G. V., M. A. Hoskin and O. Pedersen, eds. *Gregorian Reform of the Calendar: Proceedings of the Vatican Conference to Commemorate its 400th Anniversary.* Vatican City: Specola Vaticana, 1983.

Poole, Robert. *Time's Alteration: Calendar Reform in Early Modern England.* London: University College London Press, 1998.

Richards, E. G. *Mapping Time: The Calendar and Its History.* Oxford: Oxford University Press, 1998.

Periodical Articles

Gingerich, Owen. "Notes on the Gregorian Calendar Reform." *Sky and Telescope* 64 (December 1982): 530-33.

"Luigi Lilio and the Gregorian Reform of the Calendar." *Sky and Telescope* 64 (November 1982): 418-19.

Moyer, Gordon. "The Gregorian Calendar." *Scientific American* 246 (May 1982): 144-52.

Newton's Law of Universal Gravitation

Overview

In 1687 English physicist Sir Isaac Newton (1642-1727) published a law of universal gravitation in his important and influential work *Philosophiae Naturalis Principia Mathematica* (Mathematical Principles of Natural Philosophy). In its simplest form, Newton's law of universal gravitation states that bodies with mass attract each other with a force that varies directly as the product of their masses and inversely as the square of the distance between them. This mathematically elegant law, however, offered a remarkably reasoned and profound insight into the mechanics of the natural world because it revealed a cosmos bound together by the mutual gravitational attraction of its constituent particles. Moreover, along with Newton's laws of motion, the law of universal gravitation became the guiding model for the future development of physical law.

Background

Newton's law of universal gravitation was derived from German mathematician and astronomer Johannes Kepler's (1571-1630) laws of planetary motion, the concept of "action-at-a-distance," and Newton's own laws of motion. Building on Galileo's observations of falling bodies, Newton asserted that gravity is a universal property of all matter. Although the force of gravity can become infinitesimally small at increasing distances between bodies, all bodies of mass exert gravitational force on each other. Newton extrapolated that the force of gravity (later characterized by the gravitational field) extended to infinity and, in so doing, bound the universe together.

The impact of Newton's law of gravity was initially more qualitative than quantitative. Newton's law of gravitation, mathematically expressed as $F = (G)(m_1 m_2)/r^2$, stated that the gravitational attraction between two bodies with masses m_1 and m_2 was directly proportional to the masses of the bodies, and inversely proportional to the square of the distance (r) between the centers of the masses. Accordingly, a doubling of one mass resulted in a doubling of the gravitational attraction, while a doubling of the distance between masses resulted in a reduction of the gravitational force to a fourth of its former value. Nearly a century passed, however, before English physicist Henry Cavendish (1731-1810) was to determine the missing gravitational constant (G) that allowed a reasonably accurate determination of the actual gravitational force. Regardless, the parsimony of Newton's law made its quantitative application easy to translate to problems in astronomy and mechanics.

Newton admitted having no fundamental explanation for the mechanism of gravity itself. In *Principia* Newton stated, "...I have been unable to discover the cause of those properties of gravity from phenomena, and I feign no hypotheses (regarding its mechanism)." Moreover, Newton asserted, "To us it is enough that gravity does really exist, and act according to the laws which we have explained, and abundantly serves to account for the motions of the celestial bodies and our seas." In his later work, *Opticks*, Newton raised the possibility that the gravitational force might be conveyed through a medium or "ether."

Regardless, that the force of gravity had calculable and measurable effect on the universe proved vastly important to astronomers and scientists. It was useful enough to explain that gravity was the force accelerating planets in their orbits (i.e. keeping the direction of orbital motion forever changing toward the Sun). Paradoxically, the widespread acceptance of Newton's law was enhanced by being detached from an underlying mechanism. Accordingly, Newton's law became a powerful descriptive and predictive tool that could be used as a confirmation of the existence of God or, in the alternative, as proof that no divine intervention was needed to move the heavens.

Impact

Newton's law of gravitation proved to be a precise and effective tool wherever applied. A truly universal law, it could be verified by the simplest fall of an apple or measured against the most detailed observations of celestial movements. In the twentieth century Newtonian mechanics, based in part on Newton's laws of universal gravitation, still proved accurate enough to guide the navigation of spacecraft.

Although Newton's law of gravitation offered no fundamental explanatory mechanism for gravity, its usefulness of explanation lay in a higher level of cause and effect. Using the law allowed physicists and astronomers to bridge the relationship between cause and effect with regard to falling bodies and orbiting planets.

In addition, Newton's law found widespread acceptance and usage because it was a universal law that related information about bodies far removed from direct experimentation. Observation and induction based upon Newton's law yielded rich insight into the workings of the natural world, and Newton's law of universal gravitation became a powerful impetus to further the generalizations of natural laws gleaned from the rise of experimentalism during the Scientific Revolution. Newton's law became a powerful and testable verification that the cosmos could yield to inductive reasoning (i.e. where a more general case is used to reason to a more specific case).

Providing proof of the universality of Newton's law provided an impetus to eighteenth-century astronomers, including German-born English astronomer William Herschel (1738-1822). Although Herschel is most famous for his discovery of the planet Uranus in 1781, his celestial surveys not only provided an extensive star catalogue, they also provided abundant and unwavering validation of the universality of Newton's law.

As a derivation of Kepler's second law, Newton's law mathematically fulfilled all of the requirements of a force propelling planetary motion. In accord with both Kepler's laws and Newton's laws of motion, the Sun was at the focus of the elliptical planetary orbits exerting a gravitational pull that, in specific accord with Kepler's third law, resulted in the proper relation of the planets' sidereal period to their mean distance from the Sun. Because the force of gravity directly depended upon the masses of the bodies, Newton's law of universal gravitation was also in accord with Newton's own third law of motion. More importantly, Newton used the law of universal gravitation to actually correct a defect in Kepler's third law. Kepler had failed to consider the gravitational influence of the smaller planetary body on the greatly more massive Sun. Newton's refinement and advancement of a mutual gravitational force proved important to the determination of subtleties in orbital mechanics that ultimately allowed the prediction of masses for the planets and other celestial objects. Ultimately, Newton's law of universal gravitation would, in the twentieth century, provide evidence of the existence of black holes.

The Newtonian methodology of simplifying mass to a point (i.e. with regard to gravitational

Isaac Newton. *(Bettman/Corbis. Reproduced with permission.)*

fields, all of the mass of a body can be considered to lie in a center of mass without physical space) also proved a brilliant simplification that enabled the mathematical advancement of mechanics and electromagnetism.

Newton's law of universal gravitation also laid the template for the articulation of subsequent physical law. For a century after the publication of *Principia,* scientists tried to seize on Newton's law of universal gravitation to explain other at-a-distance phenomena (e.g. magnetism). Many failed in their attempts to characterize electrical and magnetic phenomena as forces analogous to gravitational force because they failed to properly integrate the effects of infinitesimal forces. By the start of the nineteenth century, however, it was discovered that electrostatic forces, the force between two charged particles, indeed was mathematically similar to Newton's law of universal gravitation.

The magnitude of the electrostatic force was found to be directly proportional to the product of the magnitudes of the charges and inversely proportional to the square of the distance between the charges. Accordingly, both the electrostatic force and the gravitational force obey Newton's third law, are characterized by magnitude and direction (i.e. forces can be added as vectors), act at a distance through seemingly empty space, and are inverse square law forces.

Although there were important differences (e.g. although gravitational force is always attractive, electrostatic forces can be both attractive and repulsive), the formulation of the electrostatics force—a force much stronger than gravity, as articulated in Coulomb's law—was built on the Newtonian formulation of gravitational force.

The lack of a fundamental explanation for the actual mechanism of gravity, and Newton's own speculations, led to the assumption that there must be some universal or cosmic ether through which gravity acted. Although the need for such an ether was dispelled by Scottish physicist James Clerk Maxwell's (1831-1879) development of a set of equations that accurately described electromagnetic phenomena and subsequently by the late-nineteenth-century ingenious experiments of Albert Michelson (1852-1931) and Edward Morley (1838-1923), the quest for the discovery of a such an ether was to consume physicists until early in the twentieth century, when the need for its existence was rendered moot by the advancement of relativity theory.

Because Newton's law of universal gravitation was so mathematically simple and precise, it strengthened the idea that all the laws describing the universe should be mathematical. Correspondingly, Newton's law also fostered belief that the universe was governed in accord with mathematical laws. In turn this led to the reassertion of a Pythagorean concept of God as the ultimate mathematician.

For theologians the predictability of the law of gravity provided comforting reassurance that the universe was governed by regular laws that provided glimpses of divine revelation. For others, the mechanistic, mathematical, and predetermined influence of gravity left no need for a God to guide the heavens, and they too relied on the predictability of Newton's law to advance their arguments. For both camps, Newton's law was simply sufficient to explain a clockwork universe.

After Newton, the appearance of comets was not to be interpreted as a direct sign from God, but rather, in accord with Newton's law of universal gravity, a natural consequence of the attraction of the Sun for a body traveling through the solar system on a highly elliptical orbit. In essence, Newton's law of universal gravitation, a marvel of scientific reasoning, swept away the supernatural and made the expanse of the universe knowable and predictable.

K. LEE LERNER

Further Reading

Bronowski, J. *The Ascent of Man*. Boston: Little, Brown, 1973.

Cragg, G.R. *Reason and Authority in the Eighteenth Century*. London: Cambridge University Press, 1964.

Deason, G.B. "Reformation Theology and the Mechanistic Conception of Nature." In *God and Nature*, ed. by Lindberg, D.C. and Numbers, R.L. Berkeley: University of California Press, 1986.

Hawking, S. *A Brief History of Time*. New York: Bantam Books, 1988.

Hoyle, F. *Astronomy*. New York: Crescent Books, 1962.

Isaac Newton's *Principia Mathematica* Greatly Influences the Scientific World and the Society Beyond It

Overview

Isaac Newton's (1642-1725) most influential writing was his *Philosophiae Naturalis Principia Mathematica* (The Mathematical Principles of Natural Philosophy), published in sections between the years 1667-86. It united two competing strands of natural philosophy—experimental induction and mathematical deduction—into the scientific method of the modern era. His emphasis on experimental observation and mathematical analysis changed the scope and possibilities of science.

Background

Throughout the medieval period European scholars had relied heavily on the teachings of Aristotle (384-322 B.C.) and the works of a few Christian philosophers. Then, in the late fifteenth century, there was a rediscovery and popularization of other ancient writers, such as Plato (427-347 B.C.), who opposed many of Aristotle's ideas. The intellectual community began to debate the works of these and other ancient writers, often challenging firmly held academic and religious beliefs. However, these debates were framed by the question of which of the ancient writers was correct. Some thinkers began to question the basis of such debate, arguing that new forms of thinking could go beyond the works of the ancients.

René Descartes (1596-1650) found the Aristotelian methods he was taught to be entirely unsatisfactory. He considered them to be based on false assumptions. The only knowledge he found certain was mathematics, and so he used mathematical deduction as the basis of his entire scientific method and philosophy. Descartes' most comprehensive work was his *Principia Philosophiae* (1644), which attempted to put the whole universe on a mathematical foundation, reducing the study of everything to that of mechanics. For Descartes, knowledge could only be gained from deduction from fundamental principles.

Deduction is the method by which consequences are derived from established premises. From the observed or established facts predictions of future events or possible consequences can be deduced. In a sense it is an educated guess. For example, from the premise that all swans are white, if the bird you observe is a swan, then deductively the bird must also be white. As long as the premise and observation are correct then the conclusion must be true. However, deduction can never prove the premise, no matter how much supporting evidence is gathered, as a single contradictory result will overturn the rule. Black swans were discovered in Australia, and so the deduction was incorrect.

Galileo Galilei (1564-1642) also championed the notion of deduction, although in his case from experimental observations. He found the academic focus on ancient knowledge to be suffocating and limiting. He used experimental deduction to show that the universe was not as he had been taught. He observed the mountains and craters of the Moon, which tradition held was a perfect sphere. He saw moons orbiting Jupiter, and observed that the Milky Way was made up of tiny stars. From these observations he deduced that Earth was not the center of the universe, that the planets were not perfect and unchanging, and that the Copernican theory (that Earth revolved around the Sun, not vice versa) was correct.

An alternative to both the reliance on ancient writings and the deductive method was proposed by Francis Bacon (1561-1626). Bacon dismissed deduction as merely the logic of argumentation. He preferred induction, which is the process of reasoning from particular events to general rules. For example, when numerous observations of swans also gave the result that all observed swans are white, then it was induced that all swans were white. As we have seen, this is not correct.

Bacon did not think that scientists should seek to prove particular theories as Galileo had done. He proposed that scientists should unselectively and objectively collect facts from experiment and observation and then organize and classify them. When enough facts had been collected, then they would be generalized to create a universal theory.

Another inductive thinker was Robert Boyle (1627-1691). His 1661 work *The Sceptical Chemist* argued against Aristotle's views on the composition of matter. His experimental work was wide and varied, and only when he had performed numerous variations of an experiment would he then induce general rules to explain the results.

Like these other scientists, Newton also found the stale debates over ancient writings to be frustrating. Initially he turned to the mathematical deduction of Descartes as an escape from Aristotle. However, the more he considered Descartes's ideas the more he disagreed. He was also influenced by the works of Nicolaus Copernicus (1473-1543), Johannes Kepler (1571-1630), and Galileo, and began to combine them all together. The work that resulted was his *Principia Mathematica*.

Newton's *Principia* was mainly a description of the laws of planetary motion. However, it also contained more universal material that was to influence the way all science developed. In effect it combined the methods of induction and deduction. Newton agreed with the inductionists that first a scientist should establish the facts by careful observation and experiment. However, he then proposed using deduction from already known principles to formulate new hypothetical principles. Then laws of nature could then be induced. These new laws could be tested by further experiment and observation, and so on.

Impact

The *Principia* received good reviews, perhaps because some were written by close friends of Newton. The book was an all-encompassing explanation of physics, starting with definitions of mass, force, and motion, providing mathematical explanations of these principles, and going on to explain planetary motion, lunar motion, the ocean tides, and many other things besides.

Descartes had stressed the importance of mathematics, and on this point Newton agreed. The *Principia* established mathematics as the language of science. Mathematics became a means of knowing about the universe. However, the *Principia* was directly, and deliberately, opposed to Descartes' philosophy. Newton fundamentally disagreed with the separation of spirit from matter that existed in Descartes' mechanical view of the world. In part Newton's work was an attempt to restore the place of God in science. Newton used his mathematical method to show that Descartes' system of mechanics was impossible.

However, despite the problems with Descartes' theories they remained popular on the European continent, particularly in France, for nearly one hundred years. There were, however, a number of non-British scientists who followed Newton's ideas. The prominent French intellectual Voltaire (1694-1778) was in England at the time of Newton's funeral, and was impressed with the scientific culture he found there. He wrote a glowing description of the British intellectual climate, but these writings were immediately banned in France.

The *Principia* provided a standard for doing scientific investigations, and with his other published works, such as *Opticks* (1704), formed the cornerstone for the modern scientific method. It offered a coherent method that seemed free of the occult and reliance on the ancients. However, Newton's influences included alchemy, unorthodox religious ideas, and a belief that God had given the ancients the true secrets of science and religion.

The *Principia*'s focus on experiment and observation seems to owe much to the ideas of Bacon. Yet Newton had been more strongly affected by alchemical philosophy, partly because of its mystical and religious elements. He found alchemy's reliance on experiment to be more solid than many other forms of study. He also preferred its description of the universe as a living force over the mechanical philosophy of Descartes. Newton wrote over a million words on his alchemical studies, but published nothing.

Newton had been careful to include God in his overall plan of the universe. Against Newton's wishes later followers of his ideas tended to reduce, or eliminate completely, the religious aspect of his theories. Later editions of the *Principia* often edited out the philosophical sections,

and emphasized Newton's mechanics. In a sense the mechanical views of Descartes were eventually triumphant, but only within the setting of Newton's theories.

Newton's success enabled him to wield great influence in British scientific affairs. He was careful to promote the careers of those who supported his ideas, and obstruct those who opposed him. A culture of Newtonianism grew, helping to spread the ideas and changing the way science was performed. Britain developed a more practical and hands-on scientific approach than the rest of Europe. The emphasis on experiments was different from the contemplative, hands-off philosophy that remained popular elsewhere.

Newton's methods lent themselves easily to everyday applications in mining, agriculture, and industry. Newtonian mechanics could be applied to drain swamps, construct bridges, and pump air into deep mines. Newton's writings helped change the perspective of his followers, and they saw the world with practical, and mechanical, eyes.

The ideas of the *Principia* were used outside of science as well. Indeed, Newtonian mechanics was applied to almost anything, including society itself. John Locke's (1632-1704) democratic philosophy, one of the sparks of the revolutionary period, used Newtonian concepts. Newton even revolutionized the Freemasons (a fraternal order who adopted the rites of ancient religious orders), who introduced new rituals modeled on his philosophy.

Newton's writings took on the form of doctrine to many later scientists, particularly in England. His findings were often held to be unshakable, even when experiments and analysis using his own method showed them to be wrong. In the eighteenth century many physicists insisted that there were only seven colors, the ones Newton had shown with his prisms. Work that contradicted the old master was discouraged or denied. In nineteenth-century England there was fierce resistance to the wave-theory of light, as it opposed Newton's corpuscular model. While the culture of Newtonianism helped spark the Industrial Revolution, it later held back research into new areas.

Newton's ideas were practical, and his method allowed predictions and discoveries to be made. Perhaps the most spectacular use of Newton's laws of planetary motion were the calculations made by U. J. J. Le Verrier (1811-1877). On the basis of slight variations from Newtonian calculations of the orbit of Uranus he predicted the existence of a new planet, Neptune. However, a similar wobble in the orbit of Mercury was shown by Albert Einstein (1879-1955) not to be caused by another planet, but rather due to effects of relativity, which was to supersede Newtonian mechanics in the twentieth century.

DAVID TULLOCH

Further Reading

Dobbs, Betty Jo Teeter, and C. Margaret Jacob. *Newton and the Culture of Newtonianism*. Atlantic Highlands, NJ: Humanities Press, 1995.

Gjertsen, Derek. *The Newton Handbook*. London: Routledge & Keegan Paul, 1986.

Hall, A. Rupert. *Isaac Newton—Adventurer in Thought*. Oxford: Blackwell, 1992.

Koyré, Alexandre. *Newtonian Studies*. London: Chapman & Hall, 1965.

Westfall, Richard S. *Never at Rest—A Biography of Isaac Newton*. Cambridge: Cambridge University Press, 1980.

From Alchemy to Chemistry

Overview

At the beginning of the seventeenth century, chemistry remained in its infancy. Scientists still had not agreed upon language to describe chemicals and had no ways of classifying them. In addition, chemistry played a role in many different fields that did not necessarily share knowledge with one another: medicine, metallurgy (the science of metals and their uses), pottery making, glass manufacturing, and alchemy. The field that had the most direct impact on the birth of modern chemistry was alchemy. Alchemy was a combination of philosophy, religion, and primitive science whose chief goal was the perfection of matter. This goal included the conversion of metals into gold and the discovery of a potion that would cure all disease. Many scientists of the time viewed chemistry as a pseudo-science much like astrology and palm reading are viewed today. The work of Robert Boyle (1627-1691) helped to change this impression and led to the establishment of chemistry as an independent, modern science.

Background

Throughout the Middle Ages, alchemists tended to cloak their written work in symbolism and secrecy. This was partly due to the religious circumstances of the times. Many faced the threat of the Inquisition if their experiments were looked on unfavorably. The Inquisition was established in the thirteenth century by the Catholic Church to try people who rebelled against religious authority. The punishments of the Inquisition could be severe and even deadly. Another reason for alchemists' secrecy was their fear that any powerful

Robert Boyle. *(Library of Congress. Reproduced with permission.)*

secrets they uncovered might lead to great evil if they fell into the wrong hands. As a result, alchemists used symbols and coded language that made it nearly impossible for an outsider to understand them. For example, mercury is referred to in their writings as green lion, venomous dragon, mother egg, or doorkeeper.

Alchemists believed that all matter was made of the same four elements (earth, air, fire, and water) in different arrangements and proportions. In other words, silver was thought to consist of earth, air, fire, and water as were frogs, bricks, and everything else in the universe. Alchemy consisted of trying to alter the proportions of these elements to make desired substances. According to this idea of matter, every chemical reaction was a kind of transmutation. Transmutation is a change in which one type of matter becomes another. For instance, alchemists believed that lead could be transmuted into gold. (It is now known that transmutation can occur only in special circumstances, such as during some types of radioactive decay.)

THE GOLDEN TOUCH

Although the central beliefs of alchemy were discredited centuries ago, people have never lost their fascination with the idea of creating gold from other substances. In fact, with the introduction of modern nuclear chemistry in the twentieth century, it seemed that such a goal might be possible. When atoms of an element are bombarded by high-speed particles, the atoms will sometimes break apart into a lighter atom and one or more particles or into two lighter atoms. As a result, the original atoms are transmuted from one element to another. In 1980, scientists at the University of California at Berkeley fired charged atoms of carbon and neon at the metal bismuth. This experiment transmuted part of the bismuth into gold. However, the sample of gold the scientists produced was so small that it was worth only one-billionth of one cent, or $0.00000000001. The experiment itself cost about $10,000. So, it appears that although it may be possible to change base metals into gold, transmutation is probably not the most practical way to make a fortune.

STACEY R. MURRAY

One of the first alchemists to break with centuries of secrecy was the German alchemist Andreas Libavius (1540?-1616). In 1597, he published what is considered by some to be the first chemistry textbook. This book summarized the knowledge of the alchemists in clear language that anyone could understand.

Another book, however, was to have an even greater impact on the traditions of alchemy. In 1661, the Irish chemist Robert Boyle (1627-1691) published *The Sceptical Chymist*. In this book, he opposed the alchemists' theory of the four elements. A central argument that the alchemists presented for this theory was the case of burning wood. The wood gave off fire as it burned. The smoke represented air, and liquid that boiled off the ends of the wood represented water. The ashes left behind were considered earth. In other words, the wood broke down into the four elements of fire, air, water, and earth. Boyle, however, argued that some substances, such as gold and silver, could not be reduced to these elements by burning. He also observed that some substances seemed to break down into more than four elements.

Alchemists saw the four elements as mystical substances whose existence could be reasoned by logic alone. Boyle, on the other hand, believed that elements were concrete substances whose existence could be only verified by experiment. He did not necessarily reject the four elements, and he did not offer a list of replacements for them; however, he wanted chemists to establish the elements based on scientific observations.

Like Libavius, Boyle opposed the mysterious language of the alchemists. He wrote, "And indeed I fear that the chief reason, why chymists have written so obscurely of their three principles, may be, that not having clear and distinct notions of them themselves, they cannot write otherwise than confusedly of what they but confusedly apprehend." He felt that the alchemists' secrecy kept true scientific advances from being made.

Impact

With his book, Boyle helped to transform alchemy into chemistry. He introduced the experimental method into chemistry that was being used in physics. Boyle helped to draw parallels between these two sciences, showing that chemistry was just as worthy of study as physics. This raised the social and intellectual status of chemists above that of second-rate magicians and reduced their tendency for secrecy.

The new science of chemistry attempted to investigate only that part of the universe that is observable. Unlike the alchemists, the new chemists did not attempt to involve religion and philosophy as a central part of their work. Chemists began to focus on chemical substances and their changes rather than the perfection of matter and humanity. Instead of taking ancient beliefs and trying to put them into practice, chemists attempted to form general rules about the natural world based on their own observations.

Prior to Boyle's work, the main method alchemists had used to analyze chemicals was with fire. Boyle believed that fire, although useful, was not a sufficient way to analyze the chemical composition of substances. Therefore, he searched for and described other methods of analysis, many of which are still used today. These included color tests, flame tests, and examination of crystal shape.

Color tests, naturally, involve color changes. One type of color test is the use of acid-base indicators. These chemicals change color when added to a solution based on the solution's acidity. For example, an indicator might turn an acid red and a base blue. This type of test could therefore be used to distinguish acids from bases. Flame tests involve wetting a chemical with hydrochloric acid and then putting a small sample in a flame. The color of the flame often indicates the composition of the chemical. For instance, copper burns with a bright green flame. Although many of these tests had been developed years earlier by other scientists, they became well known because of Boyle's writing. Only gradually, however, did chemists begin to accept other methods of analysis as being equally important as fire.

During the 1600s, scientists began to see language as a tool to express knowledge clearly and exactly, and in *The Sceptical Chymist,* Boyle helped to clarify chemical classification and naming conventions. The alchemists sometimes had dozens of names for the same chemical. This practice led to great confusion and difficulty of communication. Boyle believed that every chemical should have a single name upon which all scientists would agree. He attempted to connect the names of chemicals with their composition. He had only limited success because the composition of many chemicals was not known at that time. However, the modern naming system used in chemistry is based on composition.

Although Boyle had rejected the four elements of the alchemists, he did not offer an alternative system. Chemists, however, still needed the concept of an element and as a result, they returned to the four elements for lack of an alternative. However, Boyle's work made it possible for chemists in the 1700s to slowly increase the number of accepted elements. An element was eventually defined as any substance that could not be broken down into simpler substances by ordinary chemical means. By 1789, in his *Traité Élementaire de Chimie (Elements of Chemistry)*, the French chemist Antoine Lavoisier (1743-1794) was able to complete a fairly accurate list of the more common elements.

Gradually, the new science of chemistry began making its way into universities. Guerner Rolfinck (1599-1673) began the first university chemistry laboratory in Germany in 1641 at the University of Jena. Throughout most of the 1600s, however, chemistry was not officially recognized at most colleges, except for those in Germany. However, by 1672 Nicolas Lemery (1645-1715) was giving public chemistry lectures in Paris that drew enormous crowds. Chemistry departments began appearing in major European universities in Montpellier (France) in 1673, Oxford (England) in 1683, Utrecht (The Netherlands) in 1694, Leyden (The Netherlands) in 1702, and Cambridge (England) in 1703. Most chemistry programs were initially associated with medical schools, and many of the chemicals produced from university laboratories were tested as medicines.

A scientific revolution had taken place in the 1600s. Astronomers and physicists, such as Galileo (1564-1642) and Isaac Newton (1642-1727), rebelled against the ancient ideas of Greek scientists that had been accepted for centuries. However, a similar revolution did not really take place in the field of chemistry until the next century. During the 1700s, many chemists abandoned the mysticism of the alchemists and began to rely on precise measurements in the laboratory. Eventually, chemical theories based on speculation were replaced by theories based on experiment, and by the mid-eighteenth century, nearly all chemists and physicists had rejected alchemy and transmutation. The spark for the revolution, however, had been set by Boyle during the 1660s.

STACEY R. MURRAY

Further Reading

Cobb, Cathy and Harold Goldwhite. *Creations of Fire: Chemistry's Lively History from Alchemy to the Atomic Age.* New York: Plenum Press, 1995.

Hudson, John. *The History of Chemistry.* New York: Chapman & Hall, 1992.

Multhauf, Robert P. *The Origins of Chemistry.* New York: Franklin Watts, Inc., 1966.

Stillman, John Maxson. *The Story of Alchemy and Early Chemistry.* New York: Dover Publications, Inc., 1960.

Taylor, F. Sherwood. *The Alchemists.* New York: Arno Press, 1974.

Advances in Geological Science, 1450-1699

Overview

During this period, the study of the Earth began to change dramatically from a nearly complete reliance on religion for an explanation of the Earth's features to the beginnings of a more scientific approach. Major steps in explaining the origins of the Earth, its age, the origins of the oceans, and the Earth's geologic features were taken by Agricola (1494-1555), Leonardo da Vinci (1452-1519), Thomas Burnet (1635?-1715), Nicolas Steno (1638-1686), Robert Hooke (1635-1703), and others. Although geology could not yet be considered a science, by the end of the seventeenth century, it was nearly at that point.

Background

Through most of recorded human history attempts have been made to explain how the Earth was formed and why it looks as it does. Every culture, it seems, has developed a creation myth, and many cultures invoked supernatural powers that formed the seas, pushed up mountains, and made the other features we see on Earth.

In the Western world, the best-known creation story is found in Genesis, the first book of the Old Testament. This book is common to the Jewish, Muslim, and Christian religions and it holds that a single deity created the Earth, heavens, and ocean. It goes on to explain that a great flood (Noah's flood) at one point covered the entire Earth with water and, when it receded, the animals carried in Noah's ark repopulated the Earth. The stories of Genesis and Noah had a profound impact on early speculations regarding the formation of the Earth and the nature of its visible features, an impact that lasted for hundreds of years.

Because of the widespread acceptance of the Bible as an infallible history of the Earth, virtually all early theories of the Earth were based on trying to fit observations into a biblical framework. So, for example, the presence of clam fossils far inland was assumed to mean that water had once covered that part of the land, depositing clams that were stranded and fossilized when the waters receded. An alternate theory was that marine fossils found far inland were simply the remains of traveler's lunches that somehow became fossilized in only a century or so. These views of the Earth began to slowly change during the Renaissance.

Agricola (the latinized name of Georg Bauer) was a German mining engineer and scholar who wrote one of the first treatises on techniques of mining. He also wrote about the origins of metal ore deposits, something of great interest to his contemporaries who spent a great deal of time and effort trying to find gold, silver, mercury, lead, tin, iron, and other valuable metals. Although Agricola's theories were not tremendously accurate from a theoretical standpoint, they were a good compilation of empirical evidence he noted in a number of European mines. They also marked one of the first efforts to arrive at a systematic explanation for economically important geological phenomena.

One of the first phenomena to receive more critical inquiry was fossils. Long thought to be either rock formations or relatively recent animal remains, they were looked at in a different light by Leonardo da Vinci. Da Vinci speculated that they were, indeed, the remains of actual animals long dead. Da Vinci's work was followed by that of Nicolas Steno, who wrote a long paper on the origins of fossils, pointing out that fossils tended to be similar in rocks occupying similar positions. Although he did not explicitly suggest using fossils to organize the rock record, this observation was later confirmed as one of the principles upon which modern stratigraphy is based.

More significant than his observations of fossils were Steno's thoughts regarding the original horizontality of sedimentary rocks, their original deposition in parallel beds, and the relative ages of adjacent rock beds. These observations all continue to be taught to geology students today and are, in general, still considered accurate. With these observations, Steno helped set geology on a path towards becoming a science.

At the same time, Steno, Englishman Robert Hooke, and some others were beginning to grapple with the origins of the Earth. No longer content to simply note that the Earth was created in strict accordance with the biblical story just a few thousand years earlier, many were beginning to realize that it might be more realistic to assume a greater age for the Earth. They simply thought that 4,000 or 5,000 years was simply too little time to have formed mountains and oceans, sculpted the land, or to have accomplished all the other factors that must have gone into shaping the Earth they saw. Some, such as Burnet, saw the history of the Earth as an endless cycle of ocean, mountain building, and erosion; while others saw, instead, a finite Earth that had simply developed at a somewhat slower rate than stated in the Bible. Both of these were a step away from biblical literalism and towards a more scientific approach to the question of origins.

Impact

All of these men, and others like them, are important because they mark humanity's first attempts to explain the world around them in other than biblical or mythological terms. To be sure, it was not until the nineteenth century that the Earth's antiquity was well accepted by the scientific community. Indeed, even today there are those who believe in the literal truth of the Bible. However, it was during the Renaissance that humanity took its first steps away from religion and towards science to explain the physical phenomena around them. This, in turn, had several impacts that would reverberate for centuries, and, in some cases, until the present:

1. The beginning of the ascendancy of science to explain physical phenomena.

2. The debate between the respective place of science and religion.

3. The beginning of public interest in nonreligious explanations of natural events.

As noted above, this period of time saw the first attempts to develop explanations for natural phenomena that departed from strictly religious interpretations. In our century, with the benefit of centuries of hindsight, it sometimes seems difficult to believe that intelligent, even brilliant men and women could seriously believe in a literal interpretation of the Bible over the evidence before their eyes. However, we must remember that the world was, at that time, emerging from the Dark Ages, and little thought had gone into many of these topics for over a thousand years. The only references that were available were the writings of Aristotle (384-322 BC) and the ancient Greeks or the Bible. Consider, too, that in a deeply religious age wracked by religious wars, literal belief in the Bible was the norm. Add to that the belief that the Bible was the literal word of God, dictated to Man—not many had the courage, let alone the inclination, to second-guess an infallible source. Finally, even those who dared speculate against biblical facts had difficulty in gaining an audience because it just did not make sense to believe the word of a fallible human being when that word contradicted that of

an infallible God. From that standpoint, the first men (and they were, at that time, invariably men) to suggest the Bible might not be literally true were taking a huge chance. They must also have been extraordinarily confident in their abilities and in what their eyes and reason told them, for they had to break their own conditioning, as well as that of their contemporaries.

In any event, what they accomplished was to develop the first plausible (given the knowledge of the day) secular explanations of many phenomena in the world. Their step was tiny in terms of advancing scientific knowledge, for they made many more mistakes than accurate guesses. However, this step was intellectually huge, because it marked such a radical departure from the beliefs of the day. As we now know, this first step was followed by many more, leading to the Enlightenment and to the society we have today in which the scientific method and scientific techniques are the accepted way to investigate questions of biology, geology, medicine, astronomy, and other parts of the natural world.

During this time, too, a debate began that continues to this day over the respective roles of science and religion in education, theorizing about the world, and everyday life. At the end of the Dark Ages, the Catholic Church was the most powerful political organization in Europe, and one of the most powerful in the world. The Church had a near-monopoly on all facets of daily life in Europe, and was beginning to spread to other parts of the world, carried by missionaries and the Spanish conquistadors. The rise of nonreligious explanations for the world was seen as a direct challenge to the Church, and it reacted protectively by attacking those it perceived as threats to its power. As a result, many thinkers, Galileo (1564-1642) and Nicolaus Copernicus (1473-1543) being the most famous, faced extreme opposition from the Church.

Over the next few centuries, these turf battles eventually came to resolution until, at present, most major religions see no inherent conflict between science and religion. Instead, both sides agree that science addresses those things that can be quantified, examined, and predicted, while religion addresses other areas that are less amenable to the scientific method. So, science tries to tell us *how* the big bang started and what happened afterwards, while religion takes up the issue of *why* the universe exists.

Finally, these early musings over the origins of the Earth, fossils, and landforms caught the attention of some of the public. While the average man on the street was likely unaffected by these debates, those who were more educated were often intrigued by the findings of the nascent scientists. Science was still some time away from general popularity, but this popularity and acceptance began to stir in the days of the Renaissance.

P. ANDREW KARAM

Further Reading

Gohau, Gabriel. *A History of Geology.* Translated by Albert and Marguerite Carozzi. New Brunswick, NJ: Rutgers University Press, 1990.

Gould, Stephen Jay. *Rocks of Ages: Science and Religion in the Fullness of Life.* New York: Ballantine, 1999.

Oldroyd, David. *Thinking about the Earth: A History of Ideas in Geology.* Cambridge: Harvard University Press, 1996.

Christiaan Huygens Makes Fundamental Contributions to Mechanics, Astronomy, Horology, and Optics

Overview

In the period between the death of Galileo (1564-1642) and the rise to fame of Isaac Newton (1642-1727), Christiaan Huygens (1629-1695) stood alone as the world's greatest scientific intellect. His treatment of impact, centripetal force, and the pendulum helped clarify the ideas of mass, weight, momentum, and force, thus making it possible for dynamics and astronomy to advance beyond mere geometrical description, while his wave theory of light helped initiate modern physical optics. Beyond such specifics, Huygens exercised a profound influence on the progress of science through his use of quantitative methods.

Background

The scientific achievements of Huygens were realized under the aegis of a methodology that successfully combined empiricism and rationalism. The empiricist tradition, which found its canonical formulation in Francis Bacon's (1561-1626) *Novum Organum*, was primarily concerned with building knowledge of the world through direct observation and experimentation. The rationalist tradition, whose foremost exponent was René Descartes (1596-1650), eschewed perceptual knowledge as fallible, preferring instead to focus on the certainty attainable through *a priori* reasoning.

Descartes sought to place knowledge of the world on a secure foundation by organizing it into an axiomatic structure similar to Euclid's geometry. A few self-evident truths were proposed as axioms and used to deduce the body of existing empirical knowledge. This rational re-

construction from first principles was meant to undermine the metaphysical speculations of scholasticism and replace the architectonic of Aristotelian physics with a completely new Cartesian physics and cosmology.

A central feature of Cartesian scientific methodology was its reliance on mechanistic explanation. Accordingly, Descartes maintained that the analysis of any natural phenomenon must proceed by considering the motions and direct-contact interactions of matter's various particles. Descartes believed he had metaphysically demonstrated the truth of this principle and the other structural elements of his system, but he conceded that the details of his explanations might be modified by future mathematical and experimental advances.

Huygens accepted the need for mechanistic explanations but was dissatisfied with Descartes' limited application of mathematics to physical phenomena. Though the Cartesian program embodied the ideal of mathematization in its general structure, Descartes produced little in the way of detailed mathematical analysis of physical phenomena. This was disconcerting to Huygens, who felt the subtleties of physical phenomena could only be captured by combining mathematical laws with mechanistic explanation. He also challenged the Cartesian devaluation of experimental knowledge. Though aware of the inadequacies of naive empiricism, Huygens realized, as Galileo had before him, that the mathematical analysis of natural phenomena depends critically on the careful definition and quantification of relevant physical conditions. In this regard, experimental work plays an essential regulative role in the process of scientific discovery.

Critical of the inadequacies inherent to empiricism and rationalism, Huygens combined the best features of each to craft his own research methodology. His strong penchant for mathematization and mechanistic explanation, tempered by a deep understanding of the importance of experimental data, produced stunning scientific successes.

Impact

One of the first problems Huygens addressed was Cartesian impact theory. Descartes believed that once the clockwork mechanism of the universe was set in motion the universe would run indefinitely, requiring no divine intervention; to suppose otherwise implied God was imperfect. Consequently, Descartes maintained that the amount of motion initially imparted to the different parts of the universe must be conserved. He defined a body's motion as the product of its mass and speed. Though motion can be transferred between bodies through collisions, he claimed the total quantity of motion must remain constant. Unfortunately for Descartes, his conservation law disagreed with experiments.

In 1652 Huygens applied himself to the problem and showed that Descartes's principle holds only if speed is taken as a directed quantity—velocity. Huygens collected his results in 1656 in *De motu corporum ex percussione*. When the Royal Society began focusing on the same problem in 1666, John Wallis (1616-1703) and Christopher Wren (1632-1723) were asked to examine the problem anew, and Huygens was solicited for a report of his discovery. The results of all three scientists, obtained independently and published together in the *Philosophical Transactions* (1669), established the law of conservation of momentum.

Additionally, Huygens showed that for elastic collisions the product of the mass times the square of the velocity—called *vis viva* (living force) during the seventeenth century—is conserved. The debate over the nature of *vis viva* was one of the main threads leading to the development of the concept of energy and the conservation principle thereof.

Huygens's mathematical analysis of physical problems had immediate application in observational astronomy. Aided by his theoretical researches in optics, Huygens and his brother Constantijn developed lens-polishing techniques that reduced spherical aberration. They incorporated these lenses and other improvements into their telescopes. With their first instrument, Huygens discovered Saturn's satellite Titan and fixed the planet's period of revolution at 16 days (1655). The following year he provided a correct description of Saturn's ring. He later made the first observation of Martian surface markings and determined that planet's rotational period (1659). Huygens also invented a two-lens eyepiece and an improved micrometer.

Huygens rendered astronomy a greater service in 1657 with his invention of the pendulum clock. There had been many attempts to produce accurate pendulum clocks, and Huygens's first device was a combination of existing elements. This explains why his priority over the invention was challenged by other scientists. Nevertheless, Huygens's design was original in its application of a freely suspended pendulum whose motion was transmitted to the clockwork by means of a fork and handle. He also introduced an endless chain that allowed the clock to be wound without disturbing its progress. More importantly, he realized the pendulum is not quite tautochronous—that its period depends on the amplitude of swing. Huygens solved this problem by devising, through trial and error, fulcrum attachments that altered the arc of the pendulum bob so the period was independent of amplitude. He described his device in *Horologium* (1658).

The appearance of Huygens's clock inaugurated the era of accurate time keeping and revolutionized the art of exact astronomical measurements. Many towns in Holland quickly built tower clocks, and Jean Piccard (1620-1682) later instituted a program of regular horological measurements at the Paris Observatory.

Dissatisfied with the qualitative nature of this work, Huygens undertook a systematic mathematical analysis of the simple pendulum (1659). He quickly derived the relationship between pendulum length and period of oscillation for small amplitudes. In considering the general case, he discovered that the period of a pendulum will not be completely tautochronous unless the arc of swing is a cycloid. Huygens developed the theory of evolutes to mathematically demonstrate the correspondence between his empirically constructed fulcrum plates and those necessary to force the pendulum bob to describe a cycloidal path. By 1669 his study of the center of oscillation had yielded a general rule for determining the length of a simple pendulum equivalent to a compound pendulum. All of these results were published in *Horologium Oscillatorium* (1673), which Florian Cajori (1859-1930) ranked as the greatest work of science next to Newton's *Principia*.

Huygens's mechanistic tendencies are most evident in his studies of gravity and light. His 1659 gravitational researches presupposed and built upon Descartes's vortex theory—gravity is caused by particles of subtle matter swirling with great speed around Earth. Huygens maintained that vortex particles have a tendency (conatus) to move away from Earth's center. In realizing their conatus, vortex particles exert a force on ordinary particles of matter through direct contact, which brings about in the latter a conatus to move toward Earth's center. Thus, the centrifugal force of vortex particles produces a centripetal force in ordinary matter. Fleeing vortex particles are continually replaced, thus maintaining a constant gravitational force.

Next, Huygens established the law of centrifugal force for uniform circular motion as well as the similarity of the centrifugal and the gravitational conatus. He also distinguished between *quantitas materiae* and weight, the former being proportional to the space occupied by ordinary matter, while the latter was treated as a gravitational effect proportional to *quantitas materiae*. This is likely the earliest insight into the distinction between mass and weight. Though Huygens rejected Newton's theory of universal gravitation because it required action-at-a-distance, his own mechanistic account failed to explain satisfactorily how subtle vortical-matter transferred centripetal conatus to ordinary matter.

Huygens's application of mechanistic principles to optical phenomena culminated in his wave theory of light, published in 1690 under the title *Traité de la lumière*. He conceived of light as a disturbance propagated by mechanical means at a finite speed through a subtle medium of closely packed elastic particles. According to Huygens's principle, a vibrating particle transfers its motion to those touching it in the direction of motion. Each particle so disturbed becomes the source of a hemispherical wave-front. Where many such fronts overlap there is light.

Considerable research on the wave nature of light had been done before Huygens, but the significance of his work lies in its systematic treatment that allowed the theory to be fruitfully applied and developed. He mathematically demonstrated the rectilinear propagation of light, deduced the laws of reflection and refraction, and accounted for double refraction in the mineral known as Iceland Spar. However, he was not able to explain polarization.

Newton's corpuscular theory of light dominated eighteenth-century optical thinking, but it was eclipsed by Huygens' wave theory in the early nineteenth century. Though the two views were later synthesized in the quantum theory of light during the early years of the twentieth century, Huygens' principle remains the basis of modern physical optics.

Huygens's work fell into relative oblivion shortly after his death. Nevertheless, his achievements remain an enduring testament to the explanatory power of quantitative methods in the analysis of physical phenomena.

STEPHEN D. NORTON

Further Reading

Books

Bell, Arthur E. *Christiaan Huygens*. London: Edward Arnold & Co., 1947.

Burch, Christopher B. *Christiaan Huygens: The Development of a Scientific Research Program in the Foundations of Mechanics*. Pittsburgh, PA: University of Pittsburgh Press, 1981.

Elzinga, Aant. *On a Research Program in Early Modern Physics*. New York: Humanities Press, 1972.

Struik, Dirk J. *The Land of Stevin and Huygens*. Dordrecht, Holland: D. Reidel Publishing, 1981.

Yoder, Joella G. *Unrolling Time: Huygens and the Mathematization of Nature*. Cambridge: Cambridge University Press, 1988.

Periodical Articles

Cohen, H. F. "How Christiaan Huygens Mathematized Nature."*British Journal for the History of Science* 24 (1991): 79-84.

Erlichson, Herman. "The Young Huygens Solves the Problem of Elastic Collisions."*American Journal of Physics* 65 (1997): 149.

Kubbinga, H. H. "Christiaan Huygens and the Foundations of Optics."*Pure and Applied Optics* 4 (1993): 37-42.

The Emergence of Scientific Societies

Overview

Despite the persistent stereotype of the scientist as a solitary genius, science has always been a communal endeavor. Investigators have sought inspiration from the exchange of ideas, from collaborative experiments, and from personal rivalries within the context of a community of others who share their interests in science. The seventeenth century saw the emergence of one of the most important institutions in the history of science, the scientific society. For more than two centuries, scientific societies and the publications they supported were the primary communication networks for scientists and their work.

Background

As the pace of scientific activity increased during the fifteenth and sixteenth centuries, individuals pursued scientific discoveries within a variety of institutions. Some were employed by rulers or noblemen, others worked within monasteries or at universities, still others were independently wealthy and powerful citizens themselves. Scientific books were published in Latin—universally known to learned Europeans of the time—and sold to the relatively small number of enthusiasts who shared the authors' interests.

In 1603 the first group organized explicitly for the purpose of advancing science was formed. That year, the Accademia dei Lincei was established in Rome by Duke Federigo Cesi and included some of the most prominent scientists of the day, such as Galileo Galilei (1564-1642), Gimbattista Della Porta (1535?-1615), and Francesco Stelluti. They championed the role of experiments in advancing knowledge and the role of a group of experts such as themselves to sanction and support scientific investigation. The Lincei had as many as 32 members during its rather short existence (it was disbanded by 1630 after the death of Cesi and due to increasingly hostile pressure from the Catholic Church). The Lincei published a number of important books, including those of Galileo, each inscribed with the Academy's shield on which a lynx symbolized scientific truth in its struggle with ignorance. Galileo identified particularly strongly with the Lincei and its goals. He used the title "Lynceus" throughout his career, and in his famous dialogue presenting his revolutionary

understanding of the universe it is the "Academician" who speaks for Galileo and for science.

The Accademia dei Lincei had a direct heir in another short-lived scientific society. The Accademia del Cimento was formed by two brothers in the powerful Medici family, Grand Duke Ferdinand II and Prince Leopold. Both men, as well as some of the other members, had been pupils or acquaintances of Galileo and his disciple Evangelista Torricelli (1608-1647). From these men the Medicis had learned a keen respect and enthusiasm for systematic observation and experiment. They sought to apply these empirical methods to a wide range of natural phenomena. The meetings of the Cimento, which began formally in 1657 following several years of less-organized gatherings, took place at Leopold's home. He provided instruments and other services as well as participating in the experiments. The society had just nine working members, which made simpler their goal to work as a team in their investigations. Many of their experiments were tests of Galileo's theories, perfecting methods and techniques for the use of thermometers, pendulums, barometers, and vacuum pumps; others investigated elementary problems in electricity. They published an account of their experiments in 1667, which was subsequently translated into other languages and served as a guide to investigators and other nascent societies throughout Europe. The Cimento dissolved in 1667, when Leopold was named cardinal.

These small, private societies set an example for future scientific groups, and they helped to advance scientific knowledge and practice through their publications. Perhaps the most important paradigm for the development of scientific societies came not from an actual group, but from an imaginary one. Francis Bacon (1561-1626), an English philosopher, was a widely influential spokesman for systematic scientific inquiry. In several important books published during the early seventeenth century, Bacon argued against scholasticism and the reliance on classical texts and for experiment and what came to be called inductive science. He believed that by an iterative process of hypothesis and experimental testing, science could discover Nature's secrets and lead society toward perfection. In his book *The New Atlantis* (1626),

Bacon described a community of scientific workers who would divide the labor of science among themselves and work together to advance knowledge. The "Salomon's House" of this fable was an idealized scientific utopia, conjured to inspire actual scientists to work together in an organized manner.

Impact

From these early, largely inspirational, developments followed the establishment of the two most influential scientific societies, the Academie des Sciences in Paris and the Royal Society of London. These two organizations were institutionally very different. The Paris Academy was founded in 1660 by Louis XIV. It began as a small group of mathematicians and physicists, supported by government stipends, who worked in teams on experiments and theoretical investigations aimed at extending and propagating the work of the French physicist René Descartes (1596-1650). By century's end the Academy had increased its size and extended its activities to include all areas of scientific investigation. Election to the Academy was a great achievement for a scientist, and assured financial support as well as scholarly prestige. The Academy of Sciences became the center of scientific activity in France—perhaps in the world—for most of the eighteenth century. It published its proceedings, which foreshadowed in importance the scientific journals of later centuries, as well as historical accounts of the scientific achievements of deceased members that helped to establish the idea of what a scientific career should be, and summaries of scientific work done by investigators in other countries.

The Royal Society of London was established in 1662 with a charter from King Charles II, but without any financial support. The Royal Society grew out of many years of informal meetings among scholars in London and Oxford. They treated their independence from the government as a point of pride, although the need for the members to support not only themselves but also the activities of the Society made money troubles a recurring problem throughout the Society's first century of existence. The Royal Society was strongly criticized by the Church in its early years, and defenses of the Society written by members Thomas Sprat and Joseph Glanvill provide interesting views into its early accomplishments and ideology.

Like the Academy of Sciences, the Royal Society became a clearinghouse for scientific ideas and reports, and a central node in the developing communication network among scientists throughout Europe and the European colonies. Investigators around the world would send the results of their work or even their chance observations to the learned group in London. The members would then discuss the ideas and—most importantly—publish the accounts in their *Proceedings*. In its early decades, the *Proceedings* featured an eclectic assortment of reports and results; the scientific discoveries of Isaac Newton (1642-1727) and Benjamin Franklin (1706-1790) appear side by side with reports from country farmers about calves born with two heads. Gradually, the judgment of the Society became more discriminating, and publication of work in the *Proceedings* or a discussion of one's results at the Society was an important validation of scientific merit.

The Royal Society and the Academy of Sciences inspired many imitators. During the eighteenth century scientific societies were formed in most of the capitals of Europe and in many of the smaller provinces as well. Societies became a part of the fabric of science, providing a place for like-minded individuals to share ideas and experimental techniques. The belief that science would be advanced by the collaborative work of men and the direct encounter between man and nature was made tangible in scientific societies. The practical benefits of the existence of societies and their publications were enormous. While the primary benefit of the spread of scientific societies would seem to be to the smaller cities where scientists had new opportunities to congregate, in fact the spread of these societies helped to make an international community of science. With their publications and by welcoming traveling investigators to meetings, scientific societies gave those working on the increasingly esoteric and difficult study of nature a reliable means of connecting to anyone else in the world who might share their interests. In addition, many societies sponsored prizes to identify and honor scientific accomplishment, and competitions to focus attention on particularly pressing scientific problems.

By the middle of the nineteenth century general scientific societies and their publications began to be less important to the practice of science than new, smaller, more-specialized groups and journals devoted to particular studies such as physics or botany. Scientists continued to form groups and to network internationally, but as science itself became increasingly complex the multidisciplinary institutions such as the Royal Society and the American Association for the Advancement of Science took on more honorific roles. As scientific training became formalized and absorbed within university curricula, universities became increasingly important to maintaining networks of scientific communication and interaction. Scientific societies, mostly organized according to discipline rather than regionally, have remained important to science throughout the twentieth century and beyond. They are the backbone of the international network of communication and cooperation among scientists.

LOREN BUTLER FEFFER

Further Reading

Hahn, Roger. *The Anatomy of a Scientific Institution: The Paris Academy of Sciences 1666-1803*. Berkeley and Los Angeles: University of California Press, 1971.

Heilbron, John L. *Early Modern Physics*. Berkeley and Los Angeles: University of California Press, 1982.

Lyons, Henry. *The Royal Society, 1660-1940*. Cambridge, England: University Press, 1944.

McClellan, James. *Science Reorganized: Scientific Societies in the Eighteenth Century*. New York: Columbia University Press, 1985.

Ornstein, Martha. *The Role of Scientific Societies in the Seventeenth Century*. Chicago: University of Chicago Press, 1938.

Purver, Margery. *The Royal Society: Concept and Creation*. London: Routledge, 1967.

Pyenson, Lewis and Susan Sheets-Pyenson. *Servants of Nature: A History of Scientific Institutions, Enterprises, and Sensibilities*. New York: Norton, 1999.

Development of Stellar Astronomy

Overview

The advance in astronomy as primarily a progressive improvement in accurate plotting of the stars was reasserted about the middle of the fifteenth century, but within the next 150 years, the discoveries of physical aspects of stars would lead traditional astronomy toward the foundations of astrophysics. The rediscovery of ancient knowledge, called the Renaissance, had advanced in astronomy with improved applications of geometry and trigonometry. Traditional naked-eye sighting instruments progressed in accuracy from the mid-fifteenth century. Astronomers expanded their interest in accurate positional cataloguing by reevaluating ancient appraisal of the stars in terms of relative size and brightness. This led to discoveries of unusual traits in stars; some stars varied in brightness and appeared to pulsate. On rare occasions stars seemed to appear where none had been before. Studying and cataloguing stars by these new features expanded with the introduction and improvement of the telescope through the seventeenth century.

Background

Primitive awe of the celestial pinpoints of sparkling light, the stars, had first prompted study of the heavens as partitioner of time, a natural calendar, and as religion which resolved into ancient astrology, the study of the superior influences of the celestial on every aspect of terrestrial life. Grouping stars into familiar objects such as mythical characters was common to many ancient cultures. Plotting the accurate positions of the stars in the effort to interpret astrological effects on humanity became the basis of astronomy, the study of stars and the heavens as science. Ancient Greek thought passed down earlier conception that the stars were considered a fixed backdrop of the celestial vault that did not change. Aristotle popularized the concept of rigid and individual crystalline spheres holding the sun, planets, and the fixed stars. There was an early tradition in Greek thought that the stars were clouds of fire, fires in general, and even some celestial light shining through pinprick holes in the black cover of the night.

Important aspects of Greek astronomy would have been lost or longer in coming to Europe if not for the various medieval Muslim cultures, which preserved and translated Greek science. Yet improving on this secondhand presentation by translating Greek astronomy from the original Greek was taken up by several Europeans in the fifteenth century, among them the German astronomer Georg Peurbach (1423-1461). By the mid-fifteenth century the important mathematical aspect of Greek astronomy, using geometric and trigonometric methods (use of parallax, defined as the apparent change in position of an object when viewed from different points on earth due to orbit and rotation of the earth) to determine star positions, was revitalized in Europe by astronomers, particularly Johannes Regiomontanus

(1436-1476), a student and heir to Peurbach's efforts, who developed applications of solving problems by triangulation. Accurate charting of the stars would benefit from both mathematical method and improved sighting instruments (sextants, quadrants, and compass-like tools) and their use. Regiomontanus established a workshop for the construction of astronomical instruments and wrote detailed descriptions of these.

Making a catalogue of the positions of all stars visible to the naked eye was important to the mathematical applications of this data in astronomical almanacs and tables (called ephemerides), calendar making, referencing lunar and planetary motions, astrology, etc. The only comprehensive early star catalogue was that of Greek astronomer Hipparchus (fl. second century B.C.) with about 850 stars, which included spherical coordinates for each. He was the first to designate visual star relative brightness or dimness by magnitude (in his day, first magnitude being the brightest, the sixth being the faintest). This catalogue was preserved and revised significantly by the late Greek astronomer Claudius Ptolemy (fl. 150, d. 180) with additions bringing the total to 1,022 stars in 48 constellations. Star catalogues were also the basis for a Renaissance period of innovation to astronomy, the star atlas, and plot maps of constellations.

Impact

Into the sixteenth century, several astronomers followed Regiomontanus' impetus toward accuracy. Nicolaus Copernicus (1473-1543) himself built his own graduated torquetum (a sort of large three-arm compass), though he was not a regular observer. Gemma Frisius (1508-1555), mathematician of the Netherlands, applied his theory of land triangulation in surveying to astronomy and cartography and made several astronomical sighting instruments, including a long cross-staff with movable sights. Then, in 1572, the stimulus of Mother Nature proved the most efficient of all. What appeared to be a new star in the constellation of Cassiopea shone in early November and had everyone with any instrument attempting to measure the parallax. Among them was Englishman Thomas Digges (1546-1595) who observed the star accurately and wrote trigonometric theorems to find its parallax, which being negligible pointed to a celestial origin.

But the best known observer was the Danish noble Tycho Brahe (1546-1601), beginning to build large and accurate astronomical instru-

ments, who kept a detailed observational log of the new star using his own large sextant (graduated to one minute of angular arc). He measured the angular distances from neighboring stars in the constellation to form the sides of spherical triangles by which he plotted the angular latitude and longitude of the so-called New Star (which was a supernova or exploding star). Significantly, he noted that its position relative to the neighboring stars did not show any measurable parallax, which confirmed that this object was celestial. Thus the celestial sphere was not unchangeable as thought since ancient times and needed to be reappraised. Thinking the star remnant material of the Milky Way, Brahe was about to launch his own campaign of accurate star plotting with a progression of unusually precisely graduated, large instruments using adjustable slits as eyepiece sights to explore that need for celestial reappraisal.

A graphical means of indicating star positions for science and navigation was yet another ancient accomplishment poised for improvement. It was an efficient observational aid, but unless the star positions were actually given coordinates in a useful cartographic projection, they could not be used to locate stars precisely or figure the positions of other stars discovered near them. Ancient peoples including Greek astronomers had made globes and maps of constellations but no Greek examples survived. There had been inaccurate medieval representative depictions, sometimes without showing pertinent stars. The best depiction by the early sixteenth century was the first printed star chart (1515) by German artist Albrecht Durer which depicted two planispheres (for the northern and southern hemispheres) laid on an hourly-graduated zodiac circle (the twelve classical constellations on the ecliptic plane of the earth) with the stars depicted in numerical sequence for each constellation. A stereographic projection of Ptolemy's constellations with their principle stars was done as a single sheet in 1535 by Peter Apian (1495-1552). But the first true atlas in book form was that of Alessandro Piccolomini (1508-1578) in 1540. While he did not use a practical coordinate scale, he depicted the stars as seen from earth in sizes relative to their magnitudes and was the first to use letters to label the prominent stars in constellations, a technique adopted thereafter.

Still standing after 1400 years, the Hipparchus/Ptolemy catalogue was inaccurate and incomplete. With his collection of mammoth sighting instruments for better accuracy and fully aware of how much an accurate star catalogue was needed, Brahe worked from a base of known angular distance coordinates of 22 stars to calculate that of all other naked-eye stars between 1578 and 1591. The coordinates of the resulting 777 stars of this new catalogue (published in 1602) were never in positional error by more than four minutes of arc (amazing eye plotting accuracy). Johannes Kepler (1571-1630), who inherited Brahe's data, was able to bring the later total star count to a less accurate 1,004 (published in the later Rudolfine Tables, 1627). The first star atlas provided with useful spherical coordinates, using Ptolemy's catalogue, was that of Giovanni Paolo Gallucci (1588). Brahe's full catalogue was used in the most complete atlas to the time, the *Uranometria* (1603), compiled by lawyer turned astronomer Johann Bayer (1572-1625). It also contained the accumulated southern hemisphere stars plotted by Dutch explorer Pieter Keyser as well, for a total of over 2,000 stars, all designated with Greek letters for the first time.

Meanwhile, the supernova of 1572 had generated a keen interest in what seemed to be an inconstant field of fixed stars. Copernicus' heliocentric theory (1543), taking the earth away from the center of the universe, had already generated additional theorizing that the stars could not be fixed. Digges, for instance, the leader of the so-called English Copernicans, believed the stars varied in distance in an infinite space. On August 13, 1596, German clergyman/astronomer David Fabricius (1564-1617) noted a bright star in the constellation of Cetus. Its brightness faded into the next year, so it was thought to be another new star phenomenon. But it was a variable star, the first observed pulsating star (fluctuating in brightness). Bayer observed it in 1603 and designated it by the Greek letter omicron (Omicron Ceti) in his star atlas. There was talk of another new star (P Cygni) in Cygnus in 1600 (observed by one W.J. Blaeu), but this would later be identified as yet another irregular type of star. But in 1604, another new star or supernova did appear in the constellation Ophiuchus (the Serpent Bearer), sighted on October 10-11. Kepler, who had not believed before, saw for himself, believing it celestial but of the same Milky Way material Brahe had labeled the 1572 star. He observed it until it faded in 1606.

Mid-seventeenth century astronomers spent time on observations of the stars, detecting not only more variables but also binary or double stars and multiple star groups. There was the added impetus of the revolutionary refracting telescope (using lenses for magnification) and its power to see more stars. Observational astronomer Giambattista Riccioli (1598-1671) used the telescope to good advantage in studying the solar system and turned it on the stars to discover the first observed double star, Mizar in Ursa Major (1643). Dutch astronomer Johann Phocyclides Holwarda (1618-1651) studied Fabricius's Omicron Ceti in late 1638-39 and realized it varied in luminosity as periodic fluctuations, meaning stars must rotate. French astronomer Ismael Boulliau (1605-1694) would discover this star's actual period in 1667. Holwarda's famous compatriot Christiaan Huygens (1629-1695) telescopically found a grouping of three stars instead of the one star tagged as Theta in Orion. To be called the Trapezium, its fourth star was found by Jean Picard (1620-1682) in 1673. English scientist Robert Hooke (1635-1703) discovered the grouping Gamma Arietis in 1665. Observation in the southern hemisphere revealed the double star systems of Alpha Crucis (1685) and Alpha Centauri (1689). Italian physician Geminiano Montanari (1633-1687) discovered the variable character of Algol (Beta Persei) in 1670, though its brilliant flickering had long determined its name of the Demon or Satan's Head. Famous English astronomer Edmund Halley (1656-1742) discovered the variable Eta Cariae in 1677, and German astronomer Gottfried Kirch (1639-1710) did the same for Chi Cygni in 1687.

One of the most influential observational astronomers of the time was Pole Johannes Hevelius (Jan Heweliusz, or Hewelcke, 1611-1687), who brought innovation to the star catalogue and atlas. A committed observer, he studied Omicron Ceti from 1648 to 1662 and renamed it Mira (the Wonderful). Although Hevelius made and used telescopes, he seemed to take a step back by imitating Brahe's naked-eye sighting instruments, rather than contemporary sighting instruments applying telescopic lenses for sights. But he claimed this preference for measuring stellar positions as just as accurate. With a sophisticated rooftop observatory at his home in Gdansk (Danzig), Hevelius compiled a comprehensive catalogue of 1,564 stars, which included the 341 southern hemisphere stars plotted by Edmund Halley (published with an atlas in 1679). Hevelius' catalogue and accompanying atlas (*Uranographia*, 1690) provided wider recognition of Halley's catalogue and its accuracy over Bayer's southern plots compiled crudely by explorers. Hevelius used a graphic projection of the stars as on a celestial globe (rather than from earth as contemporaries did) and added eleven new constellations (seven of which remain in use).

English astronomer John Flamsteed (1647-1719), friend and contributor to Isaac Newton (1642-1727), rounded out the seventeenth century's array of stellar astronomers. Flamsteed, an amateur observational astronomer, was appointed the first astronomer royal at the founding of Greenwich Observatory (1675). Using measuring instruments with telescopic sights, which, contrary to Hevelius's claims, provided the reading of finer measurements, Flamsteed concentrated on observing the moon and the stars with twenty thousand observations between 1676 and 1689, especially using a large sextant accurate to 10 arc seconds. His systematic observations were completed in 1705.

The growing importance of positional astronomy's accuracy to the needs of astronomy and navigation was exemplified in the British furor over the high caliber of Flamsteed's work, which would entail the first Greenwich, star catalogue (nearly 3,000 stars). There was great pressure to publish his catalogue because of its accurate usefulness before it was completed into the early eighteenth century, which did happen against Flamsteed's wishes. His efforts had culminated a 150-year progression of instrumental improvements. But more important was the dimension of new astronomical theory enabled during this time span, lured by heliocentrism and able to systematically dissolve the geocentric cosmos in favor of Newton's mathematico-mechanical one by the late century. This was partially through physical discoveries of the stars. Improving the refracting telescope, joined by the reflecting telescope, pointed to deeper stellar discoveries in a seeming infinite universe but also elevation as a true tool with measuring adaptability. The stars would prove to be tools as well, for those interesting variables were little more than two centuries away from being used as yardsticks to measure the extension of the cosmos itself.

WILLIAM J. MCPEAK

Further Reading

Books

Bennett, James A., ed. *Astronomical Instruments, History of Astronomy An Encyclopedia.* John Lankford. New York: Garland, 1997.

Christianson, John R. *Tycho's Island: Tycho Brahe and his Assistants: 1570-1601.* New York: Cambridge University Press, 2000.

Gingerich, Owen. *The Eye of Heaven: Ptolemy, Copernicus, Kepler.* New York: American Institute of Physics, 1993.

Moore, Patrick. *Watchers of the Stars.* New York: G.P. Putnam's Sons, 1974.

North, John. *The Fontana History of Astronomy and Cosmology.* London: Fontana Press, 1994.

Wightman, W. P. D. *Science in the Renaissance.* 2 vols. Edinburgh: Oliver & Boyd, 1962, vol. 1, ch. 7.

Observing and Defining Comets

Overview

By the middle of the fifteenth century, improvements in the accuracy of astronomical observing instruments and the use of geometrical mathematics presaged a closer look at the strange fiery visitors known as comets and a new era in astronomy beyond dependence on a mix of Greek traditions and instrumental inaccuracy. The sixteenth century marked a landmark period of fortuitous comet appearances, affording a level of observation that pushed professional opinion toward the celestial origin of comets, rather than an ancient terrestrial one. Into the seventeenth century, the appearance of comets and accurate observational data prompted research into past comet records and comparisons which along with the inspiration of the elliptical orbital nature of the solar system and new mechanical theory brought the recognition that comets were celestial, had orbits, and were periodic.

Background

Comets, the transient astronomical phenomena that remain long enough to inspire awe, curiosity, and even fear had remained a puzzle for centuries. Although prior to the fifteenth century there had been diverse opinion on the phenomenon and the state of physical nature, the comprehensiveness of Aristotle's (384-322 B.C.) interpretation dominated the late Middle Ages, particularly in his interpretation of the celestial and terrestrial realms based on a logic of observation and the Greek tradition that all terrestrial matter was composed of four basic elements. These were fire, air, water, and earth, in order of density, light to heavy. Celestial space was formed of a perfect, incorruptible substance called aether. The elemental order fit the logic that the heavy Earth was at rest at the center of the universe with relative circumjacent regions or spheres of the other elements, represented by the oceans, the atmosphere, and a region of fire below the Moon. Beyond that lay the celestial spheres: the Moon, Sun, and planets each within a solid crystalline orb propounded by Aristotle to affirm the unchanging nature of the celestial which spun around the terrestrial realm in concentric circular orbits (defined as a perfect, unending motion) enveloped by a vast firmament of fixed stars. So in what realm did comets fit?

The heavens were considered superior to and the ultimate influence on the terrestrial sphere. This was the basis of astrology from which astronomy developed. The influence also came into play with phenomena within the earthly sphere. Any hot, dry, or fiery byproducts on Earth rose as "exhalations" to the sphere of fire and kindled into meteors, aurora, the Milky Way, and comets, essentially all meteorological phenomena, and all moved by the friction of and influences of celestial motion. Because these were changeable phenomena (even the Milky Way seemed to change through the course of the year) and seemed close to Earth, they were taken as terrestrial in origin. Although there was medieval debate about the possibility of the Milky Way being celestial, comets were accepted as terrestrial phenomena. And since they were engineered from heat and dryness, they were taken as portents of adverse weather conditions and impending bad fortune.

Yet as the century progressed toward the sixteenth century, some thinkers increased the height of the spheres of air and fire, even qualifying a less perfect celestial space, where the four elements might invade to account for growing suspicions that so called fiery impressions, like comets, were celestial. Comets had been observed for centuries but not systematically with naked-eye instruments as with the stars. These instruments were angle-measuring instruments (quadrants, sextants, etc.) meant to plot the locations of the stars and strange patches of dim light seen from Earth before the optical telescope at the end of the sixteenth century. Evidently, the first concerted effort at accurate instrumental

Tycho Brahe. *(Library of Congress. Reproduced with permission.)*

observation of cometary paths was by the Florentine mathematician/astronomer Paolo Toscanelli dal Pozzo (1397-1482). He has left detailed manuscript observations of the comets of 1433, 1449-1450, 1456 (this was Halley's Comet), the two comets of 1457 (May and June thru August), and 1472.

By this period, efforts at making more accurate instruments reflected the growing desire for more accurate astronomical measurement, using revitalized geometric and trigonometric techniques of the ancient Greeks. Among early proponents of the effort was German astronomer Johann Regiomontanus (1436-1476). He suggested that as with celestial objects, measurement by parallactic geometry be used to determine a comet's distance from Earth. Parallax is defined as the apparent change in position of an object when viewed from different points on Earth due to orbit and rotation of Earth. Celestial objects were so far away that there was little or no parallax. The Moon's parallax is about 1°, so a comet's should be more if it were below the Moon or less above it. Regiomontanus, who developed more accurate instruments for his observatory at Nuremberg, began careful astronomical observations which included plotting the path of the comet of January 1472. Though his parallax measurements were inaccurate and thus inconclusive, he had pointed the way to determining the origin of comets.

Impact

In the first half of the sixteenth century, comets appeared to the naked eye in 1500, 1517, 1531-1532, 1533, 1538, 1539, 1547, 1556, and 1558, all believed to be terrestrial until revised opinions brought the question of terrestrial or celestial origin to focus. Surprisingly, it was from astrology that published doubts first appeared, namely that the predicted droughts and other natural disasters of several of these comets did not occur which made their terrestrial origins and effects suspect. Netherlander mathematician Gemma Frisius (1508-1555) applied applications of trigonometry to astronomy, noting that comets had a proper motion against the background stars in his observations of those of 1533, 1538, and 1539. He along with Italian physician Girolamo Fracastoro (1483-1553) and German astronomer and cartographer Peter Apian (Bienewitz, 1495-1552) were the first to note upon observing the comet of 1531-1532 (again, Halley's) that its tail pointed away from the Sun. Apian wrote a tract and illustrated this. They saw this influence of the Sun as contrary to a strictly sublunar object, as Aristotle had defined the comet.

Several thinkers tentatively suggested comets were perhaps some sort of celestial reflection. One of the most stimulating reappraisals of the makeup and origins of comets came from the royal professor of mathematics at Paris, Jean Pena (d. 1558), who became intrigued by the fact that a comet's tail pointed away from the Sun. Perhaps recalling the celestial crystalline sphere idea, in his work on geometrical optics (1557) he suggested comets were made of a transparent crystal-like material through which the Sun's rays could be "refracted" causing an internal fire and the tail. He went on to say that optical measurements proved some comets were above the sphere of the Moon (in celestial space), and he joined a growing number of thinkers, among others Girolamo Cardano (1501-1576), who were denying the existence of a sphere of fire to spawn terrestrial comets.

The 1570s brought momentous celestial events and a crises to the supposed perfection of the celestial sphere. Seemingly, a new star (this was a super nova, an exploding star) appeared in 1572 where no star had been recorded before—impossible in the defined unchanging celestial space. The sight prompted an astronomical ca-

reer for Danish noble named Tycho Brahe (1546-1601), who had begun building large, accurately graduated astronomical instruments. Because this new star slowly faded, many thinkers convinced themselves that it was only an unusual comet. Yet, accurate measurements, particularly Brahe's, showed the object to be celestial. With royal Danish patronage Brahe was able to build and perfect several large, precision instruments set up on the island of Hven off Denmark at his extensive observatory.

When a brilliant comet appeared in 1577, it was keenly observed by many astronomers, none more than Brahe, who recorded it as being far in celestial space and heading for the Sun from a detailed tracking of its total visible path. Other astronomers' detailed observations, particularly those of German Michael Maestlin (1550-1631), prompted the same conclusion. Brahe also decided from his data that comets moved in circular orbits and bolstered this with his observations of comets in 1580 and 1585, and others in 1590, 1593, and 1596. Maestlin's friend Helisaeus Roeslin (fl. c. 1578), offering a more extensive sphere of fire as explaining the comet's celestial aspect, anticipated that comets moved in regular orbits with poles and axes (1578). Brahe's findings, literally, shattered the theory of solid crystal celestial spheres, since the comet intersected some of these supposed spheres. Still there were attempts at qualification, suggesting comets were: below the Moon but in a lesser celestial region; celestial but falling into the upper atmosphere; both celestial and terrestrial (with the elements also found in space); and unexplained miracles.

The strong evidence leaning toward the celestial origin of comets prompted all the more careful observation by astronomers into the seventeenth century. Brahe's assistant and heir to his observational data Johannes Kepler (1571-1630) observed a comet in 1607 (Halley once more), three in 1618, and several others. Though not an exacting observer, he wrote a book on his comet observations (1619), noting that the lack of parallax shift indicated they were celestial. Yet, since he preferred the continued conception of solid orbital spheres separating celestial bodies, he theorized comets moved in straight lines. Some theorized movement in parabolic arcs. Interestingly, the great Italian physicist Galilei Galileo (1564-1642) still agreed that comets were atmospheric in origin but optical phenomenon, not real, so could not be measured legitimately by parallax. René Descartes (1596-1650) argued philosophically that comets were celestial bodies traveling among many solar systems!

The three comets that appeared in 1618 were all the more significant for the new tool used to observe them—the telescope. Though the early century Dutch invention is usually astronomically associated with Galileo, it was not he but a Swiss mathematician/astronomer Johann Baptist Cysat (1586-1657) who first turned the instrument toward the comets of 1618, describing the head and tail—cometary substance still a mystery—with a published work (1619). The path of these comets, as that before in 1607 and those to follow to just after mid century (1652, 1664, 1665), remained a point of contention, because each path was only partially observed and recorded. Through the period, the path of a comet's travel became a matter of defining the comet by two interpretations, as either of permanent or of transitory character. The former defined a celestial object moving in some closed or circular orbit, close to Brahe's assessment. This stance was taken by Adrien Auzout (1622-1691, who built very long telescopes), Giovanni Alfonso Borelli (1608-1679), Giovanni Domenico Cassini (1625-1712), and Pierre Petit (before 1623-c.1677). The second interpretation defined the comet as a transitory object, perhaps of some sub-celestial material, moving in a uniform rectilinear path, the view of Gaileo, Kepler, Polish astronomer Johannes Helvelius (1611-87) who observed four comets, and seminal Dutch scientist Christiaan Huygens (1629-1695).

But in 1680 a comet appeared with a trajectory which allowed its path of travel to be observed before and after perihelion (closest approach as it orbits around the Sun), and the outcome did not fit the theory of rectilinear motion. The celestial origin of comets was now becoming widely accepted. One of the most comprehensive collections of data on this comet was that of English astronomer John Flamsteed (1646-1719). His famed compatriot physicist/mathematician Isaac Newton (1642-1727) took that data and applied his evolving mechanical principles of motion and deduced that comets were matter and would be attracted to the Sun as the planets. As also with the planets, the comet's rectilinear inertia would be drawn to the central pull of the Sun, so that the comet's path would be a conic section, a parabolic orbit which he calculated from three position observations.

It was left to astronomer Edmond Halley (1656-1742) to most effectively apply Newton's

mathematical technique. He collected the observational records of 24 comets between 1337 and 1698 and used the parabolic formula of comet orbit to find three with similar trajectories (that of 1531-1532, 1607, and the recent comet of 1682 which he had carefully observed). He also noted the comet of 1456 but that there was a lack of accurate observations on it—not aware of Toscanelli's data. The trajectories were very similar, and since they all were between 75 to 76 years apart, he theorized that this was the same comet and that its periodic character indicated a return in 1758. The comet did return, and became known as Halley's Comet.

The sixteenth century debate over the origins of comets as terrestrial or celestial had helped to redefine the makeup and extension of Earth's atmosphere and provided a first resolution of the boundaries between atmosphere and celestial space. Aristotelian chemical conceptions of elemental constituents of terrestrial and celestial regions had been significantly challenged and ultimately dismissed within the context of the cometary question. So to, the progressive opportunities to apply and record accurate measurement of cometary trajectories, bolstered proof of set laws of celestial motion, first set down by Kepler, departing from the complexities and frustrations of the stubbornly upheld Ptolemaic system. Thus by the beginning of the eighteenth century, comets were commonly accepted to be celestial bodies, moving in defined orbits which could be periodic like the planets, and obeying the mathematics of the newly defined physical laws of Newton and his contemporaries.

WILLIAM J. MCPEAK

Further Reading

Drake, Stillman, C.D. O'Malley. *The Controversy of the Comets of 1618*. Philadelphia: University of Pennsylvania Press, 1960.

Hellman, C. Doris. *The Comet of 1577: Its Place in the History of Astronomy*. New York: ANS Press, 1971.

Heninger, S.K. *A Handbook of Renaissance Meteorology*. First reprint edition. New York: Greenwood Press, 1968.

Jervis, Jane L. *Cometary Theory in Fifteenth-Century Europe*. Dordrecht: D. Reidel, 1985.

Thorndike, Lynn. *History of Magic and Experimental Science*. 8 vols. New York: Columbia University Press, 1964-66, vols. 3-8.

Wightman, W.P.D. *Science in the Renaissance*. 2 vols. Edinburgh: Oliver & Boyd, 1962, vol. 1, ch. 7.

The Rise of the Phlogiston Theory of Fire

Overview

The late seventeenth century saw the rise of the phlogiston (pronounced FLO-jis-ton) theory of fire, which sought to explain the burning of objects. First proposed by Johann Joachim Becher (1635-1682) and Georg Ernst Stahl (1660-1734), the phlogiston theory evolved into an complete theory of the chemical sciences. It was modified by later followers, giving it coherence, but also exposing its weaknesses. Phlogiston determined the direction that chemistry was to take for the next one hundred years, suggesting not only what experiments to perform, but how to interpret the results. The theory was eventually overturned by the concept of the combustion of oxygen, but only after a protracted series of debates and experiments.

Background

The nature of fire has long been a source of wonder and mystery. Aristotle (384-322 B.C.) combined fire with air, water, and earth to explain the composition of all things. For Aristotle, when wood burnt the flame was the element of fire escaping, any vapor was air, moisture was water, and the ash that remained was the element of earth. Aristotle's system dominated medieval European thought, partly due to the Church's adoption and modification of his philosophical ideas.

The European Renaissance helped revive the works of other Greek philosophers, such as Plato (427-347 B.C.), who had been Aristotle's teacher. Plato had proposed a "burnable principle" that existed within inflammable objects. This fitted well with the alchemical notions of the Renaissance, and the burnable principle became associated with sulfur, or "some vague spirit of sulfur." A new system of elements evolved, with substances explained by a combination of sulfur, mercury, and salt. So wood burned because it contained sulfur, gave off flame because it contained mercury, and left ash because it contained salt. However, Aristotle's four elements were still used, sometimes in conjunction with the new system, creating a confused mix of explanations for the nature of fire.

Johann Becher was an alchemist and adventurer. He sold a process for turning silver and sand into gold to the Dutch government, told tales of handling a stone that made him invisible, and claimed to have seen Scottish geese that lived in trees and hatched eggs with their feet. In his science Becher followed alchemical ideas, but these were a confused body of work and he attempted to order and simplify some of the common notions of his time. First, he proposed an "oily spirit" as an essential principle of matter, but he also used the alchemical principles of mercury, sulfur, and salt. Later he altered Aristotle's four elements, keeping only water, and dividing earth into three separate substances. One of these, "oily earth" (terra pinguis), would form the basis of the phlogiston theory. Becher stated that in order for a substance to burn it must contain oily earth.

Becher's ideas were not clearly explained in his writings, and it is thanks to his disciple, Georg Stahl, that the theory was popularized. Stahl's writings were still heavy reading, but he published widely. Stahl's phlogiston was a material principle that escaped from objects when they burned. In the act of escaping the phlogiston caused violent motion, the flames, sparks and explosions of various combustions. So burning wood was explained as the substance phlogiston escaping from within the wood, causing the motion of flames and sparks as it left, and leaving behind an ash with little or no phlogiston.

Impact

The phlogiston theory became popular in the German states and then spread quickly across Europe. Its acceptance was wide and lasting, dominating the chemical sciences for a hundred years. The theory was successful for a number of reasons. It was a development of known theories of combustion. It appealed to those familiar with the alchemical notions by retaining the importance of sulfur, but also explained the combustion of substances that appeared to contain no sulfur. It was based on Aristotle's four elements, and used Plato's idea of a burnable principle. In a sense phlogiston united all these competing ideas, allowing supporters of each to easily adopt the new theory. Even the name phlogiston was not new; it was derived from the Greek word *phlogistos* (to burn), and had been coined early in the seventeenth century.

The phlogiston theory was passed on to entire generations of academics by Stahl's teaching.

Stahl believed that his theory was divinely inspired, and that the common herd (which included his students) would not understand it. His intentionally uninteresting lecturing style made the details of phlogiston unclear, but his students dutifully copied and repeated his words. Stahl was also a very successful physician—he became the King of Prussia's personal doctor. The popularity of his medical text, *The True Theory of Medicine* (1708), also helped to spread the phlogiston theory.

Other theories of the nature of fire, some of them far closer to modern theories, lost out to phlogiston's popularity. There had been a strong British trend of thought that had suggested air as the supporter of combustion and life. However, the ideas of Robert Hooke (1635-1703), John Mayow (1643-1679), Stephen Hales (1677-1761), and others were not popular on the European continent. Opponents of phlogiston were smothered by the sheer weight of numbers of phlogistonists that trumpeted the new theory.

Phlogiston, although originally introduced to explain combustion, became the center of a whole system of chemistry. The wide scope of phlogiston's supposed attributes led to the theory becoming the first unifying principle of the chemical sciences. Breathing was explained as part of the process by which food was burnt in the body. Phlogiston would escape from the burning food and was expulsed by the lungs.

The theory also explained the process of metal calcination (rusting). When a metal rusted phlogiston escaped, leaving behind a lighter, more fragile, substance. Experiments were made to add phlogiston back to rusts, and were often successful. By adding a substance rich in phlogiston (charcoal) to certain rusts and heating them the metal was restored. Even more convincing was the fact that this process gave off a gas that was not capable of supporting fire or breathing. This was seen as proof that all the phlogiston had been taken out of the surrounding air and returned to the metal. Indeed, it is a simple and coherent explanation.

Phlogiston was also said to be the foundation of color. Many substances when burnt or rusted changed color. The change in color was explained by the phlogiston leaving during the process. Phlogiston was also said to be completely indestructible, nonelastic, dry, and imperceptible to all the human senses. Even thunder and lightning could be explained by phlogiston. Lightning flashes were the combustion of concentrations of phlogiston in the air, and

Johann Becher. *(Library of Congress. Reproduced with*

thunder was the result of the collapsing dispersed air as the phlogiston escaped.

While the phlogiston theory had come from Becher's mind as conjecture based on older philosophical ideas, it helped inspire practical experimentation. Stahl performed a number of key experiments that were published widely. The results were easily explained by the internal logic of the phlogiston theory. Across Europe chemists dutifully repeated Stahl's work, and guided by his writings they also came to the same "obvious" conclusions. This new trend to finding practical support for a theory, no matter how misguided, is one of phlogiston's most important legacies.

Although phlogiston was considered a real substance, it was not originally conceived as having any weight. To Becher and Stahl it was a substance as insubstantial as sunlight, but like sunlight it could still have dramatic effects even if it could not be contained or measured.

Later supporters of phlogiston began to alter the theory from the original concepts of Becher and Stahl. Often this was because the complex, vague, and sometime contradictory writings of the founders were misunderstood. By the 1730s most phlogistonists regarded their imagined substance as having an actual weight. This had dramatic consequences for the theory. If phlogiston had weight then when it left a substance during combustion or rusting the remaining material should weigh less. However, a number of experiments gave results that conflicted with this. For example, when a metal rusted (gave off phlogiston) the material left behind often weighed considerably more than the original metal.

Some scientists suggested that phlogiston had a negative weight, and so its absence made materials heavier. However, this did not seem to apply to all situations—for example, when animals breathed (expelled phlogiston) they did not appear to gain weight. The negative weight was not accepted by all supporters, and alternative versions of the phlogiston theory began to appear.

The discovery of new gases, later identified as hydrogen and oxygen, which burned brighter and more fiercely that normal air, also caused problems for the phlogiston theory. More and more alterations were made by supporters to answer the critics. Fierce debates raged between the two sides. This had the effect of making the anti-phlogistonists more rigorous in their research and experimentation. In order to gain support for his alternative theory of combustion, Antoine Lavoisier (1743-1794) produced many papers carefully describing his methods and analysis. Over time the weight of this evidence convinced more and more scientists to abandon the phlogiston theory and adopt Lavoisier's new explanation of oxygen combustion.

The phlogiston theory died a lingering death, with some supporters like Joseph Priestley (1733-1804) maintaining its truth against all opposition. The theory also went through occasional revivals as late as the nineteenth century, often for metaphysical reasons as opposed to chemical ones. While it was a mistaken path, phlogiston is often seen as a halfway stage between alchemy and modern chemistry. For while the theory limited the analysis of results it also encouraged experimentation, and was eventually overturned by the weight of printed evidence.

DAVID TULLOCH

Further Reading

Conant, James Bryant. *The Overthrow of the Phlogiston Theory: The Chemical Revolution of 1775-1789.* Cambridge, MA: Harvard University Press, 1956.

Partington, J. R. *A History of Chemistry.* 4 vols. London: Macmillan, 1961-70.

Partington, J. R. *Historical Studies on the Phlogiston Theory.* New York: Arno Press, 1981.

White, John Henry. *The History of the Phlogiston Theory.* London: E. Arnold, 1932, reprinted by AMS Press, 1973.

Seventeenth-century Experimental and Theoretical Advances Regarding the Nature of Light Lay the Foundations of Modern Optics

Overview

Johannes Kepler's (1571-1630) early seventeenth-century researches on the nature of light were the culmination of medieval developments in the science of *perspectiva* and inaugurated a century of research that laid the foundations of modern optics. Willebrord Snell (1580-1626) shortly thereafter discovered the law of refraction, which allowed mathematical-physical theories of light to be developed in earnest, while René Descartes (1596-1650) developed a mechanistic wave-theory of light that did much to define the boundaries for future optical studies. Christiaan Huygens (1629-1695) was the first to successfully mathematize the wave picture, while Isaac Newton (1642-1727) developed a corpuscular theory. The two latter views were eventually synthesized in quantum theory during the early years of the twentieth century.

Background

Theories about the nature and propagation of light in antiquity were intimately connected with theories of vision, and implicit in all theories of vision was the requirement that there be direct contact between the visual organ and objects of vision. Different accounts of how this contact occurred were promulgated and developed into opposing schools of thought in Ancient Greece.

The atomists adopted the position of Leucippus (c. 500-c. 450 B.C.), according to whom objects were thought to emit thin films or images of themselves through the intervening space to the eye. This "intromission" theory of vision was an alternative to the "extromission" theory championed by the Pythagoreans. Exponents of the extromission theory believed the eye emitted an invisible fire that "touched" objects of vision to reveal their colors and shape. Aristotle (384-322 B.C.) proposed a "mediumistic" theory whereby objects transmit their visible qualities through the intervening air to the eye.

Euclid (c. 330-c. 260 B.C.) developed the extromission theory in *Optica* (c. 300 B.C.), offering a geometrical theory of perspective in which the apparent size and shape of objects was determined by their distance and orientation with respect to an observer's line of sight. Ptolemy (c. A.D. 100-170) continued in this tradition, teaching the equality of the angles of incidence and reflection. He further maintained that the angles of the incident and refracted light rays had a constant relationship. Also during the second century, Galen (c. 130-c. 200) produced an alternate mediumistic theory. The purpose of Aristotle's mediumistic theory was to provide a physical explanation of how light was transmitted from visual objects to an observer, while Galen's was primarily designed to satisfy physiological criteria derived from the eye's anatomy.

When Greek optical works were translated into Arabic during the ninth century A.D., traditional distinctions were adopted and old arguments rehearsed. Only the work of Alhazen (c. 965-1038) decisively broke with the past. Alhazen proposed a new intromission theory that, for the first time, sought to simultaneously satisfy mathematical, physical, and physiological criteria. Exploiting the geometrical optics of Euclid and Ptolemy in conjunction with his knowledge of ocular anatomy, Alhazen explained the physical contact between an object and observer through intromitted rays. Though difficulties remained, the intromissionist character of vision was never again seriously challenged.

As Greek and Arabic texts became available in the Latin West during the twelfth and thirteenth centuries, the conflicting views of Aristotle and Alhazen held sway. Albertus Magnus (c. 1200-1280), the first great expositor of the Aristotelian corpus, defended the Aristotelian mediumistic theory, according to which light is a state of the medium that makes objects on the other side of it visible. Following Alhazen's lead, Roger Bacon (c. 1214-c. 1294) attempted to display the underlying unity of the major traditions by reconciling them. Most notably, he mathematized Robert Grosseteste's (c.1168-1253) Neoplatonic views on species transference through the medium and posited it as an explication of Aristotle's qualitative transformation of the medium. This produced the doctrine of multiplication of species.

Bacon argued that what was transferred was a series of simulacra called forth successively from the medium. He believed these likenesses were corporeal. Bacon's views circulated widely and helped establish the tradition of *perspectiva* in the West. They were developed in John Pecham's (c. 1230-1292) *Perspectiva communis* and Witelo's (c. 1230-c. 1277) *Perspectiva*. Nevertheless, Aristotelians were still in the majority by far.

Impact

New life was breathed into the perspectivist tradition during the Renaissance partly due to a growing interest in realistic painting. Kepler seized upon these ideas, summarizing and extending them in his own optical researches. In *Ad Vitellionem paralipomena* (1604) he developed a more satisfactory theory of vision—arguing the only way to establish a one-to-one correspondence between points in the visual field and points in the eye was if light rays were refracted through the eye's humors to focus on the retina as an inverted image. He also produced the first analysis of the telescope in *Dioptrice* (1611).

After Kepler it became generally accepted that light was not a modification of the transparent medium, rather that it existed as an independent thing whose properties could be inquired into. It was likewise accepted that light was emitted by luminous bodies and that it was rectilinearly propagated in rays. Furthermore, Kepler's methodological emphasis on the mathematical properties of reflection and refraction was widely adopted. In fact, it remains the basis of modern physical optics. Unfortunately, Kepler was unable to derive a mathematical law of refraction.

In 1621 Snell discovered the law of refraction. He demonstrated that the ratio of the sines of the angles of the incident and refracted rays to the normal remains constant. However, priority of publication goes to Descartes, who presented the law without proof in *Dioptrique* (1637) along with his wave theory of light.

Descartes's wave theory, despite its shortcomings, introduced a fruitful new area of study that many subsequent researchers took as the starting point for their own investigations. He asserted that light was a mechanical disturbance transmitted with infinite speed through the subtle matter filling the universe. However, in attempting to explain reflection and refraction, he posited a mechanical model that treated light as particles. When light particles strike a surface they are reflected elastically so that, in agreement with observation, the angle of incidence equals the angle of reflection. Descartes's corpuscular model also accounted for the quantitative law of refraction. However, it implied that light travels faster in denser media.

Pierre de Fermat (1601-1665) later showed that Snell's law of refraction could be deduced from the least-time principle, which implied that light travels slower in denser media. As striking as this result was, it was still generally believed that light was propagated instantaneously. In fact, Descartes stated that if light were not propagated instantaneously then he would be ready to confess that he knew "absolutely nothing." He avoided dealing with the problems raised by his corpuscular analysis by treating the model as a merely theoretical-educational device.

Newton took Descartes's model more seriously and developed a comprehensive corpuscular theory. By treating optical phenomena as a species of particle dynamics, Newton provided a plausible physical mechanism for light propagation. His was also the only seventeenth-century proposal to provide an adequate theory of colors. When his prism experiments of 1666 revealed white light was composed of different colors refracted through characteristic angles, Newton interpreted this to mean that white light was composed of streams of particles that were sorted and differently diverted to produce the spectrum of colors. After Newton published his theory in 1672, a multiyear controversy with Robert Hooke (1635-1703) ensued. In *Micrographia* (1665) Hooke had advocated a wavelike theory of light and spoke in general terms of the finite speed of light. However, he provided nothing comparable to Newton's theory of colors. It also seemed to Newton that wave theories were incapable of explaining the sharpness of shadows or the optical phenomena newly discovered by Erasmus Bartholin (1625-1698).

In 1669 Bartholin noticed that the mineral known as Iceland spar (calcite) produces double images of objects viewed through it. He assumed that light transmitted through the crystal was being refracted through different angles so as to produce

two rays. He further noticed that the resulting beams could be split again but only for certain orientations. Huygens developed a wave theory of light that, in opposition to Newton, explained the initial double refraction. Though the latter effect—polarization—could not be explained by existing wave theories, it was eventually accounted for by nineteenth-century wave theories.

Huygens, influenced by Fermat's work, adopted the finite velocity of light as a hypothesis of his theory several years before Ole Römer provided a demonstration of this fact. In 1676 Römer noticed the intervals between successive eclipses of Jupiter's satellites varied depending on Earth's positions—diminishing as Earth approached and increasing as it receded. He correctly attributed this to the time required by light to travel the Jupiter-Earth distance. Based on his own estimate of Earth's orbital diameter, Huygens exploited Römer's finding to calculate light's velocity. Though his value of 140,000 miles (225,000 kilometers) per second is about 25% too small, it represented a considerable achievement.

Huygens presented his completed wave theory before the Académie des Sciences in 1679 but waited until 1690 to publish his *Traité de la lumière*. He conceived of light as a disturbance propagated by mechanical means at a finite speed through a subtle medium of closely packed elastic particles. According to Huygens's principle, a vibrating particle transfers its motion to those touching it in the direction of motion. Each particle so disturbed becomes the source of a hemispherical wave-front. Where many such fronts overlap, light is visible. Huygens's systematic treatment allowed his wave theory to be fruitfully applied and developed. With it he mathematically demonstrated the rectilinear propagation of light and deduced the laws of reflection and refraction.

Newton's corpuscular theory dominated eighteenth-century optical thinking. It was eclipsed by Huygens's wave theory in the early nineteenth century. The two views were later synthesized in the quantum theory of light during the early years of the twentieth century.

STEPHEN D. NORTON

Further Reading

Books

Lindberg, David C. and Geoffrey Cantor. *The Discourse of Light from the Middle Ages to the Enlightenment.* Berkeley: University of California Press, 1985.

Ronchi, Vasco. *The Nature of Light.* V. Barocas, trans. Cambridge, MA: Harvard University Press, 1970.

Sabra, A. I. *Theories of Light from Descartes to Newton.* London: Oldbourne Book Company, 1967.

Articles

Burke, J. G. "Descartes on the Refraction and the Velocity of Light."*American Journal of Physics* 34 (1966): 390-400.

Cohen, I. B. "Roemer and the First Determination of the Velocity of Light."*Isis* 31 (1940): 327-379.

Lindberg, David C. "The Science of Optics." in D. Lindberg, ed., *Science in the Middle Ages.* Chicago: University of Chicago Press, 1978.

Sparberg, E. "Misinterpretation of Theories of Light." *American Journal of Physics* 34 (1966): 377-89.

The Founding of England's Royal Observatory

Overview

European exploration of the New World ushered in the transformation of the political, economic, academic, and social systems that had predominated in late medieval Europe. Mercantilism, an economic philosophy that emphasized the need for massive gold reserves and promoted trade, became the dominant economic theory of many Western European nations. The increased perception of a need for gold bullion forced England, France, Spain, Holland, Portugal, and other nations into intense rivalries for dominion over the seas and land in the New World. Also, these nations all sought ways in which to expedite and gain monopolies over trade with the nations of the Far East. The competition for wealth to stock national coffers—both in terms of plunder and trade goods—revolutionized European business. Corporations were founded to subsidize colonial and trade ventures, banks were established to handle personal reserves, and credit was levied to help provide capital for such ventures.

This increased interest and economic reliance on trade and colonization was not without risk. In the seventeenth century the marine endeavors themselves were a great risk, both in terms of money and human life. Innovations in shipbuilding, weaponry, and navigational instruments abounded, but ships continued to be stranded or lost with alarming frequency. Not only was this phenomenon a problem for the merchant fleets, but it also plagued the burgeoning fleets of military ships needed to defend the competing national shipping interests. Perhaps the greatest question surrounding safe passage on the seas was finding a way for sailors to determine their longitude—one's precise position east and west—while at sea and out of the sight of land. To address this problem, King Charles II of England established the Royal Observatory. Advised by the predominant scientists of the time, the king believed that the answer to the problem lay in finding a method by which to calculate longitude using charts of celestial observations. The problem of determining longitude at sea was eventually settled in a much more practical manner, by the invention of a portable timepiece resistant to the constant motion of a ship, but the Royal Observatory remained dedicated to advancing the science of astronomy.

Background

The Royal Observatory was founded at Greenwich, England, on June 22, 1675. The complex of buildings itself is an architectural masterpiece. Designed by renowned English architect Sir Christopher Wren (1632-1723), the Observatory featured high vaulted ceilings that could accommodate great instruments. Its earliest contributions were in the fields of observational and practical astronomy, most especially in the determination of star positions, the transit of certain planets, and the compilation and publication of astronomical charts and almanacs that were used as navigational tools. Accurate timekeeping was also a major endeavor of the Royal Observatory at Greenwich. For this purpose, *The Nautical Almanac* was published in 1767. The publication established the line of longitude (or meridian) that passed through Greenwich as a baseline for the calculation of time.

The matter of governing the Royal Observatory came into question in 1675. Charles II appointed John Flamsteed (1646-1719) the first Astronomer Royal, and charged the then-28-

An early-eighteenth-century engraving depicting the interior of the Royal Greenwich Observatory. *(Bettmann/Corbis. Reproduced with permission.)*

year-old clergyman "to apply himself with the most exact care and diligence to the rectifying the tables of the motions of the heavens, and the places of the fixed stars, so as to find out the so much-desired longitude of places for the perfecting the art of navigation." Flamsteed endeavored to find a method of calculating longitude using celestial observations and carefully calculated astronomical charts, an effort that took several voyages to the New World for which to collect data. However, he himself did not put forth any ultimate mathematical solution to the longitude problem.

Upon being granted the stewardship of the Royal Observatory, Flamsteed undertook the project of acquiring the necessary tools to collect the most accurate and scientific data then possible. The financial burden of equipping the Observatory fell almost solely upon Flamsteed himself. With the help of a small family inheritance and a few generous gifts from outside benefactors, he constructed a mural arc, a large instrument used for measuring the altitudes of stars as they passed over the meridian. He also purchased several practical tools used aboard ships for navigation in hopes of making technical advances on the instruments themselves, or incorporating their basic elements into equipment that would aid astronomical observations. The limited funding for the Observatory forced Flamsteed to take on students in order to earn

his salary. This practice was eventually helpful in maintaining the caliber of scientific inquiry at the Observatory and ensuring its continuance beyond Flamsteed's tenure.

Impact

The accomplishments of the Royal Observatory and the several noteworthy astronomers who graced its halls are numerous. The Royal Observatory is perhaps the only institution of the Royal Society that from its inception promoted active research, scholarship, and publication of modern science. The Society, which held steadfast to an exclusion of matters political and religious, was a torch-bearer of the Enlightenment-era philosophy that made secular institutions of scientific inquiry more politically, socially, and academically palatable. However, even the Royal Observatory was not completely free from its own internal politics and academic prejudices.

One of Flamsteed's several regular duties included serving on the panel, composed of scholars from the Royal Society, appointed to evaluate the claims of those who asserted different methods and theories of solving the longitude problem. However, he was rumored to have demonstrated a bias in favor of astronomical methods of solving the longitude problem. Recent historians have levied the criticism that one of Flamsteed's successors, Nevil Maskelyne (1732-1811), who was appointed Astronomer Royal in 1765 and who expanded upon Flamsteed's stellar catalog by adding lunar observations, and his colleagues may have, through deft bureaucratic maneuverings, hindered the development and final acceptance of a mechanical timepiece, now called the Harrison Timepiece after its creator, as a means of readily ascertaining longitude at sea.

Flamsteed's position as director of the Royal Observatory put him in contact with several of the foremost astronomers and scientists of the age, including Sir Edmond Halley (1656-1742) and Sir Isaac Newton (1642-1727). Indeed, it was Flamsteed's equipment at the Royal Observatory that Halley used to sight and track the comet that is his namesake. However, relations between the other scientists and Flamsteed were not always cordial. Recurring illness plagued Flamsteed and reportedly he was often ill tempered. Flamsteed spent decades compiling a remarkable atlas of stellar observations. Though he had already amassed a substantial volume, Flamsteed wished to delay publication on any part of the work until it was completed in entirety. Both Halley and Newton led the charge for the immediate publication of the atlas and gained the sponsorship of the Prince of Denmark to pay for the printing costs. The Prince died a few years later, but Halley continued to edit the volume and push for its publication. Despite Flamsteed's objections, 400 copies of the volume were published in 1712. Flamsteed managed to secure 300 copies of the newly printed atlas and burned them. The completed catalog of his observations was published in 1725 under the title *Historia Coelestis Britannica*. The compilation listed the names and positions of over 3,000 stars.

Discoveries and innovations in the field of astronomy continued throughout the entire life of the Royal Observatory. However, with the problem of longitude solved, the research of the eighteenth and nineteenth centuries at the Observatory were primarily focused on using practical astronomy to accurately measure time and distance on Earth. The reliance of international mariners on charts and catalogs from the Royal Observatory strengthened the case for the longitude of Greenwich itself to be designated the Prime Meridian, the imaginary line that marks 0° 0' 0", the longitude from which all east-west directional coordinates are measured. An international convention approved the designation of Greenwich as the Prime Meridian in 1884. The exact location of the meridian is marked by the sighting crosshairs inside of the eyepiece of a telescope inside the Observatory. The distinction of the Prime Meridian meant that not only east-west bearings, but also international time zones were measured from Greenwich—hence the advent of the designation of Greenwich Mean Time (or Universal Standard Time.)

Fleeing light pollution from nearby London, the Royal Observatory left its historic grounds at Greenwich in 1948. The institution carried on its research at Hertmonceux in Sussex until 1990, when it moved again to the grounds of Cambridge University. With university resources, Observatory research once again pioneered new fields in astronomy and particle physics—especially in efforts to further determine the dynamics of the Milky Way and the composition of stellar objects. After 300 years of pioneering research in astronomy, the Royal Observatory was disbanded. Research in progress and equipment was moved to the UK Astronomy Technology Center at the Royal Observatory Edinburgh; historic instruments, charts, catalogs, and other devices were returned to the old complex at Greenwich, now a museum. At its close in 1998, the Royal Observatory was the oldest scientific institution in the British Isles.

ADRIENNE WILMOTH LERNER

Further Reading

Sobel, Dava. *Longitude: The True Story of the Lone Genius Who Solved the Greatest Scientific Problem of His Time.* New York: Penguin, 1996.

Revival of Corpuscular Theories
During the Seventeenth Century

Overview

During the seventeenth century Scientific Revolution, the Aristotelian theory of substance which had dominated European thought for 2,000 years was progressively abandoned in favor of various "corpuscular" or particulate matter theories. While foreshadowing the classical and modern atomic theories of the nineteenth and twentieth centuries, not all corpuscular theories were atomic in nature, and various ones differed on key points. Whereas some were inspired by ideas drawn from the ancient Greek atomists, others developed out of Western and Islamic medieval matter theories. All of them, however, fundamentally challenged previously accepted notions of matter, motion, space, and substance in physics and chemistry, and generated intense philosophical and religious controversies as well.

Background

Western culture inherited two major philosophical theories about matter and motion from ancient Greece. Leucippus (fl. fifth century B.C.) and Democritus (c. 460-370 B.C.) argued that the world consists of minute, indivisible bits of matter (*atomos* means indivisible). Atoms are eternal, uncreated, and indestructible, infinite in number, and in constant motion through void space; their only properties are size, shape, and solidity. All objects and their qualities—color, taste, texture, etc.—and changes in these result from the motion, collision, combination, and separation of atoms. Combinations of atoms can vary in quantity, shape (H or V), order (HV or VH), or position (V). All atomic motions occur due to strict necessity or fate, rather than chance or free will. Later, Epicurus (c. 341-270 B.C.) and his disciple the Roman poet Lucretius (c. 99-55 B.C.) modified these views by introducing the concepts of relative weight for different types of atoms and of an indeterminate swerve in atomic motions.

Against atomism, Aristotle (384-322 B.C.) pitted his hierarchical theory of qualities, elements, and substances. In the sublunary realm, four primary qualities—hot, cold, wet, dry— combine in complementary pairs to "inform" or determine a completely qualityless, indeterminate substrate of "prime matter"—fire = hot + dry, air =

hot + wet, water = cold + wet, earth = cold + dry. All four elements are in turn combined in various proportions to form the *homoiomere*, or materially uniform, tangible substances (e.g., granite, iron, blood, bone) that are the basic physical constituents of the world. These *homoiomere* in turn combine to constitute *anhomoiomere*, or complex, higher-order parts and organisms (e.g., petals, hands, plants, human beings), which are also substances. (Whether Aristotle himself believed in prime matter, and also considered the four elements to be substances, or only was interpreted as doing so by later commentators, is now a much-debated topic.)

For Aristotle, a substance is a unitary entity having an essence, or unique and irreducible set of defining activities and properties, and a nature, or innate principle that determines and directs these. Each physical substance is a complex unity of a single form and its matter, which respectively manifest its actuality, or characteristic pattern of activity, and potentiality, or latent capacities for alternative actualities. All real changes involve substances, and consist of two types—essential (substantial) change, in which one substance becomes another substance (e.g., a caterpillar becomes a butterfly), and accidental change, in which a substance alters in size, place, or quality, while its essence remains the same (e.g., John Smith grows taller, sits down, or his hair changes from black to gray, but he still remains John Smith). Aristotle also rejected the existence of void space in favor of a concept of place as positional relations between physical objects. Finally, he distinguished four principles or causes of change—formal, material, efficient, and final—which explain not just how a substance changes but also why, offering not just a mechanism for change but a purpose and goal for it as well.

The theories of the atomists and of Aristotle were thus profoundly opposed in several ways. The atomist universe was constructed from the bottom up, out of atoms as the basic real units of the physical world. All tangible objects and their qualities are merely accidental, phenomenal aggregates of atoms, governed by necessity or chance, without an essential unity of structure, function, or purpose to guide and perpetuate their existence. Aristotle's cosmos was organized from the top down, with substances and qualities as the fundamental real entities of the physical world. The elements are conceptual constitutive principles which do not exist separately, but only in combination with one another. As essential unities, substances are structured, organized objects, whose inherent natures perpetuate their existence and direct their activities toward the realization of specific goals, thereby allowing for the existence of reason and free will in man.

Aristotle's system quickly triumphed over the atomists' ideas for several reasons. First, it more convincingly explained many natural properties and activities, especially biological ones. Second, its emphasis on essential structure and function, instead of merely accidental phenomena, attributed a sense of order to the universe that made it more explicable. Third, its provisions for purpose and free will, instead of chance or determinism, were more compatible with belief in God and personal moral responsibility, whereas Epicureanism gained a reputation for atheism and immorality (hedonism). With the subsequent rise of Christianity in Western culture and the adaptation of Aristotelian philosophy in support of Christian theology, atomism remained out of favor for over 1,500 years. Lucretius's poem *De Rerum Natura*, the only atomist text from antiquity to survive intact, would not be recovered by Western Europe until 1417.

Impact

By the late sixteenth century, severe difficulties with Aristotle's theories had become apparent. His generally qualitative approach proved unsuitable to newly developed quantitative and mathematical methods in physics. The doctrine of substance did not explain adequately how existing chemical substances combine to form new substances, or why the original substances reappear upon decomposition. The Paracelsian chemical doctrine of three immaterial principles of matter challenged Aristotle's four-element theory; new astronomical discoveries undermined his distinctions in material composition and motion between the superlunary and sublunary cosmic spheres; and barometric experiments by Evangelista Torricelli (1608-1647) proved the existence of the vacuum. Particulate matter theories, however, could explain physical phenomena in quantitative rather than qualitative terms, and avoid many problematic aspects of Aristotelian substance.

Ironically, the first early modern particulate matter theories developed out of Aristotle's philosophy, not atomism. A passage in Book I.4 of Aristotle's *Physics* was interpreted to suggest that, while matter as such is infinitely divisible, specific substances consist of *minima naturalia*, or particles of a minimum natural size, whose physical division entails destruction of their substantial natures. The great Arabic commentators, Avicenna [Ibn Sina] (980-1037) and Averroës [Ibn Rushd] (1126-1198), developed this concept and modified Aristotle's theory of substance to suggest that in chemical reactions, a substantial form of the product "supervenes" upon the forms of the constituents, but the latter persist as "virtual" forms that reappear when the product decomposes. This theory was transmitted to Europe during the thirteenth century, attracting the interest of the medieval scholastics. By the sixteenth century, commentators such as Augustine Nifo (1473-1546?), Julius Caesar Scaliger (1484-1558), and Jacopo Zabarella (1532-1589) increasingly stressed the concept of *minima naturalia* and de-emphasized that of substance.

The revival of Epicurean atomism was due primarily to the French priest and philosopher Pierre Gassendi (1592-1655), who overcame most theological objections to atomism by recasting it to conform to Christian doctrines. Instead of being infinite in number, eternal, indestructible, and in perpetual motion, atoms are finite in number, being originally created and set into motion by God, who directs their movements according to regular laws to accomplish specific ends. Atomic matter is passive and inert; the fundamental properties of atoms are size, shape, weight, and solidity (impenetrability), with all other qualities resulting from their combinations by means of interlocking sets of hooks, points, and pores. Space exists apart from matter, as a real geometric framework that contains both material bodies and vacuums that separate them. By contrast, God made angels and human souls out of an incorporeal matter that does not constitute physical bodies or occupy space. Matter, its phenomenal modes of appearance, and space thus replaced the Aristotelian categories of substance, real qualities, and place; corporeal matter became equated with physical body and void space with the absence of body; and motion (as efficient cause alone) became the only category of real change.

While Gassendi's system was the most influential, propagated in England by Walter Charleton (1620-1707) and in continental Europe through the correspondence network of Marin Mersenne (1588-1648), it competed for decades with several rivals. Some philosophers endorsed indivisible atoms, other *minima natu-ralia*; some accepted and others rejected the existence of void space apart from matter; some believed motion was inherent in atoms, others that atoms were inert and moved by external forces. Generally, physicists favored atoms, while chemists preferred *minima naturalia*. Galileo Galilei (1564-1642), in formulating the laws of physical motion, treated atoms as mathematical points instead of hard bodies occupying space. René Descartes (1598-1650) equated matter with three-dimensional spatial extension, denied the existence of the void, and posited a material plenum filled by particles of three different sizes but indeterminate shapes that formed swirling vortices in constant motion. The German physician Daniel Sennert (1572-1637) proposed a compromise system, wherein atoms constituted the four Aristotelian elements, which in turn constituted the three Paracelsian principles, whose various combinations then constituted the corpuscular *minima naturalia* of tangible substances. The British chemist Robert Boyle (1629-1698), influenced by Charleton, adopted the term *corpuscles* to avoid a commitment to either atoms or *minima*, since neither the divisibility or indivisibility of matter could be proved by experiment.

Though conclusive proofs for corpuscular theories were lacking, advocates appealed to several types of phenomena to support their position. Processes of growth, condensation, evaporation, and abrasion were explained by accretion and dispersion of microscopic particles. So too was the composition and motion of nonmaterial species such as light, heat, sound, magnetism, and (later) electricity. Observations made after the invention of the microscope in 1661 suggested that many macroscopic objects have a particulate microscopic constitution. Ultimately, the authority of Boyle's friend, Sir Isaac Newton (1642-1727), who employed atoms in his theories of gravitation and optics, established a dualistic theory of atomic matter and immaterial forces. By separating inert atomic matter from active forces as the manifest power of God in nature, Newton (like Gassendi) sought to preserve atomism from any taint of atheism or denial of human free will, against philosophical sceptics such as Thomas Hobbes (1588-1679).

The shift from substantial to corpuscular matter theories was vital to a broader conceptual change from organic to mechanical modes of explanation. In Aristotelian philosophy the activities of inanimate objects and systems, from metallic ores to planetary motions, were ex-plained by analogies to living organisms. Corpuscular theories reversed this pattern, so that biological and even mental processes were explained by analogies to inanimate machines. Instead of being conceived in holistic terms as complex, irreducible unities, entities were now viewed in reductionist terms as merely sums of their simpler parts. Scientific explanations became descriptive rather than purposive, asking and answering questions of "how" rather than "why" objects act as they do. A corresponding shift from qualitative to quantitative descriptive methods encouraged use of standard schemes of classification, and the formulation of universal natural laws in mathematical terms. This promoted higher standards of accuracy and precision, indispensable to the technological advances of the Industrial Revolution that have radically transformed everyday life since 1750. The "mechanization of the world picture," as one historian has termed it, reigned for two centuries, until the twentieth century revolutions in subatomic physics, astronomy, biochemistry, and ecology began to swing the pendulum back, with theories of fields, networks, and systems all being opposed to mechanistic reductionism and emphasizing organic and holistic concepts instead.

JAMES A. ALTENA

Further Reading

Books

Crosland, Maurice P., ed. *Science of Matter: A Historical Survey; Selected Readings.* Harmondsworth, U.K.: Penguin, 1971.

Emerton, Norma E. *The Scientific Reinterpretation of Form.* Ithaca, NY: Cornell University Press, 1984.

Kargon, Robert H. *Atomism in England from Hariot to Newton.* Oxford, U.K.: Clarendon Press, 1966.

Melsen, Andreas G. M. van. *From Atomos to Atom.* Pittsburgh, PA: Duquesne University Press, 1952.

Osler, Margaret J. *Divine Will and the Mechanical Philosophy: Gassendi and Descartes on Contingency and Necessity in the Created World.* Cambridge: Cambridge University Press, 1994.

Toulmin, Stephen, and Goodfield, Jane. *The Architecture of Matter.* New York: Harper and Row, 1962.

Articles

Boas, Marie. "The Establishment of the Mechanical Philosophy." *Osiris* 10 (1952): 412-541.

Clericuzio, Antonio. "A Redefinition of Boyle's Chemistry and Corpuscular Philosophy." *Annals of Science* 47 (1990): 561-89.

Leclerc, Ivor. "Atomism, Substance, and the Concept of Body in Seventeenth-Century Thought." *Filosofia della Scienza* 18 (1967): 761-76. Revised version reprinted as ch. 5 in Ivor Leclerc, *The Philosophy of Nature.*

Washington DC: Catholic University of America Press, 1986.

LeGrand, Homer E. "Galileo's Matter Theory." In *New Perspectives on Galileo*, ed. by Robert E. Butts and Joseph C. Pitt. Dordrecht/Boston: D. Reidel, 1978: 197-208.

Macintosh, J. J. "Robert Boyle on Epicurean Atheism and Atomism." In *Atoms, Pneuma and Tranquillity: Epicure-* *an and Stoic Themes in European Thought*, ed. by Margaret J. Osler. Cambridge: Cambridge University Press, 1991: 197-219.

Meinel, Christoph. "Early Seventeenth-Century Atomism: Theory, Epistemology, and the Insufficiency of Experiment." *Isis* 79 (1988): 68-103.

Advances in Electricity and Magnetism

Overview

Beginning in the late sixteenth century with the work of William Gilbert on magnets, an increasing number of individuals began to sort out the difference between electric and magnetic phenomena and to develop an understanding of Earth's magnetism. Gilbert, an eminent English physician, was among the first to carefully distinguish between magnetic phenomena and those associated with static electricity. He studied the properties of a magnetized solid iron sphere and showed that they matched those of Earth as a whole. Other researchers mapped Earth's magnetic field more precisely and showed that it varied in time.

Background

The basic phenomena of magnetism and static electricity were known to the ancient Greeks. The philosopher Thales (624-546 B.C.) was familiar with lodestone, a naturally occurring magnetic rock, and felt it necessary to attribute to it a soul because it was able to cause motion. The legend that Thales also knew about the electrification of amber by rubbing has been called into question, but there is no doubt that it was accurately described by the philosopher Plato (427-347 B.C.). The Greek medical writer Galen (A.D. 130-200) recommended the use of magnets for "expelling gross humors." The first practical use of electricity or magnetism outside of the medical area was the magnetic compass, which appeared in Europe in the twelfth century.

The first experimental study of magnetism is to be found in a letter of the French engineer Petrus de Marincourt, usually known as Petrus Peregrinus (1240-?), who described a number of experiments with lodestones. It was Perigrinus who first described a magnet as having two poles and studied the interaction of floating magnets. Perigrinus considered the possibility that a compass might be attracted to a point on Earth, but he rejected this idea. Instead, he concluded that the reason a magnetized needle suspended by a thread aligns in a north-south direction is that its poles are attracted by the North and South celestial poles, the points around which the stars in the sky appear to move. The contents of Peregrinus's letter were popularized by Roger Bacon (1220-c.1292), a Franciscan friar who taught at the universities of Paris and Oxford.

By the Elizabethan Age, named after Queen Elizabeth I of England (1533-1603), the compass had become an established tool for navigation. Christopher Columbus (1451-1506) had employed one in his voyage to the New World. Beliefs about magnetism still included much pure speculation; it was believed to have a wide range of curative powers, for instance. Understanding of magnetic phenomena advanced, however, with the 1600 publication of *De Magnete,* a treatise on magnetic phenomena in six short volumes by William Gilbert (1544-1603), an English physician and philosopher. Following an initial volume devoted to the history of magnetism, Gilbert devoted one volume each to five distinct motions or tendencies of magnets: magnetic attraction (which Gilbert called "coition), direction, variation, declination, and rotation.

In the second volume, dealing with the ability of magnets to attract iron and other magnets, Gilbert was careful to distinguish this property from the ability of amber charged by rubbing to attract small bits of other materials. He further argued that while magnetism was a property of lodestone only, electrification by rubbing could be induced in a wide range of materials that Gilbert termed "electrics." Gilbert declared that Earth is a magnet, and that the attractive power of the magnet arises from its "form," which resembles that of Earth. The attraction exerted by electrics, however, was described as resulting from a "material cause"—the materials resulting from the incomplete drying of humid matter in the earth from which the materials had formed, and their attractive capabilities due to "emanations" of fluid after rubbing.

EARLY DISCOVERIES ABOUT EARTH'S MAGNETIC FIELD

Sailing across the Atlantic, Christopher Columbus noticed something odd—his compass didn't always point north. In fact, the further he traveled, the more it deviated from pointing to true north. Today, we know that this is because the magnetic North Pole (in northern Canada) is some distance away from the physical North Pole, which marks the rotational axis of Earth. But in the late fifteenth century nobody knew what magnetism was, let alone did they understand how compasses worked. Put quite simply, this phenomenon puzzled a lot of people, but scientists eventually worked out tables of magnetic declination, showing what the deviation was at various points on Earth, and sailors stopped worrying about it. Some time later, in 1544, Georg Hartmann noticed that a freely floating magnetized needle didn't always stay perfectly horizontal and, in fact, dipped more and more strongly as he traveled north while becoming more closely horizontal when he moved to the south. This magnetic dip also proved troubling from the standpoint of physics, although it interfered little with navigation. Eventually, researchers in the area of magnetism found out that Earth's magnetic field had much in common with common bar magnets. By analogy, they discovered that Hartmann's dip was simply due to the changing direction of Earth's magnetic field lines as they dipped to re-enter Earth at the magnetic North Pole. This discovery, though, was still a few centuries in the future.

P. ANDREW KARAM

Much of Gilbert's work deals with his experiments with a magnetized iron sphere, which he called a *terrella*, or little Earth. Direction, or the tendency of a suspended lodestone to point north, he explained as attraction to the pole of the terrella. The variation in the direction taken by a compass needle at different points on Earth was explained in terms of the deviation of Earth

from a perfect spherical shape. By moving a compass around the globe he was able to show that the declination—the angle above or below the horizontal assumed by the needle—duplicated that taken by a compass needle at the corresponding points on Earth's surface.

In the last volume of *De Magnete*, Gilbert endorsed the idea of Peregrinus that a uniformly magnetized sphere with its poles aligned with the celestial poles would rotate; he attributed the daily motion of Earth to this effect. In doing so Gilbert challenged the notion of solid celestial spheres, a fundamental part of the Earth-centered cosmology challenged by the new Sun-centered Copernican system.

Gilbert's confirmation of the magnetic nature of Earth and the value of the compass to navigation motivated more detailed studies of Earth's magnetic field. Gilbert had expressed the hope that knowledge of the magnetic variation at different points on Earth might allow sailors to determine their longitude at sea. In 1635 an English clergyman and professor of astronomy, Henry Gellibrand (1597-1636), published a book on the magnetic variation in which he documented that the variation changed with time. Gellibrand also tried to solve the problem of determining longitudes but without practical success.

In 1629 Italian Jesuit priest and philosopher Nicholas Cabeo (1586-1650) published the results of his own investigations of electric phenomena. Cabeo had noted that when sawdust or other bits of matter made contact with a charged "electric," they would fly away as if pushed. He is thus the discoverer of the electrical repulsion between like charges. Such behavior could not be explained through Gilbert's emanation theory, and so Cabeo proposed a modified version of the theory in which the effluvium interacts with the air, creating a sort of wind that does the pushing away.

The German Otto von Guericke (1602-1686) was educated as a lawyer and an engineer and served as mayor of Magdeburg for 35 years. In 1672 von Guericke, who also had dabbled in alchemy, described how to produce a solid sulfur ball, which could be rotated on an axis and charged mechanically by rubbing. He might have chosen sulfur because of its role in alchemical theory or because its yellow color suggested a similarity with amber. Von Guericke's device was the first to produce electrical charges large enough to generate the sparks that are now considered emblematic of electrical phenomena.

Impact

Gilbert was a contemporary of Galileo (1564-1642) and like him interested in challenging the unquestioning acceptance of Aristotle's physics and cosmology that had taken hold in the Church and the universities of Europe. While scholars disagree about the extent to which Gilbert accepted the Sun-centered Copernican cosmology, it is clear that in proposing a rotating Earth and the idea that the stars are at different distances from Earth, Gilbert was coming into conflict with the Aristotelian view. Nonetheless, Gilbert had little choice but to express his ideas about attraction and rotation in terms borrowed from Aristotle's metaphysics. It is perhaps for this reason that Gilbert failed to note the repulsion of like charges that is one of the simplest electrical phenomena.

A modern-day physicist would say that the common characteristics of electric and magnetic phenomena reflect a subtle relationship between them, but that this relationship could not be fully understood before the development of the theory of electric and magnetic fields of the Scottish physicist James Clerk Maxwell (1831-1879) in the latter nineteenth century. This synthesis follows a tradition of experimentation with electromagnetic phenomena that has continued from the time of Gilbert.

Gilbert's electrics, now called dielectrics, and the phenomenon of charging by friction gave rise in the eighteenth century to a fascination with static electricity. In part, this was an entertainment for the aristocracy. Demonstrations in which a Leyden jar, which could store a substantial electric charge, was discharged through a line of monks or soldiers were conducted on several occasions before the French Royal Court. Developments were sufficiently interesting that one American, Benjamin Franklin (1706-1790), sold his printing business to devote his time to electrical experimentation. Dating from this period are the two-fluid theory of the French scientist Charles DuFay (1698-1739) and the one-fluid theory of Bejamin Franklin. Both preserve something of the electric emanations of Gilbert.

With the invention of the Voltaic pile or battery in 1800 by the Italian Count Alessandro Volta (1745-1827), a source of steady electrical current became available and the pace of electrical experimentation advanced. Motivated by a philosophical commitment to the unity of all forces, it was the Danish physicist Hans Christian Oersted (1777-1851) who demonstrated in 1820 that a current flowing in a wire would generate a magnetic field. The further discovery of magnetic forces on electric currents by the French physicist André Marie Ampère (1775-1836) and electromagnetic induction by the English physicist Michael Faraday (1791-1867) made possible the technology of motors and generators and the grand synthesis of Maxwell.

The importance of the compass as an aid to navigation is obvious. The fact that the magnetic variation changed in time as shown by Gellibrand was one of the first indications that the history of Earth was accessible to study by observational means. Determining that Earth was much older than the 6,000 years suggested by the Biblical book of Genesis set the stage for the theory of evolution proposed by Charles Darwin (1809-1882) and the later theory of plate tectonics in geology.

DONALD R. FRANCESCHETTI

Further Reading

Benjamin, Park. *A History of Electricity*. New York: Arno Press, 1975.

Kelly, Suzanne. "Gilbert, William." In Charles C. Gillispie, ed. *Dictionary of Scientific Biography*. New York: Scribner's, 1972: vol. 5, 396-401.

Magie, William F., ed. *A Sourcebook in Physics*. Cambridge, MA: Harvard University Press, 1963.

Verschuur, Gerrit L. *Hidden Attraction: The Mystery and History of Magnetism*. New York: Oxford University Press, 1993.

Biographical Sketches

Georgius Agricola
1494-1555
German Mineralogist, Metallurgist and Physician

Georgius Agricola is often referred to as the father of mineralogy. His series of treatises on the principles of geology and mineralogy were instrumental during the formative period in the development of these fields. As influential as these works were, he is best remembered for his magisterial *De re metallica,* which faithfully recorded sixteenth-century mining practices.

AGRICOLA REDISCOVERED BY HERBERT HOOVER

Agricola's *De Re Metallica* remained widely read and used by European miners and metallurgists, appearing in many editions until the late eighteenth century, when the development of a more quantitative and accurate chemistry made its description of smelting processes outmoded. The book then slipped into obscurity until 1912, when a young American mining engineer and his wife translated its Renaissance Latin into English for the first time, publishing it in *The Mining Magazine* of London, England. Her name was Lou Henry Hoover. His was Herbert Clark Hoover; he went on to be the thirty-first President of the United States (1929-33).

GLYN PARRY

Agricola, whose real name was Georg Bauer, was born on March 24, 1494, in Glauchau, Saxony. After attending various schools in Glauchau, Zwickau, and Magdeburg, he matriculated at the University of Leipzig in 1514 and received his B.A. the next year. He remained at the university as a lecturer in elementary Greek until 1517, when he accepted a position at the Municipal School in Zwickau. He became *rector extraordinarius* in 1519 but eventually returned to Leipzig, where he studied medicine under Heinrich Stromer von Auerbach. He continued these studies in Italy and later established a medical practice in the Bohemian city of St. Joachimsthal (1527).

Joachimsthal was an important mining center in the Tyrol, and Agricola was called upon to treat smelters and miners suffering from various occupational illnesses. He undertook a systematic study not only of their ailments but also of their lifestyles, working conditions, equipment, and methods. The results of his researches appeared in *Bermannus sive de re metallica dialogus* (1530). Books on politics and economics followed, and as his reputation grew so did the demands on his time. In 1534 he moved to the smaller, though still important, mining town of Chemnitz to continue his research.

Agricola had developed an interest in minerals, possibly because of the widely held belief in their supernatural and curative properties. His two most important works on mineralogy were both published in 1546. In *De ortu et causis subterraneorum* he developed the idea of a *succus lapidescens* (lapidifying juice). His *succi* can anachronistically be viewed as mineral-bearing solutions, though they are more akin to the humours of Galen (c. 130-c. 200). As he conceived them, stony matter could be condensed out of *succi* when heated, cooled, on becoming cool, or by exposure to air. Agricola was also one of the first to attempt a systematic classification of minerals. His scheme, presented in *De Natura Fossilum*, was based on the physical properties of minerals including weight, color, opacity, taste, texture, solubility, etc.

After serving as mayor of Chemnitz (1545) and then councilor to the court of Saxony, Agricola returned to his scientific work in 1548. New books appeared shortly thereafter, including *De animantibus subterraneis* (1549); and in 1550 he completed his master work *De re metallica libri XII*. This was the culmination of his researches begun over 15 years before in Joachimsthal. The work was published posthumously in 1556.

De re metallica presents a detailed and accurate account of sixteenth-century Saxon mining practices and is lavishly illustrated with 292 beautiful woodcuts. Drawing intelligently on Vannocio Biringuccio's (1480-c. 1539) *De la Pirotechnia* (1540), the first 11 sections deal exclusively with the extraction of metals, smelting, and assaying techniques of the day. The final section deals with the chemical technologies associated with metallurgical processes. *De re metallica* remained the standard text on mining and metallurgy for over four centuries.

As the black plague spread through Saxony (1552-53) Agricola's medical skills were in high demand. His ceaseless efforts to alleviate the suffering of some of the worst victims caused him great concern as it placed his own family at great risk. (Indeed, he was to lose a daughter to the plague.) His researches during this period are recorded in *De peste libri III* (1554). Agricola died in Chemnitz on November 21, 1555.

STEPHEN D. NORTON

Francis Bacon
1561-1626
British Philosopher and Lawyer

Francis Bacon's position in the history of science is still debated. To some he was the first spokesman for the new science of the modern era. Yet to others Bacon was an unoriginal philosopher who stated a false formula for science. He urged the rejection of ancient knowledge in favor of observation and experiment. His proposed method of induction stressed the role of impartial observation of the particular in order to make general rules of nature. Bacon also had a troubled political career, which was shaped by the whims of royal favor, noble patronage, and the complex politics of his era.

Francis was the fifth son of Sir Nicholas Bacon, Queen Elizabeth I's Lord Keeper of the Great Seal (a high governmental post). At the age of 12 Bacon studied at Trinity College, Cambridge, and developed his dislike of the standard academic philosophy of his day. He described his teachers as "men of sharp wits, shut up in their cells of a few authors, chiefly Aristotle, their Dictator."

In 1575 he moved to Gray's Inn, a law school that catered to the extravagant and wild lifestyles of the young nobility as well as the academic interests of the studious. Bacon was 18 when his father died, leaving him with little money. He then began to seriously study law in order to make his living, and through his talent and family connections began a career in politics, gaining a seat in parliament when he was 23.

His career suffered from competition with his relatives, especially the powerful Cecil family. The elder Cecil, Lord Burghley, quite naturally preferred to advance his son's career over Bacon's. Bacon turned to the Cecils' rival, the Earl of Essex, to promote his career. However, when Essex was involved in a plot to kidnap the Queen, Bacon had no qualms in helping to prosecute his former patron.

In 1603 James I succeeded to the English throne and Bacon's career then advanced apace. In 1617 he became the Lord Keeper of the Great Seal, the same office his father had held. He was made Lord Chancellor and created Baron in 1618, and finally given the title Viscount St. Albans in 1621. However, that year he was accused and found guilty of bribery. His public career ruined, he retired to his estate to devote his remaining years to his writings.

Bacon had already developed a literary reputation. He had penned a number of masques for Royal entertainment. In 1597 he had published a collection of essays on various topics, from gardening to the nature of good and evil. In 1605 he published *The Advancement of Learning*, a new categorization of the whole of the natural sciences. He continued this theme with the *Novum Organum* (1620), which outlined a new method of natural philosophy to replace Aristotle (384-322 B.C.).

Bacon proposed that, through his method of induction, the secrets of the universe could be unlocked and used to benefit society. His method involved the unbiased, almost random, collection of data, which would later be generalized into rules of nature. Bacon's method never became popular, but many of his other ideas proved influential. In the *New Atlantis* (1626), he described an imaginary society of scientists, which had a profound effect on many of those who founded the British Royal Society.

His death resulted from a misadventure many see as typical of Bacon's concept of science. Bacon had a sudden impulse to see whether snow would help preserve meat, and so he stopped his carriage, acquired a hen, and buried it in the snow. However, he caught a sudden chill, from which he died shortly after.

DAVID TULLOCH

Johann Bayer
1572-1625
German Astronomer and Lawyer

Johann Bayer produced the most comprehensive pre-telescopic star catalog and introduced the nomenclature still in use for designating stars visible to the naked eye. His was also the first celestial atlas to represent the stars around the South Pole and to cover the entire sky.

Bayer was born in Rhain, Bavaria, in 1572. In 1592 he matriculated at Ingolstadt University as a philosophy student and later obtained a law degree. Though a lawyer by profession, he maintained a keen interest in astronomy and after moving to Augsburg, published his comprehensive celestial atlas *Uranometria* on September 1, 1603. He dedicated the volume to two of Augsburg's leading citizens. The city council responded with an honorarium of 150 gulden. Bayer was later appointed legal advisor to the city counsel of Augsburg with an annual salary of 500 gulden.

The significance of Bayer's work lies in his innovative method for naming stars within each constellation. Though traditional constellations continued to provide a convenient means of dividing the heavens, the profusion of names for individual stars that resulted from the translation of Greek into various languages proved most cumbersome and confusing. Bayer sought to reform this situation by systematically identifying each star precisely and succinctly. He assigned to each star in a constellation one of the 24 letters of the Greek alphabet. If a constellation had more than 24 stars then additional characters were provided by the Latin alphabet. Thus, Castor and Pollux, the two brightest stars in the constellation Gemini, became Alpha Gemini and Beta Gemini respectively. Many stars in the southern skies were only carefully observed in Bayer's day and are today known by his designation, as for example with Alpha Centauri. Bayer's system is still in use, and as more and fainter stars have been identified, Roman numerals, both alone and in combination with alphabetic characters, have been resorted to.

The *Uranometria* contains over 2,000 stars, of which around 1,200 were taken from Tycho Brahe's (1546-1601) catalog. These were sorted into 49 constellation maps and two hemispheric charts that were beautifully engraved by Alexander Mair. Bayer retained the traditional 48 constellations of Ptolemy's (fl. second century A.D.) *Syntaxis*. He was also at great pains to cross-index the names of his stars with those in the catalogs of Ptolemy and others so as to facilitate their identification. In addition, the *Uranometria*'s forty-ninth constellation map contained the 12 new southern constellations that had recently been defined by the Dutch navigator Pierter Dirckszoon Keyser, also known as Petrus Theodori (d. 1596). These are the constellations Apus, Chamaeleon, Dorado, Grus, Hydrus, Indus, Musca, Pavo, Phoenix, Triangulum Australe, Tucana, and Volans.

Despite the terminological convenience of his system, certain aspects of Bayer's celestial atlas created problems. First, Bayer's left-right labeling of constellations was the reverse of that used by all previous atlases. Second, he bracketed stars of the same magnitude in each constellation but failed to include the method whereby he assigned the letters within each bracket. It was initially assumed that he ordered them according to descending magnitude, but this created considerable confusion in later work on variable stars. An alternate hypothesis was that he employed spatial criteria. However, this interpretation has faced serious objections as well.

A devout Protestant and amateur theologian, Bayer was never comfortable with the traditional heathen names assigned to the constellation. In the *Uranometria* he therefore proposed alternate names from the Bible. Constellations in the Northern Hemisphere were named for figures from the New Testament while those in the Southern Hemisphere were given names from the Old Testament. Needless to say, this suggestion failed to gain wide acceptance.

STEPHEN D. NORTON

Robert Boyle
1627-1691
Irish Chemist and Physicist

Robert Boyle never earned a college degree. As a physicist, however, he performed some of the earliest experiments with gases. As a chemist, he helped to separate the science of chemistry from its roots in alchemy and, for this reason, is sometimes known as the father of chemistry.

Boyle was born in Ireland in 1627. He was the fourteenth child of a wealthy and aristocratic family. By the time he was eight years old, he was already studying Latin and Greek. From 1639 to 1644, beginning when he was eleven, he traveled throughout Europe while being taught by a tutor. The following decade, Boyle lived partly in England and partly on his estates in Ireland. His father had died by this time, and Boyle's inherited wealth gave him the money and the freedom to pursue scientific experiments.

From 1656 to 1668, he resided at the University of Oxford. (He was neither a student nor a professor, however.) There, Boyle participated in meetings of scientists who favored experimentation over logic alone. This group was called the Invisible College. In 1663, the group was officially recognized by King Charles II and was renamed the Royal Society—the first scientific society of England. Like other members of this group, Boyle believed that all experimental results should be clearly and quickly reported so that other scientists could profit from them.

At Oxford, Boyle performed some of the first experiments on gases. One of his assistants was Robert Hooke (1635-1703), who would eventually be the first scientist to describe cells. Together, Boyle and Hooke constructed an air pump. With this pump, Boyle could produce a vacuum in a sealed container. Boyle used his vacuum chamber to discover several processes that require the presence of air. By placing a ticking clock in the chamber, for example, Boyle showed that sound does not travel through a vacuum. Instead, the sound of the clock faded away as air was withdrawn by the pump. In addition, Boyle demonstrated that a bird can live for only a short time in a vacuum, showing that air is necessary for respiration. He also showed that air is required for combustion. In this experiment, he placed a red-hot iron plate in a vacuum and dropped a piece of sulfur on it. The sulfur did not burst into flames until air was allowed into the chamber. He described his findings in a book published in 1660, titled *New Experiments Physio-Mechanicall, Touching the Spring of the Air and its Effects.*

During his experiments, Boyle found that air is compressible. In 1661, he reported what is now known as Boyle's law, which states that at a constant temperature, the volume of a gas is inversely proportional to its pressure. In other words, as the volume of a gas decreases, its pressure increases. Because of air's compressibility, Boyle came to the conclusion that air is not a continuous substance, but consists of individual particles separated by empty space.

Boyle's interests included not only physics, but also chemistry. In 1661, he published *The Sceptical Chymist.* This book helped to transform alchemy into chemistry. Alchemy was a combination of science, religion, and philosophy whose chief goals included the creation of gold from other metals.

Most alchemists accepted a theory of matter that was based on four elements (earth, air, fire, and water) or three principles (salt, sulfur, and mercury). Boyle, however, proposed a theory of matter that eventually evolved into the modern theory of chemical elements. The alchemists saw elements as mystical substances. Boyle, on the other hand, believed that elements could only be identified by experiment. To Boyle, any substance that could not be broken down into simpler substances was an element. He did not necessarily reject the alchemists' elements, but he wanted them to be established by experiment. He did believe, however, that the number of elements might be much larger than three or four. (Today, more than 100 elements have been identified.)

Boyle also suggested a method of distinguishing between acids and bases that eventually led to the use of indicators. An indicator is a chemical that changes color as the acidity of a solution changes. For example, Boyle described how blue solutions obtained from plants, such as syrup of violets, are turned red by acids and green by bases. He also noticed that some solutions did not cause syrup of violets to change color. He called these solutions neutral. (It had earlier been thought that all solutions are either acids or bases.) In 1664, Boyle published *Experimental History of Colors* in which he described his work with acid-base indicators.

In 1680, Boyle was elected president of the Royal Society although he turned down the position. He became quite well known as a scientist during his lifetime and is considered by some to have established the science of chemistry.

STACEY R. MURRAY

Tycho Brahe
1546-1601
Danish Astronomer

Tycho Brahe is considered the greatest observational astronomer of the pre-telescopic era. His observations of the 1572 nova and 1577 comet helped undermine Aristotelian cosmology, while his observations of Mars proved critical to Johannes Kepler's (1571-1630) discovery of the laws of planetary motion.

Brahe was born to aristocratic parents on December 14, 1646, at Skåne in southern Sweden (then under Danish rule). While attending the University of Copenhagen, Brahe developed an interest in astronomy after observing a partial eclipse of the Sun (1560). His newfound interest was discouraged by his paternal uncle, who sent him to study law at the University of Leipzig (1562).

Brahe secretly continued his astronomical researches, and in August 1563 he observed the conjunction of Saturn and Jupiter, noting it occurred about a month later than the *Alphonsine Tables* and a few days later than the *Prutenic Tables* predicted. He resolved to prepare more accurate tables and worked at producing improved instruments to make the necessary observations. After his uncle's death in 1565 he continued his astronomical and mathematical studies at the universities in Wittenberg, Rostock (where he lost the greater part of his nose in a dual), Basel, and Augsburg before returning to Copenhagen in 1570.

Brahe's reputation was established with the publication of *De nova stella* (1573), which details his observations of the nova of 1572. His measurements indicated the phenomenon was not part of the atmosphere nor was it attached to the sphere of a planet, but that it was located among the fixed stars. This undermined the prevailing Aristotelian notion that the heavens were perfect and unchanging.

In 1576 King Frederick II of Denmark granted Brahe the island of Hven and funds to construct and maintain an observatory. Brahe accepted and spared no expense in building the magnificent research facility Uraniborg (Castle in the Sky). In accordance with his desire to reform positional astronomy he equipped Uraniborg with instruments of unsurpassed accuracy. He invented new viewing sights, better instrument mounts, and improved methods for inscribing scales by transversals.

Brahe dealt another blow to Aristotelian cosmology with his observations of the comet of 1577. Because the erratic behavior of comets was incompatible with the immutability of the heavens, Aristotle (384-322 B.C.) maintained they were atmospheric exhalations. Brahe's parallax measurements indicated the comet of 1577 was more distant than the Moon. Furthermore, he concluded that its orbit was elongated, suggesting that it had passed through several planetary spheres.

Brahe rejected the Ptolemaic system as incompatible with his observations, but scripture and a lack of stellar parallax prevented him from accepting Copernican heliocentrism—the idea that Earth revolves around the Sun. As a compromise he advanced his Tychonic theory in which all of the planets but Earth orbit the Sun, with the Sun and its train of planets revolving about a stationary Earth. This system had the advantage over the Ptolemaic system of being able to account for the phases of Venus and gained acceptance in many quarters. Brahe's ability as a theorist was also revealed by his brilliant lunar theory.

Almost every astronomical measurement of importance was improved upon by Brahe during his years at Uraniborg. Unfortunately, difficulties with Frederick II's successor forced him to leave Hven in 1597. At the invitation of Emperor Rudolph II he moved to Prague in 1599. He was joined there by Johannes Kepler in 1600, but their collaboration was cut short by Brahe's death on October 24, 1601. Kepler eventually completed the *Rudolphine Tables* (1627) begun by Brahe, developing their theory not in accordance with Tychonism but with heliocentrism. In the process he used Brahe's data to discover the law of planetary motion.

STEPHEN D. NORTON

Giovanni Domenico Cassini
1625-1712
Italian-French Astronomer

Giovanni Domenico Cassini made the first accurate determination of the dimensions of the solar system. A gifted observationalist, he was an extremely conservative theorist, refusing to accept Nicolaus Copernicus's (1473-1543) heliocentric view and opposing Isaac Newton's (1642-1727) gravitational theory.

Cassini was born June 8, 1625, in Perinaldo, near Nice (then in Italy). His early education was completed at Genoa. He later became Professor of Astronomy at the University of Bologna (1650), where his scientific reputation was established through a series of solar and planetary observations. This prompted an invitation to France, where he became director of the Royal Observatory (1671). He observational work and duties as director ceased after he went blind in 1710. He died September 11, 1712, in Paris.

At Bologna, Cassini produced new tables of the Sun's motion (1662). In 1664 he determined Jupiter's rotational period to within a few minutes and detected bands and a red spot on the Jovian surface. After detecting markings on the surface of Mars, he determined that planet's rotational period to within three minutes of its presently accepted value (1666). In 1668 Cassini produced accurate tables of the motions of Jupiter's satellites, which were widely used for determining terrestrial longitude.

His growing reputation brought him to the attention of the French finance minister Jean Baptiste Colbert, who was working to attract prominent scientists to France. Colbert nominated Cassini for membership in the newly established Académie des Sciences and invited him to Paris to oversee the establishment of the Royal Observato-

termine the astronomical unit. His value of 87 million miles was the first fairly accurate estimate of the Earth-Sun distance.

Richer also observed that pendulums beat slower in Cayenne than in France. Newton argued this was due to a decreased gravitational attraction at the equator. This suggested, in conformity with his gravitational theory, that Earth was an oblate spheroid—bulging equator and flattened poles. Supporting the Cartesian view (named after French philosopher René Descartes) that Earth was a prolate spheroid—elongated along the polar axis—Cassini maintained that temperature differences explained the effect. In 1683 Cassini undertook to measure an arc of the meridian between the northern and southern French borders to settle the issue. Completed in 1700 and published by his son Jacques (1677-1756) in 1718, the measurements seemingly supported Descartes. However, expeditions to Peru (1734-1744) and Lapland (1736) later settled the issue decisively in Newton's favor.

STEPHEN D. NORTON

Giovanni Domenico Cassini. *(Bettmann/Corbis. Reproduced with permission.)*

ry. Cassini accepted, arriving in 1669 for what was to be but a temporary stay. When the Observatory opened in 1671 he accepted the directorship and in 1673 became a French citizen, changing his name to Jean Dominique Cassini.

In 1671 Cassini discovered a second satellite of Saturn, Iapetus, and correctly attributed variations in its brightness to always having the same face turned towards Saturn. He later discovered three more Saturnian satellites, Rhea (1672), Tethys (1684), and Dione (1684). In 1675 he drew attention to the dark gap—today referred to as the Cassini Divide—splitting Saturn's ring in two and postulated each part was composed of minute particles behaving like small satellites. This hypothesis has since been corroborated. Cassini also made extensive observations of the Moon's surface (1671-79), which culminated in his magnificent lunar map presented to the Academy in 1679. In addition, he carried out the earliest continuous observations of the zodiacal light and produced improved tables of atmospheric refraction.

Cassini's most significant work is associated with the Académie-sponsored astronomical expedition to Cayenne off the coast of French Guiana (1672-73). Measurements of Mars's opposition by Cassini and others in France in conjunction with those made by expedition leader Jean Richer (1630-1696) allowed Cassini to de-

Nicolaus Copernicus
1473-1543
Polish Astronomer, Priest, and Mathemetician

Nicolaus Copernicus founded modern astronomy by breaking with classical and theological tradition and proposing a mathematically supported heliocentric theory of planetary motion. His work also initiated the process that led to the Scientific Revolution.

Copernicus was born and spent his early life in Poland. After beginning his education at the University of Cracow, he studied mathematics, astrology, astronomy, canon and civil law, and medicine in Italy at Bolgna, Rome, and Padua. He received a doctorate in canon law from the University of Ferrara. Returning to Poland, he devoted his life to church administration as canon of Frombork Cathedral.

Copernicus became interested in problems in astronomy and mathematics when he served as an assistant to Domenico Maria de Novara during his studies in Italy. He became aware that a sun-centered rather than an Earth-centered universe had long been supported by some astronomers. This idea was initially proposed by Greek astronomer Aristarchus (310-230 B.C.) and had been supported more recently by Nicholas de Cusa (1401-1464) as fitting observable data better.

The earth-centered (geocentric) model, in which the planets and the Sun moved around Earth as the stationary center of the universe, was proposed by Greek astronomer Ptolemy (90-168) and was supported by the Church and by scholars of the day. It was thought that a Sun-centered (heliocentric) model, in which the planets, including Earth, moved around the Sun, was not in agreement with the Bible. More importantly, such a model would seem to remove Earth and humans from their unique position as God's chosen ones for whom all else was created.

Copernicus worked on his astronomical ideas in his spare time. He became convinced that the heliocentric model correctly describes the relative motion of Earth, the other planets, and the Sun. He used the principle of relativity of motion to explain why the motion of Earth is not detectable by observers who are moving with Earth and proposed that Earth rotates daily about its axis and revolves annually around the Sun. He proposed that other planets behave in the same manner, pointing out that the further a planet is located from the Sun, the longer its period of motion around the Sun will be.

Aware of the possible repercussions of proposing such a theory in direct opposition to the Church, he first wrote a short version of his ideas, entitled *Commentariolus*, in 1513 and distributed it for comment to friends and colleagues. The full work, entitled *The Revolutions of the Heavenly Spheres (De Revolutionibus Orbium Coelestium* or simply *De Revolutionibus)* was completed by 1530 but was not published until 1543. *De Revolutionibus* was not received well by the Church and scholars, even though a churchman, without Copernicus's permission, had added a preface which stated that the theory was not being proposed as representing the actual motion and position of the Earth and Sun, but merely as a mathematical model to make calculations easier. Copernicus died soon after *De Revolutionibus* appeared, thereby, escaping inevitable punishment.

Although the Copernican model successfully explained such phenomena as the retrograde motion of the planets, it was significantly undermined by his insistence, primarily for aesthetic reasons, on circular orbits for the planets. The Copernican Revolution was completed after his death by Galileo Galilei (1564-1642) and Johannes Kepler (1571-1630), who correctly adjusted the theory by making the orbits elliptical. Galileo was forced by the Inquisition to recant his support for the Copernican theory in 1633, and it was subsequently suppressed. Even though it was substantiated by the work of Isaac Newton (1642-1727) on mechanics and universal gravitation, it took another one hundred years for the theory to gain acceptance.

J. WILLIAM MONCRIEF

John Flamsteed
1646-1719
English Astronomer

John Flamsteed is remembered for his accurate and extensive work in positional astronomy. As the first British Astronomer Royal, he pioneered the systematic use of telescopic sights and produced the first great star catalog based on telescopic observations.

Flamsteed was born on August 19, 1646, at Denby, England. He attended the Derby Free School but was forced to leave when stricken by a severe rheumatic condition. So serious was the illness that his health was compromised for the rest of his life. During his extended period of convalescence Flamsteed developed an interest in astronomy, which he perused through self-study (1662-69).

In 1670 he entered his name at Jesus College, Cambridge, to pursue an M.A. degree. That same year he submitted a small ephemeris of lunar occultations to the Royal Society that was published in the *Philosophical Transactions*. This gained him some recognition and led to his acquaintance with Jonas Moore, who presented him with a micrometer and high quality telescopic lens. These instruments enabled Flamsteed to begin serious observational work that firmly established his reputation before his graduation from Cambridge (1674).

One of the major problems of this time was the accurate determination of longitude at sea. This was of particular interest to England, whose merchant fleet was quickly becoming the world's largest. Calculating longitude required comparing local time with a standard reference time. It had been suggested that the Moon's motion against the stellar background could be used to determine this standard. Flamsteed was asked by Moore to comment on a proposed method for exploiting this suggestion. Flamsteed's judgment was that no method would work until stellar and lunar positions were more accurately known. Moore used this assessment to successfully petition King Charles II to establish a national observatory at Greenwich and appoint Flamsteed the first Astronomer Royal on March 4, 1675.

John Flamsteed. *(Library of Congress. Reproduced with permission.)*

ductions were incomplete. He was eventually obliged to deposit his observations with the Royal Society, after which Halley unlawfully published a muddled and incomplete version (1712). Flamsteed publicly burned 300 of the 400 copies printed, then commenced work on his own edition. Before his death on December 31, 1719, he had completed enough work on the three-volume *Historia Coelestis Britannica* that his assistants were able to see it to completion in 1725. It contained the positions of nearly 3,000 stars and established Greenwich as one of the world's leading observatories. A set of star maps—*Atlas Coelestis* (1729)—based on the catalog also appeared posthumously.

STEPHEN D. NORTON

Galileo Galilei
1564-1642
Italian Astronomer, Mathematician, and Physicist

Galileo Galilei, best known simply as Galileo, made fundamental discoveries in mechanics and observational astronomy as well as inventing the thermometer and improving the telescope. More significant, his emphasis on direct observation and mathematization of natural phenomena in conjunction with a refusal to allow science to be guided by metaphysical speculation had a transforming impact on scientific methodology.

Greenwich Observatory was completed in 1676, but Flamsteed was given only a meager salary and no provisions for assistants or instruments. Moore eased the situation somewhat by donating 2 clocks and a 7-foot (2.1-meter) sextant. Nevertheless, Flamsteed worked alone for the first 13 years, recording over 20,000 observations. His observations were six times more accurate than Tycho Brahe's (1546-1601), and he was the first astronomer to make systematic use of clocks for taking measurements. An inheritance from his father's estate solved his financial problems and allowed him to commission a 140-degree mural arc with which he precisely determined Greenwich's latitude, the ecliptic's obliquity, and the position of the equinox. He then devised a method for observing absolute right ascensions—the celestial equivalent of terrestrial longitude—that removed all stellar positional errors from parallax, refraction, and latitude. Flamsteed also deduced a 25.25-day solar rotational period from sunspot observations as well as producing tables of atmospheric refraction, tides, and the Moon's elliptic inequality according to Kepler's second law.

The unsurpassed accuracy of Flamsteed's data placed them in high demand. Isaac Newton (1642-1727) and Edmond Halley (1656-1742) pressed for publication so they could test their theories, but Flamsteed resisted because his re-

Galileo was born on February 15, 1564, in Pisa, Italy. He enrolled at the University of Pisa to study medicine (1581), but his interests soon turned to mathematics. His first scientific discovery was made in 1582, when he realized a pendulum's period remains approximately the same regardless of the amplitude of oscillation. Galileo left Pisa in 1585 without finishing his degree. He continued his studies in Florence, where he completed his first scientific treatise, *La bilancetta* (1586), which describes his improved hydrostatic balance. As Professor of Mathematics at the University of Pisa (1589-92) he conducted experiments on falling bodies, which he published in *De motu* (1590). In 1592 Galileo assumed the chair of mathematics at the University of Padua.

In early 1609 Galileo heard reports of a device, invented in Holland the year before, consisting of two glass lenses that made objects at a distance appear closer. Based on this alone Galileo constructed his own telescope. His im-

Galileo. *(Archive Photos. Reproduced with permission.)*

provements allowed him to produce a 30X instrument by year's end, and in early January 1610 he became one of the first, if not the first, to use the telescope to observe the heavens. Galileo announced his controversial discoveries in *Sidereus nuncius* (1610).

Galileo's discovery of four satellites orbiting Jupiter contradicted the widely held belief that Earth was the center of rotation for all celestial bodies, and his observation of mountains and depressions on the lunar surface refuted the Aristotelian notion that the Moon was a perfect sphere. Galileo also resolved the Milky Way into a multitude of fixed stars. The existence of so many celestial objects invisible to the naked eye was difficult to understand if the universe had been created solely for man's benefit. Galileo later discovered the phases of Venus, which removed a serious objection to Nicolaus Copernicus's (1473-1543) heliocentrism—the idea that Earth revolves around the Sun—and independently discovered sunspots, which he realized provided evidence of solar axial rotation.

Rejecting a lifetime appointment at the University of Padua, Galileo returned to Florence in 1610 as mathematician and philosopher to the Grand Duke of Tuscany. Shortly thereafter a series of disputes with Dominican and Jesuit theologians over his support of Copernican heliocentrism brought him into conflict with the church. An edict was issued in 1616 declaring Copernicanism heretical, and Galileo was admonished not to defend Copernicanism in public. When Urban VIII became Pope in 1624, Galileo obtained permission to present an impartial discussion of the Copernican and Ptolemaic systems. The discussion, which appeared in *Dialogue Concerning Two Chief World Systems* (1632), was anything but impartial, marshaling as it did overwhelming empirical evidence in support of heliocentrism. Galileo was tried as a heretic, convicted, and sentenced to permanent house arrest.

Galileo's remaining years were spent preparing *Two New Sciences* (1638), which deals with the engineering science of strength of materials and kinematics. In the first the law of the lever is used to establish the breaking strength of materials. The second provides a mathematical treatment of motion in which Galileo introduces the idea of uniformly accelerated motion. He also established the law of free fall in a vacuum, deduced the terminal velocity for any body falling through air, and derived the parabolic trajectory of projectiles from uniform horizontal and accelerated vertical motions. Shortly after its publication Galileo went blind. He died four years later on January 8, 1642, at Arcetri, near Florence.

STEPHEN D. NORTON

Pierre Gassendi
1592-1655
French Natural Philosopher, Mathematician, and Priest

Pierre Gassendi is best known as the seventeenth-century rehabilitator of the atomism of the ancient Greek moralist and natural philosopher, Epicurus (341-270 B.C.). Gassendi found in atomism a way both to combat the extreme skepticism that had pervaded French intellectual life since the late sixteenth century and to overturn Aristoteleanism, which was the dominant philosophy and "science" of both the church and learned culture. Gassendi thus straddled a thin line between orthodox and heterodox ideas: on the one hand an ordained priest in search of certainty amid the skepticism brought about by a revival of ancient Greek Pyrrhonism and the disillusioning experience of the French Wars of Religion (1559-1598); on the other hand something of an iconoclast, whose stated goal was to overturn the metaphysical assumptions that had underpinned the church and intellectual authority since the Middle Ages.

Gassendi was born on January 22, 1592 in Champtercier, in the south of France. He became a doctor of theology in 1614 and was ordained a priest two years later. He mastered ancient Greek and Latin, mathematics, theology, and philosophy at an early age and began teaching philosophy at the University of Aix in Provence in 1617. Gassendi left Aix in 1622, as the Jesuits took over the school, and traveled throughout Europe for a number a years. Always compelled intellectually to dislodge the authority of Aristotle (384-322 B.C.), he began avidly reading Epicurus for that purpose in 1628. Gassendi worked on his Epicurean project throughout the 1630s; however it was not until 1647 that the first of his works on Epicurus was published. A more important work, the *Animadversions on Book X of Diogenes Laertius*, was brought out two years later; but not until three years after his death in October of 1655 was Gassendi's *Opera Omnia*, which contained his extensive commentaries on Epicurus, finally brought to light.

The atheist Epicurus had long been on the Church's most wanted list, so it is surprising on the surface that Gassendi was compelled to spend a lifetime resurrecting the works and theories of a heretic. Part of the explanation for this apparent contradiction is that Epicurus was an effective weapon against Aristoteleanism, which from an early age Gassendi found empty and useless. During his tenure at Aix, Gassendi taught his students that just as important as learning Aristotle was acquiring the tools needed to attack the ancient philosopher's philosophical system. Of course, one might also wonder about the piety of someone who attacked Aristotle in the early seventeenth century, but Gassendi defended himself writing that he found offensive both the blind veneration of Aristotle by learned authority and the way in which Aristotelean argumentation, dialectic, and rhetoric kept one from understanding the true reality of nature. Aristoteleans could deduce truths from their universal definitions of things in nature; for instance, from the premises that all stars are bright and the sun is a star. Aristoteleans could logically conclude that the sun was bright. But Gassendi thought such types of statements only told one about definitions; they did not necessarily reveal a truth about nature. Nature, instead, had to be understood using the senses alongside reasoning; experience, which the moderately skeptical Gassendi always recognized could deceive the sentient being, was the key to providing an instrument of proof about the real world and an antidote to radical skepticism.

Pierre Gassendi. *(Bettmann/Corbis. Reproduced with permission.)*

Yet this might lead one to wonder why the empirical Gassendi embraced atomism—after all, no one had ever actually seen an atom. Gassendi's answer was that even though one cannot see atoms, one can induce their existence through the senses. For instance, one can witness the effects of an invisible and seemingly matterless wind on the branches of a tree, or one can smell an unseen fragrance before one sees the source from which it is emanating. Such phenomena suggested to Gassendi that the world is composed of smaller particles, which he further hypothesized were solid and indivisible, varied in shape, magnitude, and weight, and, variously configured, the building blocks of all the matter in the universe. Echoing Epicurus, Gassendi thus thought the universe was in essence atoms and the void in which they were contained. Unlike Epicurus, however, Gassendi also made room in the cosmos for God and human souls, those things which Epicurus had adopted atomism specifically to reject.

Gassendi's lifelong balancing act between heterodox and orthodox ideas made him one of the most interesting intellectual figures in the seventeenth century. He rejected Cartesian deduction in favor of induction but was as much a mathematician as René Descartes (1596-1650) and thus also recognized the importance of deduction used in union with sense experience; he sought to explain the universe in mechanistic terms, but, always a moralist, he saw proof of God in nature everywhere he looked. In this first sense, Gassendi is a forebear of John Locke (1632-1704); in the second, he resembles Isaac Newton (1642-1727). Neither comparison is accidental. Gassendi's once formidable reputation may not have survived the early modern period intact—in his day he stood on even ground with Thomas Hobbes (1588-1679) and Descartes—but he nonetheless had a deep impact on the generation of philosophers and scientists who followed in his wake.

MATT KADANE

William Gilbert
1544-1603
English Geophysicist and Physician

William Gilbert earned his place in annals of science with the publication of *De magnete*. A landmark in the history of experimental science and widely influential, it records his pioneering researches on magnetism and electricity.

Nothing is known of Gilbert's early life other than that he was born on May 24, 1544, at Colchester in Essex, England. In 1558 he matriculated at St. John's College, Cambridge. After earning his A.B. in 1561, M.A. in 1564, and M.D. in 1569, he was elected a fellow of St. John's College. He established a successful medical practice in London in 1673 and became a Fellow of the Royal College of Physicians. He served as censor of that body (1582, 1584-87, 1589-99), treasurer (1587-94, 1597-99), consiliarium (1597-99), and elect (1596-97) before being elected president in 1600. That same year he was appointed physician to Queen Elizabeth I, serving in that same capacity for James I. He died in London on December 10, 1603.

Gilbert began studying magnetic phenomena sometime after leaving Cambridge. He published his findings in 1600 under the title *De magnete, magneticisque Corporibus, et de magno magnete tellure* (Concerning Magnetism, Magnetic Bodies, and the Great Magnet Earth). The work's emphasis on direct observation and rigorous experimentation earned Gilbert praise from Galileo (1564-1642), who considered him the founder of the experimental method.

De magnete begins by reviewing the history of magnetism and discussing the properties and behavior of lodestones. Gilbert experimentally refuted many superstitious beliefs concerning the

William Gilbert. *(Library of Congress. Reproduced with permission.)*

nature and curative properties of magnets such as garlic's ability to nullify magnetism. He also demonstrated how the magnetic effects of lodestone could be increased by arming them with soft-iron pole-pieces and that steel rods could be magnetized by stroking them with lodestones. He further showed that iron magnets lose their magnetism when heated to red-hot and cannot be remagnetized while remaining so. Gilbert introduced the idea of Earth as a giant spherical lodestone with magnetic poles (a term he introduced), an equator, and the ability to attract objects to itself. He also noted that compass needles align themselves along a north-south axis and argued they point toward Earth's magnetic poles.

Gilbert established the discipline of electrostatics by carefully distinguishing the amber effect from effects due to magnetism. It had been known since antiquity that amber, when rubbed, acquired the ability to attract light objects. Gilbert extended knowledge in this area with experiments that revealed substances other than amber exhibiting this same attractive force when rubbed. He referred to all such substances as "electrics" (from the Greek *elektron*, meaning amber) and believed they attracted objects through the emission of an effluvium that seized and pulled inward the particles of other bodies. This eventually led to the idea of electric charge. Gilbert's versorium—a light metallic needle turning about a vertical axis—was specifically designed to test for such effects.

Gilbert extended his magnetic theories to the cosmos, though most of his conclusions were not well supported by experimental results. He believed the "fixed" stars were in the same category as the Sun and, following Nicholas of Cusa (1401-1464), believed they were not all equally distant from Earth. He accepted the diurnal rotation of Earth but incorrectly attributed this motion to magnetic effects. He rejected the reality of solid celestial spheres and speculated that magnetic attraction maintained the planets in their orbits. Johannes Kepler (1571-1630) sought to apply Gilbert's ideas in a similar fashion, but his use of magnetic attraction as the motive force of his astronomical theory failed. Though Gilbert never championed Nicolaus Copernicus's (1473-1543) system in writing, his ideas were effectively exploited by others in arguing for Copernican heliocentrism.

STEPHEN D. NORTON

Francesco Maria Grimaldi
1618-1663
Italian Physicist

Francesco Maria Grimaldi was the first scientist to recognize the tendency of light to bend around objects, a phenomenon he named diffraction. He also constructed one of the most detailed maps of the Moon up to his time, and may have initiated the practice of naming lunar features after scientists. Today there is a crater on the Moon named after Grimaldi.

Born in Bologna, Italy, in 1618, Grimaldi came from a wealthy background. His father died when he was young, and at 14 he and his brother entered the Society of Jesus, or the Jesuit order. He studied theology and philosophy until he was 27, and also taught at the College of Santa Lucia, a school operated by the Jesuits, in Bologna. In 1645 he received his bachelor's degree, and an additional two years of study yielded his doctorate.

During his student years in the 1640s, Grimaldi had an opportunity to work as assistant to astronomer Giovanni Riccioli (1598-1671), who was also a Jesuit professor. Their early work together followed up on experiments made by Galileo (1564-1642) concerning falling weights, the speed of which they timed using a pendulum. For their astronomical work, Grimaldi developed a new and highly precise telescope, which helped him construct an extremely detailed Moon map or selenograph. Actually, the selenograph consisted of hundreds of drawings pieced together by Grimaldi and Riccioli.

Soon after earning his doctorate, Grimaldi gained an appointment as professor in the philosophy department of the College of Santa Lucia. Health problems forced him to give up this position, however, shortly afterward he was appointed to a mathematics professorship. At the age of 33 in 1651, he was ordained as a priest.

Around this time, Grimaldi began conducting his famous experiments with optics, allowing light to pass through a series of two apertures, or slits, and onto a blank screen. He noted that the area covered by the light on the screen was much wider than the last aperture, which indicated that the light had bent outward from the second opening.

Up until that time, scientists accepted the view that light traveled in the form of particles, whereas Grimaldi's research indicated that it actually came in waves, since only a wave could bend around objects. Some three centuries later, scientists would be confronted with the perplexing realization that light can travel either in waves *or* in particles, and though Grimaldi was incorrect in his conclusion that it only came in waves, his work was important for introducing the wave theory.

In choosing the word "diffraction," Grimaldi was referring to the manner in which water flowed around stones, branches, or other obstacles in its path. As he continued to study diffracted beams, he began to notice colors at the edges of the light beam, but could not figure out how they were created. The latter discovery would have to wait for Joseph von Fraunhofer (1787-1826).

As for Grimaldi, he continued to teach at Santa Lucia, the Jesuit College, for the rest of his life. He died in 1663, at a mere 45 years of age.

JUDSON KNIGHT

Otto von Guericke
1602-1686
German Physicist

Otto von Guericke, a German aristocrat and politician, made important contributions to two of the liveliest areas of physical investigation in the seventeenth century. He is credited with the invention of the air pump, a device that facilitated the study of the phenomena of vacuums. Von Guericke also constructed one of the earliest machines to produce static electricity.

Von Guericke was born in Magdeburg, a prosperous city in central Germany, into a wealthy and politically influential family. He studied law, science, and engineering at the universities of Leipzig, Helmstedt, Jena, and Leiden. Upon the completion of his studies in 1625, Guericke returned to Magdeburg where he was made an alderman, the start of a long period of public service to his native city.

Catholic troops in the service of the Hapsburg emperor set siege to Lutheran Magdeburg in 1631. Magdeburg was decimated by the attack, and Guericke left there to work for the governments of Sweden and Saxony as an engineer. He worked on behalf of Magdeburg as he traveled, serving as a foreign envoy and representative for his city. Guericke became the mayor of Magdeburg in 1646.

Settled once again in his hometown, Guericke used his leisure time to perform some remarkable scientific investigations. Like many others in the mid-seventeenth century, Guericke was interested in the philosophical problem—posed by Aristotle (384-322 B.C.) but made newly fascinating by the work of René Descartes (1596-1650) and Evangelista Torricelli (1608-1647)—of whether a vacuum could exist in nature. While Torricelli had investigated the behavior of mercury in glass to determine if a vacuum might exist there, Guericke sought to construct a device that could remove the air from a hollow vessel to test Descartes' idea that a container from which all the air was removed would collapse.

Guericke's first pump did eventually succeed in imploding a copper vessel made of two hemispheres (known later as Magdeburg hemispheres), but problems he encountered left him even more curious about the phenomena of pressure. He continued to design better equipment for more and more dramatic experiments. Guericke's most famous demonstration used a well-sealed pair of copper hemispheres that, when evacuated, could not be pulled apart even by powerful workhorses. Once air was admitted to the spheres, however, they immediately came apart—a memorable illustration of the power of the vacuum. Guericke eventually showed that sound could not travel in a vacuum (although light could) and that neither combustion nor respiration could take place. Guericke took advantage of his political position to gain attention for his scientific work. He announced his invention of the air pump at the Imperial Diet in 1654, and

Otto von Guericke. *(Library of Congress. Reproduced with permission.)*

his fellow delegates helped circulate word of Guericke's invention throughout Europe.

Guericke also investigated other areas of science. Related to his work on the vacuum were a series of experiments with barometers to study atmospheric pressure and meteorological conditions performed in 1660. A project that sought to simulate the magnetic properties of Earth by constructing a model of it out of sulphur led to another important, if unexpected, discovery. Guericke noticed that his model globe produced static electricity when rubbed, and he went on to make a primitive machine for the production of static electricity. This device fascinated onlookers as it attracted and repulsed feathers and other light objects. Christiaan Huygens (1629-1695), among others, was very interested in Guericke's sulphur globe, but they all had difficulty reproducing the effects Guericke had reported. Because of these difficulties—the sulfur globe requires very particular humidity conditions to perform—Guericke's discoveries had to be repeated in new contexts before their results were accepted as reliable electrical phenomena.

Von Guericke was made a nobleman in 1666. He retained his post as mayor of Magdeburg until 1676. He retired to Hamburg in 1681, and died there in 1686.

LOREN BUTLER FEFFER

Johannes Hevelius
1611-1687
German Astronomer

Johannes Hevelius was the last astronomer of repute to carry out major observational work without a telescope. Though he rejected the use of telescopic sights for stellar observations and positional measurements, he did use telescopes to produce accurate maps of the Moon and is considered the father of lunar topography.

Hevelius was born in Danzig (now Gdansk, Poland) on January 28, 1611. He studied jurisprudence at the University of Leiden before touring Europe. Upon his return he worked in his father's brewery while preparing to enter public service. His occasional interest in astronomy developed into a serious occupation in 1639 when he began making systematic observations. For the rest of his life his time was split between managing the family brewery, civic service, and astronomical research.

Hevelius's first observatory was a small room atop his father's house. In 1644 he added a small roofed tower and later erected a terrace with two observation enclosures to accommodate his large quadrants, sextants, and masts for supporting extremely long telescopes. Hevelius built and used these "aerial" telescopes because his experience and the optical theories of Johannes Kepler (1571-1630) and René Descartes (1596-1650) had shown that strongly curved lenses produced badly distorted images. Weak objective lenses—convex lenses with small curvatures—produced better images but required extremely long telescopes to accommodate the long focal lengths.

Using telescopes of 8.2 to 11.5 feet (2.5 to 3.5 meters) Hevelius observed the phases of Mercury in 1644 that previously had been predicted by Kepler and then observed by Ionnes Zupo in 1639. Hevelius also used these instruments to observe the Moon. These observations provided the material for his *Selenographia* (1647). The 133 color plates of this work represent the first detailed, accurate maps of the Moon's surface. Many of his names for the Moon's features were taken from Earth's geography and are still used, such as Mare Serenitatis (Pacific Ocean). However, his names for individual craters were not adopted.

Hevelius's next great work was his two-volume *Cometographia* (1668), in which he discussed the nature of comets and collected a con-

siderable body of literature on comets observed in previous centuries. He considered comets planetary exhalations and believed them responsible for sunspots. Like Giovanni Borelli (1608-1679) he suggested their orbits might be parabolic.

Interested in positional astronomy, Hevelius decided to compile a new star catalog for the Northern Hemisphere, which was to be more extensive and accurate than Tycho Brahe's (1546-1601). In taking positional measurements he preferred naked-eye observations. Some of these measurements appeared in *Cometographia*, which he sent to certain fellows of the Royal Society, including Robert Hooke (1635-1703). Hooke replied by recommending the use of telescopic sights. Their correspondence continued but Hevelius refused to yield. When the first volume of *Machina Coelestis* appeared in 1673—it contained much new observational data and a detailed description of Hevelius's observatory and instruments—Hooke publicly criticized him. In 1679 Hooke and John Flamsteed (1646-1719) persuaded Edmond Halley (1656-1742) to pay Hevelius a visit during his tour of Europe and try to convince him of the advantages of the telescope. To Halley's surprise, he found Hevelius could make consistently accurate measurements on par with the best telescopic work.

In September 1679 Hevelius's observatory burned to the ground with most of his instruments and many notes. Though he diligently rebuilt his observatory and constructed new instruments, the loss greatly affected his health. He died on January 28, 1687. His second wife and collaborator, formerly Catherina Elisabeth Koopman, published the *Uranographia* posthumously in 1690. It is his best known work, cataloging over 1,500 stars and introducing several new constellations, including Lacerta, Leo Minor, Lynx, Scutum, Sextans, and Vulpecula.

STEPHEN D. NORTON

Robert Hooke
1635-1703
English Physicist

Robert Hooke was a scientist with broad interests and accomplishments. He is best known for his research on elastic solids and for the discovery of the law that governs their behavior.

Hooke was educated at Westminster and at Christ Church of Oxford University. While at Oxford, he met chemist Robert Boyle who hired Hooke to assist him in his research on the be-

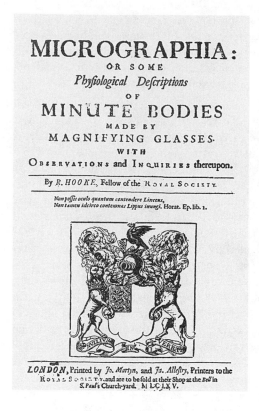

Title page from Robert Hooke's book *Micrographia*.
(Bettmann/Corbis. Reproduced with permission.)

havior of gases. In 1622, Hooke was made curator of experiments for the Royal Society of London and was elected a fellow of the Society the following year. In 1665, he was chosen to be professor of geometry at Gresham College in London, a position he held for thirty years. He also served as City Surveyor of London and was Christopher Wren's chief assistant in the effort to rebuild London after the Great Fire of 1666.

The expression *Renaissance man* is applied to a person who is able to gain more than a superficial knowledge of a wide range of intellectual areas. Robert Hooke could be used as the prototype of such a person. A man of great intellect and ability, he developed a curiosity regarding virtually all scientific fields and devoted his life to pursuing these interests. Among his early endeavors was the construction of a telescope, the first of its type, with which he made significant astronomical observations, especially concerning Jupiter and Mars, obtaining results that were useful later in establishing the properties of these planets. He also proposed that a numerical value for gravity could be ascertained by using a pendulum.

Elastic materials are defined as solids that will return to their original condition when an external force that has stretched them is re-

moved. Hooke's experiments with elastic substances led to the discovery of a fundamental relationship known, in his honor, as Hooke's Law. This principle states that the amount that an elastic material will stretch when an external stress is placed upon it is proportional to the stress. He made a number of applications of this principle, including the improvement of the accuracy of watches using balance springs.

In 1865, he gained wide recognition as a result of the publication of his book *Micrographia*, an illustrated discussion of observations he made with a reflecting microscope he built himself. His commentary included biological specimens, and he coined the word *cell* to explain the microscopic structures he observed. *Micrographia* also presented the results of his microscopic studies of crystalline solids, including snowflakes. Based on his observations, he proposed that solids form different structures as a result of different packing arrangements of microscopic spherical particles. This led to more extensive studies of crystal structure, and consequently, he is regarded as the founder of the science of crystallography. He became one of the earliest supporters of the concept of extinction of species and of biological evolution as a result of his microscopic study of fossils.

Results of his eclectic interests included the invention of a telegraph system, the discovery of the diffraction of light, a proposal of a wave theory of light, and a theory of earthquakes. He was first to assert that matter expands when heated and first to propose that air is made up of microscopic particles located at relatively large distances from each other. He stated the inverse square law of planetary motion in 1678, without mathematical proof, and informed Isaac Newton (1642-1727). Newton was able to support the postulate mathematically and later used it as a fundamental principle in his theory of gravitational attraction, without giving any credit to Hooke. This led to a prolonged acrimonious controversy between the two over credit for the discovery.

J. WILLIAM MONCRIEF

Christiaan Huygens
1629-1695
Dutch Physicist

Christiaan Huygens is famous for establishing the wave theory of light. He formulated the conservation law for elastic collisions, produced the first theorems of centripetal force, and developed the dynamical theory of oscillating systems. He also made improvements to the telescope, discovered Saturn's moon Titan, and invented the pendulum clock.

Huygens was born on April 14, 1629, in The Hague, Netherlands. His father Constantijn (1596-1687) was a diplomat and well-known Renaissance poet. The Huygens household received frequent visits from French intellectuals including René Descartes (1596-1650), who greatly influenced young Christiaan. Huygens was educated at home before entering the University of Leiden to study law and mathematics (1645-47). From 1647 to 1649 he studied law at the Collegium Arausiacum in Breda. Rejecting the idea of a diplomatic career, he returned home in 1650 to devote himself to science.

Although an outstanding mathematician, Huygens's only original mathematical contributions were his theory of evolutes, developed in connection with his work on the pendulum clock, and his probability theory, in which he introduced the concept of the expectation of a stochastic variable. His importance as a mathematician lies in his improvement and application of existing techniques to the analysis of physical problems.

In 1652 Huygens began his study of colliding bodies and by 1656 arrived at a correct solution for the case of elastic collisions. The Royal Society asked John Wallis (1616-1703) and Christopher Wren (1632-1723) to examine the theoretical aspects of this problem in 1666, and Huygens was solicited for a report of his discovery. The results of all three, obtained independently and published together in the *Philosophical Transactions* (1669), established the law of conservation of momentum.

With his brother Constantijn, Huygens developed lens-grinding techniques that reduced spherical aberration. After incorporating these lenses into their telescopes, Huygens discovered Saturn's satellite Titan (1655), correctly described Saturn's ring (1656), and first observed Martian surface markings (1659). Huygens also invented a two-lens eyepiece—the Huygens ocular—and an improved micrometer.

Huygens rendered astronomy a greater service in 1657 with his invention of the pendulum clock, which made possible accurate time measurements. Huygens demonstrated that the period of pendulum swings will not be equal unless the arc of the swing is a cycloid. He devised fulcrum attachments to produce the appropriate arc and patented his device.

When the Académie des Sciences was established in 1666, Huygens became its most prominent member and continued his research on oscillatory systems in Paris. These culminated in the publication of *Horologium Oscillatorium* (1673), which includes a mathematical analysis of the compound pendulum and derivation of the relationship between pendulum length and period of oscillation. He also included the laws of centrifugal force for uniform circular motion and an early formulation of Isaac Newton's (1642-1727) first law of motion.

In 1678 Huygens completed work on *Traité de la lumière*, which was his response to Newton's corpuscular theory of light. Publication was delayed until 1690. Huygens presented a wave construction capable of explaining light's rectilinear propagation, reflection, refraction, and certain properties of double refraction in Iceland spar. He also predicted, in opposition to Newton, that light travels slower in denser media. Newton's theory dominated eighteenth-century optical thinking but was eclipsed by Huygens's theory in the early nineteenth century. The two views were later synthesized in quantum theory during the early years of the twentieth century.

Huygens's fifteen-year residence in Paris was interrupted by two extended periods of convalescence at The Hague. When illness brought him home once again in 1681, he decided not to return to France due to the climate of growing intolerance towards Protestants. Huygens died in The Hague on July 8, 1695.

STEPHEN D. NORTON

Johannes Kepler
1571-1630
German Astronomer and Mathematician

Johannes Kepler was a mathematician and astronomer. He used observations of heavenly bodies to destroy the ancient idea that planets move in perfect circles. He also mathematically described the relationship of Sun and planets in our solar system. His three laws of planetary motion, along with the work of Nicolaus Copernicus (1473-1543) and Galileo (1564-1642), helped Isaac Newton (1642-1727) devise his law of universal gravitation.

Kepler was born in 1571 in Weil, Germany, a sickly, myopic child with a brilliant mind and intermittent double vision. He became interested in astronomy after seeing the 1577 comet when he was six years old and an eclipse of the

Moon when he was nine. He attended a seminary and the University of Tubingen intending to enter the church. His abilities as a mathematician led him instead to teach math in Graz, Austria, where he was also district mathematician. He cast official horoscopes and was court astrologer. While in Graz he cast many horoscopes for local citizens and published a calendar of astrological forecasts to augment his income. At this time he already believed that Earth moves around the Sun, not vice versa. Kepler was an original thinker and could grasp and manipulate

Johannes Kepler. *(Library of Congress. Reproduced with permission.)*

KEPLER'S MOTHER'S WITCH TRIAL

Johannes Kepler's mother, Katharina, was known in her hometown as a cantankerous old woman. During the upheavals of the Thirty Years' War she became a victim of witch-hunting. A dutiful son, Kepler helped in her defense, which was both long and costly. Katharina's accusers pointed to dozens of occasions where various aliments or actions had occurred due to her magical potions or spells. Kepler's successful defense strategy was to show how all these occurrences could be explained by natural causes. For instance, a young girl claimed Katharina had made her arm temporarily paralyzed. Kepler pointed out that she had been carrying many bricks, and the heavy load had caused her problem. A woman's sickness was revealed to be from an abortion. The schoolmaster had been injured while jumping a ditch. The butcher had lumbago. And so on. Kepler was careful never to dismiss witchcraft out of hand, as many officials believed deeply in such magic. However, he showed that for each specific event attributed to witchcraft there was a more likely natural cause that explained the result at least as well. Yet, while Katharina was acquitted, she died shortly after, a broken woman.

DAVID TULLOCH

new ideas. He was exceedingly patient and made calculations with scrupulous care.

In 1596 he published his first book corroborating Copernicus and looking for a relation between planetary bodies. In 1600 he moved to Prague, where he became an assistant to Tycho Brahe (1546-1601), Imperial Mathematician. This was a fateful association because Brahe, a Dane, was a tireless recorder of movements of heavenly bodies. When Brahe died, Kepler became Imperial Mathematician and possessor of

Brahe's observations. Brahe's family insisted that he had usurped the data, but Kepler was able to keep and use the material.

It is said that Kepler "discovered" his three laws, but it is more accurate to say he constructed them to fit the observations. Kepler explained celestial phenomena to fit these observations and believed that a mathematical relationship existed between celestial bodies. In 1602 he discarded perfect circles, which had been the model of the heavens since the Greeks. Using Brahe's data, he postulated that a planet's distance from the Sun determined its speed and calculated the exact relationship with precise accuracy. This became his second law. In 1605 he came to the conclusion that planets move in elliptical orbits, his first law. These were the first instances of discoveries made to fit observed data. In 1618 he formulated the third of his three laws, which states that the square of the time a planet takes to revolve around the Sun is proportional to the cube of its distance from the Sun.

He occasionally corresponded with Galileo, but they worked on different aspects of astronomy. After he and his family moved to Linz, Austria, Kepler published several books on astronomy and the harmony of the spheres. They were unique but read by few because they were complex and hard to understand. He continued to look for relations between planets and the Sun.

Some of his published ideas failed or were deemed absurd. His reputation was not helped by the fact that his mother was put on trial for witchcraft in 1615 and imprisoned for a time.

When he died in November 1630 in Regensburg, Germany, he was famous for his writings and revered as a careful mathematician whose work had aided the advancement of of astronomy. Kepler's work assisted Newton in formulating his 1687 theory of universal gravitation that led to modern ideas about physics and astronomy.

LYNDALL B. LANDAUER

Andreas Libavius
1540?-1616
German Alchemist

The alchemist Andreas Libavius is often considered the author of the first chemistry textbook. Contrary to the beliefs of many alchemists, he promoted the idea that the scientific discoveries should be shared openly, rather than kept secret.

Libavius was born in about 1540 in Halle, Germany, where his father was a linen weaver. At this time, children of working class parents were rarely allowed to attend universities. Usually, the only people to receive an advanced education were those who had been born into wealthy families. Despite his family's background, Libavius began attending the University of Wittenberg at the age of eighteen. This fact suggests to historians that he must have shown signs of great intelligence and been very determined to receive an education.

Libavius later attended the University of Jena, where he studied history, philosophy, and medicine. He completed his medical degree there in 1581. For several years, he taught history and poetry at the University of Jena, and then, from 1591 to 1596, he worked as a town physician. In 1605, he helped to found a school in the city of Coburg, where he remained until his death his death in 1616.

Libavius is best known, however, not as a physician or as a professor, but as an alchemist. Alchemy was a primitive science whose main goals were the discovery of the *philosopher's stone* and the *elixir of life*. The philosopher's stone would supposedly allow alchemists to create gold from other metals, and the elixir of life was thought to be a potion that would cure all diseases. Although these goals proved to be un-achievable, alchemy would eventually evolve into modern chemistry.

Libavius believed that alchemy would prove to have benefits to physicians, and most of his chemical work was concerned with its reference to medicine. He wrote books on numerous scientific topics, but his most important was titled *Alchemia* (Alchemy), which was published in 1597. This book summarized the discoveries that alchemists had made up to that date. Later editions of the book were more than 2,000 pages long and contained 200 illustrations.

Many alchemists wished to cloak their findings in mystery and secrecy. They felt their work had religious power and therefore should not fall into the wrong hands. Libavius's writing, however, was clear and meant to be easily understood. In his book, he describes alchemy as being "valuable in medicine, in metallurgy [the science of metals] and in daily life." In other words, he saw alchemy (and chemistry) as having a practical value beyond its lofty goals.

Libavius also organized his book in a systematic manner. *Alchemia* is divided into four parts. The first part describes the equipment needed for a chemistry laboratory, such as furnaces, vials, and mortars. It also includes descriptions of chemical procedures, such as distillation—a method of separating liquids based on their boiling points. The second part of *Alchemia* gives instructions, or recipes, for preparing certain chemicals. The third describes several early methods of chemical analysis—ways of determining the composition of chemicals. For example, Libavius explained that some copper-containing chemicals would turn blue when placed in a solution of ammonia. Therefore, this test could often be used to detect the presence of copper.

The first three parts of the book contain information that clearly belongs to the science of chemistry. The final part of the book, however, belongs strictly to alchemy. It discusses the theory of transmutation, a change of one type of matter directly into another. For example, it was thought that the philosopher's stone would transmute so-called base metals such as lead and tin into gold. Libavius (incorrectly) considered many types of chemical reactions to be a form of transmutation. Today, however, scientists know that transmutation can occur only in very special circumstances, such as those involving some types of radioactivity.

In addition to his writing on alchemy, Libavius performed chemical experiments of his own. He discovered methods of preparing several important chemicals, including hydrochloric acid, ammonium sulfate (now used as fertilizer), and tin tetrachloride. Because tin tetrachloride gives off fumes when it is exposed to the water vapor in air, alchemists named it "fuming liquor of Libavius."

During Libavius's life, chemistry experiments were usually conducted by private individuals using their own equipment (and often their own homes). To solve this problem, Libavius drew up plans for a building containing a series of laboratories where chemical experiments could be carried out. This "chemical house" could be considered a forerunner of the modern chemical laboratory. The plans included a storeroom, a room where chemicals would be prepared, an assistant's room, a crystallization and freezing room, a fuel room, and a wine cellar (wine was used as a source of alcohol in experiments). Libavius, however, did not live to see his chemical house, or others like it, constructed.

STACEY R. MURRAY

Sir Isaac Newton
1642-1727
English Physicist, Mathematician, and Astronomer

Isaac Newton's combination of abilities as an experimentalist, theorist, and pure mathematician have never been surpassed, and his *Philosophiae Naturalis Principia Mathematica* (1687) stands as one of the greatest, if not the greatest, scientific work ever written.

Newton was born on December 25, 1642, in Woolsthorpe, England. A premature and sickly baby born after his father's death, Newton spent his youth building mechanical contrivances including water-clocks, a mouse-powered mill, and kites bearing fiery lanterns. He entered Trinity College, Cambridge, in 1661 and earned his bachelors degree in 1665, having earlier that year discovered the binomial theorem. Soon thereafter the university closed because of the plague, and Newton retired to Woolsthorpe for the next 18 months.

This period of seclusion was among the most productive of his life. By November 1665 he had discovered differential calculus and by May of the following year integral calculus. During the intervening months his experiments with prisms had revealed that white light was composed of different colors refracted through char-

acteristic angles. Also in 1666 Newton first considered extending gravity to celestial bodies. He correctly theorized that the rate of fall was proportional to the gravitational force with the force falling off according to the square of the distance from Earth's center. However, disagreements between his calculations and observations led Newton to shelve this work for almost 15 years.

Newton returned to Cambridge in 1666 and received his M.A. in 1668. That same year he built the first reflecting telescope, important because it eliminated chromatic aberration inherent to refractors. In 1669 he became Lucasian Professor of Mathematics. (It is generally thought the incumbent Isaac Barrow [1630-1677] resigned so Newton might have it.) After exhibiting his telescope before the Royal Society, he was elected a fellow in 1672 and shortly thereafter presented his optical experiments underlying the invention. A multiyear controversy with Robert Hooke (1635-1703) ensued whereby Newton refined his corpuscular theory of light. These ideas were eventually presented in the *Optiks* (1703), which dominated optical research for the next century.

A letter from Hooke in 1679 stimulated Newton to renew his work on gravitation. Newton quickly achieved a mature understanding of the dynamical principles involved and deduced that a central gravitational force implied Kepler's equal area law. He further showed that elliptical orbits, with the center of force at a focus, implied an inverse-square law of force.

In 1684 Hooke boasted to Christopher Wren (1632-1723) and Edmond Halley (1656-1742) that he had worked out the laws governing planetary motion. Wren was unimpressed and offered a prize for the correct solution. Halley took the problem to Newton, who informed him he had already solved it. Halley encouraged Newton to renew his work on gravitation and prepare it for publication through the auspices of the Royal Society. Newton complied, but the Society's financial difficulties forced Halley to assume full financial responsibility for the *Principia*, which finally appeared in 1687.

The *Principia* represents the culmination of the Scientific Revolution. Newton synthesized Kepler's laws of planetary motion and Galileo's experimental results on falling bodies by codifying the principles of mechanics and extending their application to celestial phenomena. His three laws of motion clarified the distinction between mass and weight and how they are related under a variety of circumstances, while the law of universal gravitation explained Earth's equatorial bulge, orbital inequalities, the parabolic orbits of comets, and more.

Newton also conducted comprehensive alchemical experiments and studied the Bible throughout his life, leaving extensive manuscripts on Church history and ancient chronology. His academic output diminished in later years as he accepted various civic positions. He died in London on March 20, 1727.

STEPHEN D. NORTON

George Owen
1552-1613
Welsh Geographer, Geologist, and Cartographer

George Owen of Henllys, lord of Cemais, is best known for his work *Description of Pembrokeshire*, written in 1603 but not published until 1892. The map he made of Pembrokeshire was considered a landmark, and Owen has been referred to as the patriarch of English geologists. Interested in the lineages of local families, he also documented family histories.

Owen was born in Henllys, Pembrokeshire, Wales, in 1552. His father was a lawyer who belonged to an old Welsh family, who did much himself to improve the social standing of his family. Collecting information on genealogy, heraldry, and historic structures of Wales, Owen was also interested in the topography and geography of Wales and in geological structures of the area, including strata of limestone and coal. Owen did not attend university, but did study law at the Inns of Court in London.

When George Owen came of age he inherited the lordship of Cemais under the Earl of Pembroke, and the Newport Castle from his father. There was some conflict from the local council as to his possession of these lands, and he was even placed under arrest in his own castle. These disputes were later settled. As landlord of Cemais, he wanted to improve agricultural practices, and a paper on the use of marl as a fertilizer was written by Owen, but it was never published. He also invented a new tool for cutting marl that he considered more efficient. He became the vice admiral of the maritime counties of Pembroke and Cardigan in 1573. Owen held many positions important to his community including the Commission of the peace in Pembrokeshire, and Deputy Lieutenant of Pembrokeshire from 1587-90 and 1595-1601. In that position he was responsible for defense in Milford Haven and trained country militia. He was the sheriff of Pembrokeshire in 1587 and 1602. He married twice, first in 1573 to Elizabeth, and he had ten children from this marriage. His second marriage was to Ann.

After 1592, when Owen was commissioned by the crown to survey the property of Sir John Perrott, the Earl of Pembroke, he created a map of Milford Haven, which began his work in the field of geology and geography. Producing the map of Milford Haven under the direction of the Earl of Pembroke, Owen used his own survey and completed a genealogical catalogue of the Earl of Pembrokeshire for him as well. He produced the *Description of Milford Haven* in 1595, and the *Description of Wales* in 1602. Later in 1603 he wrote *Description of Pembrokeshire*, not published until 1892. It is considered a landmark in Welsh geography; the original manuscript can be found in the British Museum. It has since been reproduced by a descendant, Henry Owen, under the title *Owen's Pembrokeshire*. George Owen also wrote *A Cataloge and Genelogie of the Lord of Kemes, and Baronia de Kemes*, published in 1861 and 1862, and *A Treatise on the Government of Wales and Pembrock and Kemes*. Owen's interest in heraldy led him to write a collection of pedigrees, and he also wrote a story called the "Taylor's Cushion." Another notable achievement of Owen's was a detailed map of Pembrokeshire, later included in *Owen's Pembrokeshire*.

While Owen did not receive a formal education in geography or cartography, his detailed writings and maps were groundbreaking. *Description of Pembrokeshire* contained detailed writings about the area and its landscape, including geological structures. His were the first for Pembrokeshire, and served as a suitable basis for further studies in the area. George Owen died in Haverford West, Pembrokeshire, in 1613.

KYLA MASLANIEC

Jean Picard
1620-1682
French Astronomer

Jean Picard was a French astronomer who was the first to accurately measure the length of the arc of the meridian, the imaginary line running across the Equator between the North and South Pole. Picard's historic measurement allowed him to do something magnificent—compute the size of Earth. His observations helped

English physicist and mathematician Sir Isaac Newton (1642-1727) to verify his theory of gravitation.

Early in his career Picard observed stars on the meridian during the day and measured their position using cross-wires at the focus of his telescope. Since many of his colleagues had lost their standard default measurements, Picard devised a method of comparing his with the length of a simple pendulum beating seconds at Paris. His ingenuity allowed him to reproduce the standard at any time.

Shortly after, the French astronomer applied telescopes and micrometers to graduated astronomical and measuring instruments. In 1669-70 he made his historic observations using a specially designed telescope and Willebrord Snell's (1591-1626) theory of triangulation. By measuring the angles of a series of triangles extending from Paris northward, he determined *latitude*, a term used in mapping to locate a place north or south of the Equator. Latitude is expressed by angular measurements ranging from 0° at the equator and 90° at the poles. Picard was among the first to apply scientific methods during mapmaking. He produced a map of the Paris region, then went on to join a project to map France.

Picard is regarded as the founder of modern astronomy in France. He studied for the priesthood at the Jesuit college at Le Flèche and later received a masters in astronomy from the University of Paris. Ten years after observing the solar eclipse of August 1645 he became professor of astronomy at the Collège de France, Paris. In 1666 he became one of the first members of the Academy of Royal Sciences.

Shortly after joining the Academy, Picard visited the observatory of noted astronomer Tycho Brahe (1546-1601) on Hven Island in Sweden. His goal was to precisely determine the observatory's location so Brahe's astronomical observations of could be directly compared with others. Picard later visited the Paris Observatory, where he collaborated with rival Italian astronomer Jean-Dominique Cassini (1625-1712), Ole Römer (1644-1710), and, slightly later, mathematician Philippe de la Hire (1640-1718). Putting aside his own ambitions, Picard recommended Cassini to King Louis XIV for the direction of the new observatory at Paris.

One of Picard's less publicized discoveries occurred during his study at Tycho Brahe's observatory. While taking measurements on the observatory mountaintop, Picard observed what later became the single most important concept behind neon signs. The French astronomer discovered that a faint glow appeared in the mercury-filled tube (barometer) he used to measure atmospheric pressure. When he shook the tube, the glow intensified. The effect, called barometric light, caused quite a stir in the scientific community, although the actual cause of the light was not well understood.

In 1679 Picard began publication of the first national almanac, the *Connaissance des temps* (Knowledge of Time or the Celestial Motions). He authored the first five volumes, which contained tables for the crude determination of longitude for the position of celestial bodies. Since then it has been published continuously.

Picard is also credited with the introduction of telescopic sights and the introduction of the pendulum clock.

KELLI A. MILLER

Johannes Regiomontanus
1436-1476
German Astronomer and Mathematician

Johannes Regiomontanus played a key role in reforming astronomical studies in fifteenth-century Europe by emphasizing and acting on the need for new and improved observations over those of the ancients. He also introduced Arabic algebraic and trigonometric methods to Europe, thus providing a systematic basis for their further development.

Regiomontanus, whose real name was Johann Müller, was born on June 6, 1436, in Königsberg, Franconia. The name of his birthplace means "King's Mountain," and in accordance with the practice of the day his parents adopted the Latinized version of this name—Joannes de Regio monte—from whence Regiomontanus was derived. He studied dialectics at Leipzig sometime around 1448. He was then drawn to the University of Vienna by the reputation of the astronomer Georg von Purbach (1423-1461). Regiomontanus matriculated there in 1450 and after receiving his bachelor's (1452) and master's degrees (1457) joined the Vienna faculty.

European astronomical work of the Middle Ages, with the exception the efforts of Alfonso X (1223-1284) and his assistants, amounted to little more than the collection and reorganization of Arabic and ancient Greek material. No new observations of importance were undertaken,

gether with his *Tabulae directionum* it was the primary means whereby Arabic algebraic and trigonometric methods were reintroduced to Europe. In *De triangulis,* Regiomontanus developed the earliest statement of the cosine law for spherical triangles; the *Tabulae directionum,* which was primarily an astronomical work, contained a valuable table of tangents.

By 1468 Regiomontanus had returned to Vienna. He then moved to Nuremberg in 1471. There he attracted the wealthy, amateur astronomer Bernard Walther (1430-1504), who provided him with an observatory and work shop. Of the many observations Regiomontanus made while in Nuremberg his most important were of the comet of 1472—later to be known as Halley's comet. This work appears to have been the first attempt to study comets scientifically instead of viewing them merely as objects of superstition. Regiomontanus also established the first press devoted to astronomical and mathematical literature, intending to advance the work of science by providing quality texts free of scribal and printing errors.

In 1475 Regiomontanus traveled to Rome. According to some reports he was summoned by Pope Sixtus IV to assist with the reform of the Julian calendar. Whether or not this is true, nothing substantive along these lines emerged from his trip, for he died on July 8, 1476, quite probably due to the plague that spread after the Tiber overflowed earlier that year.

STEPHEN D. NORTON

Johannes Müller, Regiomontanus. *(Bettmann/Corbis. Reproduced with permission.)*

and by the mid-fifteenth century the Alfonsine Tables (1252) were sorely in need of revision. Purbach pointed out to Regiomontanus the inaccuracies of the Tables as well as the need for better translations of Greek texts. Though his knowledge of ancient Greek left much to be desired, Puerbach attempted to produce a revised and corrected Latin verion of Ptolemy's (fl. second century A.D.) Almagest. Though he did not live to finish the project, he pledged Regiomontanus to see it to completion.

In 1461 Regiomontanus traveled to Italy in search of early Greek scientific manuscripts. While there he finished the *Epitome of Astronomy*. The work contained, in addition to the Purbach-Regiomontanus translation of Ptolemy's work, critical commentary, revised computations, and additional observations. Though not printed until 1496, 20 years after Regiomontanus's death, the work was a great success and attracted the attention of a young Nicolaus Copernicus (1473-1543). Struck by the errors in Ptolemaic lunar theory revealed by Regiomontanus, Copernicus went on to develop his heliocentric view of a Sun-centered Solar System.

While in Italy, Regiomontanus completed much of *De triangulis omnimodis,* which appeared posthumously in 1533. This work represents the first systematic treatment of trigonometry presented independently of astronomy. To-

Jean Richer
1630-1696
French Physicist

Jean Richer is remembered for having determined that pendulums beat more slowly at the equator than at higher latitudes, a discovery that initiated the famous controversy between Newtonians and Cartesians over Earth's shape. Also, his observations of Mars were used by Giovanni Domenico Cassini (1625-1712) to make the first accurate estimate of the size of the solar system.

The details of Richer's early life have been lost, but when the Académie des Sciences was organized in 1666, he was admitted as an *élève astronome* (assistant astronomer). In March of 1670 he was selected to carry out astronomical measurements in the East Indies in conjunction with measurements to be made simultaneously in Paris. During this trip he was also responsible for testing Christiaan Huygens's (1629-1695) marine clocks constructed for the purpose of determining longitude at sea.

Last minute changes resulted in the expedition being diverted to New England. Bad weather early on stopped Huygens's clocks, thus preventing the collection of any useful horological data. In New England Richer took tidal measurements at different sites and accurately determined the latitude of the French fort at Penobscot Bay. This was the most precise astronomical measurement made in the Western Hemisphere to that time. The expedition was back in France by September, and Richer reported his results to the Académie in January 1671.

Notwithstanding Huygens's unjustified imputation of Richer's abilities in the failure of his marine clocks, the Académie was suitably impressed with Richer's performance to select him for their next project, an expedition to Cayenne off the coast of French Guiana. The primary goals of this expedition were to accurately determine the motions of the plants and Sun, assess existing tables of refraction, and determine the parallax of Mars. Additionally, it was hoped that the uniformity in length of the seconds pendulum at all latitudes could be established, thus making it possible to determine a universal standard of linear measurement. The expedition departed in February 1672 and arrived at Cayenne in April. Due to an illness Richer left the expedition early, departing in May 1673.

His lunar and planetary observations corroborated the accuracy of existing astronomical tables. He also carried out extensive solar observations and accurately determined the obliquity of the ecliptic and time of the solstices and equinoxes. All of these measurements helped establish the accuracy of Cassini's tables of refraction. Richer also made observations of Jupiter and its satellites that allowed him to fix the longitude of Cayenne—a result of vital importance for the proper reduction of the his measurements.

Richer took careful observations of the meridian altitude and meridian transit times of Mars and certain nearby fixed stars. Cassini and others made corresponding measurements in France. By reducing Richer's observations to the Paris meridian and comparing several sets of corresponding measurements, Cassini determined the horizontal parallax of Mars and then the astronomical unit. His value of 87 million miles (140 million km) was the first fairly accurate estimate of the mean Earth-Sun distance.

Richer also observed that pendulums beat slower in Cayenne than in France. This was interpreted as a decrease in the gravitational attraction at the equator, suggesting that points along the equator were further from Earth's center than at higher latitudes. Isaac Newton (1642-1727) argued that his gravitational theory adequately accounted for Richer's results because it predicted Earth was an oblate spheroid—bulging equator and flattened poles. This contradicted the Cartesian view that Earth is a prolate spheroid—elongated along the polar axis. Expeditions to Peru (1734-44) and Lapland (1736) later settled the issue decisively in Newton's favor.

Richer was elected to full membership in the Académie in 1679. He died in Paris in 1696.

STEPHEN D. NORTON

Ole Christensen Römer
1644-1710
Danish Astronomer and Physicist

Ole Römer is famous for demonstrating the finite velocity of light. He produced various scientific instruments including an improved micrometer, planetaria, the first transit circle, and an alcohol thermometer.

Römer was born on September 25, 1644, in Aarhus, Denmark. In 1662 he matriculated at the University of Copenhagen, where he studied astronomy and mathematics with Thomas (1616-1680) and Erasmus Bartholin (1625-1698). He lived at the home of Erasmus while a student, eventually becoming his personal assistant and marrying his daughter.

In 1671 Römer met Jean Picard (1620-1682) and accompanied him to the island of Hven to assist in redetermining the position of Tycho Brahe's (1546-1601) observatory Uraniborg. To this end they made measurements, in conjunction with those by Giovanni Domenico Cassini (1625-1725) in Paris, of a series of eclipses of Jupiter's moon Io. Römer returned to Paris with Picard and was elected to Académie membership in 1672. He was appointed tutor to the Crown Prince, responsible for making various astronomical observations, and constructed many precision instruments including an improved micrometer that was quickly adopted into wide use.

Römer's most important result was an outgrowth of his further work on occultations of Jupiter's satellite Io. Cassini had already published ephemerides for the motions of Jupiter's

light's velocity. Though his value of 140,000 miles per second (225,000 kilometers per second) is about 25% too small, it represented a considerable achievement. Full acceptance of Römer's conclusion came only after James Bradley (1693-1762) announced his discovery of stellar aberration in 1729.

Römer returned to Denmark in 1681 to become professor of mathematics at the University of Copenhagen, Astronomer Royal to King Christian V, and Director of the Royal Observatory. In 1704 he built his own observatory, the Tusculaneum, which he equipped with quality instruments of his own design, including the first transit circle. Römer also invented a thermometer with a scale based on two fixed points that influenced Daniel Fahrenheit's (1686-1736) thermometric researches.

Römer held many civic and advisory positions including master of the mint, harbor surveyor, and inspector of naval architecture. He was Copenhagen's first judiciary magistrate (1693), chief tax assessor (1694), and then mayor (1705). He was also appointed senator and then named head of the state council of the realm (1707). He died in Copenhagen on September 19, 1710.

STEPHEN D. NORTON

Ole Christensen Römer. *(Library of Congress. Reproduced with permission.)*

satellites (1668) and in 1675 discovered an inequality responsible for periodic fluctuations in the timing of Jovian satellite eclipses. The effect seemingly depended on the position of Earth relative to Jupiter. Cassini entertained and then rejected the idea that these fluctuations were due to the finite velocity of light.

Römer pursued the issue more carefully and in September 1676 announced to the Académie that the eclipse of Io expected on November 9 would occur exactly ten minutes late. Skeptical, Académie members made careful observations. They reported the eclipse took place at 45 seconds after 5:35 A.M.—exactly 10 minutes late as predicted. Two weeks later before a baffled assembly of the Académie, Römer explained the delay was due to the finite velocity of light. Scrutiny of his observations and those of Cassini had revealed that the interval between successive occultations of Io diminished as Earth approached Jupiter and increased as Earth receded. Römer correctly surmised that this was because it took a shorter or longer time respectively for light to reach Earth. By comparing the predicted eclipse times of Io with those observed at various points in Earth's orbit, Römer estimated it took 22 minutes for light to cross Earth's orbit.

Using this estimate and his own estimate of Earth's orbital diameter, Christiaan Huygens (1629-1695) made the first determination of

Willebrord Snell
1580-1626
Dutch Physicist and Mathematician

Willebrord Snell is remembered for discovering the law of refraction that bears his name. He has also been called the father of modern geodesy for perfecting the method of determining distances by trigonometric triangulation.

Snell was born in 1580 in Leiden, Netherlands. He was the son of Rudolph Snell van Royen (Latinized as Snellius), professor of mathematics at the University of Leiden. Willebrord studied law and taught mathematics at Leiden. After touring Europe (1600-04) he returned home, where he prepared a Latin translation of Simon Stevin's (c. 1548-1620) *Wisconstighe Ghedachtenissen* and worked on restoring the two existing books of Apollonius's (c. 262-c. 290 B.C.) work on plane loci. In 1608 he received his M.A. and married. After the death of Rudolph in 1613, Willebrord assumed his father's teaching duties, officially succeeding him in 1615.

It was also during 1615 that Snell set himself the task of determining the length of a de-

Willebrord Snell. *(Photo Researchers, Inc. Reproduced with permission.)*

gree of the meridian. For this purpose he chose the method of triangulation originally suggested by Gemma Frisius (1533). Starting with his house and taking the spires of nearby churches as reference points, he measured a net of triangles from Alkmaar to Bergen-op-Zoom using a huge 130-inch (210-centimeter) quadrant. This allowed him to accurately compute the distance between these towns and also calculate the length of a degree of the meridian. His results were published in *Eratosthenes batavus* (1617). Seeking to improve his work he extended the net of triangles from Bergen-op-Zoom to Mechelen. Reduction of this data occupied him throughout the rest of his life, and his findings were published posthumously by one of his students. His corrected value of 69 miles (111 kilometers) for the length of a degree of the meridian is within a few hundred meters of the presently accepted value.

In 1621, or shortly thereafter, Snell discovered the law of refraction that today bears his name. When light rays pass obliquely from a rarer to denser medium (e.g. air to water) they are bent toward the vertical. Scientists from Ptolemy (fl. second century A.D.) to Johannes Kepler (1572-1630) had searched in vain for a law to explain this phenomenon. Ptolemy thought the angles of the incident and refracted light rays maintained a constant relationship,

while Kepler had produced nothing more than approximate empirical relations. Snell's years of research revealed that it was the ratio of the sines of the angles of the incident and refracted rays to the normal that remains constant.

Though Snell never published his findings, the manuscript containing the discovery was examined by Isaacus Vossius (1618-1669) and Christiaan Huygens (1629-1695), who commented upon it in their own works. However, priority of publication goes to René Descartes (1596-1650), who presented the law without proof in his *Dioptrique* (1637). Huygens and others accused Descartes of plagiarism. Though Descartes's many visits to Leiden during Snell's life make the charge plausible, there seems to be no evidence to support it.

Snell's astronomical work includes observations of the comet of 1618. His parallax measurements clearly indicated the comet was above the sphere of the Moon. Nevertheless, his support for the Ptolemaic system remained unshaken. In *Cyclometricus* (1621) he used Van Ceulen's methods to determine the value of π to 34 decimal places. His work on navigational methods focused on the study and tabulation of Pedro Nuñez's rhumb lines (1537), which Snell referred to as *loxodromes*. This material appeared in *Tiphys batavus* (1624). *Canon triangulorum* (1626) and *Doctrina triangulorum* (1627) contain the fruits of his research on plane and spherical trigonometry. The latter unfinished work was completed and published posthumously by his student Martinus Hortensius. Snell died in Leiden on October 30, 1626.

STEPHEN D. NORTON

Nicolaus Steno
1631-1687
Danish Anatomist, Naturalist, and Physician

Nicolaus Steno was one of the great scientists of the seventeenth century. Best known today for his views on fossils and as the founder of stratigraphy, he was trained as a physician and did important work in anatomy, crystallography, geology, and paleontology.

Steno, whose real name was Niels Steensen (sometimes spelled Stensen), was born on January 1, 1631, in Copenhagen, Denmark. Raised a devout Lutheran, he entered the University of Copenhagen in 1656 to study medicine. He continued his studies at the University of Leiden (1660-1663), receiving his M.D. *in absentia* while in Paris (1664). After making important anatomical discoveries he traveled to Italy, where in 1666 he was appointed court physician to Ferdinand II, Grand Duke of Tuscany (1610-1670). His most significant geological and paleontological results were developed during his stay in Florence. He converted to Catholicism in 1667 and returned to Copenhagen in 1668 to accept the position of royal anatomist. After becoming a priest in 1675 he abandoned science completely and devoted his remaining years to the Catholic Church. He died in acute pain from gallstones at Schwerin, Germany, on November 25, 1686.

In 1660 Steno discovered the parotid gland (Steno/Stensen's duct)—the oral cavity's principal source of saliva. His further investigations led to a basic understanding of the lymphatic system as a whole. In *Observationes anatomicae* (1662) and *De musculis et glandulis* (1664) Steno presented his new discoveries, describing the structure and function of other glands including the lachrymal apparatus, which facilitates the movement and cleansing of the eye, the nasal duct, the earwax duct, ducts of the cheek glands, smaller ducts under the tongue, and the glandular ducts of the epiglottis and palate.

Beginning in 1662 Steno conducted research on muscles from which he developed a comprehensive view of their structure. He showed that muscle tissue contains arteries, veins, and nerves, and is composed of closely woven fibers. Steno described the function of the diaphragm during respiration, classified the tongue as a muscle, and advanced the then-novel idea that the heart is nothing more than a muscle with contractions controlled by muscle fiber. From 1665 to 1667 he worked on brain anatomy, embryology, and comparative anatomy. During this period he demonstrated that animals possess a gland resembling the pineal gland of humans. This undermined René Descartes' (1596-1650) claim that the pineal gland was the seat of the uniquely human soul.

The catalyst for Steno's paleontological, geological, and mineralogical discoveries was his dissection of a huge shark caught near Leghorn, Italy (1666). His examination of the shark's teeth revealed their similarity to *glossopetrae* (tongue-stones, so called because of their resemblance to petrified serpent or bird tongues) found on Malta and elsewhere. These and other fossils were widely believed to be either direct productions of nature or God. Steno challenged these views in *De Solido intra Solidum naturalites con-*

tento dissertionis Prodromus (1669), maintaining the organic origins of fossils.

Prodromus contains the outlines of modern geology. Steno argued that fossils are animal remains that had become embedded in sea-floor strata and then petrified by the concreting action of chemical forces, heat, and compression over time. He suggested that strata are deposited horizontally in layers from aqueous fluids with occasional crustal collapses accounting for diversities of topography. In explaining this stratification he established the principle of superposition—underlying sedimentary layers are older than those overlaying them—upon which modern geochronology is based. To accompany his explanations, Steno produced what are considered the earliest geological cross sections. In this same treatise, attempting to account for crystallization, he propounded Steno's law—angles formed by corresponding faces of quartz crystals remain constant for a given mineral.

STEPHEN D. NORTON

Simon Stevin
1548-1620
Flemish Mathematician and Engineer

Simon Stevin was the first to systematically develop the ideas of Archimedes on the equilibrium of solid bodies and liquids. He established the law of equilibrium for bodies on an inclined plane, explained Archimedes' law for submerged bodies, and propounded the hydrostatic paradox. He also greatly influenced the use of decimal fractions.

Stevin, known also as Stevinus, was born in 1548 at Bruges in present-day Belgium. He earned his living as a bookkeeper before leaving the southern Netherlands in 1581 for Holland. Settling in Leiden, he established himself as an engineer. As an advisor for the construction of mills, locks, and harbors he received several patents and attracted the attention of Maurice of Nassau, stadholder of Holland and commander-in-chief of the States Army. Maurice held him in high regard and regularly sought out his advice in matters of defense and navigation. Stevin was entrusted with the organization of a school for military engineers at Leiden (1600) and appointed quartermaster in the army (1604). A bachelor most of his life, he married Cartherina Cray in 1616; they had four children. He died in 1620, most likely at his home in The Hague.

Stevin wrote on a variety of subjects ranging from commerce and navigation to hydrostatics and music theory. His first book, *Tafelen van Interest* (1582), presents rules for calculating single and compound interest as well as tables for computing discounts and annuities. Such information was well known in the banking establishment but considered a trade secret. Stevin's tables quickly gained wide usage in the Netherlands. He also published a slim pamphlet persuasively arguing for the systematic use of the decimal fraction. Though the notation of *De Thiende* (1585) was awkward, Steven found a sympathetic audience. His ideas gained wider currency when John Napier (1550-1617), inventor of logarithms, championed and then greatly facilitated their use with the introduction of the decimal point.

In *De Beghinselen der Weeghconst* (1586) Stevin introduced what is perhaps his most famous discovery, the law of the inclined plane. He showed geometrically that a linked chain of spheres must remain motionless when hung over two inclined planes joined to form a triangle, in effect demonstrating that the gravitational force is inversely proportional to the length of the inclined plane. His geometric proof is the basis for the parallelogram method for analyzing forces. In *De Beghinselen des Waterwichts* (1586) he provided the first systematic development of Archimedes' hydrostatics. He explained Archimedes' displacement principle for submerged bodies and showed that the pressure exerted by a liquid on a surface depends on the height of the liquid above that surface and is independent of the shape of the vessel containing it. Also in 1586, he experimentally refuted Aristotle's (384-322 B.C.) claim that heavier bodies fall faster than lighter ones.

In 1608 he revealed himself as one of the earliest converts to Copernicanism with the publication of *De Hemelloop*. Additionally, he developed a theory of the tides and tried his hand at solving the problem of determining longitude at sea, proposing a method based on deviations of compass needles from the astronomical meridian. Stevin's corpus also includes works on military fortification, music theory, civic life, and various treatises on engineering, including two books devoted to sluices and locks that he had helped design.

Stevin lived during a period of general scientific resurgence attendant upon the commercial and industrial prosperity of the Netherlands and northern Italy during the sixteenth century. Reflecting the new spirit of confidence of the time, Stevin chose to write in the vernacular. This required his introducing new scientific terms, many of which remain part of the Dutch scientific vocabulary.

STEPHEN D. NORTON

Evangelista Torricelli
1608-1647
Italian Mathematician and Physicist

Evangelista Torricelli is best known for inventing the mercury barometer (1644) and for his fundamental results in hydrodynamics. He also made important contributions to many areas of mathematics.

Torricelli was born on October 15, 1608, in Faenza, Italy. After Torricelli demonstrated his talents at an early age, his father, a textile artisan in modest circumstances, sent him to his uncle who supervised his education. He studied mathematics and philosophy at the Jesuit school in Faenza before going to Rome in 1627 to attended Sapienza College, run by Benedetto Castelli (1577-1644), a former student of Galileo (1564-1642).

In 1641 Torricelli completed *De motu gravium*, in which he developed some of Galileo's ideas on projectile motion. He experimentally verified many new conclusions and stated what is today known as Torricelli's law—a rigid system of bodies can move spontaneously on Earth's surface only if its center of gravity descends. Useful theorems of external ballistics followed, as well as artillery firing tables. He also propounded a fundamental hydrodynamic theorem that bears his name—the efflux velocity of a jet of liquid exiting a small orifice equals the velocity of a single drop of liquid falling freely in a vacuum from the same height as the liquid level at the orifice.

Castelli showed this work to Galileo. Suitably impressed, he engaged Torricelli as his personal assistant in 1641. They developed a close friendship, and Torricelli remained with Galileo until his death early the next year. Shortly thereafter Ferdinand II, Grand Duke of Tuscany, appointed him to Galileo's post of court mathematician.

After his death Galileo's followers in Rome and Florence continued a long-standing debate over why water in suction pumps would not rise more than 29.5 feet (9 meters). Galileo had argued that pumps created a vacuum, which in turn exerted a force on the water, thus preventing it from rising. Giovanni Battista Baliani

Evangelista Torricelli. *(Corbis Corporation. Reproduced with permission.)*

(1582-1666) as early a 1630 maintained that it was the weight of air that was responsible. Torricelli undertook a series of experiments in 1643 to settle the issue.

The clearest evidence was provided by what now is readily recognized as the barometer. Torricelli filled a long glass tube with mercury. Placing his finger over the open end, he inverted the tube and inserted it in a large dish of mercury. As expected, the mercury began to drain out of the tube—all but 3 inches (76 millimeters) that is. Torricelli interpreted the result in accordance with Baliani's hypothesis, arguing that the weight of air pressing on the mercury in the dish balanced that of the mercury column. The discovery was announced in a 1644 letter to Michelangelo Ricci (1619-1692). (In fact, the letter suggests Torricelli, based on observations of various hydrostatic devices, was aware of variations in atmospheric pressure before the experiment and was more concerned with producing "an instrument that would show changes of air, now heavier and denser, now lighter and thinner.")

Torricelli also made significant contributions to the development of the calculus—a subject he possibly would have invented if he had lived long enough. He developed important results on maxima and minima, used infinitesimal methods to complete the first modern rectification of a curve (1645), and produced what is

perhaps the first graph of a logarithmic function (1647). His independent discovery of the quadrature and center of gravity of the cycloid embroiled him in a bitter priority dispute with Gilles Persone de Roberval (1602-1675), who most certainly arrived at these results first. While in the process of assembling his correspondence and notes to defend his claims, Torricelli fell violently ill. He died, possibly of typhoid fever, in Florence on October 25, 1647.

STEPHEN D. NORTON

Biographical Mentions

Henricus Cornelius Agrippa
1486-1535

German lawyer and physician, also known as Agrippa of Nettesheim, who in the sixteenth century was a leading early proponent of the serious inspection of occult philosophies as a means of understanding nature beyond conventional science. Along with talents as diplomat and historiographer, he successfully defended a woman accused of witchcraft. His *The Occult Philosophy* (c. 1510) became a major influence in central European intellectual circles and along with *The Uncertainty and Vanity of Science* (1527) focused on the logic of the occult alternative.

Ulisse Aldrovani
1522-1605

Italian naturalist who advanced work in the natural sciences through emphasis on direct study and observation of the world. Appointed professor at the University of Bologna (1560), Aldrovani established Bologna's botanical garden, created collections of minerals and fossils, and aroused interest in the systematic study of nature through his lively lectures. His collections provided the material for a 14-volume encyclopedia of living things, which remained authoritative until superseded by Georges Buffon's *Histoire Naturelle* in the eighteenth century.

Guillaume Amontons
1663-1705

French physicist who conducted the first serious research on air thermometers, showing experimentally that the pressure of a constant mass of air is directly proportional to temperature increases, and deducing from Mariotte's law that any volume of air expands by the same fraction for given temperature changes (1699). Deaf from an early age, Amontons established that friction is proportional to load, designed many instruments including a hygrometer (1687) and conical barometer (1695), and proposed water's boiling point as a thermometric fixed-point (1702).

Peter Apian
1495-1552

German astronomer and geographer best known for noting comet tails point away from the Sun (1531). Apian's first major work, *Cosmographia seu descripto totius orbis* (1524), describes the use of maps and surveying techniques, defines weather and climate, and suggests lunar distances be used for calculating longitude. In *Instrumentum sinuum sive primi mobilis* (1534) he published the first table of sines calculated for every minute. Apian also produced the first large-scale map of Europe (1534).

Adrian Auzout
1622-1691

French astronomer and physicist who, with Jean Picard, fashioned and made systematic observations with an improved micrometer (1666). Consisting of two parallel hairs whose separation could be varied by a precision screw, their instrument allowed measurements of image sizes at a telescope's focal point. They also contributed to the systematic use of telescopic sights (1667-71). By forming one vacuum inside another, Auzout also demonstrated air pressure is responsible for the rise of mercury in barometers (1647).

Erasmus Bartholin
1625-1698

Danish physician and mathematician remembered for discovering double refraction (1669). Bartholin noticed that the transparent crystal Iceland spar (calcite) produced double images of objects viewed through it. He assumed light transmitted through the crystal was being refracted through different angles so as to produce two light rays. Christiaan Huygens explained certain aspects of double refraction with his wave theory of light but full understanding awaited the work of Etienne Malus, on polarization (1809), and Augustin Fresnel (1817).

Johann Joachim Becher
1635-1682

German chemist remembered primarily for his influence on Georg Stahl. Becher's *Physicae subterraneae* (1669) attempted to adapt traditional alchemical ideas to the growing body of chemical knowledge. He divided solids into three kinds of earth—*terra vitrescible, terra fluida,* and *terra pinguis.* Becher treated the latter substance, known as "fatty earth," as the principle of inflammability similar to the alchemical sulfur. Stahl later developed the *terra pinguis* concept into the phlogiston theory.

Isaac Beeckman
1588-1637

Dutch physicist who first formulated the principle of inertia—bodies in motion continue in motion unless impeded—though applying it to circular motions as well as rectilinear (1613) ones. He also discovered the law of uniformly accelerated motion for bodies falling *in vacuo* (1618), determined that the velocity of water out-flowing from a container bottom varies as the square root of water-column height (1615), and derived the relationship between pressure and volume for measured quantities of air (1626).

Jean Beguin
c. 1550-c. 1620

French chemist credited with first mentioning acetone and producing ammonium sulfide. Beguin revealed many mysteries of iatrochemistry through public lectures on the preparation of chemical remedies, and his *Tyrocinium chymicum* (1610) focused on chemical operations for producing safe medicines. Originally published to obviate the need for dictating lectures to his students, the *Tyrocinium* was very popular and issued in many editions, remaining the authoritative chemical text until 1695, when Nicolas Lémery's *Cours de chymie* appeared.

Filippo Beroaldo, the Elder
1453-1505

Italian humanist remembered as one of great philologists and classical scholars of the Renaissance. His translations and detailed commentaries of ancient Greek authors did much to stimulate scientific work in the early Renaissance. He works include a commentary on Pliny (1476) in which he deals with earthquakes, a corrected edition of Ptolemy's *Cosmographia* (1477), and annotations on Galen (1505). Beroaldo also estimated the time from creation to Christ's birth at 3,929 years.

Vannoccio Biringuccio
1480-c. 1539

Italian engineer and metallurgist who produced *De la Pirotechnia* (1540)—the first comprehensive account of mining practices. *Pirotechnia* describes mining of metallic ores and semimetals. Also described are furnaces and methods for smelting and alloying, various casting techniques, and the manufacture of canons and gunpowder. Biringuccio's descriptions are based on his experiences running an iron mine and forge for the tyrant of Siene, casting canons for Venice, and heading the papal foundry and arsenal.

Anselmus Boetius de Boodt
c. 1550-1632

Flemish mineralogist who produced the first systematic treatise on minerals. In *Gemmarum et lapidum historia* (1609) Boodt describes and classifies over 600 minerals based on his own observations and lists over 200 more mentioned by others. He used various categories to classify minerals, dividing them into great and small, rare and common, transparent and opaque, combustible and incombustible, as well as employing a three-degree scale of hardness. Boodt also noted the crystalline forms of some minerals.

Ismael Boulliau
1605-1694

French astronomer who established the period of the variable star Mira Ceti (1667). In *Astronomia philolaica* (1645) he suggests that if a planetary motive force exists it should vary inversely as the square of the distance. Though he rejected this dynamical explanation of planetary motion, preferring instead a purely kinematical account, Boulliau's inverse-square hypothesis was praised by Isaac Newton. A Copernican and supporter of Galileo, Boulliau was one of the first to accept Johannes Kepler's elliptical orbits.

Sophie Brahe
1556-1643

Danish horticulturist and astronomer who occasionally assisted her brother Tycho Brahe (1546-1601). She frequently visited Tycho at his observatory Uraniborg on the island Hven and assisted him with observations he used to compute the December 8, 1573, lunar eclipse. Sophie was schooled in classical literature, astrology, and alchemy. She married Otto Thoft when nineteen or twenty and later became a horticulturist after his death (1588) left her to manage their property at Ericksholm, Scania.

Hennig Brand
fl. 1670

German alchemist who discovered phosphorus (c. 1669). Believing he could produce gold, Brand heated concentrated urine with sand and collected the products under water. He named the white waxy substance "phosphorus" (light bearer) because it glowed in the dark. Brand also discovered that it burst into flame when exposed to air. Robert Boyle independently discovered phosphorus in 1680, and a priority dispute ensued since Brand had not published his results.

Giordano Bruno
1548-1600

Italian philosopher and scientist who wrote on a number of topics, including memory and the effect of language on human behavior. Initially a Dominican friar, Bruno's unorthodox views caused him trouble, so he left Italy and traveled throughout Europe. He mistrusted mathematics, preferring symbols and images, which gave his works a mystical tone. He enthusiastically supported the ideas of Nicolaus Copernicus, despite Church opposition. When he returned to Italy in 1592 he was arrested for heresy and, eventually, burnt at the stake.

Niccolo Cabeo
1586-1650

Italian physicist who discovered electrostatic repulsion (c. 1620). He is also remembered for misinterpreting Giovanni Battista Baliani's experiments on falling weights. Baliani observed that different weights take approximately the same time to fall equal distances. Cabeo concluded that any two weights will fall equal distances in the same length of time regardless of the medium. Vincenzo Renieri conducted experiments that refuted Cabeo's claim. Cabeo's major works are *Philosophia magnetica* (1629) and *In quatuor libros meteorologicorum Aristotelis commentaria* (1646).

Girolamo Cardano
1501-1576

Italian mathematician and physicist whose work *Ars magna* (1545) contained Cardano's rule for solving reduced cubic equations and Tartaglia's method, obtained under oath not to reveal it, for solving general cubics. Cardano's publication initiated debates on the ethics of scientific secrecy that eventually crystallized into the belief that secrecy is of great harm to science. Cardano was a physician whose repute was second only to Vesalius, did important research in mechanics, and outlined the hydrologic cycle of rivers.

Vincenzo Cascariolo
fl. early 1600s

Italian alchemist who created the first synthetically luminiscent material (c. 1603). A cobbler by profession in Bologna, Cascariolo heated a mixture of barium sulfide with coal and found that the powder thereby obtained exhibited a temporary bluish glow at night that could be restored by exposure to sunlight. The material was called *lapis solaris* or sunstone because it was hoped it could be used for transforming base metals into gold (the symbol for gold being the Sun).

Guillaume Cassegrain
fl. 1672

French physicist who invented the reflecting telescope that bears his name. Cassegrain reflectors employ a concave primary mirror to focus and reflect light onto a convex secondary mirror. The light is then reflected back through a hole in the primary mirror to a lens. Cassegrain's design increased angular magnification over other types of reflectors. A more advantageous feature, not recognized for another century, is the elimination of spherical aberration inherent in other two-mirror reflectors.

Benedetto Castelli
1578-1643

Italian astronomer and hydrologist whose pioneering work *Della misura dell'acque correnti* (1628) marks the beginning of modern hydraulics. Castelli related river cross-sectional areas to the water volume passing through those areas and discussed the relation of velocity to head in flow through an orifice. In a celebrated series of letters to Galileo (1637-1638), he discussed absorption of radiant heat by black and white objects. A loyal supporter of Galileo, Castelli played an important role in extending and transmitting his work.

Christoph Clavius
1537-1612

German astronomer and mathematician who supported Ptolemaic geocentricism over Copernican heliocentrisim, arguing the latter was physically impossible and contradicted scripture (1581). As Papal Astronomer, Clavius's recommended improvements to the Julian calendar (1582) and confirmed Galileo's telescopic discoveries but rejected his interpretation (1611). Known as the Euclid of the sixteenth century, he attempted to prove the parallel postulate; worked on fractions, algebra, and trigonometry; and was one of the first to use parentheses for aggregating mathematical terms.

Maria Cunitz
1610-1664

German astronomer remembered for revising Johannes Kepler's Rudolphine Tables. Her *Urania propitia siva tabulae astronomicae mire faciles* (1650) greatly simplified Kepler's work. Cunitz prefaced the volume by explaining that despite being the work of a woman, the *Urania* was nevertheless accurate. Her tables proved of great utility and went through many editions. This led many to suspect her husband had computed them. Cunitz's husband wrote a preface for later editions to dispel such rumors.

Johann Baptist Cysat
1586-1657

Swiss astronomer credited with the first telescopic observations of a comet (1618). Cysat observed the comet of 1618-19 almost continuously for two months, attempting to show the object was supralunary. He proposed two models to explain its motion, both of which assumed a stationary Earth. Cysat also observed the lunar eclipse of 1620, the transit of mercury in 1631, and the Orion Nebula in 1619—which had previously been observed by Peiresc in 1610.

René Descartes
1596-1650

French mathematician and philosopher whose most significant contribution was analytic geometry. His coordinate system—subsequently called Cartesian in his honor—allowed geometric problems to be solved algebraically and thus paved the way for Isaac Newton's development of the calculus. With his new tools Descartes made the first systematic classification of curves. His vortex theory—according to which Earth is carried around the Sun in a vortex—held sway for over a hundred years on the European Continent and delayed acceptance of Newton's work in France.

Thomas Digges
c. 1546-1595

English astronomer and mathematician who was the leading proponent of Copernicanism in England. He translated portions of Copernicus's *De revolutionibus* and appended his own views on an infinite universe with fixed stars at varying distances from Earth (1576). Digges had earlier published *Alae seu scalae mathematicae* (1573), containing his observations of the 1572 supernova. The appendices to *Stratioticos* (1579, 1590) and his father Leonard's posthumously published *Pantometria* (1571, 1591) constitute the first serious English research on ballistics.

Marco Antonio de Dominis
1566-1624

Yugoslavian (Dalmatian) scientist and mathematician who wrote about lenses, telescopes, and tides, and developed a theory of rainbows.

He studied law at the University of Padua, became a Jesuit, and taught mathematics, logic, and rhetoric at Verona, Padua, and Brescia. He left the Jesuits in 1596 and became a Bishop in Dalmatia. In 1616 he fled to England to escape the Inquisition. When he returned home in 1622, he was arrested by the Inquisition and died in confinement.

Cornelius Drebbel
1572-1633

Dutch instrument maker best known as one of the possible inventors of the thermometer. His famous perpetual-motion clock—actually an astronomical clock—operated on the same principles as the air thermometer, which Drebbel certainly could have constructed first. An expert lens grinder, he produced compound microscopes as early as 1619. Drebbel also invented thermostats for regulating oven and furnace temperatures, a diving-bell-like submarine, and discovered a tin mordant for dying scarlet with cochineal.

Lazarus Ercker
c. 1530-1594

Bohemian metallurgist whose *Beschreibung allerfürnemisten mineralischen Ertzt und Berckwerksarten* (1574) was the first systematic treatment of analytical and metallurgical chemistry. In it he reviews methods for obtaining, refining, and testing alloys and minerals of silver, gold, copper, antimony, bismuth, lead, and mercury as well as acids, salts, and other compounds including methods for purifying saltpeter. He also discusses the design and construction of apparati for assaying such as the cupel, furnaces, and assaying balance.

Mikkel Pedersön Escholt
c. 1610-1669

Norwegian theologian remembered for authoring *Geologia norvegica* (1657)—the first scientific treatise printed in Norway and one of the first books published in Norwegian. In the work Escholt describes the April 24, 1657, Norwegian earthquake. He held that earthquakes were portents of divine intervention in the world, though their production by God was occasioned by physical causes—principally escaping air from Earth's interior. He also was the first to use "geology" in its modern sense of Earth science.

David Fabricus
1564-1617

German astronomer who was the first to discover a variable star. In 1596 Fabricus observed a third magnitude star in the constellation Cetus.

He assumed it was a nova since it later disappeared. Johann Bayer observed the star again (1603) and named it Omicron. Johannes Holwerda discovered Omicron Ceti's variability in 1638 and named it Mira (miraculous) Ceti. Fabricus's observations of Mars, together with Tycho Brahe's, were used by Johannes Kepler to derive the laws of planetary motion.

Ferdinand II, Grand Duke of Tuscany
1610-1670

Italian prince who invented the liquid-in-glass thermomenter (1641). When the variability of air pressure was discovered, it was realized thermometers responded to such changes as well as temperatures changes. Ferdinand's sealed instruments were designed so as not to be affected by atmospheric pressure changes. He also founded the Accademia del Cimento (Academy of Experiments) in 1657, which greatly influenced the development of experimental physics with the publication *Saggi di Naturali experenze fatte nell' Accademia del Cimento* (1668).

Girolamo Fracastoro
c. 1478-1553

Italian physician and astronomer remembered for his pioneering work in epidemiology. Syphilis, then rampant in Europe, derives its names from his poem *Syphilis sive morbus Gallicus* (1530). His work *De contagione* (1546) lists the three modes by which contagion spread—simple contact, carriers (such as cloths, bedding, etc.), and from a distance (in which some have seen his unlikely anticipation of microbes). Fracastoro was also the last to defend a theory of solid celestial spheres.

William Gascoigne
c. 1612-1644

English astronomer who introduced telescopic sights and invented the first micrometer for measuring small angular distances at a telescope's focal point. Gascoigne produced a working model by 1641 but had little time to promulgate his ideas due to his untimely death in the royalist disaster at Marston Moor during the English Civil War. The micrometer and telescopic sight only saw widespread application after Adrian Auzout and Jean Picard duplicated Gascoigne's work in the late 1660s.

Henry Gellibrand
1597-1636

English astronomer who discovered secular variations in magnetic declination—changes over time in the acute angle between magnetic and true north. Edmund Gunter's declination measurements at Limehouse in 1622 were about 5° less than William Borough's 1580 results. Gunter attributed the discrepancy to errors by Borough. Gellibrand and John Marr's 1633 measurements showed a further decrease, which Gellibrand announced in 1635. Gellibrand also made mathematical contributions to navigation including a method for determining longitude at sea.

Konrad von Gesner
1516-1565

Swiss naturalist who is considered one of the pioneers of modern animal description. Gesner's *Historiae Animalium* (Animal History) of 1551-54 is commonly regarded as the basis of modern zoology. Unlike many of his contemporaries, von Gesner supplemented knowledge of the natural world taken from antiquity with his own biological research. Furthermore, his texts are characterized by a reliance on illustrations in order to aid students: his *Opera Botanica* (Botanical Works) featured close to 1,500 plates of his own composition.

Johann Rudolph Glauber
1604-1670

German chemist who founded industrial chemistry. His *Furni novi philosophici* (1649) offered detailed descriptions of new furnaces, laboratory equipment, and experimental techniques. Glauber pioneered the synthesis of mineral acids and salts (Glauber's "salt" is sodium sulfate); produced benzene and phenol by coal distillation; and derived acetone and acetates from wood distillation. *Dass Teutschlands-Wohlfahrt* (1661) promoted the economic advancement of Germany with recipes for beer and wine concentrates, use of chemical fertilizers, and suggestions for chemical weaponry. The *Pharmacopoea spagyrica* (1668) advocated chemical medicines as substitutes for herbal treatments.

James Gregory
1638-1675

Scottish mathematician and astronomer who designed the first plausible reflecting telescope (1663), though he was unable to obtain mirrors of sufficient quality to produce a working model. Gregory also made significant contributions to the discovery of calculus. He independently discovered the general binomial expansion (1670) and introduced the terms "convergent" and "divergent." Proposition 6 of his *Geometricae pars Universalis* (1667) has often incorrectly been viewed as the first proof of the fundamental theorem of calculus.

Edmund Gunter
1581-1626

English mathematician and instrument maker who designed many instruments to simplify astronomical, nautical, and surveying calculations. These include a portable quadrant and forerunner of the modern slide rule known as "Gunter's Line." He also introduced the terms "cosine" and "cotangent." Variations in magnetic declination—the acute angle between magnetic and true north—were first observed by Gunter in 1622, but he attributed the effect to errors. Henry Gellibrand later realized the effect was real (1633).

Edmond Halley
1656-1742

English astronomer and physicist famous for predicting the 1758 return of Halley's comet and for the instrumental role he played in the publication of Isaac Newton's *Principia*. Halley established the Southern Hemisphere's first observatory and published the first telescopically based star catalog of the southern skies. He conducted research on tidal phenomena (1684-1701), proposed a core-fluid-crust model to explain the westward drift of Earth's magnetic field (1692), and was the first to detect the proper motions of stars (1718).

Thomas Harriot
c. 1560-1621

English mathematician and astronomer who was among the first to view celestial objects telescopically. He anticipated Galileo's use of the telescope for viewing the Moon's surface (1609) and observing Jupiter's moons (1610), independently discovered sunspots (1611), and made detailed studies of comets. Harriot also made many contributions to algebra including a comprehensive theory of equations and improved notation that includes the signs < for less than and > for greater than.

Johannes Hartmann
1568-1631

German chemist and physician important for having introduced medical and pharmaceutical chemistry into the university curriculum. In 1609 at the University of Marburg Hartmann initiated lectures and laboratory instruction on the chemical preparation of medicines. Later that year he was appointed professor of iatrochemistry at Marburg, the first chemistry professorship in Europe. Though he was not very successful as a physician, Hartmann's textbook on pharmaceutical chemistry, *Praxis chymiatrica*, went through many editions and was widely respected.

Johannes Baptista van Helmont
1579-1644

Flemish physician, physiologist, and chemist who first recognized the existence of more than one air-like substance. He introduced the term "gas" to refer to the various substances he isolated, which include carbon dioxide and chlorine. Helmont conducted some of the earliest quantitative chemical experiments. In his most famous experiment he grew a willow tree in a measured quantity of soil, adding only water. After five years the tree had gained 164 pounds while the soil weight had decreased negligibly. Helmont also correctly noted the role of stomach acid in digestion.

Johannes Holwerda
1618-1651

Dutch astronomer who discovered Mira Ceti's variability (1638), thus helping undermine the Aristotelian concept of the heavens' immutability. In 1596 David Fabricus noticed a third-magnitude star that faded and disappeared. Holwerda later observed a star—now called Mira Ceti—in the same place and watched as its magnitude fluctuated over 11 months. A professor of logic and astronomy at Franeker's Frisian University, Holwerda later, in *Philosophia naturalis seu physica vetus-nova* (1651), defended an Aristotelian atomism derived from Pierre Gassendi.

Jeremiah Horrocks
c. 1619-1641

English astronomer among the first to accept Johannes Kepler's theory of elliptical orbits. Having corrected Kepler's Rudolphine tables, Horrocks predicted and became the first to observe a transit of Venus (1639). He calculated an improved solar parallax value that challenged then-current estimates of the solar system's size, undertook the first continuous series of tidal observations, demonstrated the Moon's orbit is elliptical, and made tentative steps towards universal gravitation by recognizing the Sun's perturbing influence on the Moon's orbit.

Gottfried Kirch
1639-1710

German astronomer among the first to systematically search the heavens telescopically. He observed sunspots, eclipses, the 1701 Mercury transit, and discovered several comets as well as the variable star chi Cygni (1685). Kirch computed calendars and a well-known set of ephemerides, based on Kepler's Rudolpine tables,

for the years 1681-1702. He also designed a new circular micrometer and was appointed the first director of the Berlin Observatory (1700), which was not completed until after his death.

Johannes Marcus Marci von Kronland
1595-1667

Bohemian physicist who anticipated by two decades many of the observations on optics made by Sir Isaac Newton (1642-1727). Sometimes called simply Marcus Marci, Johannes served as professor and rector of Prague's Charles University, where in 1648 he published *Thaumantias liber de arcu coelesti....* In it, he discussed the colors of the rainbow, the dispersion of light beams directed through a prism, and the diffraction of light around a wire. He also observed that repeated refraction does not cause monochromatic rays to change color. In addition to his work in optics, Johannes performed a number of experiments in mechanics using a pendulum.

Johann Kunckel
1630?-1702?

German chemist who in about 1675, along with Hennig Brand and Robert Boyle, isolated and identified phosphorus, the first new element to be discovered since antiquity. He was also noted for his writings on glassmaking and the chemistry of salts, and for his detailed descriptions of laboratory equipment and experimental techniques.

Nicolas Lemery
1645-1715

French chemist, physician, and pharmacist, whose *Cours de chymie* (Course on Chemistry) of 1675 was Europe's most influential chemical textbook for over fifty years, with eleven editions in his own lifetime and over thirty editions by 1756. Influenced by René Descartes's theories of corpuscular matter and mechanical motion, Lemery explained chemical properties and reactions in terms of minute particles, which joined together by means of hooks, points, and pores of complementary shapes. He also advocated use of "wet" (solution) over "dry" (combustion) methods of chemical analysis.

Camillus Leonardus
fl. 1480s

Italian astronomer, mineralogist, and physician whose *Speculum lapidum* (1502) treats over 200 minerals. This compilation proceeds according to traditional knowledge and concepts, relying little on independent observation, although the invention of gunpowder is alluded to. His focus is primarily the occult powers of gems and images carved upon them. Leonardus also worked on astronomical tables, verified the positions of particular stars, and published astrological rules for bleeding and administering drugs.

Edward Lhwyd
1660-1709

English paleontologist and botanist whose illustrated catalog of fossils from Oxford's Ashmolean Museum (1699) contains letters to John Ray on fossil origins. Lhwyd incorrectly maintained that mists transported animal spawn and minute seeds great distances, after which they penetrated deep into the ground where they germinated and grew to complete or partial replicas in stone. Lhwyd's *Archaeologia Britannica* (1707) contained the first comparative study of Celtic languages as well as the first Gaelic dictionary.

Hans Lippershay
c. 1570-1619

Dutch optician who possibly invented the telescope. He is one of three spectacle-makers who claimed priority for the invention. It may be that the other two—Sacharias Jansen and Jacob Adriaenszoon (a.k.a. Jacob Metius)—realized they were in possession of the same device, though employing it for different purposes, when they learned of his claim. What is certain, though, is that the earliest mention of a telescope is in connection with Lippershay's 1608 patent application.

Martin Lister
1638-1712

English zoologist who is best known for his *Tractatus de Araneis,* the world's first scientific work on spiders, published in English as *Martin Lister's English Spiders* (1678). Modern entomologists study Lister's book and value both his astute observations about spiders and the precise drawings by Michael Roberts. Lister was a medical doctor, zoologist, and a Fellow of the Royal Society. He was appointed Queen Anne's physician in 1709.

Longomontanus
1562-1647

Danish astronomer, also known as Christian Severin, who was Tycho Brahe's only disciple. Upon his master's death Longomontanus assumed responsibility for selecting and integrating Tycho's data into a coherent account of planetary motions and presenting this work in a systematic treatise. His task was brought to fruition in *Astronomia danica* (1622), which was enthusiastically received by the astronomical community. The prestige attached to his work saw it through two reprintings despite the appearance of Johannes Kepler's *Tabulae Rudolphinae* in 1627.

Edmé Mariotte
?-1684

French physicist celebrated as one of the founders of French experimental physics. Mariotte produced the first comprehensive work on elastic and inelastic collisions (1673), relying heavily on the work of Christiaan Huygens, John Wallis, and Christopher Wren. In 1676 he made an important qualification to Boyle's law by noting that the inverse relationship between volume and pressure only holds if temperature remains constant. He attempted to calculate the atmosphere's height from this law. Mariotte also discovered the eye's blind spot (1668).

Simon Marius
1573-1624

German astronomer among the first to use the telescope for viewing celestial objects. He discovered Jupiter's satellites, if not before then shortly after Galileo, but waited until 1614 to publish his observations. Marius computed tables of the mean periodic motions of the Jovian satellites, directed attention to variations in their brightness, and assigned the names by which they are known today. He was the first to view the Andromeda nebula through a telescope (1612).

Michael Mästlin
1550-1631

German astronomer whose observations of the nova of 1572 demonstrated it was a new star, indicating the heavens were changeable, not fixed as many had previously thought. Mästlin also failed to detect parallax for the comets of 1577 and 1580, suggesting they were supralunar bodies. These celestial events led him to reject Aristotelian cosmology. After attending Mästlin's lectures on the superiority of Nicolaus Copernicus's cosmology, Johannes Kepler embraced Copernicanism. Mästlin published the first correct explanation of earthshine—pale illumination next to the lunar disk crescent due to reflection from the sunlit Earth.

John Mayow
1641-1679

British chemist and physiologist and early advocate of seventeenth century corpuscular theories of matter. His theory of a "nitro-aerial spirit" as a distinct reactive component of atmospheric air offered an early explanation for chemical and biological processes of combustion, fermentation, and respiration. Mayow's possible anticipation of aspects of combustion theories proposed a century later by Cavendish, Priestley, and Lavoisier remains a subject of scholarly controversy.

Geminiano Montanari
1663-1687

Italian astronomer, physicist, and engineer who discovered the star Algol's variability (1667). Though failing eyesight prevented him from discerning the regularity or period of variation, Montanari's report to the Royal Society (1671) nevertheless helped undermine the Aristotelian concept of the heavens' immutability. He later discovered another variable star in the constellation Hydra (1672). Montanari also artificially incubated chicks (1657), produced what at the time was the largest moon-map (1662), and successfully transfused blood between animals (1668).

Jean-Baptiste Morin
1583-1656

French astronomer who proposed a method for determining terrestrial longitude at sea based on observations of the Moon's motion relative to the stars. Morin's method required nautical instruments of much greater precision than existed, better mathematical solutions for spherical triangles, and refined lunar tables. Though unable to provide a practical proposal for implementation of his method, Morin did make important contributions to instrumental technique by introducing the telescopic sights and Vernier micrometers for angle measurements.

Robert Norman
fl. late 1500s

English instrument maker whose treatise *The Newe Attractive* (1581), which was one of the first systematic treatises in experimental physics, announced his discovery of the dip of compass needles. Norman measured the deviation from the horizontal of compass needles at London and speculated on whether or not the effect varied over Earth's surface. He also suggested compass needle orientation was due to turning toward, rather than attraction to, a particular point.

Blaise Pascal
1623-1662

French mathematician and physicist who proved that the mercury-column height in barometers depends on air pressure. In hydrostatics Pascal's principle (1654) provides the connection between the mechanics of fluids and mechanics of rigid bodies. His correspondence with Pierre de Fermat laid the foundations for probability theo-

ry and combinatorial analysis. Pascal also invented a calculating machine that by means of cogged wheels could add and subtract. Some of the principles involved are still used in mechanical calculators.

Nicolas Claude Fabri de Peiresc
1580-1637

French astronomer who discovered the Orion nebula (1610). Primarily interested in positional astronomy, Peiresc calculated terrestrial longitude and determined the length of the Mediterranean Sea (1635). However, Peiresc's importance to science is more appropriately gauged by his patronage. He provided for the publication of various scholarly works and sponsored much research, including work on the human circulatory system and supporting Pierre Gassendi, who lived at his estate for a few years while working on his philosophical writings.

Pierre Perrault
1611-1680

French hydrologist whose *De l'origine des fontaines* (1674) is one of the foundational works in experimental hydrology. Perrault demonstrated rainfall is sufficient to maintain the flow of rivers. He determined the drainage area for a portion of the Seine River and calculated the total precipitation. After adjusting for losses, he compared the total volume of water with the river's annual flow and showed that only one-sixth of the annual rainfall was necessary to sustain it.

Alessandro Piccolomini
1508-1578

Italian astronomer and littérateur remembered for publishing the first printed book of star charts. *De le stelle fisse libro uno* (1540) consisted of 47 maps—one for each Ptolemaic constellation except Equuleus. Piccolomini introduced a lettering system for stars, and his maps indicated magnitudes, the direction of the equatorial pole, and direction of the celestial sphere's daily rotation. His failure to provide coordinates, though, made it impossible to accurately determine stellar positions.

Giambattista della Porta
1535-1615

Italian philosopher and scientist who, while employed as a writer of dramas, also did scientific work making optical instruments. He wrote on a variety of topics: optics, cryptography, mechanics, squaring the circle, steam engines, military engineering, the camera obscura, agriculture, chem-

istry, hydraulic machines, medical cures, de-
monology, and magnetism. He formed a short-
lived scientific society, which was closed by the
Inquisition, and built a personal museum. Although della Porta joined the Jesuits, the Inquisition banned publication of many of his writings.

Petrus Ramus
1515-1572

French philosopher and mathematician, born
Pierre de la Ramee, who led the movement to reform the dogmatic adherence to Aristotelian philosophy by the intellectuals of his day (Scholastics). He laid the groundwork for the development
of a philosophical system more conducive to scientific thought. A ban on his works in 1544 was
removed in 1547. He converted to Protestantism
in 1561, suffered persecution, and was assassinated by his intellectual and religious enemies in the
St. Bartholomew's Day Massacre in 1572.

Henricus Regius
1598-1679

Dutch physician who introduced Cartesian
views into Dutch universities. Appointed
Utrecht's professor of medicine (1638), Regius
attacked Aristotelian natural philosophy, publishing numerous disputations with clear Cartesian biases. Subsequently, Cartesianism was officially condemned (1642). Dutch Cartesians afterward adopted René Descartes's views on
natural phenomena while distancing themselves
from their metaphysical foundations, which
smacked of atheism. Descartes originally defended Regius; but when Regius published *Fun-
dameta physices* (1646), Descartes publicly dis-
tanced himself to avoid any suspicion of impiety.

Erasmus Reinhold
1511-1553

German astronomer and mathematician who calculated the *Tabulae Prutenicae* (1551)—the first
practical set of planetary tables based on Nicolaus Copernicus's theory. More accurate than the
Alfonsine Tables, Reinhold's tables were widely
adopted and provided a strong argument in favor
of Copernicanism. Reinhold's focus on Copernicus's mathematical modeling and silence regarding the physical reality of heliocentrism—the notion that Earth revolved around the Sun, not vice
versa—encouraged a similar attitude in German
astronomers that persisted, long after his death
from the plague, into the late sixteenth century.

Giovanni Battista Riccioli
1598-1671

Italian astronomer who discovered the first double star (1643) and proved instrumental in undermining Aristotelian cosmology. In *Almages-
tum novum* (1651) he maintained the identity of
celestial and terrestrial matter and thus the cor-
ruptibility of the heavens. Riccioli also departed
from traditional cosmology in two important
ways: viewing the world's center as the cosmos's
noblest place and making Earth more noble and
perfect than any celestial body. Riccioli deployed
these ideas in challenging Nicolaus Copernicus's
claim that the Sun occupied the cosmos's geometric center.

Julius Caesar Scaliger
1484-1558

Italian physician and humanist who established
his intellectual reputation with an attack on
Erasmus's criticism of Cicero. After a minor military career, Scaliger studied at the University of
Padua. He traveled to France and adopted the
name Jules Cesar de l'Escale de Bordonis. He
translated a number of ancient texts into Latin,
including Aristotle's *Natural History*. He published works on grammar, poetry, literary criticism, zoology, and botany, and proposed a new
system of botanical classification based on distinctive characteristics.

Christoph Scheiner
1573-1650

German astronomer among the first to detect
sunspots telescopically (1611) and the first to
publish such observations (1612). Hoping to
preserve the incorruptibility of the Aristotelian
heavens, Scheiner interpreted the spots as small
stars orbiting the Sun. Galileo disagreed, believing them contiguous with the Sun and thus providing evidence of solar axial-rotation. Scheiner
correctly attributed the Sun's elliptical appearance near the horizon to refraction (1617) and
localized the retina as the seat of vision (1619).

Wilhelm Schickard
1592-1635

German polymath who designed and built the
first modern mechanical computer (1623).
Schickard's calculating clock performed the operations of addition and subtraction automatically and multiplication and division partially
so. He was a skilled cartographer, engraver, and
astronomer. Schickard and his family died during the plagues brought about by the Thirty
Years' War. His work lay forgotten until 1957,

when some of his letters to Johannes Kepler were discovered, which allowed his device to be reconstructed.

Daniel Sennert
1572-1637

German physician notable for his chemical theory that sought to reconcile the rival systems of the traditional four Aristotelian elements, three Paracelsian chemical principles, and newly revived corpuscularian theories. The elements constituted the principles, which in turn combined to constitute all chemical substances. The matter of these compounds consisted of *minima naturalia*, or microscopic particles which, while infinitely divisible in theory, in actuality had a minimum natural threshold of indivisibility, beyond which they underwent destruction and lost their distinctive properties.

Georg Ernst Stahl
1659?-1734

German chemist and physician who is best remembered as one of the main developers of the phlogiston theory of combustion, which dominated the chemical sciences for a hundred years. Stahl studied medicine at the University of Jena, and later lectured there. He became the court physician at Weimar, then professor of medicine at the University of Halle, and eventually served as personal physician to the King of Prussia. Stahl also produced influential medical writings and founded a short-lived chemical journal.

Leonhard Thurneysser
1531-1596

Swiss/German physician, alchemist, mining and metallurgical technologist who, though steeped in occult philosophical sympathies of nature as in his alchemical pursuits, was also a practical scientist who made contributions to mining, metallurgy, and hydrology. Thurneysser represented the transitional state of later sixteenth century central European thinkers, holding to traditional pseudo-science concepts but pursuing application as well. He was a disciple of Georg Agricola's (1494-1555) practical mineralogy, mining, and metallurgical techniques. He also made contributions to river hydrology among other ideas in *Ten Books on Mineral and Metalic Waters* (published in 1612).

Isaacus Vossius
1618-1669

Dutch physicist whose *De lucis natura proprietate* (1662) contains many bold conjectures about light, including the assertion that light is not a corporeal body, as well as commentary on Willebrord Snell's unpublished law of light refraction. Vossius also claimed comets were real bodies, not specters or illusions; that a vacuum existed above Earth's atmosphere; and that sea water could not rise through subterranean channels to form mountain rivers, maintaining instead that all rivers come from rain-water (1666).

John Wallis
1616-1703

English mathematician who, with Christiaan Huygens and Christopher Wren, established conservation of momentum (1668). In *Arithmetica Infinitorum* (1656) he presented a method for calculating areas under curves using infinite sums, which greatly influenced Isaac Newton's development of calculus. Wallis extended exponents to negative numbers and fractions, first used ∞ to symbolize infinity, represented imaginary numbers geometrically, and produced an infinite product for π. He also wrote on grammar, logic, and theology; was an expert on decipherment; and taught deaf-mutes to speak.

Gottfried Wendelin
1580-1667

Flemish astronomer who staunchly supported Copernicanism. Wendelin appears to have been the first to propose the law of the variation of the obliquity of the ecliptic. He also studied the pendulum, noting the effect of temperature on the period of oscillation as well as showing that increases in amplitude increase the period of oscillation. Known as the Ptolemy of his age, Wendelin was highly respected and his views were solicited by, among others, René Descartes, Christiaan Huygens, and Pierre Gassendi.

William Whiston
1667-1752

English mathematician whose *New Theory of the Earth* (1696) sought rapprochement between Newtonian science and the Biblical account of Creation. According to Whiston, Earth was originally a comet whose eccentric orbit was made circular by God; until the Flood, Earth possessed no diurnal rotation and its poles were perpendicular to the ecliptic. Finally, a comet guide by God penetrated Earth's surface, causing the waters of the abyss to overflow, shifting the poles, and imparting axial rotation.

Bibliography of Primary Sources

Agrippa, Henricus. *The Occult Philosophy* (c. 1510). This work became a major influence in central European intellectual circles and along with *The Uncertainty and Vanity of Science* (1527) focused on the logic of the occult alternative.

Apian, Peter. *Cosmographia seu descripto totius orbis* (1524). Described the use of maps and surveying techniques, defined weather and climate, and suggested lunar distances be used for calculating longitude.

Bacon, Francis. *The Advancement of Learning* (1605). A new categorization of the whole of the natural sciences.

Bacon, Francis. *Novum Organum* (1620). Outlined a new method of natural philosophy to replace Aristotle. Bacon proposed that, through his method of induction, the secrets of the universe could be unlocked and used to benefit society. His method involved the unbiased, almost random, collection of data, which would later be generalized into rules of nature. Bacon's method never became popular, but many of his other ideas proved influential.

Bacon, Francis. *The New Atlantis* (1626). Here Bacon described a community of scientific workers who would divide the labor of science among themselves and work together to advance knowledge. The "Salomon's House" of this fable was an idealized scientific utopia, conjured to inspire actual scientists to work together in an organized manner.

Bayer, Johann. *Uranometria* (1603). A comprehensive celestial atlas. The significance of Bayer's work lies in his innovative method for naming stars within each constellation. Though traditional constellations continued to provide a convenient means of dividing the heavens, the profusion of names for individual stars that resulted from the translation of Greek into various languages proved most cumbersome and confusing. Bayer sought to reform this situation by systematically identifying each star precisely and succinctly. He assigned to each star in a constellation one of the 24 letters of the Greek alphabet. If a constellation had more than 24 stars then additional characters were provided by the Latin alphabet.

Becher, Johann. *Physicae subterraneae* (1669). This work attempted to adapt traditional alchemical ideas to the growing body of chemical knowledge. Becher divided solids into three kinds of earth—*terra vitrescible, terra fluida,* and *terra pinguis.* Becher treated the latter substance, known as "fatty earth," as the principle of inflammability similar to the alchemical sulfur.

Beguin, Jean. *Tyrocinium chymicum* (1610). Focused on chemical operations for producing safe medicines. Originally published to obviate the need for dictating lectures to his students, the *Tyrocinium* was very popular and issued in many editions, remaining the authoritative chemical text until 1695, when Nicolas Lémery's *Cours de chymie* appeared.

Boodt, Anselmus de. *Gemmarum et lapidum historia* (1609). The first systematic treatise on minerals. Boodt described and classified over 600 minerals based on his own observations and listed over 200 more mentioned by others. He used various categories to classify minerals, dividing them into great and small, rare and common, transparent and opaque, combustible and incombustible, as well as employing a three-degree scale of hardness. Boodt also noted the crystalline forms of some minerals.

Boulliau, Ismael. *Astronomia philolaica* (1645). Here Boulliau suggested that if a planetary motive force exists it should vary inversely as the square of the distance. Though he rejected this dynamical explanation of planetary motion, preferring instead a purely kinematical account, Boulliau's inverse-square hypothesis was praised by Isaac Newton.

Boyle, Robert. *New Experiments Physio-Mechanicall, Touching the Spring of the Air and its Effects* (1660). This work described the air pump constructed by Boyle and Robert Hooke.

Boyle, Robert. *The Sceptical Chymist* (1661). This book helped to transform alchemy into chemistry.

Boyle, Robert. *Experimental History of Colors* (1664). In this work Boyle described his work with acid-base indicators.

Brahe, Tycho. *De nova stella* (1573). Detailed Brahe's observations of the nova of 1572. His measurements indicated the phenomenon was not part of the atmosphere nor was it attached to the sphere of a planet, but that it was located among the fixed stars. This undermined the prevailing Aristotelian notion that the heavens were perfect and unchanging.

Castelli, Benedetto. *Della misura dell'acque correnti* (1628). A pioneering work that marks the beginning of modern hydraulics. Castelli related river cross-sectional areas to the water volume passing through those areas and discussed the relation of velocity to head in flow through an orifice.

Copernicus, Nicolaus. *De Revolutionibus Orbium Coelestium* (The Revolutions of the Heavenly Spheres, 1543). Landmark book in which Copernicus discussed his heliocentric view of the heavens. Aware of the possible repercussions of proposing such a theory in direct opposition to the Church, he first wrote a short version of his ideas, entitled *Commentariolus,* in 1513 and distributed it for comment to friends and colleagues. The full work, *De Revolutionibus,* was not received well by the Church and scholars, even though a churchman, without Copernicus's permission, had added a preface that stated that the theory was not being proposed as representing the actual motion and position of the Earth and Sun, but merely as a mathematical model to make calculations easier. Copernicus died soon after *De Revolutionibus* appeared, thereby, escaping inevitable punishment.

Cunitz, Maria. *Urania propitia siva tabulae astronomicae mire faciles* (1650). This work greatly simplified Johannes Kepler's *Alfonsine Tables.* Cunitz prefaced the volume by explaining that despite being the work of a woman, the *Urania* was nevertheless accurate. Her tables proved of great utility and went through many editions. This led many to suspect her husband had computed them. Cunitz's husband wrote a preface for later editions to dispel such rumors.

Descartes, René. *Principia Philosophiae* (1644). Here Descartes attempted to put the whole universe on a mathematical foundation, reducing the study of everything to that of mechanics. For Descartes knowledge could only be gained from deduction from fundamental principles.

Digges, Thomas. *Alae seu scalae mathematicae* (1573). Contained the author's observations of the 1572 supernova.

Ercker, Lazarus. *Beschreibung der allerfürnemisten mineralischen Ertzt und Berckwerksarten* (1574). The first systematic treatment of analytical and metallurgical chemistry. In it Ercker reviewed methods for obtaining, refining, and testing alloys and minerals of silver, gold, copper, antimony, bismuth, lead, and mercury as well as acids, salts, and other compounds including methods for purifying saltpeter. He also discussed the design and construction of apparati for assaying such as the cupel, furnaces, and assaying balance.

Escholt, Mikkel. *Geologia norvegica* (1657). The first scientific treatise printed in Norway and one of the first books published in Norwegian. In the work Escholt described the April 24, 1657, Norwegian earthquake. He held that earthquakes were portents of divine intervention in the world, though their production by God was occasioned by physical causes—principally escaping air from Earth's interior. He was also the first to use "geology" in its modern sense of Earth science.

Flamsteed, John. *Historia Coelestis Britannica* (3 vols., 1725). Published posthumously, this work contained the positions of nearly 3,000 stars and established Greenwich as one of the world's leading observatories. A set of star maps—*Atlas Coelestis* (1729)—based on the catalog also appeared posthumously.

Galileo. *La bilancetta* (1586). This treatise described Galileo's improved hydrostatic balance.

Galileo. *De motu* (1590). Contained Galileo's experiments on falling bodies.

Galileo. *Sidereus nuncius* (1610). This work included Galileo's descriptions of his self-built telescope, which he constructed and used in January 1610, making him perhaps the first person to use a telescope to view the heavens.

Galileo. *Dialogue Concerning Two Chief World Systems* (1632). This book was ostensibly meant to present an impartial discussion of the Copernican and Ptolemaic systems. The discussion was anything but impartial, marshaling as it did overwhelming empirical evidence in support of heliocentrism. Galileo was tried as a heretic, convicted, and sentenced to permanent house arrest.

Galileo. *Two New Sciences* (1638). Dealt with the engineering science of strength of materials and kinematics. It included the law of the lever, used to establish the breaking strength of materials. It also provided a mathematical treatment of motion in which Galileo introduced the idea of uniformly accelerated motion. He also established the law of free fall in a vacuum, deduced the terminal velocity for any body falling through air, and derived the parabolic trajectory of projectiles from uniform horizontal and accelerated vertical motions.

Gilbert, William. *De magnete, magneticisque Corporibus, et de magno magnete tellure* (Concerning magnetism, magnetic bodies, and the great magnet earth, 1600).

This work's emphasis on direct observation and rigorous experimentation earned Gilbert praise from Galileo, who considered him the founder of the experimental method.

Hevelius, Johannes. *Selenographia* (1647). The 133 color plates of this work represent the first detailed, accurate maps of the Moon's surface. Many of Hevelius's names for the Moon's features were taken from Earth's geography and are still used, such as Mare Serenitatis (Pacific Ocean). However, his names for individual craters were not adopted.

Hevelius, Johannes. *Cometographia* (2 vols., 1668). Here Hevelius discussed the nature of comets and collected a considerable body of literature on comets observed in previous centuries. He considered comets planetary exhalations and believed them responsible for sunspots. Like Giovanni Borelli he suggested their orbits might be parabolic.

Hevelius, Johannes. *Uranographia* (1690). Hevelius's best known work, cataloging over 1,500 stars and introducing several new constellations, including Lacerta, Leo Minor, Lynx, Scutum, Sextans, and Vulpecula.

Hooke, Robert. *Micrographia* (1685). An illustrated discussion of observations Hooke made with a reflecting microscope that he built himself. His commentary included biological specimens, and he coined the word *cell* to explain the microscopic structures he observed. *Micrographia* also presented the results of his microscopic studies of crystalline solids, including snowflakes.

Huygens, Christiaan. *Horologium Oscillatorium* (1673), Included a mathematical analysis of the compound pendulum and derivation of the relationship between pendulum length and period of oscillation. Huygens also included the laws of centrifugal force for uniform circular motion and an early formulation of Isaac Newton's first law of motion.

Huygens, Christiaan. *Traité de la lumière* (1690). Huygens' response to Isaac Newton's corpuscular theory of light. Huygens presented a wave construction capable of explaining light's rectilinear propagation, reflection, refraction, and certain properties of double refraction in Iceland spar. He also predicted, in opposition to Newton, that light travels slower in denser media. Newton's theory dominated eighteenth-century optical thinking but was eclipsed by Huygens's theory in the early nineteenth century. The two views were later synthesized in quantum theory during the early years of the twentieth century.

Kronland, Johannes von. *Thaumantias liber de arcu coelesti....* (1648). Here Kronland discussed the colors of the rainbow, the dispersion of light beams directed through a prism, and the diffraction of light around a wire. He also observed that repeated refraction does not cause monochromatic rays to change color.

Lemery, Nicolas. *Cours de chymie* (Course on chemistry, 1675). Europe's most influential chemical textbook for over fifty years, with eleven editions in Lemery's own lifetime and over thirty editions by 1756.

Leonardus, Camillus. *Speculum lapidum* (1502). Treated over 200 minerals. This compilation proceeded according to traditional knowledge and concepts, relying little on independent observation, although the invention of gunpowder was alluded to. The author's focus was primarily the occult powers of gems and images carved upon them.

Libavius, Andreas. *Alchemia (Alchemy)* (1597). This book summarized the discoveries that alchemists had made up to the date of publication. Later editions of the book were more than 2,000 pages long and contained 200 illustrations.

Longomontanus. *Astronomia danica* (1622). Longomontanus was Tycho Brahe's only disciple. Upon his master's death Longomontanus assumed responsibility for selecting and integrating Tycho's data into a coherent account of planetary motions and presenting this work in a systematic treatise. His task was brought to fruition in this work, which was enthusiastically received by the astronomical community. The prestige attached to his work saw it through two reprintings

Newton, Isaac. *Philosophiae Naturalis Principia Mathematica* ("The Mathematical Principles of Natural Philosophy," 1667-86). Perhaps the greatest scientific work ever written. It united two competing strands of natural philosophy—experimental induction and mathematical deduction—into the scientific method of the modern era. His emphasis on experimental observation and mathematical analysis changed the scope and possibilities of science.

Norman, Robert. *The Newe Attractive* (1581). One of the first systematic treatises in experimental physics, in which Norman announced his discovery of the dip of compass needles. Norman measured the deviation from the horizontal of compass needles at London and speculated on whether or not the effect varied over Earth's surface. He also suggested compass needle orientation was due to turning toward rather than attraction to a particular point.

Owen, George. *Description of Pembrokeshire* (written 1603, published 1892). The map Owen made of Pembrokeshire was considered a landmark, and Owen has been referred to as the patriarch of English geologists.

Pena, Jean. *Euclidis Optica et Catoptrica* (1557). In his treatise on geometrical optics of lens and mirrors, Pena noted that a comet's tail pointed away from the Sun, prompting him to theorize that comets were made of some celestially transparent substance that refracts light and causes combustion and thus the tail.

Perrault, Pierre. *De l'origine des fontaines* (1674). One of the foundational works in experimental hydrology. Perrault demonstrated rainfall is sufficient to maintain the flow of rivers. He determined the drainage area for a portion of the Seine River and calculated the total precipitation. After adjusting for losses he compared the total volume of water with the river's annual flow and showed that only one-sixth of the annual rainfall was necessary to sustain it.

Regiomontanus, Johannes. *Epitome of Astronomy* (1496). The work contained, in addition to the Georg von Puerbach-Regiomontanus translation of Ptolemy's work, critical commentary, revised computations, and additional observations. Though not printed until 20 years after Regiomontanus's death, the work was a great success and attracted the attention of a young Nicolaus Copernicus.

Reinhold, Erasmus. *Tabulae Prutenicae* (1551). The first practical set of planetary tables based on Nicolaus Copernicus's theory. More accurate than the *Alfonsine Tables* of Johannes Kepler, Reinhold's tables were widely adopted and provided a strong argument in favor of Copernicanism.

Steno, Nicolaus. *De Solido intra Solidum naturalites contento dissertionis Prodromus* (1669). Here Steno maintained the organic origins of fossils. The work contains the outlines of modern geology. Steno argued that fossils are animal remains that had become embedded in sea-floor strata and then petrified by the concreting action of chemical forces, heat, and compression over time. He suggested that strata are deposited horizontally in layers from aqueous fluids with occasional crustal collapses accounting for diversities of topography. In explaining this stratification he established the principle of superposition—underlying sedimentary layers are older than those overlaying them—upon which modern geochronology is based. To accompany his explanations, Steno produced what are considered the earliest geological cross sections.

Stevin, Simon. *De Beghinselen des Waterwichts* (1586). Here Stevin provided the first systematic development of Archimedes' hydrostatics. He explained Archimedes' displacement principle for submerged bodies and showed that the pressure exerted by a liquid on a surface depends on the height of the liquid above that surface and is independent of the shape of the vessel containing it.

Torricelli, Evangelista. *De motu gravium* (1641). Here Torricelli developed some of Galileo's ideas on projectile motion. He experimentally verified many new conclusions and stated what is today known as Torricelli's law—a rigid system of bodies can move spontaneously on Earth's surface only if its center of gravity descends.

Vossius, Isaacus. *De lucis natura proprietate* (1662). Contained many bold conjectures about light, including the assertion that light is not a corporeal body, as well as commentary on Willebrord Snell's unpublished law of light refraction.

Whiston, William. *New Theory of the Earth* (1696). This work sought rapprochement between Newtonian science and the Biblical account of Creation. According to Whiston, Earth was originally a comet whose eccentric orbit was made circular by God; until the Flood, Earth possessed no diurnal rotation and its poles were perpendicular to the ecliptic. Finally, a comet guided by God penetrated Earth's surface, causing the waters of the abyss to overflow, shifting the poles, and imparting axial rotation.

NEIL SCHLAGER

Technology and Invention

Chronology

1450 Johannes Gutenberg invents a printing press with movable type, an event that will lead to an explosion of knowledge as new ideas become much easier to disseminate.

1504 German clockmaker Peter Henlein builds the first truly portable clock, a 6-inch (15-cm) high spring-driven "watch" made entirely of iron.

1521 Cesare Cesariano, student of Leonardo da Vinci, publishes his master's observations on the phenomenon called the camera obscura, forerunner of the modern photographic camera.

1525 Rifled gun barrels, which cause a bullet to spin—thus providing greater stability and accuracy for its flight—make their appearance.

1589 William Lee of England invents the first knitting machine.

1590 Some 40 years after the first simple, or single-lens, microscopes appeared, Dutch optician Sacharias Jansen, together with father Hans, invents the first compound microscope.

1610 Dutch-English inventor Cornelius Drebbel devises the first self-regulating oven, which uses a thermostat.

1617 Scottish mathematician John Napier first develops what comes to be known as Napier's bones, an early calculator that makes use of wooden rods labeled with numbers.

1622 English mathematician William Oughtred invents the slide rule, which makes possible rough but rapid multiplication and division by sliding a numbered stock between two slats.

1631 Pierre Vernier invents the vernier, a two-part scale that measures angles and lengths in small divisions; ultimately it replaces the astrolabe, and finds its optimum use in surveying.

1667 The cabriolet, a light, two-wheeled carriage that will become very popular during the eighteenth century, makes its first appearance in France.

1681 France builds the Languedoc Canal, also known as the Canal du Midi, a 150-mile (241-km) waterway considered the greatest feat of civil engineering between Roman times and the nineteenth century.

1698 English engineer Thomas Savery patents the first steam engine, a monocylinder suction pump that drains water from mines.

Overview:
Technology and Invention 1450-1699

Background

The age of humanism that followed the Medieval era built upon a revolution in science that celebrated human curiosity and its use of rational inquiry. This embrace of human discovery also affected technology with a rational approach to the material world and a growing interest in transforming that world through individual action. With the advent of mechanical printing and the resulting development of a literate technology, technological change accelerated and technological diffusion widened. As a result, the Western world associated science and technology with progress, incorporating them into the larger framework of a humanist perspective so characteristic of the Renaissance and Age of Enlightenment.

Mechanized Printing

More than any other development of this era, mechanized printing transformed the nature of technology. Johannes Gutenberg (c. 1398-1468) used his background as a metallurgist to devise a means of printing with interchangeable, standardized metal type. He achieved this process of using movable type by casting various letters in the same mold; each piece of type was identical in size and shape to every other piece, which gave him complete interchangeability. His type mold, not the printing press itself, was the critical part of this revolutionary means of printing. The result was the first information revolution with an explosion of accessible knowledge available to a wide audience. Mechanized printing made the production of books less expensive, allowed for the wide dissemination of knowledge, greatly enhanced the accuracy of description, and permitted individuals to gain knowledge and to challenge established authority through their own self-learning and exploration.

Less expensive printing fueled the growth of technical literature and created the world of handbook technology. Georg Bauer (also known as Agricola [1494-1555]) published the widely used handbook of metallurgy and mining, *De Re Metallica*. Vannoccio Biringuccio (1480-c. 1539) focused on the metallurgy of precious metals and the art of casting in his *De la pirotechnia*. These works, among many others, contained engrav-

ings as well as descriptive text so that their readers could replicate various technological methods and machinery. Ready access to the tools and processes of technology through the printed record also stimulated invention and innovation as this knowledge spread throughout the Western world in the new portable form. This age of literate technology spurred so much development that technology for the first time was divided into two categories: military engineering and everything else—civilian or civil engineering.

New Weapons of War

Gunpowder first used in China diffused to the West during the Renaissance with a significant impact on weapons development. New weapons of war such as the cannon, muskets, and pistols became hallmarks of armies as both strategic and tactical methods changed to reflect the use of these new weapons. For example, ship design was modified to include a cannon deck, allowing the ship to play a more active role in military action. In addition, land-based warfare saw the increased use of cannons, first to frighten and later to destroy communities. This new weapon made the typical Medieval walled fortress much more vulnerable because cannon fire easily could permeate those walls. Many Renaissance engineers successfully pursued careers in designing and building new city fortifications using slopped walls, placing cannons throughout the line of defense. The resulting star fort design replaced the high stone walls of the Medieval fortress with lower earthen and brick walls less susceptible to damage by cannon fire. Cannon fire was less likely to destroy an entire wall made of brick or of angled earthen surfaces.

The increased use of gunpowder also transformed the role of governments in encouraging and supporting military technology. The new weapons required skilled metallurgists, manufacturing systems, trained soldiers, and standing armies. Governments, especially those intent on expansionism and increased regional influence, began supporting a military establishment for the maximum exploitation of new weapons and new military activities. A growing partnership between governments and the military establishment made war a more professional activity.

Mechanized Time

In this era, the ways people measured time moved from the use of sun dials and water clocks to mechanical clocks. The regular effects of weights, springs, or a pendulum became a means of measuring time; time keeping moved away from natural rhythms to abstract mechanical intervals established by scientists and inventors. Christiaan Huygens (1629-1695) played a key role in developing the pendulum clock, and men such as Jean de Hautefeuille (1647-1724), Peter Henlein (1480-1542), and Robert Hooke (1635-1703) provided techniques for producing portable clocks in the form of watches that used a controlled force released by a spring to measure and display time.

The widespread use of mechanical clocks mirrored a changing attitude toward nature. Instead of seeing the world as an organic or mystical place, those involved in science and technology sought mechanical measures of their world. Based largely on the new scientific attitudes of the time, this analytical, logical, and mechanical approach provided an intellectual foundation for the rationalized technology that was essential for the industrial age. This mechanical mindset also encouraged people to observe, measure, and analyze various aspects of the world around them. Man, indeed, became the measure and measurer of all things.

Measuring Instruments and Measuring the World

The interest in measurement and the perception of a mechanical world enhanced the development and use of scientific instruments. Devices such as the telescope, the microscope, calculators, the magnetic compass, barometers, gun sights, scales, lenses, and clocks provided a technological underpinning for both scientific inquiry and practical application. Measurement was critical for astronomy, navigation, and surveying; precise observation and description became a hallmark for the world of Renaissance science. For example, Galileo's (1564-1642) use of the telescope allowed him to challenge long-standing astronomical theories and to replace them with explanations based on his personal observations; these new models contributed to the Newtonian revolution in science.

In the world of transportation technology, the bringing together of the compass, the lateen or triangular sail, and the stern-post rudder created a vastly improved sailing ship worthy of trans-oceanic travel. The results were the voyages of discovery and exploration that flourished in the seventeenth century. With these experiences explorers discovered new flora and fauna, increased the pool of human knowledge, and provided new trade routes that stimulated trade and commerce. As a consequence, the material wealth of Western Europe increased substantially, a happening that fed the materialistic thrust of the era.

Improved measuring instruments and techniques affected technology on land as well as at sea. Canal building on a vast scale attested to the civil engineering skills of Renaissance engineers. In Britain, Hugh Myddleton (c. 1560-1631) created a canal known as the New River, which remains as a means of supplying water to London. In France, Pierre-Paul Riquet de Bonrepos (1604-1680) was instrumental in the construction of the Canal du Midi (Languedoc Canal); this impressive engineering feat had a 500-foot (152-m) tunnel, several aqueducts, and a hundred locks along its 180-mile (290-km) length. The widely traveled and experienced Dutch engineer Simon Stevin (1548-1620) engaged in several projects focusing on canals, drainage, and irrigation projects throughout Western Europe. The surveying instruments and techniques necessary for canal building also played a key role in the design and construction of urban fortifications, another major contribution of Renaissance civil engineers. No less a person than Leonardo da Vinci (1452-1519) devoted most of his time and talent to various civil engineering tasks such as building canals and forts. Leonardo was typical of the many Renaissance engineers who made a living by selling their skills and experience at designing and building forts for an era of gunpowder warfare, and canals and other water control devices in an age that valued the control and distribution of water.

Conclusion

The improvements in agriculture, the growth of urban centers for trade and commerce, and the enhanced shipping technology available created a more urban-based society in the West during this era. These developments stimulated population growth, an interest in the material, and a utilitarian/mechanical perspective on the world. Clearly, the advent of mechanized printing accelerated the spread of technical knowledge and increased the literacy of engineers and the growth of technical treatises or handbooks. As Renaissance society became more secular, technology played an increasingly important role.

Technology became associated with secular progress in a world that embraced rational inquiry, careful and precise observation, and widespread experimentation. To a degree not seen before in the West, technology was more closely linked to science; the resultant scientific revolution that created the modern scientific method affected technology as well. Rationalism, a mechanical world view, and logical inquiry served both science and technology at a time when both allowed people to take the measure of themselves and the world that surrounded them. The pace of technological change and diffusion accelerated, aided in large measure by the new technologies in the fields of printing and sailing. These technological developments stimulated more invention and innovation in both civil and military engineering. The resultant Renaissance emphasis on and rewards for technological change provided a proto-industrial base from which industrialism would emerge a century later. Because individual action and analysis played an important part in understanding the world, increasingly individuals took credit or were credited with technological developments; Western history became the record of human achievement to a new degree. All of these factors contributed to a new era of humanism that stimulated both science and technology and shaped modern Western culture, a culture in which science and technology are central to human progress.

H. J. EISENMAN

The Birth of Print Culture: The Invention of the Printing Press in Western Europe

Overview

The invention of the printing press in Western Europe stands as one of the most important events in the history of human civilization. Though various printing processes had earlier appeared in Asia, it was the mechanical apparatus developed by Johannes Gutenberg (1398?-1468) that ushered in a new age of communications and comprehension with far-reaching implications that continue to shape our world and perception. The Gutenberg Bible, the first known printed book in the West, remains the marker by which 500 years of "print culture" began.

Prior to Gutenberg's invention, which involved the use of moveable type, books were expensive and relatively scarce, as each text was painstakingly copied by hand over a period of weeks or months. As a result, the transmission and preservation of knowledge relied heavily upon oral, or spoken, communication and memorization, particularly among the vast majority of people who could not read or write. For those with access to manuscript books, the lack of standardization among texts and the time-consuming process of duplication limited the accuracy and speed by which both existing and new ideas could be distributed.

Through the use of Gutenberg's mechanized instrument of reproduction, numerous exact copies of books could be generated in a short period of time, making available an unprecedented number of familiar books and encouraging the creation of new works. Though this influx of printed books was initially limited to the moneyed and literate upper and scholar classes, the growing availability and affordability of printed texts later resulted in increasing levels of mass literacy and provided for the free flow of information essential to scientific progress, capitalism, and modern democracy.

Background

The invention of printing had its origins in China, where as early as the second century A.D. printing from engraved wood blocks, or xylography, was in use. In 1041-1048 Pi Sheng, a Chinese blacksmith and alchemist, developed the process of printing with movable type; however, due to the large number of Chinese characters—some 80,000—and the frailty of his baked clay type, this method proved impractical. Later, in the fourteenth century, moveable wood type was used in China, and in the fifteenth century moveable type cast in bronze was developed in Korea. But despite the apparent success of these innovations, it remained unknown in Europe.

During the late 1430s, Gutenberg, a German goldsmith, began experimenting with

An early printing press (1511). *(Bettman/Corbis. Reproduced with permission.)*

techniques for "artificial writing," as the printing press was first known. Its development in Europe hinged upon several key factors—the new availability of paper, the casting of metal type, and increasing demand for books. While paper had existed in China since 240 B.C., paper and papermaking technology did not reach Europe until the eleventh century A.D., almost 1,200 years later. Before paper arrived in Europe via the Muslim Middle East, the pages of manuscript books were made of vellum (calfskin) or parchment (pigskin). Both materials were expensive and, furthermore, the outer surface of the skin was too rough to use, allowing only one side of the page for writing.

Gutenberg's expertise as a metallurgist enabled him to develop a method for casting interchangeable lead typefaces for printing. The use of metal typefaces, as opposed to wood or clay, improved the quality and durability of the printing process. Gutenberg also developed ink that readily adhered to his type and to the paper surface. A relatively simple machine, Gutenberg's printing press worked by arranging type—individual characters consisting of letters, numbers, spaces, or symbols—into sentences and lines, thus forming the text of a full page. The type, arranged in a type "bed," was then inked and pressed upon a sheet of paper under the force of a screw press. The resulting impression on the paper could be duplicated repeatedly by simply re-inking the type and inserting additional paper. The advantage of reusable moveable type involved its ability to be dissembled and re-arranged into new pages of text, from which additional pages could be printed and the process repeated. The printed sheets could then be arranged, folded, and bound into books.

During the fourteenth and fifteenth centuries the demand for books among European scholars and aristocrats increased dramatically. This period, known as the Renaissance, was characterized by a new scientific and artistic interest in secular—as opposed to religious—understanding of the world. Coupled with the earlier rise of the university during the eleventh and twelfth centuries, there emerged intensified pressure to produce greater numbers of textbooks, scientific treatises, and other written reports of new discoveries. Before the printing press, manuscript books were produced primarily by medieval monks in specialized workshops called scriptoria, where groups of clerics hand-copied individual texts. Later, as demand increased, professional and state-sanctioned scribes were also employed to mass produce books for sale. However, the labor-intensive process of transcribing by hand was expensive and could not meet the growing demand, thus encouraging the development of a machine to speed this process.

The Gutenberg Bible, also known as the "42-line Bible" in reference to the number of double column lines on each page, is universally hailed as a landmark technical and aesthetic achievement. Of the estimated 200 copies printed between 1455-56, fewer than 50 are known to exist. While Gutenberg's Bible at once set the standard for all subsequent books, many early printed books were designed to replicate the appearance of more expensive manuscript books, with ornamental initials and Gothic print. However, the familiar conventions of printed books were quickly established and widely adopted, including the use of title pages (previously nonexistent), page numbers, running titles, colophons indicating the publisher of the book, and easily readable Latin typefaces such as roman and italics (except in Germany, where gothic Fraktur persisted until the twentieth century). Books produced during the infancy of the printing press—the period from 1450 to 1500—are referred to as *incunabula,* Latin for "swaddling clothes" or "infancy."

Impact

Within 50 years of its invention in Germany, printing presses appeared in nearly all major European centers: Cologne (1464), Basel (1466), Rome (1467), Venice (1469), Paris (1471), Valencia and Budapest (1473), London (1480), and Stockholm (1483). Though documentary evidence is lacking, it is estimated that by 1500 more than 10,000 different titles had been published and several million printed books were already in existence. Another indication of the press's rapid spread was the proliferation of paper mills throughout Europe during this time, as the corresponding demand for paper increased manyfold. The Spanish brought the press to the New World, where the first was established in Mexico City in 1539. During the sixteenth century, European colonists and missionaries, particularly the Portuguese, introduced the printing press to other parts of the world, including China, Japan, India, and Africa. The printing press did not appear in North America until 1638, where in Cambridge, Massachusetts, Stephen Day published the first American book.

From the beginning, book publishing was a profit-motivated industry inextricably linked to the rise of capitalism in early modern Europe. Printing press operators simultaneously served as agents, editors, publishers, and booksellers, and competition was aggressive as printers vied for market share. Gutenberg was himself entangled in a legal dispute with his financial benefactor Johann Fust (1400?-1466?), who seized Gutenberg's invention in 1455 and with Peter Schoeffer (1425?-1502), Fust's son-in-law and Gutenberg's former assistant, founded the first commercial publishing company. By 1560 many of the most famous early printer-publishers had established themselves, including Anton Koberger (1445-1513) in Nuremberg, Aldus

Manutius (1450-1515) in Venice, Johann Froben (1460-1527) in Basel, William Caxton (1420?-1491) in London, Robert Estienne (1503-1559) in Paris, and Christophe Plantin (1520?-1589) in Antwerp.

The invention of the printing press exerted an immediate effect on the social and intellectual climate of Europe. As a product of the Renaissance, the press served the interests of humanists by making available many ancient Greco-Roman classics, such as the works of Homer, Plato (427?-347? B.C.), Aristotle (384-322 B.C.), Cicero, and Virgil, previously rare or unavailable in western Christendom. The rediscovery and study of scientific works by ancient writers such as Hippocrates (190?-120? B.C.), Ptolemy (100?-170?), and Galen (c. 130-200) also had a major impact on fifteenth- and sixteenth-century science. In addition to the rise of classical scholarship, one of the hallmarks of the Renaissance, printing facilitated the growth and consolidation of learning by making available the works of more recent medieval philosophers, including Albertus Magnus (1193-1280), Roger Bacon (1214?-1294), Thomas Aquinas (1227?-1274), and Arab scholars Avicenna (980-1037) and Averroes (1126-1198).

It is significant that the first book to be printed was the Bible, as one of the first discernable impacts of the printing press on society was the Reformation, a period of dramatic, often violent, religious dissent during which various sects of European Christians broke from the Roman Catholic Church to form the branches of Protestantism. This sixteenth-century religious upheaval was given impetus and international significance largely through printed texts that asserted opposing views. In 1517 Martin Luther (1483-1546) published his famous "95 Theses," followed by three additional tracts in 1520, in which he denied the supremacy of the Church. Because Luther's writings were printed and widely distributed, a local controversy was transformed into a major public event, touching off the Reformation in Germany and elsewhere. Other religious dissenters, in particular John Calvin (1509-1564) in Switzerland, published important texts that opposed the authority of the Church and further spread the Reformation throughout Europe.

One of the central arguments of the reformers was that the Bible itself—not the Church—was the supreme authority on matters of morality and religious practice. This emphasis on the *text* of the Bible, previously accessible only to those who could read Latin, prompted numerous translations into the vernacular, or common languages, of Europe. Though the first vernacular Bible appeared in German in 1466, Luther's 1522 German translation of the New Testament (followed by the Old Testament in 1534) was the most popular and influential, reprinted in numerous editions and serving as the basis for subsequent translations into Dutch, Swedish, and English.

While Latin remained the universal language for European religious and scientific texts for several centuries, the growing availability of vernacular texts fostered a larger reading audience of both men and women, including a burgeoning nonaristocratic class of wealthy merchants, artisans, and financiers—the bourgeoisie, or middle class. Though early literacy rates are uncertain, in 1533 Thomas More (1478-1535) claimed that half of the English population could read—undoubtedly an inflated figure, but telling in its optimistic presumption. In England, unlike the rest of Europe, native-language books were printed from the beginning, and English superceded Latin as the most popular language of publication. The growth of vernacular texts promoted wider readership in general, but it also had the effect of delimiting boundaries, out of which distinct national literatures arose. As Latin progressively lost favor, translations between vernacular languages became increasingly important and, particularly in the case of scientific knowledge, the absence of translations could result in isolation and unnecessarily duplicated research. For most of Europe and America, however, high levels of literacy—rather than basic competence—remained limited to the wealthy, educated, and clergy until the nineteenth century.

The new availability of books soon gave rise to the growth of libraries. While the existence of libraries as storehouses of knowledge dates back to the ancient Egyptian and Greco-Roman worlds, the advent of printing permitted far more extensive collections to be amassed, particularly as new titles and knowledge rapidly became available. Aristocrats, wealthy merchants, and humanist scholars began to assemble personal libraries that brought together the most valuable works on diverse subjects—and served to display their owner's social status and cultivation. Some of these early royal and private collections later became the core of major national libraries, such as the Bibliothèque Nationale in France and the British Museum (1759) in England. Both of these libraries required that a copy of all new printed works be donated to their holdings, establishing the concept of the depository library system, adopted by the Library of Congress (1800) in the United States.

Important university collections were also founded, notably the Bodleian Library at Oxford, established by Thomas Bodley in 1602. Public libraries began to emerge in the eighteenth century, originating first as lending libraries, such as that established by Benjamin Franklin (1706-1790) in Philadelphia, and later, during the nineteenth century, as civic institutions intended for the education and benefit of the public at large. The proliferation of books made libraries important as centers of consolidation and organization. It is perhaps not surprising that several highly influential scholars—Gottfried Wilhelm Leibniz (1646-1716), David Hume (1711-1776), and Johann Wolfgang von Goethe (1749-1832)—worked as librarians during their careers.

The impact of the printing press on science and technology was tremendous. Whereas scientists and inventors had long worked in relative isolation and had limited ability to share information, the use of the press to publish the results of their work greatly accelerated the rate of scientific discovery and progress, and many of these advances directly impacted the quality of life for large groups of people through improved medicine, domestic and agricultural technology, and transportation. The birth of the modern scientific revolution can be directly linked to early printed works by Nicolaus Copernicus (1473-1543), Johannes Kepler (1571-1630), Galileo Galilei (1564-1642), and Isaac Newton (1646-1728), with his *Principia* (1687) as the crowning achievement of this early phase. In addition to major advances in mathematics and astronomy, medicine and physiology were advanced by the published (and illustrated) works of Andreas Vesalius (1514-1564) and William Harvey (1578-1657), mineralogy and metallurgy by Georgius Agricola (1494-1555), physics and chemistry by Robert Boyle (1627-1691), microbiology by Robert Hooke (1635-1703) and Anton von Leeuwenhoek (1632-1723), entomology by Jan Swammerdam (1637-1680), plant and animal classification by Carolus Linnaeus (1707-1778), and scientific philosophy by Francis Bacon (1561-1626) and René Descartes (1596-1650). This is only a small sampling of the many new works that contributed toward advances in the sciences.

The new empirical methodology of scientific research—mainly the emphasis on observation, experimentation, and cumulative documentation—was well complemented by the advent of printing. Printed books provided a means by which knowledge could be systematically recorded, arranged, preserved, and reproduced with unprecedented accuracy, speed, and consistency. Thus scientists were increasingly able to assimilate the research and innovations of their predecessors in order to produce further advances of their own. With the rise of scientific societies during the seventeenth century also emerged the role of the scientific journal as a tool of knowledge preservation and dissemination. The first such journal was the *Philosophical Transactions* of the Royal Society of England, whose first issue appeared in 1665. Others soon followed, such as the *Mémoires* of the Academié des Sciences in France, and the standard practice of publishing one's research for the use and scrutiny of the scientific community was established. As a growing number of scientific works became available, and new and existing branches of science became increasingly complex, the trend toward specialization also emerged as a new feature of the sciences. Mastery of all branches of science by a single individual—the celebrated Renaissance man—was nearly inconceivable by the end of the eighteenth century, as so much new knowledge had been created and disseminated by the press.

The press also facilitated the publication of travel accounts that made known the voyages and expeditions of European explorers during the great age of discovery. Such reports were essential to monarchs interested in expanding colonial empires, as well as to merchants seeking lucrative new lands where raw materials and goods could be traded and obtained—another important way in which the press is linked to the rise of capitalism. Some of the most famous early accounts include "The Columbus Letter" (1493), in which Christopher Columbus (1451-1506) announced his discovery of the New World, accounts of the travels of Marco Polo (1254-1324), Ferdinand Magellan (1480?-1521), and James Cook (1728-1779), as well as important compilations by Richard Hakluyt (1552?-1616) and others. Exploration was further aided by the publication of maps, providing explorers with increasingly accurate guides. Cartography flourished in the Netherlands, due to the knowledge of Dutch mariners, and resulted in several important atlases, including those by Ortelius (1527-1598), Willem Janszoon Blaeu (1571-1638) and his son Jan, and Gerardus Mercator (1512-1594), who developed the modern projection map.

The printing press had a dramatic effect on literature and the arts in Europe. The humanism of the Renaissance fostered a new spirit of creative individuality and self-consciousness that resulted in many new works of poetry, drama, and history. Literary works such Thomas Malory's *Le Morte Darthur* (1470), Ludovico Ariosto's *Orlando furioso* (1516), and Dante Alighieri's *Divine Comedy* (1555) soon appeared in print. A vast array of devotional literature also flourished, such as the early bestseller *The Imitation of Christ* (1473), attributed to Thomas à Kempis, and later John Bunyan's allegorical work *The Pilgrim's Progress* (1678). In England the Renaissance plays of William Shakespeare (1564-1616) were published at the time of their performance, and the first collection of Shakespeare's dramatic works, the so-called *First Folio*, appeared in 1623. With printing also emerged the professional writer, whose livelihood could be earned by publishing and selling original works to the public, rather than to a wealthy patron alone.

New genres of secular writing also emerged, such as the personal essay, pioneered by Michel de Montaigne (1533-1592), and, perhaps most notably, the modern novel, whose history in the West is generally thought to begin with the publication of *Don Quixote* (Part I, 1605; Part II, 1615) by Spanish writer Miguel Cervantes. The novel became one of the most popular forms of literature during the first half of the eighteenth century, particularly in England through the works of Daniel Defoe, Samuel Richardson, and Henry Fielding. The confluence of literature, politics, and social criticism also found expression in the new genre of utopian fiction, exemplified by Thomas More's *Utopia* (1516), and satires, such as Desiderius Erasmus's *The Praise of Folly* (1511) and Francois Rabelais's masterpiece, *Gargantua and Pantagruel* (1532-64). In addition, the development of art and architecture was shaped by several important works, including Giorgio Vasari's *Lives of the Most Eminent Italian Painters, Sculptors, and Architects* (1568) and Antonio Palladio's *Four Books of Architecture* (1570).

The rapid proliferation of new and often highly controversial scientific and literary publications prompted the emergence of improved modes of control—censorship and copyright—to serve the interests of both the authorities and the author. Censorship, the official prohibition of certain books deemed scandalous, subversive, or heretical, was implemented by both the Church and ruling monarchs almost immediately after the first books came off the press. Though books had always been subject to official sanction, the printing press made their suppression more difficult, as condemned books could be quickly produced and distributed in great numbers and in relative secrecy. In 1559 the Roman Catholic Church established the *Index Librorum Prohibitorum,* a notorious list of banned books that was issued continuously until 1966. Likewise, royal decrees were regularly issued in England and other countries with the aim of controlling the content and publication of books. As a result, many great Renaissance intellectuals, most famously Galileo, were persecuted for their published writings. Resistance to such controls by printers and authors helped establish the "freedom of the press," a political and artistic principle championed by writers such as John Milton (in his *Areopagitica,* 1644), and a cornerstone of democratic society.

The rights of the author and publisher found protection in the development of publishing guilds and copyright laws. While printers sought to preserve their exclusive right to print and sell certain titles, authors demanded that their unique ideas, or intellectual property, be protected from plagiarism and unauthorized sales from which they did not receive proceeds—a problem that hardly existed before printing. In England the Stationer's Company was incorporated in 1557, taking on the role of regulating publishers by requiring that new works be registered and licensed. This concept was extended in the Copyright Act of 1709, the basis for all subsequent copyright laws, through which authors and publishers were granted legal rights to their work for a limited period of time. Unauthorized, or "pirated," editions of books were common during the first several centuries of printing, and the effort of publishers and authors to protect against infringement began early and remains an ongoing struggle in modern times, particularly as new technologies, such as the Internet, pose new challenges to such protections.

While the printing press is inextricably linked to the history of the book, the invention of the press gave rise to many other important forms of publication such as the newspaper, the magazine, the broadside, and the pamphlet. The first newspapers, or newsletters, appeared in the mercantile centers of Holland and Germany during the first decade of the seventeenth century, spreading to other parts of Europe shortly thereafter. The earliest known magazines were published during the 1660s, first in Germany and then in France and England. Other "ephemeral" formats, including the pamphlet and broadside, a hybrid of poster and newssheet, were even cheaper to produce than books and made possible the rapid dissemination of current information, such as local news, advertisements, and literary pieces, as well as religious controversies and unpopular or officially censored ideas. The diffusion of such publications attests to the expansion of literacy in early modern Europe, and the power and importance of the press as a vehicle of mass communication.

In many ways the social impact of Gutenberg's printing press culminated with the Enlightenment. This intellectual milieu during the seventeenth and eighteenth centuries, also known as the Age of Reason, was characterized by new confidence in the powers of the rational human mind to discover and order the laws of the universe and society. These lofty aspirations hinged upon the supreme importance of knowledge and freedom, both of which the press embodied. With the power of the Church diminished by humanism and the Reformation, philosophers such as Thomas Hobbes (1588-1679), John Locke (1632-1704), Voltaire (1694-1778), Jean-Jacques Rousseau (1712-1778), Immanuel Kant (1724-1804), and others took up controversial issues involving the nature of religion, human agency, and the rights of government. Such investigations, widely published and debated throughout Europe, had a profound impact on religious, political, social, and economic thought of the time, introducing the concepts of free-market capitalism and democracy, and shaping modern attitudes of skepticism, secular materialism, and individualism. These developments precipitated the American and French Revolutions, both of which were fueled by the circulation of pamphlets, such as Thomas Paine's "Common Sense" (1776) in America, that challenged the established order and galvanized public support.

In addition to fomenting social and political upheaval, one of the central roles of the press during the Enlightenment involved standardization. The ambitious effort to organize and present all knowledge based on the principles of rationalism is exemplified by the 17-volume *L'Encyclopèdie* (1751-65) compiled by Denis Diderot and Jean d'Alembert, as well as works in specific fields: in natural history, Georges Buffon's 44-volume *Historie Naturelle* (1749-1804); in lexicography, Samuel Johnson's *Dictionary of the English Language* (1755); in history, Edward Gibbon's 6-volume *Decline and Fall of the Roman Empire* (1776-88); and in law, William Blackstone's 4-volume *Commentaries on the Laws of England* (1765-69). In addition to encouraging

comprehensive scholarship, the movement toward standardization also influenced pedagogy, as a growing abundance of textbooks, guides, and reference volumes provided an established framework for the education of the young.

With only minor improvements, the technology of Gutenberg's printing press remained relatively unchanged until 1814, when the first steam-powered press, built by Friedrich König (1774?-1833), was installed to print the London *Times*. Thenceforth, during the Industrial Revolution, new and much faster powered presses became available and the printing revolution begun by Gutenberg was greatly expanded, further assisted by improved methods of transportation that aided the distribution of books and other reading materials. With superior tools of mass production, this period witnessed an astonishing proliferation of published works, beginning an overwhelming acceleration of knowledge production in all fields of study that continues to this day. During the late nineteenth and early twentieth centuries, the benefits of public education, libraries, and literacy were finally extended to the greater majority of society in the West. As the number of readers increased, so too did the uses of the printed text for mass consumption. Advertising became a became a major industry in the nineteenth century, as did the use of print for propaganda in various forms.

During the twentieth century the primacy of the printed word was challenged by several new mediums of mass communication—film, radio, television, and the computer. Though none of these technologies have effectively supplanted reading, the great popularity, efficiency, and emotional power of the modern multimedia experience has raised new questions about the future of the book and the mentality of print culture, including the meaning of authorship. While books require the active participation of a solitary reader who abides by the linear logic of the author's fixed text, video and computer presentations have altered the parameters of communication with nontangible, nonlinear montages of sound and imagery, including text, photography, and visual art, that often involve extensive creative collaborations. Some note that aspects of this new malleability and interaction seem to hark back to oral tradition, and that the interconnectivity of electronic media evokes the notion of a global village.

It may be said that the authority of the text—an authority that books once carried due to their relative expense, seeming permanence, and veneration as cultural objects—has been undermined by the use of computer technology to cheaply and instantaneously transmit vast quantities of electronic data around the world. This progressive affordability and expedience has also contributed toward a daunting over-abundance of information in the modern world, a defining feature of the so-called Information Age. Yet this same technology has also made the tools of mass communication more accessible than ever before, offering the prospect of a more free and equitable democratic society, along with the attendant risks of exclusion and manipulation. In many direct and indirect ways these freedoms—religious, political, social, intellectual, and technological—are testament to the profound and enduring impact of Gutenberg's press.

JOSH LAUER

Further Reading

Carter, John, and Percy H. Muir, eds. *Printing and the Mind of Man: A Descriptive Catalogue Illustrating the Impact of Print on the Evolution of Western Civilization During Five Centuries.* London: Cassell, 1962.

Eisenstein, Elizabeth L. *The Printing Press as an Agent of Change: Communications and Cultural Transformations in Early-Modern Europe.* 2 vols. 1979. Reprint (2 vols. in 1), Cambridge: Cambridge University Press, 1994.

Eisenstein, Elizabeth L. *The Printing Revolution in Early Modern Europe.* Cambridge: Cambridge University Press, 1996.

Febvre, Lucien, and Henri-Jean Martin. *The Coming of the Book.* Translated by David Gerard. 1958. Reprint, London: Verso, 1999.

Katz, Bill. *Dahl's History of the Book.* 3rd English ed. Metuchen, NJ: Scarecrow Press, 1995.

McLuhan, Marshall. *The Gutenberg Galaxy: The Making of Typographic Man.* 1962. Reprint, Toronto: University of Toronto Press, 1995.

Nunberg, Geoffrey, ed. *The Future of the Book.* Berkeley: University of California Press, 1996.

Steinberg, S. H. *Five Hundred Years of Printing.* 3rd ed. New York: Penguin, 1974.

The Advent of Newspapers

Overview

Before the invention of printing in the 1500s, people kept each other up to date with news and gossip through personal letters. Members of formal organizations accomplished the same function through handwritten newsletters. Following the arrival of print in the sixteenth century, news publishing went through four distinct stages before it evolved into what we would recognize as a newspaper. The very first examples were not sheets of two or four pages made up into columns but news books, or news pamphlets, looking a lot like a book, with a title page. Only in the seventeenth century did news begin to be distributed regularly and frequently.

Background

The first stage in the evolution of newspapers was publication of a single story, called a relation. This kind of publication recounted an event long after it had occurred. For example, in 1619 Nathaniel Newberry of London published a pamphlet under the title "Newes out of Holland," translated from the Dutch, that recounted details of a conspiracy.

A second genre of news publishing was known as the coranto. In England, Thomas Archer, Nicholas Bourne, and Nathaniel Butter published several such publications between 1620 and 1625. That year the government temporarily suspended publication of foreign news. After the restriction was dropped, in 1638 Bourne and Butter were granted the exclusive right to print news from abroad.

The coranto bore some resemblance to conventional newspapers in that it was published on a weekly basis, with some gaps. However, it had no single voice, and its front-page title would change from week to week. Sometimes editions were announced as a continuation of the previous week's edition. Corantos took their content from letters and foreign publications, a practice acknowledged on the title pages. For example, a coranto dated May 30, 1622 announces "Weekly Newes from Italy, Germanie, Hungaria, Bohemia, the Palatinate, France, and the Low Countries." The coranto's significance lay in its attempt to cover events the world over, and to establish itself as an authority on world affairs. During this time, short news items began to appear, called "broken stuffe."

The coranto evolved into the diurnal, which was a weekly account of several days' worth of events. In the aftermath of the Thirty Years War (1618-1648), news attention in England shifted to domestic issues. Robert Coles and Samuel Pecke had a large hand in producing English diurnals, and they tended to draw their copy from daily goings-on in Parliament. These diurnals often featured woodcut illustrations.

The fourth stage of early news publishing was also a book-like publication, called a mercury. It took its name from a Latin publication of the 1580s that covered affairs in central Europe. When civil war broke out in England, mercuries multiplied, and it was possible to buy one every day of the week, including Sunday. A contemporary publication to the mercury—slightly more official in tone—was called the intelligencer. John Thurloe (1616-1668), secretary of state under Oliver Cromwell (1599-1658), supervised such a publication during the second phase of the Civil War (1648). Its purpose was allegedly to "prevent misinformation." A singular feature of the intelligencer was that it marked a move toward coverage of a greater breadth of subject matter, and even to provide entertainment. An additional characteristic of news pamphlets published during the Civil War was that they dropped the book-style title page in favor of putting as much news as possible onto the front page.

The printing press used to generate the first regular news publications was a variation on Johannes Gutenberg's (c.1400-1468) device for moving type. This machine consisted of a wooden screw with a horizontal iron plate attached to its lower end. A page of type was placed on a movable flat bed that slid under the iron plate. Lowering the plate put enough pressure on the type to make a printed impression on a sheet of paper positioned over the type. The type was inked between each printing. In 1620, a Dutch innovation basically doubled printing capacity by modifying the machine to allow the iron plate to rise automatically after each impression. The next major advance in printing would not appear until the early 1800s.

Early printers were supervised by the state in a process called prior censorship. Prior censorship was a way authorities had of controlling what could and could not be published through licensing agreements. The degree of supervision varied from country to country and changed with time. In Italy and Germany, for example, both the Church and state were involved in censoring. In France, all publications were subjected to royal censorship up until the time of the French Revolution (1787-1799). In England, in 1538, the Royal Privy Council forbade publication of any book without its permission. A period of religious dissent in the late 1500s broke through this control. Although periodical publication of all news was banned in 1637, the beginning of the English Civil War in 1640 helped to cement the collapse of the censorship and control system.

Sporadic publication gave way rapidly to regular publication. In 1597, Samuel Dilbaum was printing a monthly newssheet. In Strasbourg in 1609, a bookseller brought out a monthly account of "noteworthy happenings." In Antwerp, Belgium, Abraham Verhoeven began publishing his *Niewe Tydinghe* in 1605. It appeared irregularly at first, but by the 1620s had reached three editions per week. Verhoeven's newssheet was influenced by the ornate style favored by the French, and featured engravings and maps. In 1610, both France and Switzerland had begun regular city papers.

The medium was adopted most enthusiastically in Holland and Germany. At first, the appearance of newspapers followed trade routes, and early papers were copycat versions of other papers. In 1650, a paper called *Einkommende Zeitung* (Incoming news) was produced—the oldest daily newspaper in the world. By 1626, Holland had 140 news publications.

In France, the physician Théophraste Renaudot (1586?-1653) was the first person to conceive of combining advertising with publishing the news. Widely traveled before accepting a post as commissioner-general for the poor, Renaudot was quick to spot the journalistic innovations of the time and to incorporate them in his experiments in communicating urgent information regarding "human life and society." In 1631, Renaudot obtained from Louis XIII (1601-1643) the sole right in France to "make, print, and to have printed and sold by those appropriate, news, gazettes and accounts of all that has happened and is happening inside and outside the Kingdom." This Renaudot did until the king died, after which an era of political instability set in that made publishing the news a hazardous occupation.

In England, the changes in attitudes wrought by the Civil War led emboldened newspaper publishers to flout legal prohibitions in recounting parliamentary affairs. Stimulated by a public appetite for domestic news, an unbridled flowering of journalism occurred in the 1640s that would not happen again for another century. Relations, corantos, diurnals, intelligencers, and mercuries proliferated, covering stories on marine monsters, battles, and the limits of permissibility in the press.

Early newspapers were not aesthetic objects. Material was crammed into eight to twelve pages, and the size of the typeface kept shrinking to accommodate more stories. Text was set in a single column, and there were few paragraph breaks and no subheads. But an increase in exports between the Restoration (1660) and the Glorious Revolution of 1688 that brought William III of Orange (1650-1702) and Mary II (1662-1694) to the throne translated into a market for both ordinary and financial information. In 1665, the first number of the *Oxford Gazette* appeared in a new format: a half-sheet printed on both sides, made up in two columns, topped by a title of a single line.

Impact

The growth of newspapers depended on a constellation of different developments. For example, at the start of the seventeenth century, a basic network of postal routes, printing capacity, and local distribution was already in place over much of Europe. An additional factor was the century's religious wars, which provided a demand for news. Moreover, permanent shifts in how families were structured, where people lived, and how aware they were of the world around them created a further desire for news that was evident even in the first corantos shipped from Holland in 1620.

Early printing was expensive in terms of human effort and the cost of making the metal letters and composing a page. But news publishing was a less precise affair than book printing, and printers depended on news to subsidize the high cost of making books.

The attitude of political authorities to the rise of newspapers was deeply ambivalent. On the one hand, so long as they maintained control over who could print what, governments could use the dissemination of news to their own ends. But the press could also turn politics upside down. The disappearance of biological succession in England resulted in a growing power of public opinion that was rooted in political factionalism. This development was helped by the rise of coffeehouses, where news was discussed and distributed. Additionally, after 1680, a penny post operated hourly within London to transport materials quickly between the suburbs and the city. By 1695, licensing was viewed no longer as an effective system of control but as an intrusion on good government. From then on, newspapers in England could expand and multiply as they liked, a freedom exceeded only by Holland as the seventeenth century drew to a close.

GISELLE WEISS

Further Reading

Hutt, Allen. *The Changing Newspaper: Typographic Trends in Britain and American, 1622-1972.* London: Gordon Fraser, 1973.

Smith, Anthony. *The Newspaper: An International History.* London: Thames and Hudson, 1979.

Advances in Firearms

Overview

The first firearms were small canons, the earliest of which was known as the *millimete canon,* dating from 1326. Public records from Florence in that same year that indicate a provision for guns in the protection of the town. Within about 30 years, there is evidence that handguns were being used for personal protection.

While early handguns had some distinct differences based on where they were made and where they were used, they had many things in common. They consisted of a wooden plank with an iron or bronze barrel at-

tached to it. The barrel was plugged at one end and a hole was drilled through the barrel, which was called a vent. The gun was loaded with powder and ammunition. The ammunition consisted of a ball, shot, or small stones. Finally, after the vent had been sprinkled with powder, the gun was ready to fire. The wood was tucked under the arm and a hot iron was applied to the vent to ignite the powder and discharge the weapon. This method proved to be unreliable in both the discharge and accuracy. In addition, there was tremendous recoil associated with the discharge, so improvements were necessary in design.

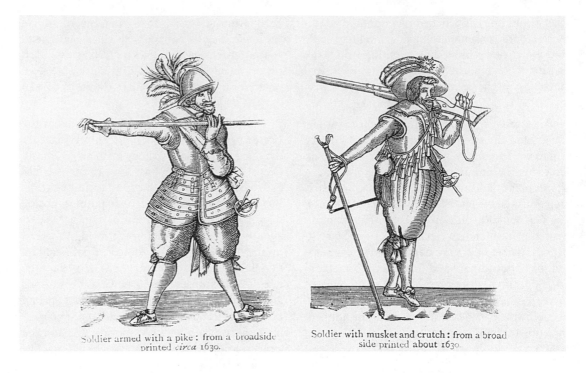

Soldier armed with a pike : from a broadside printed *circa* 1630.

Soldier with musket and crutch : from a broad side printed about 1630.

Seventeenth-century soldiers, one armed with a pike (left), the other with an early musket. *(Alan Towse; Ecoscene/Corbis. Reproduced with permission.)*

Background

The first major improvement to firearms came in the fifteenth century, when a hook was attached at a right angle to the barrel. This could be slung over a wall so that the shock from the recoil of the gun would be absorbed by the supporting structure. These weapons were given the name *hakenbuchse* (for hook), which eventually evolved into *harquebus* or *arquebus*. Further modifications markedly improved the use of the personal firearm.

An important early innovation was a simple design that allowed a lighted cord, rather than a hot iron, to ignite the powder. Initially, the cord was carried by hand, but the serpentine was added to hold the cord and act as a trigger. This freed the gunman from having to be near a hot iron and greatly increased the aim of the weapon. Improvements continued as the vent was moved to the side, then covered to protect the powder from the elements.

The *matchlock* soon replaced the early serpentine lock. This mechanism consisted of a trigger, an arm holding a smoldering match, a pin connecting the trigger and arm, and a mechanical link that opened the vent as the match descended to ignite the powder. This step proved to be an important one and opened the door to modern gun design. Another refinement

was the snap matchlock. This was a spring that drove the matchlock down to the vent. At the same time, the gunstock changed to reflect the increased ability to aim the gun. Rifle butts were enlarged, allowing them to be placed against the shoulder, which absorbed most of the recoil.

These simple firearms were inexpensive to build and operate. They did, however, have certain disadvantages. Because they always needed to be lit, the matchlock made stealth nearly impossible, and harsh weather could severely compromise its effectiveness.

The *wheel lock,* an ancestor of the cigarette lighter, was probably developed in Germany around 1515. It contained a serrated iron wheel that struck a spark to ignite powder on the vent. This happened when the iron wheel was rotated against a shard of flint or a piece of iron pyrite to form sparks. This was a fairly complicated mechanism, so despite the fact that it was an improvement over the matchlock, it saw little use in the military. It did become the preferred weapon for private citizens, however. It also allowed the firearm to be concealed. There is evidence that as early as 1518 there were laws passed in Europe to forbid carrying concealed weapons.

Since the military was a large market for guns, it was important to design one that was cheaper than the wheel lock, but better than the matchlock. The *snaphaunce* was invented for just this purpose in the late sixteenth century. It consisted of a cock (the hammer in a gun) and a frizzen (striker) that when brought together caused a spark; it then opened the firing chamber to the spark, causing the gun to fire. This design was modified in the early seventeenth century to become the *flintlock*.

The flintlock quickly became the gun of choice for the military. It was simple, reliable, strong, easy to repair, and most of all, inexpensive. These guns were also much faster to load. The flintlock had a one-piece frizzen and pan cover that when triggered, made the frizzen strike the flint, showering sparks onto the gunpowder in the priming pan; the ignited powder, in turn, fired the main charge in the bore, propelling the ammunition. This design seemed to be an adaptation of the tinderbox used at the time to start fires—the flint and the steel striking each other created the spark. A further development was the prepared cartridge. This put the bullet and powder together, premixed and ready to fire. The flintlock remained the most widely used gun until the percussion lock was developed in the early nineteenth century.

Impact

Early guns like the matchlock made modern small arms possible, with obvious ramifications for both the military and the general public. But they were not immediately adopted by the armed forces. They were expensive, unreliable, and cumbersome, weighing between 15 and 20 lbs (7 and 9 kg). As the guns evolved over time, however, their use became more widespread, and adaptations were made to improve its performance and utility.

As guns gained greater acceptance in the military, their use changed fighting tactics. While swords remained the primary weapon until the eighteenth century, soldiers with guns began to accompany swordsmen. Men riding horses started to use pistols in conjunction with swords as part of their attack. In the heat of battle, gun technology often made the difference between victory and defeat, as shown during the Age of Exploration. Much of the Spanish conquest of America was, in part, due to superior weaponry. Francisco Pizarro's (1470?-1541) army, for example, which numbered fewer than 200 men, used arquebuses to kill nearly 5,000 Inca warriors while sustaining only one casualty. This battle, which lasted less than an hour, began the fall of the Inca empire.

Guns have had profound and subtle effects on society. They afford a great deal of protection and allow people to defend themselves, but, like many forms of technology, can also be used for evil purposes. Ever since the gun was invented, people have argued over whether their benefits outweigh their risks. As early as 1588, Toyotomi Hideyoshi (1536-1598), emperor of Japan, banned citizens from possessing any type of weapon, including firearms. The military, however, were armed with guns, largely to protect the nobility against a peasant uprising. Because the government was able to regulate firearms, many people turned to makeshift weapons, substituting one weapon for another. It remains to be seen what course modern society will take.

JAMES J. HOFFMANN

Further Reading

Chant, Christopher. *The New Encyclopedia of Handguns & Small Arms*. New York: Smithmark Publishing, 1995.

Diamond, Jared. *Guns, Germs, and Steel: The Fates of Human Societies*. New York: W. W. Norton & Company, 1999.

Hogg, Ivan. *An Illustrated History of Firearms*. New York: A & W Publishers, Inc., 1980.

The Military Revolution

Overview

War is a characteristic of virtually every human society and civilization in nearly every era of human history for which some sort of records exist. However, until the fifteenth century most military conflicts were fought using largely the same weapons and tactics as those of Alexander the Great (356- 323 B.C.). Several inventions between the fifteenth and eighteenth centuries changed this, leading to a revolution in the way wars were fought. This set in motion many trends in warfare that have continued to this day, such as the concept of total war, the almost exclusive use of guns, and tactics better suited for the age of guns.

Heavy artillery at the beginning of the sixteenth century. *(Bettmann/Corbis. Reproduced with permission.)*

Background

Mankind is almost unique among animals in consciously waging war. Through all of recorded human history, accounts of warfare are common, and archeological evidence exists confirming the presence of warfare deep in prehistory. For most of human history, in fact, until the fifteenth century, warfare changed very little. Men armed with sharp objects (typically swords, arrows, or pikes) would try to kill each other. Whichever side lost so many men that it could no longer fight cohesively would lose the battle. The major innovations in warfare were the use of chariots by the Romans and the use of cavalry by many powers.

In general, battles were waged by foot soldiers, with mounted soldiers (knights, in medieval Europe) adding mobility, all under the di-

rection of a king, general, or other highly placed leader. In this brand of warfare, the rules were fairly well fixed, tactics were well known, and there was little in the way of strategic planning. That is to say, wars were fought one battle at a time, mostly without coordinating the activities of various armies, industry, transportation, and with no overall plan for arranging the downfall of the enemy beyond destroying its armies on the field of battle. And, necessarily, most fighting was done in hand-to-hand combat because archers were few and their range limited. These general rules began to change in the mid-fifteenth century, and the changes accelerated and became more profound with time.

One of the first innovations was the introduction of firearms on a large scale. Another was the use of large guns (artillery) to bombard cities, fortifications, and armies from a distance. At sea, the change from galleys (which were rowed rather than sailed) to larger and more powerful sailing ships turned naval warfare into a strategic weapon. Another innovation was mounting large guns on ships. This was first done by simply cutting holes in the sides of ships and putting guns behind them. Over time, this evolved into ships with two or three gun decks that might carry 100 guns or more. This, in turn, caused ships to become important weapons for fleet actions, raiding merchant ships, and bombarding shore facilities in support of armies ashore.

As these technological changes became entrenched, the Swedish army developed revolutionary new battlefield tactics, quickly adopted by other European powers, that maximized the utility of guns while rendering the horse-mounted knights obsolete. In fact, the Swedish army of the 1650s was perhaps the first army that used tactics significantly different from those of Alexander the Great. They were also the first army in nearly 2,000 years that could have defeated Alexander in battle.

At the same time, power was becoming more centralized, with national governments (albeit still monarchies) assuming power previously held by many smaller nobles. As this happened, larger armies became the norm, although many nations relied on mercenaries (especially the Swiss) to fight wars for them.

As a side note, Japan's retreat from the use of firearms in battle must be mentioned. The Japanese used firearms enthusiastically for over a century, achieving a proficiency that exceeded that of European powers. However, the ruling Samurai class realized that a common peasant with a gun could kill a highly trained warrior. They understood that this could threaten the Samurai's place in Japanese society and decreed that guns could only be manufactured for the government, then failed to purchase any. As a result, as late as the nineteenth century Japan had no indigenous firearms industry and relied on essentially medieval weaponry.

By the dawn of the eighteenth century, most of the innovations noted above had become incorporated into European military strategy. Although technology continued to improve, the world caught its breath for a century or so before the next round of rapid change again transformed the face of warfare.

Impact

These dramatic innovations had a significant impact on the manner in which wars were fought in Europe and, as they spread, throughout the world. In addition, they helped change the global balance of power in favor of Europe, which helped Europe financially dominate most of the world for several centuries. Finally, changes in weaponry changed the nature of nations and the way they fought wars, as well as helping to usher in the end of the feudal system in Europe.

Perhaps the most obvious impact of the innovations mentioned above involves the manner in which wars were fought. Extensive use of firearms was the final step in driving the knight and chivalry into oblivion. With this, and the advent of standing national armies, the stage was set for introducing standardized tactics, military drills, war games, and the like. This, in turn, started to turn soldiering into a profession, as it remains in many nations. A corollary of this trend was that professional (or at least highly trained conscripts) could fight more effectively and more efficiently than swarming masses of peasants who were given a pike and told to stab at anyone from the other side. Warfare again began to involve military maneuvers and more sophisticated tactics, as had the Roman legions and the army of Alexander the Great. These tactics had fallen into disuse in the intervening centuries because the large masses of untrained soldiers were simply unable to learn the techniques, and the large numbers of smaller fighting forces (consisting of the men each knight or nobleman brought with him) tended to preclude coordinated movements on the battlefield.

In addition, more highly trained armies were expensive. This not only led to somewhat smaller army sizes, but also made each soldier more valuable because of the training that had gone into making him an efficient cog in the war machine. In turn, this made commanders less willing to risk their men's lives because they knew that each casualty would require replacement, which was often a very junior, inexperienced, and poorly trained soldier. Armies did everything they could to avoid high losses.

At sea, warfare also changed dramatically, to the point that sea power became indispensable to the Dutch, Spanish, and British and was very important to the Portuguese and French. As noted above, the chief breakthrough lay in realizing that ships could hold large guns. This dictated a sturdier, stronger vessel to hold the gun and withstand its recoil when discharged. The larger ships became nearly impossible to row, making them dependent on their sails for propulsion. This, in turn, vastly increased their range, although at the cost of speed and maneuverability. Adding larger and more numerous guns required even larger ships, and it quickly became apparent that more guns resulted in a greater chance of victory in battle. With all of these developments, naval tactics changed dramatically as commanders strove to take advantage of the positive characteristics of their new platforms. With all of this, ships became increasingly capable of destroying other warships, raiding enemy merchant vessels, and bombarding enemy shore facilities. These new offensive capabilities served to turn naval war into a legitimate strategic weapon. This, in turn, helped small seagoing nations such as Britain and Holland become major powers, even though their armies were often outnumbered on the field of battle.

Europe's technological and economic advantages, developed during this period, enabled various European nations to dominate much of the rest of the world for several centuries. Superior firepower was combined with advanced tactics and an appreciation for increasingly wide-ranging strategy, a combination that was used to defeat most of the armies Europeans faced for several centuries. This, in turn, gave Europe a decided financial advantage over most of the world, an advantage that largely remains to this day. In fact, of the major economic powers, only Japan is neither European nor a former European colony.

Finally, changes in the nature of warfare ushered in by these technological innovations led to changes in the structure of nations. As noted above, new weapons led to new tactics, which led to increasing emphasis on foot soldiers at the expense of heavy cavalry. In effect, this caused much of the aristocracy to lose prominence in the army. At the same time, forming large standing armies led to the elimination of the smaller armies formerly maintained by the aristocracy. To help remedy feelings of discontent, many nations developed a practice of choosing officers entirely from the ranks of theses aristocrats. This not only provided a corps of leaders who were in positions of social authority, but also helped turn the aristocrats into paid employees of the central government instead of relatively independent leaders with private armies. By so doing, the relatively new central governments helped eliminate a potential threat to their authority.

Most of these new weapons and trends emerged in the fifteenth century and were in full swing by the eighteenth century. Once incorporated into military and political strategy, they tended to remain relatively constant for the next 150 years, until the American Civil War.

P. ANDREW KARAM

Further Reading

Dyer, Gwynne. *War.* New York: Crown, 1985.

Keegan, John. *The Price of Admiralty.* New York: Viking, 1989.

Inventing the Submarine

Overview

In 1623 Dutch inventor Cornelius Drebbel (1572-1633) invented the first submersible that could remain underwater for an extended period of time, be propelled through the water, and be steered. Although this invention was not capitalized upon for more than two centuries, Drebbel's submersible marked the first step towards submarine warfare and caused quite a stir in many circles at the time. Today, nearly four centuries later, the submarine is a powerful tool for research and a potent weapon in war.

Background

As legend has it, in 332 B.C. Alexander the Great descended to the bottom of the sea in a glass diving bell, accompanied by two companions and lunch. Although likely a legend only, this is the first record of anyone entering the water for longer than they could hold their breath, and it was not to be repeated (or at least, not written about) for nearly 2,000 years.

The next mentions of submarines (or, more properly, submersibles) was not until Leonardo da Vinci (1451-1519) mentioned a military diving system in the late fifteenth century, although he gave no details because of "the evil nature of men who practice assassination at the bottom of the sea." Following the passage of a few more centuries, William Bourne (1535-1583) described the principles by which a ship could operate submerged, although he did not propose building such a vessel or provide any drawings for one.

After Bourne, only another 40 years were to pass before Drebbel's invention made its appearance.

Drebbel made a few significant advances, some of which are unfortunately lost to us because of his penchant for guarding his secrets. Perhaps the most important of these was his apparent discovery of a method for replenishing the atmosphere of his tiny vessel while remaining underwater. This was important on a few levels. For starters, Drebbel was one of the first to realize that part of the air is necessary for life and the rest is not. We now know that about 20% of air is comprised of oxygen, termed the "quintessence" by Drebbel. It is unlikely that Drebbel actually succeeded in separating oxygen from air; this was not to be accomplished for another century. However, it is possible that he stumbled upon a method of scavenging carbon dioxide from the air; prolonging the time a vessel could remain submerged. Unfortunately, while there seems little doubt that Drebbel did make use of a chemical reaction to do so, exactly which chemicals he used is not known. Many modern "scrubbers" use either lithium hydroxide or complex chemicals for this task; however, these did not exist in the seventeenth century and could not have been used by Drebbel.

There is some doubt, too, that Drebbel's craft actually submerged fully, as a modern submarine does. The scheme described for submerging the boat, by changing the volume of water contained within goatskin sacks inside the boat, seems implausible at best. It is more likely that Drebbel designed the craft to float with the upper deck just awash, counting on the vessel's forward momentum to carry it beneath the water, in a manner similar to that used by many submarines today. Yet another innovation was the ability to propel and steer the craft. This was successful to the point that, according to contemporary accounts, it "could be rowed and navigated under water from Westminster to Greenwich, the distance of two Dutch miles: or even five or six miles, or as far as one pleased." It was demonstrated to a number of people, and Drebbel may have even built additional craft. And, invariably, one of the first thoughts was about the military advantage that could be enjoyed by any nation building a fleet of submarines.

Impact

Since the first voyage of Drebbel's machine over three centuries ago, the submarine has exerted three primary impacts on science and society. These involve military advantage, scientific exploration and discovery, and inciting the public imagination.

"Assassination at the bottom of the sea" was probably the first and most important outcome of developing a successful submarine. The immediate impact of Drebbel's invention was to spur a flurry of activity among submarine designers, both serious and crackpot, each trying to find some way to exploit this invention for purposes of warfare or exploration. One of the first recorded uses of submarines in warfare was David Bushnell's (1742?-1824) boat, the *Turtle,* which made a total of three unsuccessful attacks on British warships during the American Revolution.

In subsequent years, many nations experimented with submarines, including the French, Germans, British, and Americans. All nations shared several problems: keeping the air fresh, navigating underwater, and affixing explosives to the hulls of ships (or finding some other way of causing damage). And all struggled with these problems for similar reasons: because the potential military value of a warship that could approach and sink a vessel undetected was so great that it was worth the investment in time and money.

Eventually, around the beginning of the twentieth century, these problems were either solved or sidestepped. The German navy took a terrible toll on Allied shipping during the First World War, forcing the Allies to learn some lessons, although they were largely forgotten by the start of the next great war. However, the ability of German submarine wolfpacks to interdict Allied shipping during the Second World War, and the similar impact that U.S. submarines had on Japanese shipping during the same war, forced military strategists to make future plans with submarines in mind. The further refinement of submarines by the addition of nuclear reactors, nuclear weapons (in some cases), air regenerating equipment, and advanced diesel engines (in some cases) turned them into one of the most formidable—and revolutionary—naval and strategic weapons of the twentieth century.

Submarines and submersibles have also seen much use in oceanographic and marine biological explorations for at least 200 years. Edmund Halley (1656-1742) described some excursions to the sea floor in a diving bell in the latter part of the eighteenth century, and the use of bathyspheres and bathyscaphs also came into practice. Submarines have played, and continue to play, a vitally important role in the exploration of the ocean and its inhabitants.

This exploration, in turn, has led to a vastly increased understanding of the Earth, the oceans, and the role played by the oceans and their inhabitants. For example, the theory of plate tectonics was developed based on data from both the land and seas. However, some of the most interesting information confirming and elaborating on this theory depends on marine data. Research submersibles have returned with photos of "black smokers," where the oceanic crust is being formed. These same submersibles have brought back observations of extensive colonies of bacteria, tube worms, and other organisms living completely divorced from the surface—the first ecology found on Earth that did not rely, even indirectly, on solar energy for sustenance. Other submersibles have returned with countless geologic and biologic specimens, archeological data, sunken treasure, photos of the *Titanic,* and more.

Submarines do not seem to have appeared in the consciousness of the general public until several centuries later, when Jules Verne wrote *20,000 Leagues Under the Sea* (1869-1870), but this and the use of unrestricted submarine warfare by the German navy in the First World War quickly brought them to the attention of the public. In fact, in the world of the seventeenth century, Drebbel's submarine was the talk of the town, assuming the "town" consisted of those with enough spare time to watch or talk about it, or those who were sufficiently literate to read the accounts. Although those who were informed (or could read) found it important and fascinating, they comprised a small minority of the total population. Most people lived a bookless life of hard work, narrow perspectives, and little or no formal education. In addition, in spite of the stir caused by Drebbel's submarine in Europe, Europe is only a small part of the world. In Asia, Africa, Australia, and the Americas, the inhabitants (largely tribal aboriginal populations, with the exception of parts of Asia) were oblivious to happenings in Europe. Drebbel's submarine caused a stir among a small group of the educated elite in a few of the nations of Europe, and had absolutely no impact whatsoever on the rest of the world.

As time went on, however, technology caught up with the sporadic military interest in submarines. Once a successful military weapon was produced and used, the public could not help but pay attention. The romance and mystery of submarines, which existed primarily because they could attack unseen from beneath the waves, caught the attention of an increasingly literate and engaged public. This interest was enhanced during World War II, with the successes of the German and, later, American submarines. The publicity surrounding the development of nuclear submarines and, later, ballistic missile submarines further encouraged public interest in these ships, as did popular books and movies, such as *Ice Station Zebra, Run Silent, Run Deep, The Hunt for Red October,* and others. Although slow in starting (over 300 years passed between Drebbel's craft and *Ice Station Zebra*), the public seems fascinated with submarines and their uses, both military and scientific.

It is likely that Drebbel would have expected submarines to have enormous military value, although he would probably not have imagined the missions they have turned out to have. However, it is likely that he would be amazed or disbelieving at the range of scientific applications developed for them, and he would likely be even more amazed at the level of interest on the part of the general public.

P. ANDREW KARAM

Further Reading

Harris, Brayton. *The Navy Times Book of Submarines.* New York: Berkeley Books, 1997.

The Invention of Spectacles

Overview

The human body has evolved in a rough and ready response to environmental demands. The eye is no exception to this rule, for very few people have perfect vision. The many with imperfect vision fall into two general groups, the farsighted who have trouble focusing on near objects (presbyopes) and the nearsighted who have poor vision beyond a very short distance (myopes). Presbyopes require spectacles with convex lenses that curve outwards on both surfaces, myopes need lenses that are concave, or curve inwards. Attempts to correct human vision are as old as human society. The oldest known lens, of polished rock crystal, was found in the ruins of ancient Nineveh. The classical Roman writer Seneca is said to have read all the books in Rome by using a glass globe of water to enlarge the handwritten letters. However, spectacles in some way shaped to sit on the face, (as opposed to eyeglasses, which are held to the eye but are not on the face for long periods), seem to have been an invention of medieval Europe. Over the period from 1450 to 1700 they helped to change the way in which Europeans both perceived the world and operated within it.

Background

The first spectacles contained convex lenses and were only effective for presbyopia or farsightedness, which normally occurs around the age of 40 as the crystalline lens of the eye hardens. They may have developed from "reading stones," segments of glass spheres used by presbyopic monks to read manuscripts by holding the glass against the letters. Convex spectacles seem to have evolved by chance, not through optical theory, although medieval Europe had acquired some scientific knowledge of optics from the Muslim world. The Muslim mathematician and natural philosopher Ibn al-Haitham (c.965-1039), called Alhazen by Europeans, wrote about the properties of lenses in a work translated from Arabic into Latin in 1266. In 1267 the English monk and scientist Roger Bacon (c. 1214-1294) wrote about his experiments in using convex lenses to correct vision, advocating their use to help old people.

Convex spectacles seem to have been invented around 1285. The first reference to spectacles dates from 1289, in a manuscript written about the Popozo family from Tuscany, Italy. The Dominican Friar Giordano da Rivalto of Pisa, in a sermon delivered in Florence in 1305 or 1306, claimed that spectacles had been invented nearly 20 years earlier, and that he had met the unnamed inventor. Historians differ over whether Florence or Pisa was the site of this invention, though it has been considered so important that for centuries patriotic historians of Italian cities have altered manuscripts and invented evidence to claim the prestige of the invention for their city. Venice became an early center of the mass production of lenses. The glass blowers of the Venetian suburb of Murano produced thicker and clearer glass, better for grinding high-quality lenses, than elsewhere in Europe, just as they

had produced the best "reading stones." The Venetian crystalworkers' guild laid down regulations for producing "glass discs for the eyes" in 1301, and by c. 1320 a guild of spectacle-makers existed there.

These earliest spectacles had no sidepieces, or temples, the invention of which in 1725-1730 is attributed to Edward Scarlatt of London. Medieval spectacles were riveted at the center and had leather grips to hold on to the bridge of the nose. Some pictures show readers holding spectacles on the face by hand, and some frames were made of leather to reduce their weight. By the 1360s the early Renaissance writer Petrarch could refer to spectacles for the elderly as if they were commonplace in Florence, and in paintings of this period and the fifteenth century they are often included in portraits of saints and scholars to signify piety and learning. By the late fifteenth century their use had spread so far outside the elite that artists increasingly used them to signify folly or senility.

Making concave lenses was more difficult than convex ones, which evolved naturally from the magnification observed through convex reading stones. Any convex lens will magnify, and even inaccurate medieval lenses helped presbyopes. However, concave lenses have an important relationship with literacy, since they enable shortsighted myopes to read even small letters and to write more clearly themselves. The invention of concave lenses is sometimes attributed to Nicholas Krebs (1400?-1464), better known as Nicholas of Cusa, a Cardinal, senior politician, and diplomat of the Roman Catholic Church, who wrote on philosophy, theology, and science. In a treatise of 1450, *De beryllo* (Concerning Beryl), Nicholas described that semi-precious, sometimes transparent stone, "to which a concave as well as a convex form is given; by looking through it you reach what was previously invisible." Even though *beryllo* was a Latin word used at the time for spectacles, the stone itself was used as a magnifying glass rather than in spectacles, for which cheaper and lighter concave glass lenses were by then available.

Nicholas, however, may have been aware of the recent development of concave lenses in Florence, through his Florentine friend Paolo Toscanelli (1397-1482). The need for literacy among the merchant and political elite of Florence probably led to the discovery of concave lenses by an anonymous inventor before 1450. Certainly by 1451 Florence had overtaken Venice as the manufacturing center for high-quality spectacles, including concave lenses for myopes, which are first mentioned in a letter of August 25 that year. On October 21, 1462, the Duke of Milan ordered three dozen spectacles from his resident ambassador at Florence, including a dozen with concave lenses for improving distance vision. Although the order was filled within days, suggesting a large-scale manufacture and a stock of lenses ready to be ground by several specialist shops, these were not prescription lenses in the modern sense. However, the Florentines understood that vision declines with age and made convex lenses in different strengths for five-year age groups from 30 years old, and con-

in the early fifteenth century of the principles governing linear perspective, by the Florentine artists Filippo Brunelleschi (1377-1446) and Leon Battista Alberti (1407-1472). "Perspective" at this period meant mostly "optics," a branch of mathematics, and Brunelleschi and Alberti used geometry to create the illusion of space and distance on a flat or curved surface. This is also connected with map-making, the perspective projection of the curved surface of Earth onto a flat surface, practiced by Paolo Toscanelli, who was perhaps the source of Nicholas of Cusa's knowledge of Florentine optics.

LATER DEVELOPMENTS IN SPECTACLES

The development of sidepieces, or "temples," by 1730 allowed a further change to spectacles in the 1780s. The human eye as it ages adjusts its focus less rapidly when moving from close work to distance vision, and may need a combination of both concave and convex lenses to see distant and near objects properly. However, the eye can adjust to using such lenses, called bifocals, when they are held at sufficient distance from the eyes along the nose by the "temples." The first craftsman to perceive the need for bifocals and to find a solution to their construction was typical of Western civilization in that he was an inveterate scientific experimenter and tinkerer with machinery, always seeking ways to improve upon received knowledge. Realizing that a combination of convex and concave lenses in one spectacle frame would enable his aging eyes both to work close up with machinery and adjust to seeing at a distance, he worked out a way of grinding such lenses on a lathe and putting them into a frame. His name was Benjamin Franklin.

GLYN PARRY

Impact

The invention of concave and convex spectacles aided the revival of learning in the early Renaissance, adding years of eyesight to older readers. The increasing refinement of spectacles at the moment when printing was invented in 1450 may be no coincidence. The rapid spread of printing increased the supply of books and increased the number of readers, enlarging the market for spectacles, encouraging cost cutting and the production of specialized lenses. In turn the availability of spectacles allowed more readers to read small pages of small type, allowing printers to reduce the cost of their books by using less paper. The spread of printing accompanied the spread of spectacles, which raised standards of book and manuscript production by placing more emphasis on fine workmanship, accuracy, neatness, and detail, for example in miniature illustrations in manuscripts.

This carried over into other aspects of European society. Effectively, spectacles doubled the skilled craft workforce by doubling the working life of skilled craftsmen, especially those who did fine jobs—scribes and readers, instrument and toolmakers, metalworkers, and close weavers. The demand for precision-ground lenses encouraged improvements to the basic lathe, as instrument makers required precision parts. So conversely, spectacles pushed Europe towards the invention of precision instruments found nowhere else. Europeans further developed the crude instruments borrowed from other societies to create a range of gauges, micrometers, and many other tools linked to precision measurement and control. This process led to the establishment of a machine-tool industry, of machines to make other machines, and continues in the precisely fitted parts of modern articulated machines. Thus Europe could move to skilled replication, batch and then mass production of identical goods that did

cave in two strengths. Lenses bought by the Duke of Milan were ground for distant, near, and normal sight. The latter suggests that spectacles had become a fashion accessory at the Milanese court, perhaps to make courtiers look more intellectual, and perhaps also because the elderly Duke and his young Duchess both wore glasses with concave lenses for their myopia.

Venice challenged Florence in the large-scale production of quality spectacles, which also spread to Germany and other parts of Europe. However, Florence may have gained its advantage in optical research through the rediscovery

not rely on skills learned over long apprenticeship, but on finely made machines.

Spectacles magnified the authority of the eye and allowed people to challenge received authority on the basis that seeing was believing, after overcoming some initial resistance. George Bartisch in the first book on eye diseases, entitled *Ophthalmodouleia* (The service of the eye, 1583), could not imagine how an imperfect eye could see better through a lens. Contemporary theory held that the eye emitted a visual spirit, which created vision when reflected back from objects. Only after Galileo's (1564-1642) improved telescope of 1609 encouraged Johannes Kepler (1571-1630) to formulate modern optical theory in his book *Dioptrice* (Concerning measuring altitude, 1611) did the eye become accepted as an optical instrument that received light rays reflected from objects. Thereafter, the cosmic world seen through telescopes, and the microcosmic world seen through microscopes, would be used to develop entirely new systems of scientific knowledge in the seventeenth century.

GLYN PARRY

Further Reading

Books

Derry, T.K., and Trevor I. Williams. *A Short History of Technology*. Oxford: Oxford University Press, 1960.

Dreyfus, John. *Into Print: Selected Writings on Printing History, Typography and Book Production*. London: The British Library, 1994.

Landes, David S. *The Wealth and Poverty of Nations: Why Some Are So Rich and Some So Poor*. New York: W.W. Norton, 1998.

Mumford, Lewis. *Technics and Civilization*. London, 1934.

Singer, C., E.J. Hall, A.R. Holmyard, and Trevor I. Williams. *A History of Technology*. 8 vols. Oxford: Oxford University Press, 1954-84, vol. 3 (1957).

Periodical Articles

Ilardi, Vincent. "Eyeglasses and Concave Lenses in Fifteenth-Century Florence and Milan: New Documents." *Renaissance Quarterly 29*, no. 4 (Winter 1976): 341-60.

Rosen, Edward. "The Invention of Eyeglasses." *Journal of the History of Medicine 2*, no. 1 (January 1956): 13-46 and no. 2 (April 1956): 183-218.

Camera Obscura: Ancestor of Modern Photography

Overview

Capturing an image from life was long ago the sole proprietorship of the skilled artist, whose brushstrokes precisely recreated portraits of man and landscape on canvas. That art is now shared by anyone who cares to peer through a camera's viewfinder and snap the shutter. Modern photography had its start in the 1800s, with the invention of the Dageurrotype and Englishman W. H. Talbot's (1800-1877) negative-positive development process. But its roots can be traced to centuries earlier, to a simple mechanism—a dark room, in which light passed through a pinhole, projecting an image on the opposite wall. It was called *camera obscura*.

Background

More than 2,000 years before the invention of the camera obscura, its earliest predecessor came to light in ancient Greece. In 500 B.C., the philosopher Aristotle (384-322 B.C.) discovered that by passing sunlight through a pinhole, he could create a reversed image of the Sun on the ground. He used this device as a means for viewing an eclipse without having to stare directly into the Sun.

Aristotle's experiments went no further. He could not explain why the image was created, or why it was in reverse. In 1035 an Egyptian scientist named Ibn al-Haitham (965-1039) continued Aristotle's work, by first devoting himself to the understanding of what makes up light. He had a theory that light traveled in straight lines, called rays, and set out to prove his theory by arranging a line of candles on a table, lighting them, then standing behind a screen that separated him from the candles' light.

Al-Haitham pricked a hole in the screen and watched as the light rays passed through it. The image from a candle placed at the left side of the hole would pass through and strike the wall to its right, and vice versa. Light from the top of the candle flame would strike downward of the hole. Thus, he proved that light rays move in

A depiction of the camera obscura's use. *(Library of Congress. Reproduced with permission.)*

straight lines. He recorded his findings in a book titled *Kitab al-Manazir*. Centuries later, the book found its way to Europe.

In Italy, inventor and artist Leonardo da Vinci (1452-1519) caught word of al-Haitham's work and decided to try his own experimentation using mirrors, lenses, and pinholes. In around 1510 he wrote in his notebooks:

> *When the images of illuminated objects pass through a small round hole into a very dark room...you will see on the paper all those objects in their natural shapes and colors. They will be reduced in size, and upside down, owing to the intersection of the rays at the aperture.*

In Italian, the name for darkened room, *camera obscura*, became synonymous with the projection of light through a small hole. In 1521 one of da Vinci's students, Cesare Cesariano (fl. 1520s), published the first description of the camera obscura, but it was not widely read. The public did not gain knowledge of this new device until an account was written more than 30 years later by Italian nobleman Giovanni Battista della Porta (1535-1615). He described the process for assembling the camera obscura in his book *Magiae Naturalis* (Natural magic, 1558):

> *You must shut all the Chamber windows, and it will do well to shut up all holes be-*

> *sides, lest any light breaking in should spoil all. Only make one hole...as great as your little finger....*

When the Sun shone through the hole, an image would appear "and what is right will be the left, and all things changed."

Impact

Della Porta was the first to be able to manipulate the image, through the use of a camera obscura fitted with convex lenses and mirrors. A concave mirror would enable the device to reflect an image right side up. He suggested that artists could project scenes from nature on a piece of paper to assist in the rendering of their works.

Thrilled by his new invention, della Porta summoned his friends and important members of Naples society to his home for a demonstration. Instead of sharing his excitement, the group was appalled when they saw real human images displayed on the wall, believing it to be the work of witchcraft. The Catholic Church got wind of della Porta's demonstration and promptly charged him with sorcery. His work was banned for six years.

This did not usher in the end for the camera obscura, however. The *Magiae Naturalis* was disseminated throughout Europe, and the device became a novelty item across the continent.

Meanwhile, an interest in optics exploded throughout the European scientific community. In 1604 German astronomer Johannes Kepler (1571-1630) studied the mathematical laws governing mirror reflection, and seven years later worked out the theory of lenses. Scientists were unknowingly providing an invaluable service to the artistic community.

Throughout the next century, several improvements were made to the camera obscura. A German mathematics professor, Daniel Schwenter, discovered that by drilling a hole through a wooden ball and placing a lens at either end, he could focus images in any direction upon a wall. Scientist Friedrich Risner (d. 1580) invented a portable version of the camera, housed in a collapsible tent. The only problem was, the operator still had to climb inside the clunky device.

Johann Zahn (fl. 1680s), a German monk, solved that dilemma by inventing a camera obscura that was just 9 inches (22.86 cm) high and 24 inches (61 cm) long. Inside the box was a mirror placed at a 45-degree angle to the lens. The mirror reflected the image to the top of the box, where he had placed a sheet of frosted glass. The glass was covered with tracing paper, allowing images to be easily copied by an artist. Zahn's design would remain in use for nearly 200 years.

Further assisting the artist was the invention of the *camera lucida* by William Hyde Wollaston (1766-1828) in 1806. This was no actual camera, but rather a glass prism suspended at eye level from a brass rod. This device enabled even the most untrained artist to trace an image on a piece of paper, by allowing him to view his subject and the paper at the same time. But even the camera lucida required the skilled hand of an artist, and many more people clamored for the chance to capture reality.

The camera obscura had made it possible to successfully project an image—the next challenge was to make that image permanent. The first attempts at freezing an image were made by Tom Wedgwood (son of the famous pottery maker Josiah Wedgwood) in 1800. Drawing on his understanding of the camera obscura, as well as the discovery by German natural philosopher Johann Heinrich Schulze (1687-1744) that silver salts are light sensitive, he began experimenting by coating paper with silver nitrate, then placing it in a camera obscura and exposing it to the Sun's rays. His experiments were unsuccessful—the image refused to hold on the paper. Wedgwood abandoned his work due to poor health.

In the early 1800s French printer Joseph Niepce (1765-1833) achieved a greater degree of success. He was convinced that if he could coat a surface with light-sensitive chemicals, he could recreate an image by exposing that surface to the Sun. Niepce placed an engraving between a plate coated with a substance called bitumen of Judaea, and a sheet of clear glass. The resulting images of the engravings he called *heliographs.*

In 1824 Niepce switched from copper and zinc plates to a pewter plate, which he coated with a special light-sensitive varnish and placed into a camera obscura. He set the camera in the window of his home and opened the aperture. He left it in the Sun for a full eight hours.

When Niepce returned, he removed the pewter plate and dipped it into a chemical bath. The softest portions of the varnish softened, while the hardest sections remained. While indistinct, the image revealed was unmistakably the view out of his window in Saint-Loup-de-Varennes. He could see the outlines of buildings, roofs, and chimneys. It was the world's first permanent photograph.

In 1829 French scene artist Louis Jacques Mandé Daguerre (1787-1851) formed a partnership with Niepce, believing that he could improve upon the latter's heliographic process even further. Daguerre discovered that he could create an image on plates coated with silver iodide, which could then be "developed," that is, permanently affixed to the plates, with the use of sodium chloride (salt). Exposure time quickly dropped from eight hours to a mere (at the time) 30 minutes. His photographic process, which he called "daguerreotype," proved the next major step in the evolution of modern photography.

The final piece of the puzzle that launched modern photography was discovered by Englishman William Henry Fox Talbot. In 1834 Talbot invented the first light-sensitive paper by soaking it in a salt solution, then coating it with silver nitrate. A year later, he wrote: "In the Photogenic...process, if the paper is transparent, the first drawing may serve as an object to produce a second drawing, in which the lights and shadows would be reversed."

The image that Talbot captured was a "negative"—that is, the light objects appeared dark on the paper, and vice versa. He realized that by placing this negative on top of a second sheet of paper and exposing both to sunlight, the process would repeat itself, forming a "positive," or true image.

By the mid-1800s photography became the rage throughout the world. People everywhere were thrilled at the sight of their own visage, preserved forever on Daguerre's patented plates, and later on film. From purely artistic application, photography spread into the news arena, capturing scenes of horrific violence during the American Civil War through the lens of Matthew Brady (c. 1823-1896).

From a simple projection of light through a pinhole eventually emerged an entire industry. The expression "a picture is worth a thousand words" has proven true, with photographs cap-turing our emotions, encouraging us to purchase a wide array of products, and showing us a world that mere words could never reveal.

STEPHANIE WATSON

Further Reading

Newhall, Beaumont. *The History of Photography*. Boston: Little, Brown and Company, 1988.

Pollack, Peter. *The Picture History of Photography*. New York: Harry N. Abrams, Inc., Publishers, 1958.

Steffens, Bradley. *Photography: Preserving the Past*. San Diego, CA: Lucent Books, Inc., 1991.

Antonio Neri Reveals the Secrets of Glassmaking and Helps Make High Quality Glass Available to the World

Overview

The second oldest of all manmade materials (after bronze) is glass, which for thousands of years has served important building, decorative, and utilitarian purposes. Because of its value, glass became a prized possession of the wealthy and powerful. By the 1600s the world's finest glasses were produced in Venice, Italy, whose leaders sought to protect their dominance by imposing a strict ban on the sharing of knowledge about glassmaking methods and techniques. In 1612 a Florentine priest and chemist, Antonio Neri (1576-1614) published a book, *L'artra vetraria* (The art of glass) that revealed the glassmaker's secrets. Making those secrets available to the public not only made possible the duplication of Venetian glassmaking, but also provided a basis for further innovation, hastening improvements in glassmaking and spurring the shift in glassmaking from artisan's craft to mass-manufactured material.

Background

Among the most important of the advances in the material sciences, the invention of glass, and the development of techniques for manufacturing it, go back at least as far as Egypt in 3000 B.C. Indeed, the only older manmade substance is bronze. Naturally occurring glass, also known as obsidian, the product of volcanoes, was used as a tool by early humans long before the emergence of social organizations resembling ancient civilization.

It was with the Egyptians, though, that the actual manufacture of glass was first undertaken, with glass beads produced between approximately 3000 and 2500 B.C. Within a thousand years the art of glass manufacture had progressed to the point where the Egyptians were producing small glass containers, although many of these were carved or ground from glass blocks, rather than being molded or shaped into finished form during the glass-manufacturing process. Another early method of forming glass into desired shapes involved forming glass around clay or metal molds.

That process itself is both simple and complex. At it simplest, glass is the product of applying high heat to a mixture of silica (sand), soda (generally wood- or plant-ash), and limestone. These ingredients—along with smaller amounts of other materials that yield different qualities to the glass—are heated at temperatures of approximately 1500°C. The ingredients melt and combine; when cooled the resultant material is glass.

The complexity enters the formula through the various additional ingredients that can be used to alter the qualities of the finished glass. The addition of lead, for example, was found to increase the clarity of the finished glass. Variances in the amount of silica also produced differences in the finished quality of the glass, as did alterations in the amount of heat applied to the mix of materials. Of particular importance to the development of glass manufacture was the develop-

Seventeenth-century glassmakers. *(Historical Picture Archive/Corbis. Reproduced with permission.)*

ment, by at least 2500 years B.C., of simple but effective furnaces that allowed for greater control of heat, resulting in a more pure final product.

The most dramatic early refinement in glass manufacture occurred around A.D. 1 with the Roman development of a hollow iron pipe used for blowing glass—using blown air to create desired glass shapes and consistencies from the molten glass adhering to the far end of the pipe. The Romans also made large advances in etching, coloring, molding, cutting, and engraving glass.

Between A.D. 200 and 1200 the fundamental nature of glassmaking changed little—refinements in furnaces and in methods for adding color to glass were among the few innovations in the process. Despite that stagnation in further development, glassmaking remained an important undertaking, and by the thirteenth century had spread throughout Europe, with Venice becoming the acknowledged center of both the glassmaking art and industry.

That centrality resulted in great wealth and influence for Venice. As a result, Venice's governing body, the Grand Council, relocated all glassmaking enterprises to the island of Murano in an effort to keep secret the mastery of glassmaking and related technologies. Any who violated the code of secrecy faced penalties including execution.

Protectiveness about glassmaking techniques was nothing new. The Mesopotamians, more than 3,000 years ago, identified their instructions for glassmaking as secret. Yet the value of glass and the public's desire for the material resulted in almost constant attempts to wrest the secrets away from those who controlled them.

One who knew the secrets was Italian priest and chemist (and alchemist) Antonio Neri. A citizen of Florence, Neri was the son of a physician. From an early age Neri applied himself to the study of glassmaking, and by 1612 had introduced at least one major innovation. This was the addition of a minute amount of gold to the molten mixture of glassmaking materials. Chemical reactions instigated by the gold produced a glass of a brilliant ruby color. The glass—which to this day cannot be mass-produced and is thus more of an artisan's product than an industrial one—came to be known as cranberry glass or gold ruby glass.

For all of his technical skills, it was Neri's ability—and boldness—as a writer that earned him his greatest fame. In defiance of the edict against sharing the secrets of glassmaking, Neri wrote a small book called *L'arte vetraria* (The art of glass) in 1612, the same year as his discovery of cranberry glass.

In *L'arte*, Neri presented virtually all of the body of knowledge that surrounded the manufacture of glass. He addressed the nature of furnaces, proper temperatures and melting times, mixtures of materials that resulted in different types and qualities of glass, and more.

Impact

Neri's little book created a revolution—the elements required for high level glassmaking became widely known, and the industry spread rapidly throughout Europe, where most previous glass manufacturing undertakings had failed to approach to the quality of Venetian glasses.

Neri's book hastened the decline of Venice as the world's glass capital, although Venetian domination of the industry was already threatened by the immigration of glassmakers who took their secretes with them. In 1622, for example, an English company sent six Italian glassmaking artisans to the Jamestown colony in the New World.

More importantly, by consolidating in one volume the essence of glassmaking, Neri provided all glassmakers with knowledge that had previously been restricted only to a few. In short, he made available not only an existing body of knowledge but also, and ultimately of more importance, he provided glassmakers with a foundation upon which they could build improvements. By having fundamental principles at their disposal, glassmakers were better and more easily able to experiment, refine, and improve the formulas and mixtures that were used to produce glass. Better and more diverse types of glass, both practical and decorative, were developed over the course of the 1600s. Antonio Neri had opened the windows of the world.

With the ability to make glass of high quality more widespread, more glass was produced. What had been a material restricted to the upper classes began to spread throughout society, transforming the nature of everyday life. Windows became common for all buildings, and this brought an unexpected but highly beneficial social and cultural change. Because windows allowed more light into buildings and homes than was previously common, the occupants of those structures became more aware of their surroundings, particularly of dirt and grime. Transparent windows led to cleaner homes, and increased cleanliness—hygiene—led to a general improvement in the quality of life. Neri's revolt against secrecy made a large contribution to both individual and social wellbeing.

KEITH FERRELL

Further Reading

Brock, William H. *The Norton History of Chemistry*. New York: W.W. Norton, 1993.

Mumford, Lewis. *Technics and Civilization*. New York: Harcourt, Brace and Company, 1934.

Sass, Stephen L. *The Substance of Civilization: Materials and Human History from the Stone Age to the Age of Silicon*. New York: Arcade, 1998.

The Origins and Development of the Magic Lantern

Overview

More than twenty-five hundred years ago, intellectuals from many cultures began to experiment with image projection in their attempts to understand the relationship between the mechanics of the human eye and the physical principles of light. Consequently, they discovered the value of image projection for religious, educational, and entertainment purposes. In the fifth century B.C., Chinese philosopher Mo Ti described a "collecting place" or "locked treasure room" where an inverted image appeared on a screen when light passed through a pinhole. During the fourth century B.C., Aristotle (384-322 B.C.) similarly described what would eventually be called the *camera obscura*; Chinese inventor Ting Huan is said to have perfected a device for projecting moving images in c. 207 B.C. Others attribute the earliest description of sequential animation to Titus Lucretius Carus, a Roman poet and philosopher, in c. 65 B.C. Around the tenth century A.D., the Arab philosopher Ibn al Haytham (965-1038) described the mechanics of the pinhole camera. It is now thought that al Haytham was describing optic principles already well known throughout the world. By the Renaissance, ground glass lenses had made possible the improvement and widespread use of spectacles and telescopes, as well as improved focus in pinhole boxes or "cameras obscuras," from which the magic lantern derives.

In the seventeenth century, the optical principles of the *camera obscura* were applied to display multiple images painted on glass plates. This "magic lantern" technology incorporated the same basic components as modern optical projection systems—a subject in a holder, a light source, a projection lens, a condenser, or lens to redirect as much light as possible through the projection lens; and a ground, or viewing screen. The combined processes of controlled light projection through specialized lenses and intermittent movement of painted images in the magic lantern lay the foundation for modern slide and movie projectors.

Background

The invention and first public demonstration of the magic lantern has long been a subject of de-bate. In 1646 Athanasius Kircher (c. 1602-1680), a German Jesuit priest, detailed improvements in "mirror writing" in *Ars Magna lucis et umbrae*. This method of projecting images etched into highly polished metal plates had earlier been described by Giambattista della Porta (c. 1535-1615) of Naples, Italy, in his *Magiae naturalis libri viginti* (1589). Kircher's modifications included the addition of a bi-convex lens arrangement, using either candle or sun as a light source. The mirror was replaced with images painted on a water-filled glass container. A few years after the initial publication of *Ars Magna*, Kircher's pupil, missionary Martin Martini, reportedly began touring Europe, illustrating his trip to China by projecting images from glass slides, a contraption Martini attributed to Kircher. In 1657, and again in 1671, Gaspar Schott (1608-1666), in his *Magica Optica,* described every type of magic lantern then known, and gave Kircher credit for originating the technology. In Kircher's expanded second edition of *Ars Magna* (1671), he illustrated a projection lantern, and took credit for its invention. Kircher's drawing indicated a lamp as light source, a lens, a mirror, slides, and an image projected on a wall. Interestingly, the slides were shown mounted in the inverted position in order to provide an upright presentation, and thus indicated his understanding of optics. Although the illustration contained some technical contradictions, it indicates Kircher's objective of projecting a series of images successively onto a large screen.

Some historians maintain that it was the Dutch physicist Christiaan Huygens (1629-1695) who in around 1659 constructed the first working magic lantern. This view is supported by a letter that Huygens wrote to his brother Ludwig in which he described the lantern, and by reports of a slide presentation of Martini's trip by Huygens's friend, the Jesuit priest Andreas Tacquet. Other historians have argued that reports of magic lanterns predate Huygens's description, and that Huygens's model never advanced beyond the experimental stage. Yet Schott wrote that Tacquet used Kircher's method of projection to illustrate Martini's journey from China back to the Netherlands.

In the second edition of *Ars Magna*, Kircher himself attributed another magic lantern prototype to "the learned Dane who came to Lyons in 1665." It is thought that the reference was to Thomas Rasmussen Walgensten, a professor of mathematics who had begun touring Europe with an improved version of the magic lantern in the mid 1660's. Francesco Eschinardi, an Italian priest who has been credited with coining the term, "magic lantern," referred to a description of Walgensten's device, the scare lantern in his *Centuriae Optical Pars Altera* (1668). A similar reference was made by Claude François Milliet de Chales (1621-1678) a French mathematician. In the second edition of his *Cursus seu mundus mathematicus*, (1690), he included illustrations of Walgensten's version of the lantern. Milliet de Chales is another one of the people credited with the invention. In his book, he advanced lantern technology by addressing issues of focus and illumination, and by suggesting the procedure of moving a series of glass slides from side to side through the lantern apparatus. Nevertheless, he disclaimed any credit for inventing the process.

Despite the controversy over who actually produced the first lantern projector, by the mid-seventeenth century descriptions of prototypes and reports of demonstrations were being published throughout Europe. A London optician and acquaintance of Huygens named John Reeves began producing his own lanterns In 1663. In 1668, British scientist Robert Hooke described a universal projector for both transparent and opaque slides, which could use either sun or candle as a light source. In the same year, Francesco Eschinardi published *Centuriae opticae pars altera seu dialogi optici pars tertia*, which included a detailed description of the construction of the magic lantern. In 1672, Johann Christoph Sturm, professor of mathematics at the University of Nurnberg, introduced the lantern into Germany where he gave experimental lectures supposedly using Walgensten's model lantern and slides.

Twenty years later, William Molyneux (1656-1698), a professor at Trinity College, Dublin, published *Dioptrica Nova* (1692), in which he devoted a whole section to a description of a metal lantern with adjustable focusing lenses. Molyneux pointed out in the introduction to his book that there had previously been nothing written on mathematics in English, making his description of the lantern also the first work in English on that topic. Molyneux's illustration of the lantern indicated a candle as the light source, and a condensing lens (absent from the lantern constructed by Reeves in the 1660's). The illustration also indicates telescopic effects, the ability to project at a distance through a combination of lenses, and the inversion principle with the subject upside down, and the image right-side up. At the end of Molyneux' book was an advertisement for the lantern and components "made and sold by John Yarwell at the Archimedes and Three Golden Prospects. . .London."

As with many new inventions, the magic lantern was first used to entertain. However, Milliet de Chales noted that the lantern had value for scientific instruction, as it could be used to show enlarged images of insects and other specimens. Johann Zahn (fl. 1600s) advanced the idea of slide presentations for anatomical lessons, and in his *Oculus artificialis teledioptricus sive telescopium* (1685/1686) he suggested tracing book illustrations onto glass plates for that purpose. Zahn included illustrations of both cameras obscuras and magic lanterns, and suggested hiding the magic lantern out of sight of the audience for effect.

By the end of the seventeenth century, the popularity of the magic lantern motivated the development of prepared slides for sale to educators and to the public. For example, in 1705 Johann Conrad Creiling, a professor of natural history in Tübbingen proposed the production of educational magic lantern slides to accompany lectures in subjects such as natural history, biblical studies, geography, and mathematics. In 1713, B. H. Ehrenberger of Hildburghausen produced such slides, as described in his *Novum et curiosum laternae magicae augmentum* (1713). Ehrenberger's teacher, Hamberger, credited the origin of prepared slides to Erhard Weigel, a physicist who he said made them before 1700.

Impact

Improved projection technology had an immense impact on society in the following centuries. In many ways, because of the magic lantern's descendants—slide shows and motion pictures—people gained a new perspective about themselves and the world around them. Public moving picture exhibitions made everything larger than life, literally reflecting back upon society both its cultural and technological accomplishments and its social ills. But it also provided working people and a

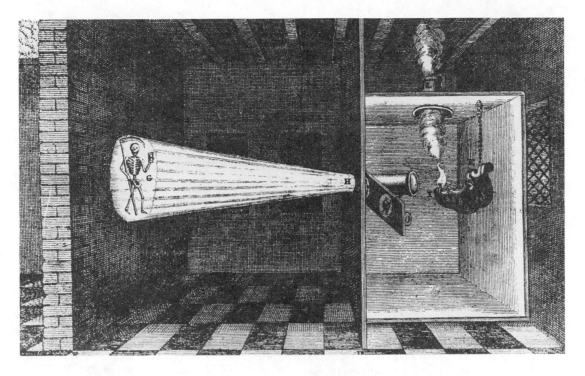

Athanasius Kircher's projection device. *(Christel Gerstenberg/Corbis. Reproduced with permission.)*

growing middle class with a new social outlet for entertainment.

The value of projecting multiple images became obvious to both scientists wishing to share new discoveries about natural phenomena and entrepreneurs anxious to make money entertaining the public. By the late eighteenth century, public exhibitions were already widespread. One of the most popular of the magic lantern shows were "Phantasmagoria" demonstrations, in which ghostly apparitions seemed to appear and vanish in the dark before viewers' eyes. Demand for such performances was so great that in 1798 E. G. Robertson took his London phantasmagoria show on the road to Paris, Vienna, and St. Petersburg.

In the nineteenth century, a variety of "optical" toys and gadgets were introduced concurrently with improvements in magic lantern technology. Among them was the "thaumatrope," developed by astronomer John Herschel and popularized by British doctor, John Ayrton Paris. As a cardboard disk with a different picture on either side is rotated rapidly on a string, the optical principle of persistence of vision takes over, and the viewer sees an optical illusion—one combined image. More elaborate devices were developed that worked on the same principle. The Daedalum, introduced by W.H. Horner in the mid-nineteenth century

and refined by French inventor Pierre Desvignes under the name Zoetrope, was later patented in the United States by William F. Lincoln and in England by Milton Bradley. The device consisted of a drum with evenly spaced slits that spun on a pedestal. As the drum was turned, individual images drawn in a narrative sequence and wrapped around the inside of the drum would appear as one smooth action when viewed through the slits. The Zoetrope had become a popular toy by the late 1800s. In 1877, Emile Reynaud introduced the Praxinoscope, a modified version of the Zoetrope in which the image was viewed on mirrors placed in the center of the drum instead of directly through the slits.

By the late 1800s, larger, more complex versions of these devices were being developed for commercial use. More streamlined methods of producing and displaying slides were introduced, advancing from painted glass strips to flexible gelatin or film. In 1892, Reynaud gave the first public exhibition of his commercial Praxinoscope, using long strips of hand-painted frames. The jerky effect of Reynaud's method would be corrected by the likes of the Lumiere brothers in France and Thomas Edison (1847-1931) in the United States with the introduction of moving pictures. Movie projection uses the same principle of persistence of vision as the thaumatrope toy: Multiple frames of individual pictures snapped in rapid succession give the illusion of continuous motion when projected at rapid speed. In 1895 the Lumiere brothers exhibited the first short motion pictures in France using their cinematograph. Quickly, others with new versions of moving picture lanterns followed, including Edison and Georges Melies, who became famous for his special effects.

Carousel and tray slide projectors became a mainstay of educational, corporate, and industrial facilities by the middle of the twentieth century. In recent years, innovations in digital technology have broadened the scope of displayed images far beyond that introduced by Kircher and his contemporaries. Today, close-up images of a ball game or concert can be projected to thousands of people on giant screens. Enormous curved screen IMAX movie theaters, originally installed in museums, have made their way to shopping malls and now include 3-D effects. Early projections of mystifying images have modern parallels in holographic presentations and laser light shows. In medical centers, expectant parents watch their developing off-spring through sonograms projected on lab monitors. Millions of people around the world can view the same images at once through satellite and internet broadcasts, making what was once larger than life seem smaller and more accessible.

LISA NOCKS

Further Reading

Books

Chadwick, W. J. *The Magic Lantern Manual.* London: Frederick Warne, 1878.

Hammond, John H. *The Camera Obscura: A Chronicle.* Bristol: Adam Hilger, 1981.

Liesegang, Franz Paul. *Dates and Sources: A contribution to the History of the Art of Projection and to Cinematography.* Hermann Hecht, ed/trans. London: The Magic Lantern Society of Great Britain, 1986.

Lindberg, David. *Studies in the History of Medieval Optics.* London: Variorum Reprints, 1983.

Weiss, Richard J. *A Brief History of Light and Those That Lit the Way.* River Edge, NJ: World Scientific, 1995.

Internet Sites

Burns, Paul T. "The Complete History of the Discovery of Cinematography." www.precinemahistory.com.

William Lee and the Stocking Knitting Frame: Micro- and Macroinventions

Overview

Some technological inventions can be explained by the societal conditions in which they emerge, since they clearly meet an existing societal need. Historians call these gradual changes in techniques *microinventions*, trial and error tinkering that usually resulted from an intentional search for improvements and can be explained by economic theories—as ways to improve production, lower costs, or use less labor. However, *macroinventions* are more difficult to explain, because they seem to be the result of individual genius and luck more than economic forces. They seem to come out of nowhere, because they bring to bear seemingly unrelated ideas that just happen to produce a totally new solution. The invention of the stocking knitting frame by William Lee is one such invention.

Background

Stockings, or hose, were universally worn in Tudor England instead of trousers or pants, being tied to the bottom of the pantaloons (hence the word pants), as we can see from any picture of Tudor men's clothing. Tudor women wore stockings under their long skirts. The vast majority wore woolen stockings, either cut from cloth and seamed up the back of the leg, or knitted by hand as with modern hand-knitted sweaters. The rich wore silk stockings knitted this way. Despite the huge demand for stockings, no one knows why William Lee (c. 1550-1610) decided to design a machine or frame to knit them. He was the son of prosperous farmers in Calverton, Nottinghamshire, England. His education at Cambridge University, where he entered Christ's College in May 1579, and later moved to St. John's College to take his B.A. in 1583, was in traditional scholastic subjects such as Latin grammar and rhetoric. There is evidence that he graduated with an M.A. in 1586, but again this degree provided no training in practical skills. Lee invented the stocking-frame in his spare time as a church minister back in Calverton in 1589, and legend

A spinning wheel. *(Leonard de Selva/Corbis. Reproduced with permission.)*

claims that he did so because a young woman whom he wished to marry would always pay more attention to her hand-knitting than to him when he came courting.

Like many macroinventions, Lee's frame-knitting machine combined two previously unrelated ideas into a dramatic new development. Knitting had been done on circular or rectangular frames for many centuries. Wooden or bone pegs were fitted at regular intervals all round the top of the frame. Thinner pegs placed closer together made finer knitting. Yarn was tied to the first peg and then wound counter-clockwise around each peg until every peg had a crossed loop of yarn lying at its base. More yarn was then wound around the pegs in the same way to make a second set of loops above the first. The first set was then drawn over the second set of loops. By Lee's time a hooked implement rather like a crochet needle was used to raise one loop at a time over the other by hand. Continually repeating this process created a tube or hose of knitted fabric in crossed stocking-stitch, hanging down from the frame. Lee's very ingenious machine had a series of rigid hooks on the frame, and a second series of moving hooks at right angles to them. The first stitches were wound on to the series of rigid hooks in the traditional way, but then the knitter used a simple mechanism to move all the moveable hooks into the stitches looped around the rigid hooks. The yarn was then laid horizontally under the rigid hooks, and the stitches were drawn over it by the moveable hooks. Repeating the process created stockings much faster than before.

Impact

The long-term impact of Lee's development has been enormous. This simple action for knitting has never been bettered, and is basic to all the machines used in the machine-knitting industry throughout the modern world.

However, its immediate impact was limited by sixteenth-century social and political beliefs. Although individuals in Tudor England could rise or fall in social status through their abilities, most people believed that God had ordained the social hierarchy and everyone's place within it, and that there should be no social change. This meant that each person knew their responsibilities to others in the hierarchy, for example to obey superiors and protect inferiors, but also that they strongly believed in their just rights, for example to be employed when they offered their labor to support themselves, and not to have their accustomed work taken away. Politicians from local magistrates to Queen Elizabeth shared these beliefs, not just because the emphasis on obedience kept the political system stable, but because they accepted their God-given obligation to protect the rights of those lower in the social hierarchy to work and feed themselves and their families. Lee's invention threatened this social order, as it offered the promise of producing the same amount of stockings with fewer workers. Therefore, when Lee first began to use his frame to make stockings in a small workshop at Calverton, employing his brother and others, he met widespread opposition from the local hand-knitters, who recognized the threat that his fast, cheap production of stockings represented to their livelihood.

Perhaps because of this resistance, in 1591 he moved his machine and workers to Bunhill Fields outside the walls of London, and sought the patronage of Lord Hunsdon, the Queen's cousin and a leading courtier. Lee's aim was to obtain from Elizabeth a royal patent of monopoly for the machine, forbidding anyone to copy it and perhaps demanding that all hose should be knitted on similar machines—for a fee payable to Lee. At that time Elizabeth was raising money for her wars against Spain by selling such monopoly control over many everyday commodities. However, when Elizabeth saw Lee's machine in action she was disappointed by the coarseness of the ordinary stockings it produced—she had been hoping that he would make the fine silk stockings she preferred. Even though Lee began to improve the machine with microinventions, so that by 1598 he could make the frame smaller to produce silk stockings, and presented a pair to her, she still refused his monopoly. Members of the House of Commons had begun to complain about the hardship monopolies caused by driving up prices, and a series of bad harvests had created economic unrest, unemployment, and political instability in the countryside, which was already feeling the burden of continuing war. As well as the social and political assumptions outlined above, conditions were too dangerous for Elizabeth to risk making hand-knitters unemployed by encouraging Lee to bring the frame into general use. Unemployed workers created problems of law and order in a country without a police force, and weakened the state in its struggles with Spain, as well as destabilizing the social hierarchy that everyone assumed should be preserved. For similar reasons, when James I succeeded Elizabeth in 1603 he also refused to support Lee.

What happened next shows how Europe's political system gave it an advantage over other parts of the world in developing and applying new technology. In a highly centralized empire such as Ming China, for example, one person's political decision could prevent any technological changes from being introduced, or even cease

the application of established technology, in order to maintain a stable and controllable social and political environment. However, Europe was politically decentralized, containing many states who competed along technological and economic as well as political lines. When technological change was discouraged in one state, it often found encouragement from another ruler who saw its economic and political advantages. Historians call this *technology transfer.* Thus Henry IV of France invited Lee to settle in France, promising him great rewards, and Lee established a workshop at Rouen, where he manufactured stockings under the king's protection. During the political unrest that followed the assassination of Henry IV in 1610, Lee died in Paris. His son and seven of his workmen returned to England, and together with one Aston of Calverton, one of Lee's former apprentices, they laid the foundation of an important new household industry in the East Midlands of England around Nottingham. People began to make stockings in large numbers, working on the frame at home, and its use spread throughout Europe in the first half of the seventeenth century.

Lee's knitting frame remained at the heart of stocking-frame technology until late into the nineteenth century, with only one major change in 1758, when Jedediah Strutt (1726-1797) created an attachment for knitting ribbed fabrics. The main problem in applying mechanical power to the knitting industry was that it was difficult to adapt the fine mechanism that varied the frame width, on which the quality of the stocking depended, to machinery more powerful than human muscle power. Thus for centuries one of Europe's basic and most important industries depended on an invention of genius by an obscure Nottinghamshire clergyman.

GLYN PARRY

Further Reading

Derry, T.K., and Trevor I. Williams. *A Short History of Technology*. Oxford: Oxford University Press, 1960.

Kranzberg, M., and C. W. Pursell, Jr., eds. *Technology in Western Civilization*. New York: Oxford University Press, 1967.

Mokyr, Joel. *The Lever of Riches: Technological Creativity and Economic Progress*. New York: Oxford University Press, 1990.

Singer, C., E.J. Holmyard, A.R. Hall, and Trevor I. Williams. *A History of Technology*. 8 vols. Oxford: Oxford University Press, 1954-84, vol. 3 (1957).

"William Lee." In *The Dictionary of National Biography*. 22 vols. London: Oxford University Press, 1921-22.

Overview

In Europe, more than any other part of the world, industrial manufacturing and technology has developed from metallurgy, the mining and smelting of metals. Advances in metallurgy have been at once the cause and effect of European technological superiority. In the Renaissance the extraction and smelting of ore was a strongly traditional industry, and Vanoccio Biringuccio's book *De la pirotechnia libri X* (Ten Books of a Work in Fire, 1540), like George Agricola's *De Re Metallica Libri XII* (Twelve Books on Metals, 1556), did not announce any dramatic inventions. However, their books described with exceptional clarity craft processes that trial and error had gradually improved over centuries. Their works demonstrate how printing helped to systematize knowledge and helped the spread of mechanization. In the eighteenth century the techniques required for profitable mining could be applied to developing the new steam technology that powered the Industrial Revolution in Britain and throughout the world.

Background

In the High Renaissance of the early sixteenth century, the Mediterranean region, and especially Italy, dominated European civilization. Italy led in many production techniques, including the most skillful metalwork, described by the Italian Biringuccio (1480-1539). He learned his trade working in foundries throughout the metallurgical regions of Italy and southern Germany, where profitable mining centers had developed using abundant ores and water-powered machinery.

A self-taught expert who wrote in vernacular Italian for nonscholars, Biringuccio's practical manual described metallurgical processes—assaying, smelting, alloying, and casting—that he had seen or practiced. In many cases his was the first written description of processes for smelting and casting gold, silver, copper, lead, tin, iron, steel, and brass. He also described the casting of medallions, fine art objects, type for printing, and cannon, which in typical Renaissance fashion he insisted should be ornamented to make them beautiful. Biringuccio's book went through many editions for practical metallurgists. His expertise explains his appointment in 1538 as head of the papal foundry at Rome. Biringuccio modeled his *Pirotechnia* on the scholarly Agricola's earliest book on metallurgy, *Bermannus* (1530). In return Agricola translated some passages from *Pirotechnia* into Latin for his *De re metallica*, thus paying tribute to Biringuccio's expert knowledge.

Like Biringuccio, German mineralogist George Agricola (1494-1555)—Agricola is the Latin version of his original name, Bauer—was an empiricist, not a scientist. He did not start from mining theory but described practical solutions, which had been developed to a uniquely high level in the southern German mining region, to such problems as flooding, vertical haulage, and the blasting of rocks. His systematic study provides much of our knowledge of mining techniques in the sixteenth century. He is often called the Father of Mineralogy as he was the first to describe minerals in terms of their observable, physical properties, rather than their supposed magical or philosophical powers.

Agricola sought a different readership from Biringuccio by writing in Latin according to the standards of Renaissance scholarship, which is why he translated his name into Latin. Like many Renaissance intellectuals he pursued several careers—as a diplomat, historian, editor, physician, and apothecary. Following his education in Greek and Latin at Leipzig University, he first published a book of Latin grammar and then studied medicine between 1524 and 1526 at Bologna, Italy. His classical education made him very receptive to Italian Renaissance culture, and he emulated the Italian architect-engineers of the period who had enhanced the social status of their profession by creating a learned literature about it. Coming from the mining area of Saxony, he believed that by writing about mining techniques in good Latin, full of terms ransacked from classical Greek and Roman literature, he could raise a practical art into a learned subject. He succeeded so well that the great humanist scholar Erasmus (1469-1536) wrote the preface to Agricola's *Bermannus* in 1530.

He could use classical Latin to describe the methods of mining at the rock-face because those methods had remained unchanged since classical antiquity. Yet Agricola, because he was a doctor and a humanist, also used analytical observation and precise Latin to describe the new machinery developed since c. 1450 to solve increasing problems of drainage, ventilation, and

A sixteenth-century depiction of metallurgical processes. *(Corbis Corporation. Reproduced with permission.)*

haulage as Saxon mining went deeper in search of precious ores.

Cleverly organized, Agricola's *De re metallica* discussed all aspects of mining from prospecting through the production of gold, silver, lead, salt, soda, alum, sulfuric acid, sulfur, bitumen, and glass. One of the great monuments of technology, for two centuries it remained through many reprints the best mining textbook in Europe, because of its comprehensive coverage and brilliantly clear illustrations of machinery, powered by horses or water wheels, for pumping water, hoisting spoil, ventilating shafts, crushing and grinding ore, assaying and sampling for quality, stirring and mixing ore in tubs, blowing smelting furnaces, and moving heavy hammers to work wrought iron. Agricola gave two or three variant types for each machine, all of their massive

wheels and shafts built from heavy timber, using iron only for ties and bearings to reduce costs.

Agricola paid much attention to the pumping and ventilation machinery that were vital for extending mining deeper once easily accessible ores were exhausted. He described suction pumps and pistons, whose principles he had found described in ancient classical works, and chain and rag pumps, in which each carefully made rag ball on a chain moved up a tightly fitting tube, pushing water in front of it to the next level.

Impact

The fact that Biringuccio and Agricola began the large-scale publishing of books dealing with technology does not prove that technology in this or other fields began to change principally because the highly educated class now found its study socially acceptable. Agricola had to defend mining against charges that it caused diseases, and deflect modern-sounding complaints that it polluted the soil and water-courses, and that its huge demand for wood as fuel and building materials caused deforestation, destroying the eco-systems of birds and animals eaten by humans.

More practical limitations also slowed the impact of this technology. Agricola himself acknowledged that only in the largest mines over 100 feet (30.5 m) deep was water-powered machinery cheaper and more efficient than horsepower. The machinery itself was poorly designed and inefficient, its power output proportional to its size, and therefore its useful limits were soon reached. Also a large gap existed between this best practice and standard techniques, which limited the impact of these expensive, complex machines on industrial practice. They were difficult to adapt to different environments, being dependent on regular supplies of water to the wheels.

Many German craftsmen also found Agricola's elegant Latin an obstacle to understanding, and some inaccurate German translations gave very confused versions of his processes. In the second half of the sixteenth century the German mining industry also declined because the technology he described reached its limits in dealing with growing technical problems, and succumbed to competition from Spanish mines in the New World.

However, German mining's great demand for capital and the return it gave in silver did create early capitalist enterprises. All the major banking firms of the period, such as the Fuggers and the Welsers, were heavily involved in mining and became lenders to Renaissance rulers who required increasingly vast sums of money to wage war with the expensive new gunpowder weapons of cannons and muskets. The increasing frequency of such wars and the mechanization of gunpowder armies increased the consumption of iron for weapons and projectiles to the extent that by the end of the sixteenth century rapid deforestation was leading Europe to the limits of the charcoal resources required for smelting iron.

The rising price of this energy source encouraged attempts to substitute coal in its place. In England, Dud Dudley (1599-1684) claimed in his *Mettalum Martis* (1665) to have successfully smelted iron with coal as early as 1619. However, this was part of an unsuccessful publicity campaign to secure a royal patent for the process, and his claims are generally rejected. Raw coal contains sulfur and other substances that make cast iron brittle, and the only way of using coal for smelting was to reduce it to coke first, since coking eliminates the sulfur content. Abraham Darby (1678?-1717) achieved the first successful smelting of iron ore with coke in 1709, by chance using coal with a naturally low sulfur content.

While Biringuccio and Agricola gave much more space to gold and silver than to describing the production of iron, the machinery they described did make an eventual contribution to technological development in the eighteenth century. Smelting iron with coke released for

ORIGIN OF THE DOLLAR

Georgius Agricola spent part of his career as a doctor at the copper and silver mines at Joachimstal in Saxony, now in Germany. The silver produced at Joachimstal was so pure that it was used to make a coin accepted throughout Europe for its high precious metal content. This coin was known as the *Joachimsthaler*, after the mine. Later coins were also named *thalers* to suggest their high silver content, and from the German word *thaler* evolved the word "dollar," which continues the Renaissance tradition of its earlier namesake by being accepted as a hard currency.

GLYN PARRY

Britain the immense accumulated potential energy of her coalfields, instead of relying on the current available energy of wood and water. Although cheaper coke-smelted iron proved inferior in quality to charcoal-smelted iron, economic pressure lowered standards and made it acceptable, particularly when precisely engineered. Precision engineering, like the increasing knowledge of metallurgy, chemistry, mechanics, and civil engineering, can be traced back to sixteenth-century mining.

As Agricola's book demonstrated, mining sought out desirable minerals even in the face of increasing problems with water drainage that pushed its pumping technology to the limits. The further refinement of this technology required increasingly accurate boring machines and machine tools to make efficient pumps. In the eighteenth century the earliest coal-burning steam engines required precise engineering to be efficient. Their designers could turn to the manufacturers of mining pumps for the means of making the precisely engineered vessels that powered the early Industrial Revolution in Britain and throughout the world. Therefore, the metallurgical industry described by Biringuccio and Agricola had a direct impact on the early Industrial Revolution and helped to solve its most difficult problem, that of harnessing the stored energy of coal.

GLYN PARRY

Further Reading

Cardwell, D.S.L. *Technology, Science and Society*. London: Heinemann, 1972.

Derry, T.K., and Trevor I. Williams. *A Short History of Technology*. Oxford: Oxford University Press, 1960.

Klemm, F. *A History of Western Technology*. Trans. by D.W. Singer. New York: Scribner, 1959.

Kranzberg, M., and C. W. Pursell Jr., eds. *Technology in Western Civilization*. New York: Oxford University Press, 1967.

Mokyr, Joel. *The Lever of Riches: Technological Creativity and Economic Progress*. New York: Oxford University Press, 1990.

Mumford, Lewis. *Technics and Civilization*. New York: Harcourt, Brace, 1934.

Pacey, Arnold. *The Maze of Ingenuity*. New York: Holmes & Meier, 1975.

Singer, C., E.J. Holmyard, A.R. Hall, and Trevor I. Williams. *A History of Technology*. 8 vols. Oxford: Oxford University Press, 1954-84, vol. 2 (1956), vol. 3 (1957).

Development of the Horse-Drawn Coach

Overview

Although carriages were used in continental Europe as early as 1294, vehicles to carry passengers first appeared in England in 1555. That they did not appear earlier was due to the appalling condition of English roads, which were little more than cattle tracks and water courses. Winter was an especially treacherous time for wheeled transport. In England, in the twelfth century, wagons were used by distinguished persons for travel. Because they were comparatively more comfortable, litters supported by two horses (one in back, one in front) carried ladies of rank, the sick, and also the dead.

Background

The earliest surviving carriages (from the 1500s) were four-wheeled, with an arched tilt (covering) of leather or fabric over a bent-wood hooped frame. Although the wooden body and tilt framework from earlier carriages also survive, the undercarriage and wheels are gone. These carriages are long, and were mainly used by aristocratic ladies. From the beginning of the sixteenth century, a new type of body—a box slung on wheels, or coach—was invented.

Passengers in early carriages could look forward to a jerky ride. Queen Elizabeth I (1533-1603) suffered so much from her first experi-

ence riding to the opening of Parliament in 1571 that she never used that particular vehicle again. No one knows when exactly builders first used springs to soften the jolting caused by the rough roads the carriages had to travel on. But already in the mid-1400s, there is evidence to suggest that coach bodies were being hung on leather straps or braces connected to a wooden frame to take some of the dead weight of the coach body off the undercarriage.

The first coach to be made in England was made by Walter Rippon of Holland for the Earl of Rutland. It had a covered body and a pivoted front axle, unlike the rigid-axle carriages of earlier times, and was driven by a pair of horses. Queen Elizabeth preferred another coach brought out of Holland by William Boonen, who was made and remained her coachman to the end of the century. This coach had four wheels with seven spokes each. Each wheel had a thick wooden rim bound round it and was secured to the wheels with pegs.

The most common model of the first generation of coaches had a seat, called the boot, projecting outwards at either side, between the wheels. These seats were usually occupied by pages, grooms, or ladies in attendance. The boot was an uncomfortable seat because it had no covering of any kind and would have exposed those sitting in it to the wet and cold. The boot remained a feature of coaches until they became enclosed and supplied with glass windows.

The first coaches were drawn by two horses, but as coach travel over country roads became more frequent, additional horses were required to deal with the demands of the road surface. More horses also meant that the vehicles could travel at faster speeds, since the horses had to work less and were thus able to trot or to gallop.

In 1605, the first hackney coaches came into use. These were four-wheeled coaches drawn by two horses that could accommodate six people and were used for hire to transport people about the city. At first, hackneys remained in their owners' yards until they were sent for. However, by 1634 hackney stands had appeared in London, where drivers in uniforms called livery would wait for fares. A year later, there were so many hackney coaches on the streets in London, creating a nuisance, that Charles I (1600-1649) issued a proclamation prohibiting their use for journeys under three miles. In France, hackney-like vehicles were called fiacres, and they performed a similar function.

Carriages with glass windows first appeared in 1599 in Paris, where they created a scandal at the court of Louis XIII (1601-1643). Glass was first used in the upper panels of the doors, but soon covered all the upper half of the sides and the front of the body. Although in England glass windows were common in houses before 1650, the kind of plate glass needed to withstand the rigors of carriage travel had to be imported from France. From 1670, it was also made in England.

The stagecoach came into vogue in England in about 1640. These coaches were constructed like a hackney coach but on a larger scale, and were intended to take passengers between London and towns between 20 and 40 miles away. Journeys to further towns such as York, Chester, and Exeter took four days and were accommodated by so-called flying coaches. Stagecoaches carried eight passengers inside, and provided a large basket behind, over the axle, for baggage and as many passengers as could fit in what space was left. Passengers inside were protected from the rain and cold by leather curtains.

As early as 1625, Edward Knapp was granted a patent for suspending the bodies of carriages on steel springs. Steel springs were hard to make, and his design failed, but 40 years later, others took up the problem more successfully. Soon after its founding in the mid-1600s,

the Royal Society, too, took up the question of improvements in carriage design.

Ever-increasing road traffic led to a demand for smaller vehicles for general use. The gig, a light, two-wheeled carriage, was invented in France in 1667. A later version was called the cabriolet, and proved enormously popular there, and in England. The gig had a curved seat set on two long bending shafts that were placed in front on the back of the horse and behind on the two wheels. Like other carriages of the period, its springs were constructed of leather straps.

Impact

Although carriages were important to the Romans, as is evident from their excellent roads, with the fall of the Roman Empire, carriage technology suffered. And without the incentive to maintain repairs for the passage of vehicles (horseman required less well maintained roads), the roads disintegrated. In Western Europe, accounts from England and France describe roads damaged, decayed, and hindered by brooks, stones, brambles, and trees.

Carriage building enjoyed a renaissance in the sixteenth century, owing to the growth of trade, and increasing mobility among people. But the technology was not embraced wholeheartedly. In the thirteenth century, for example, as part of a bid to stamp out luxury, Philip the Fair of France forbade the wives of citizens to ride in carriages—though at this time, carriages were little more than baggage carts. By the fourteenth century, more luxurious carriages had evolved in England from the four-wheeled wagon, and were used to transport wealthy ladies (the men rode behind). In England, as well, by 1580 coaches were so usual among the wealthy classes that they became associated with degeneracy. Those who chose to ride in coaches instead of actively riding on horseback were seen as lazy. Critics called the carriages "upstart four-wheeled tortoises." Nonetheless, so popular did coaches become that in 1601 a bill was passed in Parliament to "restrain the excessive use of coaches." The bill was never enforced, and in any event, coaches were little used outside of London and large towns, owing to the bad condition of country roads.

The success of hackney coaches had a dramatic effect on the livelihood of the watermen who until then had monopolized passenger traffic across the Thames River. Some chalked passenger preference up to a taste for novelty, but in fact, at the time, crossing the Thames in a boat was a risky proposition. One waterman complained of seeing his fares dwindle from eight or ten in a morning to two for the entire day.

Just as hackney coachman were accused of taking bread from the mouths of Thames watermen—and in fact, the profession of waterman was scarcely heard of after 1662—so stagecoaches were accused of robbing licensed hackney coachmen of their livelihood. Moreover, stagecoaches were believed to be destroying the breed of good horses, the profession of watermen, and lessening royal revenues formerly brought by saddle horses. They were also accused of encouraging simple folk to make idle visits to London, where they would be exposed to vice. On the positive side, stagecoaches were believed to have provided the first incentive to improve country roads, though the truth was the other way around. As long-distance coach travel flourished, prohibitions were put into place to prevent existing highways from being ploughed up by carriage wheels carrying heavy loads.

Early carriages often had to be driven across fields and through ditches. It was symptomatic of the attitudes of the times that between 1684 and 1792, 10 patents were granted for devices to keep carriages from overturning, though few thought to work on improving roads so they did not cause

upsets in the first place. In 1663, the first turnpike gate was erected on the Great North Road to collect tolls to repair the highway in the surrounding region, but it proved so unpopular that it took a hundred years to erect the next one. Repairs to highways were made by forced labor, only when the roads absolutely required it.

In 1677, Charles II (1630-1685) founded the Company of Coach and Coach Harness Makers, illustrating the importance and favor coaches had taken on at the time. When England was at war with France in 1694, a new system of taxing hackney coaches provided revenues for defense. Although roads and highways were slow to improve, the diversity of vehicles ensured that Europe would dominate technological development in transportation until the eighteenth century.

GISELLE WEISS

Further Reading

Gilbey, Walter. *Early Carriages and Roads.* London: Vinton, 1903.

Piggott, Stuart. *Wagon, Chariot, and Carriage: Symbol and Status in the History of Transport.* London: Thames and Hudson, 1992.

Straus, Ralph. *Carriages and Coaches: Their History and Their Evolution.* London: Marin Secker, 1912.

Systematic Crop Rotation Transforms Agriculture

Overview

The French landowner and lawyer Olivier de Serres (1539-1619) published in 1600 his book *Théatre d'agriculture*, which described systematic crop rotation for the first time. His ideas were developed further in England by Sir Richard Weston (1591-1652) in his book *Discourse of Husbandry Used in Brabant and Flanders, Showing the Wonderful Improvement of Land There, and Serving as a Pattern for Our Practice in This Commonwealth* (1650). Neither man invented the ideas they collected in their books. However, their descriptions helped to spread the efficient farming practices that had developed in some European regions in the sixteenth century to meet the demands of a rising population. As such they demonstrate the importance of microinventions,

in this case those small changes that over centuries gradually improved farming technology and productivity, and of the dissemination of these best practices throughout Europe by the printing press, perhaps the key technological invention of the period because it helped to spread knowledge of other inventions.

Background

Europe before the late eighteenth century was a subsistence society. Its agricultural productivity was so low that in some regions up to 90% of the population had to labor in the fields to ensure that crops and farm animals produced enough food for the whole population, and to plant the next harvest and breed the next gener-

Seventeenth-century etching of farmers rolling fields. *(Christel Gerstenberg/Corbis. Reproduced with permission.)*

ation of animals. This was because at any one time up to one third of the fertile arable land was left fallow, unplanted and used for rough grazing. Villagers practiced an ancient crop rotation that divided their lands into three large fields, each of which was successively planted in winter wheat (planted in the fall, harvested in early summer), spring wheat (planted in spring, harvested in late summer), and then left fallow, that is, allowed to grow rough grass. So every year one field in three remained fallow, in order to manure it by feeding cattle on it. The great problem was that the fallow did not provide enough food to sustain the cattle through winter, so that many had to be slaughtered in the fall. Therefore cattle could not be improved by careful breeding, and farmers also missed out on their future manure and on using their muscle power to work farm implements.

Before de Serres and Weston, many writers on farming had recommended the cultivation of fodder crops, such as clover, turnips, sainfoin, and buckwheat, to increase animal winter food stocks. This would use the ground formerly left fallow more productively and in turn produce more manure, increasing crop production and allowing more animal energy to be applied to farming. However, it is now impossible to determine where fodder crops were first grown. Clover may have been first used in northern Italy, and from there spread to Flanders (now in

Belgium) and from Flanders to Germany and England after the mid-seventeenth century.

The chief importance of de Serres and Weston is that they described in detail agricultural innovations that had actually been proven to increase agricultural productivity over time. By publishing their books, they distributed this best practice all over Europe and influenced their readers to apply the ideas themselves.

De Serres was a Calvinist lawyer who lived all his life on the small family estate at Villeneuve de Berg in the Vivarais region of France, where he tested the innovations that he proposed in his *Théatre d'agriculture* (1600). This very popular work appeared in several editions throughout the seventeenth century. Serres surveyed all aspects of agriculture, starting with advice on running a religious Calvinist household. He discussed how to domesticate and cultivate all the plants and animals he knew. He enthusiastically advocated using irrigation to improve meadows, carefully draining land, and conserving water. He supported the sowing of "artificial grasses"—that is, nonnative fodder crops—and their use in a rotation of fields that avoided leaving them fallow. He introduced hops to France, vital for the development of the brewing industry because they preserved beer. He was the first agricultural writer to describe and encourage the cultivation of maize and potatoes, newly imported from the Americas. Eventually these new crops improved the diets of many French peasants because they were cheap and nutritious.

Around the time that he published his book, dedicated to King Henry IV of France, de Serres successfully lobbied Henry to expand sericulture, the cultivation of silkworms and the mulberry tree on whose leaves they fed. Beginning at Henry's Tuilleries palace, de Serres planted mulberry trees in many other areas of France. This laid the basis for the important French silk industry. It is little wonder that de Serres is often called the father of French agriculture.

Because France was so geographically varied and traditional localism was so strong, few of these innovations were adopted widely by French farmers. Until the French Revolution of 1789, most farmers persisted in using medieval methods, leaving a third of their land fallow, and not breeding their cattle selectively because they lacked sufficient feed to keep them through the winter. The improvement of French agricultural techniques began only just before the Revolution, and in imitation of what had been achieved in England.

England had increased its agricultural productivity by the later eighteenth century largely because writers like Sir Richard Weston had a greater impact on English society, which was vigorously developing its farming before 1650. Many earlier books on farming, such as Fitzherbert's *Boke of Husbandry* (1523) and Thomas Tusser's *Hundreth Good Pointes of Husbandrie* (1557), summarized changes in England such as the floating of water-meadows to better feed cattle, the grafting and planting of trees, the cultivation of hops, and the management of poultry and cattle, but they also imitated the methods used in the Low Countries, now the Netherlands and Belgium. In this way the printing press helped to transfer new technology across Europe.

The Low Country farmers had to be very careful in their farming methods because this was the most densely populated area in Europe, and they also grew many industrial crops such as flax for the linen industry, red madder and blue woad for dyeing, barley and hops for brewing, hemp for ropes, and tobacco, recently introduced from North America. Farmers specialized in market-gardening of vegetables and fruit growing, which also required extreme care and the use of fertilizers from cattle-raising areas and human waste from the towns. They also changed the ancient three-field system by introducing artificial grasses. Even in the relatively

small area of the Low Countries, differences in the soils and the level of the water-table required different systems. However, their treatment of the light sandy soils astonished foreign visitors like Weston, and this sandy soil system played a great part in the development of modern farming in Britain, which has similar soils, as well as other countries of the world.

Weston came to know the Low Country farming methods because he fought for the losing side in the English Civil War. A devoted supporter of Charles I, his estates were seized by Parliament in 1644, and he was forced into exile in the Low Countries until 1649. He published his *Discourse of Husbandry* in 1650 to spread the knowledge of the "ley" farming techniques that he saw between Antwerp and Ghent while in exile. Ley farming emphasized the careful accumulation of manure from animals hand-fed in stalls during summer with green and root crops, such as hay, clover, turnips, or flax. When the ley lands were ploughed under, the roots left in them fertilized the soil. This created a rising cycle of production, because better-manured crops produced more food for animals, which in turn produced more manure. Animals could also be kept through the winter on the reserves of larger crops, enabling them to be selectively bred and preserving their muscle power and manure.

Weston tried out this system in Surrey, England, after his return from exile. He planted flax, turnips, and oats mixed with clover, the latter mowed three times in the second year, and then left as improved grazing for four or five years before being ploughed under. More than any other individual Weston was responsible for the crop rotation system that spread over large areas of Britain after 1650.

The traditional system had used large communal open fields divided into scattered personal strips of arable land, with communal grazing rights over the fallow. This new system stimulated the conversion of the open fields into individually owned enclosed fields, because no one could grow the new fodder crops on his share of the fallow fields where cattle traditionally grazed, for they would be eaten by other people's animals as well as his own. Nor could good cattle be kept apart from the common herd that bred promiscuously and passed on diseases. Therefore in the long run traditional English society based on the communal use of arable and pasture land gave way to private farming by individuals. In the process some suffered and others prospered, increasing social inequality and leading to conflict between those who wanted to preserve traditional communal rights and those who wanted to pursue their own self-interest.

On the other hand, the greater productivity of the new kind of agriculture meant that fewer workers were needed to ensure an adequate food supply. So more labor was available for the growing requirements of industry, as the increasing demand of a rising population of generally more prosperous, better-fed people stimulated the early Industrial Revolution in Britain.

Impact

Overall, the modern world could not be possible without the innovations practiced and advocated by de Serres and Weston to increase the productivity of the soil and of the animal resources of Europe. Our modern industrial society, with its highly specialized workforce, depends on the application of technology to ensure that the tiny minority who work in agriculture can produce an adequate food supply for our growing population. Without the new rotations that allowed European farmers to bring more of their land into production and to keep more and better animals alive for their fertilizing and energy resources, the industrialized nations would still be living in a subsistence society comparable to many third-world countries today.

GLYN PARRY

Further Reading

Derry, T.K., and Trevor I. Williams. *A Short History of Technology*. Oxford: Oxford University Press, 1960.

Mokyr, Joel. *The Lever of Riches: Technological Creativity and Economic Progress*. New York: Oxford University Press, 1990.

Pacey, Arnold. *The Maze of Ingenuity*. New York: Holmes & Meier, 1975.

"Richard Weston." In *The Dictionary of National Biography*. 22 vols. London: Oxford University Press, 1921-22.

Singer, C., E.J. Holmyard, A.R. Hall, and Trevor I. Williams. *A History of Technology*. 8 vols. Oxford: Oxford University Press, 1954-84, vol. 3 (1957).

The Development of
Key Instruments for Science

Overview

The unaided human eye can see individual objects as small as a few tens of microns, can detect single photons (when dark-adapted), and can see objects millions of light-years away in space. Our fingertips can feel differences in texture resulting from features less than a thousandth of an inch high, and our other senses can detect similarly small differences in molecular concentrations (taste and smell) and vibration (hearing). Yet, our eyes are poor compared to a hawk's, we cannot hear or smell as well as most dogs, and we cannot begin to duplicate a salmon's ability to taste the waters of its home stream. In order to explore and understand our world and universe, we must extend our senses further still. So we have learned to make telescopes that can see nearly to the beginning of time and microscopes that can see individual atoms. And whatever we can see, we have learned to measure. Our ability to understand our world is limited by our senses, as they have been augmented by our scientific instruments, descendents of the first telescope, the first microscope, and other devices for observing and measuring the world and the universe in which we live. The growth, then, of ever-more sophisticated scientific instruments has had a significant impact on our view of the world, the universe, and ourselves.

Background

There is no doubt that for as long as humanity has existed, we have squinted at objects impossibly tiny or off in the distance. We have done so to try to understand the world in which we find ourselves, or simply out of wonder and curiosity. And where our senses were unable to take us, our imagination picked up, resulting in fanciful hypotheses in areas too numerous to mention. Our early astronomical observatories date to the time of Stonehenge or before, and the Babylonians, Egyptians, Mayans, Chinese, and most other cultures have some history of naked-eye astronomical observations.

At the same time, the small has fascinated us, too. Unable to see individual cells, we wondered what the human body was made of, the biology of fertilization, and where disease came from. We felt that we should be able to measure anything we could detect, no matter how large or small, while the use of ever-more-complex calculations made early scientists yearn for a way to reduce the drudgery of endless arithmetic.

In the sixteenth century humanity began to overcome some of these limitations. The Renaissance saw the invention of the telescope, the microscope, precision measuring devices, and rudimentary calculating devices. All of these helped scientists to see things previously only speculated about, and in many cases, what was seen was completely unexpected. As a result, humanity's notions of how the world worked were turned upside down.

Impact

The impact of these new instruments on science and on society was profound and nearly universal. Although a great many instruments were developed during this time, we will concentrate on the scientific and societal impacts of only four: the telescope, the microscope, the vernier caliper, and calculating devices.

The Telescope

Although there is some evidence of lenses going back over 2,000 years, the first telescope was not consciously made until sometime in the late sixteenth or early seventeenth centuries. The standard story is that a Dutch spectacle maker by the name of Hans Lippershay (c. 1570-1619) invented the first telescope, but it was almost certainly invented earlier, and Lippershay was simply the first to apply for a patent (later denied) in 1608. By 1609 Galileo (1564-1642) had heard about the device and made one of his own that turned out to be a significant improvement over Lippershay's. Over the next century the telescope was improved by a number of people, including Galileo and Isaac Newton (1642-1727), coming to roughly modern form (in design, if not in size) relatively quickly.

The telescope was a boon to scientists and the bane of the Church and tradition. Looking through a relatively simple and inexpensive device, astronomers found that the Moon's face was cratered and mountainous, the Sun's face was mottled with dark spots, the Milky Way was actually a collection of uncountable stars, Jupiter had both bands and moons, and much more. Galileo was actually threatened with excommu-

nication from the Catholic Church unless he renounced his unsettling discoveries. However, within a century, scientists had accepted the fact that Earth was not the center of the universe, that the Sun and the Moon were not perfect and unblemished spheres, and that other planets had their own satellite systems.

These discoveries had the result of displacing Earth from a special place in the cosmos. Once Earth's motion around the Sun was confirmed, we could no longer claim to live at the center of the universe. We were simply one planet of many around a star that later observations were to show was really nothing special, either. Until Galileo's observations, and those that were to follow, man could claim to be the favored species of God, living at the center of Creation. Ultimately, developments in new and different telescopes have given us "eyes" above the atmosphere and in wavelengths not even known to exist by Newton. We can now see back in time almost to the big bang (at least, to within about a million years of the big bang) and can look at our universe in wavelengths ranging from gamma rays to radio waves. And with each new improvement, we continue to find explanations, new phenomena, and still more questions.

The Microscope

If the telescope gave us the ability to peer outwards into the universe, the microscope has given us the ability to gaze within; within ourselves and within just about any other object we care to examine. The microscope may have been invented by Lippershay, too, although Sacharias Jansen (1588-c. 1628) is thought to have performed important work as well, and Robert Hooke (1635-1703) was among the first to make microscopic discoveries widely known.

Just as the telescope gave us an entirely new way to look at the heavens, the microscope gave us an unprecedented ability to look at ourselves and the details of our world. Cells were discovered in cork, showing us for the first time what our bodies were made of. Pond water was shown to be swarming with microscopic life, leading eventually to the germ theory of disease upon which so much of public health is based. More discoveries followed, and each of them showed more detail about how the living world worked.

These discoveries, in turn, helped to show again that humans did not differ significantly from other animals. Like them, we were comprised of cells, and most of these cells looked like their counterparts in the animal world. This was another indication that, although gifted with the power of abstract thought, mankind was also a member of the animal kingdom, and was likely not placed on a pedestal. At the same time, biological and medical research performed with these microscopes has led to an immeasurable increase in human medical knowledge, and it's safe to say that much of the success of modern medicine is based in whole or in part on the microscopic examination of tissues, germs, and more.

Over the intervening centuries, the microscope, too, has been expanded in capability and in utility, now finding uses in geology, materials science, and more. Modern microscopes include the confocal microscope, able to focus on very small slices of individual cells; the electron microscope, which can image viruses and smaller objects; and the atomic force microscope, which can "see" and manipulate individual atoms.

The Vernier Caliper

Although often overlooked or taken for granted, the ability to make precise and accurate measurements is crucial to both science and technology. Manufacturing a steam engine, for example, requires one to measure with sufficient accuracy that a piston will slide freely in a cylinder, neither binding to the sides nor leaving such a gap that steam leaks out. Similarly, making a telescope requires grinding lenses and fitting them to a tube with some degree of precision. These tasks are both dependent on the ability to make dimensional measurements that are precise and accurate.

One of the first precision measuring devices was invented in the early seventeenth century by French mathematician Pierre Vernier (1584-1638). This device, still called a "vernier," consists of two scales, a main scale and a sliding scale. Using this, one can measure dimensions with a fairly high degree of precision and accuracy. In fact, vernier calipers are still in common use in machine shops, scientific laboratories, instrument shops, and elsewhere because of their simplicity and accuracy.

The vernier was, of course, only the first precision measuring instrument. In the intervening centuries, measuring technology has expanded greatly to the point where scientists can directly weigh individual molecules, can measure lengths to a matter of angstroms, or can measure distances to the edge of the visible universe. In many ways, science and technology can only advance to the limits of what we can measure, because what we cannot measure, we cannot scientifically understand. And with these increased powers of measurement, our ability to understand our place in the universe and our ability to manipulate our world have increased substantially, contributing directly to the development of engines, aircraft, and modern electronics, among others.

Calculating Devices

The last of the devices this essay will address is the calculating device. Until John Napier (1550-1617) developed the first slide rule in 1617, all calculation was performed by hand. As anyone who has labored through a difficult series of calculations can attest, working through involved calculations by hand greatly increases the opportunity to make mistakes, often requiring each set of calculations to be repeated several times to ensure everything was done properly. To get an idea of the work involved, try to picture one of the lengthy tables of logarithms or trigonometric functions often still found in the back of mathematics textbooks. Now, try to picture calculating all of those numbers individually, by hand, without the benefit of a calculator or computer. Yet this is precisely how such tables were originally constructed.

The first slide rule simplified these calculations enormously by making the basics of multiplication and division mechanical rather than mental. Assuming the slide rule was made properly, all one had to do was to line up the appropriate numbers and read out the answer according to some simple rules. Not only was much of the labor taken out of these calculations, but the chances of error were reduced substantially. Scientists, engineers, and mathematicians became more accurate as well as more productive.

As with the other devices mentioned here, the process of calculation has advanced beyond Napier's dreams in the intervening centuries. Surprisingly, the slide rule survived almost unchanged for nearly 350 years before being supplanted by the hand calculator in the 1970s. Since that time, electronic calculators have become more powerful than the earlier computers, while computers and computer-aided computation have gained the ability to perform more calculations in a few minutes than an early mathematician could have performed in a lifetime.

The impact of calculating devices on society and the ability to perform calculations rapidly and accurately has been a tremendous boon to science and technology, as well as to the worlds of finance, insurance, business, and more.

creased greatly. This, in turn, helped humanity to gain a better understanding of our place in the universe.

P. ANDREW KARAM

Summary

The last centuries of the Renaissance saw the development of the first of what became a centuries-long succession of scientific instruments and devices. These devices helped to reduce much of the tedium of scientific calculation, made possible the development of precision machinery, and extended the range of human senses immeasurably. With all of this, our ability to understand and manipulate our world, our universe, and ourselves also in-

Further Reading

Brecht, Bertholt. *Galileo*. New York: Grove Press, 1966.

King, Henry C. *The History of the Telescope*. Cambridge, MA: Sky Publishing, 1955.

Maor, Eli. *E: The Story of a Number*. Princeton, NJ: Princeton University Press, 1994.

Morrison, Philip, and Phylis Morrison. *Powers of Ten: A Book about the Relative Size of Things in the Universe and the Effect of Adding Another Zero*. Redding, CT: Scientific American Books, 1982.

The Measure of Time

Overview

Beginning with the designs of Leonardo da Vinci (1452-1519) and Galileo Galilei (1564-1642), pendulums set in motion an evolution in accuracy and utility of clocks and watches. Various escapements managed motion into regular intervals. Balance wheels made miniaturization possible. Reliability became the holy grail of clockmakers, as temperature, torque, and friction were countered. Measurements became precise enough to justify adding a second hand, and clocks and watches became standard instruments for navigation and scientific experimentation. Ultimately, accurate clocks initiated changes beyond the measurement of hours, minutes, and seconds, creating mechanical devices, new philosophical concepts, and a new view of time itself.

Background

There is only one measurement most people make every day—that of time. Clocks and watches are both widely used and highly personal devices. For timepieces—and a mechanical view of time—to permeate our culture, they had to become accurate and mobile. The first breakthrough came with a simple mechanism for dividing time into equal intervals—the pendulum. The person who had this insight was Galileo.

In Galileo's time, clocks were imprecise and used primarily for religious purposes. Mechani-

Christiaan Huygens. *(Library of Congress. Reproduced with permission.)*

cal clocks had existed since about A.D. 1000, powered by falling weights and controlled by verge escapements (bars with points at each end that rocked back and forth, allowing the energy of the falling weight to escape bit by bit by limiting the turning of a toothed gear). These clocks, at best, lost 15 minutes a day. The hours of the day were not yet standardized (a practice which came about with another invention, the train), and the modern sense of timekeeping, absolute, scientific, and independent of nature and tradition, did not yet exist.

But careful measurement had begun to come into its own during the Renaissance as observation and the scientific method took root. Nicolaus Copernicus (1473-1543) tracked the planets; Albrecht Dürer (1471-1528) used devices to add perspective to his drawings; explorers like Christopher Columbus (1451-1506) and Ferdinand Magellan (1480?-1521) took careful notes of winds, currents, and stars. Leonardo da Vinci, a keen observer with exceptional mechanical intuition, even sketched out a design of a pendulum clock. But, as was typical of Leonardo, the design remained hidden in his notebooks for generations.

Galileo was a major proponent of making science more quantitative, and taking advantage of the power of experimentation. When he first turned his attention to the pendulum in 1582, the device was the subject of the experiment, not the means for its measure. He used his own pulse to time the intervals of the swings of a chandelier and discovered that they were exceedingly regular, even when the angles were varied. He and his students observed oscillations of pendulums throughout an entire day to confirm that the swing of a pendulum has a constant period (now known not to be completely true). From this point on, Galileo used pendulums to measure short periods of times in his experiment, but he did not succeed in creating the first pendulum clock. Though weight power, gears, and escapements were familiar to him, Galileo was defeated in his attempts to build a pendulum clock by the difficulty in transferring the energy from the pendulum to a cog-wheel, and by problems with keeping the pendulum from slowing up and stopping.

The honor of building the first practical pendulum clock goes to the Dutch physicist and astronomer Christiaan Huygens (1629-1695). He completed his first model in 1656, using gravity not just to power the mechanism, but also to regularly give a "kick" to the pendulum to keep it going. By 1657 he had created a clock that was accurate to less than a minute per day, many times better than any clocks that preceded it. Huygens's triumph was not immediately celebrated in his day. In fact, there quickly were charges of plagiarism against the young, un-

known scientist. It was only when Huygens revealed his detection of the ring of Saturn that he was accepted as an original who was extending, rather that stealing, the legacy of Galileo.

As valuable as the regulating quality of the pendulum was, it wasn't a very portable mechanism. In about 1500 Peter Henlein (1480-1542), a German blacksmith, created the first watch, powered by a spring. Henlein's use of a spring as a portable power could replace weights, but how could a pendulum be made equally portable? Here, Englishman Robert Hooke (1635-1703) is given credit for a clever leap of imagination. Hooke, an accomplished scientist who discovered refraction, first used the word "cell" as a biological term, and stated the inverse square law to explain the motion of planets, was fascinated by the physics of springs. He had discovered the principle (now known as Hooke's Law) that the stretching of a solid body is proportional to the force applied to it, and saw that a spring could be used to perform the task of a pendulum. His spiral balance spring (1660) was fixed to the movement at the outer end and to a friction-held collar at the other. The spring winds and unwinds, according to the balance of the mechanism, oscillating in exact analogy with the working of a pendulum. (As with the pendulum, the balance wheel must get a regular "kick," provided by the mainspring, to keep going.) By 1674, Huygens was making watches that included balance wheels and spring assemblies (with what was probably the first useful instance of the spiral balance spring) that were correct within 10 minutes over the period of a day.

About that same time, Englishman Thomas Tompion (1639-1714) also began making watches that used balance springs. He went on to become the most famous English clockmaker of his time, and introduced a number of improvements that made pendulum clocks more accurate and portable. Together with Edward Barlow and William Houghton, he also patented the cylinder escapement (1695). Here, the wheel's teeth alternately ride on the inside of the cylinder, then on the outside. This created a compact mechanism for regulating spring power, which allowed the making of flat watches. Jean de Hautefeuille (1647-1724), a French physicist interested in acoustics and tidal phenomena, claimed priority for the invention of the spiral balance spring, but is not generally credited with it (though he is credited with the invention of the virgule escapement for watches in 1670).

In the 1660s, William Clement invented an escapement that was so effective that it justified the use of a second hand. The anchor escapement (so-called because of its shape) rocks back and forth in the same plane as the toothed wheel, providing a highly controlled braking of the motion of the mechanism.

Impact

Over a 250-year period, timekeeping went from the village clocktower, accurate to the hour, to timepieces that were personal, portable, reliable, and accurate to the second. The very concept of time and its use changed during this period. Noon came not to mean when the Sun was at its zenith (apparent time), but the average of the times it was at its zenith (mean time), and humans began dividing their days by the rule of a mechanism rather than nature.

The ever more exact measurement of time was part of a general trend of measurement brought about by the successes of the scientific method. Numbers and standards were being applied to mass, volume, distance, heat, and other physical properties, inevitably moving toward the adoption of the metric system. Accurate measurement became essential to the progress of science and, on a more fundamental level, measurement allowed scientific concepts to be developed and expressed with equations, where pictorial and geometric methods had previously been dominant. The clock's contribution to science initially centered on astronomy, allowing an ever more detailed understanding of celestial dynamics. However, physics and engineering were not far behind.

The Age of Exploration was nearly over before the clock made a contribution. John Harrison's (1693-1776) chronometer allowed accurate determination of longitude, but Columbus, Magellan, and Vasco da Gama (1460?-1524) had depended on luck and "sailing the parallel" to avoid getting lost at sea. While this may have been adequate for adventurers, it was not sufficient for naval operations and was particularly unacceptable for commerce. Thus, the chronometer became a key tool in establishing sea power and facilitating worldwide trade.

Accurate clocks also made a commercial impact in factories by regulating work and synchronizing activity. This allowed more complex processes and greater efficiencies. In addition, the development of precision parts for clockmaking made at least as large a contribution as accurate timekeeping did. As clocks came to represent shared, community time, rather that personal time, they especially needed regularity, consistency, and standardization. Accomplishing this required new tools and new skills. With this new capability, springs, gears, and other devices for managing energy found their way into other devices, including toys, automata, weapons, and industrial equipment. The success of these devices and the mechanical view they engendered led to a cultural change. Clockmakers began forming their own craft guilds as early as 1544, with the Guild of Clockmakers in Paris. These people shared a point of view. They actively sought to use their skills to solve other problems. It is no coincidence that inventors John Fitch (1743-1798), John Whitehurst, David Rittenhouse (1732-1796), Eli Terry (1772-1852), Alexander Bain (1818-1903), Benjamin Huntsman (1704-1776), and John Kay (1704-1780?), as well as French economist and politician Pierre du Pont de Nemours (1739-1817), were all clockmakers. The image of the clockmaker-inventor even entered literature (for example, Drosselmeyer in *The Nutcracker*) and became the precursor for nutty inventors and mad scientists that are prevalent in popular culture.

The clock exemplified a mechanical view that, with the rise of science, moved inexorably into the philosophical arena. With the development of Isaac Newton's (1642-1727) laws, the idea of a clockmaker god, who built the universe then stood aside to listen to it tick, took hold. By the 1700s this concept was an inspiration to Deists, Rationalists, and Materialists, and continues to be a lively subject of debate.

Western culture is obsessed with time (time is money), creating ever more accurate electronic digital clocks, as well as watches that are the hallmarks of fashion and beauty. Our machines also have an obsession with time. Every computer includes a clock, which is essential to its operation. And the watch, in particular, has been a model for interactive devices of the information age.

Accoding to historian Lewis Mumford: "The clock, not the steam engine, is the key machine of the modern industrial age. In its relationship to determinable quantities of energy, to standardization, to automatic action and finally to its own special product, accurate timing, the clock has been the foremost machine in modern technics; and at each period it has remained in the lead: it marks a perfection toward which other machines aspire."

PETER J. ANDREWS

Further Reading

Asimov, Isaac. *Isaac Asimov's Biographical Encyclopedia of Science & Technology.* New York: Doubleday, 1976.

Barnett, Jo Ellen. *Time's Pendulum: From Sundials to Atomic Clocks, the Fascinating History of Timekeeping and How Our Discoveries Changed the World.* Chestnut Hill, MA: Harvest Books, 1999.

Landes, David S. *Revolution in Time: Clocks and the Making of the Modern World,* Cambridge, MA: Harvard University Press, 2000.

Yoder, Joella G. *Unrolling Time: Christiaan Huygens and the Mathematization of Nature.* Cambridge: Cambridge University Press, 1989.

Development of the Self-Regulating Oven

Overview

At the start of the seventeenth century, there was no way to measure heat. Although it was known that air expanded as it was heated, and compressed as it was cooled, no one had thought to assign numbers to the degrees of hot and cold. Cornelis Drebbel (1572-1633) was one of a small group of European practical and learned men who worked on developing air thermometers that included numerical scales. Their inventions and innovations illustrate the seventeenth-century trend toward quantifying natural phenomena. But Drebbel's greatest invention was the thermostat.

Background

Drebbel began his career in Holland as an engraver, but turned to mechanical invention in 1598. He had a special interest in how temperature and pressure cause a volume of air to vary. For example, Drebbel observed that heating air and water causes them to expand, whereas cooling compresses them. He described experiments in which he hung an empty glass laboratory vessel, called a retort, with its mouth in a container of water and its bulb toward a flame. As the air in the glass warmed, air bubbled out of the mouth of the retort into the water. But removing the flame from the fire cooled the air and compressed it, drawing water up into the glass.

Drebbel first applied the results of his observations to a kind of perpetual motion machine that he called the *perpetuum mobile.* The *perpetuum mobile* was an elaborate toy that Drebbel delighted in showing off to the Emperor Rudolf II (1552-1612) in Prague in 1610. Drebbel divided his working life between England and continental Europe, and his initial fame was probably due to this invention.

Basically, the *perpetuum mobile* consisted of a glass bulb filled with air and connected with a glass spiral that contained a small amount of liquid. Changes in temperature caused the air in the bulb to increase by day and decrease during the night. These variations caused the liquid in the spiral to move back and forth in a continuous motion, mimicking the flow of the tides. To many, the motions of the liquid in the tube were mysterious; but other of Drebbel's contemporaries realized that what he had built was an air thermometer.

Drebbel later used the principle of the *perpetuum mobile* in constructing his most notable invention: an oven with an automatic temperature-regulating device that we now call a thermostat. So that he could hatch duck and chicken eggs all year round, Drebbel built an apparatus consisting of a closed pool of alcohol connected to a tube containing mercury. As the temperature rose in the oven, the alcohol expanded, and pushed the mercury up the tube, which opened a valve that admitted cold air. The air then cooled the oven, and as the temperature fell, the volume of alcohol decreased. This caused the mercury column to drop and thereby shut the valve, so that the heat of the oven would again begin to raise the temperature.

Drebbel wrote little, and left no records concerning his thermostatic furnaces. What we know of his work, we know from his contemporaries. After Drebbel's death, the executors of his estate filed an English patent on his furnaces on behalf of his descendants. His inventions did not become widely known until years later. At a meeting of the Royal Society in 1662, in response to a problem proposed by Sir Robert Moray, a member suggested Drebbel's method of regulating a furnace using a mercury thermometer. But the proceedings in which Moray's remarks are quoted provide no technical details. Drebbel's incubators for hatching eggs were mentioned in the Royal Society in 1668. In fact it is Drebbel's son-in-law Johan Sibertus Kuffler (1595-1677) that we have to thank for building several of Drebbel's furnaces and supplying technical information about the temperature regulators.

Drebbel was not alone in proposing ideas for keeping heat at a constant degree. In 1677 the physicist Robert Hooke (1635-1703) described how several oil lamps could be made to burn evenly. And in 1680, Johann Joachim Becher (1635-1682) claimed to have invented temperature regulators. But Becher failed to supply technical details and illustrations.

Although the idea of temperature regulation remained undeveloped in England for the next century, a nobleman from Lyon named Balthasar de Monconys (1611-1665) had visited researchers and learned of Drebbel's thermostatic furnace in conversation with them. Johan Sibertus Kuffler had even shown it to him. Monconys published his observations in a book that appeared in several editions between 1677 and 1697. Thus, though neglected in England, the temperature regulator became common knowledge in continental Europe.

Impact

The thermostatic principle of Drebbel's self-regulating furnaces applies to self-regulation of all kinds, in engineering and industry, as well as biological systems such as body temperature. We use the term "feedback control" to describe the mechanisms that drive these systems. Drebbel's thermostatic furnace has been called the first feedback system invented since antiquity.

Feedback control functions to make unstable systems stable. It's what allows quantities such as pressure, temperature, velocity, and thickness to be maintained at a constant level. Feedback control systems have three components: they measure output, compare the output with a desired value, and then make corrections to achieve that value. For example, in a steel rolling mill a feedback mechanism controls the thickness of the billets, in a robot it controls the position of the end of the robot's arm, and in a tank it controls the pressure of the contents of the tank.

As you produce a product, the accumulation of the product slows or stops further production. When something is being heated in the oven, as the temperature of the oven increases, the process of heating decreases. A modern room thermostat has a sensing element that responds to the air temperature in the room, and a control element that regulates the heating process by making a correction whenever the air deviates

from the desired temperature. In Drebbel's time, the oven was going constantly. For him, the feedback mechanism didn't reduce the production of heat, as it does in a modern thermostat where the furnace goes off. Instead, for Drebbel, as the heat increased, the feedback mechanism dissipated the heat by applying cold air. Drebbel applied a principle that we realize is a much more significant one than just hatching eggs. It led to the self-control of mechanical devices.

The question persists whether Drebbel can also claim credit for having independently invented the air thermometer. His perpetual motion machine is usually interpreted as one, although Drebbel did not call it that. Certainly, contemporary accounts appear to indicate that he was widely thought to have invented it. A letter written to the philosopher and astronomer Galileo Galilei (1564-1642) mentions that the machine's tube was marked with equidistant diagonal lines. But although it is clear that Drebbel understood the principles involved and could have constructed a thermometer if he had wanted to, the actual inventor remains unknown. Along with Drebbel, the other likely candidates are Galileo; Santorio Santorre (1561-1611), a professor of medicine at Padua in Italy; and the Welsh mystic Robert Fludd (1574-1637).

Drebbel lived in troubled times. The Reformation, a religious revolution that began in 1517, had raised questions that were still not resolved in the early 1600s. Artists and inventors depended for their survival on the favor of kings, and they were often caught up in a web of religious and political machinations. Drebbel himself was arrested and released several times in various countries. The established order of science was still governed by the Aristotelian view of the universe, and by rigidly held ideas about the constitution of matter that had no basis in the observable world.

In his inventions, Drebbel displayed understanding of established scientific principles. Moreover, his research in chemistry and physics may well have contained significant discoveries. But he was also a showman, and sometimes mixed his results with magic to exhibit to an admiring public. What science there was would have been hard to discern behind the hocus-pocus. What he was, through and through, was an inventor who had little concern for publicizing his inventions.

Although no one disputes Drebbel's invention of the temperature regulator, he did not himself document his invention, nor did he attempt to produce it on any scale. It was others who in the eighteenth century turned his improvised laboratory device into a carefully designed practical appliance. It is one of the mysteries of the history of technology why some inventions appear and fail to develop, even when the science and know-how to develop them are at hand. One argument, aside from Drebbel's own peculiar methods, is that interest in feedback mechanisms had to await inventions that required them, such as the steam engine or float regulators for domestic water supplies.

Mystery notwithstanding, the importance of the idea of feedback control from the eighteenth century forward cannot be underestimated. Applied to biology, it is known as homeostasis, and refers to the ability of organisms to remain stable while adjusting to conditions they need to survive. It was also adopted in fields beyond engineering, for example, in Adam Smith's (1723-1790) economic postulate of laissez-faire, by which, given the right conditions, economies will automatically swing into equilibrium.

GISELLE WEISS

Further Reading

Books

Mayr, Otto. *The Origins of Feedback Control.* Cambridge, MA: MIT Press, 1970.

Middleton, W. E. Knowles. *A History of the Thermometer and Its Use in Meteorology.* Baltimore: Johns Hopkins Press, 1966.

Tierie, G. *Cornelis Drebbel (1572-1633).* Amsterdam: H. J. Paris, 1932.

Thorndike, Lynn. *The Thermometer: History of Magic and Experimental Science.* Vol. 7: *Seventeenth Century.* New York: Columbia University Press, 1958.

Internet Sites

http://es.rice.edu/ES/humsoc/Galileo/Things/thermometer.htm

Denis Papin Invents the Pressure Cooker

Overview

A pressure cooker is a vessel that uses steam under high pressure for cooking food. It offers a number of benefits, including fast, often low-fat cooking that preserves the minerals—and even the coloration—of fruits, vegetables, and meats. For Americans born in the twentieth century, pressure cookers were a familiar part of home life, so much so that they had come to seem positively old-fashioned by the 1990s, when they began making a comeback among health-conscious consumers. In fact the genesis of the pressure cooker dates back to a 1679 invention by French physicist Denis Papin (1647-1712), though it would be many centuries before a modified version of his "steam digester" would be adapted for household use.

Background

Many of the features associated with the modern kitchen and dining room are more recent in origin than most modern people would imagine: the plate, for instance, did not make its debut among the common people until the sixteenth or seventeenth century. Much of a cook's work was done by methods that had persisted more or less unchanged since prehistoric times. Until the stove appeared in the 1600s, the vast majority of cooking was done over an open fire, or more rarely, in a rudimentary oven. Even in the latter case, fire provided the heat.

Fire would also be the source of heat in the steam digester, though the significance of Papin's innovation was that it introduced a new medium between the flame and the food it cooked: steam. In fact the road to Papin's invention was not a direct one, because his chief concern was not cooking but steam pressure and power.

The last real progress in steam power prior to Papin's time dated back some 1,500 years, to Hero of Alexandria (fl. first century A.D.). Hero's aeliophile consisted of a sphere resting on two hollow tubes, themselves connected to a steam-producing boiler. The heated steam escaped through the hollow tubes, which then caused the sphere to whirl. It was an interesting invention, but really nothing more than a curiosity—rather like the wheels created by the Olmec of ancient Mesoamerica, who used them only in making children's toys, and failed to grasp their use in providing traction for transportation, agriculture, and building projects.

Papin himself began his career working not on steam, but on air pumps, as an assistant to physicists Christiaan Huygens (1629-1695) and Robert Boyle (1627-1691). He worked with Huygens in Paris from 1671 to 1675, then with Boyle in London until 1679, the year he presented his steam digester to the Royal Society. No doubt the heightened atmosphere of curiosity and experimentation associated with the Royal Society, one of the most influential scientific institutions in world history, had an effect on his thinking. In any case, it was during this period that he developed the idea of using steam as a form of power.

Papin first demonstrated his "New Digester for Softening Bones" before the Royal Society on May 22, 1679. To modern eyes, it was an ungainly looking contraption, a far cry from the compact pressure cookers that would later appear in modern kitchens. Shaped rather like a potbellied stove, the digester consisted of a raised metal cylinder containing a glass vessel. Papin filled the cylinder with water equal to the difference in volume between the glass container and the cylinder itself, then placed some meat into the container, added liquid to it, and sealed it. He then applied heat from a fire beneath the cylinder, using a built-in safety valve to release excess steam once the interior of the container had reached the necessary pressure.

The gentlemen of the Royal Society were pleased to discover that the steam digester cooked the meat quickly and thoroughly, producing a dish far more tender and tasty than a similar item cooked over a fire. The key element was the steam, which increased the pressure and—because of the relationship between pressure and temperature—heated the meat more quickly than a mere fire could have. As a result of cooking faster, the food retained more of its flavor and (though this was a fact far beyond the knowledge of scientists in the seventeenth century) its nutritional content.

Another aspect of scientific knowledge yet to make its appearance at the time was an accurate means of measuring temperature. That would have to wait for two men who were not even born at the time: Daniel Fahrenheit (1686-1736) and Anders Celsius (1701-1744), who produced their respective temperature scales in the following century. Because the pressure cooker did its work so quickly, it was important to have some idea when the meat was cooked, so as to avoid overcooking—but for safety, this had to be done without lifting the lid.

Papin devised an ingenious method for overcoming this challenge. He had a depression built into the top of the pressure cooker, and into this he placed a drop of water. He could then see when the water boiled, and using a three-foot pendulum, he was able to time the interval necessary for boiling as well as for evaporation. He also weighed the coal necessary to operate the pressure cooker, as a means of measuring the instrument's efficiency.

Impact

Though Papin's invention would ultimately have a great impact in the kitchen, its influence was still greater, because in fact what he had created was a forerunner of the piston and cylinder mechanism later incorporated into engines. When the cold water inside the cylinder was heated, this raised the "piston"—i.e., the cooking vessel. This in turn created a partial vacuum, and as a result, the outside air pressure forced the piston downward. Of course in the case of the pressure cooker, the purpose was not to force the cooking vessel toward the bottom of the cylinder, but simply to produce the pressure necessary for cooking; nonetheless, the principle of the active piston stroke had made its appearance.

Sir Christopher Wren (1632-1723) of the Royal Society commissioned Papin to write a booklet concerning his invention, and Papin went on to conduct experiments with the idea of a steam engine. In 1690 he produced an atmospheric engine using a tube, three inches (7.6 cm) in diameter and sealed at one end, that contained a movable piston. The tube was filled with cold water that, when heated, converted to steam. This caused the tube to rise; then as the steam cooled and condensed, the atmospheric pressure forced the piston to move downward to its original position.

In 1698 English inventor Thomas Savery (1650-1715) developed a pump used for removing water from flooded mines. Papin studied Savery's pump, which did not include a piston, and concluded that the addition of the latter would make a more effective engine. He also improved on Savery's idea of a boat propelled by side paddles, which though it seemed impractical to many at the time, would obviously offer great advantages over either human or wind power.

Savery himself, failing to see the implications of steam in this instance, had intended to use muscle power for operating the paddles, whereas Papin's paddle wheel—prefiguring the idea of a steamboat—used steam. Unfortunately, the vessel was destroyed, apparently by river boatmen who feared a challenge to their means of livelihood. The full realization of the steam engine itself would have to wait for Thomas Newcomen (1663-1729), who produced his in 1712, the same year that Papin died.

Ironically, the man who created the pressure cooker, and played a major role in the use of steam power, died poor and largely forgotten. The pressure cooker, too, lingered in obscurity for many years, and when it finally did receive attention, it was not used for the application most commonly associated with it today. Rather, as scientists in the centuries that followed came to recognize the value of sterilization in the medical environment, the pressure cooker entered service as a sterilizer for metal instruments.

The first notable culinary application of the pressure cooker after Papin's time came in 1810, when French chef and confectioner Nicolas François Appert used it for boiling sealed containers of food. Thus was born the canning industry, which made it possible to preserve foods indefinitely, and to transport them anywhere.

At the beginning of the twentieth century, when other innovations such as the harnessing of electrical power served to facilitate the use of the pressure cooker for home use, the appliance finally began to appear in the kitchen. The term "pressure cooker" itself first appeared in print in 1915, and by the 1920s pressure cookers themselves had made their way into the homes of wealthy and middle-class consumers. At the World's Fair in 1939, National Presto Industries presented what became the first widely popular commercial pressure cooker.

World War II proved to be a boon for the pressure cooker, when wartime shortages of energy forced American homemakers to seek more energy-efficient means of cooking. As a result, baby boomers grew up on meals cooked in the appliance, which became as much a fixture of 1950s and 1960s households as television sets and frost-free refrigerators. Pressure cookers were particularly useful for people living in high-altitude areas, where lowered atmospheric pressure required longer cooking times for foods prepared with a traditional oven or stove.

The pressure cooker proved so popular, in fact, that it had come to seem rather passé by the 1970s and the 1980s. Fear of dangers associated with high-pressure cooking may also have had something to do with its waning use: an episode of the highly popular *I Love Lucy* television show during the 1950s, for instance, featured a pressure cooker that exploded, sending hot food flying all over star Lucille Ball's kitchen. Pressure-cooker manufacturers responded by adding safety features such as locks, pressure regulators, and low-pressure fryers. These measures, coupled with a growing interest in healthy eating, spurred a resurgence of interest in the pressure cooker during the 1990s.

JUDSON KNIGHT

Further Reading

Books
Bluestein, Barry. *Express Cooking: Making Healthy Meals Fast in Today's Quiet, Safe Pressure Cookers.* New York: HP Books, 2000.

Larson, Egan. *The Pegasus Book of Inventors.* London: D. Dobson, 1965.

Internet Sites
"Papin Engine Animation." http://www.geocities.com/ Athens/Acropolis/6914/pappe.htm (July 12, 2000).

Andrea Palladio and Developments in Western Architecture

Overview

At the height of his popularity and influence in 1570, Italian architect Andrea Palladio (1508-1580) published his masterpiece, a treatise titled *I Quattro Libri Dell' Architettura* (*Four Books of Architecture*). The book solidified his standing as one of the greatest architects in history. *Quattro Libri* allowed Palladio's contemporaries and future generations of architects to examine his philosophies on the design of houses, bridges, civic and public buildings, and ancient temples. *Quattro Libri* is widely regarded as the finest architectural textbook ever produced.

Background

Palladio's spiritual mentor was ancient Roman architect Vitruvius, whose work *Ten Books on Ar-*

chitecture, was the first attempt to outline the theoretical principles of the field. Although Vitruvius's efforts gave architecture an intellectual footing it lacked prior to his study, he still was unable to cover the discipline completely. Nevertheless, Vitruvius still supplied the best blueprint to date and his work was regarded as the exemplary text for hundreds of years.

Vitruvius's influence was so great that Leon Battista Alberti (1404-1472) imitated his style in his own *De Re Aedificatoria,* even though he was critical of the Roman master. Alberti incorporated numerous literary sources into his work, including Plato (427?-347 B.C.) and Aristotle (384-322 B.C.), to create a sociology of architecture.

The advent of mass printing allowed both authors to gain wider audiences. Alberti was first published in 1485, while publishers reprinted Vitruvius's treatise a year later. Architectural books increased dramatically in the sixteenth century and were translated into Italian, thus expanding readership and the use of the works as teaching tools. Translators were also able to illustrate the books with woodcuts. In 1522 an Italian version of Vitruvius appeared with more than 100 illustrations.

Another important architect and author, Sebastiano Serlio (1475-1554), published a multi-volume handbook beginning in 1537. His five books on architecture were the first to deal with the subject visually as well as in theory. Serlio's treatise served as a blueprint for Palladio's later publications. Palladio relied heavily on drawings in his own work and paired them with logical, concise narrative descriptions of his theories.

Impact

In his late twenties and trained as a stonemason, Palladio met Gian Giorgio Trissino (1478-1550), Vicenza's leading intellectual and humanist. Trissino was rebuilding a villa in nearby Cricoli in classical ancient Roman style and set up an academy there to provide young aristocrats with a traditional education. Palladio worked on the renovation project and his natural design skills led Trissino to invite him to join the academy. Trissino then directed the young man's initial forays into architecture and renamed him Palladio (he was born Andrea di Pietro della Gondola), a frequent occurrence in humanist study.

Palladio's focused architectural education was unusual in this period; most students were steered toward a general course of study. Under Trissino's guidance, Palladio adopted his men-

tor's ideas regarding symmetrical layout, with a large central room and flanking towers. During his studies, Palladio copied Serlio's woodcuts and drawings of Roman monuments. Humanistic architecture study centered on the writings and remains of classical antiquity.

Architecture historians are not sure why Palladio was drawn to publication, but it seemed to be on his mind from the start of his career. Many of his early villa and palace drawings look as if he intended them to be published. Palladio's career as an author seems to stem from his early trips to Rome in the 1540s. He acquired many blueprints for ancient buildings and added them to his own expanding portfolio.

Architects in the sixteenth century, like Serlio, commented on the difficulty and high cost associated with studying ancient buildings. Thus, publishing was a way to preserve their work. In the dedication of the first two books of *Quattro Libri,* Palladio echoes this sentiment. Also, in 1555, writer Anton Francesco Doni discussed Palladio's prodigious output, saying, "The man came into the world to put architecture to rights. His book has no title, but from its contents it could be described as a guide to good architecture."

With his publication of *Quattro Libri,* Palladio drew upon the various strands of architectural writing, but then improved on them by presenting them in a more concise and fluent narrative. In addition, Palladio's own intensely productive building career, spanning more than two decades, contributed to the book's success. He aspired to publish a complete guide to architecture that encompassed everything from the foundation to the roof. Evidence exists suggesting that Palladio hoped to publish more than four volumes and planned them to be the first in a series of books, along the lines of his spiritual mentor Vitruvius.

Quattro Libri, in essence, became Palladio's summary of his lifelong study of classical architecture. The first two books outline Palladio's principles of building materials and his designs for town and country villas. The third volume illustrates his thinking about bridges, town planning, and basilicas, which were oblong public halls. The final book deals with the reconstruction of ancient Roman temples.

Quattro Libri secured Palladio's great importance in architectural history. The treatise popularized classical design and his own innovative style. It served as a veritable blueprint for design worldwide, reaching its zenith in the eighteenth

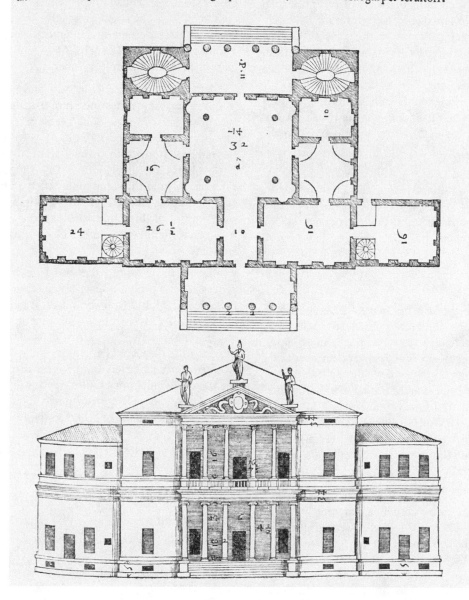

An illustration from Palladio's *I Quattro Libri Dell' Architettura* (1570). *(Corbis Corporation. Reproduced with permission.)*

century. In the eyes of many scholars, the work stands as the clearest and best-organized textbook on architecture ever produced.

Palladio's treatise revealed the link between humanistic education and architecture, and by doing so changed the way architecture was perceived. His status as an international bestseller spread his theories far beyond his native Italy, and inspired the label Palladianism, considered the search for classical beauty in architecture.

Palladianism spread beyond the grand architect's death in 1580. Palladio's most brilliant pupil, Vincenzo Scamozzi (1552-1616), continued his mentor's work as the founder of Neoclassicism. Because of his stature as Palladio's student, Scamozzi received a jumpstart on his career, which then took off even further after Palladio's death. Although Scamozzi did attempt to distance himself somewhat from Palladio, the master's influence spread.

Palladio stayed in the public's eye through subsequent editions of *Quattro Libri*, published in multiple languages. An important Palladian revival began in Venice in the early 1700s. Architects there returned to the classical work of Palladio in reaction to the ornate style of the late Baroque period. The earliest Palladians were Domenico Rossi and Andrea Tirali. The most prominent neo-Palladian was Tommaso Temanza, who provided an intellectual basis to the movement.

England faced a similar resistance to the excessiveness of the Baroque era. English tourists discovered Venice and a market developed for mementos from Italy. English consul Joseph Smith funneled drawings, paintings, and sculptures from Venice to England. Soon, the artists Smith patronized visited England and were subsidized by English nobles.

The greatest proponent of Palladianism in England was Inigo Jones (1573-1652). He visited Vicenza and the aging Scamozzi in 1614. Although Scamozzi was secretive and irritable, Jones obtained original drawings by Palladio. Up to that time, most architectural works from Italy were poorly translated or unavailable all together. Sir Henry Wotton, England's ambassador in Venice, also collected original drawings from Palladio and Scamozzi. Jones's collection supplemented, and in some ways, surpassed the woodcuts in the *Quattro Libri*.

Under Jones, Palladio's drawings gained a wider audience in the English-speaking world. He passed them among leading architects of the day, which continued for more than a century, providing a constant source of information and inspiration. Later, increasingly impressive translations of Palladio's *Quattro Libri* appeared in English and French.

In young America, Palladio's influence could be found in the aristocratic circles of plantation Virginia. Because the region's weather mimicked that of northern Italy, Palladio's villa was an ideal model for plantation owners to endure the hot humid summers. In fact, many Virginia plantation homes were so closely imitated that Palladio and later Palladians seemed to simply provide a pattern book.

Peter Harrison, a native of England but transplanted to Rhode Island, was the first true Palladian in America. He designed the Redwood Library in Newport, Rhode Island, and King's Chapel in Boston based on Palladian principles. Harrison died in 1775, at the beginning of the Revolutionary War, but another, more famous person—Thomas Jefferson (1743-1826)—emerged to grasp the Palladian mantle in America.

Jefferson's intellectual curiosity and humanist education almost destined him to adopt Palladianism. Jefferson's lifelong interest in architecture began as he started to build his own home, the famous Monticello (from Italian, meaning Little Mountain), overlooking Charlottesville, Virginia. Around the same time, Jefferson headed a committee charged with designing the new state capitol of Richmond. In this role he also trumpeted architecture in the classical style of antiquity.

In 1816, giving advice to a friend in the process of building a house, Jefferson declared Palladio, "the Bible. You should get it and stick close to it." As president, the prominent Virginian also implemented aspects of Palladianism in the design of the Capitol and along Pennsylvania Avenue. Perhaps Jefferson's greatest display of the use of Palladian principles lies in his design of the University of Virginia. Interestingly, Jefferson's use of Palladianism reflected the humanist political theories of the day and his dedication to republican principles.

BOB BATCHELOR

Further Reading

Boucher, Bruce. *Andrea Palladio: The Architect in His Times.* New York: Abbeville Press, 1994.

Constant, Caroline. *The Palladio Guide.* London: Architectural Press, 1987.

Holberton, Paul. *Palladio's Villas: Life in the Renaissance Countryside.* London: John Murray, 1990.

Tavernor, Robert. *Palladio and Palladianism.* London: Thames and Hudson, 1991.

Wundram, Manfred, and Thomas Pape. *Andrea Palladio, 1508-1580: Architect Between the Renaissance and Baroque.* Cologne, Germany: Benedikt Taschen, 1993.

The Palace of Versailles

Overview

Louis XIV, France's Sun King, demanded a palace symbolic of his power and authority, a place that would awe the nobility of Europe. He turned to noted French architects Louis Le Vau (1612-1670) and Jules Hardouin-Mansart (1646-1708) to transform a modest hunting lodge southwest of Paris into the greatest palace ever known. For nearly two decades, the men carved the Palace of Versailles out of the thick wooded marshes and achieved Louis's vision of enormity and splendor.

Background

Originally built in 1624, Versailles served as a hunting lodge for Louis XIII, but was transformed into the palace of his son, Louis XIV, over many decades. Since the elder king died when his son was only five years old, his mother and prime minister Cardinal Mazarin ruled France. The monarchy survived a civil war, but instilled in the young Louis XIV a lifelong distrust of Paris and the nobility. The boy witnessed the invasion of the royal palace by enemy forces and the memory remained permanently etched in his mind.

After Mazarin's death in 1661, Louis XIV ascended to his birthright and ruled France. Because of his boyhood memory of revolution, he decided to establish a palace far from Paris. He chose Versailles, a sight he first hunted at in 1651 and visited on several occasions until his marriage in 1660. After he chose the site, Louis XIV became obsessed with it and spent an increasing amount of time concerned with its design. Over the objections of his advisers, the young king disregarded the amount of money or time he spent at Versailles.

Major modifications of the original hunting chateau began in 1661. Louis Le Vau served as the principal designer. Transforming a chateau into the world's greatest palace was an arduous undertaking. The first step was draining the swamps and leveling the land in the palace location. Thousands of laborers paid for this work with their lives, dying from fever and pneumonia. Slowly, however, the change took shape.

One of Le Vau's first additions was a menagerie and orangery that would become the initial piece of a grand entry court. Gradually, Louis XIV decided that his entire court would be moved from Paris to Versailles. In preparation for the move, a second building phase started in 1668 that totally redesigned the chateau, essentially transforming it into an entirely new building.

Le Vau extended the palace overlooking the gardens by wrapping the area in stone. He then built two symmetrical apartments, the north end reserved for the king, while the south portion went to the queen. In the central area stood a reception area and a terrace overlooking the gardens.

Le Vau worked with noted gardener Andre Le Notre and painter Charles Le Brun to design the landscape and gardens outside the palace and the walls within. Le Vau was a student of classical architecture and combined his theories with dedication to enormous scale, in line with the Sun King's grandeur. Le Vau's interiors, including the Ambassador's Staircase (destroyed in 1752), unite a vision of largesse and baroque influences that was well beyond anything that existed at the time.

The excavation of the Grand Canal began in 1667. Louis XIV worried about having enough water for the gardens and decided the Seine River should be used as the main source. Workers used immense hydraulic machines driven by the current of the river to deliver water. The Grand Canal became a giant artificial pond of more than 70 acres with a circumference of 5 miles (8 km). An intricate network of underground reservoirs and aqueducts supplied the palace fountains and waterfalls.

In fact, Le Vau was responsible for many of the palace's remarkable gardens and its park-like setting, centralized on the Grand Canal. The palace of Versailles holds numerous monuments, sculptures, and secondary buildings, including the Grand Trianon and the Petit Trianon (a favorite of the infamous Marie Antoinette). In total, there are more than 300 statues and multiple ponds in the gardens. Interestingly, the layout of the gardens remains the same in current times as when Louis XIV originally planned them.

After Le Vau's death in 1670, French architect Jules Hardouin-Mansart became royal architect in 1675. He undertook a massive campaign that would last for more than a decade. Initiated in 1678 and involving more than 30,000 labor-

A view of the Palace of Versailles. This shows a statue of Louis XIV in the Court de Marbre. *(Adam Woolfitt/Corbis. Reproduced with permission.)*

ers and craftsmen, Hardouin-Mansart designed the north and south wings of the palace. Included in his designs was Versailles's most enduring symbol, the Hall of Mirrors, completed in 1684. He is noted for introducing French materials (rather than Italian) into the building process, including mirrors and pink marble.

Hardouin-Mansart had a special skill at designing buildings and monuments that expressed Louis XIV's wealth and power. His trademarks were buildings with enormous scale, yet possessing an understated simplicity. At the end of his career, Hardouin-Mansart fell victim to rumors and innuendo spread by jealous architects. They claimed that he stole designs, but this charge lacks credibility. Hardouin-Mansart's grand vision of Versailles is seen throughout the palace.

The building project did not stop, even after Louis XIV transferred the court and the seat of government to the Palace of Versailles in 1682. The largest court in Europe, Versailles housed 20,000 nobles.

Hardouin-Mansart began construction of the palace chapel in 1699, but it was not completed until two years after his death in 1710. The architect's brother-in-law, Robert de Cotte, finished the chapel in his place. Construction of the grand chapel even outlived the king himself, who died in 1715.

Louis XV continued an aggressive building campaign at Versailles. Jacques-Ange Gabriel designed an opera house for the palace that was completed in 1748. It took the French Revolution (1789-1799) to halt further building at Versailles. After the reign of Louis XVI ended with the revolt, the furnishings of the palace were sold and the building was turned into a museum.

Impact

In addition to being one of the world's most beautiful buildings, the palace of Versailles was also one of the most expensive. In no small measure, Versailles can be blamed for most of the economic problems the country endured in the generations leading up to the French Revolution. Louis XIV's disregard for the costs associated with the castle placed an economic burden on the people of France that eventually festered into revolt.

Moving the nobility from Paris to Versailles also undercut their power, which Louis XIV realized. After 1682 the nobility no longer had power to affect French politics. Instead, life at Versailles centered on etiquette and serving the king's every whim and fancy.

By the time Louis XVI assumed the reigns of France, the national debt exceeded 4,000 million livres. Even under intense economic pressure, Louis XVI continued adding to the palace.

In the nineteenth and twentieth centuries, architects and designers restored Versailles to its original glory. Many of the palace's stunning decorations and furnishings were located and put back in place. The refurbished areas included the Salon of Hercules and Marie Antoinette's bedroom.

Living up to Louis XIV's original vision, the Hall of Mirrors in Versailles served as an important meeting place for gathering diplomats. The Hall of Mirrors was the site where a united Germany declared Wilhelm I its emperor after the victory over France in the Franco-Prussian War (1870-1871). The palace had been used as a Prussian military headquarters during the war.

While Versailles witnessed the rise of the German Empire, it also saw its fall nearly 50 years later. After defeating Germany in World War I (1914-1918), the Allied powers held their diplomatic meeting at Versailles. On one hand, the meeting set the stage for the League of Na-

tions, but on the other, it outlined harsh reparations against Germany that would remain a rallying point for German nationalists, leading to the rise of Adolph Hitler and the Third Reich.

Versailles attracts more than 30,000 tourists a day, an average of nearly 10 million a year. The long history and beautiful grounds and buildings draw visitors from every corner of the globe. Perhaps the most lasting tribute to the Sun King is that the palace of Versailles continues to leave visitors awestruck, just as he planned.

BOB BATCHELOR

Further Reading

Ballon, Hilary. *Louis Le Vau: Mazarin's College, Colbert's Revenge.* Princeton: Princeton University Press, 1999.

Constans, Claire. *Versailles: Absolutism and Harmony.* New York: Vendome Press, 1998.

Hibbert, Christopher. *Versailles.* New York: Newsweek, 1972.

Van der Kemp, Gerald. *Versailles.* New York: Vendome Press, 1978.

Development of the Midi Canal

Overview

Civilizations depend on water supply, and since ancient times people have sought ways of building channels to hold water that would circumvent natural barriers both to the supply of water and to navigation. The Romans and Charlemagne dreamed of a way to transport merchandise that would avoid having to detour around the coast of Spain or risking attack by Barbary pirates. The solution appeared in the seventeenth century in the form of the Midi Canal.

Background

The Briare Canal in France, constructed by Adam de Crappone (1526-1576), was the first important watershed canal in the West. A watershed is a ridge of high land that divides areas drained by different river systems. On his return to France in 1516, Francis I (1494-1547) discussed with Leonardo da Vinci (1452-1519) a watershed canal between the Mediterranean Sea and the Atlantic Ocean. The idea was to join the Garonne and Aude Rivers—the latter leading south to the sea—via a canal that was later

called the Languedoc or Midi Canal. Francis died in 1547, and it was not until 1598 that Henri IV (1553-1610) resuscitated the plans for the canal. But at the time, he had several more pressing canal projects, and his first matter of business was to build the Briare Canal. Henri died before he could return to the Midi Canal, and it was Louis the XIV (1638-1715) under whose aegis the Midi Canal was realized.

The person hired to carry out the project was a French public official and self-made engineer named Pierre-Paul Riquet (1604-1680). Riquet had himself become interested in how to construct a shortcut from the Bay of Biscay to the Mediterranean Sea. In 1662, already past middle age, he wrote a canny letter to the controller-general of finances outlining his plans for a canal that would help to cut in on the King of Spain's profits from the Straits of Gibraltar. The letter was brought to the attention of the king, who appointed a royal commission to approve Riquet's project. Because the canal would have to cross a watershed, the royal commission immediately pointed out the difficulty of supplying water to

A view of the Canal du Midi in France. *(Michael Busselle/Corbis. Reproduced with permission.)*

the canal's summit. Riquet proposed to solve the problem using two feeders, one 26 miles (42 km) long. It would be one of the canal's most notable technical achievements. Work began on the water supply in 1665, and construction began on the canal soon after. Ultimately, 8,000 workers were put to work on building the canal, including several hundred women. It took 14 years to complete, and opened in 1681. Riquet did not live to see it finished.

The Midi Canal stretched 150 miles (240 km) between the City of Toulouse and the Mediterranean Sea, and had 101 locks—devices that allow a vessel to negotiate changes in altitude by raising or lowering the water level. Until the early 1800s, most locks used in river canalizations were flash locks. The flash lock was a barrier that acted as a dam. Water would build up behind the dam, and when the reach behind the dam was full, a line of waiting boats would be let in through narrow lift gates in the dam. The process circumvented some of the problems of natural rivers, but it was slow, wasteful of water, and expensive.

For the Midi Canal, Riquet used pound locks, which are enclosed chambers about the same size as the boats using them, with gates at both ends. These locks first appeared in the Netherlands and northern Italy during the fourteenth and fifteenth centuries. Although he did

not invent them, Leonardo da Vinci is credited with having brought the idea to France in 1515. But for the next 200-odd years, they were used chiefly in building watershed canals. Riquet constructed his system of locks arranged like a staircase to negotiate the 206-foot (63-m) rise to the summit, and 620 feet (189 m) down.

Near the southern French town of Beziers, he encountered a rocky rise. The problem with the rise was that it was made of alluvial stone, liable to collapse, and permeable to water. In the face of considerable opposition, Riquet used black powder to blast a 515-foot (157-m) tunnel through the rise, afterward leading an astounded commission through it by candlelight. Riquet's was the first canal tunnel built in this way, and he was the first to use explosives in underground construction. He also constructed three large aqueducts to carry the canal over rivers. In one of these aqueducts, the Repudre aqueduct, the bed of the canal served as a single-arch bridge across the river.

Without sufficient water, a canal is no more reliable than the natural waterways it seeks to improve. The Midi Canal is most renowned for its water supply system, which included the world's first known artificial reservoir for canal water supply. The reservoir was an earth-filled dam with masonry walls across the Laudot River, and it took four years to build. As a gravity dam, it relied on its weight for stability, and it was the first in Europe of its size. Water could be let out from the reservoir by means of two underground vaults. Riquet and the military engineer Sebastien Vauban (1633-1707) built feeder channels to supplement the reservoir. Additional dams were built in the eighteenth century to supply still more water to the canal.

Impact

French canal builders of what is called the old regime focused their technical abilities on specific problems of artificial waterways. For ages, people had developed ways of moving water for drainage, irrigation, and water supply. But for all their remarkable achievement, these efforts at improving or creating navigable waterways represented a minor effort in the ancient world.

The Midi Canal has been called Europe's finest seventeenth-century engineering work. Apart from its architectural interest, it served as a model of technical ambition and excellence. For many years, leaders had sought an all-water link between the two seas as a triumph of military strategy over Spain. Landowners and merchants in the region saw the canal as a means to commercial development that bypassed bad roads.

But the Midi Canal also illustrates the contradictions between high and low technological systems that existed in France prior to the eighteenth century. Until the time of the French Revolution (1787-1799), France lacked a coherent transport policy. What this meant in practical terms was that no one had ever thought through a way of transporting goods and people that would maximize the benefits and minimize the inefficiencies of both roads and rivers. The result was confusion, and a disproportionate emphasis—for military and political reasons—on a system of highways linking Paris to other parts of the country, even though water was clearly preferable for shipping heavy, low-value goods. Another reason roads were preferred was that they were cheap because they were built and maintained largely by unpaid labor.

Until 1750, therefore, when the government began to take a greater interest in them, waterways were largely funded by private means. But the state did not abdicate control over them. The state chose the engineers to build the canals, approved the building specifications, and provided a portion of funding. A disadvantage to this hands-on, hands-off system was that it was political: the entire process from conception through

construction was vulnerable to the fortunes and caprices of a particular government. An advantage for the Midi Canal was that the 14 years it took to build it fell within a peaceful and prosperous era in French history. By contrast, during the building of the Briare Canal, the king was assassinated, the treasurer was forced to resign, and a commission of inquiry was appointed.

By suggesting that the commercial success of the canal would enhance the king's revenues, Riquet was able to obtain construction subsidies from Louis XIV. Moreover, owing to the administration's affection for Riquet, royal support continued even after Riquet had spent the last penny of his personal funds. Asked why he had sunk all the money set aside for his daughters' dowries into the canal, Riquet answered, "The canal is my dearest child."

Economically speaking, Riquet's technological prowess was ahead of its time. Despite the improvements made to the waterways, including a handful of canals, of which the Midi Canal was one, navigation conditions progressed very little up to the time of the Revolution. Towpaths for hauling barges along canals were not well maintained, and in other places, farmers and millers encroached on them. Boatmen's corporations rose up that sought to establish their members' monopolies over carrying goods along a waterway. These monopolies were vigorously defended, using political, economic, and sometimes physical means. Moreover, tolls, heavy to begin with and often arbitrary from river to river, contributed to the high cost of water transport, which limited its appeal.

For all that the achievements of a few canal builders failed to fulfill the promises of canal enthusiasts, France was still the world leader in artificial waterways almost up to the end of the 1700s. Despite impressive waterway networks, some countries, like Holland, did not have to cope with the same technical problems posed by the French countryside. In other countries such as the United States and Germany, programs of waterway improvement had hardly begun.

GISELLE WEISS

Further Reading

Books

Geiger, Reed G. *Planning the French Canals: Bureaucracy, Politics, and Enterprise under the Restoration.* Newark: University of Delaware Press, 1994.

Hadfield, Charles. *World Canals: Inland Navigation Past and Present.* New York: Facts on File, 1986.

Payne, Robert. *The Canal Builders: The Story of Canal Engineers through the Ages.* New York: Macmillan, 1959.

Internet Sites
"History of the Canal des Deux Mers." http://www.canal-du-midi.org/history.htm

Biographical Sketches

Vannoccio Biringuccio
1480-1539
Italian Metallurgist

Although he had a tempestuous career in which he found himself embroiled in the political intrigue that characterized Renaissance Italy, Vannoccio Biringuccio is known chiefly for a book that appeared only after his death. This was *De la pirotechnica,* an encyclopedia of metallurgical knowledge focused more on technique and observation than on scientific generalization.

Biringuccio was born in Siena in 1480, and grew up amid the turmoil of the era. During this period Italy, which had not been united under a single government since Roman times—and would not be again until the 1800s—was dominated by powerful families of warlords and tycoons, the most famous of which were the Borgias and the Medici. The first family of Siena were the Petrucci, and from the beginning of his career Biringuccio was associated with them.

Under the patronage of Pandolfo Petrucci, young Biringuccio traveled throughout Italy and Germany, beginning to compile the information that would go into his life's work, the *Pirotechnica.* Pandolfo later appointed him director of mines in a town near Siena, and following the elder Petrucci's death in 1512, Biringuccio aligned himself with Borghese, Pandolfo's son.

Borghese appointed Biringuccio to a post in the Siena armory, but in 1515 Biringuccio found himself caught up in a controversy that had at its root a rival political faction's antipathy toward the Petrucci. On claims that Borghese had ordered Biringuccio and mint director Francesco Castori to debase the currency by adding base metal to the silver and gold, all three men were forced to leave the city along with their families and many others.

During the period of his exile, which lasted until 1523, Biringuccio traveled around Italy and Sicily. Finally Pope Clement VII became involved in the Siena conflict, and through his efforts the Petrucci family were restored to power under the leadership of Fabio, Borghese's younger brother. Biringuccio, too, was returned to his former position and property, and in 1524 was granted a monopoly over the saltpeter trade, valuable for its uses in making munitions.

Just two years later, however, the Petrucci were again thrown out, this time for good. Biringuccio lost all his property, but was fortunate enough to be in Florence when the crisis broke out in Siena. During his second exile, from 1526 to 1529, he traveled again to Germany; then, with the restoration of peace in Siena in 1530, he returned to his hometown.

During the early 1530s, Biringuccio held a number of key posts in Siena, and in 1538 was appointed director of the papal foundry and the papal munitions in Rome. He did not hold these positions for long, however: in early 1539, he died in Rome at the age of 59.

The following year saw the publication of the *Pirotechnica,* a book containing a career's worth of knowledge on mining ores and extracting metals from them. Because Biringuccio was a practical metallurgist and not a scientist, the book is little concerned with theory or speculation; nonetheless, it made an invaluable handbook for students of the subject, and went into nine subsequent editions over the next 138 years.

JUDSON KNIGHT

William Caxton
1421?-1491
British Printer

In the late fifteenth century, as printing presses on the continent were gaining prominence, one man, William Caxton, had the foresight to bring printed works to England. Although his career began in textiles, Caxton retired from the textile business before learning the art of printing. He set up a printing business in Bruges in 1474, the same year he printed the first known book in the English language, *Recuyell of the His-*

tories of Troie, which he translated from the French. In 1476 Caxton returned to England and set up his printing and publishing business near Westminster Abbey. In the ensuing years his press introduced many of the literary masterpieces of his day, including Chaucer's *Canterbury Tales* (1478) and Malory's *Morte d'Arthur* (1485).

Born around 1421 in Kent, England, William Caxton received schooling before entering the Mercers' Company, an influential London guild, and being apprenticed to Robert Large, a leading textile merchant, in 1438. He learned the export trade in textiles and, around 1441, moved to Bruges, Belgium (modern Brussels), where he developed a successful trade business. In 1462 Caxton was appointed Governor of the English Nation at Bruges, an appointment for an organization created by the Mercers and the Merchant Adventurers. After some time devoted to diplomatic missions for this organization, Caxton retired from commerce and became secretary of the household of Princess Margaret of York, the Duchess of Burgundy and sister of King Edward IV of England. The Duchess was a noted scholar of literature, and she encouraged Caxton to begin producing fine manuscripts, which he copied by hand, making translations from the French.

In 1471 Caxton traveled to Cologne to learn the art of printing. He returned to Bruges and in 1474 set up a printing business with partner Colard Mansion, calligrapher and bookseller, whom it is thought Caxton taught the art of printing. The same year, the first known book published in the English language, *Recuyell of the Histories of Troie* by Raoul le Fevre, was produced. The duo also printed, in 1475, *The Game and Playe of the Chesse Moralised*, before Caxton moved his printing and publishing business to England.

In the vicinity of Westminster Abbey, conveniently near the court and members of Parliament he expected to serve, Caxton established his printing and publishing business in 1476. In December he produced the first piece of printing done in England, a Letter of Indulgence (a collection of rules showing how to deal with the concurrence of religious festivals). In November 1477 Caxton produced the first dated book printed in England, *The Dictes or Sayengis of the Philosophhres*, translated from French, which had been translated from Latin. In 1481 Caxton's press also produced the first illustrated book in England, *The Mirrour of the World*, which included 27 crude woodcuts. Caxton did much to promote English literature, producing works from

William Caxton, from an egraving c. 1700. *(Public Domain. Reproduced with Permission.)*

Chaucer, such as *The Canterbury Tales* (which he published in 1478 and, in second edition, in 1484) and *Troilus and Creseide*, Gower and Lydgate, Malory, and others.

Before his death in 1491, when Caxton left his press to his former apprentice and current foreman, the publisher produced about 100 printed works, including 74 books, of which 20 were his own translations from Latin, French, and Dutch (he even published a French-English dictionary). Because he adopted the language of London and the court, Caxton had a tremendous impact on fixing a permanent standard for written English. The products of his press, which included many of the first editions of the literary masterpieces of the Middle Ages, hold an eternal place of honor in English literature. Caxton's scholarly vision, as well as his anticipation of the importance of the printing press, made him very influential in the history of the written word.

ANN T. MARSDEN

Giambattista della Porta
1535-1615
Italian Natural Philosopher and Scientist

Giambattista della Porta was a natural philosopher whose ground-breaking development and research in optics was undermined by his belief in miracles and magic.

Home educated, della Porta grew up in an environment where he was mostly self-taught. It is likely that his maternal uncle supervised his education. His education was augmented by the discussion of scientific topics with the learned society that frequented the della Porta home.

Della Porta's mother was from the aristocratic Spadafora family, part of the ancient nobility of Salerno. His father, Nardo Antonio served Emperor Charles V, Holy Roman Emperor and King of Spain. It was this service that transformed the modest fortunes of the Porta family, into a sum that would allow his son to devote himself entirely to study.

Della Porta was a traveler at heart, exploring Italy, France, and Spain extensively. However, his work always led him back to his estate near Naples, which afforded him the solitude he preferred to pursue his studies.

In 1579, della Porta moved to Rome to serve Luigi, cardinal d'Este. During this time, he functioned as a dramatist who wrote comedies for his patron, alongside Torquato Tasso, a well-known Italian poet of the Renaissance.

He also began making optical instruments for the Cardinal. It is this experience that later led to his work *De refractione, optices part* (1593), an expansive study of refraction. Using this publication as substantiation, della Porta claimed that he—not Galileo—was the inventor of the telescope. However, there is no evidence that della Porta constructed the apparatus before Galileo did.

Della Porta's first book, *Magiae naturalis* (1558), constituted the foundation of an expanded 20-book edition of the *Magia naturalis* (1589). The document explores the natural world, claiming it can be manipulated through theoretical and practical experimentation. *Magiae naturalis* is della Porta's most recognized work and the basis of his reputation.

The Accademia dei Segreti, dedicated to studying nature, was founded by della Porta himself some time before 1580. It met in his house in Naples, however, despite his status as a devoted Catholic, the Inquisition closed down the academy around 1578. By 1585 he joined the Jesuit Order, however, in 1594 the Inquisition banned any further publication of his works, and did not permit them again until 1598.

Interestingly, in addition to all his other accomplishments, della Porta was seen as something of an alchemist. It is said that his patron earlier in life, the Cardinal, saw him as such and consequently saved him from the Inquisition. Later, Rudolf II sent his chaplain to Naples to contact della Porta, hoping to procure some alchemical secrets. Della Porta himself spoke of favors he received from Rudolf and to him he intended to dedicate his Taumatologia.

Della Porta published additional works, including *Villae* (1583-1592), an agricultural encyclopedia, and *De distillatione* (1609), a description of his pursuits in chemistry. He wrote on many topics: Astrology, cryptography, fortification, horticulture, mathematics, meteorology, mnemonics, optics, physics, and physiognomy. His other accomplishments include conceptualizing a steam engine before it was a reality, recognizing the heating effect of light rays, and adding a convex lens to the camera obscura (the prototype of the camera).

When he died at the age of 80, he was writing a dissertation in support of his claim of inventing the telescope.

AMY MARQUIS

Cornelius Drebbel
1572-1633
Dutch Alchemist, Inventor, and Engineer

Cornelius Drebbel is best known as one of the possible inventors of the thermometer. A celebrated wonder-worker in his day, Drebbel's experimental investigations in chemistry and physics may have been of real significance, but his penchant for secrecy and mysticism meant that many of his ideas died with him and had to be rediscovered.

Drebbel was born in 1572 at Alkmaar in West Friesland of the Netherlands. His father, Jacob Jansz, was a burgher of Alkmaar. Drebbel received only an elementary education and was apprenticed to the famous Haarlem engraver Hendrik Goltzius. In 1595 he married Sophia Jansdocther, a younger sister of Goltzius, and settled in Alkmaar, where he established himself as an engraver and mapmaker. Shortly thereafter he devoted himself to developing his mechanical inventions: receiving patents for a pump and a perpetual-motion clock in 1598; building a fountain for the town of Middelburg in Zeeland in 1601; and designing a new chimney for which he received a patent in 1602.

Drebbel's perpetual-motion clock has been interpreted by many as the first thermometer.

The device was in fact an astronomical clock whose motive power was the expansion and contraction of the air contained within. Though Drebbel well understood the principles involved, this was not an air thermometer. However, a distinctive type of air thermometer, which came into use in the low countries no later than 1625, was widely attributed to Drebbel. The evidence regarding the inventor's identity remains inconclusive, but Drebbel had sufficient knowledge to construct such a device if it had so occurred to him.

Around 1605 Drebbel moved to England. His inventions so impressed the monarch James I that he was given an annuity and lodgings at Eltham Palace. He was there occupied primarily with constructing machinery for stage performances. In 1610 he moved to Prague at the behest of Emperor Rudolph II. During this period he devoted himself to alchemical studies and developing a perpetual motion machine and mining pumps. Unfortunately, Rudolph was deposed by his brother Matthias in 1611 and Drebbel imprisoned. Through the intervention of Henry, Prince of Whales, he was released in 1613 and allowed to return to England with his family.

Upon returning to London, Drebbel began manufacturing microscopes, producing compound devices as early as 1619. In 1620 he built what has been called a submarine but is more accurately described as a diving bell. The apparatus consisted of two chambers. The upper was above water and occupied by rowers. The lower chamber was completely sealed from the upper and was below water. According to eyewitnesses, the "submarine" carried a number of passengers from Westminster to Greenwich; and Robert Boyle (1627-1691) claimed Drebbel had developed a method for purifying the air within.

Around this time Drebbel made the acquaintance of the four Kuffler brothers who became his disciples and promoters (and two became sons-in-law). Their activities revolved around the distribution and sale of his microscopes and other instruments and the exploitation of his discovery of a tin mordant for dyeing scarlet with cochineal. Through their activities his dyeing method spread throughout Europe.

After the death of his benefactor James I in 1625 Drebbel was engaged by the British navy to produce explosives and construct fire-ships to be used in the failed expedition to raise the French siege of the Huguenot stronghold of La Rochelle. After leaving the employ of the navy Drebbel's spendthrift ways left him broke; and despite involvement in efforts to drain the fenlands of eastern England, he was forced to earn a living by running an alehouse. He died in London in 1633.

STEPHEN D. NORTON

Robert Estienne
1503-1559
French Printer and Scholar

Born to a family of printers, Robert Estienne made his mark not only by printing the first editions of many Greek and Roman works but by using the distinct symbol of the olive tree to represent his enterprises. An outspoken humanist, hostility from the theologians of the University of Paris ultimately drove him out of his native France.

Estienne's father Henri founded a printing firm in Paris around 1502. During his reign, the firm produced more than 100 books. Upon his death, Foreman Simon de Colines not only succeeded him but married his widow. At the age of 23, Robert took over the family business from his stepfather.

In 1526 Robert began running the firm and devoted himself to more scholarly pursuits. His focus included his first achievement, a Latin Bible issued in 1527-1528 as well as his most influential work, *Dictionarium sue linguae latinae thesaurus* (1531), a Latin dictionary. He was also responsible for preparing the first printed editions of several Greek and Roman classics, many of which he edited himself. In 1539 Francis I of France appointed him the king's printer for Hebrew and Latin works. A year later he took on the responsibility of printing Greek texts for the royal library.

Estienne paid special attention to the quality of his printing, not only by using the olive tree insignia designed by Proofreader Geogroy Tory but with types designed by Claude Garamond specifically for the printing house. It was Robert's attention to typography during this time that was unparalleled by his successors. Estienne embraced humanism, which focused on the importance of reason as opposed to faith. It was the antithesis of theology during the Renaissance. These beliefs were what made him a target of Sorbonne faculty. In 1550 he fled to Geneva, Switzerland to escape increasing pressures. It was here that he set up a press and produced a Greek New Testament (1551) that revealed the division of the text into verses for the first time in history.

His brother, Charles, took over the Parisian establishment that same year. Charles, a writer himself, chose to move his attentions toward medical and agricultural subjects, which were more in line with his own interests. However, some of his works were given to Robert to print.

Robert's son, named for Henri Estienne, eventually inherited his father's press under the condition that it would not be moved from Geneva. Following in Robert's footsteps, the second Henri was the family's greatest scholar and a humanist as well. His incumbency marks the height of the family's prowess. His accomplishments include many editions of Greek and Latin works, known for their accuracy and textural criticism. Some of his most well known pursuits are *Thesaurus Graecae linguae* (1572) and *La Precellence du language français* (1579), arguably his most important work. His outspoken *Apologie pour Herodote* (1566) caused trouble with the Consistory of Geneva. Consequently, Henri fled to France to escape punishment. Upon his return to Geneva he was briefly imprisoned and afterward became a wandering scholar. The family's printing firm survived five generations, maintaining its prominent status until the late seventeenth century.

AMY MARQUIS

Johann Fust
c. 1400-1466
German Printer

Though he is much less famous than his sometime associate Johannes Gutenberg (c. 1395-1468), Johann Fust was also a pioneer of printing. His lesser stature is deserved, since it was Gutenberg's technology, seized in a legal action, that Fust used; nonetheless, Fust was responsible for a number of firsts in the history of the printed word.

Fust, whose surname is sometimes rendered as Faust, was born in Mainz, the town Gutenberg would later make famous. He was initially a goldsmith, and with the proceeds from that trade also became a moneylender. It was in this capacity that he first came into contact with Gutenberg, who had been experimenting with the idea of a movable-type printing machine for eight years before he moved to Mainz in about 1446.

The two men met, and Gutenberg so intrigued Fust with his idea that Fust agreed to loan him money to finance the realization of his dream. But when nine years passed and Gutenberg had still not paid back his debts to Fust, the latter brought suit against him. In the resulting settlement, Gutenberg was forced to hand over to Fust all claims to the invention, along with all tangible work he had put into it to that point—including his famous 42-line Bible.

Some scholars believe that it was Fust and not Gutenberg who printed the famous Gutenberg Bible in 1455. In any case, it is unquestioned that Fust printed the second book in history: Gutenberg's Psalter, or Book of Psalms, in 1457. The latter was the first printed book that included both a publication date and a colophon, a symbol to identify the printer.

By then Fust had gone into partnership with his son-in-law Peter Schöffer (1425?-1502), and the two had set up a printing office in Mainz. Among the other works they published were the *Constitutiones* of Pope Clement V in 1460 and, five years later, *De officiis* by the Roman orator Cicero. The latter was the first printed classic, and the first printed book to contain Greek letters.

JUDSON KNIGHT

Johannes Gutenberg
1398?-1468
German Inventor, Craftsman, and Printer

Johannes Gutenberg was a German craftsman whose invention of the moveable type printing process allowed for the first mass production of books, letters, and other written documents. His technique would survive virtually intact until the twentieth century.

Little is known of Gutenberg's early years; only that he was born in Mainz, the son of a wealthy aristocrat named Friele Gänsfleisch, whose lineage dates back to the thirteenth century. Gutenberg's last name was derived from the name of his father's ancestral home, *zu Laden, zu Gutenberg*. A craftsman by trade, he was exiled from Mainz during a feud between the patricians and tradesmen of the city around 1430. Gutenberg moved to Strassburg (now Strasbourg, France), and joined the goldsmith's guild. There, he also taught various crafts, including gem polishing, the manufacture of looking glasses, and the art of printing.

At the time, printed materials were reproduced via a lengthy process, hand written by scribes one at a time or printed page by page with

Johannes Gutenberg. *(Library of Congress. Reproduced with permission.)*

the use of a hand carved wooden block. Gutenberg had the idea to use the tools of metalworking, such as casting, punch-cutting, and stamping, to mass-produce books. He created a font consisting of 300 individually cast characters, which replicated the ornate scroll of handwritten letters. These characters were cut onto small stem rods called patrices, and the dies made were impressed upon a soft metal, such as copper.

Gutenberg blended lead, antimony, and tin to make a variable-width mold that would accommodate his many fonts. Printers could then mix-and-match the various letters to create multiple pages. Some of Gutenberg's earliest printed works included "Poem of the Last Judgment" and the "Calendar for 1448."

In 1450, he returned to Mainz to continue his work on the press. He convinced wealthy financier Johann Fust to lend him 800 guilders—a large sum of money at the time—with which to complete his invention. Fust later added another 800 guilders to his investment and became a full partner in the endeavor.

The oldest surviving printed work in the Western world, *The Bible of 42 lines*, now known as the Gutenberg Bible, was completed in 1456. Shortly before the Bible was finished, Fust initiated a lawsuit against his debtor for failure to repay the loan plus interest, and eventually gained control of Gutenberg's shop and machinery. It was he

who completed printing of the Bible. Fust later carried on the inventor's work with the assistance of his son-in-law, Peter Schöffer.

Moveable-type printing presses soon became abundant throughout Europe. Many of the earliest printers learned their skill in Mainz, taught by Gutenberg himself. In the fifteenth century, there were as many as 1,000 printers actively working, most of them of German origin. By 1500 an estimated 30,000 titles had been published.

In addition to the Bible, a number of other printings were attributed to Gutenberg, including a *Türkenkalender*, which was a warning against Turkish invasion, printed in 1454, and the *Catholicon* of Johannes de Janua, a more than 700-page encyclopedia.

In January 1465 Gutenberg took a position as courtier to the Archbishop of Mainz, who provided him with a food and clothing stipend that carried him through his final years. Gutenberg passed away in his hometown in 1468.

Gutenberg's printing method is now considered among the greatest inventions of all time. Whereas reading was once only available to the elite classes, with the advent of the press, books became widely available to the general public. The printing press allowed literacy to flourish, contributed to the rapid development of science, and made knowledge and education available to all.

STEPHANIE WATSON

Peter Henlein
1480-1542
German Clockmaker

The invention of the portable timepiece or, as we know it today, the watch, is attributed to Peter Henlein, a locksmith from the city of Nuremburg, Germany. He introduced the mainspring as a replacement for weights, enabling the small size and portability of the watch.

During Henlein's time the role of locksmith extended well past locks. Such a locksmith was also an expert mechanic, similar to a modern toolmaker. The medieval locksmith, like the medieval blacksmith, was involved in producing complex and detailed devices. As a result, many locksmiths and blacksmiths were involved in the development and construction of time-keeping devices.

Around 1500, Henlein began to make small clocks that were driven by a spring. These were the first portable timepieces and, designed to be carried by hand, were frequently circular or oval in shape. Because of this oval shape, and a mistranslation of the German word Ueurlein (little clocks) for Eierlein (little eggs), these timepieces were called Nuremburg eggs. The dials of these clocks were placed on top of the device and featured only an hour hand. A record of 1511 indicates that Henlein's watches included iron movements that were mounted in musk balls. This ball was a decorated and perforated sphere in which musk was placed.

The watches invented by Henlein were both desirable, fashionable ornaments and devices indicative of an increased societal reliance on technology. As Henlein's contemporary Johannes Coeulus wrote in 1511, "every day produces more ingenious inventions. A clever and comparatively young man—Peter Henlein—creates works that are the admiration of leading mathematicians, for, out of a little iron he constructs clocks with numerous wheels, which, without any impulse and in any position, indicate time for forty hours and strike, and which can be carried in the purse as well as in the pocket." These devices were also indicative of a new kind of future.

However, despite the fascination that these devices elicited from mathematicians and intellectuals, the Nuremburg eggs were far from accurate or reliable. In essence, they could not be moved and still keep accurate time. Likewise, they kept time unevenly: the force of the mainspring was greater when fully wound than when it was nearly run down. As the design possibilities of the Nuremburg egg increased, so did hostility to the inaccuracy of the device. In Shakespeare's *Love's Labour's Lost,* for example, Shakespeare compares an inconstant woman to "A German clock / Still a-repairing, ever out of frame, / And never going aright, being a watch."

Around 1525, however, Jacob Zech, a Swiss mechanic who lived in the city of Prague (in what is now the Czech Republic), began to study the problem of equalizing the pressure of the mainspring. Zech developed the fusee, a cone-shaped grooved pulley that was used together with a barrel containing the mainspring. Because the mainspring rotates the barrel in which it is housed, the leverage of the mainspring is progressively increased as it runs down. This device improved the accuracy of the watch, and ensured its continuation and development beyond the failings of Peter Henlein's Nuremburg egg.

DEAN SWINFORD

Sacharias Jansen
1588-c. 1628
Dutch Optician

Sacharias Jansen is generally credited with inventing the first compound microscope and may possibly have invented the telescope. A traveling merchant as well as optician, Jansen was a bit of a rogue, involved as he was in various counterfeiting schemes.

Jansen was born in 1588 in The Hague, Netherlands. His father, Hans, was a lens grinder based in Middelburg, the flourishing capital of Zeeland. Hans died four years after his son's birth, and Sacharias's mother taught her son the skills necessary for managing the family business. Jansen married in 1610, and his son Johannes Sachariassen was born the next year.

It is generally believed that Jansen built the first compound microscope around 1595. Since Sacharias was so young at the time it may have been Hans who actually invented the device, with his son merely having assumed production after his father's death. Jansen's first microscopes had a maximum magnification of only 9X, with images being somewhat blurry. Though not very useful as a scientific tool, knowledge of the principles involved spread quickly and within a few years instrument makers throughout Europe were producing improved devices.

Jansen's role in the invention of the telescope is more controversial. In early October 1608 Hans Lippershay (c.1570-1619) filed a patent claim for the telescope. Shortly thereafter Jansen testified before the Committee of Councillors of Zeeland that he knew the art of making such glasses. A third claimant was Jacob Adriaenszoon of Alkmaar, known as Jacob Metius, who after learning of Lippershay's claim advanced his own. No patent was granted since it was determined the technology was too readily available and easily copied.

Metius claimed to have been perfecting his telescope over the previous two years. Evidence for Jansen's priority is mainly derived from the statements of his son. In 1634 Johannes Sachariassen claimed his father, in 1604, had copied an instrument in the possession of an Italian. In 1655 he claimed his father invented the device in 1590. The latter statement was made during an official investigation into the origins of the instrument and was clearly a self-serving prevarication since Jansen would have been two at the time. The former statement has more to recommended itself, as it was made during casual conversation. However, even if Jansen did invent the telescope in 1604, the question stands as to why it remained a secret.

The most likely explanation is that while Jansen probably did copy an instrument in 1604 —consisting of a concave and convex lens in a tube that provided a slight magnification—it was not used as a telescope. Giambattista della Porta (1538-1615) and others seem to have possessed similar instruments that they used to improve faulty vision. This was the typical use for such devices.

A likely scenario is that after Jansen and Metius learned of Lippershay's instrument they realized they were in possession of the same device, which they had been using for other purposes. They then laid claim to the invention as their own. While it may never be known who first realized such a device could be used to enhance normal vision for viewing objects at a distance, it remains certain that the earliest mention of a telescope is in connection with Lippershay's patent application.

On April 22, 1613, Jansen was fined for counterfeiting copper coins. He moved to Arnemuiden, where he expanded his operation to include gold and silver coins. After being apprehended in 1618 he escaped to Middelburg to avoid a death sentence. Financial difficulties followed, leading to bankruptcy in 1628 and the subsequent sale of property to settle his debts. He died sometime before 1632.

STEPHEN D. NORTON

Anton Koberger
1445-1513
German Publisher and Printer

One of the outstanding publishers of the fifteenth century was Anton Koberger, reputed to have operated 24 presses, employed 100 printers and craftsmen, and had agencies in most of the principal cities of Europe for the sale of his books and for manuscript acquisition. Before his death in 1513, Koberger's presses issued 236 products, most of which were theological in character, which is logical considering that reading skills were not commonplace in his lifetime and theologians were the scholars of his time. (Reportedly, his only oversight was in turning down Martin Luther's request to become his publisher.) Three of the most influential illustrated early German printed books were pro-

duced by the Koberger press—in 1488 the *Lives of the Saints* by Voragine; in 1491 the *Schatzbehalter*, a religious treatise by Stefan Fridolin; and in 1493 his most famous work, the *Weltchronik*, also known as the *Liber Chronicarum* and, more commonly, as the *Nuremberg Chronicle*, by Hartmann Schedel.

Born in Nuremberg, Germany, in 1445 to an old Nuremberg family of craftsmen, Anton Koberger's early life is shrouded in mystery, but it is known that he began his career as a goldsmith before becoming a printer. The first dated book produced by his press was Alcinous's *Disciplinarum Platonis Epitome*, printed in November 1472. This early work was indicative of many of Koberger's books for it contained nearly 100 illustrations. In the early days of his press, Koberger had the foresight to enlist the services of two distinguished artists, Michel Wolgemuth, master and teacher of Albrecht Dürer (who was Koberger's godson), and Wolgemuth's stepson Wilhelm Pleydenwurff. Their woodcut illustrations were used in many of Koberger's publications.

The scope of the works published by Koberger is phenomenal considering the technical complexities of printing in the late fifteenth century. In 1474 Koberger published *Pantheologia* by Rainerius de Pisi, which contained 865 leaves. In 1481 he surpassed that total with the *Postillae super Biblia* by Nicolaus de Lyra, a two-volume set that included 939 leaves. Koberger also made good use of ornament and illustration, including the 1491 religious treatise entitled *Schatzbehalter der wahren Reichtümer des Heils*, which contained 96 full-page illustrations. His firm published an illustrated German bible in 1483 along with numerous editions of Bibles in Latin (the first in 1475).

Of the hundreds of books and products published by Anton Koberger, none are more famous than the *Nuremberg Chronicle*, published in Latin and German versions in July and December 1493, respectively. A monument of book illustration, the *Chronicle* used 645 different woodcut illustrations, some used more than once (up to ten times for some of the ornaments) to produce 1,809 total illustrations depicting the full pictorial life of Christ, episodes in the lives of many saints, portraits of prophets, kings, emperors, popes, heroes and great men of history, genealogical tress, nature's wonders, maps, and panoramic views of cities. The *Chronicle* was a compendium of history (the text of the book is a full chronicle of the world's history from its creation up to the year the book was printed), geography, and the natural wonders of the world. The German translation by Georg Alt was published in a shorter version of only 297 leaves, but the Latin edition contained over 326 leaves and 596 pages. There was also a limited edition with hand-colored illustrations.

Koberger's presses and his agencies in cities such as Paris, Lyon, Strasbourg, Milan, Como, Florence, Venice, Augsburg, Leipzig, Prague, and Budapest thrived during his lifetime, when more than 230 publications were issued. Although it is alleged that he fathered no less than 25 children, Koberger was succeeded in business by his nephew (two sons later came into control of the publishing house). None of his family was successful in continuing his publishing and printing company, and it was no longer in business after 1540.

ANN T. MARSDEN

William Lee
1550?-1610?
English Inventor

According to legend, William Lee invented the first knitting machine because the woman he was in love with spent more time knitting than she did with him. Whatever the cause, his 1589 invention would create an economic revolution, and would establish principles of operation still used in modern textile equipment.

Lee was born in about 1550 in the town of Calverton in Nottinghamshire, England. He later became a minister in the town, and though he may indeed have begun his work on the knitting machine out of romantic frustration, the actual facts are not known. Perhaps Lee married the preoccupied woman in the end.

The initial machine of Lee's design was made to produce coarse wool for knitting stockings. As was the law in England at that time, he presented it to Queen Elizabeth I for a patent, but she refused it. Not daunted, Lee went back to the drawing table and refined his creation to produce a machine capable of knitting silk. Again he presented it to the queen, and again she denied him his patent.

In fact Elizabeth's refusal had nothing to do with concerns over the quality of the machine, or of the garments it produced. Rather, she was motivated by what would be called protectionism in modern trade parlance—that is, the suppression of free economic competition as a means of preserving jobs. The fact was that England had a strong hand-knitting industry on which many of the queen's subjects depended for their livelihood, and she was not about to endanger the status quo. Therefore Lee did something that, while not particularly patriotic, made plenty of sense from an economic standpoint: he took his idea to England's most hated rival.

In France, Lee found a warm reception from King Henry IV, who arranged for him to set up a factory in the town of Rouen. There Lee began manufacturing stockings, and he prospered for many years under the patronage of Henry. Then in 1610, Henry—who had made peace with the Protestants by signing the Edict of Nantes 12 years earlier—was assassinated by a religious fanatic named Ravaillac. Lee died around this time, though whether his death had anything to do with that of his patron is not known.

Later, Lee's brother, who had been working with him in France, returned to England with plans of establishing the knitting industry there. Elizabeth was long gone, but the opposition from hand-knitters was as strong as ever, and the brother initially faced great challenges. In time, however, Lee's methods took over the market, helping to spawn a revolution in industry that would transform England's economy.

JUDSON KNIGHT

Leonardo da Vinci
1452-1519
Italian Painter and Inventor

Leonardo da Vinci was the quintessential figure of the Italian Renaissance, and one of the most versatile geniuses who ever lived. His artistic accomplishments alone, including some of the most famous paintings in the world, would have made his name immortal. Yet he was also an inventor whose ideas were hundreds of years before his time. His technical drawings and careful scientific observations were preserved in notebooks that give a fascinating glimpse into one of history's most creative minds.

Leonardo was born near Florence, Italy, on April 15, 1452. He was the illegitimate son of a peasant girl named Caterina and a prosperous notary, Ser Piero da Vinci, who came from a well-to-do family. Leonardo was brought up by his father from the age of five, but the circumstances of his birth meant that many careers considered prestigious at the time, including his

ing detailed observations on both form and function. He kept a workshop where he tinkered with such inventions as a diving suit and a flying machine with wings that flapped like a bird's.

Forced to flee Milan when his sponsor was overthrown, Leonardo returned to Florence. There he painted the *Mona Lisa*, also called *La Gioconda*. The model was Lisa del Giocondo, the wife of a local merchant, and her enigmatic smile has been famous for 500 years. The painting is now displayed at the Louvre.

During 1502 Leonardo worked as an engineer for the infamous warlord Cesare Borgia as he marched with his armies on the region of Romagna. He traveled with the soldiers, prepared maps, and built barracks, a fort, and war machinery. Yet his notebooks of this period ignored the bloodshed around him; they describe such points of interest as an attractive fountain in Rimini and local farming methods. Perhaps the situation eventually became intolerable for the sensitive Leonardo, who refused to eat meat, and sometimes bought caged birds at market so he could set them free. In any case, he again retreated to Florence.

Leonardo did not inherit any of his father's fortune because his parents had not been married, so he was dependent on wealthy patrons all his life. These included Giuliano de Medici, brother of Pope Leo X, who brought Leonardo with him to Rome. But Leonardo was more troublesome to the Vatican than his colleagues among Renaissance artists; he was a scientist as well, and difficult to restrain in his investigations. Leo X was an art lover, and willing to put up with Leonardo, but after the pope's death, the support of the Vatican was no longer forthcoming.

In 1517 Leonardo settled near Tours, France, at the invitation of King Francis I, an admirer who provided him with a comfortable chateau and a generous stipend. He spent his last years continuing to draft designs for a number of ingenious machines. He died at his home in Cloux on May 2, 1519.

While he was arguably as close to a universal genius as humanity has yet produced, Leonardo did have his blind spots. He had little interest in history, literature, or religion. Perhaps more important, he never got around to organizing or publishing his work. His notebooks were largely an awe-inspiring jumble of thoughts and ideas put down as they occurred to him. For this reason, many of his inventions were not widely

Leonardo da Vinci. *(Library of Congress. Reproduced with permission.)*

father's profession, were closed to him. However, he showed great talent in drawing, and painting and sculpture were considered "mechanical arts," suitable for a boy of his station. In due course he was apprenticed to the artist Andrea del Verrocchio. Verrocchio was a painter of the early Renaissance style, but Leonardo's paintings, with their shadows and soft edges, portended the beginning of high Renaissance style.

After his apprenticeship and a few years of maintaining his own studio in Florence, Leonardo went to Milan as court artist and engineer for the duke Lodovico Sforza. There he spent the next 17 years, and painted *The Last Supper* on the wall of the refectory of the monastery of Santa Maria delle Grazie. Unfortunately, the experimental compound he intended to protect the painting had the opposite effect. The painting began to flake away almost as soon as he had finished it, and continues to do so to this day.

As an engineer, Leonardo provided the duke with designs for artillery, folding bridges, and armored vehicles, and plans for diverting rivers. He enlivened parties and pageants with such marvels as a mechanical planetarium. He could also provide musical entertainment, singing and playing musical instruments, some of which were of his own construction. As a scientist, he undertook a careful study of human anatomy, dissecting both human and animal corpses and record-

known until after they had been built by others hundreds of years later.

SHERRI CHASIN CALVO

Hans Lippershay
c. 1570-1619
Dutch Instrument Maker

Hans Lippershay is one of several Dutch spectacle-makers who claimed to have invented the telescope. Though the issue of priority for the invention remains clouded, the earliest mention of such a device was clearly in reference to Lippershay's 1608 patent claim.

Though the exact date of Lippershay's birth remains uncertain, it is known that he was born in Wesel around 1570. He settled in Middelburg, the provincial seat of government for Zeeland. This was the most important commercial and manufacturing center in the southwestern portion of the Netherlands and was home to the oldest glass factory in the northern provinces (established 1581). Lippershay was married in 1594 and became a citizen of Middelburg in 1602.

In September 1608, Lippershay appeared before the Committee of Councillors of Zeeland, where he demonstrated his "device by means of which all things at a very large distance can be seen as if they were nearby." He received a letter of introduction to the Zeeland delegate to the States-General in The Hague requesting an audience be arranged with Maurice of Nassau (1567-1625). The meeting was a great success and Lippershay formally applied for a patent. The States-General formed a commission to investigate Lippershay's instrument and priority claims as well as to negotiate for delivery of six binocular instrument within a year.

On October 5, 1608, Lippershay was given an advance of 300 guilders to produce one instrument built to their specifications. Within two weeks it was discovered that others possessed the art of making telescopic devices, including Sacharias Jansen (1588-c. 1628), also of Middelburg, and Jacob Adriaenszoon of Alkmaar, usually referred to as Jacob Metius. After the committee examined Lippershay's binocular scope on December 15, they pronounced it satisfactory but rejected his patent claim since knowledge of its construction was clearly possessed by others. Nevertheless, Lippershay was paid an additional 300 guilders to produce two more instruments. He delivered these on February 13, 1609, receiving a final payment of 300 guilders.

Metius claimed to have been perfecting his telescope for the previous two years. Evidence today for Jansen's priority is mainly derived from statements his son made decades after the fact. In 1634 Johannes Sachariassen claimed that in 1604 his father had copied an instrument in the possession of an Italian. In 1655 he claimed his father invented the device in 1590. The latter statement was made during an official investigation into the origins of the instrument and was clearly a self-serving prevarication since Jansen would have been two at the time. The former statement has more to recommended itself as it was made during casual conversation. However, even if he Jansen did invent the telescope in 1604, the question stands as to why it remained a secret.

The most likely explanation is that while Jansen in 1604 probably did copy an instrument—consisting of a concave and convex lens in a tube that provided a slight magnification—it was not used as a telescope. Giambattista della Porta (1538-1615) and others seem to have possessed similar instruments that they used to improve faulty vision. This was the typical use for such devices.

A likely scenario is that after Jansen and Metius learned of Lippershay's patent claim they realized they were in possession of the same device, which they had been using for other purposes. They then laid claim to the invention as their own. While it may never be known who first realized that such a device could be used to enhance normal vision for viewing objects at a distance, it remains certain that the earliest mention of a telescope is in connection with Lippershay's patent application.

STEPHEN D. NORTON

Aldus Manutius
1449-1515
Italian Printer and Scholar

A leader in the printing industry, Aldus Manutius was also a humanist scholar. He was responsible not only for establishing several publishing houses but for creating the first Greek alphabet italic fonts as well. He also produced a small, inexpensive collection of Greek and Roman classics for scholars.

Manutius was born Teobaldo Mannuci at Sermoneta in the Papal States. Between 1467 and 1473 he was a student in the Faculty of Arts at the University of Rome. In the late 1470s he attended the University of Ferrara, where he studied Greek under the distinguished humanist and educator Battista Guarino (1435-1505). In 1480 he was employed as tutor to the children of the Duke of Carpi, near Ferrara.

In 1489, Manutius abandoned teaching for the publishing world and moved to Venice. He formed a partnership with established printer Andrea Torresano (1451-1529), who provided both expertise and material resources to the fledgling company. Manutius later married the daughter of his partner in 1505.

In 1490, the Aldine Press opened its doors in Venice. One of Manutius's main goals was to produce the best quality books at the lowest possible prices. The firm's staff consisted of Greek scholars and compositors. The official language at work was Greek and became the same in Manutius's home. Manutius's duties included managing the printing shop, selecting the texts to be published, making editorial decisions, and marketing of the books. It is likely that he owned only 10% of the firm during his lifetime, although his marriage to Maria Torresano seemingly increased his holdings.

In March of 1495 came Manutius' first dated book, the *Erotemata* of Constantine Lascaris. Over the next three years, he printed five volumes of Aristotle. He published editions of many Greek classics including works from authors such as Aristophanes, Euripides, Herodotus, Plutarch, Sopocles, Thucydides, and Xenophon.

Manutius's most famous pursuit came in 1499. That year the *Hypnerotomachia Poliphili* was published, exhibiting exceptional woodcuts by an unknown artist.

His firm made several innovations in the printing domain, one of which was the revolutionary development of the pocket-sized book. This new, portable article provided convenience for traveling scholars of the day. In addition, an italic typeface was created by punchcutter Francesco Griffo, who was said to have imitated the cancellaresco script of calligrapher Bartolomeo Sanvito. Under Manutius's leadership, a Greek alphabet font was born as well. Throughout his reign, Manutius marked his work with the symbol of a dolphin and an anchor.

Manutius established the New Academy in 1500. This school, dedicated to the promotion of Greek studies, was first mentioned in a written work in 1502. The academy, founded by Manutius, was composed of scholars who devoted their time to editing classical texts.

The press stopped its production during the war of the League of Cambrai against Venice but resumed in 1513 by publishing works by Plato, Pindar, and Anthenaeus. When Manutius died in 1515, his brothers-in-law ran the business until his third son, Paulus, took over in 1533. Paulus left the press to his son, Aldus Manutius the Younger in 1561. During the Aldine family's reign between the years of 1495 and 1595, it is likely that the firm produced 1,000 editions.

AMY MARQUIS

Michelangelo di Lodovico Buonarroti Simoni
1475-1564
Italian Artist and Architect

Michelangelo was born on March 6 in the Republic of Florence. His father was a minor government official who at the time of Michelangelo's birth was administrator of the small town of Caprese. When Michelangelo was still very young, the family returned to its permanent residence in Florence. Michelangelo's father wished for him to pursue a career in banking, a longstanding family tradition. Despite his father's initial objections, Michelangelo was eventually permitted, at the age of 13, to become an artist-apprentice. This first apprenticeship, under prominent Florence painter Domenico Ghirlandajo, lasted only one year. After devoting his short apprenticeship to copying the works of earlier Florentine artists, such as Giotto, Michelangelo declared himself fully studied.

Already regarded in Florence as a gifted artist, Michelangelo was shortly thereafter granted patronage by Lorenzo de Medici. The Medici family was famous not only for their wealth and political power, but also for their generous sponsorship of artists, poets, and philosophers. Michelangelo used this period of Medici patronage to study their vast art collection. Especially impressed by the numerous pieces of ancient Roman statuary, Michelangelo dedicated himself to the mastery of marble sculpture rather than the more popular Renaissance medium of bronze.

Michelangelo's detailed research and study of the human figure was reflective of the artistic and even scientific spirit of the European Renaissance—the rebirth of Classical philosophy and

A painting of Michelangelo by Volterra.

art. However, the Renaissance was not simply the rediscovery of ancient ideas, rather it was revived exploration of those ideas combined with a pursuit of new scientific inquiry. Some Renaissance artists, such as Leonardo da Vinci (1452-1519), chose to explore the human form by conducting surgical research on cadavers and recording their findings in detailed sketches of the skeletal and muscle systems of the body. Michelangelo was certainly influenced by this new artistic philosophy, but his sculpture and painting reflect the Classical ideal more than a Renaissance understanding of anatomy.

The patronage of the Medici, however, proved to be short-lived. In 1494, the Medici were overthrown. Before their ousting, Michelangelo had sensed the increasingly tense political environment in Florence and fled the city. In Bologna, Michelangelo's first commission was to finish the remaining marble statuary for the tomb of St. Dominic. Though his reputation as a fine sculptor was forged in Bologna, Michelangelo returned to Florence. In 1501, he received a commission to design a marble statue for the cathedral; the work, *David,* was perhaps Michelangelo's most renowned work as a sculptor.

Called by Pope Julius II to create 40 statues for his tomb, Michelangelo moved to Rome. The expense of the project soon became overwhelming for the Papacy who was devoting its re-

sources to the military conquest of the Italian states, and Michelangelo was sent to work on a less costly project. As a result, he received one of his most famous assignments in 1508, the painting of the Sistine Chapel ceiling. The frescoes took four years to complete and required the assistance of a series of apprentices. Construction of the frescoes was stopped for a period of about a year, perhaps because of a dispute over payment. Julius II died in 1513 and his successor sent Michelangelo to work on projects in Florence. There he continued his painting and sculpting work, and also pursued architecture, constructing a library.

In 1530, Michelangelo briefly put aside artistic pursuits to serve as the designer of fortifications during the siege of Florence. The sketches that survive not only demonstrate his engineering know-how, but also are some of the only surviving depictions of such early modern fortifications and siege weaponry. Michelangelo designed low and thick-walled fortifications, with several pointed wall junctions, meant to withstand blasts from the relatively new cannon.

After the siege, Michelangelo remained in Florence to work on the tombs of the newly restored Medici. Though the project was incomplete, Michelangelo left Florence for Rome for the last time in 1534. His return to Rome also marked his return to the Sistine Chapel. There, Michelangelo painted the fresco of the *Last Judgment* in 1534, completing the interior of the church.

In the later years of his life, Michelangelo focused on architecture—perhaps because it was less physically demanding for the artist. His most recognized architectural work is probably St. Peter's Basilica, the center of the Vatican. The Pope commissioned Michelangelo to design the cathedral and pieta (or square) in 1557 following the death of the original project architect, Bramante. Michelangelo followed the original architect's general design, but embellished the project with his own designs. The result was an artistic and engineering masterpiece. The most striking examples of these additions are the keyhole design of the columned promenade that flanks the square and the dome that was commissioned to top the basilica. More subtle embellishments, such as window trims, gave the building a Classical character reminiscent of Michelangelo's statuary.

The work was not completed before his death, but many historians believe that the end result did not differ greatly from Michelangelo's original design. The dome atop the cathedral, however, may have suffered more modifications at the hands of his successors. The dome that was finally constructed was more pointed and steep than Michelangelo's intended perfect hemisphere. Whether the change was done for stylistic or engineering reasons remains the subject of debate.

Michelangelo sought to capture the full scope of the human experience, and succeeded brilliantly. His manifold artistic triumphs earned him fame in his own time. Not only were his works admired, but Michelangelo himself was the subject of great curiosity. Three editions of his biography were written while he was still alive. Contemporaries such as Raphael were credited with beginning whole schools of art, but the work of Michelangelo has, over the span of hundreds of years, rarely been emulated in an artistically literal manner. Traces of Michelangelo's style can be found from seventeenth-century Baroque painting to nineteenth-century sculpture, but there is no singular artistic movement that binds these works to the Renaissance master. Regardless, his frescoes, sculptures, and architecture are prized as most perfectly representing the Renaissance ideal. Michelangelo remains one of the most prolific figures in the history of western art.

ADRIENNE WILMOTH LERNER

Antonio Neri
1576-1614
Italian Glassmaker and Alchemist

Antonio Neri was a glassmaker who had an interest in chemical technology and alchemy. Best known for his book, *L'arte Vetraria* (The Art of Glass), written in 1612, Neri revealed many of the secrets of glassmaking. The book, colorful and detailed, has served as a basic reference for other essays on the subject ever since.

Antonio Neri was born in Florence, Italy, on February 29, 1576. The name of his father is not known, but it is certain that he was a physician. Little else is known about Neri's family background. Studying glassmaking and other chemical arts throughout his life, Antonio Neri did not attend university. Some suggest he learned glassmaking at Murano, a famous and influential force in glassworks, but this fact has been disputed. It is not really known how Neri supported himself, but he did become an ordained priest before 1601. It is thought that Antonio

Neri traveled in Italy and Holland, at one point working in Florence and Pisa where he experimented with glass. From 1604-1611 he stayed in Antwerp with Emanuel Ziminer, a Portuguese noble who was keenly interested in Neri and his studies in alchemy, for which he was well known during the seventeenth century.

Throughout his life Antonio Neri was active in chemical technology, alchemy, and iatrochemistry, considering himself a practicer of the spagyrical art. Difficult to define, alchemy is a many-faceted art, subject to broad interpretations and applications. Some explored the mystical aspects of the practice, while others conducted chemical experiments in hope of providing medical and spiritual benefits. Alchemy centered on the search for the philosopher's stone, which alchemists believed held the key to their art; in particular, they believed it would have the ability to change lead, or other mundane substances, into gold. Those who were interested in applying alchemy and medicine through chemistry were called iatrochemists. Using the principles of alchemy, spagyrists tried to use chemical processes to break down and purify substances and then reunite them with the intention of providing spiritual benefits. Neri was fascinated with alchemy and spent some time working with Don Antonio Medici, a fellow alchemist.

The Art of Glass was a notable work, revealing many secrets of the glassmaking world. It was rich and detailed and a major influence on other books on the subject. Neri's book contained descriptions on the coloring and making of glass, as well as information on how to imitate precious stones. Cranberry glass, a beautiful, red-colored glass, has been attributed to Neri. He added gold during the melting process, which produced a ruby-colored glass. Cranberry glass became highly collectible in nineteenth-century England and is still a prized collectible today. Neri may have studied in Murano, where glassmaking dates back to the first millennium. The glass coming out of Murano set the standard for fine glass, making Murano the glassmaking capital of the world. Many of the techniques and tools described in Neri's book remain in use.

Neri's book was first published in 1612, and later in 1661 and 1817. Dr. Merret translated it into Latin in 1662; it has subsequently been translated into many languages, including French and German. Much of what is known today about glassmaking comes from Neri's book, and it remains an important work. Although Neri is not considered a major figure in the history of alchemy, he did promote its cause. Sometimes considered to be a product of medieval superstitions, alchemy did contribute to the development of sciences like medicine and chemistry. The scientific methods developed by alchemists served as the basis for modern scientific experimentation. Neri spent the last years of his life in northern Italy and died in 1614.

KYLA MASLANIEC

Andrea Palladio
1508-1580
Italian Architect

Italian architect Andrea Palladio is one of the most important figures in the history of Western architecture. He initially gained fame for his design of palaces and villas and theories linking current thinking about architecture with classical Roman style. These ideas have been imitated again and again for more than 400 years. Palladio's influence was seen most notably in eighteenth-century America, England, and Italy. While the architect's work stands as a lasting tribute, Palladio's four-volume treatise, *I Quattro Libri dell'Architecttura* (1570), or *Four Books of Architecture,* established his enduring reputation worldwide. The collected work was an international bestseller for more than two centuries.

Born Andrea di Pietro in 1508, Palladio was the son of a grain mill worker in Padua, a major city in the Venetian Republic. He apprenticed to a stonemason at 13, but broke his contract three years later and moved to his adopted hometown of Vicenza in northern Italy. Palladio joined the mason's guild and joined the workshop of Giacomo da Porlezza, the city's most important architect at that point.

In his late twenties, Palladio met Gian Giorgio Trissino (1478-1550), the city's leading intellectual and humanist. Trissino was rebuilding a villa in nearby Cricoli in classical ancient Roman style and set up an academy there to provide young aristocrats with a traditional education. Through his association with Porlezza, Palladio worked on the renovation project, and his natural design skills led Trissino to invite him to join the academy. Trissino then directed the young man's initial forays into architecture and renamed him Palladio, a frequent occurrence in humanist study at the time.

Palladio's focused architectural education was unusual in this period; most students were steered toward a general course of study. Under Trissino's guidance, Palladio adopted his mentor's ideas regarding symmetrical layout, with a large central room and flanking towers. Trissino also introduced his student to the works of Sebastiano Serlio (1475-1554), whose five books on architecture were the first to deal with the subject visually as well as in theory. Serlio's treatise served as a blueprint for Palladio's later publications. Palladio studied ancient Roman architect and theorist Vitruvius, who he labeled as his master and guide.

In 1540 Palladio designed his first villa and first palace. These works incorporated the teachings of Trissino with Palladio's own innovations based on his study of ancient Roman buildings. Over the next several decades Palladio made many trips to Rome, which greatly enhanced and solidified his theories about architecture. He applied these ideas as he set out creating palaces for Vicenza's elite, including many of his fellow students from Trissino's academy.

While in Rome from 1554-56, Palladio published *Le Antichita di Roma* (The Antiquities of Rome), which remained the standard guidebook to Rome for 200 years. He also collaborated with a classical scholar in reconstructing Roman buildings for a new edition of Vitruvius's *De Architectura,* or *On Architecture.* Over a 20-year period of intense construction, Palladio became the first architect to systematize the plan of a house and use the Greco-Roman temple front as a roofed porch.

More important, in 1570 Palladio published his four-volume treatise *I Quattro.* The work was a summary of his lifelong study of classical architecture. The first two books outline Palladio's principles of building materials and his designs for town and country villas. The third volume illustrates his thinking about bridges, town planning, and basilicas, which were oblong public halls. The final book deals with reconstruction of ancient Roman temples.

I Quattro cemented Palladio's place in architectural history. The treatise popularized classical design and his innovative style, subsequently became a veritable blueprint for design worldwide, reaching its zenith in the eighteenth century. In the eyes of many scholars, the work stands as the clearest and best-organized textbook on architecture ever produced.

BOB BATCHELOR

The Villa Rotonda, built by Andrea Palladio in Vicenza, Italy (1567-70). *(Bettmann/Corbis. Reproduced with permission.)*

Denis Papin
1647-1712
**French-born British Inventor,
Engineer, and Physicist**

French-born engineer, physicist, and inventor Denis Papin was responsible for inventing the pressure cooker as well as other innovations. His most important contribution was developing the concept of a steam engine, which was the first step toward the Industrial Revolution.

Born to a Huguenot family, Papin's father, also named Denis, was a government official. It is known that his title was Receiver General of the Domaine de Blois, however the family's specific financial status remains uncertain. It is clear that, throughout his life, Papin had little financial stability and floated from one patron to another.

Although Papin originally left France voluntarily, it is likely that, because of his religious beliefs, the Edict of Nantes kept him in exile. He was educated at the University of Angers. His first pursuits were accomplished in the early 1670s while working with Dutch Physicist Christiaan Huygens (1629-1695) at the Royal Library in Paris. Their work on air pump experiments involved using gunpowder to create a vacuum under a piston, allowing pressure from the outside air to force the piston down. His time with Huygens lasted until 1674.

Beginning in 1675, he moved to England to become a tutor to the sons of an unknown member of the aristocracy. Soon, he made the transition to Assistant to Physicist Robert Boyle (1627-1691) in London, where he spent the next four years. However it wasn't until 1681 that Papin achieved independent notoriety by publishing a paper, dedicated to the Royal Society, on the pressure cooker. This closed vessel with a tight-fitting lid kept steam trapped within until pressure forced the boiling point of water to rise considerably. A safety valve prevented explosions. The pressure cooker, which Papin called the steam digester, made faster cooking possible. It was this creation, combined with his earlier work under Huygens, which would later lead to his idea of using steam to drive a piston in a cylinder.

Papin was appointed the Temporary Curator of Experiments at the Royal Society from 1684 until 1687, when he became Professor of Mathematics at the University of Marburg. His next big idea was the steam engine. It used steam to generate pressure. The engine itself was a tube made of metal, closed at one end with a piston inside. Under the piston, a small quantity of water would be heated until it converted into steam, forcing the piston to rise to the edge of the cylinder.

Because of his lack of stable income, Papin was always in pursuit of developing advanta-

Denis Papin. *(Library of Congress. Reproduced with permission.)*

geous relationships. He attempted this by creating inventions that would make a spectacle. One example was a steam engine that Papin called "the Machine of the Elector." A demonstration was performed in order to impress Charles-Auguste, Landgrave of Hesse-Kassel, and therefore elicit monetary support. The apparatus successfully pumped water into a tank at the top of a palace in order to run the fountains in the gardens below. The Landgrave, who spent most of his funds on wars, was center-stage for this and many other demonstrations. While Papin was able to get some funding from him, the Landgrave lost interest regularly.

In 1705, it was Thomas Savery's sketch of the first practical steam engine that inspired Papin's paper on the topic of steam engines, "Ars Nova ad Aquam Ignis Adminiculo Efficacissime Elevandam" ("The New Art of Pumping Water by Using Steam," 1707). Papin continued his developments with a man-powered paddle-wheel boat in 1709. Again he created a successful venture. This one demonstrated the efficiency of using a paddle wheel in place of oars to move steam-driven ships.

Throughout his lifetime, Papin worked on many innovations including a grenade launcher during the War of the Spanish Succession, a mine ventilator, and he even worked with food preservation. Papin moved back to London in

1709 and lived in obscurity until his death three years later.

AMY MARQUIS

Baron Pierre-Paul Riquet de Bonrepos
1604-1680
French Engineer

French engineer Baron Pierre-Paul Riquet de Bonrepos designed and built the Languedoc Canal, sometimes called the Canal du Midi. The latter, which connects the Mediterranean Sea and the Atlantic Ocean, has been called the greatest civil engineering feat between Roman times and the nineteenth century.

Riquet worked as a tax collector under King Louis XIV, during whose long reign he became interested in a problem that had long perplexed French civil engineers: how to construct a means of travelling by boat from the Bay of Biscay to the Mediterranean. The relatively narrow width of the land between the two bodies of water made this a tempting prospect, as did the fact that two rivers at either end—the Garonne and the Aude—already provided part of a waterway. Without the canal, vessels from southern France had to sail all the way around the Iberian peninsula, where they faced a number of dangers, to reach the country's western coast; with the canal, boats could travel between coasts without ever leaving France.

In 1662 Riquet, already 58 years old, went to French finance minister Jean-Baptiste Colbert with a proposal for the building of the canal. He quickly won Colbert to his side, and the powerful minister then went to work on his behalf with Louis and other figures whose support and permission would be required. Work finally began in 1665.

The remaining 15 years of Riquet's life would go into the building of the canal, which by any standards was an awe-inspiring creation. On the way to its highest point, 26 locks raised it some 206 feet (63 m); then at the summit, it ran for 3 miles (5 km) before beginning a descent of 620 feet (189 m) over the course of 114 miles (183.5 km). Other features included a reservoir to provide water for the summit during the dry season, as well as the 515-foot (157-m) Malpas Tunnel, which was 22 feet (6.7 m) wide.

In digging the latter, necessary in order to breach a rocky promontory near Bézier, Riquet became the first engineer in history to use an explosive—specifically, black powder—to blast away rock. Malpas was also the first canal tunnel ever built, and given the fact that explosives had never before been used for blasting, Riquet was exceedingly daring to do so.

The building of the canal took a considerable toll on its creator, and by 1680 his body had given out. The canal was nearing completion, and he was in the middle of work on the harbour of Cette (now called Sète) on the Mediterranean when he died on October 1. Workers continued putting on the finishing touches during the course of more than a year: thus the canal opened in 1691, but was not fully completed until 1692.

JUDSON KNIGHT

Thomas Savery
1650?-1715
English Engineer

Thomas Savery was a military engineer who is known for the invention of the Savery pump. This machine was designed to use steam in order to pump floodwaters from coal mines. While not a steam engine in the modern sense, the Savery pump was the first machine that used steam to provide mechanical power.

Savery was interested in devising mechanisms for practical applications. His significance lies in the extent to which his work serves as a transition from the laboratory-centered experiments of figures such as German physicist Otto von Guericke (1602-1686) and Irish chemist Robert Boyle (1627-1691) to the large-scale mechanical applications that precipitated the Industrial Revolution. Guericke and Boyle developed air pumps and were interested in the properties of gases and the specifics of atomic theory. The enduring image of two teams of horses straining to pull apart Guericke's vacuum-sealed metal hemispheres suggests the quest for the triumph of knowledge over physical force.

Savery, on the other hand, wanted to use scientific knowledge to accomplish tasks beyond the means of brute force. He spent his free time performing mechanical experiments. Indeed, Savery is responsible for inventing a device capable of polishing plate glass as well as a machine that used paddle wheels to move ships stuck in the open water.

His most important invention, which was patented in 1698, was the Savery pump. He designed this machine to lift quantities of water in order to keep mines dry or to supply towns with water. The pump operated according to similar principles that guided the use of vacuums. Savery used steam, and not an air pump, to form a vacuum. Consider, for instance, a chamber filled with steam. If you cooled the outside of the chamber with cold water, the steam within the chamber would condense, leaving only a few drops of water. A vacuum would exist in the place of the steam. If this chamber featured a movable wall, exterior air pressure would drive that wall into the chamber. However, this wall could also be pushed outward again if steam filled the chamber. Then, this wall could be pushed inward if the steam was condensed again. Such a movable wall is, in effect, a piston of the sort that could be used to run a pump. Savery's steam engine relied on high-pressured steam to effect this process. The vacuum then sucked water into a container and out of the coal mine.

Savery's machine was quite unstable, however. The boilers, pipes, and containers were tin soldered and unable to sustain the high pressures necessary to pull water from deep mine shafts. Another Englishman, Thomas Newcomen (1663-1729), was able to devise a steam engine that operated on more stable, low-pressure steam. This new design emerged as the safest and most efficient at the time.

The success of Newcomen and the Scottish engineer James Watt (1736-1819) helped to turn scientific developments into practical achievements. These achievements, and the increasing ease and safety with which they were accomplished, allowed for the transformations engendered by the Industrial Revolution.

DEAN SWINFORD

Peter Schoeffer
1425-1502
German Printer

Peter Schoeffer was the principal workman of Johannes Gutenberg (1398?-1468), the inventor of the movable type printing press. He helped to form the firm of Fust and Schoeffer after Johann Fust (1400?-1466) foreclosed his mortgage on Gutenberg's printing outfit in 1455. Besides being a founding member of the first publishing firm, Schoeffer was also responsible for many printing innovations such as: dating books, introducing Greek characters in print, developing the art of type-founding, and printing in colors. Furthermore, his success as an entrepreneur helped to solidify the importance of printing in late-medieval Europe.

Little is known of Peter Schoeffer's early history. He was educated at the University of Paris and lived in the city of Mainz in Germany. By the time he began to work for Johannes Gutenberg, Gutenberg had already worked through many of the preliminary stages in his development of the printing press. He began these experiments as early as 1436 while living in the city of Strasbourg. Gutenberg was forced to exile himself in this city due to an anti-aristocratic uprising led by the tradesmen and craftsmen of Mainz.

While Gutenberg came from a wealthy family (his father was an aristocrat and one of the four master accountants of the city of Mainz), Gutenberg needed capital in order to master the art of printing. As a result, he was involved in several lawsuits through the course of his life. The first of these was brought against Gutenberg in 1439 because of a partnership Gutenberg had established between a man named Dritzehn and several others. In return for the payment of a sum of money, Gutenberg agreed to teach these men the secrets of a new art. When Dritzehn died, his brother sued so that he would be admitted to the partnership in place of Dritzehn. While Gutenberg won this case, it provides an example of the extreme interest generated by Gutenberg's early experiments with printing.

A second, and much more important, lawsuit occurred in 1455, well after Gutenberg had established his new art. Johann Fust, who had lent money to Gutenberg, sued because the printer had made no attempt to repay either the interest or the principal. Gutenberg could not pay, and Fust foreclosed, taking over as much of the printing plant as possible. Furthermore, he persuaded Peter Schoeffer, who was, by that point, a skilled craftsman, to enter into a partnership.

Fust the businessman and Schoeffer the craftsman quickly surpassed Gutenberg's successes. The first book published by the firm was a Psalter, or choir book. This was produced in two editions in 1457 and is famous as the first book printed with a date. A Bible appeared in 1462, becoming the first with a publication date included.

After Fust's death in 1466, Schoeffer continued the business under his own name. He retained close ties with the Fust family, however. He was in partnership with Fust's sons, and was married to Fust's daughter, Christine. In 1467 he

printed a version of Thomas Aquinas under his own name, thus marking the beginning of his leadership of the firm. Furthermore, in 1470, he issued the first bookseller's advertisement of available printed books.

Schoeffer's business continuously expanded. While he continued to print books in Mainz, and to experiment with printing techniques, he also devoted considerable energy to extending his business throughout Europe. Because of Schoeffer, Gutenberg's technology was widely applied and practiced, and the products of the press widely distributed and sold.

DEAN SWINFORD

Thomas Tompion
1639-1713
English Clockmaker

Thomas Tompion was one of the greatest clock and watchmakers. He is often regarded as the father of English clockmaking. He made some of the first watches with balance-springs, helped to develop the quadrant, and produced clocks for prestigious clients. He is renowned both for the accuracy of his timepieces and for their beauty.

Thomas Tompion was born in Ickfield Green, a hamlet in the parish of Northill, Bedfordshire, in 1639. He was the eldest son of Thomas Tompion senior, a blacksmith, and his wife Margaret. Very little is known of Tompion's upbringing. However, the facts that are known allow us to trace Tompion's development as a maker of the exquisite clocks that ornamented the spacious salons of the English upper classes.

Tompion's grandfather was a blacksmith of limited means who guided his son's entry into the blacksmith trade. Tompion's father, however, was able to profit considerably from this profession. He not only amassed acres of land in the regions surrounding Ickfield Green, but also owned several houses in the neighboring parishes of Biggleswade and Caldecote. As a young blacksmith, Thomas was instructed by his father in tasks such as making hinges and bolts for doors, forging horse shoes, and mending the clappers of church bells.

This background helped to secure the legend of Tompion. As the poet Matthew Prior (1664-1721) remarked, "when you next set Your Watch, remember that Tompion was a farrier, and began his great Knowledge in the Equation of Time by regulating the wheels of a common

Jack, to roast meat." Tompion moved from the more plebeian role of blacksmith into a profession that, during his life, was closely connected to the ideas and theories of the greatest thinkers.

While the details of Tompion's movement from blacksmith to clockmaker are unknown, it can be surmised that Tompion was apprenticed to a blacksmith, finished his term at around the age of 21, and spent the next 11 years in a provincial town. It was during this period that Tompion became a blacksmith as well as a clockmaker of church or turret clocks.

The Court Minute Book of the Clockmakers' Company lists Tompion's admission as a Brother on September 4, 1671. This book describes Tompion as a "Great Clockmaker," indicating that he was recognized as a Master blacksmith-clockmaker who specialized in large iron clocks intended for churches. At this point Tompion began to make his fortune in London. There, he made the acquaintance of and worked with Robert Hooke (1635-1703), considered by many to be the greatest experimental physicist of the seventeenth century, whose discoveries and inventions transformed nearly all of the natural sciences.

In field of horology, Hooke was responsible for two inventions that improved the accuracy of watches and clocks: the anchor escapement that allowed a long pendulum as a regulator and the use of a spring to regulate watches. At this point Hooke was famous while Tompion was only beginning to establish himself. Hooke commissioned Tompion with the construction of a new device, the quadrant, which Hooke had invented. This device was a new kind of astrological instrument intended to improve the accuracy of measurements of celestial bodies.

After this commission, Hooke and Tompion worked on a hand-held watch that could maintain its accuracy even on a rolling ship. King Charles II requested a copy of this invention, helping Tompion's reputation to grow. Indeed, when the Royal Observatory at Greenwich was established in 1676, Tompion was chosen to make two clocks that were to be wound only once a year. These were more accurate at keeping time than those available at other observatories.

Likewise, Tompion's workshop, The Dial and Three Crowns, began to turn out a considerable number of clocks and watches in the 1670s and 1680s. Tompion produced clocks for English and European buyers. In fact, many of his early clocks betray a distinct Dutch influence. His elegance and restraint in the design of clock cases, combined with his enormous productivity, helped to make him the most famous of English clockmakers.

DEAN SWINFORD

Biographical Mentions

Georgius Agricola
1494-1555

German mineralogist and metallurgist, born Georg Bauer, whose *De re metallica* (1556) remained the authoritative text on mining and metallurgy for over four centuries. Lavishly illustrated with 292 woodcuts, this work presented the first detailed, accurate account of sixteenth-century mining practices. His series of treatises on geology and mineralogy proved influential during the formative period of these disciplines. Known as the father of mineralogy, Agricola in *De Natura Fossilum* (1546) attempted the first systematic classification of minerals.

Leone Battista Alberti
1404-1472

Italian architect and mathematician whose *Della Pittura* (1435), which contains the first general account of the laws of perspective, initiated the classical Renaissance style in art. Known as the Florentine Vitruvius, his *De re aedificatoria* (promulgated 1452, published 1485) became the bible of Renaissance architecture, incorporating as it did advances in engineering and aesthetic theory. Alberti also produced a treatise on geography that set forth the rules for surveying and mapping and wrote the first book on cryptography.

Giovanni Battista Benedetti
1530-1590

Italian mathematician who studied astronomy and optics, and designed sundials. Born in Venice, he never attended university but his father tutored him in music, philosophy, and mathematics. Benedetti was court mathematician from 1558 to 1566; the Duke of Savoy then appointed him ducal mathematician and philosopher. He also taught at the University of Turin.

Nicolas van Benschoten
fl. 1680s

Dutch inventor who created the modern thimble. Bronze thimbles had existed since Roman times, and were found in the ruins of Pompeii and Herculaneum. Benschoten, however, was the first to produce a thimble of iron that could be manufactured in large quantities.

Jacques Besson
1540-1576

French mathematician and inventor who was Leonardo da Vinci's successor as engineer to the French court. Around 1568, Besson devised his best-known invention, the first useable screw cutting lathe. In 1569, the inventor published a book on mechanical arts entitled *Theatrum Instrumentorum et Machinarum*, which included illustrations of his various machines. Besson also developed an improved vertical water mill, the principles of which were applied to a power turbine installed in the mid-1800s in waterworks throughout Europe.

William Bourne
fl. 1570s

English mathematician who published the first detailed description of a submarine. In *Inventions or Devices* (1578), Bourne provided a design for an enclosed boat that could be submerged and rowed underwater. Made of a wooden framework encased in waterproofed leather, Bourne's planned submarine was to be lowered by means of hand vices, which would contract the sides and reduce its volume. Although a craft similar to Bourne's design appeared in 1605, it sank due to errors in construction. The first builder of a workable submarine was Cornelius Drebbel (1572-1633) in 1620.

Donato Bramante
1444-1514

Italian architect who launched the High Renaissance style in architecture. Bramante was born in Monte Asdruald (now Fermignano), near Urbino. Details of his early life are sketchy, but it is known that at an early age he studied painting under the Italian masters Andrea Mantegna (1431-1506) and Piero della Francesca (1420?-1492). Bramante later relocated to Milan, where he is believed to have shared discussions on architectural style with Leonardo da Vinci (1452-1519). Bramante completed several structures in Milan before moving to Rome in 1499. His architecture was characterized by its use of illusion, which was more commonplace in painting than in building design. In 1503, Bramante entered into the service of Pope Julius II (1443-1513) and two years later began work on his greatest achievement, the Basilica of St. Peter in Rome. His other major projects included the Belvedere courtyard in the Vatican (begun c. 1505), and the choir of Santa Maria del Popolo.

Giovanni Branca
1571-1640

Italian physicist who provided the first known description of a steam turbine. In 1629, Branca published *La Machine,* a gazette of machinery that included 77 woodcuts illustrating various types of equipment. Among those depicted was a steam turbine, along with Branca's notes describing the means for turning a wheel by shooting jets of steam against vanes attached to the outer rim of the wheel. At the end of a turbine shaft were pestles for pounding materials.

Timothy Bright
1551?-1615

English physician and cleric who is credited with developing the first modern shorthand system, which he initially used to transcribe an epistle of St. Paul. Bright's characters were arranged in 18 vertical rows, similar to the method of Chinese. After Queen Elizabeth granted him a patent for his method in 1588, Bright published a short work explaining his "charactery" as the art of short, swift, and secret writing.

Jean Carre
?-1572

French glassmaker who revitalized the English glass-making industry. Carre was operating as a merchant in Antwerp in 1567, when he received a 21-year license from Queen Elizabeth to make "glass for glazing such as is made in France, Burgundy, and Lorraine." He brought to England glassmakers from Italy and the French regions of Burgundy and Lorraine, where the art had long thrived.

Salomon de Caus
1576-1626

French engineer and architect, who pioneered the use of the steam engine. In *Les Raisons des forces mouvantes* (1615), Caus described a steam pump in which water was heated in a vessel and pushed out by the resulting steam. Because he was a Protestant, Caus was exiled from his homeland, and spent much of his career in England and Germany. He used steam to power a number of small devices.

Jacques Androuet du Cerceau
c. 1520-c. 1585

French engineer and architect whose designs included the Pont Neuf bridge over the Seine in Paris. Cerceau designed a number of palaces for the French royal family, and wrote several books on architectural and decorative engravings. His plans for the Pont Neuf, for which building began in 1578, incorporated elements from two other famous bridges: the flat arches of the Ponte Vecchio in Florence and the cow's horns of the Pont Notre Dame in Paris. Cerceau was the patriarch of a distinguished family of architects that included sons Baptiste (1545-1590) and Jacques II (c. 1550-1614), and grandson Jean I (1585-1649).

Cesare Cesariano
fl. 1520s

Student of Leonardo da Vinci (1452-1519) who published his teacher's observations on the camera obscura. The latter, a forerunner of the modern camera, was an enclosed chamber in which Leonardo conducted experiments with light and darkness. In 1521, Cesariano also published *De Architectura* by the Roman architect Vitruvius (first century A.D.), the first edition of this highly significant treatise on Greco-Roman architecture and engineering to appear in a modern language.

Ch'en Yuan-lung
fl. 1600s

Chinese scholar who in the seventeenth century published a book concerning Chinese inventions. Among the latter were papermaking, printing (China had its own form of movable-type printing that dated back to the eleventh century), gunpowder, the kite, and the compass.

Chu Tsai-Yü
fl. 1580s

Chinese Ming Dynasty prince who calculated a musical system of 12-tone equal temperament. The problem of the 12-tone scale had long perplexed mathematicians and scientists, because to create a system of equal tones that would translate from instrument to instrument required computations with numbers containing as many as 108 zeroes. One had to then find the twelfth root of such a number, an operation performed by calculating the square root twice, then the cube root. In 1596 (some sources say 1584), Prince Chu published a book containing highly accurate calculations of the string length required to produce 12-tone equal temperament on a lute. His findings preceded those of Marin Mersenne (1588-1648) in France by a number of years.

Bernabé Cobo
1582-1657

Spanish missionary to the New World whose *Historia del nuevo mondo* contains a number of valuable observations, including the first recorded mention of rubber extraction. A Jesuit, Cobo went to Peru in 1615, and spent the remainder of his life in the Americas. During that time, he wrote a number of works, including the *Historia* or *History of the New World*. The work contains detailed commentaries on animal and plant life, along with a passage discussing the extraction of liquid resin from a rubber tree.

Humphrey Cole
1530?-1591

British engraver and goldsmith who was considered the most renowned scientific instrument maker of sixteenth century England. Cole's work at the Royal Mint as an engraver and his experience with metalworking led him to work as an instrument maker. He created a wide variety of mathematical instruments. Cole's masterpiece was a 2-foot (61 cm) diameter astrolabe, dated 1574, but he also designed intricate sundials, including a ring dial (c. 1575) and unusual quadrant dials (1574).

François d'Aguilon
1567-1617

Belgian optician and mathematician who coined the term "stereoscopic." A Jesuit priest whose family was of Spanish descent, d'Aguilon studied persistence of vision and visual illusions, and in 1613 published his *Opticorum*. This work contained a discussion of stereoscopic projection, the phenomenon whereby a person's two eyes capture and integrate images viewed from slightly different angles, thus giving the image greater depth. Knowledge of stereoscopy—which had been recognized but unnamed since the time of Greek astronomer Hipparchus (c. 190-c. 120 B.C.)—would prove useful in the development of binoculars and, much later, photography.

John Davis
1550?-1605

English navigator and explorer who went in search of the Northwest Passage from Europe to the Pacific. Davis was born near Dartmouth, in Devon. He fell in love with the sea as a child, and as an adult became convinced that he could navigate around North America to reach the Far East from Europe. He persuaded the British monarchy, under Queen Elizabeth I (1533-1603), to sponsor his journey. They agreed, and in 1585 Davis began his first expedition. He made three unsuccessful attempts to locate the Northwest Passage, in 1585, 1586, and 1587. On his third voyage, Davis attempted to navigate the Strait of Magellan, but was prevented from doing so by bad weather. On his way back to England in

1592 he discovered the Falkland Islands. In addition to his noteworthy seamanship, Davis authored several books on navigation, including *The Seaman's Secrets* (1594) and *The World's Hydrographical Description* (1595). He also invented the back-staff and double quadrant, or Davis's quadrant, which was used for navigation until the eighteenth century. Davis was killed in 1605 by Japanese pirates near Sumatra.

Stephen Daye
1594-1668

English-American locksmith and printer credited with printing the first product of a North American printing press, *The Freeman's Oath* (1639), a broadside of which no known copy exists. The press, set up in Cambridge, Massachusetts, belonged to the widow of Rev. Jose Glover, who had died on the journey to the colonies from England. Daye was entrusted with the working of the press from which he and his son Matthew produced 1,700 copies (only 11 survived) of the first book printed in the colonies, *The Whole Booke of Psalmes*, commonly known as the *Bay Psalm Book*, in 1640.

Leonard Digges
1520-c. 1559

English mathematician known for his applied work in navigation, surveying, and ballistics. His almanac (1555) contained much useful information for sailors. In 1556 he published *Tectonicon*, a manual of elementary surveying techniques. *Pantometria*, published posthumously by his son Thomas in 1571, was a manual of practical mathematics that contained more advanced and up-to-date surveying techniques. Digges also invented the theodolite, a portable instrument for measuring horizontal angles.

Dionysius and Pietro Domenico

Italian inventors who are said to have designed the first double-gate locks for a canal, built on the Brenta, near Padua, in 1481. Some authorities claim that this date is too late for the invention of multiple locks, but an extensive 1778 work on navigation attributed the invention of the canal lock, which had a water opening built into the wooden gates, to the Domenico brothers.

Dud Dudley
1599-1684

English ironmaster who was the first to smelt (or fuse) iron ore with coal. Dudley was born the fourth of eleven children to Edward Lord Dudley in the county of Worcester. As a child, Dudley was fascinated with his father's iron-works, where the youngster learned the trade of iron manufacture. Iron smelting had previously been done with the use of charcoal. Dudley began experimenting with what he called pit-coal, and his experiments were heartily endorsed by the English government, which was concerned that the widespread use of charcoal, which is made from wood, was depleting the country's forests. Dudley received a patent for his invention in 1621 and began iron production at his father's ironworks.

Lazarus Ercker
1530-1594

German metallurgist who wrote the first descriptive review of metallurgical chemistry. Ercker studied at the University of Wittenberg before holding several governmental positions in Saxony. He later became a control tester for coins near Prague, Czechoslovakia. In 1574, he wrote *Beschreibung allerfürnemisten mineralischen Ertzt und Berckwercksarten* (Description of Leading Ore Processing and Mining Methods), which reviewed current techniques for testing alloys and minerals of silver, gold, copper, antimony, mercury, bismuth, and lead, and for refining those minerals.

Erhard Etzlaub
fl. c. 1500

German cartographer who produced the first European road map. Based in the German town of Nüremberg, Etzlaub in 1500 produced the Romweg Map, which provided a guide for pilgrims travelling to Rome from parts of central and western Europe. In the map, Etzlaub applied principles of stereographic projection, which would later be improved by Gerhard Mercator (1512-1594), to render the curved surface of the Earth on a flat plane. Etzlaub's Compass Map (1501) offered a conformal projection—that is, a representation of small areas free of the distortions often necessary to render a mathematically accurate, if less useful, map.

Domenico Fontana
1543-1607

Italian architect and engineer who designed some of Rome's most famous structures, including the Vatican library and St. Peter's Basilica. Fontana was born in Melide in 1543. He traveled to Rome in 1563, where he was hired by Cardinal Montalto (1521-1590) (who would later become Pope Sixtus V) to design a chapel in the church of St. Maria Maggiore (1585). Assisting the Pope in his plan to modernize Rome, Fontana designed the Vatican library (1587-1590), the Lateran Palace

(1587), and collaborated with Italian architect Giacomo della Porta (1541-1604) on the completion of St. Peter's dome (1588-1590), following the plans left by the great artist Michelangelo (1475-1564). Fontana was most famous for moving the Egyptian obelisk from the Vatican to the front of St. Peter's.

Regnier Gemma Frisius
1508-1555

Dutch mathematician and mentor of Gerhard Mercator (1512-1594) who provided the first published illustration of a camera obscura, and who advanced attempts at solving the longitude problem. At that time, navigators were still many years away from finding a means of easily and accurately measuring longitude, a challenge that literally posed a life-and-death problem to sailors at sea. Frisius's *De principiis astronomiae cosmographicae* (1530) discussed a method for finding longitude using a clock and astrolabe. During the 1530s, he trained Mercator, and in 1545 published *De radio astronomico et geometrico,* which discusses his observations of a solar eclipse the preceding January. The book included a drawing of the camera obscura, a dark, enclosed chamber that was a forerunner of the modern camera.

Claude Garamond
1490-1561

French typefounder and craftsman who was the first to specialize in type design, punch cutting and type founding as a service to publishers and printers. From the late 1520s, Garamond was commissioned to cut types for the publishing firm of the scholar-printer Robert Estienne. His first roman font was used in the 1530 edition of *Paraphrasis in Elegantiarum Libros Laurentii Vallae* by Erasmus. Following the success of Garamond's roman font, King François I of France commissioned a Greek font, now known as the *Grecs du Roi,* for his exclusive use (c. 1451). In 1545, Garamond also began publishing his own type designs including a new italic font.

Fra Giovanni Giocondo
1433?-1515

Italian architect and engineer who pioneered the high Renaissance style. The young priest began his career teaching Latin and Greek in Verona, but soon put his background in archaeology and draftsmanship to work in Rome, where he aptly sketched its noble buildings. Giocondo later returned to his home, taking a position as an architectural engineer and overseeing the construction of several bridges. At the request of

Emperor Maximilian, Giocondo designed the Palazzo del Consiglio, one of Verona's most stately buildings. From 1496-1499, he was invited to France to assist in the design of several chateaus and to supervise the construction of the Notre-Dame bridge over the Seine in Paris. His work helped introduce the Italian Renaissance to French architectural styles. After returning to Italy, Giocondo assisted with the construction of St. Peter's in the Vatican.

Francesco Griffo
fl. c. 1500

Italian typographer who designed and produced the first italic type. Employed in the shop of Aldus Mantius (1449-1515), Griffo developed the type by modeling it on the informal style of handwriting. *Epistola devotissime da Sancta Catharina da Siena* (Devotional Epistles of Catherine of Siena), published by Aldus in September 1500, contained a woodcut showing St. Catherine holding a book. On the book itself were the first 18 italic characters in history. Griffo's creation would prove to have a massive impact: printers soon dropped the heavy black letter Gothic type in widespread use up to that point, and Roman typefaces designed during the next three centuries reflected Griffo's influence.

Frederico Grisone
fl. 1532-1550

Italian riding master and pioneer of equestrian arts. In 1532, Grisone founded a riding school near Naples, which became a center of equestrian education for the next century. The first known riding teacher to write on the sport, Grisone in 1550 published *Gil ordine di cavalcare,* a manual of horsemanship.

Otto von Guericke
1602-1686

German physicist and engineer who invented the air pump, which he used to study how air was used in respiration and combustion. Guericke completed his university education with degrees in law, mathematics, and mechanics. In 1631, he was hired as an engineer in the army of Gustavus II Adolphus of Sweden. From 1646-1681, he served as mayor of Magdeburg. In 1650, Guericke invented the air pump, which revealed that air travels through a vacuum, while sound does not. Through several experiments, he also discovered the force exerted by air pressure. In 1663 he invented the first electrical generating machine, creating static electricity by briskly rubbing a revolving ball of sulfur.

Jules Hardouin-Mansart
1646-1708

French architect who completed the Palace of Versailles, begun by Louis Le Vau (1612-1670). Grandnephew of the architect François Mansart (1598-1666), Hardouin-Mansart—he changed his surname in 1668—worked as building superintendent for King Louis XIV from 1675. In addition to his work on Versailles, a renovation rather than a building project, Hardouin-Mansart was responsible for the enlargement of the Château de Saint-Germain and the designs of the Place Vendôme, the Place des Victoires, and the dome of the Hôtel des Invalides. Architect of the Château de Clagny (1676-1680), Hardouin-Mansart designed a number of other châteaux, including de Dampierre, de Luneville, and de Sagonne.

John Harington
1561-1612

British writer and inventor credited with the original concept and construction of the valve-operated water closet (1586). Harington designed and installed his first water closet in his home near Bath, England. In 1596, he published a satire entitled *A New Discourse upon a Stale Subject Called the Metamorphosis of Ajax*, which described the water closet in his home. That same year, he installed his flushing toilet for his godmother, Queen Elizabeth I of England (1533-1603) at Richmond Palace. Harington's invention was not very sanitary and had poor drainage and venting, so it was not put into general use.

Jean de Hautefeuille
1647-1724

French physicist and watchmaker who made significant advances in the development of timepieces and invented the forerunner of the internal-combustion engine. Hautefeuille was fascinated by mechanics as a youth. He focused much of his attention on watches, to which he made many improvements, most notably replacing the pendulum used at the time with a spiral spring. Hautefeuille also invented a thalassameter, which was used to register the movement of tides, and was credited with inventing the first internal-combustion engine, which was designed to operate a pump.

Robert Hooke
1635-1703

English physicist with broad scientific interests and accomplishments, who is best known for his discovery of the law that governs the behavior of elastic materials, known as Hooke's law in his honor. He also coined the word *cell* for biological systems, discovered the diffraction of light, proposed a wave theory of light, invented a telegraph system, made significant astronomical observations of Jupiter and Mars, and is regarded as the founder of the science of crystallography.

Christiaan Huygens
1629-1695

Dutch physicist best known for inventing the wave theory of light. Working with his brother Constantijn he invented a two-lens eyepiece, which dramatically reduced chromatic aberration, and he constructed improved telescopes with which he discovered Saturn's largest satellite Titan (1655), correctly described Saturn's rings (1655), and first observed markings on Mars (1659). Huygens also invented the first successful pendulum clock, so useful for accurate time keeping in astronomical measurements, and independently discovered conservation of momentum.

Athanasius Kircher
1601-1680

German scholar who was renowned for his prolific writings on a variety of academic subjects. Kircher was the youngest of nine children, the son of a doctor of divinity. He narrowly escaped death several times during childhood, and felt that he was spared because he was predestined for a special purpose in life. Kircher studied Greek, Hebrew, the humanities, natural science, and mathematics at various institutions, and was ordained in 1628. Three years later, he fled to France to escape the Thirty Years' War; while in France he taught mathematics, philosophy, and oriental languages. After a few years, he began a journey through Italy and eventually settled in Rome. Once there, he began writing on a wide variety of subjects, documenting his vast knowledge of geography, astronomy, theology, language, and medicine. He eventually completed approximately 44 books in addition to several thousand manuscripts and letters. Kircher was also something of an inventor. He described several innovations, including a graduated aerometer and a method of measuring temperature by the buoyancy of small balls. Throughout his lifetime he collected a vast array of historical materials, which were eventually displayed in a museum bearing his name, the Museo Kircheriano in Rome.

Nicholas Krebs
1401-1464

German scholar and humanist who created the first modern map of Germany in 1491. Also known as Nicholas of Cusa and Nicholas Cusanus, Krebs entered the priesthood in 1433 and rose to the position of cardinal in 1448. He wrote and studied widely in a variety of disciplines, including geography, medicine, mathematics, the arts, philosophy, law, and theology. In addition, Krebs wrote commentaries on classical writers, and helped to resurrect a number of writings from antiquity.

Louis Le Vau
1612-1670

French architect who was one of the chief designers of the Palace of Versailles, particularly the central portion of the garden facade. Born into a family of architects and builders, Le Vau achieved an impressive reputation as a designer of private estates and was famous for his opulent interior designs and majestic proportions. Le Vau was also commissioned to design part of the Louvre in Paris.

Jean de Locquenghien
1518-1574

Flemish engineer, sometimes known as Jan van Locquenghien, who designed the most significant canal in Belgium. In 1561, work was completed on Locquenghien's Willebroek Canal, which gave Brussels access to the North Sea by way of Antwerp, some 30 miles (48 km) to the north.

Francesco di Giorgio Martini
1439-1502

Italian engineer and architect whose *Trattato di architettura civile e militare* (c. 1482) was one of the most significant books on architectural theory during the Renaissance. Initially employed as a military architect, first by the city-state of Siena and later by both Lorenzo de Medici and the duke of Urbino, Martini went on to influence civilian architecture throughout much of Italy. He designed parts of the ducal palace at Urbino, and created a model of the dome for the cathedral at Milan. He was also an accomplished sculptor and painter. As a theorist, he was greatly influenced by the ideas of the Roman architect Vitruvius (first century A.D.)

Marin Mersenne
1588-1648

French mathematician, philosopher, and theologian whose Mersenne numbers pioneered the effort to discover a formula representing all prime numbers. Mersenne spent five years at the Jesuit College at La Fleche and two years studying theology at the Sorbonne in Paris. He joined the Roman Catholic order of the Minims in 1611,

and taught philosophy at its convent in Nevers from 1614-1618. The following year, he returned to Paris and became a member of the city's academic circle. Mersenne performed significant research in mathematics, arriving at the formula 2^p-1, in which p represents a prime number. While this formula, called the Mersenne numbers, was not applicable to all prime numbers, it did stimulate interest in their study. Mersenne also investigated cycloids (types of geometric curves), and proposed the use of a pendulum as a sort of clock.

William Molyneux
1656-1698

Irish mathematician and astronomer who in 1692 published the first work in the British Isles on the subject of optics, *Dioptrica nova,* which contained a discussion of the "magic lantern." The latter was a primitive type of projector, considered a forerunner of modern-day motion-picture projectors. Molyneux also designed a telescopic sundial and a new gyroscope, and with his father, Samuel Molyneux, conducted experiments in gunnery.

Joseph Moxon
1627-1700

British hydrographer, publisher, and instrument-maker who created and published the most comprehensive book on the practice of type-cutting, founding, and setting before the invention of the power press. Around 1660, Moxon was appointed hydrographer to King Charles II of England. From 1667 to 1679, he published 38 papers for the professional instruction of skilled artisans in the metal and woodworking trades. In 1683, Moxon combined 24 of these papers into his famous book entitled *Mechanick Exercises on the Whole Art of Printing*, the earliest practical manual of printing in any language.

Thomas Newcomen
1663-1729

English engineer who invented the first atmospheric steam engine. Newcomen was born in Dartmouth, and spent many years as an ironmonger there. Finding out that pumping water out of mines was at the time accomplished through a labor-intensive process, using horses, he spent 10 years trying to create an engine that would complete the task mechanically. Newcomen's engine was comprised of a piston within a vertical cylinder and a huge beam which connected to the mine pumps. The engine was groundbreaking in that it used atmospheric pressure, so as not to be limited by the pressure of steam. The first Newcomen engine was used in a South Staffordshire Colliery in 1712, and within a few years the invention was put into use in mines throughout the country. In 1765, however, Newcomen's engine was overshadowed by that of Scottish mechanical engineer James Watt (1736-1819), who is now considered the true inventor of the steam engine.

Bertola da Novate
c. 1410-1475

Italian engineer and pioneer of canal-building. Novate built the Bereguardo Canal in Italy, completed in 1458, which was the first lateral canal (that is, paralleling a river) to use locks as a means of controlling its steep elevation changes. In 1470, Novate completed work on the Martesana Canal, which connected Milan with the River Trezzo and used only two locks over a distance of 24 miles (38.4 km). He is also credited with introducing the mitre gate, one in which the two gates meet at an angle. The latter is much more effective at resisting the stress placed on it, and makes it possible to build a wider canal lock.

Juan Pablos
fl. 1539-1560

Italian printer, known by his Spanish name, who established the first printing house in the New World. Pablos worked for Johann Cromberger, a German printer in Seville, who in 1539 sent him to Mexico City. In 1543, Pablos published the first surviving book printed in the Americas, *Doctrina breve muy provechosa* by Bishop Zumárraga. Pablos was joined in 1550 by Antonio de Espinosa, a type founder and die cutter whose Roman and italic types, the first in the New World, soon replaced the Gothic forms used by Pablos up to that point.

Bernard Palissy
1509?-1589

French potter, glass-painter, and writer who pioneered lead-based, enameled ceramics called majolica and popularized a rustic form of ceramic art featuring coiled snakes, scaly fish, and slinking lizards set in high relief and painted as found in nature. Around 1548, after many years of chemical experimentation, Palissy, an avid naturalist, discovered the secret of producing Italian majolica, which he combined with his interest in nature to create a unique style of ceramic art that remained popular for over 400 years.

Dom Pierre Perignon
1638-1715

French Benedictine monk who is renowned for his contribution to winemaking. Prior to the seventeenth century, abbeys produced only still wines, although sparkling wines were known in France and England. As cellar master at the Abbey of Hautvillers in Champagne, France, Dom Perignon converted ordinary sparkling wine into the exceptional wine called champagne. Dom Perignon's sparkling wine was expensive, ordered only by royalty and the nobility, and made him famous in his own lifetime.

Ottaviano dei Petrucci
1466-1539

Italian printer who in 1498 devised a method of printing musical notation using movable type. This innovation greatly spurred the spread of composers' work, and helped lead to the standardized system of notation in use today. A Venetian, Petrucci established a paper mill that remained in operation until the nineteenth century.

Sir William Petty
1632-1687

English-Irish economist and physician who invented the first modern cataraman, a boat with two hulls. A prominent figure in his time, Petty served in a number of important positions with the English government. He was a founding member of the Royal Society, and in 1662 published one of the first books on vital statistics. Also in 1662, Petty built his catamaran, the *Experiment,* in Ireland.

Hugh Platt
fl. 1600-1610

English scientist who discovered coke. In 1603, Platt heated a quantity of coal without burning it, which distilled the coal and left a residue. The latter was coke, which proved to be a highly useful fuel for heating and industry. Platt was also one of the first English writers to mention molasses, in his *Delights for Ladies* (1609).

Christopher Polhem
1661-1751

Swedish inventor who designed a number of machines and made great improvements to his country's mining industries. One of the first notable scientists from Sweden, Polhem created and built lathes, clocks, tools, and other devices. He also held a key position with the Royal Board of Mines, where in 1716 Emanuel Swedenborg (1688-1772), destined to become a famous scientist, mystic, and philosopher, came to work as his assistant. The two men worked together for a number of decades, during which time they made mineralogical studies and sought to enhance the operations of Swedish mines.

Agostino Ramelli
1531-1590

Italian military engineer and author of *Le diverse et artificiose machine* (Diverse and artifactitious machines, 1588), one of the most important works on machinery written during the Renaissance. Ramelli spent much of his career in the service of Henry of Anjou, who later became French king under the title Henry III. He was wounded while taking part in the 1572 military action against the Huguenots at La Rochelle, and afterward devoted himself to his writings. His great work on machines, written in both Italian and French, contained 194 plates depicting a variety of devices, some of them fanciful and imaginary creations. These included water pumps, mills, cranes, a water wheel, fountains, bridges, catapults, and what Ramelli called a "book wheel." The latter was a rotating fixture that made it possible to keep several books open and read from them at the same time. In the view of some scientists, this was a precursor to the idea of hypertext, as used today on the Internet.

Erhard Ratdolt
1447-1528

German printer and type-cutter who published the first book using more than two different colors of ink on one page. In 1482, Ratdolt printed an edition of the *Elements* by Euclid (c. 325-c. 250 B.C.), which was the first printed book illustrated with geometric figures. Three years later, in 1485, he published *De sphaera* by English mathematician Johannes de Sacrobosco (a.k.a. John of Holywood, c. 1200-1256), which contained pages using more than two colors. Ratdolt was also the originator of the decorated title page.

George Ravenscroft
1618-1681

English glass maker who developed lead glass, also known as flint glass. For several centuries, Venice had dominated the glass industry; in 1674, however, Ravenscroft discovered that by adding lead, he could create a glass that was not only more durable, but more brilliant and aesthetically pleasing than Venetian glass. Today Ravenscroft's discovery is known as lead crystal.

Anton Maria Schyrlaeus de Rheita
1597-1660

Bohemian astronomer who produced the first lunar map to represent the Moon as seen through an inverting telescope—with its southern-most features at the top. Rheita's only scientific treatise of value is *Oculus Enoch et Eliae, opus theologiae, philosophiae, et verbi dei praeconibus utile et iucundum* (1645), in which he describes an eyepiece, of his own invention, that re-inverted the inverted images of Keplerian refracting telescopes (1645). Rheita coined the terms "ocular" and "objective."

Friedrich Risner
?-1580

German mathematician who first suggested the idea of a portable camera obscura. The latter is a dark chamber, used for experiments in optics and generally considered to be a precursor to the modern camera. In 1572, Risner suggested that rather than a darkened room, a wooden hut would make a more useful camera obscura since it could be moved. He also edited a highly influential edition of writings by Alhazen (965-1039) and Witelo (c. 1230-c. 1275) on optics.

John Rolfe
1585-1622

English colonist who discovered a method for curing tobacco, and was the first known settler in the New World to cultivate the crop. Rolfe arrived in Jamestown, Virginia, in 1610, and by 1616 he had a successful tobacco crop using seeds probably obtained from the West Indies. Tobacco would have an impact on the economies of Virginia and neighboring North Carolina that is still felt today. Rolfe is also famous as the husband of Pocahontas; ironically, it is believed that he was killed by Native Americans near his home.

Giovanni Ventura Rosetti
fl. 1540s

Italian textile-maker who published the first book on dyeing fibers and fabrics, *Plictho dell'arte de tentori* (1540). This coincided with the dawning of the commercial textile industry, as the process for applying natural dyes—synthetic dyes would not appear for several centuries—came to be standardized.

Anna Rügerin

German printing press owner and operator who was the first recorded female printer. In June 1484, the Augsburg press owned and worked by Rügerin printed the *Sachsenspiegel* of Eike von

Repgow. It is probable that Johann Schönsperger, who had been associated with printer Thomas Rüger, assisted his widow with the project.

Olivier de Serres
1539-1619

French agronomist, thought to be the first to practice systematic crop rotation. Sometimes referred to as the father of French agriculture, Serres presented King Henry IV with plans for the expansion of sericulture. His advocacy was largely responsible for the widespread planting of mulberry trees in France at the time, and his agricultural manual, *Théatre d'agriculture et mesnage de champs* (1600), proved highly influential. In addition to crop rotation, Serres advocated new irrigation methods, water conservation, and the sowing of artificial grasses.

Ludwig von Siegen
1609-c. 1680

German engraver and painter who in 1642 invented the mezzotint process for printing in graduated tones. Siegen, who described his method as one of engraving by dots rather than by lines, used a small roulette or fine-toothed wheel to achieve the desired effect. For years he kept his method a secret, before revealing it to Prince Rupert of the Palatinate, a patron, in 1654. Eventually the process made its way to engravers in England, who adopted it to a much greater extent than their counterparts in other countries.

Edward Somerset
1601-1667

British inventor and Army officer who designed one of the earliest known steam engines, a steam-operated pump for raising water. Somerset was granted a patent for the device in 1663, the same year he published a book written in 1655 entitled *A Century of the Names and Scantlings of Such Inventions as at Present I Can Call to Mind to have Tried and Perfected*, which included a description of his steam pump as "an admirable and most forcible way to drive up water by fire."

Frederick Staedtler
fl. 1660s

German manufacturer who produced the first commercially made pencils. In Nüremberg, Germany, in 1662, Staedtler established a factory for making pencils by fitting a thin strip of graphite into a grooved piece of wood, then covering the groove and sealing it in place with a glued wooden strip.

Tartaglia
1499-1557

Italian mathematician, born Nicolò Fontana, whose *La nuova scientia* (1537) was the first book in history on the science of ballistics. The latter had come to be increasingly important to military forces armed with guns and cannons, and Tartaglia's work was highly influential; however, he incorrectly stated that a ball falls straight downward after being propelled forward from a cannon. As a mathematician, Tartaglia is best known for his partial solution to the problem of cubic equations, a discovery that later led him into a dispute with Girolamo Cardano (1501-1576) and Cardano's assistant Ludovico Ferrari (1522-1565). Interestingly, Cardano published his own work on ballistics and other subjects, *De subtilitate,* seven years after Tartaglia's.

Giacomo Torelli
1608-1678

Italian stage designer and engineer who created a number of innovations for the theatre. In 1641, Torelli designed and built the Teatro Novissimo in Venice, for which he developed number of machines, in particular a revolving stage. Invited to Paris by King Louis XIV in 1645, he designed for the Théatre du Petit-Bourbon the first effective machinery for making rapid changes to large sets. His set designs, most notably for Pierre Corneille's *Andromède* in 1650, won him praise. After returning to Italy in 1661, he designed the Teatro della Fortuna in Fano (1677).

Geoffroy Tory
c. 1480-c. 1533

French printer and engraver whose designs in *Champ fleury* (1529) provided the model for styles of book decoration in the French Renaissance. Typographers long struggled with the problem of developing a formula for the design of capital Roman letters, a question Tory effectively addressed in his seminal work. He advocated a number of other reforms, including the use of accents, the cedilla, the apostrophe, and other punctuation marks, that exerted a profound influence not only on French orthography, but on the French language as it is written today. *Champ fleury* won such great acclaim that King Francis I appointed him to the position of Imprimeur du Roy in 1530.

Robertus Valturius
1413-1484

Italian military engineer whose *De re militari Libri XII* included the earliest printed technical illustrations. The book, published in Verona in 1472, featured depictions of the latest military technology, as well as scenes illustrating various military activities and warfare.

Sébastien Le Prestre de Vauban
1633-1707

French military engineer whose tactics helped win a string of victories for King Louis XIV. Three years after joining the French engineer corps, Vauban served as engineer-in-chief at the successful siege of Gravelines in 1658. He went on to great acclaim for his work in enabling victories during 12 military engagements from 1667 to 1703. Vauban designed fortifications at Strasbourg and a number of other towns, and invented the socket bayonet. His *De l'attaque et de la défense des places,* published posthumously in 1737, was a highly influential study of fortifications and siege technology.

Sir Cornelius Vermuyden
1595?-1683

Dutch-born English engineer responsible for a number of drainage and land reclamation projects. Vermuyden went to England in 1621 (he later became a naturalized citizen) to repair the Thames embankments. In 1626, he received a commission from King Charles I to drain Hatfield Chase in Yorkshire, and spent the better part of the 1630s draining the Great Fens in Cambridgeshire. He later reclaimed the area after much of it was flooded during the English Civil War.

Pierre Vernier
1584-1638

French military engineer who invented the Vernier caliper or Vernier scale for measuring small angles and lengths. The instrument, sometimes simply called the vernier, consisted of a large stationary scale for measuring whole numbers, along with a smaller, movable scale for measuring fractions. Originally used in astronomy, it replaced the much more complicated astrolabe; ultimately, however, the vernier would gain even wider use among surveyors.

Bernard Walther
1430-1504

German astronomer who was one of the first to use weight-driven clocks. Walther was a patron of Johann Müller Regiomontanus (1436-1476), and contributed funds toward the building of the latter's observatory in Nuremberg in 1471. After Regiomontanus's death in 1476, Walther continued his associate's work in measuring star

positions. He first made use of the weight-driven clock in 1484.

Wan Hu
d. c. 1500

Chinese rocketry pioneer. Wan Hu was reportedly a government official, though some sources identify him as a legendary or semi-legendary figure. In about 1500, he attached some 47 "fire-arrow rockets" or fireworks to two large kites, which were in turn attached to a chair. He then sat in the chair, and on his command, the 47 rockets were lit. The result was an explosion that took his life.

Richard Weston
1591-1652

English agriculturalist who was first to describe a system of crop rotation using no fallow break—that is, a period in which no crops are planted on the land. An advocate of growing root crops, Weston published his *Discours of Husbandrie* in 1650.

Henry Winstanley
1644-1703

English engineer who designed the first Eddystone Lighthouse, which greatly reduced shipping accidents off the Plymouth coast. An inventor known for his many bizarre contraptions, Winstanley owned two ships that in 1696 sank on the dangerous Eddystone Rocks 14 miles (22.5 km) from shore. He then resolved to build a lighthouse. Given its distance from inhabitable land—the Eddystone Rocks were, as their name indicated, mere rocks jutting from the sea—this was a formidable task. Simply digging the foundation took five months, but finally the lighthouse was completed in 1698. Five years later, Winstanley died in a violent storm that destroyed the lighthouse.

Hannah Woolley
1623-1684?

British schoolteacher and writer, whose name is well known to collectors of books on cookery and the domestic arts as the first woman to publish a cookbook. Left an orphan at a young age, Woolley found work as a schoolmistress and governess before attaining the post of stewardess and secretary for an unnamed woman. She eventually married and was widowed, about which time she published her first book, *The Ladies' Directory* (1661). Her four subsequent books (and their later editions) were a curious mixture of etiquette rules and behavior in society, the art of letter-writing, cooking recipes, first aid instructions, and education advice for girls' governesses and others seeking domestic positions.

Sir Christopher Wren
1632-1723

English polymath who, after the great fire of London (1666), designed the new St. Paul's Cathedral and over 50 churches. He also designed the Greenwich Observatory. Wren was a charter member of the Royal Society, becoming that body's president in 1680. He produced the first lunar globe (1661), determined the arc length and center of gravity for the cycloid (1658), independently established conservation of momentum (1668), and proposed a unit of length based on pendulum oscillations (1673).

Yi Sun-shin
?-1598

Korean admiral who built the first ironclad warships. In 1592, when the Japanese invaded Korea with the aim of conquering the peninsula and moving on to China, Yi ordered the building of 12 *kobukson,* or "tortoise ships." These low-decked, armed galleys, covered with an iron-plated dome, faced a vastly larger Japanese force—133 ships—and won an overwhelming victory, sinking 31 vessels and routing the remainder. Despite this victory, however, the Koreans did not continue building ironclad warships, and the technique did not appear in the West until the nineteenth century.

Johann Zahn
fl. 1680s

German inventor whose box-sized camera obscuras provided a prototype for the box and reflex cameras that appeared at the advent of photography in the nineteenth century. Originally the camera obscura consisted of an entire room in which all light was shut off, except for a small hole at one end; thanks to innovations beginning with Friedrich Risner (d. 1580), however, they were small and portable by Zahn's time. Zahn, a monk from Würzburg, developed lenses of varying length for viewing scenes at a greater or lesser distance.

Jakob Zech
fl. 1520s

German clockmaker, sometimes known as Jacob the Czech, who built the first clock using a balance wheel, which made it more regular and therefore more accurate. Zech's wheel, called a fusee, represented the first successful attempt to regulate the power of the watch spring.

Bibliography of Primary Sources

Agricola, Georgius. *De re Metallica* (1556). This cleverly organized work discussed all aspects of mining from prospecting through the production of gold, silver, lead, salt, soda, alum, sulfuric acid, sulfur, bitumen, and glass. One of the great monuments of technology, for two centuries it remained through many reprints the best mining textbook in Europe, because of its comprehensive coverage and brilliantly clear illustrations of machinery, powered by horses or water wheels, for pumping water, hoisting spoil, ventilating shafts, crushing and grinding ore, assaying and sampling for quality, stirring and mixing ore in tubs, blowing smelting furnaces, and moving heavy hammers to work wrought iron.

Alberti, Leone Battista. *Della Pittura* (1435). This work contained the first general account of the laws of perspective.

Alberti, Leone Battista. *De re aedificatoria* (promulgated 1452, published 1485). This book became the bible of Renaissance architecture, incorporating as it did advances in engineering and aesthetic theory.

Aldus Mantius. *Epistola devotissime da Sancta Catharina da Siena* (Devotional epistles of Catherine of Siena, 1500). This work contained a woodcut showing St. Catherine holding a book. On the book itself were the first 18 italic characters in history, created by Francesco Griffo. Griffo's creation would prove to have a massive impact: printers soon dropped the heavy black letter Gothic type in widespread use up to that point, and Roman typefaces designed during the next three centuries reflected Griffo's influence.

Besson, Jacques. *Theatrum Instrumentorum et Machinarum* (1569). A book on mechanical arts that included illustrations of Besson's various machines.

Biringuccio, Vannoccio. *De la pirotechnia libri X* (Ten books of a work in fire, 1540). Biringuccio's practical manual, written in vernacular Italian, described metallurgical processes—assaying, smelting, alloying, and casting—that he had seen or practiced. In many cases his was the first written description of processes for smelting and casting gold, silver, copper, lead, tin, iron, steel, and brass. He also described the casting of medallions, fine art objects, type for printing, and cannon, which in typical Renaissance fashion he insisted should be ornamented to make them beautiful.

Bourne, William. *Inventions or Devices* (1578). The first detailed description of a submarine. In the work Bourne provided a design for an enclosed boat that could be submerged and rowed underwater. Made of a wooden framework encased in waterproofed leather, Bourne's planned submarine was to be lowered by means of hand vices, which would contract the sides and reduce its volume. Although a craft similar to Bourne's design appeared in 1605, it sank due to errors in construction.

Branca, Giovanni. *La Machine* (1629). A gazette of machinery that included 77 woodcuts illustrating various types of equipment. Among those depicted was a steam turbine, along with Branca's notes describing the means for turning a wheel by shooting jets of steam against vanes attached to the outer rim of the wheel. At the end of a turbine shaft were pestles for pounding materials.

Caus, Salomon de. *Les Raisons des forcesmouvantes* (1615). Here Caus described a steam pump in which water was heated in a vessel and pushed out by the resulting steam.

d'Aguilon, François. *Opticorum* (1613). This work contained a discussion of stereoscopic projection, the phenomenon whereby a person's two eyes capture and integrate images viewed from slightly different angles, thus giving the image greater depth.

della Porta, Giambattista. *Magiae naturalis* (1558). This book constituted the foundation of an expanded 20-book edition of the *Magia naturalis* (1589). The document explores the natural world, claiming it can be manipulated through theoretical and practical experimentation. *Magiae naturalis* is della Porta's most recognized work and the basis of his reputation.

Ercker, Lazarus. *Beschreibung allerfürnemisten mineralischen Ertzt und Berckwercksarten* (Description of leading ore processing and mining methods, 1574). This work reviewed current techniques for testing alloys and minerals of silver, gold, copper, antimony, mercury, bismuth, and lead, and for refining those minerals.

Fevre, Raoul le. *Recuyell of the Histories of Troie* (1474). This was the first known book published in the English language.

Frisius, Regnier. *De principiis astronomiae cosmographicae* (1530). Discussed a method for finding longitude using a clock and astrolabe.

Frisius, Regnier. *De radio astronomico et geometrico* (1545). This work discussed Frisius's observations of a solar eclipse the preceding January. The book included a drawing of the camera obscura, a dark, enclosed chamber that was a forerunner of the modern camera.

Grisone, Frederico. *Gil ordine di cavalcare* (1550). A manual of horsemanship.

The Gutenberg Bible (1456). The oldest surviving printed work in the Western world was originally known as *The Bible of 42 Lines.*

Molyneux, William. *Dioptrica nova,* (1692). The first work in the British Isles on the subject of optics, this book contained a discussion of the "magic lantern." The latter was a primitive type of projector, considered a forerunner of modern-day motion-picture projectors.

Moxon, Joseph. *Mechanick Exercises on the Whole Art of Printing* (1683). Moxon combined 24 papers to create this famous book, the earliest practical manual of printing in any language.

Neri, Antonio. *L'arte Vetraria* (The art of glass, 1612). In this influential work Neri revealed many of the secrets of glassmaking, which had previously been closely guarded among Italian glassmakers. The book, colorful and detailed, has served as a basic reference on the subject ever since.

Palladio, Andrea. *I Quattro Libri dell'Architecttura* (Four books of architecture, 1570). Four-volume treatise that popularized classical design and Palladio's innovative style, and became a veritable blueprint for design worldwide, reaching its zenith in the eighteenth

century. In the eyes of many scholars, the work stands as the clearest and best-organized textbook on architecture ever produced. The first two books outline Palladio's principles of building materials and his designs for town and country villas. The third volume illustrates his thinking about bridges, town planning, and basilicas, which were oblong public halls. The final book deals with reconstruction of ancient Roman temples.

Papin, Denis. *Ars Nova ad Aquam Ignis Adminiculo Efficacissime Elevandam* (The new art of pumping water by using steam, 1707). Thomas Savery's sketch of the first practical steam engine inspired Papin's paper on the topic of steam engines.

Ramelli, Agostino. *Le diverseet artificiose machine* (Diverse and artifactitious machines, 1588). One of the most important works on machinery written during the Renaissance. Written in both Italian and French, the work contained 194 plates depicting a variety of devices, some of them fanciful and imaginary creations. These included water pumps, mills, cranes, a water wheel, fountains, bridges, catapults, and what Ramelli called a "book wheel." The latter was a rotating fixture that made it possible to keep several books open and read from them at the same time. In the view of some scientists, this was a precursor to the idea of hypertext, as used today on the Internet.

Rheita, Anton de. *Oculus Enoch et Eliae, opus theologiae, philosophiae, et verbi dei praeconibus utile et iucundum* (1645). Here Rheita described an eyepiece of his own invention that re-inverted the inverted images of Keplerian refracting telescopes.

Rosetti, Giovanni. *Plictho dell'arte de tentori* (1540). The first book on dyeing fibers and fabrics, This coincided with the dawning of the commercial textile industry, as the process for applying natural dyes—synthetic dyes would not appear for several centuries—came to be standardized.

Serres, Olivier de. *Théatre d'agriculture* (1600). Described systematic crop rotation for the first time. This very popular work appeared in several editions throughout the seventeenth century. Serres surveyed all aspects of agriculture, starting with advice on running a religious Calvinist household.

Somerset, Edward. *A Century of the Names and Scantlings of Such Inventions as at Present I Can Call to Mind to have Tried and Perfected* (1663). This book included a description of Somerset's pioneering steam pump as "uoan admirable and most forcible way to drive up water by fire."

Tartaglia. *La nuova scientia* (1537). The first book in history on the science of ballistics. The field had come to be increasingly important to military forces armed with guns and cannons, and Tartaglia's work was highly influential; however, he incorrectly stated that a ball falls straight downward after being propelled forward from a cannon.

Tory, Geoffroy. *Champ fleury* (1529). The designs in this work provided the model for styles of book decoration in the French Renaissance. Typographers long struggled with the problem of developing a formula for the design of capital Roman letters, a question Tory effectively addressed in his seminal work. He advocated a number of other reforms, including the use of accents, the cedilla, the apostrophe, and other punctuation marks, that exerted a profound influence not only on French orthography, but on the French language as it is written today. *Champ fleury* won such great acclaim that King Francis I appointed Tory to the position of Imprimeur du Roy in 1530.

Valturius, Robertus. *De re militari Libri XII* (1472). Included the earliest printed technical illustrations. The book, published in Verona, Italy, featured depictions of the latest military technology as well as scenes illustrating various military activities and warfare.

Vauban, Sébastien. *De l'attaque et de la défensedes places* (1737). This work, published posthumously, was a highly influential study of fortifications and siege technology.

Weston, Richard. *Discourse of Husbandry Used in Brabant and Flanders, Showing the Wonderful Improvement of Land There, and Serving as a Pattern for Our Practice in This Commonwealth* (1650). Weston published this work to spread the knowledge of the "ley" farming techniques that he saw between Antwerp and Ghent while in exile. Ley farming emphasized the careful accumulation of manure from animals hand-fed in stalls during summer with green and root crops, such as hay, clover, turnips, or flax. When the ley lands were ploughed under, the roots left in them fertilized the soil. This created a rising cycle of production, because better-manured crops produced more food for animals, which in turn produced more manure. Animals could also be kept through the winter on the reserves of larger crops, enabling them to be selectively bred and preserving their muscle power and manure.

NEIL SCHLAGER

General Bibliography

Armitage, A. *Copernicus, the Founder of Modern Astronomy.* London: Allen & Unwin, 1938.

Asimov, Isaac. *Adding a Dimension: Seventeen Essays on the History of Science.* Garden City, NY: Doubleday, 1964.

Basalla, George. *The Evolution of Technology.* New York: Cambridge University Press, 1988.

Benson, Don S. *Man and the Wheel.* London: Priory Press, 1973.

Beretta, Marco. *The Enlightenment of Matter: The Definition of Chemistry from Agricola to Lavoisier.* Canton, MA: Science History Publications, 1993.

Black, Jeremy. *The Cambridge Illustrated Atlas, Warfare: Revolution, 1492- 1792.* New York: Cambridge University Press, 1996.

Boorstin, Daniel J. *The Discoverers.* New York: Random House, 1983.

Bowler, Peter J. *The Norton History of the Environmental Sciences.* New York: W. W. Norton, 1993.

Brock, W. H. *The Norton History of Chemistry.* New York: W. W. Norton, 1993.

Bruno, Leonard C. *Science and Technology Firsts.* Edited by Donna Olendorf, guest foreword by Daniel J. Boorstin. Detroit: Gale, 1997.

Buchwald, Jed Z. and I. Bernard Cohen, editors. *Isaac Newton's Natural Philosophy.* Cambridge, MA: MIT Press, 2000.

Bud, Robert and Deborah Jean Warner, editors. *Instruments of Science: An Historical Encyclopedia.* New York: Garland, 1998.

Butterfield, Herbert. *The Origins of Modern Science, 1300- 1800.* New York: Macmillan, 1951.

Bynum, W. F., et al., editors. *Dictionary of the History of Science.* Princeton, NJ: Princeton University Press, 1981.

Carnegie Library of Pittsburgh. *Science and Technology Desk Reference: 1,500 Frequently Asked or Difficult-to-Answer Questions.* Washington, DC: Gale, 1993.

Christianson, Gale E. *Isaac Newton and the Scientific Revolution.* New York: Oxford University Press, 1996.

Crombie, Alistair Cameron. *Medieval and Early Modern Science.* Garden City, NY: Doubleday, 1959.

Crone, G. R. *Man the Explorer.* London: Priory Press, 1973.

Debus, Allen G. *Chemistry, Alchemy, and the New Philosophy, 1550-1700: Studies in the History of Science and Medicine.* London: Variorum Reprints, 1987.

Debus, Allen G. *Man and Nature in the Renaissance.* New York: Cambridge University Press, 1978.

Debus, Allen G. *Science and Education in the Seventeenth Century: The Webster-Ward Debate.* New York: American Elsevier, 1970.

Ellis, Keith. *Man and Measurement.* London: Priory Press, 1973.

Finocchiaro, Maurice A., translator and editor. *The Galileo Affair: A Documentary History.* Berkeley: University of California Press, 1989.

Gascoigne, Robert Mortimer. *A Chronology of the History of Science, 1450-1900.* New York: Garland, 1987.

Gaukroger, Stephen, editor. *Descartes: Philosophy, Mathematics, and Physics.* Totowa, NJ: Barnes & Noble Books, 1980.

Good, Gregory A., editor. *Sciences of the Earth: An Encyclopedia of Events, People, and Phenomena.* New York: Garland, 1998.

Grafton, Anthony and Nancy Siraisi, editors. *Natural Particulars: Nature and the Disciplines in Renaissance Europe.* Cambridge, MA: MIT Press, 1999.

Grattan-Guiness, Ivor. *The Norton History of the Mathematical Sciences: The Rainbow of Mathematics.* New York: W. W. Norton, 1998.

Gullberg, Jan. *Mathematics: From the Birth of Numbers.* Technical illustrations by Pär Gullberg. New York: W. W. Norton, 1997.

Hellemans, Alexander and Bryan Bunch. *The Timetables of Science: A Chronology of the Most Important People and Events in the History of Science.* New York: Simon and Schuster, 1988.

Hellyer, Brian. *Man the Timekeeper.* London: Priory Press, 1974.

Holmes, Edward and Christopher Maynard. *Great Men of Science.* Edited by Jennifer L. Justice. New York: Warwick Press, 1979.

Hoskin, Michael. *The Cambridge Illustrated History of Astronomy.* New York: Cambridge University Press, 1997.

Hunter, Michael Cyril William. *The Royal Society and Its Fellows, 1660-1700: The Morphology of an Early Scientific Institution.* Chalfont St. Giles, England: British Society for the History of Science, 1982.

Kelly, John T. *Practical Astronomy during the Seventeenth Century: Almanac-Makers in America and England.* New York: Garland, 1991.

King, Lester S. *The Road to Medical Enlightenment, 1650-1695.* New York: American Elsevier, 1970.

Lankford, John, editor. *History of Astronomy: An Encyclopedia.* New York: Garland, 1997.

Levere, Trevor H. and William R. Shea, editors. *Nature, Experiment, and the Sciences: Essays on Galileo and the History of Science.* Boston: Kluwer Academic Publishers, 1990.

Lincoln, Roger J. and G. A. Boxshall. *The Cambridge Illustrated Dictionary of Natural History.* Illustrations by Roberta Smith. New York: Cambridge University Press, 1987.

Lindberg, David C. *Theories of Vision from al-Kindi to Kepler.* Chicago: University of Chicago Press, 1976.

Moran, Bruce T., editor. *Patronage and Institutions: Science, Technology, and Medicine at the European Court, 1500-1750.* Rochester, NY: Boydell Press, 1991.

Porter, Roy. *The Cambridge Illustrated History of Medicine.* New York: Cambridge University Press, 1996.

Reeds, Karen. *Botany in Medieval and Renaissance Universities.* New York: Garland, 1991.

Sarton, George. *Introduction to the History of Science.* Huntington, NY: R. E. Krieger Publishing Company, 1975.

Singer, Charles. *A History of Biology to About the Year 1900: A General Introduction to the Study of Living Things.* Ames: Iowa State University Press, 1989.

Singer, Charles. *A Short History of Science to the Nineteenth Century.* Mineola, NY: Dover Publications, 1997.

Smith, Roger. *The Norton History of the Human Sciences.* New York: W. W. Norton, 1997.

Spangenburg, Ray. *The History of Science from the Ancient Greeks to the Scientific Revolution.* New York: Facts on File, 1993.

Stiffler, Lee Ann. *Science Rediscovered: A Daily Chronicle of Highlights in the History of Science.* Durham, NC: Carolina Academic Press, 1995.

Stwertka, Albert and Eve Stwertka. *Physics: From Newton to the Big Bang.* New York: F. Watts, 1986.

Travers, Bridget, editor. *The Gale Encyclopedia of Science.* Detroit: Gale, 1996.

Westfall, Richard S. *Force in Newton's Physics: The Science of Dynamics in the Seventeenth Century.* New York: American Elsevier, 1971.

Whitteridge, Gweneth. *William Harvey and the Circulation of the Blood.* New York: American Elsevier, 1971.

Wolf, Abraham. *A History of Science, Technology, and Philosophy in the Sixteenth and Seventeenth Centuries.* London: Allen & Unwin, 1950.

Whitehead, Alfred North. *Science and the Modern World: Lowell Lectures, 1925.* New York: The Free Press, 1953.

World of Scientific Discovery. Detroit: Gale, 1994.

Young, Robyn V., editor. *Notable Mathematicians: From Ancient Times to the Present.* Detroit: Gale, 1998.

JUDSON KNIGHT

Index

*Numbers in bold refer to
main biographical entries*